INTRODUCTION TO MACROMOLECULAR SCIENCE

SECOND EDITION

INTRODUCTION TO MACROMOLECULAR SCIENCE

Second Edition

PETR MUNK
TEJRAJ M. AMINABHAVI

⟨W⟩WILEY-INTERSCIENCE

A JOHN WILEY & SONS, INC., PUBLICATION

This book is printed on acid-free paper. ♾

Copyright © 2002 by John Wiley & Sons, Inc., New York. All rights reserved.

Published simultaneously in Canada.

No part of this publication may be reproduced, stored in a retrieval system or transmitted in any form or by any means, electronic, mechanical, photocopying, recording, scanning or otherwise, except as permitted under Sections 107 or 108 of the 1976 United States Copyright Act, without either the prior written permission of the Publisher, or authorization through payment of the appropriate per-copy fee to the Copyright Clearance Center, 222 Rosewood Drive, Danvers, MA 01923, (978) 750-8400, fax (978) 750-4744. Requests to the Publisher for permission should be addressed to the Permissions Department, John Wiley & Sons, Inc., 605 Third Avenue, New York, NY 10158-0012, (212) 850-6011, fax (212) 850-6008, E-Mail: PERMREQ@WILEY.COM.

For ordering and customer service, call 1-800-CALL-WILEY.

Library of Congress Cataloging-in-Publication Data:

Munk, Petr.
    Introduction to macromolecular science.—2nd ed. / Petr Munk and Tejraj M. Aminabhavi.
        p.   cm.
    Includes bibliographical references and index.
    ISBN 0-471-41716-5 (cloth : alk.paper)
    1. Macromolecules.   I. Aminabhavi, Tejraj Malleshappa.   II. Title.

    QD381 .M85   2001
    547′.7—dc21   2001045583

Printed in the United States of America.
10 9 8 7 6 5 4 3 2 1

To our wives, Zdenka and Devika

CONTENTS

Preface   xv

Preface to the First Edition   xvii

Prologue: Macromolecules Around Us   xix

    Natural Macromolecules, xx
    Man-Made Macromolecules, xx

1   Structure of Macromolecules   1

  1.1   Concept of a Macromolecule, 1
  1.2   Primary Structure—Covalent Bonds, 2
      1.2.1   Linear Polymers, 2
          1.2.1.1   Polymers with All-Carbon Backbones, 3
          1.2.1.2   Polymers with Heteroatoms in the Backbone, 9
          1.2.1.3   Polymers with Inorganic Backbones, 12
      1.2.2   Cyclolinear Polymers, 14
      1.2.3   Branched Polymers, 18
      1.2.4   Copolymers, 21
          1.2.4.1   Types of Copolymers, 22
          1.2.4.2   Some Important Copolymers, 23
      1.2.5   Isomerism of Polymer Chains and Tacticity, 24
      1.2.6   Nomenclature of Polymers, 28
      1.2.7   Macromolecular Networks, 31
          1.2.7.1   Loose Networks, 31

1.2.7.2   Dense Networks—Thermosets, 32

1.2.7.3   Two-Dimensional Networks, 35

1.2.8   Natural Macromolecules, 36

1.2.8.1   Polysaccharides, 37

1.2.8.2   Proteins and Polypeptides, 39

1.2.8.3   Nucleic Acids, 42

1.3   Higher Structure—Conformations, 45

1.3.1   Random Macromolecular Coils, 46

1.3.1.1   Short-Range Interactions, 49

1.3.1.2   Statistical Coils, 52

1.3.1.3   Long-Range Interactions, 53

1.3.1.4   Polyelectrolyte Effect, 56

1.3.2   Secondary Structure—Regular Conformations, 58

1.3.3   Tertiary Structure—Arrangement of Larger Segments, 61

1.3.4   Enzymes—Tertiary Conformations in Action, 62

1.4   Multimolecular Arrangements—Quaternary Structure, 64

1.4.1   Multiunit Proteins, 64

1.4.2   Double Helix of Nucleic Acids, 66

1.4.3   Genetic Coding and Reproduction, 68

1.4.4   Protein Synthesis, 69

1.4.5   Natural Supportive Structures, 71

1.5   Aggregates of Small Molecules, 71

1.5.1   Colloids, 72

1.5.2   Micelles, 75

1.5.3   Block Copolymer Micelles, 79

1.6   Molecular Weight, 80

1.6.1   Molecular Weight Averages, 81

1.6.2   Distribution of Molecular Weights, 84

1.A   Suggestions for Further Reading, 90

1.B   Review Questions, 91

1.C   Derivations, 92

1.D   Numerical Problems, 93

2   Techniques for Synthesis of Polymers   **95**

2.1   Polycondensation, 96

2.1.1   Carboxylic Acid Derivatives, 97

2.1.1.1   Reaction Mechanisms, 97

2.1.1.2   Cyclization Reactions, 101

2.1.1.3   Kinetics of Polycondensation, 102

2.1.1.4   Equilibrium; Distribution of Molecular Weights, 105

2.1.1.5   Gel Point; Three-Dimensional Structures, 108

2.1.1.6   Dendrimers and Hyperbranched Polymers, 110

2.1.2   Isocyanates, 115

2.1.3   Aldehydes, 119

2.1.4   Condensations Forming Cycles, 124
2.1.5   Siloxanes, 127
2.1.6   Epoxy Resins, 130
2.1.7   Miscellaneous Polycondensations, 131
2.2   Radical Polymerization, 135
2.2.1   Production of Radicals, Initiators, 135
2.2.2   Reactions of Radicals, 139
2.2.3   Reactivity of Radicals, 142
2.2.4   Kinetics of Radical Polymerization, 145
    2.2.4.1   Initiation, Propagation, and Termination, 145
    2.2.4.2   Rate of Polymerization and Molecular Weight, 149
    2.2.4.3   Chain Transfer, 157
    2.2.4.4   Inhibition and Retardation, 162
    2.2.4.5   Kinetics at High Conversions—Gel Effect, 164
2.2.5   Radical Copolymerization, 166
2.2.6   Thermodynamics of Radical Polymerization, 172
2.2.7   Living Radical Polymerization, 175
    2.2.7.1   Atom Transfer Radical Polymerization (ATRP), 176
    2.2.7.2   Nitroxide-Mediated Radical Polymerization, 177
2.2.8   Industrial Polymerizations, 178
    2.2.8.1   Suspension Polymerization, 179
    2.2.8.2   Precipitation Polymerization, 179
    2.2.8.3   Emulsion Polymerization, 180
    2.2.8.4   Cross-Linked Structures, 181
2.3   Ionic and Coordination Polymerization, 184
2.3.1   Anionic Polymerization, 184
    2.3.1.1   Ions in Nonpolar Media, 185
    2.3.1.2   Anionic Initiators, 187
    2.3.1.3   Initiation and Propagation, 189
    2.3.1.4   Living Polymers and Block Copolymers, 193
    2.3.1.5   Ring-Opening Polymerizations, 195
2.3.2   Cationic Polymerization, 198
    2.3.2.1   Cationic Initiators, 199
    2.3.2.2   Polymerization of Olefins, 200
    2.3.2.3   Living Cationic Polymerization, 203
    2.3.2.4   Polymerization of Aldehydes and Cyclic Monomers, 204
2.3.3   Coordination Polymerization, 205
    2.3.3.1   Ziegler–Natta Initiators, 206
    2.3.3.2   Metallocene Catalysts, 208
    2.3.3.3   Metathesis Polymerization, 214
    2.3.3.4   Group Transfer Polymerization (GTP), 218
    2.3.3.5   Miscellaneous Coordination Initiators, 220
2.4   Reactions on Macromolecules, 220
2.4.1   Reactions of Side Groups, 221
2.4.2   Cyclization Reactions, 222

2.4.3   Modifications of Cellulose, 224

2.4.4   Cross-Linking and Vulcanization, 226

2.4.5   Polymer Grafting, 228

2.4.6   Polymer Degradation and Stabilization, 230

2.A   Suggestions for Further Reading, 234

2.B   Review Questions and Derivations, 235

2.C   Numerical Problems, 238

3   **Macromolecules in Solution**   **241**

3.1   Thermodynamics of Macromolecular Solutions, 242

    3.1.1   Flory–Huggins Model, 244

        3.1.1.1   Low-Molecular-Weight Mixtures, 244

        3.1.1.2   Macromolecular Solutions, 248

        3.1.1.3   Chemical Potentials, 250

    3.1.2   Excluded-Volume Theories, 254

        3.1.2.1   Compact Molecules, 254

        3.1.2.2   Macromolecular Coils, 256

    3.1.3   Equation-of-State Theories, 261

    3.1.4   Phenomenological Approach, 269

    3.1.5   Anisotropic Solutions and Liquid Crystals, 271

        3.1.5.1   Types of Liquid Crystals, 272

        3.1.5.2   Liquid Crystalline Polymers, 275

3.2   Equilibrium Methods for the Study of Macromolecules
in Solution, 277

    3.2.1   Osmometry, 279

        3.2.1.1   Macromolecular Solutions in Mixed Solvents, 284

        3.2.1.2   Osmometry of Polyelectrolytes, 288

        3.2.1.3   Technical Aspects of Osmometry, 291

    3.2.2   Equilibria in the Ultracentrifuge, 293

        3.2.2.1   The Ultracentrifuge, 295

        3.2.2.2   Sedimentation Equilibrium, 298

        3.2.2.3   Equilibrium in a Density Gradient, 305

    3.2.3   Phase Equilibria, 308

        3.2.3.1   Fractionation of Polymers, 318

3.3   Hydrodynamics of Macromolecular Solutions, 321

    3.3.1   Viscous Flow of Liquids, 322

    3.3.2   Particles Moving Through a Liquid—Frictional
Coefficients, 326

    3.3.3   Particles Suspended in a Flowing Liquid—Viscosity
Increase, 330

    3.3.4   Hydrodynamic Interactions, 332

    3.3.5   Hydrodynamics of Macromolecular Coils, 335

    3.3.6   Concentration Effects in Macromolecular Hydrodynamics, 340

    3.3.7   Orientation and Deformation of Particles in
a Flowing Liquid, 343

3.4 Hydrodynamic Methods for the Study of Macromolecules
    in Solution, 346

    3.4.1 Diffusion, 346

        3.4.1.1 Experimental Diffusimetry, 349

        3.4.1.2 Interpretation of Diffusion Coefficients, 351

    3.4.2 Sedimentation Velocity, 352

        3.4.2.1 Homogeneous Solutes, 352

        3.4.2.2 Heterogeneous Solutes, 358

        3.4.2.3 Archibald Method, 359

    3.4.3 Viscometry, 360

        3.4.3.1 Viscometers, 361

        3.4.3.2 Intrinsic Viscosity, 364

        3.4.3.3 Molecular Weight and Coil Dimensions, 365

        3.4.3.4 Unperturbed Dimensions, Thermodynamic
             Parameters, 367

        3.4.3.5 Branched Chains, 370

        3.4.3.6 Polydisperse Polymers, 370

        3.4.3.7 Concentration Dependence of Viscosity, 371

        3.4.3.8 Non-Newtonian Viscosity, 372

    3.4.4 Flow Birefringence, 374

        3.4.4.1 Optical Properties of Dielectrics, 374

        3.4.4.2 Molecular Anisotropy, 376

        3.4.4.3 Birefringence of Systems Oriented by Flow, 377

        3.4.4.4 Birefringence and Stress, 382

        3.4.4.5 Experimental Arrangements, 383

        3.4.4.6 Interpretation of Flow Birefringence Data, 385

3.5 Light Scattering, 386

    3.5.1 Scattering by a Single Small Isotropic Particle, 386

    3.5.2 Light Scattered by an Anisotropic Particle, 390

    3.5.3 Interference of Light Waves, 392

    3.5.4 Scattering by Large Particles, 393

    3.5.5 Scattering by Macroscopic Systems, 398

        3.5.5.1 Theory of Fluctuations, 400

        3.5.5.2 Scattering by Gases and Liquids, 401

    3.5.6 Light Scattering by Polymer Solutions, 403

        3.5.6.1 Measurement of Molecular Weight and Size, 403

        3.5.6.2 Effect of Polydispersity, 406

        3.5.6.3 Polymers in Mixed Solvents, 406

        3.5.6.4 Turbidity, 408

    3.5.7 Quasi-Elastic Light Scattering, 409

    3.5.8 Experimental Arrangements, 411

    3.5.9 Small-Angle X-Ray Scattering (SAXS), 413

3.6 Spectral Methods, 417

    3.6.1 Ultraviolet Spectrophotometry, 417

    3.6.2 Fluorescence, 418

    3.6.3 Infrared Spectra, 419

3.6.4   Nuclear Magnetic Resonance (NMR), 421
3.6.5   Mass Spectrometry, 425
3.7   Separation Techniques, 426
      3.7.1   Electrophoresis, 428
              3.7.1.1   Free Electrophoresis, 429
              3.7.1.2   Paper Electrophoresis, 430
              3.7.1.3   Gel Electrophoresis, 432
              3.7.1.4   Capillary Electrophoresis, 433
              3.7.1.5   Electrofocusing, 434
              3.7.1.6   SDS Electrophoresis, 435
      3.7.2   Gel Permeation Chromatography (GPC), 435
              3.7.2.1   Principles of GPC, 438
              *3.7.2.2   Molecular Weight and Universal Calibration
                         Curve, 440*
              3.7.2.3   Molecular Weight Distribution, 443
              3.7.2.4   Critical Point GPC, 443
              3.7.2.5   Selection of Columns, 444
              3.7.2.6   Detectors Used in GPC, 445
      3.7.3   Field Flow Fractionation, 447
      3.7.4   Supercritical Fluid Chromatography, 449
3.8   Techniques for the Study of the Structure of Nucleic Acids, 449
      3.8.1   Polymerase Chain Reaction, 450
      3.8.2   Separation of DNA Fragments, 452
              3.8.2.1   Pulsed Field Gel Electrophoresis (PFGE), 452
              3.8.2.2   Southern Transfer Technique, 453
              3.8.2.3   DNA Profile Analysis, 454
      3.8.3   Sequencing of DNA Fragments, 455
3.A   Suggestions for Further Reading, 456
      3.A.1   General Reading, 456
      3.A.2   Thermodynamics, 457
      3.A.3   Equilibrium Methods (Osmometry, Sedimentation Equilibrium,
              and Phase Equilibria), 458
      3.A.4   Hydrodynamics and Hydrodynamic Methods, 458
      3.A.5   Light Scattering and Spectral Methods, 458
      3.A.6   Separation Techniques, 459
      3.A.7   Techniques to Study the Structure of Nucleic Acids, 459
3.B   Thermodynamics, 459
      3.B.1   Review Questions, 459
      3.B.2   Derivations and Numerical Problems, 460
3.C   Osmometry, Sedimentation Equilibrium, and Phase Equilibria, 461
      3.C.1   Review Questions, 461
      3.C.2   Derivations and Numerical Problems, 462
3.D   Hydrodynamics and Hydrodynamic Methods, 464
      3.D.1   Review Questions, 464
      3.D.2   Derivations and Numerical Problems, 465
3.E   Light Scattering, 467

3.E.1　Review Questions, 467

3.E.2　Derivations and Numerical Problems, 468

3.F　Spectral Methods, 470

　　3.F.1　Review Questions, 470

3.G　Separation Methods, 471

　　3.G.1　Review Questions, 471

　　3.G.2　Derivations and Numerical Problems, 472

3.H　Techniques to Study the Structure of Nucleic Acid, 473

4　Bulk Polymers　　　　　　　　　　　　474

4.1　Properties of Bulk Polymers, 474

　　4.1.1　Stress and Strain, 476

　　4.1.2　Glassy Polymers, 479

　　4.1.3　Elastic Networks, 481

　　　　4.1.3.1　*Theory of Rubber Elasticity,* 482

　　　　4.1.3.2　*Swelling of Gels,* 488

　　4.1.4　Polymer Melts, 489

　　4.1.5　Viscoelastic Materials, 492

　　　　4.1.5.1　*Creep and Stress Relaxation,* 493

　　　　4.1.5.2　*Dynamic Experiments,* 498

　　4.1.6　Crystalline Polymers, 504

　　　　4.1.6.1　*Morphology of Crystalline Polymers,* 506

　　　　4.1.6.2　*Mechanical Properties of Crystalline Polymers,* 511

　　4.1.7　Multicomponent and Multiphase Materials, 512

　　　　4.1.7.1　*Plasticization of Polymers,* 512

　　　　4.1.7.2　*Polymer Blends,* 514

　　　　4.1.7.3　*Heterophase Materials,* 516

　　4.1.8　Electrical and Optical Properties of Polymers, 519

　　4.1.9　Transport Through Polymers, 522

4.2　Techniques for the Study of Bulk Polymers, 525

　　4.2.1　Mechanical Methods, 525

　　4.2.2　Differential Scanning Calorimetry (DSC), 528

　　4.2.3　Inverse Gas Chromatography (IGC), 531

　　4.2.4　X-Ray Diffraction, 535

　　4.2.5　Neutron Scattering, 539

4.3　Techniques for the Study of Polymer Surfaces, 542

　　4.3.1　Optical Microscopy, 542

　　4.3.2　Electron Microscopy, 543

　　4.3.3　Atomic Force Microscopy (AFM) and Scanning Tunneling Microscopy (STM), 545

　　4.3.4　Electron Spectroscopy for Chemical Analysis (ESCA), 547

4.A　Suggestions for Further Reading, 550

4.B　Study Questions, 552

4.C　Numerical Problems, 554

5   Technology of Polymeric Materials   **557**

  5.1   Fabrication of Polymers, 557
        5.1.1   Compounding and Mixing, 558
        5.1.2   Casting, 559
        5.1.3   Extrusion, 559
        5.1.4   Bubble Blown Film Extrusion, 561
        5.1.5   Cast Film Extrusion, 563
        5.1.6   Coating, 563
        5.1.7   Fiber Spinning, 564
        5.1.8   Calendering, 567
        5.1.9   Molding, 567
        5.1.10 Foam Fabrication, 570
  5.2   Testing of Polymers, 571
        5.2.1   Mechanical Testing, 571
        5.2.2   Thermal Testing, 573
  5.3   Barrier Properties of Polymers, 574
        5.3.1   Membrane Types, 575
        5.3.2   Membrane Preparations, 576
                5.3.2.1   Symmetric Porous Membranes, 576
                5.3.2.2   Asymmetric Membranes, 577
                5.3.2.3   Ion-Exchange Membranes, 579
        5.3.3   Membrane-Based Separation Processes, 580
                5.3.3.1   Reverse Osmosis and Filtration Techniques, 580
                5.3.3.2   Electrodialysis, 583
                5.3.3.3   Gas Separation, 583
                5.3.3.4   Pervaporation, 585
        5.3.4   Polymeric Devices for Drug Delivery, 588
  5.A   Suggestions for Further Reading, 590
  5.B   Review Questions, 591

Epilogue: Literature About Macromolecules   **593**

  A. Textbooks About Macromolecules, 593
  B. Monograph Series and Encyclopedias, 594
  C. Handbooks and Reference Sources, 594
  D. Macromolecular Journals, 595

Index   **597**

PREFACE

More than a decade passed since the publication of the first edition of this textbook. During this period the importance of macromolecular science has ever been increasing both in academia and industry. Courses on polymer science are newly offered at many universities, both in departments of chemistry and chemical engineering. Biochemistry and biology curricula offer courses covering important parts of macromolecular science. Thus, the demand for good textbooks in the macromolecular field is compellingly manifesting itself.

When the senior author was writing the first edition, he emphasized the necessity of a balanced approach to the field. His idea of a balanced textbook called for equal concern for the structure of macromolecules, their synthesis, methods for studying them, their properties, and technological aspects of their utilization. We believe that this approach is advantageous and we will employ it in this edition as well. We are again focusing on the basic understanding of the phenomena involved. That leaves many sections of the book unchanged.

It was, of course, necessary to include in the second edition a host of new topics that gained importance in the last decade. Thus, we are adding sections about newer synthetic procedures for polymers: living radical and cationic polymerizations, group transfer polymerization, polymerizations using metallocene and metathesis catalysts, syntheses leading to dendrimers. The treatment of separation techniques was greatly expanded. We are also adding a short introduction to techniques that are used for studying the structure of nucleic acids. The treatment of technological aspects of polymer processing was expanded significantly. New sections deal with techniques for studying polymer surfaces and polymeric membranes. Topics like polymer nomenclature, liquid crystalline polymers, block copolymer micelles, etc., are also included.

The first edition did not give the teacher enough tools for helping the students to digest the new material. To remedy this situation we are attaching at the end of each chapter an expanded list of references including the most recent ones, a set of review questions, and a list of suitable theoretical derivations and numerical problems.

The co-author, Tejraj M. Aminabhavi wishes to place on record the encouragement of his brother (A.M. Aminabhavi) and the assistance he received from Dr. R. H. Balundgi and Ph. D. students (particularly Dr. H.G. Naik) while preparing this edition. Dr. S. Rame Gowda (former Vice Chancellor) and Dr. P.E. Cassidy (Southwest Texas State University) have always been the guiding spirits in the academic excellence of the co-author. The Karnatak University authorities (Vice Chancellor, Dr. A. M. Pathan) have been very kind to sanction the leave of absence during which time Aminabhavi planned the book with his mentor Professor Petr Munk. Finally, a word of thanks to his family members (wife, Devika, and children, Bharathi, Hema, and Sneha) who have been very cooperative and patient while he was working on the book.

PETR MUNK
TEJRAJ M. AMINABHAVI

PREFACE TO THE FIRST EDITION

I have been teaching macromolecular science under its various names at the University of Texas at Austin since 1969. During this time I have been perpetually searching for an ideal textbook: a textbook that would give a fair introduction into the macromolecular field to a student who is conversant in organic chemistry and in physical chemistry but has never been exposed to the concepts of macromolecular science.

In my eyes, the ideal textbook should present the problems in their full complexity, yet their physical and chemical aspects should not be drowned in excessive mathematics. It should recognize that the term *macromolecule* encompasses synthetic materials as well as important biochemical entities and that these two types of materials have much more in common than is usually acknowledged. An ideal textbook should be equally concerned with the structure of macromolecules, with the way they are made, with the methods used for studying them, with their useful (as well as annoying) properties, and finally with the procedures by means of which they are put to practical use.

Alas, the quest for an ideal is never-ending. The older masterpieces were getting – well older and older. Some of the textbooks dwelled too much on the synthesis and neglected the physicochemical aspects, others made just the opposite mistake. Some put too much emphasis on the industrial viewpoint; others neglected it altogether. Some overestimated the relative significance of biological polymers; some did not mention them at all. Some textbooks were too short and did not cover the field extensively enough; others were so thorough that the reader suffocated under the amount of information.

Inevitably I came to a painful conclusion: if I want to use an ideal textbook, I must write it myself. Some five years later, I present here the result of that revelation. Part of the responsibility for my endeavor must be shared by Professor Herbert Morawetz of the Polytechnic University of New York, who gave me early encouragement as well as many valuable suggestions, and by my wife, Zdenka, who patiently supported me during the evenings, weekends, and holidays spent with the manuscript.

No doubt, it will be found that in this book I have replaced biases of other authors by my own. Modesty (or is it common sense?) leads me to admit that the ideal is as elusive as before. But it is too late: the book has been written, and here it is without further apologies.

PETR MUNK

Austin, Texas
June 1989

PROLOGUE
MACROMOLECULES AROUND US

Although macromolecular materials comprise a major portion of our world, macromolecular science has existed for not more than about seventy years. In fact, the existence of molecules with very high molecular weight was not firmly established until the late 1920s. This late development is no doubt related to the fact that macromolecular materials exhibit properties that appear to defy classical rules of chemistry and physics. We will see again and again that what was a small correction in a theory of small molecules becomes a dominant term in macromolecular systems. Another characteristic of macromolecular science is the relative paucity of suitable experimental methods. More often than not, the experimental values are related to the desired quantities in a very indirect way: quite elaborate theories are sometimes needed to bridge the gap. In such a situation, misinterpretations are not surprising and indeed occur rather frequently.

Notwithstanding these difficulties, macromolecular science is experiencing an almost explosive growth. This is due to the immensely important role macromolecular materials and the knowledge of their properties play in modern technology. Our world would be quite different if technology did not encompass the present level of understanding and mastery of these materials. Knowing that this understanding is still rather limited, we can easily see the necessity for scientists and engineers of all inclinations to understand macromolecular science and develop it further.

In this section, we would like to impress readers with the ubiquity of macromolecules and with their significance for life in general and for human society in particular.

NATURAL MACROMOLECULES

The proper heading for this chapter would read "Macromolecules Around and Within Us!" Indeed, a living organism is a complex machine, most functions of which are performed by macromolecules. The informational package including blueprints for the composition, synthesis, and control of all parts and functions of the organism is composed of deoxyribonucleic acids (DNA). The synthesis of proteins is carried out by ribonucleic acids (RNA) and ribosomes. Proteins are the workhorses of every organism. Structural proteins (together with bones) provide the support, muscle proteins, the movement; blood proteins (hemoglobin, albumin) transport all needed substances. Enzymes are extremely specific and extremely efficient catalysts for every chemical reaction needed by the organism. While the organisms of the plant kingdom lack muscles, they developed cellulose (wood, cotton) as an all-purpose building material. Starches serve as energy reservoirs for plants as well as for the animals who eat them. To make the story short: DNA, RNA, ribosomes, proteins, cellulose, starches, and a host of other biologically important substances are all macromolecules.

From the very beginning of civilization, humans made technological use of natural macromolecules. Structural proteins in the form of leather were used to make sandals, belts, shields, and (later on) parchment. When another fiberlike protein was still attached to the leather ("fur" is the name used for these composite materials), it made an excellent material for clothing, especially in cold climates. Later on these protein fibers were separated (usually sheared from a living animal) and used alone; known as "wool" they were the basis of the textile industry.

However, cellulose proved to be the most versatile macromolecular material. Its technological importance is still unmatched. As wood, it has served as an excellent building material; the U.S. homebuilding industry still relies on it almost exclusively. In the form of natural fiber (cotton, flax) it helped the textile industry to make lighter-weight fabrics. Later, when a process was developed to make paper (an inexpensive substitute for parchment) from cellulose, writing could become widespread, with the resulting boom in printing and information industries (and the hides of many donkeys were saved in the process).

Another important macromolecular material is natural rubber. Its significance was discovered relatively late; nevertheless, the early automobile industry would be unthinkable if natural rubber had not been available for tire manufacturing.

MAN-MADE MACROMOLECULES

The first man-made macromolecular materials were made without knowledge of their chemical nature. They were shellacs used by old Chinese artists for their elaborate carvings and by Renaissance masters to coat their paintings. The first modern macromolecular material was Celluloid – modified cellulose that made possible early cinematography. After the turn of the century, new materials emerged more frequently. The first was Bakelite – a hard resin easily formed into light switches, ashtrays, and hundreds of other articles.

The arrival of modern macromolecules is marked in our book by the introduction of a new term. Many of these new materials were prepared by polymerization of some simple molecules; accordingly, these materials were called *polymers.* It will become obvious that all polymers are composed of *macromolecules,* and the two terms are used almost interchangeably in the literature and in the rest of this book. However, we will keep a slight distinction: While all polymers are macromolecules, the polymeric nature of many macromolecules (especially the natural ones) is not obvious; for them we will use the broader term *macromolecules.* Two other common names frequently used for polymers are derived from the properties and/or appearance of these materials in solid form: *plastics and resins.*

Nowadays, much of the world around us is made from polymers. Our houses are built from wood, their walls are painted with polyurethane latexes, and similar materials are used as varnishes and other paints. Carpets are made from various polymeric fibers; other fiber-forming polymers are used for fabrics in upholstery, clothing, and car interiors. The supports for photographic films (including movies) and for magnetic tape are polymeric, as well as most plumbing, wire insulation, printed materials, and sporting and recreational equipment. (Yes, your fishing boat and fishing line too!) In the automobile industry, polymers are used not only for tires and interiors, but also for many structural parts. The packaging industry is switching fast from paper to lighter and cheaper plastics.

It is tempting to extrapolate a little and to predict what may be the future of polymers in our world. However, we will leave this to you, after you have read this *whole* book as well as many others. To help you with their selection, we are listing in the Epilogue some of the more important textbooks that cover the macromolecular field in its entirety. Books that cover narrower topics are listed at the end of individual chapters.

1

STRUCTURE OF MACROMOLECULES

We will talk a lot about macromolecules in this book. It is therefore appropriate to explore what macromolecules are: their structure on the atomic level, the mutual relations of different parts of the same macromolecule, the way they form larger structures, and the proper measures of their size. These are the major topics of this chapter.

1.1. CONCEPT OF A MACROMOLECULE

Although the term *macromolecule* itself suggests that macromolecules are large molecules, it is useful to review the meaning of the term *molecule* and to discuss how large a molecule must be to qualify as a macromolecule.

For typical organic molecules, the definition of a molecule is easy: It is a set of atoms bound together by covalent bonds. Fortunately, this definition will be satisfactory even for macromolecules in most cases. We will call such macromolecules *true macromolecules*. However, in some materials, several or many true molecules are held together by multiple secondary forces, and the resulting particle behaves in most physicochemical aspects as a molecule. Several classes of materials belong to this category:

1. Double-stranded nucleic acids, in which the two strands are held together mainly by hydrogen bonds.
2. Colloidal solutions, the fine dispersions of usually crystalline materials in a solvent in which they are not soluble; they show no tendency to dissociate into individual molecules.

3. Detergent micelles, in which the interplay of forces is very subtle, but is nevertheless sufficient to keep the micelles together.

In all these instances it is practical to consider each particle as a molecule. The criterion is kinematic. A molecule (read "particle") is a set of atoms that move together and do not dissociate under given experimental conditions. Obviously, this definition is useful only for liquid systems and gases (aerosols); it fails for solids.

Some macromolecular materials (polyelectrolytes) can in aqueous or similar solvents dissociate into ions, of which one is macromolecular; the others are small counterions. Other macromolecules can be "solvated" by some small molecules that are bound to them more or less firmly. Should the counterions and the solvating molecules be counted as part of the macromolecule? This question becomes important when measurements of molecular weight are interpreted on the stoichiometric level. Actually, the answer is a matter of convention, and both possible answers are used occasionally. Obviously, it is necessary to know what convention has been used in any particular case, and the evaluation must be performed accordingly. Any ambiguity has its origin in sloppy experimental or reporting practice.

The question, "At what size does a molecule become a macromolecule?" is essentially a semantic question. Usually, a material is considered to be macromolecular when it exhibits properties typical of high-molecular-weight substances. This usually happens when the molecule is composed of at least several hundred atoms, that is, has a molecular weight in the range of at least 1000–10,000.

1.2. PRIMARY STRUCTURE—COVALENT BONDS

The basic way to describe a macromolecule is the same as for small molecules: The molecules are described by their structural chemical formula, which specifies all covalent bonds. With the huge number of atoms constituting a macromolecule, such a description may seem an impossible task. Fortunately, most macromolecules exhibit a rather regular structure, and their description is relatively easy.

In this section we will describe the chemical structure of some of the more important polymers. At this opportunity we will also familiarize ourselves with the nomenclature of polymers, with their basic properties, and, for some of the most important polymers, with the raw materials from which they are made and with their basic uses. However, the synthesis of polymeric materials will be left to Chapter 2.

1.2.1. Linear Polymers

Many important polymers have a uniquely simple structure: A large number of constituent atoms are linked together to form a chain that is called the *backbone*. The backbone is linear in the topological sense; its actual shape is very complex (see Section 1.3), and the backbone bonds exhibit more or less typical valence angles. The side groups on the backbone, if any, are small in the sense that their size corresponds to small molecules and not to macromolecules. Moreover, for many simple polymers (sometimes called *homopolymers,* as opposed to the copolymers of Section 1.2.4) the molecular chain could be mentally divided into identical units, which are frequently called *monomeric units.* Thus, the structure of linear homopolymers can be written as Y—M_n—Y, where M represents a monomeric unit and Y is an end group. For a typical polymer, n is a rather large number called the *degree of polymerization* (DP). If n is small, say 2–20, the molecule is called an *oligomer.* When DP is large, the nature of the end groups becomes more or less irrelevant and is frequently not even known. In such a case, the formula is written simply as —M_n—. It should be noted that it is virtually impossible to prepare a polymeric material that has all molecules with the same value of n. (Only nature can do it in some natural macromolecules; see Section 1.2.8.) A mixture of molecules with the same —M—, but different values of n is still considered to be a single substance—a *polydisperse polymer* (as opposed to the virtually unknown *monodisperse polymer*). For polydisperse polymers it is convenient to know the distribution of the DP values or at least some average of them. We will treat this problem in some detail in Section 1.6.

Frequently the monomeric unit —M— can be identified with some small molecule (monomer) from which the polymer was (or could have been) prepared. This identification leads to the common nomenclature of polymers: The name of the polymer is the name of the monomer with the prefix poly-. Thus, if —M— is —$CH(CH_3)$—CH_2—, the polymer is called *polypropylene* (the more proper name, *polypropene,* is virtually never used). If —M— is —NH—CH_2—CH_2—CH_2—CH_2—CH_2—CO—, we are talking about polycaprolactam; poly(methyl acrylate) is a polymer with —M— equal to —$CH(COOCH_3)$—CH_2—. In the last example we see how the parentheses are used in names of polymers derived from monomers whose names are composed of two or more words. Similarly, a polymer having —M— equal to —$CH(OH)$—CH_2— is called poly(vinyl alcohol) because it could have been made from vinyl alcohol if the latter compound were known. It is actually prepared indirectly by hydrolysis of poly(vinyl acetate). A more detailed treatment of nomenclature will be presented in Section 1.2.6.

In the following sections we will catalog the more common polymers according to their chemical structure.

1.2.1.1. Polymers with All-Carbon Backbones.   Polymers with all-carbon backbones are usually prepared from monomers having carbon-carbon double bonds by the chain-polymerization methods described in Sections 2.2 and 2.3. We will now list and briefly describe these homopolymers according to the structure of their monomers.

A. *Polymeric Hydrocarbons*

A.1. The simplest possible repeating unit is —CH_2—CH_2—, and the polymer is called *polyethylene.* Obviously, the same material may be described as having the still simpler repeating group —CH_2— and may be called *polymethylene.* Actually, the former name is used almost exclusively, because the polymer is made from ethylene. Industrially, ethylene is an inexpensive monomer; it is made from the C_2 fraction of natural gas and oil. However, ethylene also can be used for the synthesis of ethanol and ethylene glycol; this increases its synthetic value and its price.

The first polymerization process for ethylene used radical polymerization initiated by traces of oxygen at high temperatures and high pressures. This process leads to many defects in the structure of the polymer. There are a fair number of side groups present, ranging from methyl to butyl, that interfere with the crystallization of the polymer and lower its density. Accordingly, this product is called *high-pressure polyethylene* or *low-density polyethylene*. It is a clear material, rather inert chemically (it is an alkane), and is easily formed into sheets. It is an excellent electrical insulator (it played a significant role in the development of radar during World War II) and a cheap, good packaging material (produce bags in grocery stores, for example, are made of polyethylene). Its mechanical properties are good enough for it to be made into labware and many other light-duty articles.

Ethylene can also be polymerized using coordination polymerization (Section 2.3.3). The polymerization proceeds at ambient temperatures and pressures, and the product has almost no defects. It is highly crystalline and dense and is called *low-pressure polyethylene* or *high-density polyethylene*. It is translucent and mechanically very strong—thinner sheets can do the work. It is used similarly to the low-density material, especially where greater strength is needed.

A.2. Polypropylene [monomeric unit —CH(CH$_3$)—CH$_2$—] is made by polymerization of propylene. This monomer is made from the C$_3$ fraction of natural gas and oil. Its polymerization is the most important use of the C$_3$ fraction (except as a fuel); accordingly, the monomer is rather inexpensive. The polymerization by coordination methods (Section 2.3.3) is the only one feasible for propylene. In polypropylene, the mutual arrangement of the methyl side groups may be either random (such a polymer is said to be *atactic*) or regular (*isotactic*). (A more detailed treatment of tacticity will be presented in Section 1.2.5.). According to the details of the polymerization process, the polymer may be isotactic, atactic, or a mixture of both; they can be separated by extraction with heptane. The atactic polymer is a soft material with inferior properties. However, the crystalline isotactic polymer has very useful properties and is one of the fastest-growing industrial products. It makes excellent fibers, especially useful for outdoors:—carpeting, webbing for garden furniture, upholstery, etc. Being strong and having the lowest density of all plastics, it is used whenever weight is a consideration. It is easily molded and is an excellent electrical insulator.

Polymers based on longer *n*-alkenes yield polymers of little significance.

A.3. Polyisobutylene, with monomeric unit —C(CH$_3$)$_2$—CH$_2$—, is made by cationic polymerization (Section 2.3.2) of isobutylene. It is a rubberlike polymer with good resistance against oxidation, acids, and alkalis. It remains rubberlike even at subzero temperatures. It is used whenever these properties are important.

A.4. When butadiene is polymerized, either 1,2 monomeric units **1a** or 1,4 units may be incorporated into the *polybutadiene* chain. In the 1,4 units the configuration of the remaining double bond may be either *cis* **1b** or *trans* **1c**. The actual polymer contains all three forms; the proportion of these is governed by the details of the polymerization procedure. A polymer having mostly *cis* units is an excellent rubberlike material.

Accordingly, the producers of this rubber strive to find synthetic methods leading to the highest *cis* content. We shall see in Section 4.1.3 that in rubbers the individual polymer chains must be cross-linked to maintain the desired shape of the article. The unreacted double bond in the monomeric unit of polybutadiene provides a convenient group for the cross-linking reaction (vulcanization, Section 2.4.4).

$$
\begin{array}{ccc}
-\text{CH}-\text{CH}_2- & \underset{-\text{CH}_2}{\overset{H}{\diagdown}}\text{C}=\text{C}\underset{\text{CH}_2-}{\overset{H}{\diagup}} & \underset{-\text{CH}_2}{\overset{H}{\diagdown}}\text{C}=\text{C}\underset{H}{\overset{\text{CH}_2-}{\diagup}} \\
\overset{|}{\text{CH}}=\text{CH}_2 & & \\
\textbf{1a} & \textbf{1b} & \textbf{1c}
\end{array}
$$

A.5. Polyisoprene, $-\text{CH}_2-\text{C}(\text{CH}_3)=\text{CH}-\text{CH}_2-$, has properties very similar to those of polybutadiene. However, the leading manufacturer of both forms of this polymer is nature. The pure *cis* form is produced by several plants growing mainly in the tropical regions; it is *caoutchouc,* or *natural rubber.* In fact, it was the threatened cutoff of the United States from the sources of natural rubber during World War II that caused the explosive growth of polymer science in the 1940s. The all-*trans* polymer, called *gutta-percha,* is also produced by some trees. It is hard, not rubberlike, and its utilization is limited (insulation, golf balls).

A.6. Polystyrene **2** is a leading representative of polymers that are made with a vinyl group on an aromatic nucleus. Styrene is easily polymerized by all chain polymerization techniques; its solutions exhibit very few anomalies. Consequently, it is the polymer of the physical chemist; it has served as a model in the vast majority of physicochemical studies. It is a clear, hard, and somewhat brittle plastic that is used for making small articles of all kinds (knobs, pencils, trays, for example). It can be formed into an extremely lightweight foam that serves as an excellent thermal insulator (homes, ice chests) as well as a packaging material for fragile goods. The monomer is made by alkylation of benzene followed by dehydrogenation. This longer synthesis makes it more expensive than simple olefins. When a substituent (methyl, halogen, etc.) is introduced on the aromatic nucleus or the α-carbon, the properties of the polymer may be modified.

$$
-\text{CH}-\text{CH}_2-
$$

2

A.7. Diacetylenes, $R_1-C\equiv C-C\equiv C-R_2$, are easily polymerized in crystals; the resulting polymer conforms to the monomeric crystalline lattice. *Poly(diacetylenes),* $=CR_1-C\equiv C-R_2=$, have a conjugated enyne structure and display a deep red color. We will see later that their ease of polymerization in oriented form can be exploited in the preparation of models of biological structures—vesicles. This feature is a result of the fact that only a small displacement of atoms accompanies the solid-state polymerization of diacetylenes [equation (1.2.1)]

$$R-C\equiv C-C\equiv C-R \longrightarrow \overset{\displaystyle \underset{R}{\diagdown}}{C}-C\equiv C-\overset{\displaystyle \underset{\diagdown\diagdown}{\overset{R}{\diagup}}}{C} \qquad (1.2.1)$$

B. Halogen-Containing Polymers   Halogen substitution on polymers usually increases the polymer's toughness and its resistance to hydrocarbon solvents.

B.1. Poly(vinyl chloride) (PVC), —CHCl—CH$_2$—, is the most important polymer in this group. The monomer is prepared by the addition of hydrogen chloride on acetylene; it is polymerized radically, usually in emulsion (Section 2.2.8.3). PVC is a very versatile polymer, easily molded into various shapes. The pure polymer is very tough; it is used for pipes and a number of structural components. When plasticized by moderate amounts of low-molecular-weight substances (typically esters, e.g., dioctyl phthalate or tricresyl phosphate), it becomes pliable and can be molded into flexible items such as water hoses, luggage, seat covers, tent floors, and protective clothing. When plasticized further, it is used in the form of soft sheets for making raincoats and covers of all types. PVC was the first major plastic developed, and it still accounts for a large portion of polymer production.

B.2. Polytetrafluoroethylene, —CF$_2$—CF$_2$—, is better known under its trade name, Teflon. It is one of the most inert substances known—it is attacked only by hot alkali metals. It is highly crystalline and does not dissolve or swell in any solvent. It is also extremely stable thermally. The unique properties of Teflon follow from the unusually small dispersion forces exhibited by all polyfluorinated compounds. (This behavior is manifested in perfluorinated small molecules by their very low boiling point and very low refractive index.) Thus Teflon, even while crystalline, is very soft and easily stretches into very thin films that are used for sealing metal joints, etc. It does not stick to other materials; when properly bonded to the supporting metal it is the basis for nonsticking cooking utensils. (Its thermal stability also comes in handy in this case.) It is also used for making or lining dishes when extreme inertness is needed.

B.3. Polytrifluorochloroethylene, —CClF—CF$_2$—, combines the properties of PVC and Teflon. It is tougher than Teflon, while retaining most of its inertness. It is highly resistant to organic acids and oils.

B.4. Vinyl acetylene adds hydrogen chloride to yield chloroprene (2-chlorobutadiene), which can be polymerized to *polychloroprene,* —CH$_2$—CCl=CH—CH$_2$—. This polymer exhibits rubberlike properties similar to those of other diene polymers but is more resistant to hydrocarbons. During World War II, the Germans used it as a substitute for natural rubber. Its negligible permeability to seawater makes it a useful outer covering material for nuclear submarines.

C. Polymers with Polar Side Groups   In many important polymers the polar side group is an ester group; however, such polymers are *not* called polyesters. The latter term is reserved for polymers having the ester group as part of the backbone (see Section 1.2.1.2). The properties of the polymers in our present group are governed mainly by the interactions among the side groups. The stronger interactions make a harder (and sometimes more brittle) polymer.

C.1. Poly(methyl acrylate) **3** is a rubbery polymer with limited significance. This is also true of acrylates with larger alkyls. All of them are used mainly as comonomers in the production of copolymers. The goal of this operation is to modify the properties of the other comonomer. It should be mentioned that the acrylic group $CH_2{=}CH{—}$ $CO{—}$ has a combination of kinetic constants in radical polymerization (Section 2.2.4) that are conducive to the formation of polymers with very high molecular weights (in the million range).

$$
\begin{array}{ccc}
 & CH_3 & CH_3 \\
 & | & | \\
-CH-CH_2- & -C-CH_2- & -C-CH_2- \\
| & | & | \\
COOCH_3 & COOCH_3 & COOCH_2CH_2OH \\
\mathbf{3} & \mathbf{4} & \mathbf{5}
\end{array}
$$

C.2. Poly(methyl methacrylate) **4** is one of the most important polymers. The monomer is made by methanolysis of an adduct of hydrogen cyanide on acetone. The polymer is hard but not brittle and can be prepared as very large sheets or other objects with exceptional optical clarity. Thus, it serves under a variety of trade names (Plexiglas, Lucite) as an organic glass (airplane windshields, store windows, etc.). It also has reasonable resistance to alkalis and acids, so it finds many uses in laboratories (especially in biochemistry). Polymethacrylates with a longer alkyl group have much inferior mechanical properties and are of very limited use—mainly as components of copolymers.

C.3. Poly(hydroxyethyl methacrylate) **5,** or poly-HEMA, is a polymer that swells appreciably in water. The resulting gel contains about 35% water and is very elastic and strong. It is used for making soft contact lenses in ophthalmology as well as for other biomedical applications. The monomer is made from ethylene oxide and methacrylic acid.

C.4. Vinyl acetate and vinyl butyrate are prepared by addition of the corresponding acid on acetylene. The polymers *poly(vinyl acetate)* **6a** and *poly(vinyl butyrate)* **6b** (as well as their copolymers) form clear, flexible films that are used as backing for photographic films and for similar purposes.

$$
\begin{array}{cc}
-CH-CH_2- & -CH-CH_2- \\
| & | \\
OCOCH_3 & OCOCH_2CH_2CH_3 \\
\mathbf{6a} & \mathbf{6b}
\end{array}
$$

C.5. Acrylonitrile is made by the addition of hydrogen cyanide on acetylene or, more recently, by catalytic oxidation of a mixture of propylene and ammonia (Sohio process). Its polymerization leads to *polyacrylonitrile* (PAN) **7,** a strongly polar polymer that is insoluble in most solvents. However, PAN is too intractable; acrylonitrile is therefore copolymerized with small amounts of acrylamide, which facilitates its tractability, dye absorption, etc. The modified polymer forms fibers highly appreciated by the textile industry. These are either called *acrylic fibers* or go under a number of trade names such as Orlon.

$$
\begin{array}{ccc}
 & C\equiv N & \\
 & | & \\
C\equiv N & -C-CH_2- & O=C-NH_2 \\
| & | & | \\
-CH-CH_2- & COOCH_2CH_3 & -CH-CH_2- \\
\mathbf{7} & \mathbf{8} & \mathbf{9}
\end{array}
$$

C.6. *Poly(ethyl cyanoacrylate)* **8** is the product of the polymerization of ethyl cyanoacrylate. This monomer can be polymerized by anionic initiators at an amazing rate; the resulting polymer binds well to many materials. Accordingly, the monomer is used as the main component of the "miracle glues." The polymer is used as a bonding agent. Recent studies have demonstrated the potential of poly(alkyl cyanoacrylate) as a colloidal carrier of drugs and as a surgical glue because of its bioresorbable properties.

C.7. Partial hydrolysis of acrylonitrile yields acrylamide, which can be polymerized to *polyacrylamide* **9,** a polymer that frequently has an extremely high molecular weight (up to tens of millions). It dissolves easily in water, enhancing its viscosity considerably. Consequently, like some other water-soluble polymers, it is used as a thickening agent. Because of the enormity of the viscosity increase, polyacrylamide is being explored as a thickening agent in tertiary oil recovery. It is also used for flocculation of colloidal solutions.

C.8. Polymerization of the monomeric acids yields *poly(acrylic acid)* **10** and *poly(methacrylic acid)* **11.** The acidic properties of the carboxyl are essentially the same whether it is part of a small molecule or a macromolecule. Accordingly, both of these polyacids are easily soluble in aqueous solutions at higher pH values as the acidic hydrogen dissociates. Under these circumstances, there are a large number of negatively charged groups along the polymer backbone that repel each other. This leads to a number of anomalies observed in polyelectrolyte solution. We will learn more about them in Section 1.3.1.4.

$$
\begin{array}{cc}
 & CH_3 \\
 & | \\
-CH-CH_2- & -C-CH_2- \\
| & | \\
COOH & COOH \\
\mathbf{10} & \mathbf{11}
\end{array}
$$

C.9. As already mentioned, *poly(vinyl alcohol)*, —CH(OH)—CH$_2$—, is prepared by hydrolysis of poly(vinyl acetate). It is a water-soluble polymer often used as a thickening agent. Its water solution changes properties over time; this process is called *aging* and is probably caused by reactions among hydroxyl groups.

C.10. A number of compounds with a vinyl group attached to a nitrogen atom can be polymerized. Examples of the resulting polymers are *poly(vinyl pyrrolidone)* **12a** and *poly(vinyl carbazole)* **12b**. The former is easily soluble in water and fully biocompatible. When cross-linked, it forms a gel useful for manufacturing contact lenses with a high water content. Poly(vinyl carbazole) is used in xerography.

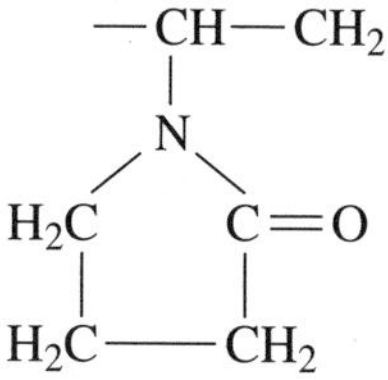

$$—CH—CH_2—$$

12a 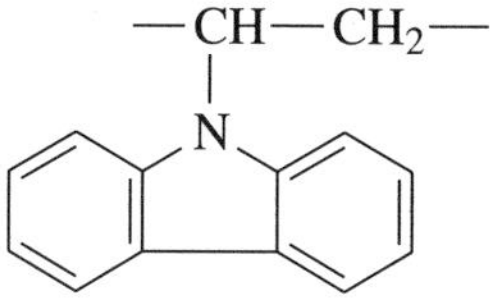   **12b**

1.2.1.2. Polymers with Heteroatoms in the Backbone.

In this section we will describe polymers that, in addition to carbon atoms, have some other atoms (mainly oxygen, nitrogen, sulfur) in their backbones. These heteroatoms are actually a part of some functional groups well known from organic chemistry: ethers, esters, anhydrides, carbonates, amides, urethanes, ureas, sulfides, sulfoxides, sulfones, etc. We will arrange our descriptions according to the above classification. The polymers in this group are usually prepared either by polycondensation techniques (Section 2.1) or by chain polymerization of appropriate heterocyclic monomers.

A. Polyethers

A.1. Polyoxymethylene or polyformaldehyde, —CH$_2$—O—, is the simplest polymer in this group. It is easily prepared by cationic polymerization of either formaldehyde or its cyclic trimer trioxane. The polymer is quite unstable unless the hydroxyl groups on the ends of the chain are stabilized by conversion to acetates or to methyl ethers. The stabilized polymer is a highly crystalline material, very strong, with good thermal and dimensional stability. It replaces metals in light- to medium-duty applications, especially where weight is a consideration.

A.2. Poly(ethylene glycol) or poly(ethylene oxide), —CH$_2$—CH$_2$—O—, is easily prepared by either anionic or cationic polymerization of ethylene oxide. The polymer is water-soluble. It is used as a thickening agent; it also finds several uses in biochemical research.

A.3. Epichlorohydrin is polymerized by some proprietary processes to *polyepichlorohydrin* **13**, a water- and oil-resistant elastomer, which is used in specialty rubbers.

$$—CH—CH_2—O—$$
$$|$$
$$CH_2Cl$$

13

A.4. Polymers prepared from other oxygen heterocycles, such as tetrahydrofuran or propylene epoxide, have only limited significance.

B. Polyesters   There are two groups of polyesters. In one group, the repeating unit is a hydroxy acid (or lacton), —O—(CH$_2$)$_a$—CO—, for example, *polycaprolacton* ($a = 5$); in the other group, the repeating unit is a combination of a diacid and a diol, —O—(CH$_2$)$_a$—O—CO—(CH$_2$)$_b$—CO—, for example, *poly(ethylene adipate)* ($a = 2$; $b = 4$). Polyesters are known for most combinations of diols and diacids. They are prepared by condensation of the appropriate diol either with the diacid under acid catalysis or with its dimethyl ester.

Most polyesters that are based on aliphatic components have similar properties. At ambient temperatures they are usually sticky substances with little direct significance. However, they are often used as part of more complicated materials to which they lend appreciable toughness. Polyesters with maleic (or fumaric) acid have carbon-carbon double bonds, which may be utilized for further reactions, usually for the cross-linking of the polymer. If a part of the diol is replaced by a triol, usually glycerol, the resulting product is more or less extensively branched. We will return to all these materials in Sections 1.2.3 and 1.2.7.

The most important polyester is *poly(ethylene terephthalate)* **14**. Strictly speaking, it belongs to Section 1.2.2—it is a cyclolinear polymer. Its high crystallinity and high melting point are responsible for its toughness and its excellent fiber-forming properties. A vast majority of the so-called polyester textiles are based on this polymer.

$$-O-CH_2-CH_2-O-CO-\!\!\left\langle\!\!\bigcirc\!\!\right\rangle\!\!-CO-$$

14

C. Polyanhydrides   Polyanhydrides, with a typical formula —CO—(CH$_2$)$_a$—CO—O—, can be prepared by reaction of acetanhydride with the appropriate diacid. They are tough polymers that form strong fibers. However, they are hydrolyzed so easily that they are virtually worthless.

D. Polycarbonates   Polycarbonates with a repeating unit —O—R—O—CO—, where R is any divalent group, are usually prepared by polycondensation of diols with phosgene or by a reesterification reaction between diols and organic carbonates. Their properties are similar to those of polyesters. Again, the more important members of this group are based on aromatic diols and are actually cyclolinear polymers. The polymer prepared from the so-called bisphenol A **15** (condensation product of phenol and acetone) is a clear, tough polymer used for organic lenses, safety glass, beverage containers, and similar applications.

$$HO-\!\!\left\langle\!\!\bigcirc\!\!\right\rangle\!\!-\overset{\overset{\displaystyle CH_3}{|}}{\underset{\underset{\displaystyle CH_3}{|}}{C}}-\!\!\left\langle\!\!\bigcirc\!\!\right\rangle\!\!-OH$$

15

E. Polyamides   Similarly to polyester classification, only those polymers that have the amidic function as part of the backbone are classified as polyamides, not those with amidic function in the side chain. There are two classes of polyamides: one is based on ω-amino acids (or lactams), —NH—$(CH_2)_a$—CO—, for example, *polycaprolactam* ($a = 5$); the other is formed by a polycondensation of a diamine and a diacid, —NH—$(CH_2)_a$—NH—CO—$(CH_2)_b$—CO—, for example, *poly(hexamethylene adipamide)* ($a = 6; b = 4$). Polymers of lactams are prepared either by polycondensation or by an alkali-catalyzed polymerization. The latter polyamides may have quite high molecular weights.

Aliphatic polyamides have another nomenclature. They are called *nylons,* with one or two numbers indicating the number of carbon atoms in the monomeric unit. One number designates polylactams (nylon 6 is a polycaprolactam). Two numbers belong to the other class of polyamides [nylon 6, 10 is poly(hexamethylene sebacamide)].

Because of the presence of a hydrogen atom on the amidic nitrogen, polyamides form strong intrachain and interchain hydrogen bonds between two amidic groups. Consequently, the polymers are usually highly crystalline and very tough. Nylon 66 was the first fully synthetic textile fiber, and it still successfully defends its position in the fiber and textile industries.

When an aromatic diacid and/or aromatic diamine are used in polymerization, the product (an aramid polymer) is especially tough and highly temperature-resistant, as are most members of the cyclolinear family. Industrially important polymers include that of terephthalic acid and p-phenylenediamine (trade name Kevlar) and the polymer of isophthalic acid and m-phenylenediamine (Nomex).

It should be mentioned that proteins and peptides are actually polymers of α-amino acids. Along with synthetic polypeptides, we will discuss them in Section 1.2.8.2.

F. Polyurethanes   Polyurethanes have a repeating unit, —NH—R_1—NH—CO—O—R_2—O—CO—, where R_1 and R_2 are divalent groups. They can be prepared by reaction of bis(chloroformates) with diamines. However, the reaction of diols with diisocyanates is used almost exclusively. Polyurethanes are very versatile polymers, and a large number of different monomer combinations are used to obtain materials with desired properties.

Polyurethanes in which R_1 and R_2 are aliphatic groups are useful as elastic fibers. When one or both of these groups are aromatic, the polymer gains rigidity; such polymers are used in lacquers and varnishes as well as for structural elements. Frequently, a short polymer with two or more hydroxyl groups is used instead of the diol. The properties of the product obviously depend very strongly on the nature of this polymeric diol. Valuable elastomers result when "hard" and "soft" segments are combined within the same chain. Polyurethanes easily form foams that are used extensively as packaging and insulating materials. One of their major uses is in upholstery, as padding for mattresses, pillows, and so on. Linear segmented polyurethanes are the choice materials for construction of cardiac valve materials.

G. Polyureas   Polyureas, —NH—R_1—NH—CO—NH—R_2—NH—CO—, exhibit properties similar to those of the corresponding polyamides. They are slightly more hydrophilic. Polyureas are prepared by reacting diamines with diisocyanates.

H. Polyisocyanates Isocyanates can be polymerized by an anionic mechanism to yield *polyisocyanates* **16**. These polymers have an unusually stiff backbone, which leads to a number of interesting properties (e.g., in solution they form helices). However, they are normally unstable and consequently are of no commercial significance.

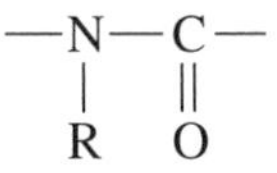

$$-N-C-$$
$$\begin{array}{cc} | & \| \\ R & O \end{array}$$

16

I. Polymers with Sulfur in the Backbone

I.1. Poly(alkylene polysulfides), $-(CH_2)_a S_b-$, are most conveniently prepared by polycondensation of dichloroalkanes with sodium polysulfide. In solutions of sodium polysulfide, molecules with different numbers of sulfur atoms are present at the same time. This feature is retained by the polymer; the subsequent polysulfide groups in the chain have randomly distributed numbers of sulfur atoms. The polysulfides were first prepared in the 1930s. They serve as rubbers with good flexibility and chemical and solvent stability.

I.2. Polysulfones are tough, thermally stable materials that are good electrical insulators. They also serve as precision molding materials. Polysulfones are usually prepared by reactions that do not involve the sulfone group. One of the more significant commercial polysulfones has structure **17**. Its rigidity and stability are caused not only by the sulfone groups but also by its cyclolinear structure.

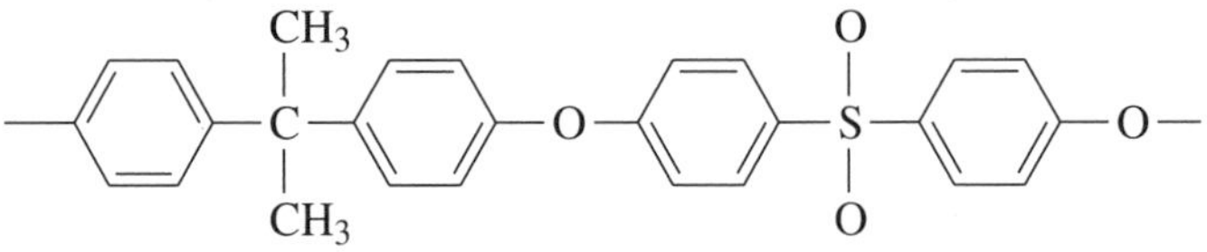

17

Polysulfone membranes are useful in gas separation studies.

1.2.1.3. Polymers with Inorganic Backbones. Many elements from the second and higher rows of the periodic table are capable of making strong covalent bonds either among themselves or with oxygen and nitrogen. Actually, materials with a very large number of atoms bound together by such bonds form a major part of inorganic chemistry. We will name only a few: silicates, aluminosilicates, borosilicates, some phosphates, titanates, molybdates, and nitrides. Although such materials satisfy our definition of macromolecules (they would fit into Section 1.2.7.2), we will follow the usual practice of not including them in the macromolecular science.

However, there exists another group of materials that are linear and possess an inorganic backbone. The side groups (if any) are either inorganic or organic. This section is devoted to such materials.

A. Polymeric Sulfur   —S—S—S—S—, also called *plastic sulfur,* is easily prepared by heating crystalline sulfur (consisting of S_8 rings) to about 170°C and then cooling it quickly. It exhibits typical polymer behavior with good elastic properties above its glass transition. However, it slowly reverts to the cyclic octamer and loses its polymer properties. Nevertheless, short sulfur chains confer good properties to polymers, as we have seen in the case of poly(alkylene polysulfides) and as we will see in Section 2.4.4, dealing with vulcanization.

B. Polysiloxanes   **18,** which have alternating oxygen and silicon atoms as a back-bone, are presently the most important polymers with inorganic backbones. The sub-stituents R_1 and R_2 may be alkyl groups, cyanoalkyls, perfluoroalkyls, or phenyls. However, the most useful polysiloxane is *poly(dimethyl siloxane)*, in which both substituents are methyls; it has the most flexible chains among all polymers. Accord-ingly, the materials with high molecular weight, when cross-linked, form excellent rubbers that perform well between about −30°C and 200°C. They have good sta-bility against oxidation, but at about 250°C they undergo decomposition to a cyclic tetramer.

$$
\begin{array}{cc}
\overset{\displaystyle R_1}{\underset{\displaystyle R_2}{\overset{|}{\underset{|}{-\text{Si}-\text{O}-}}}} & \overset{\displaystyle R_1}{\underset{\displaystyle R_2}{\overset{|}{\underset{|}{-\text{Si}-}}}} \\
\textbf{18} & \textbf{19}
\end{array}
$$

Polymers with lower molecular weight serve as very inert oils; their viscosity de-pends on their molecular weight. A unique property of poly(dimethyl siloxane) is its unusually high permeability for oxygen. Actually, the transport of oxygen is faster through this polymer than through water. The introduction of phenyl or other more polar substituents into polysiloxanes increases their toughness and thermal stability.

Just like polyurethanes, polysiloxanes are choice materials in biomedical applica-tions. Siloxane rubber is widely used in making artificial membrane lung because of its high permeability to respiratory gases and its blood compatibility.

Polysiloxanes are also known as *silicones.* This incorrect name was originally coined because their empirical formulas correspond to those of ketones.

C. Polysilanes   **19** represent a relatively recent entry into the world of polymers. Their silicon-silicon bonds are light-sensitive, a property that makes them useful as a component of photoresists. Polysilanes are currently used as electroactive components in light-emitting diodes.

D. Polyphosphazenes   Phosphorus pentachloride reacts with ammonium chloride to form a cyclic trimer **20a,** a rather stable inorganic heterocycle with pentacovalent phosphorus. This substance polymerizes with heating to *poly(dichlorophosphazene)* **20b,** a very stable polymer with typical rubberlike properties.

$$
\begin{array}{c}
\text{PCl}_2 \\
\nearrow \quad \diagdown \\
N \qquad N \\
| \qquad \parallel \\
Cl_2P\diagdown \quad \diagup PCl_2 \\
N
\end{array}
\qquad \longrightarrow \qquad
-N\!\!=\!\!\underset{\underset{Cl}{|}}{\overset{\overset{Cl}{|}}{P}}-
$$

20a **20b**

Unfortunately, the phosphorus-chlorine bonds easily undergo hydrolysis, which destroys the polymer. However, it is possible to subject the polymer to (otherwise standard) substitution reactions and to replace the chlorine atoms by alkoxy, fluoro-alkoxy, alkylamino, and other groups. The resulting polymers have many useful properties, which, of course, depend on the substituents. Generally, they show good thermal stability and chemical resistance. The phosphazene fluoroelastomers are currently used in military, commercial, and biomedical applications. The elastomers prepared from polyphosphazene show excellent potential for applications such as flexible foams, coatings, and wire covering. The large number of different pendant groups with widely varying chemical functionality that can be attached to the P-N backbone demonstrate the unusual molecular design potential for this class of polymers. Materials with outstanding flame resistance can be obtained by grafting polymers onto polyphosphazene structures. The ease with which their structure can be modified to suit any particular purpose promises them a bright future.

E. Organo-Inorganic Polymers   The ongoing search for newer and better polymeric materials led to the synthesis of numerous polymers with many metallic and nonmetallic elements, exhibiting covalent bonds, coordination, and chelating bonds, etc. So far, none of these materials has reached commercial significance. We consider their description to be beyond the scope of this introductory textbook.

1.2.2. Cyclolinear Polymers

In the previous section we were mainly concerned with polymers with a backbone formed by a single chain of covalently bonded atoms. Many important polymers have cyclic structures incorporated into the backbone. We will consider several cases:

A. The rings are connected by a single-stranded chain.
B. The rings are connected directly to each other.
C. The rings are fused together, forming a double-stranded "ladder" structure.

The introduction of rings into the backbone has multiple effects. First, the backbone is less flexible, and consequently the polymer is more rigid. Second, many rings, especially the aromatic ones, are thermally very stable and confer this property to the polymers. The extent of rigidity is controlled by the proportion of double-stranded segments (i.e., rings) to single-stranded ones. A fully double-stranded structure often leads to intractable polymeric materials.

A. Polymers with Alternating Ring-Chain Backbones   We have already met several polymers belonging to this group: poly(ethylene terephthalate), polyamides derived from aromatic diacids, polyurethanes derived from phenylene diisocyanates and similar diisocyanates, and polysulfones. All these polymers are tougher and more thermally stable than their single-chain analogs. It should be mentioned that polymers in which the benzene rings are replaced by cyclohexane rings display even more rigidity (steric effects reduce the flexibility of the chain even further); however, the thermal resistance is reduced. Several more polymers of this group deserve mention.

A.1. Poly(p-xylylene) **21** is made when *p*-xylylene **22** condenses on a cool surface; the process is used for coating surfaces to make them inert. The highly reactive monomer exists only in a vacuum at high temperatures; it is prepared by oxidative pyrolysis of *p*-xylene.

$$H_2C = \underset{\textbf{22}}{\underset{\big|}{\bigcirc}} = CH_2 \quad \longrightarrow \quad -CH_2 - \underset{\textbf{21}}{\underset{\big|}{\bigcirc}} - CH_2 -$$

A.2. Poly(m-phenylene oxide) **23** is prepared by self-condensation of potassium *m*-bromophenolate.

A.3. The polymer that is routinely called *poly(phenylene oxide)* is actually *poly(dimethylphenylene oxide)* **24.** It is prepared by copper-catalyzed oxidation of 2,6-dimethylphenol. As we can expect from its structure, it is a thermally stable hard polymer useful for machined parts. However, excessively high temperatures are required for its processing. It is therefore blended with polystyrene, which reduces its processing temperature and its price.

$$-O-\underset{\textbf{23}}{\bigcirc} \qquad \overset{CH_3}{\underset{\underset{\textbf{24}}{CH_3}}{-\bigcirc-O-}}$$

A.4. Cellulose and *amylose* are also cyclolinear polymers belonging to this group. However, we will treat them together with other natural macromolecules in Section 1.2.8.1.

Some linear polymers may be subject to cyclization reactions; two neighboring side groups may react together and/or with some reagent to form a cycle. Generally, it is possible that an unreacted side group is surrounded by two reacted pairs—the reaction cannot go to completion, and the linear portion of the chain is left between two cyclic portions. Such materials are included in this section. Let us consider a few examples.

A.5. Poly(vinyl alcohol) can be reacted with aldehydes to form *poly(vinyl acetals)* **25.** Useful acetals are *poly(vinyl formal)* (R = H) and *poly(vinyl butyral)* (R = C_3H_7).

The latter polymer is used in manufacturing safety glass (two sheets of glass bound by a thin film of polymer).

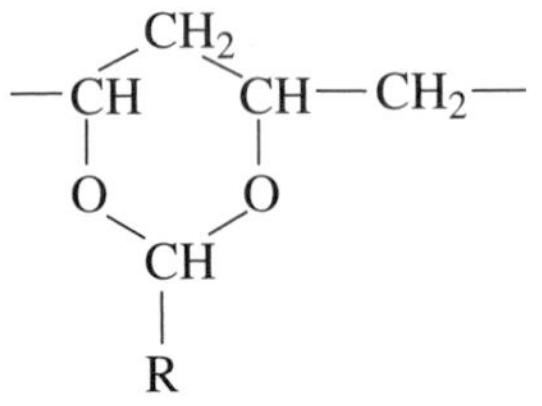

$$
\begin{array}{c}
\quad\quad CH_2 \\
-CH \quad\quad CH-CH_2- \\
| \quad\quad\quad\quad | \\
O \quad\quad O \\
\backslash\quad/ \\
CH \\
| \\
R
\end{array}
$$

25

A.6. Natural rubber and other polydienes under Lewis acid catalysis react further through their remaining double bonds and yield highly cyclic structures. The exact mechanism and structure of the product are not known. Of course, this procedure converts a highly elastic material into a hard plastic.

A.7. Poly(methyl vinyl ketone) **26** undergoes aldol condensation to yield a cyclic structure **27.**

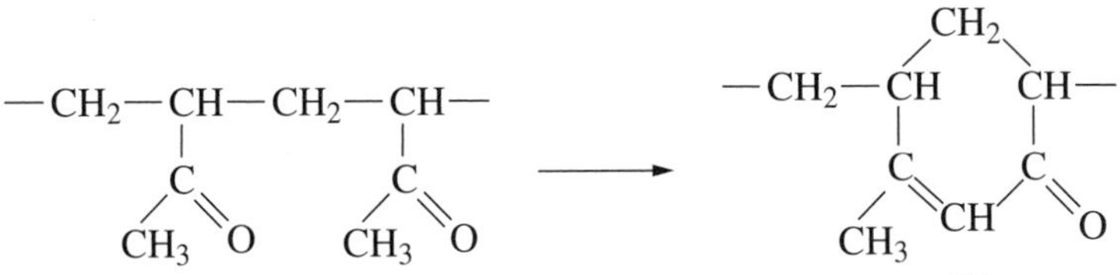

$$
\begin{array}{c}
-CH_2-CH-CH_2-CH- \\
| \quad\quad\quad\quad | \\
C \quad\quad\quad C \\
/ \backslash\backslash \quad / \backslash\backslash \\
CH_3 \ O \quad CH_3 \ O
\end{array}
\quad \longrightarrow \quad
\begin{array}{c}
\quad\quad CH_2 \\
-CH_2-CH \quad\quad CH- \\
| \quad\quad\quad\quad | \\
C \quad\quad\quad C \\
/ \backslash\backslash CH / \backslash\backslash \\
CH_3 \quad\quad O
\end{array}
$$

26   **27**

B. Polymers with Connected Rings

B.1. Poly(p-phenylene) **28** is the simplest representative of this group. This insoluble polymer can be prepared from benzene and Lewis acids in the presence of oxidation agents.

$$ -\langle\!\!\!\bigcirc\!\!\!\rangle - $$

28

An important group of thermally resistant polymers is prepared by polycondensations when the rings (usually heterocycles) are formed during the reaction. We will consider all such polymers in this section even if (because of the structure of the nonreactive part of the monomer) they could belong to the previous or following section.

B.2. Polyimides are prepared by reacting aromatic dianhydrides with diamines (usually also aromatic). For example, reaction of pyromellitic dianhydride with *p*-phenylenediamine yields a very rigid, high-melting, and thermally stable polymer with repeating unit **29.** When *p*-phenylenediamine is replaced by some long diamine with a more flexible structure, the polymer exhibits a little more flexibility.

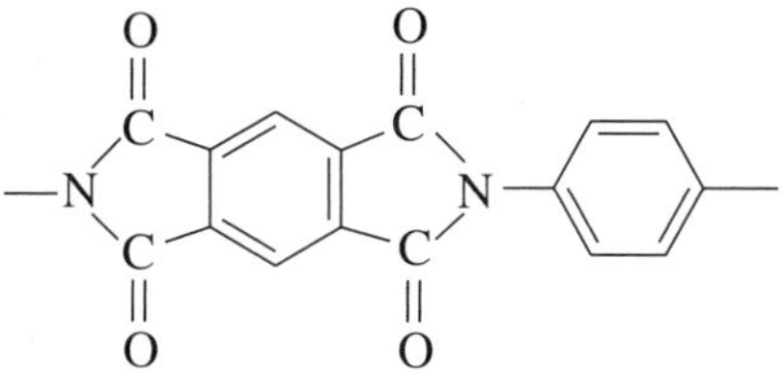

29

This is one of the earliest thermally stable commercial polymers, originally synthesized by duPont and marketed under the trade names H-film and, later, Kapton as a yellow film. Polyimides are available in precured films and fibers, curable enamels, adhesives, and resins for composite materials. The resistance of polyimides to acidic hydrolysis and bacterial attack makes them good candidates as reverse osmosis membranes.

B.3. Diacids or their diphenyl esters can be reacted with tetramines (usually diaminobenzidine). The reaction leads to the formation of benzimidazole rings; the resulting *polybenzimidazole* **30** is again a rigid, thermally stable polymer. The nature of R in formula **30** depends on the diacid; both aliphatic and aromatic diacids have been used.

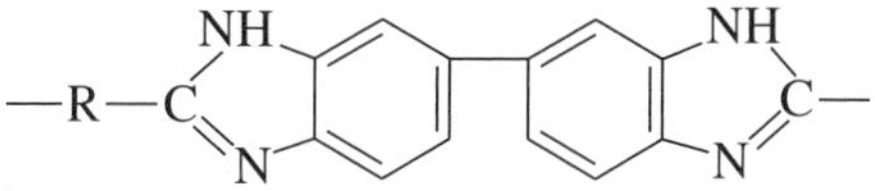

30

Polybenzimidazoles have demonstrated a good adhesion as films when cast from solution onto glass plates. This quality led to their use in glass composites, laminates, filament-wound structures, and space suits.

B.4. Many synthetic reactions leading to aromatic heterocycles can be adapted to yield high-molecular-weight polymers, such as *polyoxadiazoles, polytriazols and polythiazols.* However, these interesting polymers have not yet attained any commercial significance. A more detailed discussion of them can be found in specialized books.

C. Polymers with Fused Rings (***Ladder Polymers***)   Macromolecules in this group can be viewed either as very extended polycyclic structures (such a view is preferable for fused aromatic rings) or as two linear macromolecular chains regularly bridged (preferable for loose structures). The best-known examples of the latter structure are nucleic acids; the bridges are actually formed by multiple hydrogen bonds. Nucleic acids in the form of double helix are not "true macromolecules." In any case, they will be described together with other natural macromolecules in Section 1.2.8.

C.1. When trichlorophenylsilane is hydrolyzed under carefully controlled conditions, it yields *poly(phenyl sesquisiloxane)* **31** with eight-member rings, which are typical of cyclic siloxanes. This is a well-behaved polymer that is soluble in most organic solvents. However, because of its more rigid ladder structure it does not display the elastic behavior that is characteristic of other siloxanes.

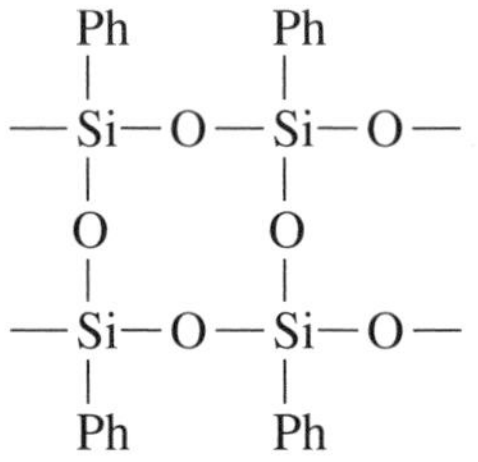

31

C.2. We have already seen that polycondensation reactions can lead to the formation of heterocyclic repeating units in a polymer backbone. It is possible to choose the reaction components in such a way as to produce a full ladder structure. We will give only one example: Condensation of pyromellitic dianhydride with tetraaminobenzene yields polyimidazopyrrolone **32.**

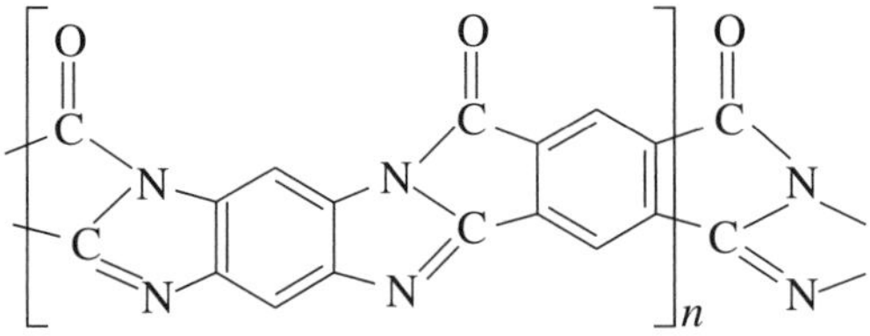

32

C.3. As we have mentioned earlier, some linear polymers [polydienes, poly(methyl vinyl ketone), for example] can be converted into cyclolinear structures. In the case of polyacrylonitrile the procedure can be quantitative and lead to a full ladder structure. The pendant nitrile groups can be induced to polymerize by higher temperatures. Still higher temperatures and the presence of oxygen (air) will then lead to oxidative aromatization and yield a polymer with structure **33.** This polymer is a very tough material; it serves as an intermediate for the synthesis of carbon fibers (by pyrolysis at extremely high temperatures).

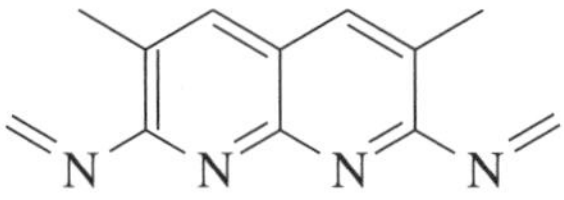

33

1.2.3. Branched Polymers

Many polymers are not strictly linear; the backbone may be branched. In most cases the branching is caused by imperfections in the polymerization mechanism,

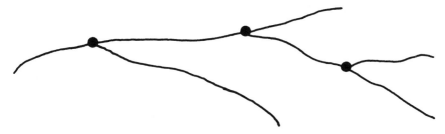

Figure 1.1. A randomly branched polymer molecule.

but sometimes it is introduced intentionally. Topologically, the branched polymers form several groups.

Most common are randomly branched polymers (Fig. 1.1). For these polymers, the branching points are randomly distributed along the backbone. In fact, the branches may be branched further; for a heavily branched molecule it may not even be possible to specify the main backbone. Sometimes branching produces cycles as depicted in Fig. 1.2. Branched polymers are produced during chain polymerization when a deactivated chain ("dead" polymer) is reactivated in the middle of the backbone either by chain transfer (Section 2.2.4.3) or by an attack by some reagent (e.g., vulcanization, Section 2.4.4). If the newly introduced growing chains terminate by recombination (Section 2.2.4.1), the cyclic structures of Figure 1.2 may be produced. Otherwise, such a polymer has the structure of Figure 1.1.

Another source of branches is the addition of small amounts of a monomeric component that has more functional groups than are needed for producing a linear polymer. For example, branching of polyesters may be achieved by the addition of glycerol. Similarly, the addition of a divinyl compound during chain polymerization of vinyl monomers may cause branching. Such an addition may be intentional or unintentional; the additives may actually be impurities. For example, it is extremely difficult to prepare pure hydroxyethyl methacrylate; it is almost always contaminated by the diester, ethylene glycol dimethacrylate, that acts as a branching agent.

Figure 1.2. A branched polymer molecule with internal loops.

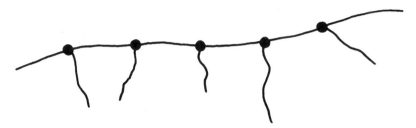

Figure 1.3. A comblike polymer

There is a subtle but important consideration for branched macromolecules: What are the relative probabilities that a representative branch of a polymer molecule is either further branched or terminated? If the probability of further branching prevails, at least part of the polymeric materials will form an infinite network (on the molecular scale). Such materials exhibit properties quite different from those of either branched or linear polymers; we will consider them in Section 1.2.7. On the other hand, when the probability of chain termination prevails, the branched molecules will be of finite size and will behave similarly to their parent linear polymers.

During polymerization of low-density polyethylene the growing radical frequently transfers about four carbons back along the chain (backbiting, Section 2.2.4.3). The result is a polymer with many short side chains. Similar polymers (especially when the number of branches is large) are called *comblike polymers* (Fig. 1.3).

Comblike structures are also produced either by attaching a large number of short polymeric chains to the original backbone or by a multiple reactivation of the backbone with the subsequent growth of many polymer side chains. However, these strategies are usually used for attachment of a different polymer. In such a case, the result is a graft copolymer belonging to Section 1.2.4.1. The comblike structure is manifested also by some linear polymers with very long side chains, for example, *poly (hexadecyl methacrylate)*. Here, as in other comblike polymers, the side groups are extremely crowded, forcing the backbone into a quite stretched conformation.

When studying conformational properties of polymer molecules, physical chemists must have available not only linear molecules but also molecules with well-controlled modes of branching. Special types of such molecules are star macromolecules with three or more branches (Fig. 1.4). Such polymers can be prepared either using a special initiator (usually anionic) with multiple initiation sites or, alternatively, by an attack of a living polymer (Section 2.3.1.4) on some multifunctional substrate.

Figure 1.4. Star macromolecules

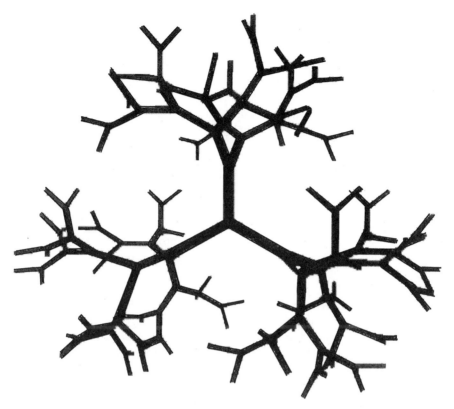

Figure 1.5. Structure of a dendrimer

Currently, a new class of branched polymers called *dendrimers* is gaining significance. (The name was coined from *dendron,* Greek for "tree") (Fig. 1.5) These highly
branched polymers, also called *dendritic* or *hyperbranched,* have a unique structure
that may result in unusual properties and lead to novel applications. These molecules
with almost spherical shape have a starlike center. Each arm of the star is branched at
a very small distance from the center. Thus an onionlike layer is formed at the surface
that has multiple terminal groups. These groups are then utilized as new branching
points for building the next layer. This procedure is repeated several times, eventually
producing a spherical molecule with high concentration of surface terminal groups.
Dendrimers have very different properties compared with their linear analogs. For
instance, they are extremely soluble in various organic solvents and exhibit low intrinsic viscosity. We will treat the syntheses of dendrimers in some detail in Section
2.1.1.6.

1.2.4. Copolymers

Industry keeps demanding materials fitting ever more closely to the myriad of intended
uses. However, the number of polymerizable structures is limited, and the number of

useful monomers is increasing only slowly—it cannot satisfy the demand. Quite often, the requested properties are somewhere in between those of two known polymers. Unfortunately, two polymers are incompatible more often than not; thus, blends of polymers are only rarely an answer. One way to force two monomers to stay mixed is to force them to be parts of the same macromolecule. Such molecules are called *copolymers*. Although the original idea was to combine the useful properties of both homopolymers, sometimes new materials have been developed with very valuable properties that are not just a plain combination of properties of the corresponding homopolymers.

The number of available materials increased tremendously with the introduction of copolymers. If the number of existing homopolymers is n, then the number of possible binary copolymers is $n(n-1)/2$. We leave it to the reader to go through the counting. On top of that, the same binary copolymer can be prepared with different ratios of the two components, and they can be combined in different ways. Whereas for a great many monomer pairs the techniques for production of the copolymer are either unknown or impractical, the number of possible combinations is still quite respectable.

1.2.4.1. Types of Copolymers. To describe a molecule of a linear homopolymer it is sufficient to specify the number of monomeric units forming the chain. For a copolymer, it is also necessary to specify the ratio of the numbers of the two monomeric units (composition of the copolymer) and the rule of sequencing the monomeric units **A** and **B**. This rule is a result of the way the copolymer was synthesized. Let us first analyze a scheme in which monomeric units are attached to the chain one at a time (this scheme corresponds to chain polymerization, Sections 2.2 and 2.3). The probability that the unit being added is a unit of **A** depends on (1) the fraction of **A** units in the monomer mixture and (2) the nature of the end group of the chain to which it is attached.

In the simplest case of *random copolymer*, the nature of the end group plays no role. In this case, the sequence corresponds to a sequence of outcomes of random events. If the ratio of monomers is 1:1, the sequence is the same as the one obtained in series of tosses of a coin. Obviously, sequences of a few heads or tails (read **A** or **B** units) are quite frequent; however, their frequency is smaller and smaller for longer sequences. If the ratio of monomeric units is different from 1:1, the same similarity holds; however, we should now consider a false coin, which when tossed gives more heads than tails. Copolymers that follow the above rules are called *random* or *statistical copolymers*.

However, the end group usually influences the probability. In many cases (when the double bonds of the two monomers have opposite polarity), attachment to the other monomer is preferred. The sequences of the same units become shorter, and the copolymer exhibits an alternating tendency. In the limiting case (and when **A:B** = 1:1), the copolymer is an *alternating copolymer*.

The opposite case, when attachment to the same unit is preferred, is only hypothetical. In such a situation, the sequences of the same unit would grow longer, and eventually the two monomers would be fully separated. If there remained one or a few **A–B** sequences, a *block copolymer* would result. In practice, special procedures must be designed for their synthesis. Typically, living polymers (see Section 2.3.1.4) are used for this purpose. Briefly, in anionic polymerization, it is possible to prepare a solution of growing chains with an active end. Once the monomer is exhausted, the chains can no longer grow. When another monomer is added, they resume growth, forming a block copolymer. This procedure can be repeated, yielding copolymers with several blocks. Copolymers with two blocks are often called *AB block copolymers;* those with three blocks are *ABA copolymers.* Some of the latter are becoming important in modern rubbers.

Another stratagem for synthesis of block copolymers uses the principle of condensation or polycondensation. However, in this case, longer or shorter polymer chains are utilized instead of monomers. These chains should have very active end groups. For example, one polymer may be a living polymer (organometallic end) while the other is halogenated. One or both ends may be active. The combination of two polymers with a single active end yields an **AB** block copolymer; a polymer with one active end combined with another having two active ends yields an **ABA** copolymer. Finally, when both polymers have both ends active, a *multiblock copolymer,* **ABABAB . . . ,** results.

Graft copolymers are materials in which chains (usually short ones) of one polymer are attached more or less randomly to another preformed homopolymer. Most graft copolymers have a comblike structure, in which the backbone is formed by one monomer and the side chains by the other. Usually, the second monomer is polymerized in the presence of the backbone polymer under conditions favorable for grafting. Most methods rely on activation of the backbone; the grafted chain grows on the activated site. When radical polymerization is employed, radicals have to be formed on the backbone. The strategies used are similar to those employed in the radical polymerization (Section 2.2): Radicals are formed either by irradiation or as a result of an attack by another radical (usually by the primary radical from the initiator). Alternatively, the backbone polymer may first be modified to carry labile groups (usually by means of some reaction leading to peroxides), which under polymerization conditions act as an initiator. An interesting variation of this method is preradiation of a solid polymer: The radicals are trapped in the rigid material and react after the material is swollen by the other monomer.

The activation principle also is used when preparing graft copolymers by anionic or coordination mechanism. A suitable group on the backbone is converted into the catalyst (e.g., into a metallo-organic compound), which starts growing after addition of a monomer.

Graft copolymers are sometimes prepared from two polymers. A living anionic polymer (an organometallic compound) may react with a susceptible group on the backbone of the other polymer (e.g., an ester group of polyacrylates). Grafting also can be achieved mechanically. Mastication of two polymers may break the chemical bonds, which then re-form, connecting the chains of the two homopolymers in a random way.

1.2.4.2. Some Important Copolymers.

The rubber industry has utilized the principle of copolymerization extensively. The majority of rubberlike materials are based on polydienes. We have seen that this class of polymers exhibits very high elasticity as well as other desirable properties. However, they are rather soft and have poor resistance to solvents. Greater toughness is demanded, especially by the tire industry, which is the largest user of rubbers. The combination of choice proved to be the copolymer of styrene and butadiene, with styrene providing the necessary toughness. When resistance to hydrocarbon solvents is required, butadiene is frequently copolymerized with acrylonitrile. Finally, the terpolymer of butadiene, styrene, and acrylonitrile exhibits both toughness and inertness. The triblock copolymer styrene-butadiene-styrene forms in the solid state a domain structure in which the polystyrene blocks segregate and form cross-links, which can be reversibly dissolved by increasing the temperature. It is hoped that similar materials will lead to a rubber that can be recycled.

The use of copolymers is indicated whenever the crystallinity of a polymer must be modified. We will learn that a prerequisite for crystallinity is a very regular structure of the polymer chain. This regularity is very effectively disrupted by the presence of a comonomer—even in relatively small amounts. For example, most polyamides are highly crystalline. However, polyamides synthesized using a mixture of two diacids are virtually amorphous. The same idea is applied in ethylene-propylene copolymers. As another example, the crystallinity of some textile fibers is the basis of their fiber-forming properties, yet it interferes with their dyeability. The problem is solved by the addition of appropriately chosen amounts of various comonomers.

Some polymers have a tendency to depolymerize at high temperatures. For example, polydimethylsiloxane decomposes to its cyclic tetramer. This process can be suppressed if the siloxane chain is interrupted by other monomers. In carborane-siloxane copolymers, the carborane cages are connected by polydimethylsiloxane segments. The result is an interesting combination of rigid and thermally stable carborane with flexible (and now also thermally stable) siloxane. This copolymer is used as an extremely thermally stable inert stationary phase for gas chromatography.

In another stratagem, the hydrophilicity of water-soluble polymers such as polyacrylamide is modified by partial hydrolysis, which leads to polyelectrolytic behavior. The resulting material is actually *poly(acrylamide-acrylic acid)* copolymer.

At least two more copolymers deserve mention. *Poly(ethylene-vinyl acetate)* copolymers are used for insulation and for the manufacture of syringes, toys, etc. *Poly(vinyl chloride-vinyl acetate)* copolymer is the preferred material for phonograph records.

1.2.5. Isomerism of Polymer Chains and Tacticity

Molecules that have the same summary chemical formula may still differ in the spatial arrangement of their constituent atoms. When a molecule changes from one spatial arrangement to another during ordinary thermal movement, both arrangements are considered to belong to the same molecular species; they are *conformations* of the same molecule. When, on the other hand, the different spatial arrangements cannot change from one to the other or when such a change requires a more drastic molecular event than a thermal movement (for example, at least momentary breaking of chemical bonds), then we are talking about *isomers*. Different isomers have different

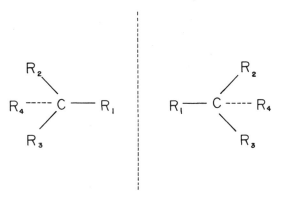

Figure 1.6. Mirror symmetry of two antipodes of molecules that have four different substituents on an asymmetric carbon.

configurations. In the rest of this book, we will always clearly distinguish between molecular conformations and configurations.

Trivial examples of isomeric polymers are poly(methyl acrylate), poly(vinyl acetate), and polybutyrolactone. However, in this section we will be concerned with a more subtle type of isomerism, which is displayed by *stereoisomers.* In two stereoisomers, corresponding atoms are bound to the same neighbors, yet the two molecules cannot be interconverted by thermal movement. In organic molecules, this phenomenon is usually related to the tetrahedral arrangement of the four single bonds of carbon atoms. The simplest case is that of molecules that have four different substituents on one carbon atom *(asymmetric carbon)*—two different arrangements are possible that are mirror images of each other (Fig. 1.6). These arrangements correspond to two different chemical species; they are called *optical antipodes.* They behave identically in most respects, except that they rotate the plane of polarized light in opposite directions. A more thorough symmetry analysis shows that (1) two antipodes must be mirror images of each other and (2) none of their conformations may possess either a plane or center of symmetry.

The situation becomes more involved when the stereoisomeric molecules have more than one asymmetric carbon. In the middle of the nineteenth century, Pasteur studied tartaric acid, the first example of a substance with two asymmetric carbons. Tartaric acid exists in three isomers (Fig. 1.7a, b, c). Two of them (a, b) are optical antipodes; their mixture is called *racemic* tartaric acid. The third form has a plane of symmetry; it is called *meso*-tartaric acid, and it has different chemical properties from those of the racemic acid; it is not optically active. Later, the same nomenclature was adopted for the description of any pair of asymmetric carbons (whether they are direct or more distant neighbors). The form with a plane of symmetry for these two atoms is called the *meso* form; the other, without symmetry, is the *racemic* form.

Polymers with asymmetric carbons usually have a very large number of them. We need to understand their arrangement in the molecule. We will distinguish two

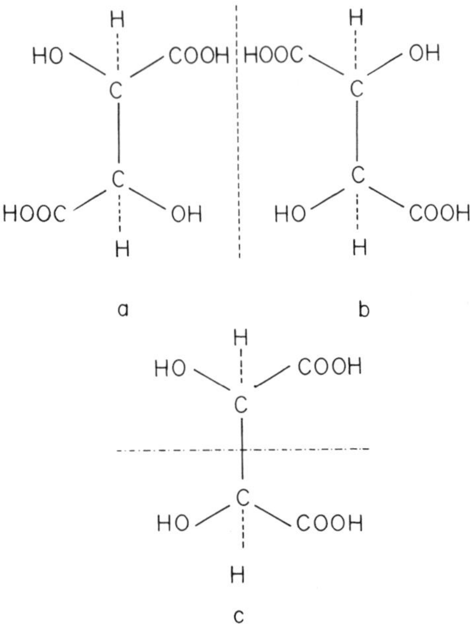

Figure 1.7. Three isomers of tartaric acid. (a, b) Optical antipodes; (c) optically inactive mesotartaric acid.

cases. First, the asymmetry and optical activity could have been present already in the monomer and could have survived the polymerization (e.g., optically active alcohol within the ester group of methacrylates, or polyamides prepared from optically active amino acids). In this case, the polymers are optically active whether the asymmetry resides in the backbone or in the side chain. (Optical activity itself, of course, depends on the neighborhood of the asymmetric center; the neighborhood could have changed during the polymerization.)

In the other case, the asymmetric center can be created by the polymerization process itself, for example, by polymerization of vinyl monomers according to reaction (1.2.2). During this reaction, a new asymmetric center is created on the carbon carrying the side group R:

$$CH_2{=}CHR \longrightarrow -CH_2-\underset{\underset{H}{|}}{\overset{\overset{R}{|}}{C}}- \tag{1.2.2}$$

These new centers do not lead to optical activity. The four substituents on the new asymmetric carbon are not really different, because the two portions of the backbone adjacent to this center are identical—at least in the vicinity of the center. However, the relative configuration on the neighboring centers is still important (the *pseudo-asymmetric centers* are now located at every other carbon). The two possible

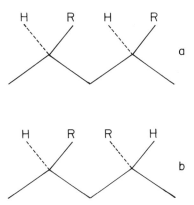

Figure 1.8. Two possible arrangements of two neighboring pseudoasymmetric centers on vinyl polymer chains. (a) An isotactic diad with a plane of symmetry; (b) a syndiotactic diad without any symmetry element.

configurations of two neighboring monomeric units (a *diad*) are depicted in Figure 1.8. Configuration a possesses a plane of symmetry; it is therefore called a *meso-* or, more often, an *isotactic diad*. Configuration b lacks the symmetry elements; it is called *racemic* or, more often, a *syndiotactic diad*. When all possible diads (note that each monomeric unit is a part of *two* diads) are isotactic, we have an isotactic polymer. Similarly, a syndiotactic polymer consists of syndiotactic diads only. Of course, quite often both kinds of diads are present in the same chain. When their ratio is 1:1 and they are distributed randomly, the polymer is *atactic*. The word *atactic* is frequently used also for polymers in which the ratio of the diads differs from 1:1 but the polymer is far from approaching the pure iso- or syndiotactic form. The ratio of the two types of diads can be measured by nuclear magnetic resonance (NMR).

Let us consider so-called *stereoblock* polymers—polymers in which one part of the chain is isotactic and the other is syndiotactic. Obviously, the structure is quite different from that of an atactic polymer, yet the ratio of diads is the same. In such cases it is useful to consider *triads* (Fig. 1.9). An *isotactic triad* (Fig. 1.9a) is composed of two isotactic diads. A *syndiotactic triad* (Fig. 1.9b) is composed of two syndiotactic diads. A *heterotactic triad* (Fig. 1.9c) is composed of one isotactic and one syndiotactic diad. The proportion of triads can be measured by NMR. Obviously, a stereoblock polymer has only one or a few heterotactic triads; an atactic polymer has the number of heterotactic triads established by random statistics (50%).

We will see in later chapters that the stereoregularity of polymers depends mainly on the technique of their preparation. Radical polymerization usually leads to atactic polymers (sometimes the syndiotactic diads prevail, but are far from totality); the stereoregular forms (isotactic, syndiotactic, stereoblock) are prepared by anionic and coordination polymerization techniques.

In solution, different stereo forms of the same polymer behave very similarly. There is, however, a big difference in their behavior in the solid phase. Both stereoregular forms may fit into crystalline lattices; these solid polymers exhibit all the valuable properties of crystalline polymers: fiber forming, great toughness, etc. On the other

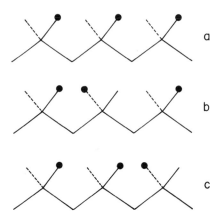

Figure 1.9. Three possible arrangements of three neighboring pseudoasymmetric centers on vinyl polymer chains. (a) An isotactic triad; (b) a syndiotactic triad; (c) a heterotactic triad.

hand, the irregular structure of an atactic polymer precludes any crystallinity, and atactic polymers are generally amorphous in the solid state. (Some stereoregular polymers do not crystallize either.)

1.2.6. Nomenclature of Polymers

In the previous sections we introduced the nomenclature of polymers in bits and pieces. Now it is time to present it in a more adequate form. The International Union of Pure and Applied Chemistry (IUPAC) developed rules for naming polymers as it did previously for organic and inorganic compounds. Two naming systems are recognized. One is *source-based;* it names polymers according to monomers from which they were prepared. This is the nomenclature we were using so far. IUPAC recommended and expected that this system would die out and be replaced by the *structure-based* nomenclature. This did not happen, and the source-based nomenclature is presently used almost exclusively.

The *structure-based* nomenclature gives a unique, unambiguous name to every possible linear polymer. However, it is rarely used because it often becomes very cumbersome. As an example, we present two systematically created names for two common polymers. Poly(methyl methacrylate) **4** can be called poly[1-(methoxycarbonyl)-1-methyl-ethylene]. Similarly, poly(vinyl butyral) **34** can carry the name poly[(2-propyl-1,3-dioxane-4,6-diyl)-methylene].

$$-\left(-CH_2-CH-CH_2-CH-\right)_n-$$

34

The rules of structure-based nomenclature are many, they are complicated, and their description requires a large space. We have decided not to include them in this text, and we refer the reader to the original literature. (W. V. Metanomski, Ed., *Compendium of Macromolecular Nomenclature,* Blackwell, London, 1991.)

We will now review the source-based naming system that is used in most literature even if it breaks down occasionally.

The name of the polymer is created when the monomer name is attached to the prefix "poly." When the monomer name consists of two or more words they are placed in parentheses. Thus we have polystyrene **2** and poly(vinyl pyrrolidone) **12a.**

Sometimes a polymer may be synthesized from several different monomers using different synthetic procedures. In such a case the simplest possible name is usually used without respect to its actual synthesis. For instance, the name poly(ethylene oxide) is derived from its monomer, ethylene oxide. However, the same name could be used even if the polymer were synthesized from ethylene glycol ($HOCH_2CH_2OH$), ethylene chlorohydrin ($ClCH_2CH_2OH$), or bis-chloromethyl ether ($ClCH_2OCH_2Cl$). Similarly, the name polycaprolactam is used for the polymers made from its monomer, caprolactam, or from its parent amino acid, ε-aminocaproic acid (H_2N—$(CH_2)_5$—COOH). Other alternative names will suggest themselves, but they are not used in practice. Thus, polyacrylonitrile **7** could be called poly(vinyl cyanide).

A few polymers have names based on the repeating unit without reference to any parent monomer. For example, silicones possess the repeating unit **18** and the most common silicone fluids are based on poly(dimethyl siloxane).

Tacticity of the polymers is described by putting the word isotactic/syndiotactic/atactic in front of the polymer name. When it is desirable to specify the end groups, it is done by using prefixes α and ω in front of the polymer name together with the name of the polymer. Thus, we may have α-benzoyl-ω-methoxy-polystyrene.

The nomenclature for copolymers includes the names of the monomers separated by an interfix. Different interfixes specify different types of copolymers. The interfix -*co*- may be used for any kind of copolymer. The names poly(methyl methacrylate-*co*-styrene) and poly(styrene-*co*-methyl methacrylate) both describe any unspecified copolymer of styrene and methyl methacrylate. Usually the first monomer name refers to the major component, if there is one. Other interfixes define specific types of copolymers. Thus the interfix -*alt*- specifies an alternating copolymer, for example poly(styrene-*alt*-maleic anhydride). Interfixes -*block*- or -*b*- describe block copolymers as poly(styrene-*b*-butadiene-*b*-styrene), a triblock polymer **35.** In this case the polymers are named from an end of the species. However, in the commercial market the monomer initials usually designate such polymers, and **35** is named SBS block copolymer.

35

Graft copolymers (Section 1.2.4.1) are named as poly(A-*g*-B) with the backbone polymer mentioned before the branch polymer. Examples are poly(vinyl alcohol-*g*-acrylamide) or starch-*g*-polystyrene. IUPAC has authorized other interfixes: -*stat*- designated for polymers obeying statistical laws; -*ran*- for polymers obeying the more restricted Bernoullian random statistics; and -*per*- for special chains in which three or more monomeric units are periodically repeated. However, these interfixes are used extremely rarely.

In instances where a repeating unit of a linear polymer contains other atoms in addition to carbon, the polymer is frequently named from the linking group between hydrocarbon portions. Thus the names polyester, polyamide, polyurethane, polyurea, polyisocyanate (**16**), and polysulfone (**17**) have been given (Section 1.2.1.2). Many of these polymers are prepared by reacting two monomers with the elimination of a smaller molecule and are called condensation polymers (Section 2.1.1.1). Such polymers (polyamides, polyesters, etc.) are not named as copolymers because the chemical structure of the joining linkage in each case shows that the parent monomers must alternate and the copolymer nomenclature would therefore be redundant.

Condensation polymers of this type are named by analogy with the lower-molecular-weight amides, esters, and so on. Thus the names of all esters end with the suffix -*ate* attached to the parent diacid molecule. Similarly to the nomenclature of low-molecular-weight esters (e.g., ethyl acetate), the glycol (alcohol) part of the ester is named only as the hydrocarbon residue. Therefore, structure **14** is named poly(ethylene terephthalate) and structure **36** is poly(tetramethylene terephthalate) or poly(butylene terephthalate).

$$-O-CH_2-CH_2-CH_2-CH_2-O-\underset{O}{\overset{\|}{C}}-\!\!\left\langle\!\!\bigcirc\!\!\right\rangle\!\!-\underset{O}{\overset{\|}{C}}-$$

36

Similar naming principles are used for polyamides. Thus the polymer, —NH—$(CH_2)_6$—NH—CO—$(CH_2)_4$—CO—, is poly(hexamethylene adipamide). However, this naming principle does not work for polyesters and polyamides for which the carboxylic function and the hydroxy or amino function are part of the same monomer molecule. Naming of such polymers reverts to pure source-based names such as poly(lactic acid) **37,** polycaprolactone **38,** and polycaprolactam **39.**

$$\begin{array}{cc} CH_3 & O \\ | & \| \\ \!\!\!\!-\!\!\left(\!-O-CH-C-\right)_{\!n} \end{array} \qquad \begin{array}{c} O \\ \| \\ \!\!\!\!-\!\!\left(\!-O-CH_2-CH_2-CH_2-CH_2-CH_2-C-\right)_{\!n} \end{array}$$

37   **38**

$$\begin{array}{c} O \\ \| \\ \!\!\!\!-\!\!\left(\!-NH-CH_2-CH_2-CH_2-CH_2-CH_2-C-\right)_{\!n} \end{array}$$

39

For polyamides, another widely accepted nomenclature was coined. They are called nylons and are named according to the number of carbons in the diamine and carboxylic acid reactants (monomers) used in their syntheses. For example, the nylon produced by the condensation of 1,6-hexamethylenediamine (6 carbon atoms) and sebacic acid (10 carbon atoms) is called nylon-6,10. Similarly, the polymer formed from 1,6-hexamethylenediamine and adipic acid (both having 6 carbon atoms) is called nylon-6,6 or nylon 66. The nylon from the single reactant caprolactam (6 carbon atoms) is called nylon-6.

The source-based nomenclature has other shortcomings: It is difficult to name polyurethanes or polysulfones **17**. For some polymers the accepted name conveys relatively little information about the repeating unit structure, for example, polycarbonate **15**, polysulfone **17**, etc.

Technical literature designates the more important polymers and copolymers by abbreviations. The list of recommended abbreviations is long, and we will not reproduce it. Examples are PVC for polyvinylchloride and ABS for acrylonitrile/butadiene/styrene copolymer.

From the above it is apparent that the actually used polymer nomenclature is a somewhat haphazard collection of various naming principles. However, after a little practice polymer nomenclature will be felt to be much simpler than the standard nomenclature of organic or inorganic chemistry.

1.2.7. Macromolecular Networks

We have already seen that branching of polymers can become so extensive that a single, highly branched macromolecule may be as large as the macroscopic sample itself. In such a case, we are talking about infinite macromolecular networks. A large number of very important materials belong in this category. It is useful to review the concept of *branching*. The branching point is a point from which the macromolecular backbone continues in at least three directions. The portion of the chain between two branching points (or between a branching point and an end of a chain) is an *elementary chain*. The branching points either are a result of the process by which the polymer was synthesized or they are formed later by reactions that connect two separate chains by covalent bonds. In the latter case, they are called *cross-links*. It is convenient to consider separately polymers with long elementary chains (5–100 or more monomeric units) and materials with very short chains, which may have even lost the appearance of a chain. We will call these materials *loose* and *dense networks*, respectively.

From a topological viewpoint, linear polymers are one-dimensional and polymer networks are three-dimensional. Polymeric materials with a two-dimensional structure are rare; however, several of them have been synthesized recently.

1.2.7.1. Loose Networks.
When polymeric materials are subjected to deformation, molecular chains may slip along each other and the deformation may become permanent. Such behavior is technologically very undesirable, especially for materials that are expected to be deformed a lot, such as rubbers. Permanent deformation can be

avoided when the chains are connected in an infinite network. However, to allow for a large deformation, the elementary chains must be long enough. We will treat this situation in more detail in Chapter 4.

In the rubber industry, light cross-linking is usually produced after the original polymer has been synthesized and the final rubber article has been formed. This process is called *vulcanization*. Different strategies are employed for vulcanization. Polymers with residual double bonds (polydienes) are frequently vulcanized by introducing sulfur bridges. Sometimes, when a polymer has no double bonds, a few are introduced into it (e.g., a small amount of isoprene is copolymerized with isobutylene) and are subsequently employed for cross-linking. It is also possible to convert some sensitive bonds (e.g., tertiary carbon-hydrogen bonds) into radicals either by treatment with peroxides or by radiation; the radicals will form cross-links by recombination.

Many monomers, when polymerized in bulk, will yield polymers that are sufficiently cross-linked to form a loose network. Sometimes, similar cross-linking is introduced deliberately by adding a cross-linking agent to the monomer. In the case of vinyl polymers, this agent is usually a divinyl compound; for example, polystyrene is frequently cross-linked by adding divinylbenzene to the polymerizing mixture. A similar effect is achieved in polycondensation polymers such as polyesters when a small amount of multifunctional acid or polyol is copolymerized with the principal monomers. When an attempt is made to dissolve a lightly cross-linked polymer in a solvent that would dissolve the linear polymer, the cross-linked material will swell and form a gel. Some gels can contain a very large proportion of the solvent and still give the appearance of a solid; they usually display rubberlike behavior. So-called soft contact lenses in ophthalmology are examples of these materials; they are based on cross-linked poly(hydroxyethyl methacrylate), on cross-linked poly(vinyl pyrrolidone), or on similar materials. Lightly cross-linked polyacrylamide will absorb a very large amount of water; it can be used for soaking up water spills and in such items as diapers. Highly swollen polyacrylamide is also used in biochemistry as the column material for gel electrophoresis—a method of separating proteins. Other gel-like materials are used in gel permeation chromatography (see Sections 3.7.1 and 3.7.2).

1.2.7.2. Dense Networks—Thermosets.   Inorganic glasses are prototypes of dense networks. For example, the standard silicate glasses have silicon-oxygen chains branched at almost every silicon atom. Organic dense networks have similar properties: They are very rigid, inflexible and sometimes brittle; they cannot be deformed unless the temperature reaches values at which the chemical structure starts to be destroyed. The heavy branching, is usually completed at high temperatures. Accordingly, the dense network materials are called *thermosets* in contrast to linear polymers, which are called *thermoplastics*. The latter materials soften and can be fabricated at relatively low temperatures, whereas the former do not soften at all.

Dense networks can be prepared by procedures similar to those used for lightly cross-linked gels, except that the amount of the branching agent is increased substantially. For example, copolymers of styrene with large amounts of divinyl-benzene are hard materials that are insoluble and almost nonswelling in any solvent. These materials have reactivities similar to those of substituted benzenes. They are easily derivatized (sulfonated, nitrated, etc.) and converted to ion-exchange resins: both catexes and anexes. Methacrylic monomers are converted to dense networks by copolymerization with ethylene glycol dimethacrylate or similar cross-linking agents.

In polycondensation, heavy cross-linking is achieved by using multifunctional components. Many important resins of this class are polyesters; a typical example is a polyester made from glycerol and phthalic acid. The degree of cross-linking (and consequently the brittleness) can be modified by addition of other acids or hy-droxy compounds. These polyesters are important in lacquers and in many compos-ite materials (laminates). Another strategy for forming thermosets from polyesters consists of using maleic or fumaric acid as a comonomer. In the second step of this synthesis, a vinyl monomer (e.g., styrene) is added and subjected to radical polymerization; the maleic moieties are built into the polystyrene chain, leading to cross-linking. Polyesters of all the above types are known as *alkyd polyester resins.*

The synthesis just described is an example of a two-step process frequently used for the preparation of thermosets. First, a *prepolymer* is prepared that has the easy workability of a low-molecular-weight polymer. In the second step, the prepolymer is formed into the final object and appropriately cured.

The high reactivity of epoxides toward alcohols, amines, acids, and anhydrides is exploited in the synthesis of *epoxy resins.* In a typical first step, bisphenol A **15** is reacted with an excess of epichlorohydrin **13** to form a prepolymer with epoxy end groups.

$$(n+1)\,HO-R-OH \; + \; (n+2)\,ClCH_2-\overset{\displaystyle O}{\overset{\displaystyle \diagup\ \diagdown}{CH}}-CH_2 \longrightarrow$$

$$\overset{\displaystyle O}{\overset{\displaystyle \diagup\ \diagdown}{CH_2}}-CH-CH_2-\!\!\left(\!O-R-O-CH_2-\overset{\displaystyle OH}{\overset{\displaystyle |}{CH}}-CH_2-O\right)_{\!\!n}\!\!R-O-CH_2-\overset{\displaystyle O}{\overset{\displaystyle \diagup\ \diagdown}{CH}}-CH_2$$

$$(1.2.3)$$

where

$$R = -\!\!\left\langle\!\!\bigcirc\!\!\right\rangle\!-\overset{\displaystyle CH_3}{\underset{\displaystyle CH_3}{\overset{\displaystyle |}{\underset{\displaystyle |}{C}}}}-\!\!\left\langle\!\!\bigcirc\!\!\right\rangle\!-$$

The prepolymer is then cured by reacting it with any of the above-mentioned com-pounds (they have to have at least two functional groups). Frequently, during the curing reaction some epoxy groups react with the hydroxy groups along the prepoly-mer chain, greatly enhancing the degree of cross-linking. Epoxy resins are used as strong adhesives. Another major use is in the field of *laminated materials.* An epoxy laminate with fabrics made from glass fibers is especially strong and durable. It is used for making hulls of pleasure boats and as lightweight roofing materials; it finds increasing use in the automobile industry.

Another major group of thermosets is based on the high reactivity of formaldehyde with respect to phenols and amines. Formaldehyde reaction with phenol yields methylol groups in the ortho and para positions of the phenol molecule. Up to 3 mol of formaldehyde can react with 1 mol of phenol. In a further reaction, the methylol group condenses with another molecule of phenol to form a methylene bridge.

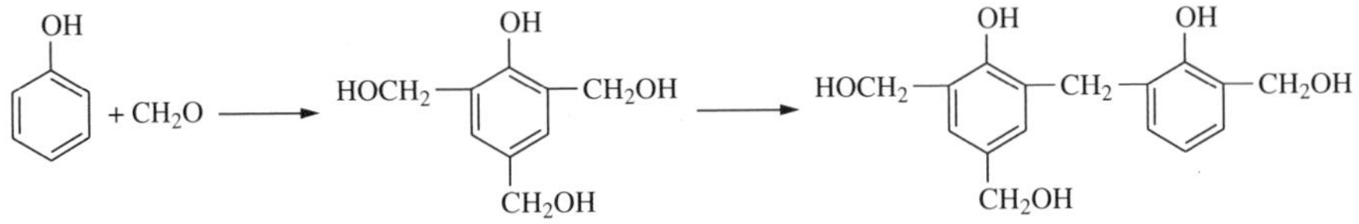

$$\text{(1.2.4)}$$

In these reactions, phenol acts as a three-functional monomer; consequently, very high cross-linking may be achieved. In practice, a prepolymer is prepared first; the final curing is done simultaneously with the forming of the article. The prepolymers prepared using basic catalysis are called *resoles;* when acidic catalysis is employed, they are called *novolacs.* The final insoluble and infusible *phenol-formaldehyde resins* are frequently called *Bakelite.* The final curing of these resins is done at rather elevated temperatures. Under these conditions, a small fraction of methylol-phenol moieties may dehydrate to form a quinone methide. Such quinoid structures are probably responsible for the dark, almost black color of Bakelite.

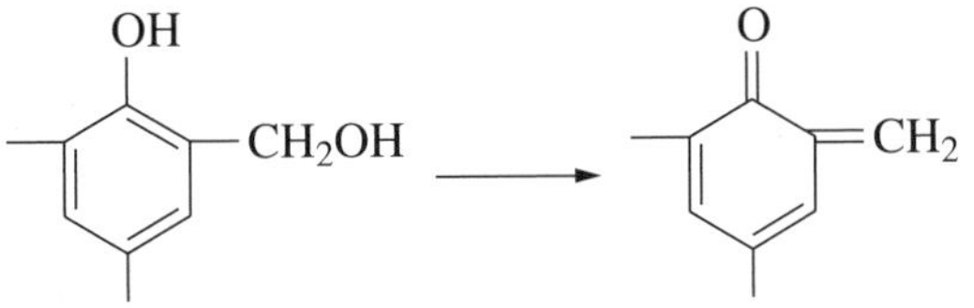

$$\text{(1.2.5)}$$

Phenol-formaldehyde resins are good electrical insulators. They are used for manufacturing parts and boxes for small electrical instruments, switches, telephones, heaters, etc. These resins are frequently filled with some inexpensive materials such as kaolin or sawdust to make them cheaper to produce.

When a compound with an amino group is reacted with formaldehyde, the amino group is hydroxymethylated. The hydroxymethyl group can further react with another amino group to form a methylene bridge.

$$R\!-\!NH_2 + CH_2O \longrightarrow R\!-\!NH\!-\!CH_2\!-\!OH \longrightarrow R\!-\!NH\!-\!CH_2\!-\!NH\!-\!R$$
$$\text{(1.2.6)}$$

Obviously, compounds with two or more amino groups form polymers with formaldehyde; each amino group can participate in two methylene bridges. These polymers are sometimes called *aminoplasts.*

Urea reacted in the above manner yields *urea-formaldehyde resins.* The reaction is usually carried in two steps as for most thermosetting materials. It is believed that formation of six-member rings from three amino groups and three methylene bridges plays a major role in the cross-linking of the resin. Thus, the cured product probably contains many structures of the following type:

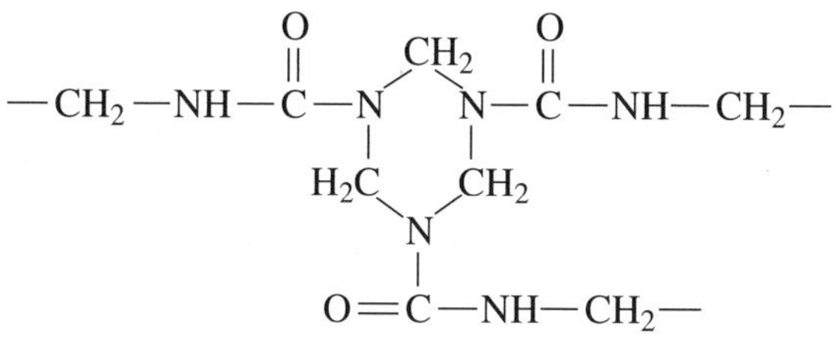

$$-CH_2-NH-\overset{\overset{\displaystyle O}{\|}}{C}-N\overset{CH_2}{\diagup}N-\overset{\overset{\displaystyle O}{\|}}{C}-NH-CH_2-$$

Urea-formaldehyde resins are tough, colorless materials that are especially useful whenever a light color is desired, as in the manufacturing of dinnerware. Otherwise, they are used for the same purposes as Bakelites. When a cotton fabric is lightly impregnated with these resins it becomes wrinkle-resistant.

Melamine **40** is another amino compound extensively used in such polycondensations. The intermediate hydroxymethylated materials are sometimes stabilized by ether-forming reaction with alcohols (usually butanol). During the final curing, some of these alkoxy groups may remain in the resin; this provides a valuable way to modify the resin properties. The uses of *melamine-formaldehyde resins* are similar to those of their urea analogs.

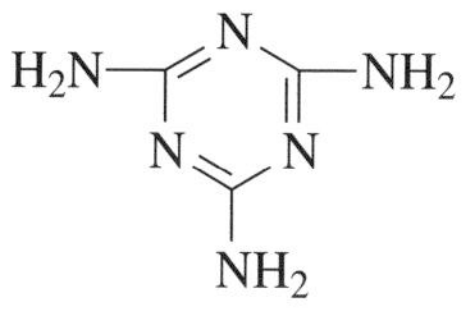

$$H_2N-\underset{N\diagdown N}{\overset{N}{\diagup \diagdown}}-NH_2$$

40

1.2.7.3. Two-Dimensional Networks. Two-dimensional structures are known mainly from inorganic chemistry. Graphite is a typical example. Cells of living organisms have many two-dimensional structures—membranes. A large number of important biochemical processes are localized on them. However, these membranes are made from smaller molecules (cholesterol and phospholipids are major components) and are held together by secondary forces: Nature seems to prefer a more flexible structure that can be modified easily (i.e., without breaking too many chemical bonds). However, physical chemists studying such membranes and their models prefer more stable structures. A very useful model is a thin capsule enclosing a small volume of a solvent (water, buffer, etc). Such capsules are called *vesicles;* they are prepared easily by sonication of some convenient detergent with the solvent. A vesicle consists of a detergent bilayer. The hydrophilic heads of the detergent point toward the solvent (both inside and outside of the vesicle); the hydrophobic tails hold the vesicle together. However, these vesicles are not stable enough to allow extended observation. The structures are stabilized if the detergent molecule contains a polymerizable group that is easily polymerized after the vesicle is formed by sonication. A typical example of a polymerizable detergent is the methacryloyl compound **41**.

$$\underset{\text{O}}{\overset{\text{CH}_3}{\underset{\|}{\text{H}_2\text{C}=\text{C}-\text{C}}}}-\text{O(CH}_2)_{11}\text{CO}_2(\text{CH}_2)_6$$

$$\text{CH}_3(\text{CH}_2)_{17}\overset{\text{CH}_3}{\underset{\text{CH}_3}{\diagdown\text{N}^+\diagup}}\quad\text{Br}^-$$

41

It is obvious from structure **41** that the polymer stabilizing the vesicle is still linear polymer with long side groups (a comblike structure). However, detergents have been synthesized with two long side groups, each carrying a polymerizable moiety. Polymerization of such vesicles may lead to "true" two-dimensional structures, because occasionally the two groups on the same detergent molecule are incorporated into different polymer backbones.

Diacetylene moieties are useful polymerization groups for this purpose. They are easily polymerized when aligned in a crystalline lattice or in another oriented structure (detergent bilayer); otherwise, their polymerization is more difficult. This feature biases the polymerization toward a more regular two-dimensional structure. Bisdiacetylene **42** is a typical example of this class of detergents.

$$\text{CH}_3-(\text{CH}_2)_{12}-\text{C}\equiv\text{C}-\text{C}\equiv\text{C}-(\text{CH}_2)_8-\text{CO}-\text{O}-(\text{CH}_2)_2$$
$$\text{CH}_3-(\text{CH}_2)_{12}-\text{C}\equiv\text{C}-\text{C}\equiv\text{C}-(\text{CH}_2)_8-\text{CO}-\text{O}-(\text{CH}_2)_2$$
$$\text{NH}^+(\text{CH}_2)_2\text{SO}_3^-$$

42

1.2.8. Natural Macromolecules

In the prologue, we pointed out the large variety of macromolecular materials, of biological origin and their significance both for life and for industry. In this section, we continue in the approach adopted in previous sections: We have not considered all the polymers of the various types discussed—not even all the important polymers. Instead, we have concentrated on the principles governing the structure of manmade polymers and on the descriptions of the main structural types. Similarly, in this section, we will discuss the basic types of materials chosen by the grand master of engineers—nature—to accomplish the necessary tasks of living organisms.

Surprisingly, the number of types is quite limited. Most natural macromolecules belong to one of four groups: polysaccharides, proteins, nucleic acids (or the combinations of them), and polymeric hydrocarbons. The first three groups will be described in this section. The last group is represented mainly by polymers of isoprene and was included in Section 1.2.1.1. It is worth mentioning that isoprene, the monomeric unit of natural rubber and gutta-percha, is also the building block of substances like terpenes and sterols (e.g., cholesterol). This gives us the first insight into the strategy of nature: Once a useful structure is developed, it is used in innumerable variations for many purposes. We will see still more striking examples of this strategy in the rest of this section.

1.2.8.1. Polysaccharides.   Polysaccharides are polymers of low-molecular-weight polyhydroxy compounds known as *sugars*. Whereas different sugars are present in natural macromolecules, most of the more important polysaccharides are polymers of glucose or its derivatives. Glucose, $C_6H_{12}O_6$, is a cyclic hemiacetal with two tautomeric structures; the α form, **43**, and the β form, **44**.

These two structures have different configurations at the hemiacetal carbon C-1. In solution, they easily change from one to the other. However, in most polyglucosides the hydroxyl on C-1 forms an ether bond with a hydroxyl on another molecule of glucose. This ether bond converts the hemiacetal into a stable acetal with the α or β structure.

Thus, *cellulose* is made of D-glucose residues linked by β (C-1 → C-4) glycosidic bonds. By its nature, cellulose is a cyclolinear polymer, and its backbone is quite rigid. Moreover, it still has three free hydroxyl groups, which can form intermolecular hydrogen bonds. In fact, cellulose is predisposed to exist in a crystalline structure with a large number of hydrogen bonds holding the chains together. The cellulose crystallites are rigid, strong, and, in the natural environment, almost indestructible. They are the universal structural material of plants. Some plants, including cotton, hemp, and flax, manufacture cellulose in a form of fiber with such good textile properties that in some respects they are still unmatched by synthetic fibers. Their value results from a combination of excellent strength and water absorption; a fiber with 30% water content still feels dry! The high crystallinity of cellulose (especially of cotton) is responsible for one undesirable effect: Small molecules do not penetrate cellulose easily; therefore, pure cotton is difficult to dye. The quest for a fiber with better dyeability led to many modifications of cellulose. Although cellulose is virtually insoluble in all solvents (the only exception being some copper-ammonia and cadmium-ethylenediamine complexes), it swells in concentrated alkali because of the formation of alcoholate groups, —ONa. The process of *mercerization* consists of swelling cellulose in alkali and subsequently regenerating it with acid. The treated product has reduced crystallinity and slightly lower strength, but better dyeability and luster.

The major source of cellulose is wood, from which it is isolated by dissolving the other major component of wood, lignin, in sodium sulfite. This cellulose is not in the form of long fibers. For production of fibers it is treated with alkali and carbon disulfide to produce *cellulose xanthate* (the hydroxy groups are converted into —OCS$_2$Na moieties). The resulting solution is very viscous and is appropriately called *viscose*. It can be spun into an acid medium, regenerating a cellulose fiber known as a *viscose fiber* or *rayon*.

When treated with appropriate reagents (see Section 2.4.3), the three free hydroxyls of each glucose ring can be derivatized. Nitration produces *cellulose nitrate* (incorrectly but commonly called *nitrocellulose*), which is remarkably stable toward hydrolysis. When prepared from cotton it is called *gun cotton*. Plasticized with camphor it forms *Celluloid*—the material of old rulers and french curves as well as the original movie film support. Other commercially important esters are acetate and butyrate; *cellulose acetate* forms valuable textile fibers sometimes called simply *acetate* fibers. Other important derivatives are the methyl ethers—*methyl cellulose* (—OCH$_3$) and the condensation product with chloroacetic acid *carboxymethyl cellulose* (—OCH$_2$CO$_2$H). The latter product has valuable cation-exchange properties.

The glucose units in 1,4-polyglucose also may be bound in a different manner; C-1 may have an α conformation. The result is a polymer with strikingly different properties: *amylose*. Because of its different steric arrangement, amylose tends to form hydrogen bonds intramolecularly, resulting in a loose helical structure; it is not crystalline. Amylose together with another polysaccharide, *amylopectin,* exists in plants as starch, a material that plants use for storing energy and glucose units. It is the most significant component of grains, potatoes, and a myriad of other agricultural products. Amylopectin is another polymer of glucose. However, it is a branched polymer; the branches are connected to the backbone through ether linkages between carbons 1 and 6.

Animals have developed another storage material, *glycogen,* a polymer of glucose that is branched similarly to amylopectin but in which the degree of branching is higher. Some bacteria produce *dextran;* a linear 1,6-polyglucose. Dextran preparations with molecular weights similar to those of plasma proteins have been used in blood transfusions as plasma substitutes that are not spoiled by long storage.

A number of important polymers use a modified glucose unit as a monomeric unit. For example, *chitin,* a structural material that serves insects in the same way as cellulose serves plants, is a homopolymer of β (C-1 $\rightarrow$ C-4)-linked *N*-acetylglucosamine residues (the hydroxyl group on C-2 of glucose is replaced by the *N*-acetylamino group, —NHCOCH$_3$).

Another interesting polymer is *hyaluronic acid,* which is an important component of synovial fluid (the fluid of high viscosity found inside the joints of the body) and of the vitreous humor of the eye. It is an alternating copolymer of two derivatives of glucose: *N*-acetylglucosamine is linked through C-1 and C-3 to the C-1 and C-4 of

D-glucuronic acid (which is glucose with its —CH_2OH group converted to carboxyl). Hyaluronic acid is a polyelectrolyte; this property undoubtedly contributes to the high viscosity of the synovial fluid.

1.2.8.2. Proteins and Polypeptides.

As mentioned in the prologue, proteins carry on a large number of different functions in living organisms. This variety notwithstanding, the primary structures of all proteins are remarkably similar. We will see in Section 1.3 that it is the higher-order conformational structure of these macromolecules that determines their specific properties. Yet the conformational structure is necessarily a result of the underlying primary chemical structure. The primary structure of the protein is engineered *in vivo* with the utmost sophistication to achieve the final result. Compared with this sophistication, the attempts of polymer chemists to modify polymer materials by changing their structure seem quite crude indeed. Proteins give us proof that the possibilities of macromolecular engineering (read "synthesis") are virtually unlimited and remind us not to set our goals too low.

All the variety in proteins is achieved by rather simple means. All proteins are polyamides derived from α-amino acids, or, more precisely, they are copolymers of the 20 L-amino acids that are listed in Table 1.1 together with their chemical structures. The macromolecular backbone is formed by amidic bonds between carboxyls and α-amino groups, —NH—CHR—CO—, where R is the group residing on the α-carbon. Only one carboxyl and its α-amino group may be part of the backbone; other carboxyls or amino groups (such as those in aspartic and glutamic acids or in lysine) are never part of the backbone. Every protein has its specific sequence of amino acids and a well-defined number of them. Thus, proteins, unlike synthetic polymers, have as definite chemical formulas as low-molecular-weight substances; in polymer terminology, they are *monodisperse*.

Each of the 20 amino acids conveys to proteins some characteristic properties that influence the conformation of the protein on the one hand and its "external" properties on the other. The first five amino acids in Table 1.1 give the protein a mostly hydrophobic character; the hydrophobicity increases with the size of the side group. The shape of the side group plays a role in the folding of the molecule. The hydroxyamino acids serine and threonine increase the hydrophilicity of the protein and may participate in the formation of hydrogen bonds. Still stronger hydrogen bonds could be formed by amidic side groups in asparagine and glutamine. Aspartic acid and glutamic acid contribute acidic properties, whereas lysine, arginine, and, to a lesser degree, histidine, are basic. The acidic and basic groups as well as the net charge at a given pH determine the polyelectrolytic properties of the protein.

Aromatic groups in phenylalanine, tyrosine, tryptophan, and histidine have characteristic electron donor-acceptor properties, which play a major role in the catalytic sites of many enzymes. The same applies to the unique chemical properties of sulfur in cysteine and methionine.

Two amino acids, cysteine and proline, have a special role with respect to the conformation of the protein molecule. Two molecules of cysteine are easily and reversibly oxidized to a dimer with a disulfide bridge, CyS—SCy. This dimer is sometimes considered to be a separate amino acid, cystine. Two halves of the cystine

TABLE 1.1   Amino Acids—Building Blocks of Proteins

1. Glycine H_2N-CH_2-COOH

2. Alanine
$$\begin{array}{c} CH_3 \\ | \\ H_2N-CH-COOH \end{array}$$

3. Valine
$$\begin{array}{c} CH(CH_3)_2 \\ | \\ H_2N-CH-COOH \end{array}$$

4. Leucine
$$\begin{array}{c} CH_2CH(CH_3)_2 \\ | \\ H_2N-CH-COOH \end{array}$$

5. Isoleucine
$$\begin{array}{c} CH(CH_3)CH_2CH_3 \\ | \\ H_2N-CH-COOH \end{array}$$

6. Serine
$$\begin{array}{c} CH_2OH \\ | \\ H_2N-CH-COOH \end{array}$$

7. Threonine
$$\begin{array}{c} CH(CH_3)OH \\ | \\ H_2N-CH-COOH \end{array}$$

8. Asparagine
$$\begin{array}{c} CH_2CONH_2 \\ | \\ H_2N-CH-COOH \end{array}$$

9. Glutamine
$$\begin{array}{c} CH_2CH_2CONH_2 \\ | \\ H_2N-CH-COOH \end{array}$$

10. Aspartic Acid
$$\begin{array}{c} CH_2COOH \\ | \\ H_2N-CH-COOH \end{array}$$

11. Glutamic Acid
$$\begin{array}{c} CH_2CH_2COOH \\ | \\ H_2N-CH-COOH \end{array}$$

12. Lysine
$$\begin{array}{c} CH_2CH_2CH_2CH_2NH_2 \\ | \\ H_2N-CH-COOH \end{array}$$

13. Arginine
$$\begin{array}{c} NH \\ \| \\ CH_2CH_2CH_2NH-C-NH_2 \\ | \\ H_2N-CH-COOH \end{array}$$

14. Phenylalanine
$$\begin{array}{c} CH_2-C_6H_5 \\ | \\ H_2N-CH-COOH \end{array}$$

15. Tyrosine
$$\begin{array}{c} CH_2-C_6H_4-OH \\ | \\ H_2N-CH-COOH \end{array}$$

16. Tryptophan
$$\begin{array}{c} CH_2-C \\ | \\ H_2N-CH-COOH \end{array}$$

17. Histidine
$$\begin{array}{c} CH \\ HN \qquad N \\ | \qquad \| \\ CH_2-C=CH \\ | \\ H_2N-CH-COOH \end{array}$$

18. Cysteine
$$\begin{array}{c} CH_2SH \\ | \\ H_2N-CH-COOH \end{array}$$

19. Methionine
$$\begin{array}{c} CH_2CH_2SCH_3 \\ | \\ H_2N-CH-COOH \end{array}$$

20. Proline
$$\begin{array}{c} CH_2 \\ H_2C \qquad CH_2 \\ | \qquad | \\ HN-CH-COOH \end{array}$$

molecule may reside in parts of the chain that are topologically quite distant, bringing these chain segments into close proximity. In some cases, the cystine disulfide (—S—S—) bridges may covalently join two chains that are otherwise unconnected. The role of proline is more subtle. As we will see later, the protein backbone has a tendency to form a helix. However, proline cannot fit into this helix; its presence is one of the means nature uses to limit the extent of the helical properties of the chain.

Amino acids may form chains of various lengths. Short chains with 2–10 amino acid residues are called *peptides.* Peptides occur in living organisms; some of them are potent hormones (oxytocin, vasopressin, enkephalin). Peptides result also from incomplete hydrolysis of proteins. Peptide chains containing 10–30 amino acids are called *polypeptides;* there is no clear demarcation between peptides and polypeptides. The latter name is also frequently used for synthetic polymers of α-amino acids or their modified derivatives. These substances have much simpler structures than the natural polymers (proteins).

Morphologically, proteins can be divided into two groups: fibrous and globular. Fibrous proteins perform mainly the structural and protective functions of organisms. We will consider only three basic types: α-keratins, β-keratins, and collagen.

α-Keratins are insoluble materials forming hair, wool, nails, horns, feathers, etc. α-Keratin in the form of hair can be stretched to almost double its length when steamed. However, at ambient temperatures the α-keratins are quite rigid. Chemically, they are especially rich in amino acids that have hydrophobic character: phenylalanine, isoleucine, valine, methionine, and alanine. They are rich in cysteine residues that form disulfide bridges among adjacent chains. These bridges increase the insolubility and rigidity, the two main biological properties of α-keratins.

β-Keratins do not stretch, even when heated. The best-known representative of β-keratins is *fibroin*—the polymeric material of silk. Fibroin has a relatively simple chemical structure (i.e., simple among proteins). Long stretches of oligopeptides comprised of six residue repeats (glycine-serine-glycine-alanine-glycine-alanine) form crystallizable sequences. Fibroin also contains some sequences that are not crystallizable.

Another fibrous protein, *collagen,* is the most abundant protein in the body. It is distributed in all kinds of connective tissues; including skin, bones, tendons, and cartilage. The main amino acids of collagen are glycine, alanine, proline, and 4-hydroxyproline **45;** hydroxylysine **46** is also present. Of the proteins known, 4-hydroxyproline and 5-hydroxylysine are found only in collagen. Every third position along the collagen chain is occupied by glycine. Denaturation (boiling) converts collagen into a sticky substance known as *gelatin.*

$$HO-CH-CH_2$$
$$\qquad | \qquad |$$
$$\qquad CH_2 \quad CH-COOH$$
$$\qquad \backslash \quad /$$
$$\qquad NH$$
$$\qquad \textbf{45}$$

$$\qquad\qquad\qquad\qquad\qquad NH_2$$
$$\qquad\qquad\qquad\qquad\qquad |$$
$$H_2N-CH_2-CH-CH_2-CH_2-CH$$
$$\qquad\qquad\qquad |\qquad\qquad\qquad\qquad |$$
$$\qquad\qquad\qquad OH\qquad\qquad\qquad COOH$$
$$\qquad\qquad\qquad\qquad\textbf{46}$$

Globular proteins in their native state have their chains rather tightly packed into more or less rigid particles ranging in shape from almost spherical to very elongated. Globular proteins comprise most classes of proteins including: *enzymes* (pepsin, trypsin, and a myriad of others), *transport proteins* (hemoglobins, serum albumin), *storage proteins* (ovalbumin, casein), *muscle proteins* (myosin, actin), *receptor proteins* (*T* cell receptor), and *antibodies* (γ-globulins). It is very tempting to present at least an overview of the better-known proteins and their biological and biochemical functions. However, this domain is a proper subject of biochemistry, and we do not believe we can give it the attention it deserves without devoting disproportionate space to it in this introductory book. We will therefore restrict ourselves to a description of the main principles governing protein structure, without going into biochemical details (see Sections 1.3 and 1.4).

Synthetic polypeptides corresponding to natural proteins have been prepared by the traditional methods of organic chemistry by adding one amino acid residue to the chain at a time. Alternatively, the cell's own synthetic apparatus for proteins has been reconstituted in a test tube and used for the production of custom-made polypeptides. Polypeptides *in vitro* can also be prepared by anionic polymerization techniques from *N*-carboxy anhydrides, which in turn are made from amino acids and phosgene.

$$R-CH-COOH + COCl_2 \longrightarrow R-CH-CO \quad (1.2.7)$$
$$\underset{NH_2}{|} \qquad \underset{NH \quad O}{|} \quad \underset{\underset{CO}{\diagdown \diagup}}{}$$

Fibers made from some of these polypeptides are used as sutures in surgery; they are absorbed by the organism in due time.

Polypeptides also can be prepared from amino acids that do not occur in nature. For instance, *poly(benzyl glutamate)* **47** exhibits very interesting solution behavior. It has been used extensively as a model for studying the conformational properties of macromolecules, especially the formation of helices (see Section 1.3). Moderately concentrated solutions of this polymer have liquid crystal properties.

$$-NH-CH-CO-$$
$$\underset{CH_2CH_2CO-OCH_2-}{|}$$

47

1.2.8.3. Nucleic Acids.   Nucleic acids are involved in most important functions of living organisms. They carry the full genetic information for every organism, duplicate the information during reproduction, and govern the synthesis of proteins by coding for the specific amino acid sequences. They are also intimately involved in the actual synthesis of proteins.

Several billions of years before military experts started using ciphers for their messages and computer experts developed their binary and hexadecimal descriptions of the world, nature perfected the art of coding by inventing the quaternary code of nucleic acids. An ingenious engineering twist used the same molecular arrangement for reproduction of the coded information. The four letters of the code are heterocyclic bases: two purines–adenine **48** and guanine **49**—and three pyrimidines—cytosine **50** and either thymine **51** or uracil **52**. (We will see that one version of the code uses thymine; the other uses uracil.)

To form a code the bases must be arranged in a one-dimensional sequence; that is, they have to be built into a polymeric structure. Nature attached them to two 5-carbon sugars: D-ribose **53** and deoxy-D-ribose **54**.

The amino groups identified by arrow in formulas **48–52** replace the hemiacetalic hydroxyl of the sugar molecule (also indicated by an arrow in formulas **53, 54**). The combined base-sugar molecules are called *nucleosides*. When any (or several) of the hydroxyls on the sugar moieties of nucleosides are esterified with phosphoric acid, the resulting molecule is called a *nucleotide*. Many nucleotides have pivotal functions in metabolism, especially in biochemical pathways for utilization of sugars and other foodstuffs for maintenance of the organism. The same molecule of phosphoric acid can be esterified by two hydroxyl groups belonging to two different nucleosides. Nucleic acids are long sequences of nucleotides bound in this manner through C-3 and C-5 of the sugar molecule—see the schematic representation in Figure 1.10.

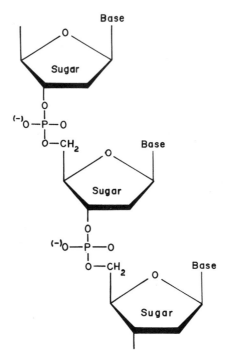

Figure 1.10. Binding of nucleotides in nucleic acids

The third acidic hydrogen of phosphoric acid gives nucleic acids their acidic character. At physiologic values of pH, it is dissociated, making nucleic acids clear-cut polyelectrolytes. *Deoxyribonucleic acids* (DNA) employ deoxyribose as their sugar; ribose is the sugar of *ribonucleic acids* (RNA).

DNA resides in the nucleus of the living cell and serves as a repository of genetic information; it usually has a very high molecular weight, on the order of 10 million and probably even higher. Although many functions of DNA are still being investigated, its role in the synthesis of enzymes and other proteins is now well understood. DNA contains four bases: adenine (A), guanine (G), thymine (T), and cytosine (C). It is the sequence of these four bases that codes the sequence of amino acid residues in proteins. Each group of three nucleotides is called a *codon;* four types of nucleotides can form 64 different codons. Most codons code for an individual amino acid; some amino acids are coded by several different codons. There also exist codons for the start (initiation codon) and the end (termination codon) of the polypeptide chain. However, there are sections of DNA whose function is still not fully understood.

A section of the nucleic acid that codes for a particular protein or other function is called a *gene*. (Geneticists have used this term for more than a hundred years for a hypothetical unit of heredity. This hypothetical unit was recently identified with a very real piece of a macromolecule.) The obviously immense importance of understanding the structure, properties, and function of the genes led to very intense research. Techniques were developed for fast sequencing of the bases in DNA. In fact, scientists have now completed sequencing the entire human genome. Experiments are in progress in many laboratories toward the treatment of diseased individuals by gene therapy. Drug companies are striving hard to emerge with appropriate drugs that match the genetic profiles of patients. We will return to these topics in Section 3.8.

Ribonucleic acids (RNA) are directly involved in the synthesis of proteins. There are three distinct forms of RNA:

1. *Messenger RNA* (mRNA), which carries the genetic code from the nucleus (DNA) of the cell to ribosomes, cellular structures where the actual synthesis of proteins takes place.
2. *Ribosomal RNA* (rRNA), which is the main component of the ribosomes.
3. *Transfer RNA* (tRNA), which "reads" the code and brings the needed specific amino acid to the place of synthesis.

RNAs employ a slightly different set of bases than DNAs—adenine, guanine, cytosine, and uracil.

All these processes are intimately related to very special conformations that are adopted by nucleic acids. We will return to these fascinating processes in the next sections after reviewing some basic aspects of the conformations of macromolecules.

1.3. HIGHER STRUCTURE—CONFORMATIONS

The actual positions of atoms within molecules must conform to a number of rules. For example, the lengths of all single carbon-carbon bonds are virtually the same; the four bonds from a carbon atom point to the corners of a tetrahedron; and all four substituents on a carbon-carbon double bond are in the same plane. For many small molecules, such rules are sufficient for determining all atomic positions. However, parts of some molecules may rotate more or less freely around some bonds—notably single bonds. Thus, for molecules with many such bonds, a large number of conformations are possible. The distribution of molecules among available conformations is governed by the rules of thermodynamics and statistical mechanics. The conformations with lower energy are preferred by the Boltzmann factor, $\exp(-E/kT)$, where E is the energy of the molecule in the particular conformation. The energies are calculated with respect to some arbitrary zero-reference energy. T is absolute temperature, and k is the well-known Boltzmann constant.

The energies of molecules depend on interactions between nonbonded atoms and/or atomic groups, both intramolecular and intermolecular. Repulsive interactions occur when two atoms approach each other too closely; in this case, the molecular energy rises sharply and the Boltzmann factor rules out such conformations. Nonpolar atomic groups are attracted to each other; this lowers the energy, and if possible the two groups will settle at a distance corresponding to the minimum energy, that is, at the position where the attractive and repulsive forces balance each other. These attractive forces hold molecules of liquid together; the larger the molecule, the more difficult it is for it to escape from the liquid (i.e., to evaporate). The attraction between polar groups is even stronger than between nonpolar groups; the largest reduction in energy is achieved whenever hydrogen bonds are formed.

The actual conformation of the molecule (or the distribution of conformations) depends on the interplay of all the intra- and intermolecular interactions. The simplest case is an isolated molecule in dilute gas. Obviously, this case is not usual for polymers; it can be achieved when a very dilute macromolecular solution is sprayed through an "atomizing" nozzle and the solvent evaporates completely from the droplets. For a more usual example, a vapor of very high-molecular-weight alkane can be considered a model of a linear macromolecule. Under these conditions the molecule will try to maximize the number of intramolecular contacts; it will collapse to a more or less spherical particle.

Two other situations are more important in practice: (1) a macromolecule surrounded by molecules of solvent, as in polymer solutions, and (2) a macromolecule surrounded by other macromolecules, as in solid polymers or polymer melts. In both cases, the intermolecular contacts lower the energy of the molecule appreciably; it is no longer the collapsed form of macromolecule that necessarily exhibits the lowest energy. More detailed analysis of the interaction energies is needed for an understanding of the conformations of macromolecules.

We will consider two cases. In Section 1.3.1 we will study situations in which the energies of various polymer conformations are sufficiently close to each other so that many different conformations can exist simultaneously. In such a case the macromolecules form so-called molecular coils. In Section 1.3.2 we will treat situations in which some conformation of the macromolecule has such a low energy as to virtually rule out any other conformation. Such macromolecules are said to have a regular secondary structure.

1.3.1. Random Macromolecular Coils

When a long macromolecule is dissolved in a solvent or is a part of an amorphous macromolecular system, it usually exists as a loose coil. We can visualize such a coil as a very long noodle suspended in water. Actually, the factors affecting the shape of the noodle and the shape of the macromolecule are essentially the same. Although each molecule has a different conformation and different outside shape, it is useful to talk about a representative molecule that has properties that are an average of the properties of all molecules. The end-to-end distance of a linear coil has a very prominent status among these properties; it is relatively easily estimated theoretically and is directly related to a number of important experimental characteristics of polymer materials—for example, to intrinsic viscosity, diffusion and sedimentation coefficients, and light-scattering properties. There is a very extensive literature dealing with a number of approaches to the calculation of the average end-to-end distance. However, before we consider them more closely we need to review their background—the conformational rules of organic molecules. In the following paragraphs we will consider only polymers with a saturated all-carbon backbone. Other polymers can be treated in a completely analogous manner.

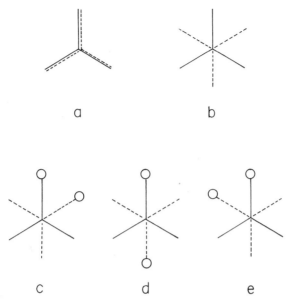

Figure 1.11. Conformations of ethane and butane. (a) Eclipsed and (b) staggered conformations of ethane, (c) *gauche,* (d) *trans,* and (c) *gauche prime* conformations of butane.

Saturated hydrocarbons can rotate more or less freely around single carbon-carbon bonds. For the simplest case of ethane it is convenient to depict the conformations using a projection parallel to the carbon-carbon bond, which in this projection is reduced to a point (Figure 1.11). The three substituents at one carbon are shown as full lines at 120° angles; the broken lines refer to substituents on the other carbon. The angle ϕ between these two sets of three lines may adopt any value. However, two conformations are prominent: the eclipsed conformation with $\phi = 0°$ (or 120° or 240°) (Fig. 1.11a) and the staggered conformation with $\phi = 60°$ (180°, 300°) (Fig. 1.11b). The six hydrogens of the ethane molecule interact with each other, and the energy of the molecule is a function of the angle ϕ (Fig. 1.12a). The energy of the eclipsed conformation is higher by about 2.8 kcal than the energy of the staggered form. The conventional wisdom recognizes that the Boltzmann factor virtually eliminates the eclipsed form and claims that all molecules are in the staggered form. However, the fact that there is an energy minimum at the staggered form implies that conformations that deviate from the staggered form by only a few degrees have energy only slightly higher and are therefore quite acceptable. This may be especially significant for larger molecules, which cannot adopt the staggered form because of some distant steric hindrance; a small deviation may relieve the problem.

Further insight is gained from consideration of the conformations of butane; these are also depicted in Figure 1.11. In this case the projection is along the bond between the two central carbon atoms; the end methyl groups are shown as circles. Only the staggered conformations need to be considered. In two of them ($\phi = 60°$ and $\phi = 300°$) the two methyl groups are relatively close together and mutually interfere.

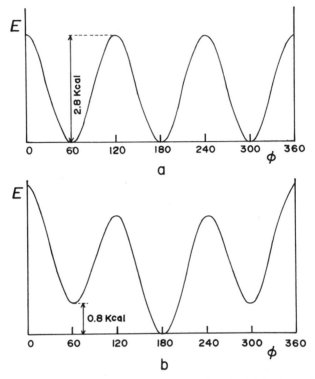

Figure 1.12. Conformational energy as a function of angle ϕ. (a) Ethane; (b) butane.

These conformations are called *gauche* and *gauche prime* (g, g'). In the *trans* (t) conformation ($\phi = 180°$) the interference is absent. Accordingly, the energy diagram (Fig. 1.12b) has one deeper minimum at 180° and two shallower minima at 60° and 300°. The difference in energy is about 800 cal. At room temperature the *trans* conformations prevail, but the fraction of the *gauche* conformations is still appreciable; actually, it is even enhanced by their degeneracy, which is equal to 2.

The conformation of higher alkanes can be represented as a sequence of *trans* and *gauche* groupings, for example, $ttgtg't$.... The sequence is random; the relative frequency of the three forms is governed by their Boltzmann factors. One more restriction exists: If a sequence gg' is attempted, a severe overlap between the first and fifth methylene group occurs. This effect is observed first for the pentane molecule and is called the "pentane effect." Undoubtedly, similar restrictions exist for longer chains too, but their mathematical treatment follows a different route (see long-range interactions in Section 1.3.1.3).

Let us return now to the conformation of a long polymer chain. Seemingly, the average conformation can be found by statistical means after all possible conformations have been analyzed and assigned their Boltzmann factors. However, this is possible only for the shortest chains because of the enormity of the number of possible conformations. If we consider a chain with 1000 links (which corresponds to a molecule of polyethylene with a modest molecular weight of 14,000) and consider three possible conformations for each link (t, g, g'), we find that the number of possible conformations is about 3^{1000}, or about 10^{477}. Compare this number to the estimated number of atoms in our galaxy, which is about 10^{70} (give or take several orders of magnitude). Even if some of the conformations need not be considered because of the pentane effect and similar effects, the number of remaining conformations is still meaninglessly high.

Obviously, a different approach is needed. Most theories use (implicitly) an assumption that the conformations of macromolecules in a given sample represent fairly all possible conformations (taken together with their Boltzmann factors). More specifically, they postulate that the average properties calculated for all possible conformations and the averages for properties of actual molecules in a sample (and quite often also the averages for a small sample of judiciously chosen conformations) are identical with an extremely high degree of accuracy.

Most theories assign some conformational behavior (e.g., the distribution of the t, g, g' conformations) to short sections of the chain and assume that this behavior is maintained along the whole chain in a random way. The main problem with this approach is the fact that distant sections of the chain may approach each other and interact, severely compromising the assumption of randomness. The theories differ in the way the elementary conformational behavior is assigned and in the way the interaction of distant chain sections is handled. A common but by no means universal approach consists of separating the interactions into *short-range* interactions dependent on the local structure of the polymer chain and *long-range* interactions related to the encounters of distant sections of the chain. The short-range and long-range interactions are handled using quite different physicochemical concepts. Obviously, there is a conceptual difficulty in the decision as to how far along the chain the two groups should be for their interaction to qualify as a long-range interaction. In the following we briefly describe some of the more important theories.

1.3.1.1. Short-Range Interactions. In most theories the linear macromolecule is modeled as a long sequence of segments. The spatial orientation of the segments is governed by rules that are more or less closely derived from the behavior of real molecules. When a study of short-range interactions is desired, individual backbone bonds are selected as segments. The actual valence bond angle between segments is required (the tetrahedral bond angle 109.5° is used for most but not all calculations).

The simplest model assumes free rotation around all bonds. It is instructive to outline the calculation of the end-to-end distance for this model. Individual segments in any particular conformation have not only a length but also a direction. They are conveniently described as vectors $\mathbf{a}$; the length a of all vectors is assumed to be the same. The end-to-end vector $\mathbf{h}$ is then simply the sum of all segment vectors,

$$\mathbf{h} = \sum_{i=1}^{N} \mathbf{a}_i \qquad (1.3.1)$$

where N is the number of segments. To eliminate vectors from the calculation we multiply vector $\mathbf{h}$ by itself (scalar product).

$$h^2 = \mathbf{h} \cdot \mathbf{h} = \Sigma \mathbf{a}_i \cdot \Sigma \mathbf{a}_j \qquad (1.3.2)$$

The scalar product of two vectors is equal to the product of one of them and the projection of the other onto the first. For two neighboring segments, the product is equal to $a^2 \cos \theta'$, where θ' is a complement to the valence angle θ.

The computation now calls for averaging equation (1.3.2) for all possible conformations. A bar over the symbols will denote the averaging. On the right-hand side of equation (1.3.2) we will do the averaging first and the summation over segments later. Now, according to our model, there is free rotation around all the bonds; their directions are independent of each other; only the valence angles have to be observed. Let us consider a particular segment $\mathbf{a}_i$. We calculate its product with all other segments. The product with itself is equal to a^2. The product with a neighboring segment is $a^2 \cos \theta'$; there are two such segments. The product with the second neighbor must be averaged for all possible directions of this freely rotating bond. Let us decompose this vector into its projection onto the first neighbor (equal to $a \cos \theta'$) and a vector perpendicular to it. The *average* value of the latter vector is zero because of the assumption of free rotation; therefore, it does not contribute to the product with our reference vector $\mathbf{a}_i$. However, the vector projected onto the first neighbor does contribute $a^2 \cos^2 \theta'$ to the scalar product with the reference segment. Again, there are two second neighbors. In an analogous way, we calculate the contribution of two third neighbors as $2a^2 \cos^3 \theta'$, and so on. Putting all the terms together we have

$$\mathbf{a}_i \cdot \sum_j \mathbf{a}_j = a^2 + 2a^2 \cos \theta' + 2a^2 \cos^2 \theta' + 2a^2 \cos^3 \theta' + \cdots\cdots\cdots \qquad (1.3.3)$$

Actually, because of the finite length of the chain, the coefficient 2 drops to 1 at some distance from the reference segment (i.e., at some value of the exponent of $\cos \theta'$) and much later it drops to 0. However, in most cases the drops occur at such high values of the exponent that the missing terms (high powers of $\cos \theta'$) are completely negligible and equation (1.3.3) can be treated as an infinite series. Its summation yields

$$\mathbf{a}_i \cdot \sum_j \mathbf{a}_j = a^2 \frac{1 + \cos \theta'}{1 - \cos \theta'} \qquad (1.3.4)$$

Finally, as there are N segments in the molecule, each of them contributing the same amount to the sum in equation (1.3.2), we obtain

$$\overline{h^2} = Na^2 \frac{1 + \cos \theta'}{1 - \cos \theta'} \qquad (1.3.5)$$

Finally, the $\cos\theta'$ is converted to the cosine of the valence angle θ, and the final expression reads

$$\overline{h^2} = Na^2\frac{1-\cos\theta}{1+\cos\theta} \tag{1.3.6}$$

The form of equation (1.3.6) is very significant: The average square of the end-to-end distance is proportional to the number of segments in the chain and therefore to the molecular weight of the polymer in its first power. This proportionality, of course, does not apply to very short chains but holds in an asymptotic limit, which is rather rapidly attained. It is strictly preserved for all models that are based on random addition of segments. However, the randomness needs to apply only with respect to distant segments; it actually means an absence of any long-range interaction. There may be any regularity along short sections of the chain, with the provision only that the effect of one segment on the orientations of other segments gradually disappears with increasing distance (measured along the backbone). It follows that any refinement of this model can change only the value of the proportionality constant between the square of the end-to-end distance and molecular weight.

Nevertheless, extensive refinements have been made. First, the freedom of rotation was replaced by a restricted rotation, either allowing only the three staggered forms (t, g, g') with appropriate Boltzmann factors or allowing all angles but weighting them according to the energy curves as in Figure 1.12. In both cases, the final result was found to be

$$\overline{h^2} = Na^2\frac{1-\cos\theta}{1+\cos\theta}\left(\frac{1+\eta}{1-\eta}\right) \tag{1.3.7}$$

where the parameter η is a measure of the steric hindrance. It is a function of temperature and can be calculated from the energy curves or from the relative energies of the *trans* and *gauche* conformations.

The above considerations were presented for a polyethylene chain. They are equally applicable to all vinyl polymers; the main difference is in the value of the parameter η. For macromolecules with a different type of backbone, the details of the analysis must take into account the peculiarities of all the bonds and atomic groups involved. Nevertheless, the main result of our analysis is preserved: The average square of the end-to-end distance is proportional to the number of bonds (or monomeric units) in the polymer backbone. All peculiarities of different polymers are absorbed into the proportionality constant.

Flory and co-workers performed the most elaborate calculations. Their method of *statistical weight matrices* takes into account simultaneously the conformations of two neighboring backbone bonds as well as a detailed analysis of the interactions of the side groups. The Flory method can be used for a very close prediction of the size of molecular coils provided (1) very detailed information is available about bond length and angles (an assumption of tetrahedral angles is usually insufficient); (2) the short-range forces are known with a good accuracy; and (3) a method is available for the assessment and/or elimination of long-range interactions.

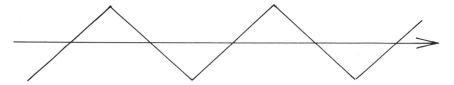

Figure 1.13. Zigzag conformation of an all-carbon polymer backbone.

1.3.1.2. Statistical Coils.   The Flory method of statistical weight matrices (developed in the 1970s) described the short-range interactions and the average end-to-end distance reasonably well. However, the older methods based on the valence-bond model were much less successful. Recognizing the difficulties of this model as early as the 1930s, Kuhn abandoned the idea of a detailed description of a coil and replaced it with a model using *statistical segments*. The basis of his model is the following consideration: In a real coil (or in a valence-bond model) the direction of the chain (which is taken as a straight line following the all-*trans* zigzag, conformation, Fig. 1.13) is gradually lost because of the occasional presence of *gauche* conformations. This gradual loss of direction is modeled by a piece of chain (a statistical segment) where the linearity is strictly preserved. At the end of the segment the continuity of direction is lost completely. The direction of the next statistical segment has no correlation with the preceding one. Thus the average projection of any statistical segment onto any other is zero, and only the products of segment vectors with themselves contribute to the averaged sum in equation (1.3.2) Consequently, the average end-to-end distance is given by

$$\overline{h^2} = NA^2 \qquad (1.3.8)$$

where N is the number of statistical segments and A is the length of a single segment. Equation (1.3.8) allows any value of A to represent a polymer and its end-to-end distance; only the value of N needs to be adjusted to satisfy equation (1.3.8). Such N values will still be proportional to the molecular weight M of the polymer. In his later work, Kuhn tried to remove this arbitrariness. He imposed another restriction on the quantities N and A, namely that they satisfy the relation

$$L = NA \qquad (1.3.9)$$

where the *contour length* L is the length of the fully stretched coil in the zigzag (all *trans*) conformation. The value of L is calculated from the known valence lengths and angles in the polymer backbone and the polymerization degree (or molecular weight) of the polymer. Thus the quantity $\overline{h^2}$ (obtained from a suitable experiment) and L (from the known structure of the polymer) determine uniquely the length of the statistical segment A, which is then considered as characteristic of the particular polymer.

However, the condition of equation (1.3.9) is not applied universally; neither is the concept of a statistical segment as a straight portion of the chain. Any portion of the chain can serve as a segment provided that the segments are chosen large enough that there is no correlation between the neighboring segments. A frequently employed model is the model of *Gaussian subchains.* Here, each segment contains enough monomer units to behave as a statistical coil in its own right. The fixed segment length A is replaced by a distribution of the segment end-to-end vectors r_0 and its average square value r_0^2. This model is frequently used in theories of long-range interaction and in theories of the behavior of polymer molecules in force (usually hydrodynamic) fields. For the latter application the subchain is considered to behave as a spring. The retractive force of the spring is entropic in origin; we will study it more closely in the section on rubber elasticity.

The model of subchains is especially useful for very long chains. For the other end of the scale—very short or stiff polymers—a *wormlike model* is sometimes employed. In this model the direction of the chain is changing gradually, that is, the chain has a constant curvature in the tangent plane. However, the orientation of the tangent plane is changing randomly, and the chain resembles a worm. The average projection of the end-to-end vector of the model onto the initial tangent of the curved chain approaches a finite (and relatively small) limit as the contour length approaches infinity. This limit is called *persistence length;* it is the characteristic quantity of the wormlike model. When the contour length is much larger than the persistence length, the particle is coiled and the model behaves essentially like the statistical segment model described above. In such a case, the persistence length is equal to one-half of the segment length A.

1.3.1.3. Long-Range Interactions.

All the above models are variations on the theme of a *random walk,* which considers the shape of a path of random steps. There may be some correlation between neighboring steps, but it is completely lost after some number of steps. The random walk approach has been used (first by Albert Einstein) very successfully for the treatment of diffusion. If we imagine that the diffusing particle leaves a material trace, we get a model of a macromolecular coil that belongs to the class of models described in the preceding section. These models are often called *Gaussian* because the average radial distribution of the segment densities is a Gaussian (error) function. However, there is a basic difference between a diffusing particle and a macromolecular chain. The former may freely cross its own path, whereas the latter cannot pass through volume elements already occupied by other segments. Moreover, the statistics of the walk is influenced even in cases when two distant segments along the chain come into close contact. When the two segments interact—that is, exhibit either net attraction or repulsion—the random walk statistics is distorted by the Boltzmann factors, which increase or decrease the statistical significance of conformations with such contacts. In fact, the effect of mutual segment exclusion and the effect of segment interaction influence the conformational statistics in a complementary way and may even cancel each other. The situation merits consideration in some detail.

First we will treat the problem in a qualitative way. Let us imagine an ensemble of hypothetical conformations obeying the random walk statistics. The average distribution of segment densities will be Gaussian, and the average end-to-end distance will be given by equation (1.3.8). Obviously, there will be many conformations with overlapping segments. Such overlapping conformations will be more frequent among dense coils (short end-to-end distances) than among loose coils (long end-to-end distances). The overlapping conformations are impossible for a real coil; let us eliminate them from the ensemble. It is apparent that most of the dense coils will be eliminated, and the average end-to-end distance of the remaining conformations will be larger than in the original ensemble. Let us now consider a case in which formation of a contact between two segments lowers the free energy of the system. (We will see later that this happens for most polymer-solvent pairs.) The conformations with many segment-segment contacts will exhibit larger Boltzmann factors than coils with no or few contacts. The former will be represented in the ensemble more than the latter. Obviously, the denser conformations have on average more internal contacts than the looser ones. The modified statistics produces shorter average end-to-end distances.

In some polymer-solvent systems, the effect of segment exclusion prevails; the end-to-end distance is larger than for the starting Gaussian coil and increases with increasing molecular weight faster than its square root. In such a case the solvent is said to be a *good solvent* for a particular polymer. For other systems, the effect of segment contacts prevails; these are the *poor solvent* systems. The end-to-end distance is smaller than for the Gaussian coil. Actually, this situation can be observed only for a narrow range of conditions. If the segment-segment contacts are preferred to too high a degree, the coil will collapse and the polymer will precipitate out of solution.

At some borderline conditions the effects of segment exclusion and segment-segment contacts on the average end-to-end distance cancel each other exactly and the latter obeys equation (1.3.8). Such solutions are called *pseudo-ideal*, and the coils are said to be *unperturbed*. Many physicochemical dependencies become especially simple under these circumstances. The Boltzmann factors influencing the statistics depend on temperature. For a given (poor) polymer-solvent system, the exact balancing of the effects will occur at a characteristic *theta temperature*. We will then talk about *theta solvents* and *theta conditions*.

It should be noted that for pseudo-ideal systems the relations for the original Gaussian coils are restored only for some properties (end-to-end distance, radius of gyration, virial coefficient). For some other properties (notably for short-range distribution of intersegmental distances), the statistics will remain distinctly non-Gaussian.

The situation described above is frequently treated by so-called perturbation theories in a quantitative way. These theories are quite involved; their main ideas are presented in Section 3.1.2.2. Although the results of the perturbation treatment may vary considerably from one version of the theory to the next, most theories agree on the following description of the dependence of $\log h^2$ on $\log N$. The dependence is curved, and its slope in the region of low N (low molecular weight) is 1.0 (the same as for the Gaussian coil). The limiting slope in the region of a very large N is equal to 1.2.

Different versions of perturbation theory predict (within the above general description) different forms of the dependence discussed above. It would be nice to be able to follow the conformations of an ensemble of polymer coils experimentally and to calculate the relevant properties as ensemble averages. Although such an experiment cannot be performed on a real solution, it can be modeled by computer. The modeling procedure is usually based on a selection of a long series of random numbers, hence the term *Monte Carlo methods.*

Because of the space-filling properties of molecules in the liquid state, there is a definite short-range order in liquids: Each molecule has a more or less well-defined number of neighbors, and these neighbors are located at a more or less defined distance from the central molecule. Thus the short-range order is very similar in liquids and in crystals. Crystals, of course, have a long-range order that is missing in liquids. However, in the study of problems in which only the short-range order is important, the liquid is often modeled as a crystal; that is, the molecules are placed onto a crystalline (usually called *pseudocrystalline*) lattice. The introduction of long-range order into the model usually has no physical consequences, but the computational procedures are immensely simplified. The Monte Carlo modeling of a macromolecular coil is usually done on such a lattice.

Typically the computer models a polymer chain by placing a string of segments onto adjoining lattice points; the chain "grows" in the computer. When the next segment is to be placed onto the lattice, it is placed onto a lattice point that is randomly selected from the neighbors of the preceding segment. In the simplest case, it is permissible to place two or more segments on the same lattice point; the chain may intersect itself. In this case, an ensemble of chains of a given length is easily generated and the ensemble averages are calculated. The results follow closely the predictions for the Gaussian unperturbed coil. In a more interesting procedure, the chains are not allowed to intersect themselves. If the random procedure calls for a chain to intersect itself, the procedure is discontinued and a new chain is started. Thus only a small fraction of the chains started reach the desired length. This decrease is called *attrition* and is especially severe when long chains are to be generated. Nevertheless, if enough computer time is available or if some computational tricks are used, it is possible to generate a sufficiently large ensemble for any required chain length. For such nonintersecting chains, the average square end-to-end distance increases with the number of segments N raised to the 1.2 power. This is a result identical with the predictions of perturbation theories in the limit of very high values of N.

When generating an ensemble of nonintersecting chains, it is also possible to keep track of the number of intersegmental contacts for each molecule. Furthermore, interaction energy can be assigned to every contact and Boltzmann factors can be computed for each molecule according to the number of contacts. The ensemble averages can then be recalculated using such Boltzmann factors. The entire procedure can be repeated using different values for the interaction energy. As the (negative) interaction energy increases, the Boltzmann factors increase also. The more compact chains with many contacts contribute more to the ensemble averages and h^2 decreases. When the calculation is repeated for a number of chain lengths, h^2 is still proportional to a power of N. However, the exponent, which was 1.2 for zero interaction energy (unweighted averages), decreases with increasing interaction energy and reaches the value 1.0 for some specific value of the energy. Although this result corresponds to the unperturbed coil as far as h^2 is concerned, the statistics remains quite non-Gaussian for the short intersegmental distances. Chains obeying this statistics are frequently called *pseudo-ideal.*

For chains with values for interaction energy between zero and the pseudo-ideal value, the plot of log $\overline{h^2}$ versus log N is a straight line with a slope between 1.0 and 1.2. This is frequently expressed as

$$\overline{h^2} \sim N^{1+2\varepsilon} \sim M^{1+2\varepsilon} \qquad (1.3.10)$$

where the parameter ε (which varies between 0.0 and 0.1) is related to the interaction energy. Relation (1.3.10) is used extensively for many calculations relating to end-to-end distance. However, it is at variance with the perturbation theories that predict that the plot is curved and that the slope of the tangent is continuously increasing from 1.0 to 1.2. The situation remains controversial, and both approaches are being used.

Before leaving the problem of the size of the polymer coil, we need to recognize that the concept of end-to-end distance is meaningful only for strictly linear molecules. Moreover, end-to-end distances cannot be measured experimentally. *Radius of gyration* r_G is another measure of molecular size; it is defined as the square of the average distance of a mass element from the center of mass of the particle, and it can be measured by light-scattering methods (Section 3.5).

$$r_G^2 \equiv \frac{\sum_i m_i r_i^2}{\sum_i m_i} \qquad (1.3.11)$$

Here, m_i is the mass of the ith element and r_i is its distance from the center of mass.

For homogeneous particles, r_G is easily calculated from their geometry. For Gaussian coils, it is easy to show that radius of gyration and end-to-end distance are related as

$$r_G^2 = \overline{h^2}/6 \qquad (1.3.12)$$

This relation is a good approximation even for non-Gaussian coils.

1.3.1.4. Polyelectrolyte Effect.

The perturbations of the polymer coils described in the previous section were caused by the van der Waals forces, which are generally short-range, that is, they act only between closest neighbors. (The term *long-range interactions,* as used in the preceding section, referred to the distances along the chain; in the physical space, the distances are short.) The other intermolecular forces—electrostatic interactions between ions—decrease only with the square of the interionic distance (the potential of these forces decreases only with the first power of this distance) and consequently can act over a distance of many molecular diameters. Thus they have quite a profound effect on the conformation of the polymer coils.

The basic rules of ionic interactions are the same in polyelectrolyte solutions and in solutions of low-molecular-weight electrolytes:

1. Oppositely charged ions attract each other; ions with the same charge repel each other.

2. The electrical forces are inversely proportional to the dielectric constant of the solvent. Thus they are relatively smaller in water, which has one of the highest dielectric constants.

3. The charges are mutually shielded from each other by the Debye–Hückel diffuse ionic atmosphere, characterized by its effective radius.

4. In solvents with low dielectric constant, oppositely charged ions form firmly bonded pairs, which are not easily separated.

Let us consider several types of polyelectrolytes in pure water and in salt solutions (buffers). We will restrict ourselves to linear molecules forming random coils in solution. Our first example will be a salt of some polyacid [e.g., polyacrylic acid or poly(styrene sulfonate)] in water. At very high dilution, a sizable fraction of small counterions will be in the intercoil region. The macromolecule itself will carry many charges (negative in this case), which will repel each other. The macromolecule will adopt a very loose conformation; in the case of short chains, a fully stretched rodlike conformation may result. Thus the end-to-end distance will be large, as well as the volume occupied by the coil; this in turn will influence strongly all the solution properties that depend on molecular size, including the diffusion coefficient, sedimentation coefficient, and intrinsic viscosity.

When the concentration of the polysalt is increased, more and more counterions remain inside the coil and shield the negative charges from each other. The situation could be treated as an increase in ionic strength resulting from the increased polymer concentration and as a decrease in the radius of the Debye–Hückel ionic atmosphere. As a consequence, the end-to-end distance and the coil volume decrease sharply with increasing concentration. This leads to a dependence of many polymer properties (diffusion coefficient, etc.) on concentration that is opposite to the one observed for uncharged polymers. However, even at moderately large concentrations, the size of a polyelectrolyte molecule in water is substantially greater than that of an otherwise identical uncharged molecule.

Let us now make a highly dilute solution of our polysalt in a solution of low-molecular-weight electrolyte. This electrolyte will penetrate the coil, lead to the Debye–Hückel shielding effect, and suppress the extreme coil expansion. When the polymer concentration is increased under such circumstances, it does not change the ionic interactions considerably; the dependence of the diffusion coefficient and other properties on concentration will follow the usual rules. Such a manipulation of the polyelectrolyte effect is used routinely in the study of biochemical materials—the ionic strength of buffers is kept moderately high (0.1 and more) to suppress undesirable concentration effects.

Our second example will be a solution of a weak polyacid (e.g., polyacrylic acid) in water. At very high dilution, a large fraction of the carboxyls will be ionized, and the solution will display a strong polyelectrolyte effect. With increasing concentration, not only will the counterions shield the charges, but also (because of the laws of dissociation of weak electrolytes) the degree of dissociation will decrease dramatically. Thus the polyelectrolyte effects of polyacids (in the absence of pH-controlling buffers) will decrease with increasing concentration even faster than in the case of the polysalt.

We would like to introduce a special case. The amide group has a weakly basic character; consequently, nylons are soluble in concentrated formic and sulfuric acids. Formic acid is a weak acid; in the nylon solutions, it is ionized only to the extent needed for protonation of the amidic groups. Thus nylon solutions in formic acid exhibit very strong polyelectrolyte effects, which may be partially suppressed either by increasing the concentration of the polymer or by adding sodium formate. On the other hand, concentrated sulfuric acid contains a large amount of ionic species (HSO_4^-, $H_3SO_4^+$, etc.), that effectively shield the polyelectrolyte effect. The concentration dependence corresponds to nonpolar polymers.

In media of low dielectric constant, polyelectrolytes (they are often called *ionomers* under these circumstances) behave quite differently. The counterions remain closely associated with the charged groups of the polymer, and the coil adopts a standard coil conformation. If the number of ion pairs is large, they attract each other. This leads to a contraction of chains at high dilution and to polymer association at higher concentration. Poly(acrylic acid) in methanol (a solvent with an intermediate dielectric constant) first expands and then collapses during neutralization.

1.3.2. Secondary Structure—Regular Conformations

Macromolecules assume lower energy conformations frequently, and such arrangements are usually quite regular. Let us consider first a similar situation in low-molecular-weight systems. The small molecules play the role of macromolecular sections, and interaction with other molecules causes the lowering of energy. In this case, the regular arrangement is a crystal.

In macromolecular systems, there are more ways to achieve this advantageous arrangement.

1. Many polymer molecules may interact and form a crystalline structure. We will treat such structures in Section 4.1.6 when we deal with crystalline polymers.

2. The regular arrangement may include only a few molecules. Section 1.4 is devoted to such systems.

3. Because of the repetitive character of the polymer chain, the macromolecule may interact with itself, usually forming a *helical structure*—the subject of the present section.

The secondary forces holding the macromolecule in a specific secondary conformation must be strong enough to overcome the tendency of the macromolecule to form a statistical coil; the coiling would increase the entropy considerably. It should be remembered that even in the coiled state many of the secondary forces would be able to play their role; the regular structure has the advantage of only a few extra contacts of low energy. It follows that the energy advantage per single contact and/or conformation should be rather large to overcome the entropy effect; with some exceptions, only the hydrogen-bonded moieties qualify.

Proteins and nucleic acids most prominently display regular secondary structures. In proteins, the hydrogen of an amidic group can form a rather strong hydrogen bond with a carbonyl of another group.

$$-NH-\underset{\underset{R}{|}}{CH}-\overset{\overset{|}{}}{C}=O\cdots H-\overset{\overset{|}{}}{N}-\underset{\underset{R}{|}}{CH}-\overset{\overset{|}{}}{C}=O$$

Not all helical structures of a polypeptide chain with intramolecular hydrogen bonds are stable enough to actually occur as the dominant conformation. A stable helical structure of a polypeptide must involve every amidic group in two hydrogen bonds: one through the nitrogen and the other through the carbonyl oxygen. In addition, the structure must satisfy the conformational rules for the bonds involved. The peptide group must be planar; the hydrogen atom involved in the hydrogen bond must lie not more than 30° from the vector connecting the oxygen and nitrogen atoms involved; the arrangements of atoms around single bonds must correspond to the energy minimum. The α-helix is the only helix satisfying all the criteria (Fig. 1.14). There are 3.6 amino acid residues per turn of the helix; that is, subsequent residues are rotated 100° with respect to the axis of the helix. The *pitch*, or spacing between successive turns, is 5.4 Å. The translation along the helical axis is 1.5 Å per amino acid residue. The side groups of the amino acid residues stick out away from the helical axis; the α-helix can accommodate all amino acids except proline (proline cannot form two hydrogen bonds either). The structure of the α-helix is very rigid; it forms a very inflexible cylinder.

Many synthetic polypeptides [notably poly(γ-benzyl glutamate)], when dissolved in solvents incapable of forming hydrogen bonds, such as dimethylformamide, adopt the α-helical structure in solution. However, hydrogen-bonded solvents such as dichloroacetic acid compete successfully with the intramolecular hydrogen bonds. In such solvents, these polypeptides form statistical coils.

Among natural proteins, α-keratins have an almost pure α-helical conformation. Many globular proteins have shorter or longer helical sequences comprising up to 60% of the residues. The helical segments are frequently terminated by proline residues.

Two other conformations of polypeptide chains satisfy the stability conditions: parallel and antiparallel "pleated sheet" structures. In both structures a number of chains are connected by lateral hydrogen bonds—the terms *parallel* and *antiparallel* specify the relation of the directions of neighboring chains (the direction is understood to point from the carboxyl to the amino end of the protein). The energy of the pleated sheet structure is even lower (the stability is higher) than the energy of the α-helix conformation. In the sheet structures every other side group is sticking out of the sheet plane; the remaining side groups are in the sheet plane and interfere with the close packing. Consequently, only proteins having glycine as every other residue can form extended structures of this type; β-keratins (fibroin, silk) are prime examples. Strictly

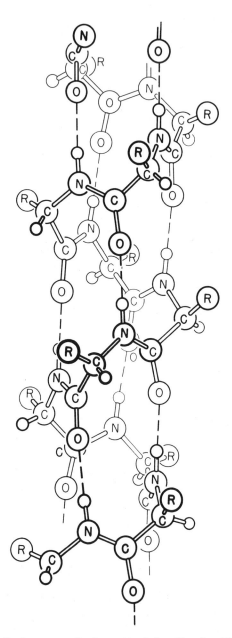

Figure 1.14. The α-helical structure of polypeptide chains. (Reprinted from C. Tanford, *Physical Chemistry of Macromolecules.* Copyright © 1961 by John Wiley & Sons, New York. Used by permission of the publisher.)

speaking, the pleated sheet structure of fibroin should be treated in Section 1.4.5 dealing with multimolecular structures. However, in many globular proteins several sections of the same polypeptide chain may run parallel to each other, forming a rudimentary pleated sheet within the molecule (the glycine rule is relaxed in this case for the outermost chains of the sheet).

Polysaccharides have a large number of hydroxyl groups capable of forming strong hydrogen bonds. However, for most polysaccharides (a typical example is cellulose) the hydrogen bonds are intermolecular, and the resulting structure is crystalline. One exception is amylose; when dissolved in water, amylose forms a random coil that slowly twists into a helical conformation with six glucose units per turn. The helix has a channel in its middle into which some smaller molecules can fit. The complex with iodine is deep blue. This coloration is so intense that it can serve as a qualitative analytical technique for either iodine or amylose.

The well-known double helix of DNA consists of two molecules and therefore belongs to Section 1.4.

1.3.3. Tertiary Structure—Arrangement of Larger Segments

Many natural macromolecules, and notably proteins in their native form, exhibit only a single conformation, which represents the tertiary structure of the protein and is quite compact (hence the term *globular proteins*). Usually, several α-helical segments are connected by either loose loops or straight segments participating in pleated sheet structures. The side chains are engineered in such a way as to form interlocking surfaces between distant parts of the chain. The desired conformation is achieved by employing a whole array of secondary forces. In addition to the forces that are instrumental in the helical and sheet segments, clusters of nonpolar segments are held together by so-called hydrophobic forces; ionized carboxyl groups of aspartic and glutamic acids can interact with the ionized amino groups of lysine and arginine; amidic groups of asparagine and glutamine as well as hydroxyls of threonine and serine can participate in additional hydrogen bonds. The sulfur-sulfur bonds holding together the two halves of the cystine residues are another important contribution to the stabilization of the native structure of proteins. In water-soluble proteins, most nonpolar side chains reside inside the structure; the polar ones are mainly on the surface.

The unique conformation of a protein molecule may be destroyed, and the protein may convert into a huge ensemble of more random conformations—in some cases into random coils. This process, called *denaturation,* can be achieved either by increasing the temperature or by appropriate chemical means. The driving force for denaturation is the increase in entropy that accompanies the transition of a single conformation into an ensemble of random ones. With increasing temperature the contribution of this entropy increase becomes more significant, and at some temperature it overcomes the energy effect—the protein is *heat denatured.* It is interesting to consider possible intermediate structures. The early unlocking of the tertiary structure deletes a large number of the bonds holding the structure together but increases the randomness only insignificantly. The later stages of denaturation lead to larger increases in entropy. Thus the intermediate states are relatively unstable, and the heat denaturation is often an *all-or-none* phenomenon. The unfolding of the protein exposes the inside nonpolar amino acid residues; their intermolecular clustering leads to aggregation of the denatured protein (remember boiling egg white). Consequently, heat denaturation is essentially irreversible.

In *chemical denaturations* the secondary bonds holding the protein segments together are disrupted by some chemicals capable of forming equally strong or stronger bonds with the groups holding the conformation together. For disrupting the hydrogen bonds, urea or guanidinium chloride are used most frequently. Under these conditions, the denatured molecules remain in solution and may revert into native molecules if the denaturing agent is slowly dialyzed away. Other powerful denaturing agents are detergents (e.g., sodium dodecyl sulfate), which disrupt both hydrophobic bonds and hydrogen bonds and effectively solvate the denatured molecule. Denaturation can also be achieved by a high or low pH or by the addition of organic solvents.

Ribonucleic acids, notably tRNA (transfer RNA) and rRNA (ribosomal RNA), frequently exhibit unique tertiary structures resembling those of proteins.

1.3.4. Enzymes—Tertiary Conformations in Action

Enzymes are proteins designed for a single purpose: to catalyze chemical reactions needed during the metabolic processes in living organisms. Nature developed the art of enzymatic catalysis to a degree that makes the most astonishing feats of modern chemistry pale in comparison. On the one hand, the catalysis is frequently extremely specific; it may be very efficient for a given substrate while the catalytic effect is completely absent for its isomer or homolog. (*Substrate* is a compound that forms a complex with enzyme that subsequently decomposes to products and enzymes.) Frequently, the substrate has several chemical groups of the same type (e.g., carboxyls, amino groups), yet the desired reaction takes place exclusively on one of them. A synthetic chemist would have a hard time protecting the other groups and removing the protection after the reaction. Enzymatic synthesis circumvents these problems by its specificity. On the other hand, enzymatic catalysis is extremely fast; the rate constants are enhanced by many orders of magnitude—often by seven orders or more. In fact, the rate-limiting factor is often the diffusion of substrates toward and away from the active catalytic sites.

As in all catalytic reactions, the presence of the catalyst does not shift the reaction equilibrium. It only speeds up both the forward and reverse reactions. This tendency of reactions to go toward equilibrium has an unpleasant consequence: No amount of catalysis can run any reaction against the dictates of thermodynamics. Both nature and synthetic chemists circumvent this obstacle in a similar way; they prepare a highly energetic derivative of the substrate and design a series of reactions going thermodynamically downhill. In synthetic chemistry such derivatives may be acyl chlorides, anhydrides, epoxides, Grignard reagents, etc. In enzyme chemistry, the

energetic derivatives are usually phosphates. Adenosine triphosphate (ATP) participates in many biochemical reactions.

ATP

The highly energetic triphosphate chain in ATP is synthesized by a series of reactions
that gain their energy from the energetic source for the organism, for example, from
glucose oxidation. The transfer of the terminal phosphate group to other hydroxy
compounds is accompanied by the liberation of a substantial amount of free energy,
which is used as the thermodynamic driving force for the desired reaction. In thermodynamic terms, the hydrolysis of ATP and the desired reaction are *coupled;* the
coupled reactions are now thermodynamically feasible.

Both the specificity of the enzymes and their catalytic activity are results of the tertiary structure. The enzymatic reaction takes place at a particular part of the molecule
of the enzyme called the *active site.* The active site is usually at the bottom of some
crevice on the surface of the enzyme. The walls of the crevice are designed in such a
way as to accommodate exactly the appropriate substrates. The degree of exactness
required determines the specificity of the enzyme. Thus, among *proteolytic enzymes*
(which as a group are not extremely specific), *papain* will hydrolyze almost all peptide
bonds as well as many amides, esters, and so on. *Chymotrypsin* has the highest activity for hydrolysis of peptide bonds where the carboxylic function is donated by
an aromatic amino acid; its activity on the peptide bonds contributed by other amino
acids is much smaller. *Trypsin* primarily attacks peptide bonds where the carboxyl belongs to a basic amino acid, arginine or lysine. *Carboxypeptidase* digests the proteins
sequentially from the carboxylic end of the chain.

The *specificity* of the crevice walls toward the appropriate substrates is achieved
both by sculpturing the shape of the crevice by employing amino acids with convenient
side chains at the right places in the molecule and by placing amino acids with groups
capable of appropriate secondary bonding (hydrogen bonds, ionic bonds, dipole-
dipole interactions) in strategic locations within the crevice. When the fit between the
enzyme and the substrate is rather tight and involves large sections of the substrate
molecule, the enzyme is very specific. Otherwise, the enzyme may catalyze reactions
on a whole class of substrates.

The most important function of the active site is the catalysis itself. Let us review the general principles of catalysis. A chemical reaction is a transition of reactants characterized by certain arrangements of chemical bonds to products with

different arrangements. During that transition the molecules must pass through some arrangement with energy higher than the energy of reactants (*activated state*).The higher the energy difference (*activation energy*), the slower the reaction. The function of the catalyst is to help to create a transition path that does not exhibit any high maximum, that is, one that has low activation energy. For example, if some bond has to be broken during the reaction, the catalyst may interact with neighboring atoms in a way that weakens the bond in question. This could be achieved by polarizing the bond by the removal or addition of a proton, by induction or coordination effects, etc. The more "smoothed" the reaction path, the more effective the catalyst.

Let us present just one example of enzymatic catalysis. The active sites of chymotrypsin and trypsin are almost identical. Chymotrypsin has a "hydrophobic pocket," made from several hydrophobic residues, which accommodates the aromatic amino acids. In trypsin, there is an ionized carboxyl present in this pocket, which interacts with the basic groups of trypsin substrates. The catalytic site itself is constituted by a histidine residue and a serine residue that are positioned in such a way that the serine hydroxyl is hydrogen bonded to the aromatic nitrogen of histidine. (The two residues are not neighbors in the polypeptide chain; in fact, they are separated by about 140 other residues.) The catalytic serine-histidine hydrogen bond is a part of an intricate network of about six hydrogen bonds between amino acid residues from all sections of the chain; they are held in their respective places by the unique structure of the rest of the molecule. A possible scenario of the catalytic action has the hydrogen on serine momentarily removed by histidine; then the (now charged) oxygen attacks the amide bond to be cleaved and is acylated by the liberated carbonyl. Next, histidine donates the hydrogen to the amino group of the cleaved peptide and serine is deacylated again. The network of the hydrogen bonds serves as a *charge-relay system,* reducing the energy changes associated with individual steps of the scheme.

1.4. MULTIMOLECULAR ARRANGEMENTS— QUATERNARY STRUCTURE

In living organisms the macromolecules are frequently assembled into higher structural units. These assemblies exist on several levels, the highest one being the whole organism. This book treats only those structures that in their physicochemical behavior resemble single particles. We also mention some crystalline and quasi-crystalline structures (cellulose, keratins).

1.4.1. Multiunit Proteins

A large number of proteins are composed of more than one polypeptide chain. These chains are held together by the same forces that are operative for establishing the secondary and tertiary structure: hydrogen bonds, salt bonds, hydrophobic interactions. We will distinguish two cases. In some proteins two or more chains are intimately intertwined. For example, myosin consists of two α-helical chains wound around each other. Hydrophobic bonds between nonpolar amino acids in every seventh position stabilize this structure. Fibrinogen is a complex of three chains; its shape is quite unique—three spherical beads connected by thin cylindrical rods. Three chains of collagen form a helical structure, which is stabilized by hydrogen bonds between glycine residues in every third position.

In another family of proteins and protein complexes, the individual chains form compact subunits, which are assembled into larger structures. The surfaces of the subunits sticking together are sculptured in an almost precisely complementary way. The situation merits closer scrutiny. The amino acid residues and other formations within a protein molecule are generally asymmetric. It follows that two identical portions of the molecular surface cannot be complementary and cannot stick together; the complementary part to any surface part must be located elsewhere on the surface. Thus, once the contact is realized, "sticky" portions of the surface still remain. In the simplest case, two units are joined by two pairs of complementary areas; the dimer exhibits twofold rotational symmetry. If the two complementary portions have an appropriate spatial relation, cyclic oligomers—trimers, tetramers, pentamers, etc.—may be formed. Frequently, the molecules possess more than one pair of complementary surfaces; this may lead to more complex structures. For example, two dimers may stick together, forming a tetramer with either square or tetrahedral arrangement. Structures with much higher symmetry are possible. The enzyme *lipoyl transacetylase* (LTA) from *Escherichia coli* is composed of 24 subunits that are arranged with cubic symmetry; a group of three subunits is located at each vertex of the cube. This arrangement is very stable; strong denaturing agents are needed for its destruction. Another enzyme with an identical biochemical function, beef LTA, is composed of 60 subunits forming an icosahedron.

Many proteins are composed of subunits of different kinds. For example, hemoglobin has two subunits designated α and two subunits that are β—it is an $\alpha_2\beta_2$ tetramer. The α and β subunits are quite similar in their secondary and tertiary structures but have several differences in their primary structure (amino acid sequences). One α subunit is bound to a β subunit very firmly, forming an $\alpha\beta$ dimer. This dimer is bound a little bit less firmly to a second dimer. The formation of tetramers in hemoglobin as well as the dimerization of many enzymes has a definite (even if not fully understood) role in the function of the protein. In the case of hemoglobin, the tetrameric structure helps to establish the isotherm for binding oxygen that is the most advantageous for the organism.

In designing biochemical pathways, nature has broken them into a large number of steps and developed an enzyme for catalyzing each of the steps. Sometimes it was convenient to assemble enzymes into larger multienzyme complexes catalyzing several successive steps. We will illustrate this strategy on an enzyme complex called *pyruvate dehydrogenase*. Dehydrogenation of pyruvic acid (it is actually oxidative decarboxylation) is one of the more important reactions in the degradation of many substances derived from food (glucose, amino acids, fatty acids). The goal of this process is to generate energy-carrying compounds, typically ATP. The reaction to be performed by pyruvate dehydrogenase complex is to split off carbon dioxide from the pyruvate anion and attach the resulting acetyl group to a reactive —SH group of a complex organic molecule called coenzyme A (CoA):

$$CH_3COCOO^- + CoASH \longrightarrow CO_2 + CoASCOCH_3$$

The pyruvate dehydrogenase complex (PDC) is composed of three types of subunits. Its core is the lipoyl transacetylase (LTA) mentioned previously. Each subunit of LTA is characterized by a long, flexible arm consisting of the side chain of a lysine residue, which through its ω-amino group forms an amide with *lipoic acid:*

$$-\text{lysine}-NH-CO-(CH_2)_4-\underset{\underset{S}{|}}{CH}\overset{\displaystyle CH_2}{\underset{\underset{S}{|}}{\diagdown CH_2}}$$

The sulfur-sulfur bond of lipoic acid can be easily and reversibly reduced to two —SH groups; lipoic acid is a useful oxidizing agent. The LTA subunit also has a bonding site for CoA. The second enzyme of PDC is *pyruvate decarboxylase,* which exists in the form of dimers. Twelve of these dimers are associated with the edges of the LTA cube. The third enzyme is also dimeric; it is *dihydrolipoyl dehydrogenase,* which is capable of oxidizing the reduced form of lipoic acid to its cyclic form. Six of these dimers are located at the faces of the cube.

The function of the complex is believed to consist of the following steps. First, the pyruvic acid binds to the decarboxylase. The lipoyl arm of LTA then reaches to the pyruvate and oxidizes it, releasing CO_2. The acetyl is bound to the —SH group of the reduced lipoic acid. Next, the arm swings to the CoA attached to the LTA; the acetyl group is transferred to the CoA, and the acetyl-CoA leaves the complex. The arm, now carrying the reduced form of lipoic acid, then swings to the dihydrolipoyl dehydrogenase, where it is oxidized back to the lipoic acid. The dihydrolipoyl dehydrogenase is then oxidized back by another coenzyme, nicotinamide adenine dinucleotide (NAD^+).

It is obvious that the special arrangement of the three components of PDC greatly facilitates the entire decarboxylation sequence. We have dwelled on this example extensively to show how details of the higher structure (in this case the quaternary structure) may be intimately related to the function of the biochemical moiety.

1.4.2. Double Helix of Nucleic Acids

Most of the important functions of nucleic acids are related to their ability to form a double helix, a structure in which two chains of nucleic acid are connected by multiple hydrogen bonds between the heterocyclic bases. The grand design is actually rather simple. Guanine is always paired with cytosine, forming three hydrogen bonds; adenine forms two hydrogen bonds either with thymine (in DNA) or with uracil (in RNA).

adenine

$X = CH_3$ **thymine**
$X = H$ **uracil**

guanine   **cytosine**

The neighboring pairs of the bases are stacked parallel to each other; the distance between the neighboring base planes is about 3.5 Å. The two backbones (sugar-phosphate chains) are twisted into a double helix. Several slightly different forms of the double helix have been suggested; actually, all of them may exist under appropriate circumstances (varying humidity, counterions). The most widely accepted form has 10 residues per turn; the base planes are perpendicular to the axis of the helix; and the diameter of the helix is roughly 22 Å. There are two helical grooves between the two strands of nucleic acid; one deeper (major) and one shallower (minor) groove. This helical structure is very stiff, but free molecules of DNA in solution are not perfect rods. In fact, the slightest possible bending of the helices must eventually lead to a coil structure. However, the length of the statistical segment for DNA is about 1000 Å; in typical synthetic polymers, segments are about 10 Å long.

The structure described above is typical for DNA; the double-helical strand may have hundreds of thousands of residues. RNA can also form a helix with two strands but does so only rarely. More often, two complementary sections of the same chain form a double helix in a hairpin fashion—sometimes with an intervening loop. Such structures are especially important in tRNAs and mRNAs. The mRNA can also form a hybrid double helix with DNA.

The main feature of all nucleic acid double helices is the fact that within the helical region the two strands must be strictly complementary. Thus when one strand has any special sequence of the four bases—a *code*—the other strand must have an exactly complementary sequence—an *anticode*. This complementarity is then used both for the reproduction of the code and for transfer of the code to the protein manufacturing plant, the ribosome.

For proper understanding of these two processes it is necessary to study first another phenomenon—*denaturation of DNA.* The denaturation of DNA has many similarities to the denaturation of proteins. Again, the decrease in enthalpy, which is caused by the extensive formation of hydrogen bonds in the paired bases, is paid for by a decrease in entropy due to the uniqueness of the double-helical structure. With increasing temperature, the entropy contribution dominates: the double-helical structure disintegrates, and the two chains separate, forming two random coils. For DNAs in which the adenine-thymine pairs (with two hydrogen bonds) prevail, the denaturation temperature is lower than for DNAs rich in guanine-cytosine (with three hydrogen bonds). The denaturation is best studied by UV absorption; the breaking of the hydrogen bond is accompanied by an increase in absorption. For homogeneous samples of DNA the transition is rather sharp. However, most samples of DNA (with the exception of some viral DNAs) are heterogeneous (composed of many different species). The heterogeneity of the base composition can be judged from the width of the temperature interval over which the transition takes place. The denaturation of DNA gives us a measure for the time scale of macromolecular processes. The separation of the two chains requires an unwinding of a helix with up to tens of thousands of turns, yet it is accomplished within seconds.

Even more remarkable than denaturation of DNA is its renaturation. When a denatured solution is slowly cooled, a substantial portion of the molecules return to the double-helical structure. This implies that the two complementary chains must find each other and match the appropriate sections of their strands. Despite apparently very low odds for this to happen, the renaturation process proceeds quite readily even for quite heterogeneous samples of DNA in which a very large number of different types of DNA are present.

Within the living cell, most of the DNA is concentrated in the nucleus. The highly acidic DNA forms complexes with a family of highly basic proteins called *histones.* A globule of eight histones with a section of the DNA wound around it forms a *nucleosome.* The "pearl necklace" of these nucleosomes is *chromatin,* the material from which heredity-carrying material—*chromosomes*—are made.

1.4.3. Genetic Coding and Reproduction

As we mentioned in Section 1.2.8.3, the sequence of the heterocyclic bases in DNA forms a code describing all the information needed by the organism throughout its lifetime. Successful function of an organism requires dependable *safekeeping* of the information, its *reproduction* during cell division, *transfer* of the information to the place where it is needed, and its *decoding* when it is actually needed. Most of these functions are intimately related to the double-strand concept of DNA. Let us look at these processes in some detail.

The whole operation closely corresponds to the operation of a manufacturing plant (one that is extremely well organized). In the plant, safekeeping is accomplished by keeping the crucial manufacturing recipes in a compact form at a safe place in more than one copy. In the cell, DNA is kept in tightly compacted form in the cell nucleus. The two complementary strands are the two copies of the information. On top of that, several enzymes are able to repair damaged DNA. For example, *DNA ligase* can join two pieces of DNA to form a continuous chain. Other enzymes can repair other kinds of damage.

When a plant establishes a subsidiary, copies of the recipes must be made. Living organisms during cell division must replicate the DNA. During this process, the double helix is unwound, and an enzyme called *DNA polymerase* synthesizes the DNA chains, using the original chain as a *template* and joining the appropriate nucleotide triphosphates into complementary chains (with release of inorganic phosphate). The two chains are not duplicated from one end to the other; it would take too long a time, not to mention the difficulties with fast rotation accompanying the unwinding of the helix. The actual duplication procedure involves nicking one of the chains at many places, unwinding short segments of DNA, synthesizing the daughter chains within these *replication forks,* and finally reconnecting the gaps using DNA ligase. During the replication about 50 nucleotides are added per second; however, thanks to the large number of replication forks, a DNA molecule with hundreds of thousands of base pairs can be replicated within minutes.

1.4.4. Protein Synthesis

Let us continue with our plant analogy. A prudent manager would not lend the original recipes for manufacturing to the workers on the assembly line. He would make copies of the master documents and send them to the plant whenever needed. Similarly, when some substrate in the cell is running out of supply, a message is sent to the headquarters (cell nucleus) that an enzyme (or set of enzymes) needed for production of this substrate should be manufactured (the exact nature of the message is not yet fully understood). The building plans (read "genetic code") for this enzyme are located in a section of the DNA molecule called a *gene.* The entire length of the DNA molecules is a linear sequence of genes; however, portions of the sequence have other not fully understood functions related to gene regulation. In any case, when the gene is activated, an enzyme called *RNA polymerase* or *transcriptase* synthesizes a single-strand RNA chain exactly complementary to the encoded message of the gene. Only the section of DNA that corresponds to the protein to be manufactured is copied. This copy of the genetic code is appropriately called *messenger RNA* (mRNA).

As the long chain of mRNA being synthesized emerges from the cell nucleus, it is immediately seized by *ribosomes*—the assembly plants of proteins. A typical cell contains about 15,000 ribosomes, which make about one-fourth of the total mass of the cell. About two-thirds of the ribosomes' mass is made of rRNA; the rest is proteins. The structure and function of ribosomes have been studied extensively. We will present only a short overview here.

Let us recall that each group of three nucleotides is a codon. Successive codons are not separated. It is vitally important to start reading the code at the beginning of a codon and not in the middle; otherwise the message will be completely garbled. The necessary steps involve capturing the mRNA molecule at its end, finding the initiation codon, bringing the initiation amino acid, moving the mRNA by one codon, bringing the next amino acid, synthesizing the peptide bond, and continuing this process until a termination codon is reached. The cell is fully equipped to carry out these functions.

The ribosomes are easily and reversibly dissociated into two subunits of unequal size. The smaller subunit contains a molecule of rRNA and 21 different proteins. The larger subunit has one large and one very small molecule of rRNA and 34 different proteins. The amino acid sequence of many of these proteins is known as well as many details of the sequence of rRNA. Moreover, the quaternary arrangement of these molecules is also partially known. The smaller subunit of the ribosome captures and transposes the mRNA; the larger unit is the place of protein synthesis.

The messenger RNA carries near its end a recognition sequence of bases that forms a complex with a complementary sequence near the end of the rRNA in the smaller ribosomal subunit. Soon after this recognition sequence follows the initiation codon signaling the start of synthesis. The mRNA molecule is situated on the ribosome in such a way that the codon that is about to be translated is exposed to the surroundings of the ribosome, where a family of *transfer RNAs* is swimming. These special nucleic acids all have a very similar cruciform structure. At one end they carry an *anticodon,* which matches only codons coding for a particular amino acid. (There is an economy involved here. We have seen that several codons code for the same amino acid, yet some of them are recognized by the same tRNA.) The ribosome-mRNA complex will interact only with appropriate tRNA. The other end of the tRNA interacts reversibly with the amino acid that it transfers; it is *charged* by the amino acid. The charging alters the tertiary structure of the tRNA; the charged molecule interacts strongly with the codon, the uncharged molecule is easily released.

Once the initiation codon is recognized at the mRNA, an involved series of reactions starts. It leads to the formation of a complex of a special tRNA molecule carrying the initiating amino acid (in *E. coli* it is always *N-formyl methionine*) with a special *P-site* (peptidyl site) on the larger subunit of the ribosome. The next codon on the mRNA complex with the appropriately charged tRNA resides at another special *A-site* (amino acid site). The next reaction is the *transfer* of the peptide chain from the P-site to the A-site and formation of a new peptide bond. Finally, in a *translocation* step, the growing chain with the last tRNA still attached is transferred from the A-site to the P-site, displacing the "spent" tRNA at the P-site. Simultaneously, the mRNA chain moves ahead by one codon and is ready to complex with the next tRNA. Obviously, the whole process is quite complicated and requires the cooperation of all the proteins and rRNAs of the ribosome as well as several initiation and termination factors, and so on, from outside the ribosome.

The newly synthesized protein chain folds spontaneously to its globular form. This is essentially denaturation in reverse. The spontaneous folding is, of course, possible only when it leads to a structure with lower energy; this is indeed the case with globular proteins.

In the above description, we have divided the process of production of proteins into several distinct steps. In fact, the process is frequently speeded up. For example, it was observed in bacteria that the same gene is transcribed into mRNA simultaneously by several transcriptase molecules; several mRNA chains are simultaneously branching off the gene. As the mRNA chain emerges from the nucleus it is seized by a ribosome, and the protein synthesis begins. The beginning portion of the mRNA chain, after leaving the first ribosome, is seized by the next one, and the next, and the next.

Frequently, a whole string of pearl-like ribosomes is attached to the same mRNA molecule, and many enzyme molecules are produced almost simultaneously. Under these circumstances, the original shortage of the enzyme is relieved very quickly and its production should stop. This is the function of the enzyme *ribonuclease,* which digests RNA. The typical lifetime of an mRNA molecule is only a few minutes.

1.4.5. Natural Supportive Structures

The principles of intramolecular and intermolecular bonding that led to the elaborate structure of protein complexes, nucleic acids, and so on are also used by nature for building three-dimensional structures intended as support for the organism. In this section we will discuss the quaternary structure of keratins, collagen, and cellulose.

We have already mentioned that the cysteine-rich protein α-keratin forms long α-helices. In hair and other natural materials, two or three α-keratin helices are twisted together to form a "supercoiled" rope sometimes called a *microfibril.* These ropes are embedded in another essentially amorphous protein with very high cysteine content. An extensive system of sulfur-sulfur bridges holds the whole structure together. Softening of the structure usually requires reducing agents capable of cleaving the sulfur bonds.

The protein of connective tissue, collagen, contains a large proportion of two amino acids, proline and hydroxyproline. These amino acids cannot fit into the α-helix form preferred by other amino acids. However, they form another helical structure composed of three intertwined chains. After the collagen molecules are transported to their final destination in the body, the hydroxyl groups of hydroxyproline and hydroxylysine as well as aldehydic groups formed by oxidation of lysine participate in extensive cross-linking of the collagen, transforming it into insoluble fibers.

In some materials, a large number of molecules are bound together by secondary forces in a quite regular manner, usually forming a fiber. These fibers actually exhibit a crystalline structure. For example, when subjected to X-ray diffraction they yield patterns less distinctive than true crystals but more detailed than powder diagrams. We have already mentioned silk fibroin, in which many polypeptide chains form antiparallel pleated sheets. Cellulose molecules, because of their β-configuration (do not confuse the β-configuration of polysaccharides with the β-conformation of polypeptides), exist as almost inflexible stretched chains. Their side hydroxyl groups are oriented in a way that allows them to form multiple intermolecular hydrogen bonds. The result is a highly crystalline fiber that is insoluble in most solvents (polar, nonpolar, hydrogen bonding).

1.5. AGGREGATES OF SMALL MOLECULES

In the middle of the nineteenth century, some solutions were discovered that did not display the usual solution properties. They exhibited only very low osmotic pressure and did not diffuse through membranes. These solutions were named *colloidal solutions* or simply *colloids.* Two types of materials formed such solutions. One type was well known substances—inorganic salts, gold particles, etc. These materials were obviously composed of small molecules; the peculiar behavior of their colloidal solutions was correctly ascribed to their state of dispersion. Unfortunately, this explanation was extended also to the other type of materials—those that today we call *macromolecules*. At that time, the only known materials of this class—proteins, gums, glues, starches, etc.—were poorly defined and even more poorly understood. For a long time, these true macromolecules were considered aggregates of small subunits; this misconception undoubtedly slowed down the progress of macromolecular science considerably. It was not until the 1920s that the macromolecular nature of these materials was clearly recognized. In the meantime, whatever meager progress was achieved in the macromolecular field went under the name of colloid science.

Two classes of colloids were recognized. *Hydrophobic colloids* were substances usually insoluble in the solvents used; they could be precipitated by the addition of small amounts of coagulation agents (quite often electrolytes). *Hydrophilic colloids* were materials known today as macromolecules. The reason that the scientists were misled into considering both classes of materials together was that both types of solutions contained *particles* that were much larger than ordinary molecules.

Today, when macromolecular science has stolen most of the glamour of colloid science, one basic fact remains the same—the colloidal particles (now we read "aggregates of small molecules") display in solution properties very similar to properties of macromolecular solutions. Consequently, we find it appropriate to give a short overview of their physical nature. We will describe two types of materials: *colloids*—materials basically insoluble in the solvent but kept in solution by some stabilizing factor—and *detergents*—materials having a special structure for which the state of the lowest free energy is in the aggregated form.

1.5.1. Colloids

Colloids are crystalline or liquid particles dispersed in a solvent in which they are not soluble. There is no sharp dividing line between colloids and suspensions or emulsions. The main difference is in the size of the particles: Colloids are usually submicroscopic. It is a well-known fact in surface chemistry that a material forming very small particles has higher chemical potential than the same material under the same conditions but forming larger particles. Hence, the small particles are thermodynamically unstable; the particles should increase in size, and the two phases should eventually separate. (Indeed, such a separation is observed under most circumstances, e.g., after shaking together oil and water.) Two mechanisms of separation are common: (1) two particles after collision may coalesce; and (2) the material from the small particle may enter the continuous phase due to its enhanced chemical potential; it will then join a larger particle with lower chemical potential. Eventually, all small particles will disappear, and only the larger ones will remain. This phenomenon is familiar from analytical chemistry: Many insoluble salts are first precipitated in a very fine form that is difficult to filter. After heating to moderate temperatures, the precipitate becomes coarser and can be filtered easily.

Then what materials form stable colloids (or emulsions)? These are materials for which both mechanisms of particle growth are prevented in some manner. Coalescence is impossible when the particles cannot approach each other, either because of electrostatic repulsion or because of protection by an adsorbed third material. Electrostatic repulsion is common in colloidal solutions of metals or insoluble salts in water solutions; usually the ion present both in the salt and (in excess) in the solution adsorbs on the crystal and gives it its charge. Now, all the crystals have the same charge and repel each other. The effect can be suppressed by increasing the ionic strength of the solution; the electric double layer shrinks, and the particles may stick to each other. However, they may still retain their identity: when the ionic strength is lowered (e.g., by dialysis), they go back into colloid solution (they are *peptized*).

The other transport mechanism, the migration of individual molecules through the solvent, may be inoperative when the solubility of the colloid material is extremely small, as for metals in water. Consequently, some colloid solutions are remarkably stable.

Colloidal solutions proper, in which the particles are separated from each other and undergo independent thermal movement, are called *sols*. Colloidal particles can also be dispersed in gases, usually in air; such systems are called *aerosols*.

Sols exhibit the basic property of fluids: They flow under the influence of the smallest shear stress. Sometimes the colloid particles can touch each other and form semipermanent bonds. When the particles form an extended network permeated by the solvent, we are talking about *gels*. In gels, the kinetic independence of the particles is at least partially lost: Small shear stresses may produce deformation, but a larger stress is needed for a sustained flow. These gels exhibit *yield* behavior. It should be mentioned that the appearance of colloidal gels and many aspects of their behavior are very similar to those exhibited by the swollen loose macromolecular networks described in Section 1.2.7.1.

Colloidal gels are formed from the corresponding sols when the agent stabilizing the sol is rendered ineffective, for example, when the ionic strength is increased and the electrostatic repulsion is suppressed. Colloidal solutions can also display gel-like behavior at higher concentrations when particles form multiple contacts because of steric factors. We should also note that some solutions of true macromolecules can develop a network of semipermanent secondary bonds (frequently hydrogen bonds) and display typical gel behavior. A prime example is gelatin, which gave its name to gels and the process of gelation.

Hydrophobic colloids (or better, lyophobic, when the solvent is different from water) are prepared by either dispersion or condensation methods. The dispersion methods are mainly mechanical or ultrasonic; depending on the size of the resulting particles, either colloidal solutions or suspensions and emulsions are obtained. During the process of dispersion usually a convenient stabilizing agent—an appropriate electrolyte, detergent, or some soluble macromolecule that will adsorb on the dispersed particles—is added.

The condensation methods of colloid preparation are based on the principle of supersaturation. The process of phase separation of metastable solutions is not simple.

Before a new phase separates out, a small *nucleus* of it should appear because of thermal fluctuations; it will then grow to the new phase. However, this small particle (nucleus) has an excess chemical potential and may dissolve again. Thus only a nucleus that is sufficiently large can grow, but the formation of a large nucleus by fluctuations is statistically improbable. Hence, slightly supersaturated solutions may be quite stable as long as some other nucleation center (seed) is not present. In the absence of a seed, and with increasing concentration of the supersaturated solution, a situation will eventually be reached when the statistical probability of the formation of sufficiently large nuclei becomes finite (this state is called *critical supersaturation*), and a number of nuclei will appear more or less simultaneously. After that, the supersaturation will be relaxed by deposition of the solute on the nuclei. If more of the solute is now brought into the solution, it will deposit directly on the growing particles; no supersaturation will be present, and no new nuclei will be formed. In practice, the concentration of the solute is increased by some mechanism (see below) until the critical concentration is reached. When the increase in concentration is fast, high supersaturation may be reached, leading to the formation of a large number of nuclei. The result of the whole procedure will be a large number of small particles. On the other hand, when the increase in concentration is slow, a smaller number of nuclei will be formed at lower supersaturation, leading to a smaller number of larger particles. This phenomenon is similar to well-known procedures in crystallization: Slow cooling produces large crystals; fast cooling small ones. Under carefully controlled conditions, each particle will grow for the same time, and the resulting particles may be remarkably homogeneous in size.

Let us consider a few examples that lead to colloidal solution by some version of the above mechanism:

1. When two highly dilute salts producing a very insoluble precipitate are mixed (e.g., a barium salt and a sulfate), the insoluble salt may form very small colloidal particles. In an extreme case, the particles may be so small that the solution seems to be quite clear; the particles can still be detected by appropriate means.

2. A metallic salt may be reduced in solution to a metal, which may form a colloid. The best-known example is purple of Cassius (known since 1685), a gold sol prepared by the reduction of gold salts with stannous chloride. The gold particles, which give the purple color to the solution, are stabilized by the by-product of the reaction, stannic acid. This colloidal solution is extremely stable.

3. Most metals can be evaporated by an electric arc under liquid—either water or organic solvent. When the conditions are selected properly, a colloidal solution may result.

4. The decomposition of the thiosulfate ion $S_2O_3^{2-}$ by hydrochloric acid is a slow reaction that is easily controlled. It has been reported that under some experimental conditions the nucleation occurs within 60–62 min after the components of the reaction mixture are mixed. The particles of colloidal sulfur are spherical and grow in a very regular manner. The reaction is stopped

and the colloid stabilized by addition of iodine. It has been demonstrated that the heterogeneity of the particle sizes is much less than 2% in particle diameter.

Aerosols originate by nucleation in the gas phase in a completely analogous manner. In nature, water droplets or ice crystals nucleate in the high atmosphere and may be quite homogeneous in size, as demonstrated by some optical phenomena (e.g., ring around the moon). In agricultural practice, insecticides (especially of the DDT type) are evaporated and condensed to form aerosols, which have much higher activity because of the ease with which they reach their target.

1.5.2. Micelles

Micelles are aggregates of small molecules, which may assemble spontaneously. Thus, in contrast to colloidal solutions, micellar solutions are often thermodynamically completely stable. The phenomenon of micellar solutions is intimately related to a special class of small molecules called *detergents*.

To understand the properties of detergents, we must look more closely into the nature of intermolecular interactions. We have already seen that the energy of molecules is lowered when they form contacts with other molecules. The amount of this energy reduction depends on the nature of the two groups in contact. Polar groups (i.e., groups displaying a significant dipole) interact strongly with each other. Still greater interaction accompanies the formation of hydrogen bonds. On the other hand, the interaction of nonpolar groups (hydrocarbons, perfluorocarbons) among themselves is much weaker; the same applies to interactions of nonpolar groups with polar groups or hydrogen-bond-forming groups. It follows that the system will try to maximize the number of the stronger polar and hydrogen-bond interactions. As it happens, many polar groups (not all of them) also may participate in hydrogen bonds (hydroxyls, carbonyls, amino groups, carboxyls, charged groups, etc.). Water is both polar and hydrogen bonded. Thus polar molecules interact strongly with water and usually mix well with it. These molecules are said to be *hydrophilic,* and their interactions with water are called *hydrophilic bonds.* The less polar molecules would need to disrupt a number of these strong hydrophilic bonds to dissolve in water or in similar solvents. Weak bonds would replace stronger ones. Consequently, the nonpolar molecules have only very limited solubility in water; they are segregated and form a separate phase. The nonpolar groups and molecules are sometimes called *hydrophobic,* and the interactions holding them together are called *hydrophobic bonds* (actually, this is not a special type of bond, it is just the general van der Waals interaction, which is not accompanied by the more special hydrophilic interactions). The result is well known: Nonpolar molecules are soluble in other hydrophobic oils; hydrophilic molecules are soluble in water, and water does not mix with oil.

However, there exist molecules that have a strong hydrophilic group attached to a long hydrophobic tail. These molecules form substances called *detergents.* The examples known for the longest time are soaps, salts of long aliphatic acids. Here, the ionized carboxyl interacts strongly with water, forming the hydrophilic head of the molecule. In modern detergents, the carboxyl may be replaced by the strongly acidic sulfonic group, SO_3^-, or by the strongly basic trimethylammonium group, $N^+(CH_3)_3$. There also exist many detergents whose hydrophilic part is nonionic. Typical examples are phenols that have a long alkyl substituent on the ring (e.g., a nonyl) and their hydroxyl is extended into an oligomer of ethylene oxide.

The molecules of detergents are not comfortable in either oils or water. In both cases, they form many undesirable polar-nonpolar contacts. Actually, the best location for them is on the surface between water and oil—the hydrophilic heads are in the water phase, the hydrophobic tails in the oil phase. However, the water-oil surface can accommodate only a limited number of detergent molecules. Thus the detergent will welcome any opportunity to increase the available surface. If the oil is dispersed in water (for example, mechanically or sonically) the detergent molecules will occupy the new surface and prevent the oil droplets from coalescing; an *emulsion* is formed. (It is also possible to make an emulsion of water in oil.)

We should note that the surface energy needed for formation of a new water-oil surface was lowered by the amount of energy gained by the expulsion of additional amounts of detergent from the bulk phases to the newly created surface. Proper thermodynamic analysis would show that the above effect means that the surface energy (equal to the surface tension) of the water-oil interface is lowered by the surface adsorption of the detergent.

If a mixture of water and detergent has no oil available to expel the detergent to, the detergent has to go to another location: the water-air interface. At this interface the molecules of detergent will align parallel to each other, their heads swimming in water and their tails sticking out into the air (see Fig. 1.15). This surface adsorption again substantially lowers the surface tension of the water-air interface. And again, the detergent will welcome any opportunity to increase the available water-air interface. Such a process forms a *foam*. *Bubbles* are made from a thin layer of water sandwiched between two layers of detergent molecules. Again, the hydrophilic heads are in water and the hydrophobic tails stick outside (Fig. 1.16).

If the detergent does not have enough surface available, it has to find other stratagems to avoid contacts between hydrophilic and hydrophobic groups. In this case, the hydrophobic groups interact with themselves and form a *micelle:* a spherical entity with the hydrophilic heads sticking into the water phase and the hydrophobic

Air

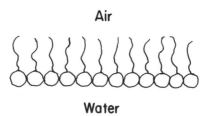

Water

Figure 1.15. Detergent molecules at water-air interface.

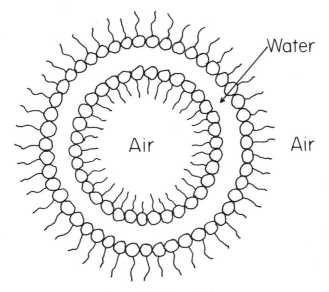

Figure 1.16. A soap bubble.

tails filling the interior of the sphere (Fig. 1.17). The size of the micelles of a given detergent can vary only within narrow limits. The surface-to-volume ratio of a micelle is determined by the cross section of the head and volume of the tail. (Nonionic detergents can also form oil-soluble micelles.)

The entropy of any two-component dispersion is maximized when the number of dispersed particles is maximized. Thus, at extreme dilutions, all dispersions must be molecular. At higher concentrations, individual molecules are still present and are in equilibrium with the larger particles—micelles in the present case. Let us calculate the concentration of molecularly dispersed detergent, [A], in equilibrium with

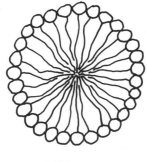

Water

Figure 1.17. A detergent micelle.

micelles A_N containing N molecules, where N is 20 or more and K is an equilibrium constant.

$$N A \rightleftharpoons A_N \tag{1.5.1}$$

$$K = \frac{[A_N]}{[A]^N} \quad \text{or} \quad [A] = \left(\frac{[A_N]}{K} \right)^{1/N} \tag{1.5.2}$$

It is easy to see that [A] is a very insensitive function of $[A_N]$ when N is a large number. In other words, the concentration of molecularly dispersed detergent is almost independent of the concentration of micelles. (An extreme case of this situation is a molecular solution in equilibrium with a pure phase. Its concentration is independent of the size of the continuous phase or of the number of macroscopic parts into which it is dispersed.) This concentration is called the *critical micelle concentration* (CMC). Thus, when the total detergent concentration is below the CMC, no micelles are present; at higher concentrations, monomer molecules are present and have just the critical concentration. We should mention that CMC depends on temperature and on the presence of other components in the system: dissolved salts reduce it; alcohols, etc., increase it. As already mentioned, the size of the micelles depends mainly on the geometry of the individual detergent molecules. However, if another nonpolar component is present in the system, it may enter the micelle and increase its size. This solubilization of nonpolar materials is the principle of detergent action: Nonpolar dirt is removed from fabrics and incorporated into micelles. Macroscopic particles, of course, adsorb the detergent molecules on their surface and are brought into solution at the same time.

At higher concentration of detergent, the micelles may not represent the lowest free energy of the system. The morphology may resemble any number of layered structures (e.g., Fig. 1.18), some of them liquid crystalline (see Section 3.1.5). The same applies also to pure detergent, in which layers of heads and tails alternate.

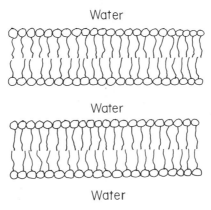

Figure 1.18. A layered structure of a concentrated detergent solution.

The morphology of these systems depends sensitively on the molecular architecture of the detergent molecules.

1.5.3. Block Copolymer Micelles

Another class of substances—block copolymers—shares the characteristic property of detergents of having two profoundly different parts. In both cases, the two parts of the molecule try to separate themselves sterically. The tendency of block copolymers to settle on a phase boundary could be utilized in compatibilization of polymer blends. In bulk, block copolymers (similarly to detergents) adopt morphologies in which the two blocks are separated in various manners. We will return to this topic in Section 4.1.7.3.

In solution, the behavior of block copolymers depends on the nature of the thermodynamic interaction of the solvent with the copolymer blocks. When the solvent is good with respect to both blocks the solution is molecular. When the solvent is good for one block but poor for the other (a selective solvent), block copolymer micelles represent the thermodynamically stable system. The insoluble blocks are segregated in a more or less spherical *core* and the soluble blocks form a diffuse outside *shell* (*corona*).

Most literature dealing with block copolymer micelles presumes that the micellar solutions are at equilibrium and that their behavior closely resembles the behavior of detergent micelles. Thus the micelles are in equilibrium with molecularly dissolved molecules (so-called *unimers*); there is exchange of unimers among micelles and between micelles and the outside solution. The *aggregation number* (average number of unimers per micelle) is again determined by the condition of the minimum free energy of the system. It depends primarily on the size of the blocks; it increases with the increasing size of the insoluble block and decreases with increasing size of the soluble block. Typical particle weight of block copolymer micelles is in millions to tens of millions. Figure 1.19 displays the formation of block copolymer micelles.

The equilibrium description of block copolymer micelles is appropriate for solvents that are good for one block and moderately poor for the other. Typical examples are solutions of Kratons, block copolymers of polystyrene and polydiene (or hydrogenated polydiene) in various organic selective solvents. In heptane, the aliphatic blocks are soluble and form the shell whereas the polystyrene blocks are segregated in the core. Heptane is only a moderate nonsolvent for polystyrene: it swells the core, allowing a reasonable mobility of the unimers through the system. Hence, such solutions are at equilibrium. A similar situation exists in butanone solutions. Butanone is a good solvent for polystyrene, which now forms the shell, and a poor solvent for the aliphatic chains in the core. Again, in this system the unimers have enough mobility to maintain the system at equilibrium.

When the solvent and the polymer are strongly incompatible (e.g., polystyrene and water), the micellar core is not swollen at all (in case of polystyrene it becomes glassy) and the mobility of unimers is totally suppressed. A typical example are micelles of poly(styrene-*b*-methacrylic acid) in aqueous buffers. Their aggregation number is fixed when they are formed and does not change during manipulation of the solvent conditions. They even survive freeze-drying and redispersion.

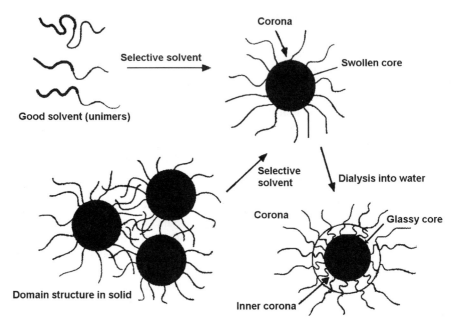

Figure 1.19. Block copolymer micelles prepared by adding a precipitating solvent for one block or by a direct dissolution of the dry polymer into an appropriate mixed solvent. (Reprinted from P. Munk, K. Prochazka, Z. Tuzar and S.E. Webber, CHEMTECH, Vol. 28, #10, p. 21, 1998 © by American Chemical Society, Washington, DC. Used by permission of the publisher.)

Block copolymer micelles are prepared by dissolving the copolymer in a good solvent and by changing the conditions to make it a selective one (by cooling or heating, by addition of a precipitating nonsolvent, by changes in pH). Alternately, they may be prepared by a direct dissolution in the selective solvent. Of course, this is possible only when such a solvent is allowing the mobility of unimers as described above. Micelles with strongly segregated cores must be prepared indirectly. The preferred procedure consists of preparing the micelles in a solvent mixture producing equilibrium micelles and then dialyzing the solution in a stepwise manner toward the solvent of choice.

1.6. MOLECULAR WEIGHT

Molecular weight is one of the more basic properties of materials. In the chemistry of small molecules, the molecular weight of a compound is obvious from its chemical formula, if the latter is known; it is not even necessary to measure it. Of course, during the analysis of an unknown substance, molecular weight is one of the clues for finding the chemical formula.

The situation is more complicated for macromolecular substances. We saw in Section 1.1 that besides true macromolecules we must also consider some multimolecular entities as individual molecules (better, particles) as long as they behave kinematically as individual particles. Fortunately, this does not pose problems; we will see later that the methods for measuring molecular weight actually measure the particle weight, directly yielding the desired quantity. (In all frankness, this fact led to our definition of a molecule in the first place.) Thus the above-described concept of molecule and molecular weight is always satisfactory and noncontroversial except in cases of ionized materials. We will return to this problem when we talk about polyelectrolytes and their molecular weights.

Strictly speaking, the term molecular weight should be replaced by the term *molecular mass*. However, the latter term is used very rarely, and we will adhere to the commonly used "molecular weight." What is more important is the dimension of molecular weight. Originally, molecular weight was a dimensionless number stating the ratio of the mass of the molecule in question to a molecular mass unit. In modern literature, molecular weight has the dimension of mass per mole and represents the mass of 1 mole, that is, of Avogadro's number of molecules. The argument is, of course, circular. Avogadro's number is a number selected in such a way that the molecular weights defined in both ways have the same numerical value.

1.6.1. Molecular Weight Averages

In the chemistry of small molecules, a chemical compound is a material composed of molecules that all have the same chemical formula and the same configuration. In the macromolecular field, this definition applies only to some materials of biological origin. Other macromolecular materials are made up of a large variety of different molecular species. Of course, these different species have many things in common: the nature of the monomer units, the configuration of the chain, and so on. The points of difference are mainly the number of monomer units, the nature of the end groups, branching, configurational details, and the composition of a copolymer. A typical polymer may contain thousands of different molecular species, all of them having very similar chemical properties. Usually, it is neither possible nor desirable to characterize each species individually. In the first approximation, the nature of the monomers is given. All other properties (molecular weight, composition of copolymers, branching, configurations, etc.) are described by their average values. In the second approximation, the distribution of properties is described by suitable functions. In this section, we will deal with averages of molecular weights; the distributions will be dealt with in Section 1.6.2.

When an average of some quantity is to be calculated, the values of the quantity for individual members of the ensemble are inserted into the averaging formula. There are a number of different formulas, leading to different types of averages. For example, algebra routinely deals with arithmetic and geometric averages. Similarly, macromolecular science uses a number of different averages and averaging methods. The reason for this is very pragmatic. When some method of measurement of molecular weight (or of another property) is applied to a polydisperse polymer, it yields an average value. The type of average depends on the nature of the method; different methods yield different averages. Let us review the more common averaging methods. In the following, we will consider polymer with a large but finite number of species; all summations will be over all these species. N_i, w_i, and M_i will be number of molecules, mass, and molecular weight, respectively, of the ith species. X_i will designate any arbitrary property of the species. Obviously,

$$w_i = N_i M_i / N_{Av} \tag{1.6.1}$$

where N_{Av}, is Avogadro's number.

One method of averaging weighs the property by the number of molecules having that property. Accordingly, the *number averages* (a bar over the symbol represents the average) are identified by the subscript n. Thus,

$$\overline{X}_n \equiv \frac{\Sigma_i N_i X_i}{\Sigma_i N_i} \tag{1.6.2}$$

$$\overline{M}_n \equiv \frac{\Sigma_i N_i M_i}{\Sigma_i N_i} \tag{1.6.3}$$

where $\overline{X}_n$ and $\overline{M}_n$ are the *number-average value* of property X and *number-average molecular weight,* respectively. In these formulas, N_i can be replaced by any quantity that is proportional to number of molecules, for example, by number of moles, molarity, molality, or molar fraction. (When using molar fractions for macromolecular solutions, the solvent may be considered when the molar fractions are calculated, or it may not. Both procedures are legitimate provided that the same procedure is applied to the sums in both the numerator and the denominator.)

Substituting equation (1.6.1) into equations (1.6.2) and (1.6.3), we obtain an alternative definition of the number averages as

$$\overline{X}_n \equiv \frac{\Sigma_i w_i X_i / M_i}{\Sigma_i w_i / M_i} \tag{1.6.4}$$

$$\overline{M}_n \equiv \frac{\Sigma_i w_i}{\Sigma_i w_i / M_i} \tag{1.6.5}$$

Again, in these relations w_i may be replaced by any quantity proportional to mass, such as mass fraction or mass per volume concentration. It should be noted that for calculations in which the solvent is not considered in calculating fractions, the sum of molar fractions as well as the sum of mass fractions are equal to unity; these sums may then be omitted from the corresponding formulas.

Number-average molecular weight is a result of those measurements that depend on the number of molecules present. Typically, these are the colligative methods: freezing point depression, boiling point elevation, change in vapor pressure, and osmometry. The latter method is rather important for macromolecular systems.

Another method of averaging weighs the property by the mass (usually called weight) of the material having that property. These averages are called *weight averages* and are identified by the subscript w. They are defined as

$$\overline{X}_w \equiv \frac{\Sigma_i w_i X_i}{\Sigma_i w_i} \tag{1.6.6}$$

and

$$\overline{M}_w \equiv \frac{\Sigma_i w_i M_i}{\Sigma_i w_i} \tag{1.6.7}$$

or, alternatively,

$$\overline{X}_w \equiv \frac{\Sigma_i N_i M_i X_i}{\Sigma_i N_i M_i} \tag{1.6.8}$$

and

$$\overline{M}_w \equiv \frac{\Sigma_i N_i M_i^2}{\Sigma_i N_i M_i} \tag{1.6.9}$$

The weight-average molecular weight $\overline{M}_w$ results from measurements of light scattering and sedimentation equilibrium.

Sedimentation equilibrium measured in a centrifuge may yield still another average of molecular weight if the experiment is evaluated using a different approach. The average is called the z average (z stands for centrifuge using the German spelling *zentrifuge*) and is defined as

$$\overline{M}_z \equiv \frac{\Sigma_i N_i M_i^3}{\Sigma_i N_i M_i^2} \equiv \frac{\Sigma_i w_i M_i^2}{\Sigma_i w_i M_i} \tag{1.6.10}$$

and

$$\overline{X}_z \equiv \frac{\Sigma_i N_i M_i^2 X_i}{\Sigma_i N_i M_i^2} \equiv \frac{\Sigma_i w_i M_i X_i}{\Sigma_i w_i M_i} \tag{1.6.11}$$

A large number of other averages may also be defined. We will mention only one of them, the *viscosity-average molecular weight* $\overline{M}_\eta$, defined as

$$\overline{M}_\eta \equiv \left(\frac{\Sigma_i N_i M_i^{1+a}}{\Sigma_i N_i M_i} \right)^{1/a} \equiv \left(\frac{\Sigma_i w_i M_i^a}{\Sigma_i w_i} \right)^{1/a} \tag{1.6.12}$$

This average molecular weight is useful when molecular weight is calculated from intrinsic viscosity (see Section 3.4.3.3). The parameter a is a constant (usually between 0.5 and 0.8) characterizing the dependence of intrinsic viscosity on molecular weight.

It can be shown mathematically that for any distribution of molecular weights the following inequality must hold:

$$\overline{M}_n \leq \overline{M}_w \leq \overline{M}_z \tag{1.6.13}$$

Here the equality sign holds only for monodisperse polymers.

The ratios of different molecular weight averages $\overline{M}_w/\overline{M}_n$ and $\overline{M}_z/\overline{M}_w$ are often used as characteristics of polymer polydispersity. However, the comparison of two polymer samples according to their ratio $\overline{M}_w/\overline{M}_n$ is meaningful only when there is a reasonable expectation that the two samples have a similar distribution of molecular weights.

1.6.2. Distribution of Molecular Weights

In the previous section, no restrictions were applied to the values of N_i. However, there is frequently a definite relationship between N_i and M_i; this relationship may follow from the way the polymer sample was prepared.

Let us study an example: a linear polymer during the preparation of which every monomeric unit had the same probability p of being followed by another monomeric unit. The probability of not being followed (i.e., of being a terminal unit) is $1-p$. We will see later that this situation occurs in some polycondensation reactions when the system is in chemical equilibrium before the reaction is quenched. The same situation also occurs for radical polymerization under some circumstances. The corresponding polymers are said to have the *most probable distribution* of molecular weights.

For this example, it is convenient to introduce a new quantity, *degree of polymerization* (DP), representing the number of monomeric units in a molecule of polymer. The degree of polymerization is related to molecular weight as

$$M = M_0 \, (\text{DP}) \tag{1.6.14}$$

where M_0 is the molecular weight of the monomeric unit. The number of molecules N_i having DP $= i$ must be proportional to the probability P_i that the molecule has i units. The latter probability is equal to the probability that the first $i-1$ monomeric units are followed by another unit, whereas the last unit is not. Consequently,

$$N_i = N_{\text{tot}} P_i = N_{\text{tot}} \, p^{i-1} \, (1-p) \tag{1.6.15}$$

where N_{tot} is the total number of polymer molecules. Substituting equation (1.6.15) into equation (1.6.2) and recognizing that DP $= i$ and $M_i = iM_o$, we can calculate the number-average degree of polymerization $\overline{\text{DP}}_n$ as

$$\overline{\text{DP}}_n = \frac{\sum_{i=1}^{\infty} i(1-p)p^{i-1}}{\sum_{i=1}^{\infty} (1-p)p^{i-1}} \tag{1.6.16}$$

Similarly, substitution into equation (1.6.8) yields for the weight-average degree of polymerization

$$\overline{\text{DP}}_w = \frac{\sum_{i=1}^{\infty} i^2(1-p)p^{i-1}}{\sum_{i=1}^{\infty} i(1-p)p^{i-1}} \tag{1.6.17}$$

Equations (1.6.16) and (1.6.17) are easily simplified when we recall the algebraic identities

$$\sum_{i=1}^{\infty} p^{i-1} \equiv (1-p)^{-1} \tag{1.6.18}$$

$$\sum_{i=1}^{\infty} i p^{i-1} \equiv (1-p)^2 \tag{1.6.19}$$

$$\sum_{i=1}^{\infty} i^2 p^{i-1} \equiv (1+p)(1-p)^3 \tag{1.6.20}$$

The resulting relations read

$$\overline{DP}_n = (1-p)^{-1} \tag{1.6.21}$$

$$\overline{DP}_w = (1+p)(1-p)^{-1} \tag{1.6.22}$$

$$\overline{DP}_w/\overline{DP}_n = (1+p) \tag{1.6.23}$$

By definition, polymers are materials with a high degree of polymerization. From equation (1.6.21), we see that p must be only slightly less than unity. Thus the ratio of averages $\overline{DP}_w/\overline{DP}_n$ is very close to 2 for this distribution.

For most polymeric materials, the molecular weight is rather high and the number of species with different values of M is extremely high. For such cases, it is usually convenient to define a continuous distribution function of molecular weights $f_n(M)$ and replace summations by integrations. The basic relation reads

$$dN = N_{tot} f_n(M) dM \tag{1.6.24}$$

where dN is the number of molecules having a molecular weight between M and $M + dM$. $f_n(M)$ is a normalized function; that is, its integral between zero and infinity is equal to unity. Definitions of various molecular weight averages for continuous distributions have the following forms:

$$\overline{M}_n = \int_0^{\infty} M f_n(M) dM \tag{1.6.25}$$

$$\overline{M}_w = \frac{\int_0^{\infty} M^2 f_n(M) dM}{\int_0^{\infty} M f_n(M) dM} \tag{1.6.26}$$

$$\overline{M}_z = \frac{\int_0^{\infty} M^3 f_n(M) dM}{\int_0^{\infty} M^2 f_n(M) dM} \tag{1.6.27}$$

The averages of other properties X are defined in a similar way. Sometimes it is preferable to express the amount of the material using mass instead of number of molecules. For this purpose, another normalized distribution function of molecular weights $f_w(M)$ is defined as

$$dw = w_{tot} f_w(M)dM \qquad (1.6.28)$$

where dw is a mass of material having molecular weight between M and $M + dM$ and w_{tot} is the total mass of the sample. The expressions for molecular weight averages now read

$$\overline{M}_n = \frac{1}{\int_0^\infty M^{-1} f_w(M)dM} \qquad (1.6.29)$$

$$\overline{M}_w = \int_0^\infty M \, f_w(M)dM \qquad (1.6.30)$$

$$\overline{M}_z = \frac{\int_0^\infty M^2 f_w(M)dM}{\int_0^\infty M f_w(M)dM} \qquad (1.6.31)$$

The functions $f_n(M)$ and $f_w(M)$ are related to each other. We leave it to the reader to prove that the relation is

$$f_w(M) = (M/\overline{M}_n) f_n(M) \qquad (1.6.32)$$

Introduction of the distribution functions is especially advantageous when the polymer sample has (or is assumed to have) a distribution function that can be described by an analytical function. For narrow and moderately broad distributions, the two-parameter *Schulz–Zimm distribution* is employed rather often. It reads

$$f_w(M) = \frac{b^{z+1}}{\Gamma(z+1)} M^z e^{-bM} \qquad (1.6.33)$$

where

$$b = z/\overline{M}_n \qquad (1.6.34)$$

Here, $\Gamma(z+1)$ is the gamma function, which is equal to the factorial function $z!$ for integer values of the parameter z. An easy calculation shows that the ratios of molecular weight averages for the Schulz–Zimm distribution are

$$\frac{\overline{M}_w}{\overline{M}_n} = \frac{z+1}{z} \qquad (1.6.35)$$

$$\frac{\overline{M}_z}{\overline{M}_w} = \frac{z+2}{z+1} \qquad (1.6.36)$$

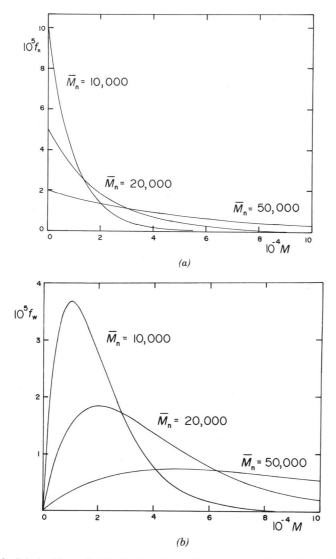

Figure 1.20. Schulz–Zimm distribution functions of molecular weight with $z = 1$. Values of $\overline{M}_n$ as noted: (a) function $f_n(M)$; (b) function $f_w(M)$.

The Schulz–Zimm distribution function with $z = 1$ is actually the most probable distribution [equation (1.6.15)] expressed by means of a continuous function. In Figure 1.20 the functions $f_n(M)$ and $f_w(M)$ are plotted for $z = 1$ and several values of $\overline{M}_n$. It is apparent that the most probable distribution is actually very broad, comprising molecules rather different in size. It is worth noting that the function $f_n(M)$ has its maximum value at the limit $M = 0$: the monomer is the most abundant species! This fact was, of course, obvious from equation (1.6.15).

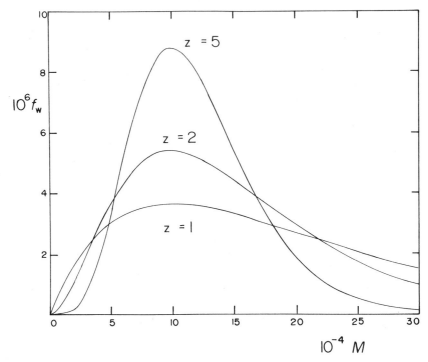

Figure 1.21. The function $f_w(M)$ for Schulz–Zimm distribution and for several values of z. $\overline{M}_n = 100,000$ for all distributions.

Let us study a hypothetical process. From a large number of molecules having the most probable distribution, we randomly select sets of z' members each and we fuse each set into a single molecule. It is easy to show that the resulting ensemble of z'-mers has the distribution function of molecular weights that can be described by equations (1.6.33) and (1.6.34). However, the parameter z changes its value from $z = 1$ applicable for the original distribution to a new value $z = z'$. Simultaneously, $\overline{M}_n$ increases z'-fold in value.

As the z value increases, the distribution becomes narrower. An infinite z corresponds to a monodisperse polymer. In Figure 1.21 the function $f_w(M)$ is plotted for several values of z; $\overline{M}_n$ is the same for all curves.

For polymers with a narrow distribution of molecular weights, the *Poisson distribution function* is frequently applicable. In fact, this distribution is theoretically predicted for polymers prepared by chain polymerization if the propagation of all chains is simultaneous and all chains are terminated simultaneously. The Poisson distribution function is usually written for the discrete variable [similar to equation (1.6.15)] as

$$N_i = N_{tot}e^{-\lambda}\lambda^i / i! \qquad (1.6.37)$$

$$\lambda = \overline{DP}_n \qquad (1.6.38)$$

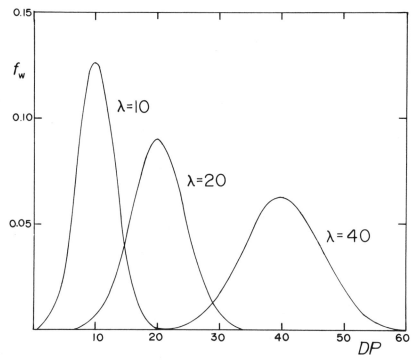

Figure 1.22. The function $f_n(DP)$ for Poisson distribution and for several values of $\overline{DP}_n$.

The Poisson distribution gets very narrow as $\overline{DP}_n$ increases. For large values of $\overline{DP}_n$ the distribution becomes close to a (continuous) Gaussian distribution, for which the average value is equal to its variance:

$$f_n(x) = (2\pi\lambda)^{-1/2} \exp\left[\frac{-(x-\lambda)^2}{2\lambda}\right] \tag{1.6.39}$$

Here x stands for the degree of polymerization DP. Alternatively, x may represent M; in this case $\lambda = \overline{M}_n$. Figure 1.22 represents the Poisson distribution [equation (1.6.39)] for several values of $\lambda = \overline{DP}_n$.

For some strongly skewed distributions, the *logarithmic normal distribution* [symmetrical in a plot of $f_w(M)$ vs. $\ln M$] is frequently useful. It reads

$$f_w(M) = A \exp\left[-\left(\frac{1}{z}\ln\frac{M}{M_M}\right)^2\right] \tag{1.6.40}$$

where A is a normalizing constant, M_M corresponds to the maximum of the $f_w(M)$ function, and z is a parameter characterizing the width of the distribution. For this distribution, the ratios of molecular weight averages are related to z as

$$\frac{\overline{M}_w}{\overline{M}_n} = \frac{\overline{M}_z}{\overline{M}_w} = \exp\left(\frac{z^2}{2}\right) \tag{1.6.41}$$

All the distribution functions mentioned above are called *differential distribution functions*. Another type of distribution functions are *integral distribution functions*, $I_n(M)$ and $I_w(M)$, which can be obtained experimentally when polymer samples are fractionated. They are related to the differential functions as

$$I_n(M) = \int_0^M f_n(M')dM' \tag{1.6.42}$$

$$I_w(M) = \int_0^M f_w(M')dM' \tag{1.6.43}$$

The differential functions are obtained from the integral functions by differentiation (usually graphical).

1.A. SUGGESTIONS FOR FURTHER READING

Alexandridis, P., and B. Lindman, *Amphiphilic Block Copolymers: Self-Assembly and Applications,* Elsevier Science, New York, 2000.

Bovey, F.A., *Polymer Conformation and Configuration,* Academic, New York, 1969.

Cantor, C.R., and P.R. Schimmel, *Biophysical Chemistry,* Part I: *The Conformation of Biological Macromolecules,* Freeman, San Francisco, 1980.

Coleman, M.M., and P.C. Painter, *Fundamentals of Polymer Science: An Introductory Text,* Technomic, Lancaster, PA, 1997.

Elias, H.-G., *An Introduction to Polymers,* Wiley, New York, 1997.

Fitch, R.M., Ed., *Polymer Colloids: A Comprehensive Introduction,* Academic, New York, 1997.

Flory, P.J., *Statistical Mechanics of Chain Molecules,* Wiley, New York, 1969.

Furukawa, Y., *Inventing Polymer Science: Staudinger, Carothers, and the Emergence of Macromolecular Chemistry (Chemical Sciences in Society Series),* University of Pennsylvania Press, PA, 1998.

Grosberg, A.I., A.R. Khokhlov, and A.Y. Grosberg, *Giant Molecules: Here, There, and Everywhere,* Academic, New York, 1997.

Havelka, K.O., and C.L. McCormick, Eds., *Specialty Monomers and Polymers: Synthesis, Properties, and Applications (ACS Symposium Series, No. 755),* Washington, DC, 2000.

Hopfinger, A.J., *Conformational Properties of Macromolecules,* Academic, New York, 1973.

IUPAC *Macromolecular Div. Comm. Macromolecular Nomenclature,* Pure and Applied Chemistry, 1985, 57, 1427

Koenig, J.L., *Chemical Microstructure of Polymer Chains,* Wiley, New York, 1980.

Lowry, G.G., *Markov Chains and Monte Carlo Calculations in Polymer Science,* Dekker, New York, 1979.

Peebles, L.H., Jr., *Molecular Weight Distribution in Polymers,* Wiley-Interscience, New York, 1971.

Rudin, A., *The Elements of Polymer Science and Engineering: An Introductory Text and Reference for Engineers and Chemists,* 2nd ed., Academic, New York, 1998.

Shaw, D.J., *Introduction to Colloids and Surface Chemistry,* 2nd ed., Butterworth, Stoneham, MA, 1978.

Sperling, L.H., *Polymeric Multicomponent Materials: An Introduction,* Wiley, New York, 1997.

Stevens, M.P., *Polymer Chemistry: An Introduction,* Oxford University Press, London, 1998.

Tanford, C., *The Hydrophobic Effect,* Wiley, New York, 1981.

Volkenstein, M.V., *Configurational Statistics of Polymeric Chains,* Interscience, New York, 1963.

1.B. REVIEW QUESTIONS

1. What do you know about polyoxymethylene? What would be another name for it?
2. Discuss the possible conformations of 1,2-dichloroethane.
3. What is the molecular weight of a block copolymer with 125 monomeric units of ethyl methacrylate and 200 monomeric units of methacrylic acid?
4. Explain the differences between a branched polymer, a star polymer, and a dendrimer.
5. Explain how tacticity of a polymer influences its properties.
6. Discuss the usefulness of the random walk problem for statistics of macromolecular coils.
7. Consider polypropylene, polyacrylonitrile, and cellulose. Predict the magnitude of intermolecular forces in these polymers from their known chemical structures.
8. Polyethylene (PE) is manufactured as high-pressure PE and low-pressure PE. Describe the difference in properties of these materials. Show the molecular basis of the difference and explain why the molecular structures are different.
9. Explain the origin of attractive intermolecular forces.
10. What is the difference between the configuration and conformation of a polymer?
11. How does the average end-to-end distance of a macromolecular coil depend on polymer-solvent interaction? Discuss the problem and give a useful relation.
12. Write the relation between end-to-end distance and radius of gyration of a linear macromolecular coil.
13. Explain why it is possible for some real polymers to adopt a distribution of end-to-end distances that closely approximates the end-to-end distance distribution of an ideal Gaussian coil.
14. Discuss the effect of the difference between the energy of *trans* and *gauche* conformations on the average end-to-end distance of a polymer coil in a

Gaussian unperturbed approximation. Include the discussion of temperature dependence.

15. Which of the following is a monodisperse polymer? (a) cellulose from cotton, (b) casein from milk, (c) high-density polyethylene, (d) poly(vinyl chloride), (e) DNA, (f) nylon-66, and (g) β-keratin.

16. Discuss the primary, secondary, and tertiary structure of a macromolecule. What is meant by quaternary structure of proteins?

17. What are the basic factors that influence the two types (short range and long range) of interactions in a long chain molecule?

18. How does the conformation of polyelectrolytes differ from that of uncharged solvated polymers and why?

19. How many different pentapeptides can be made from the 20 amino acids present in proteins: (a) allowing repetitions and (b) where each pentapeptide has 5 different amino acids?

20. Proteins and polypeptides can form helical structures. Such structures are frequently observed in their solutions. Isotactic poly(α-olefins) can also form helical structures, but these are almost never observed in their solutions. Explain this on the basis of thermodynamic arguments.

21. It has been observed that the critical micellar concentration decreases with increasing concentration of the added low-molecular-weight salts. Explain this behavior.

22. What are the characteristic structural features of a typical enzyme?

23. Discuss the process of nucleation of colloids.

24. What are block copolymer micelles? Describe how they are formed.

1.C. DERIVATIONS

1. Consider the random flight model of a polymer. Because the direction of each bond is assumed to be completely arbitrary and random, this model is often called the freely jointed chain model. To treat the mathematics of a random flight, it is convenient to treat first a simple one-dimensional version of this process. Now consider a process (a walk) in which a particle (or a random walker) starts at the origin and moves to the left or to the right at successive intervals of time with equal probability $p = 1/2$. After N such steps, the particle could be found at the points $-N, -N + 1, \ldots -1, 0, 1, 2, \ldots N$. Calculate $W(m, N)$, the probability that the particle is at the position m after N number of steps.

2. A Gaussian macromolecular coil is modeled by N statistical segments; the length of each of them is A. Derive an expression for the average end-to-end distance, h^2.

3. Show that the radius of gyration r_G of a uniform cylinder of radius r and length A is given by

$$r_G^2 = \frac{r^2}{2} + \frac{A^2}{12}$$

4. The distribution function of end-to-end distances of a real polymer chain is not Gaussian because of the excluded volume effect. A proposed modification of the distribution function was given by Mazur (see *J. Res. Natl. Bur. Standards,* 69A, 355, 1965) as

$$w(h)dh = [\Gamma(3/t)]^{-1} t\alpha^{3/t}h^2 \exp(-\alpha h^t)dh$$

where $t = 3.2$ and α is an adjustable parameter. First show that this distribution is normalized, and then show that

$$\overline{h^2} = \Gamma(5/t)/\alpha^{2/t}\Gamma(3/t)$$

1.D. NUMERICAL PROBLEMS

1. Calculate the radius of gyration of a homogeneous disk having a diameter of 4 cm and a thickness of 1 cm.
2. Calculate the radius of gyration of a cylindrical rod 5 mm long with an axial radius of 2 mm.
3. Consider a solution of polystyrene in cyclohexane at 34°C. If polystyrene has a molecular weight of 3.2×10^6 and a density of 0.90 g/cm^3, calculate its radius of gyration by assuming that the molecule is a compact sphere. Compare this to the experimental value of 49.4 nm (see W.R. Krigbaum and K.D. Carpenter, *J. Phys. Chem.,* 59, 1166, 1955).
4. Polyethylene with a molecular weight of 140,000 is composed of methylene (CH$_2$) groups with a C—C bond length of 1.54 Å. Assuming that the chain can rotate freely around every bond, calculate the end-to-end distance using C—C—C angle $\theta = 109°3'$.
5. Calculate the end-to-end distance of a polystyrene sample of molecular weight 520,000 by assuming freely rotating bonds and backbone carbon-carbon bonds that are in a strictly tetrahedral configuration. The C—C bond length is 1.54 Å.
6. Calculate the radius of gyration of a macromolecular coil having a spherical symmetry with the density function (i.e., its density ρ depends on its radius h) given by the Flory model:

$$\rho(h) = A \exp(-bh^2)$$

Assume the values for $A = 0.05$ g/cm^3 and $b = 0.1$ nm^{-2}. Using the above data, calculate also the end-to-end distance. Explain any inherent assumptions.
7. Calculate $\overline{M}_n$, $\overline{M}_w$, $\overline{M}_z$, $\overline{M}_w/\overline{M}_n$, and $\overline{M}_z/\overline{M}_w$ for a polymer sample with the Schulz–Zimm distribution given in equation (1.6.33) of the text for values of $z = 4$ and $b = 4.75 \times 10^{-7}$.
8. Calculate $\overline{M}_n$, $\overline{M}_w$, and $\overline{M}_z$ for the following mixture of macromolecular species with molecular weights M_i and mole fractions X_i.

$M_A = 350,500;$ $X_A = 0.125;$
$M_B = 896,300;$ $X_B = 0.600;$
$M_C = 6,786,800;$ $X_C = 0.275$

9. If equal weights of polymer A and polymer B are mixed, calculate $\overline{M}_w$ and $\overline{M}_n$ of the mixture.

   Polymer A: $\overline{M}_n = 35,000$, $\overline{M}_w = 90,000$;
   Polymer B: $\overline{M}_n = 150,000$, $\overline{M}_w = 300,000$.

10. If equal weights of monodisperse polymers with molecular weights 5000 and 50,000 are mixed, what is $\overline{M}_z$ of the mixture?

11. The distribution function of molecular weights for some polymer is given by the relation :

$$f_N(M) = AM^2 \exp(-BM)$$

   Calculate the molecular weight averages $\overline{M}_n$, $\overline{M}_w$, $\overline{M}_z$, and $\overline{M}_w/\overline{M}_n$ in terms of the coefficients A and B.

12. Calculate $\overline{M}_n$, $\overline{M}_w$, $\overline{M}_z$, $\overline{M}_w/\overline{M}_n$, and $\overline{M}_z/\overline{M}_w$ for the mixtures: (a) one mole of component A with molecular weight 2000 and one mole of component B with molecular weight 20,000; (b) 10 mg of a component with molecular weight 100 and 10 g of a component with molecular weight 10,000 (This is an example of contamination with low-molecular-weight material.); and (c) 1g of material with molecular weight 20,000 and 1mg of a component with molecular weight 1,000,000 (This is an example of an effect of small amounts of large aggregates.).

13. Use the data given in the Table for the fractionated polyester made from sebacic acid and 1,6-hexanediol (H. Batzer, *Macromol. Chem.*, 5, 5, 1950) to calculate $\overline{M}_n$, $\overline{M}_w$, and $\overline{M}_z$.

Fraction #	1	2	3	4	5	6	7	8	9
Mass (g)	1.15	0.73	0.42	0.35	0.51	0.34	1.78	0.10	0.94
$M \times 10^{-4}$ (g/mol)	1.25	2.05	2.40	3.20	3.90	4.50	6.35	8.10	9.40

Assume that individual fractions may be treated as monodisperse.

2

TECHNIQUES FOR SYNTHESIS
OF POLYMERS

Our approach to the description of synthetic methods in the field of macromolecules is similar to the approach adopted in Chapter 1. It is selective, not exhaustive; the emphasis is on the development of general principles and synthetic strategies, not on details of actual synthetic procedures. The outline of the chapter is a conventional one. Sections 2.1–2.3 are devoted to reactions forming macromolecules from small molecules; Section 2.4 describes methods for changing the structure of existing macromolecules.

Historically, the reactions leading to macromolecules were classified as *polycondensation* reactions if some low-molecular-weight by-product (typically water) was produced together with the polymer. A process that linked monomeric units together without a by-product was called a *chain polymerization* or simply *polymerization*. The two types of reactions were distinctly different from each other. However, some polymerizations (typically of cyclic monomers) did not produce any by-products yet had some of the other characteristics of the polycondensation reactions. They were classified as a third type of polymer-forming reactions—*polyadditions*. Their existence made the whole classification scheme somewhat unsatisfactory.

More recently, another scheme based on the mechanism of reaction has been gaining broader acceptance. In one class of these reactions, all monomers participate in the reaction from the beginning. Usually, the size of molecules is increasing slowly and the concentration of monomers decreases to rather low values before the degree of polymerization increases appreciably. Most significantly, two shorter chains may join to form a longer one. Such reactions are called *stepwise reactions;* most polycondensations and some polyadditions belong to this class.

Another group of reactions is characterized by the presence of a small number of active centers. These centers add monomers in such a way that the center is shifted along the growing chain and can add another monomer. Thus during the reaction some molecules are growing rapidly and reaching considerable size while a large portion of the monomer is still present. Such reactions are called *chain reactions* and are classified according to the nature of the active centers. The main types are radical, ionic, and coordination polymerizations.

In the following sections, we will explore the more important features of all these types of reactions.

2.1. POLYCONDENSATION

Any chemical reaction in which two smaller molecules join to form a larger one and which can proceed to high conversion can be used to synthesize polymers in a polycondensation reaction. For now, we will mention only a few examples: the reaction of carboxylic acids with alcohols, leading to esters; Diels–Alder reaction of dienes with olefins, leading to substituted cyclohexenes; the coupling of two molecules of alkyl halides by metal, leading to longer hydrocarbons; and the reaction of isocyanates with amines, leading to substituted ureas. Typically, the reactions involve two reactive groups and lead to structures incapable of further condensation. For such reactions to be used for polymer synthesis, each molecule must possess at least two reactive groups. When two such molecules are joined together, the resulting molecule still has at least two reactive groups and can participate in further reactions.

Most polycondensations involve several monomers. The number of reactive groups in the molecule of a monomer is called its *functionality*. The macromolecular nature of the resulting polymer is governed primarily by the functionality of the monomers. When only bifunctional monomers are present, the product is a strictly linear polymer. To achieve high molecular weight, the number of molecules—or better, the number of molecular ends or, still better, the number of unreacted reactive groups—must be very small. In other words, the yield of the polycondensation reaction must be very high (typically more than 99%), and no side reactions leading to nonreactive end groups are allowed. Thus, for these polymers, the yield and purity of the reaction mechanism become the overriding concerns; a relatively small number of reactions qualify.

When monofunctional monomers are incorporated into polymer, they generate an unreactive end of the molecule and severely restrict the degree of polymerization. Obviously, only very small amounts of such monomers can be tolerated in conjunction with bifunctional monomers.

Trifunctional monomers lead to branched structures. If the degree of branching exceeds some well-defined limit, the polymer may form an infinite network. Polycondensations using trifunctional (or still higher functional) monomers have a very important distinction from reactions using strictly bifunctional monomers: The growing polymer molecule has more than two growing ends; a waste of some of them (because of either an incomplete reaction or a side reaction) does *not* prevent the molecule from reaching very high molecular weight or from forming an infinite network. Thus, for such systems, the strict purity rules for the reaction can be relaxed.

In many typical polycondensations, the elementary reaction leads to the formation of a low-molecular-weight by-product. For example, when an ester is formed from an acid and an alcohol, water is the by-product. This water may again hydrolyze the ester function; that is, the ester formation is reversible. To achieve high conversion (which is necessary for the synthesis of high-molecular-weight polymers), the by-product must be removed from the reaction mixture. Thus the removal of by-products becomes one of the dominant features of the synthesis of polymers by polycondensation and frequently dictates the reaction conditions.

In the following sections, we will describe the mechanisms of the more common polycondensation reactions. The actual synthetic procedures for a few more important classes of polymers will be briefly outlined.

2.1.1. Carboxylic Acid Derivatives

We will start by considering those polycondensation reactions whose mechanisms are understood best. These reactions are the addition-elimination reactions of carboxylic acid derivatives and lead to the formation of esters, amides, and so on. Essentially, these reactions convert one derivative of carboxylic acid into another; many of them are reversible.

2.1.1.1. Reaction Mechanisms.   Generally, we may write the derivatives of carboxylic acids as $R \cdot CO \cdot X$, where X is a group carrying a free electron pair. Thus for acids, $X = OH$; for amides, $X = NH_2$, NHR', or $NR'R''$; for esters, $X = OR'$; for acyl chlorides, $X = Cl$; for anhydrides $X = O \cdot CO \cdot R'$; etc. The carboxylic carbonyl is polarized; the electron density on carbon is low. When another molecule carrying a group of the same class X (let us call it Y) approaches the carbonyl, the free electron pair of Y may attach itself to the carbon (thus completing the polarization of the carbonyl). This attachment is greatly facilitated if some catalyst increases the polarization of the carbonyl. Typically, acids perform this catalytic function by protonating the oxygen of the carbonyl. However, organic salts of some metals may have the same effect (e.g., acetates of calcium, zinc, cadmium, or manganese). In the next reaction step, either Y is eliminated again, restoring the reactants, or X is eliminated, completing the reaction. Thus the overall reaction scheme reads

$$
\underset{\displaystyle R-\overset{\displaystyle O}{\overset{\|}{C}}-X}{} \;\underset{\longleftarrow}{\overset{H^+}{\rightleftharpoons}}\; R-\overset{\displaystyle OH}{\overset{|}{\underset{+}{C}}}-X \;\underset{\longleftarrow}{\overset{|Y}{\rightleftharpoons}}\; R-\overset{\displaystyle OH}{\underset{\underset{+}{Y}}{\overset{|}{\underset{|}{C}}}}-X \;\underset{\longleftarrow}{\overset{-H^+-X|}{\rightleftharpoons}}\; R-\overset{\displaystyle O}{\overset{\|}{C}}-Y
$$

$$(2.1.1)$$

When the starting reactants are bifunctional, this reaction can lead to the gradual buildup of a polymer chain.

For polyesters, several synthetic strategies can he employed. In the simplest case, a diacid is reacted with dihydroxy a compound, usually a glycol. The esterification reaction is reversible, and equilibrium is established at about 70% conversion. To achieve the high conversions needed for high-molecular-weight polymers, it is necessary to remove the reaction water (usually using higher temperature and vacuum).

The reaction of an alcohol with an acid is usually slower than other reactions involving carboxyls. Ester interchange reaction (transesterification) is much easier and is often used for manufacturing polyesters. The reaction can be schematically written as

$$RCOOR' + R''OH \rightleftharpoons RCOOR'' + R'OH \qquad (2.1.2)$$

The commercial procedure for manufacturing the most important polyester, poly(ethylene terephthalate), uses the ester interchange repeatedly in a two-step reaction. In the first step, dimethyl terephthalate is reacted with an excess of ethylene glycol, yielding bis(hydroxyethyl) terephthalate **1** while the by-product, methanol, is continuously distilled off.

$$CH_3O-\overset{\overset{\textstyle O}{\|}}{C}-\!\!\left\langle\!\!\bigcirc\!\!\right\rangle\!\!-\overset{\overset{\textstyle O}{\|}}{C}-OCH_3 + 2\ HOCH_2CH_2OH \longrightarrow$$

$$HOCH_2CH_2O-\overset{\overset{\textstyle O}{\|}}{C}-\!\!\left\langle\!\!\bigcirc\!\!\right\rangle\!\!-\overset{\overset{\textstyle O}{\|}}{C}-OCH_2CH_2OH + 2\ CH_3OH \qquad (2.1.3)$$

1

In the second step, the intermediate product **1** is heated further. The interchange reaction between the ester function and the free hydroxyl liberates glycol, which is continuously distilled off. Obviously, in the later stages of the reaction, a molecule of glycol is liberated only when the hydroxyl reacts with the ultimate ester bond of the polymer. In this case the removal of the excess glycol by distillation alleviates the need to have the original reactants exactly balanced.

The polyesterification reaction may be performed either by using a glycol and a dicarboxylic acid (or its ester) or by self-condensation of a hydroxy acid. In a hydroxy acid, the balance of reactive groups is achieved automatically. Alternatively, a cyclic ester of a hydroxy acid (a lacton) may be used instead of the acid itself. Two mechanisms are possible: either the reaction water hydrolyzes the lacton, which then condenses with the growing chain, or the lacton is attached to the chain directly by an ester interchange. In either case, no reaction by-product is formed, and the reaction can be formally classified as *polyaddition*.

The polyamidation reaction leading to aliphatic polyamides is formally very similar to polyesterification. A mixture of a diamine with a diacid, or an amino acid, or its cyclic amide (a lactam) is used. However, there are a number of significant differences.

1. Diamines form crystalline salts with diacids; these salts have an exact balance of the amino and carboxylic groups as required for formation of high polymers.

2. The reaction equilibrium is shifted toward the formation of amides, creating less demanding conditions for the removal of water.

3. No catalyst is needed for the reaction.

4. The amide interchange reaction is slow, preventing, during later stages of the reaction, the splitting off the diamine and its loss in vacuum. (This step was important in polyesterification for the removal of excess glycol.)

The actual manufacturing of nylons is relatively simple. The monomeric mixture in the presence of water is heated rapidly to 220°C and later to temperatures above the melting point of the polymer (which is about 264°C for nylon 66). The mixture is kept under pressure to keep the diamine in the system during the early stages of the reaction. Only late in the reaction is the pressure reduced to atmospheric. Application of vacuum is not needed in this process.

The amide interchange reaction (transamidation) is a slow reaction but not negligible. When an aliphatic polyamide is heated in a melt together with an aromatic polyamide, a few interchange reactions take place and a block copolymer is formed, which has valuable mechanical properties.

The reactivity of organic carbonates is very similar to the reactivity of esters. Thus an ester interchanges of diphenyl carbonate **2** with bisphenol **A 3** leads to polycarbonates with the liberation of phenol.

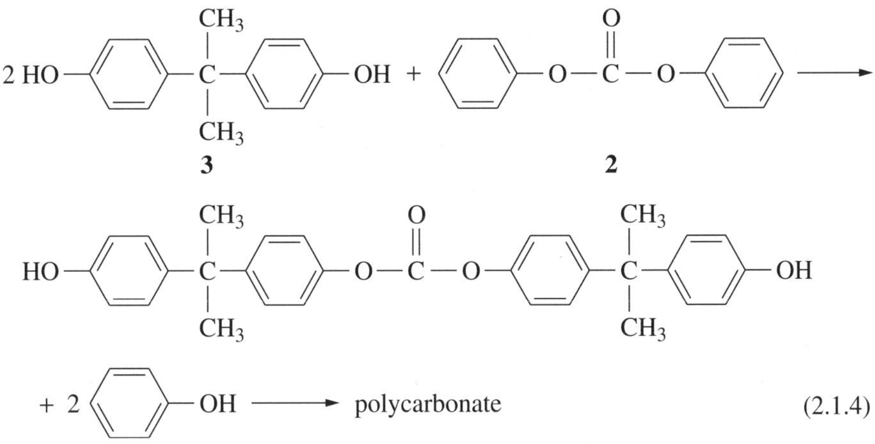

$$(2.1.4)$$

Acyl chlorides RCOCl are among the most reactive derivatives of carboxylic acids. They react vigorously with alcohols to form esters, with amines to form amides, with mercaptans to form sulfides, and so on (Schotten–Baumann reaction). Similarly reactive are sulfonyl chlorides, RSO_2Cl, and phosgene, $COCl_2$, which is actually a dichloride of carbonic acid. Reaction of dichlorides with dihydroxy compounds, diamines, etc., leads easily to high polymers.

Actually, the reaction is so vigorous that it is often necessary to keep the two reactants in two different phases. The dichloride is dissolved in an organic solvent, whereas the reactive hydrogen monomer (diamine) is dissolved in an aqueous phase that also contains a base (e.g., KOH) for neutralization of the liberated hydrogen chloride. As the diamine diffuses into the organic phase, it reacts with the dichloride while still very close to the interface. Later, diamine molecules are captured at the interface by the growing chains, and the reaction becomes diffusion-controlled as both monomers diffuse toward the interface. Very few monomers have a chance to reach an opposite monomer and start a new chain; they are incorporated into existing chains instead. Thus, unlike in most polycondensation reactions, large molecules are present from the very beginning of the reaction. Also, the molecular weight of the polymer is higher than in other linear polycondensations. This procedure is particularly useful for the low-temperature preparation of polymers with very high melting points.

Although these *interfacial polycondensations* could be performed in emulsion, the process can be arranged to yield fibers directly. In a beaker (or a reactor), one of the phases is layered on top of the other, and a continuous monofilament is drawn from the interface.

We have seen that the Schotten–Baumann synthesis is much easier and can be performed under milder conditions than reactions starting with diacids or their esters. However, the diacyl chlorides are much more expensive reagents than the diacids. Consequently, the Schotten–Baumann synthesis is used only (1) when the product cannot be prepared directly; (2) for specialty polymers, when higher molecular weight or special composition is worth the price; or (3) when the dichloride is cheap, as it is when phosgene is used. In practice, polycarbonates are prepared by coupling bisphenol **A 3** with phosgene.

Another useful reaction is the preparation of aromatic polyamides, such as poly(ethylene terephthalamide) by reacting ethylenediamine with terephthaloyl chloride **4**:

$$2\ NH_2CH_2CH_2NH_2\ +\ Cl-\overset{\displaystyle O}{\overset{\displaystyle \|}{C}}\!\!-\!\!\left\langle\!\!\bigcirc\!\!\right\rangle\!\!-\!\!\overset{\displaystyle O}{\overset{\displaystyle \|}{C}}\!-Cl\ \longrightarrow$$

4

$$NH_2CH_2CH_2NH-\overset{\displaystyle O}{\overset{\displaystyle \|}{C}}\!\!-\!\!\left\langle\!\!\bigcirc\!\!\right\rangle\!\!-\!\!\overset{\displaystyle O}{\overset{\displaystyle \|}{C}}\!-NHCH_2CH_2NH_2\ \longrightarrow\ polymer$$

(2.1.5)

In some polycondensation reactions, a reactive intermediate is prepared in situ by a *condensing agent;* it is then converted to product by continuing the reaction. A typical example is the preparation of polyanhydrides from dicarboxylic acids using acetanhydride as a condensing agent. During the first steps, acetanhydride in an anhydride exchange reaction reacts with a carboxylic group to form a mixed anhydride **5** and acetic acid. Acetic acid is the lowest-boiling component of the mixture and is distilled off. In the next step [reaction (2.1.7)], another anhydride exchange forms a polyanhydride bond, liberating another molecule of acetic acid.

$$HOOC-R-COOH \ + \ (CH_3CO)_2O \ \longrightarrow$$

$$HOOC-R-COOCOCH_3 \ + \ CH_3COOH \qquad (2.1.6)$$
$$\mathbf{5}$$

$$\mathbf{5} + HOOC-R-COOH \ \longrightarrow$$

$$HOOC-R-COOCO-R-COOH \ + \ CH_3COOH \qquad (2.1.7)$$

In this sense, the reaction of bisphenol A with phosgene can also be considered a reaction with a condensing agent (phosgene). However, in this case, a part of the agent (a carbonyl group) remains in the polymer backbone.

2.1.1.2. Cyclization Reactions. There is an important reaction competing with chain growth during the polycondensation processes: formation of cyclic compounds. For example, hydroxy acids may form lactons and amino acids may form lactams. In other reactions, two or more molecules may form a cyclic compound. For example, aminoacetic acid (glycine) may dimerize to diketopiperazine **6**.

$$2\ NH_2CH_2COOH \ \longrightarrow \ CO \overset{\displaystyle CH_2-NH}{\underset{\displaystyle NH-CH_2}{\diagup \diagdown}} CO \qquad (2.1.8)$$
$$\mathbf{6}$$

We will see other, similar reactions when talking about reactions of aldehydes, etc.

As most of the polycondensation reactions are reversible, the linear and cyclic polycondensation products can be interconverted. In fact, equilibrium between these forms is frequently established. It is very important to understand the nature of these equilibria: when the equilibrium is shifted toward cyclic molecules, the polycondensation does not lead to the desired polymer. As in all equilibria, the equilibrium constant depends on the change of the Gibbs energy ΔG^0 accompanying the equilibrium reaction. In the present case, the equilibrium is considered to be between a section of the polymer chain and a cyclic molecule, which could be formed from this section (the section may comprise one or more monomeric units). The change in Gibbs energy ΔG^0 has an enthalpic part ΔH^0 and an entropic part $T\Delta S^0$

$$\Delta G^0 = \Delta H^0 - T\Delta S^0 \qquad (2.1.9)$$

Let us study first the enthalpy change. The chemical bonds in the cyclic and linear forms are the same. The enthalpies are therefore similar; their difference is mainly related to the strain in the ring. The strain is appreciable in three- and four-member rings, and the number of such rings at equilibrium is negligible. Five-, six-, and seven-member rings have very little strain. In fact, formation of five- or six-member rings may sometimes alleviate steric crowding in a linear chain, and such rings may be even enthalpically preferred to open chains. Eight- to thirteen-member rings (especially those that have mostly methylene groups) are known to be sterically very crowded and to be absent in chain-ring equilibria. However, this does not apply to many molecules with numerous single atoms in a ring, for example, cyclic siloxanes or cyclic oligomers of formaldehyde. Large rings are usually strain-free; that is, the conformation of bonds is essentially the same as in straight-chain molecules.

The entropic term has two parts. When a number of small molecules are joined together, translational degrees of freedom are lost and entropy decreases. Thus this part of the entropic effect favors the formation of rings, especially small rings. On the other hand, the conformational entropy of a chain section is decreased when the section forms a ring. Out of all possible conformations of the chain section in the open-chain form, only those conformations are possible in the ring form in which the original section ends are neighbors. Thus this part of the entropy term acts against ring formation, especially against large rings. We will return to this topic in Section 2.1.1.4.

In summary, the three- and four-member rings are not stable with respect to linear chains because of the ring strain. Large rings are not favored because of the conformational entropy. In most cases five- and six-member rings are favored. That explains why in our list of monomers such seemingly good candidates as valerolacton, valerolactam, and diketopiperazine **6** are missing. Seven-member rings (caprolactam, nylon 6) are present at equilibrium in a small but not negligible quantity. Their proportion increases with increasing temperature.

In the case of nylon 6, the presence of the monomer is most unwelcome. It worsens the mechanical properties of the polymer and may be slowly released from the finished product. It therefore must be removed from the polymer before its final use. Moreover, it is formed again when the polymer is brought to temperatures that allow reestablishment of the equilibrium. For this reason, nylon 66 is sometimes preferred to nylon 6 even if the monomers are more expensive.

The ring formation also quite often has a negative effect on polymer stability. Polyformaldehyde decomposes spontaneously unless the chain ends are blocked, which prevents splitting of the cyclic oligomers from the ends. Polydimethylsiloxane polymers are thermally exceptionally stable with respect to breaking chemical bonds. However, they decompose at temperatures above 300°C, forming mainly the cyclic tetramer.

2.1.1.3. Kinetics of Polycondensation.

To be able to predict the progress of polycondensation reactions, we must understand their kinetics. From a formal viewpoint, there is a difference between conventional chemical kinetics and kinetics of reactions involving polymers. On cursory inspection, the kinetics of polycondensation looks formidable: There are thousands of different molecular species, and each species may react with any other species. However, the situation simplifies enormously when we consider reactions between chemical groups instead of reactions between molecules. For example, in the synthesis of polyamides, we will consider only the reaction between an amino group and a carboxyl. The key postulate of polycondensation kinetics states that, at given reaction conditions, the reactivity of a chemical group (and, consequently, the values of the kinetic constants of its reactions) depends only on its nature and, to a lesser extent, on the nature of the molecular segment to which it is attached, and not on the size of the molecule or on the nature of distant molecular segments.

According to this postulate, the kinetic constant for, say, a polycondensation of an amino acid will change moderately during the early stages of the reaction when oligomers are formed and the immediate neighborhood of the group is slightly changed. However, in later stages of the reaction, it will not change any more, despite the fact that the viscosity of the reaction mixture increases by several orders of magnitude and the diffusivity of the reactive groups decreases by about the same factor. This behavior has indeed been observed experimentally.

To explain this apparent paradox, we need to analyze in some detail the elementary events during a reaction in liquid systems. An obvious necessity for a reaction to occur is a collision between the reactive groups. When these groups collide, an extremely small probability p exists for the reaction to occur. (If the probability were not extremely small, all groups would react within a very short time and we would have an explosion.) The macroscopic rate of the reaction will be proportional to the number of collisions per unit time and to the probability p. The number of collisions is proportional to the overall concentration of the reactive groups, but it is independent of the viscosity of the medium. Thus the kinetic constant is proportional to the probability p. How is it possible that the number of collisions is independent of the viscosity and of the diffusivity of the reactive groups?

In liquids, a given molecule (or a molecular segment or a reactive group) is surrounded by a given set of neighboring molecules for a relatively long time. The neighboring molecules form a *cage*. Because of the thermal motion, the molecule oscillates in the cage, suffering a series of collisions with its neighbors before being able to escape from the cage. The rate of escape is proportional to the diffusivity (inversely proportional to viscosity). If two reactive molecules enter the same cage, we call it an *encounter*. During the length of the encounter the two molecules collide repeatedly; each collision carries a probability p of the occurrence of the reaction. Let us say that there are on average n collisions per encounter; n is proportional to viscosity. The time a molecule spends between two encounters is also proportional to viscosity; hence the number of encounters N per unit time is inversely proportional to viscosity and the product nN is independent of viscosity. It is easy to show for small values of p that a probability P of the reaction happening during an encounter is given as

$$P = 1 - e^{-pn} \qquad (2.1.10)$$

The kinetic constant is proportional to the product NP. Let us consider two situations. First, let us assume that pn is still much less than unity. Then

$$NP = N(1 - e^{-pn}) \approx N[1 - (1 - pn)] = pnN \qquad (2.1.11)$$

Thus, in this case, the kinetic constant is indeed independent of viscosity, proving our postulate.

In the other case, we will assume that the viscosity is so high that pn is large. In this case, P equals unity; that is, each encounter leads to a reaction. The kinetic constant is now proportional to N; that is, it is inversely proportional to viscosity. The reaction becomes diffusion-controlled, and its nature changes. Thus our postulate breaks down for extremely viscous systems, especially for glasslike systems. However, under conditions useful for synthetic applications, the postulate of the kinetic constant being independent of viscosity (read "conversion") holds extremely well.

As an example, we will calculate the rate of polycondensation (i.e., the rate of disappearance of reactive groups) for a reaction of a diacid with a diamine. We will assume that the molar concentrations of both components at the beginning of the reaction are the same.

Let us adopt the following notation: The original total number of molecules of both monomers is N_0; N_0 is also the original number of carboxyls and of amino groups. The original concentration of carboxyls is $[C]_0$, of amino groups $[A]_0$; during the reaction, these concentrations decrease to $[C] = [A]$. We will treat the polycondensation as a second-order reaction and write

$$\frac{d[C]}{dt} = -k_1[C][A] = -k_1[C]^2 \qquad (2.1.12)$$

The integrated form of this relation reads

$$1/[C] - 1/[C]_0 = k_1 t \qquad (2.1.13)$$

Let us define the polymerization degree DP in the present case as the number of residues of both monomers in the molecule of polymer; it is twice the number of repeating units (there are a molecule of diacid and a molecule of diamine per repeating unit). The number-average degree of polymerization $\overline{DP}_n$ is the ratio of the number of residues (i.e., original number of molecules) to the number of molecules at time t.

$$\overline{DP}_n = N_0/N = [C]_0/[C] \qquad (2.1.14)$$

Substitution of $[C]$ from equation (2.1.13) to equation (2.1.14) yields

$$\overline{DP}_n = 1 + [C]_0 k_1 t \qquad (2.1.15)$$

Thus the degree of polymerization increases linearly with time. It takes 100 times as long to achieve $\overline{DP}_n = 100$ than to reach $\overline{DP}_n = 2$, yet the latter $\overline{DP}_n$ represents a reaction one-half completed. From a practical viewpoint, the reaction needs to be accelerated in its later stages. This is achieved by gradually increasing the temperature.

In the above scheme, we have not considered the reversibility of the reaction. It would slow down the later stages of the reaction even more and would lead to equilibrium at rather low values of $\overline{DP}_n$. In practice, the reaction water is removed as it is formed, suppressing the reverse reaction but not appreciably influencing the forward one.

The kinetics of polyesterification follows the same rules as the kinetics of polyamidation except for the need of an acidic catalyst. In the presence of a catalyst (e.g., sulfuric acid), the kinetic constant k_1 is proportional to its concentration. In the absence of a catalyst, the carboxylic groups of the diacid take over the catalytic function. However, as their concentration decreases, the kinetic constant decreases, too. In fact, the reaction becomes kinetically a reaction of the second order with respect to acid and of the third order overall. The completion of the reaction is even slower in this case.

2.1.1.4. Equilibrium; Distribution of Molecular Weights.

In most chemical reactions, the presence or absence of chemical equilibrium in the final stages of the reaction determines the final composition of the reaction mixture. In polycondensations involving carboxylic derivatives, the ease of transesterification and similar reactions as well as high temperatures during the final stages of the process usually lead to equilibrated products. As always, the equilibrium composition depends only on the overall stoichiometry, reaction temperature, and pressure but is independent of the exact nature of the original reactants, kinetics of the process, and sequence of operations.

For polycondensations involving only bifunctional monomers with balanced numbers of reactive groups, two types of equilibria are important: (1) equilibrium between cyclic and linear molecules and (2) equilibrium between linear chains of different lengths.

Let us treat the second type first. The important stoichiometric quantity in this case is the extent of reaction p, which is defined as the fraction of reactive groups that have actually reacted or, alternatively, as the probability that a given group has reacted. It is related to quantities of equation (2.1.14) as

$$p = 1 - N/N_0 \tag{2.1.16}$$

$$\overline{DP}_n = 1/(1-p) \tag{2.1.17}$$

Equation (2.1.17) is the well-known Carothers equation, and it quantifies the relation between the extent of the reaction and the average degree of polymerization. When the extent of reaction approaches unity, then the degree of polymerization may grow without limits.

However, if the two types of reactive groups are unbalanced, the highest possible $\overline{DP}_n$ is finite. It is achieved when all groups of the less frequent type have reacted and the excess groups of the other type occupy all chain ends. For unbalanced components, the Carothers equation reads

$$\overline{DP}_n = \frac{1+r}{2r(1-p)+1-r} \tag{2.1.18}$$

Here p is the extent of reaction for the minor component, and r is the ratio of moles of the minor component to moles of the major component.

A similar restriction of molecular weight occurs when some monofunctional compound (e.g., any fatty acid) is added to the reaction mixture. Upon completion of the reaction this compound occupies all chain ends. For example, when in an equimolecular mixture of two bifunctional components A and B a small fraction of molecules B is replaced by monofunctional molecules C (two molecules of C replace one molecule of B), then the limiting $\overline{DP}_n$ as p approaches 1 is given as

$$\overline{DP}_n = 4/s \tag{2.1.19}$$

where s is the ratio of number of moles of C and A: $s = N_C/N_A$.

The probability p that a given group has reacted is, for a strictly equimolar mixture of two components, equal to the probability p of Section 1.6.2. (There it was a probability that a given monomeric unit is followed by another one, i.e., that it is not the terminal unit.) It follows that polycondensation leads to polymers with the most probable distribution of molecular weights as described in Section 1.6.2 by equations (1.6.15)–(1.6.23). Actually, the same distribution of molecular weights prevails in those linear polycondensation polymers that were prepared from unbalanced monomers or included monofunctional units. In this case, the probability p must be calculated from equation (2.1.17), in which the actual $\overline{DP}_n$ was used [e.g., from equation (2.1.18) or (2.1.19), p is different from the extent of reaction in such a case.

Up to this point, we have considered only equilibria involving linear chains. We need to consider also the equilibria involving ring molecules. At equilibrium, a polycondensation product is composed of linear molecules of all possible chain lengths as well as from cyclic molecules of all possible ring sizes; all the species are in mutual equilibria. Fortunately, a very basic law of chemical equilibrium states that, at full equilibrium, each possible reaction in the system must be at equilibrium without respect to any other possible reaction. This law enormously simplifies the analysis. In our case, the law implies that the equilibria among the linear chains, (and, consequently, the distribution of lengths of these linear chains) are not influenced by the presence of rings; the distribution is still given by equation (1.6.15). The only difference is in the interpretation of the probability p. It now represents the probability that a given monomer unit *within a linear chain* is followed by another unit; because of the presence of rings (where every unit is followed by another one), this probability is slightly less than the extent of reaction.

For calculation of the equilibrium concentration of a cyclic polymer of any size, it is sufficient to study any equilibrium reaction in which this ring is involved. It is convenient to select the following reaction:

$$L\text{-}P_{j+n} \rightleftharpoons L\text{-}P_j + C - P_n \qquad (2.1.20)$$

where the letters L and C represent linear and cyclic polymers, respectively. The subscripts j and n represent the number of repeating units. A repeating unit is either one monomeric unit of the A−B type or a pair of units of A−A and B−B types.

Before applying the thermodynamic rules to reaction (2.1.20), we must specify the thermodynamic reference states. In a polycondensation mixture, each species is present at very low concentration. It is therefore convenient to use the ideally dilute state as a reference state and equate activities with molar concentrations. In addition, we will assume that the activity of any component is influenced neither by the distribution of other species nor by the concentration of a diluent.

The equilibrium of reaction (2.1.20) can now be expressed as

$$\frac{[L\text{-}P_j][C\text{-}P_n]}{[L\text{-}P_{j+n}]} = K_{C_n} \qquad (2.1.21)$$

where K_{C_n} is a *cyclization constant* for a ring with n repeating units.

Replacing the molar concentrations of linear polymers by respective numbers of moles N_i as given in equation (1.6.15), we can transform equation (2.1.21) into

$$[C\text{-}P_n] = p^n K_{C_n} \qquad (2.1.22)$$

At high conversions, p is close to unity. Thus the *molar* concentration of the rings is equal to the cyclization constant and therefore independent of the dilution of the system. However, the fraction of the polymeric material present in the rings increases with increasing dilution. If the reactive groups are balanced and the reaction is driven far enough toward completion and the overall concentration of repeating units is lower than $\Sigma_n n K_{C_n}$, $[C\text{-}P_n]$ cannot reach K_{C_n} and almost no linear chains are present. Most of the material is present in the form of rings.

It remains to evaluate the cyclization constant K_{C_n}. It is related to ΔG^0 of equation (2.1.9) as

$$K_{C_n} = \exp(-\Delta G^0/RT) \qquad (2.1.23)$$

As we have seen, for small rings ΔH^0 dominates ΔG^0. For rings with 3, 4, or 8–13 atoms, ΔH^0 is positive (ring strain) and K_{C_n} is negligibly small. For rings with 5 and 6 atoms, ΔH^0 may become negative, and these small rings may comprise most of the mixture. For 7-atom rings and for very large rings, ΔH^0 is becoming negligible and ΔS^0 dominates ΔG^0.

We will show now that ΔS^0 must depend on the ring size. Let us imagine that reaction (2.1.20) proceeds in two steps. In the first step a linear segment is cut off the end of the long chain; in the second step it cyclizes.

$$L\text{-}P_{j+n} \rightleftharpoons L\text{-}P_j + L\text{-}P_n \rightleftharpoons L\text{-}P_j + C\text{-}P_n \qquad (2.1.24)$$

The entropy increase for the first step is related to the fact that the $L\text{-}P_n$ chain, which was originally constrained by being attached to the $L\text{-}P_j$ chain, is now free to move throughout the solution. This increase is the same for all values of n. During the second step (cyclization), the entropy decreases. The end of the chain, which in the linear case was free to roam throughout the volume of its coil, is confined in the ring to the vicinity of the beginning of the chain. Because the volume of the coil (for sufficiently long chains) is proportional to $n^{3/2}$, this constraint contributes $R \ln (An^{-3/2})$ to ΔS^0. Here, A is a constant decreasing with increasing length of the repeating unit and with increasing chain stiffness. Moreover, for a ring with n repeating units, there are n indistinguishable conformations of the ring given by its rotation. This contributes another $R \ln(n^{-1})$ to ΔS^0.

We can now sum these two contributions to get ΔS^0. With ΔH^0 being negligible, $-T\Delta S^0$ is equal to ΔG^0; substitution of this value into equation (2.1.23) leads to the final result:

$$K_{C_n} = An^{-5/2} \qquad (2.1.25)$$

This relation, together with equation (2.1.22), shows that at equilibrium the molar concentration of rings decreases sharply with increasing ring size (i.e., it is proportional to $n^{-5/2}$). Even the mass concentration is proportional to $n^{-3/2}$. This steep decline in the distribution function of molecular weights $f_w(M)$ [cf. relation (1.6.28)] with increasing molecular weight starting from very small rings is in sharp contrast with $f_w(M)$ of linear polymers, which goes through a maximum for equilibrated systems.

When monomers with functionality higher than 2 are present in the reaction mixture, extensively branched polymers or an infinite network are usually present at higher conversions; we will discuss such systems in the next sections. However, some monomer mixtures cannot lead to infinite networks even in the presence of multifunctional units. Such a mixture is, for example, a multifunctional unit $RA_f(f \geq 3)$ combined with a monomer of the A−B type, where the A and B groups can join together, but no condensation occurs between two A groups or two B groups. Tricarballylic acid **7** and ε-aminocaproic acid **8** form such a system.

$$CH_2COOH$$
$$|$$
$$CHCOOH \qquad NH_2CH_2CH_2CH_2CH_2CH_2COOH$$
$$| \qquad\qquad\qquad\qquad\qquad \mathbf{8}$$
$$CH_2COOH$$
$$\mathbf{7}$$

All branched molecules in these systems must have the formula $R[-A(B-A)_y]_f$, where y, the number of bifunctional units in a branch of the molecule, may differ for each of the branches (zero is a permissible value of y). Thus all the branched molecules have all branches terminated by an A group, which can react with a B group of the A−B monomer or with a B end group of a linear chain $A(-B-A)_n$ —B but not with any other branched molecule. It is easy to show that the distribution of the lengths of individual branches is again the most probable distribution, equation (1.6.15). The branched molecules can be considered a result of fusing f chains into a single molecule as described earlier in Section 1.6.2. Thus the distribution function will be of the Schulz–Zimm type and will be given by equations (1.6.33) through (1.6.36), where z will be equal to the number of branches f; it will be narrower than the distribution for linear chains. It is noteworthy that even a bifunctional unit of the RA_2 type (e.g., succinic acid) added to an A−B monomer leads to a narrower distribution ($z = f = 2$) even if all molecules are linear.

2.1.1.5. Gel Point; Three-Dimensional Structures. The presence of trifunctional monomers in a polycondensation system alleviates the strict requirements for manufacturing high polymers that were necessary for bifunctional monomers. If enough trifunctional molecules are present, eventually a huge macromolecular network will permeate the whole sample. A typical example is polycondensation of phthalic acid with glycerol in the molar ratio 3:2. Replacing part of the glycerol by any glycol can easily control the degree of branching. The resulting polymer is a glyptal polyester.

During polycondensation of this type, several new phenomena play important roles. Even if the reactivity of all reactive groups is the same, molecules having in their skeleton a trifunctional unit now have three growing ends and grow faster than molecules with just two ends. They have a better chance of joining with other molecules with three ends; they will then have four ends and grow even faster. Thus the large molecules grow faster than the small ones, and the polydispersity is increasing. There will occur a moment when the big molecules join together into an effectively infinite network and form a gel-like structure. Accordingly, this moment is called a *gel point*.

At the gel point, the mixture loses its ability to flow. Instead of being viscous, it becomes elastic. This is manifested most clearly by the following phenomenon. The small bubbles of gas, which were slowly rising through the mixture before the gel point was reached, are suddenly stopped. This effect is the most sensitive measure of the gel point.

After gelling, smaller polymer molecules unattached to the infinite network are still present in the system. In fact, the mass fraction of the network at the gel point is vanishingly small, and the number-average molecular weight of the polymer, $\bar{M}_n$, is rather low. Of course, the weight-average molecular weight $\bar{M}_w$ reaches infinity at this point.

Let us follow the development of the system after the gel point. The system now has two interpenetrating parts: *gel,* which is the infinite network, and *sol,* the part composed of molecules of finite size. The polydispersity ratio $\bar{M}_w/\bar{M}_n$ of the sol is extremely high at the gel point, and $\bar{M}_w$ of the sol approaches infinity. As the reaction progresses beyond the gel point, individual molecules join the gel, and its mass fraction is growing fast. However, the largest, multiply branched molecules with many reactive end groups have more chances to attach themselves to the network; the sol is depleted of them, and both its molecular weight and polydispersity are *decreasing*. At high conversions, only a small amount of sol is present, and it has rather low molecular weight.

When manufacturing cross-linked polymers, it is important to know at what conversion the system gels. It happens when a randomly selected monomer unit has a finite chance of being a part of an infinite network. This chance is intimately related to the value of the *branching coefficient* α, which is defined as the probability that a given functional group of a multifunctional monomer unit leads through a chain of bifunctional units to another multifunctional unit. This probability is equal to the probability that the selected unit is followed by branches of the "second generation," that is, branches beyond the nearest branch point. The average number of branches of the second generation is then $\alpha(f-1)$, where f is the functionality of the multifunctional branching unit. The average number of branches of the third generation is $[\alpha(f-1)]^2$; for the nth generation it is $[\alpha(f-1)]^{n-1}$.

A finite probability that any given unit is a part of an infinite network implies that the probable number of branches of *any* generation (n growing without limits) is not vanishingly small. This happens only for

$$\alpha \geq \frac{1}{f-1} \tag{2.1.26}$$

The equality sign represents the gel point; larger values of α correspond to increasing amounts of gel.

The branching coefficient is a function of the composition of the monomer mixture and of the extent of reaction. We will present its evaluation only for the simplest case; however, it is easily evaluated even for more complicated systems.

Let us consider a mixture of bifunctional monomers A—A and B—B with a multifunctional monomer RA_f. The concentrations are adjusted so that A and B groups are exactly balanced. The fraction of A groups belonging to RA_f monomer is ρ. All the groups have the same reactivity. The extent of reaction p is equal to the probability that a given group has reacted. First, we calculate the probability of the starting RA_f molecule being part of structure **9**.

$$A_{f-1}\text{—R—A—(B—B—A—A)}_i\text{—B—B—A—R—}A_{f-1}$$
$$\mathbf{9}$$

Clearly, the probability that a B—B group follows in the chain is p; the probability that A—A follows is $p(1-\rho)$; for RA_f it is $p\rho$. The overall probability for structure **9** is $p[p^2(1-\rho)]^i\, p\rho$. The branching coefficient α is the sum of these probabilities for all possible chain lengths.

$$\alpha = \sum_{i=0}^{\infty} p[p^2(1-\rho)]^i\, p\rho = \frac{p^2\rho}{1 - p^2(1-\rho)} \tag{2.1.27}$$

In the extreme case, when A—A units are absent, $\rho = 1$ and $\alpha = p^2$. Thus, for polycondensation of a dibasic acid with glycerol ($f = 3$), the gel point should occur when α reaches $1/2$; this corresponds to conversion $p = 0.707$.

In actual experiments, the critical value α_c is found to be higher than the predicted one. For example, the polycondensations using trifunctional components reach the gel point at $\alpha_c \approx 0.6$ instead of the predicted $\alpha_c = 0.5$. The reason lies in the oversimplification inherent in the above theory. We have implied that the number of chain ends in a branched molecule either remains the same (on addition of a bifunctional molecule) or increases (on addition of a branched molecule) during any reaction between reactive groups. Actually, two ends of the same molecule may react together, decreasing the number of reactive ends capable of creating branching points. Thus the average number of branches of higher generations decreases: A higher value of α is needed at the gel point to compensate for this loss of molecular ends.

2.1.1.6. Dendrimers and Hyperbranched Polymers.

For many years, most of the technically important polymeric materials either consisted of linear macromolecules or formed infinite three-dimensional structures. Branched polymers that were produced mainly because of some less than ideal polymerization mechanisms were considered as a nuisance. Starlike polymers were a mere curiosity. However, it was found recently that heavily branched macromolecules possess valuable properties of new types. This section is devoted to such polymers.

We have already seen that some monomer mixtures containing multifunctional monomers are incapable of forming infinite structures. A special class of such monomers can be described by a generic formula AB_2 where the A group can react with the B group. It is obvious that any polymer molecule produced from such monomers can have only one terminal A group. (It can have none if this single A group has reacted with some B end group of its own molecule.) It is equally obvious that larger molecules of this polymer are heavily branched. (The degree of branching may be reduced if a compound of the AB type is added as a comonomer.) *Hyperbranched polymers* are materials produced by this and similar procedures. In solution, their segment density is much higher than the density of corresponding macromolecular coils. Thus even high-molecular-weight materials exhibit relatively low viscosity in solution. More importantly, they have a large number of terminal B groups that may have (or may be converted to some group having) some desirable chemical property—either enhanced solubility in some solvent or some catalytic property, etc.

The main advantage of hyperbranched polymers is the ease of their essentially one-step synthesis. The main disadvantage is their extensive polydispersity and somewhat unpredictable shape. Polymers with a more uniform structure were demanded. They were found in dendrimers.

Dendrimers are well-defined globular macromolecules constructed around a star-like core unit. Each arm of the star is branched at a very small distance from the center. Thus an onionlike layer is formed at the surface that has multiple terminal groups. These groups are then utilized as new branching points for building the next layer. This procedure is repeated several times, eventually producing a spherical molecule with a high concentration of surface terminal groups. Thus dendrimers have a *treelike* structure (Fig. 1.5) composed from several layers (generations) of branched monomer units. (The word dendrimer derives from *dendron,* the Greek word for "tree".) They are prepared in a stepwise fashion from simple branched monomer units, the nature and functionality of which can be easily controlled and varied. Theoretically, synthetic procedures can lead to dendrimers that are perfectly monodisperse with all molecules having exactly the same structure, composition, and molecular weight. However, their synthesis requires numerous steps and the final products often contain defects.

One of the primary differences between the star-branched polymers described in Section 1.2.3 and dendritic polymers is the distribution of chain segment density. For star-branched polymers, the density is highest at the core and decreases as the distance from the core increases. In contrast, for dendritic polymers the density of chain segments increases as the distance from the core increases. In fact, only some maximum number of dendrimer generations can be formed beyond which only defect structures are generated.

Dendrimer synthesis is a stepwise process involving a repetitious alternation of several growth and activation reactions. The growth of successive generations of the dendrimer radially outward from the central core is called *divergent synthesis*. As the dendrimer grows larger, the end groups on the surface of the globule become more densely packed and eventually the dendrimer reaches its upper generation limit. This is known as the *starburst effect*.

In recent years, the research of the dendrimers experienced an explosive growth and dozens of ingenious approaches were developed for their synthesis. We will present only one divergent reaction scheme that was used for an early entrant in the dendrimer field—the polyamidoamine dendrimer known as PAMAM dendrimer. The stepwise synthetic scheme for PAMAM dendrimer is shown below.

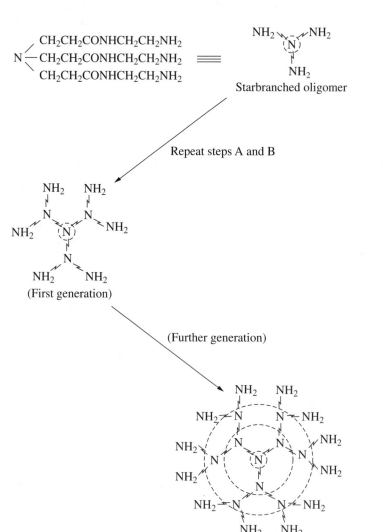

Starbranched oligomer

Repeat steps A and B

(First generation)

(Further generation)

In the first step, an initiator core such as ammonia or ethylene diamine (EDA) is reacted with methyl acrylate—the so-called Michael addition—to form a triester (or tetraester when EDA is used). In the second step, the triester is exhaustively amidated with a large excess of EDA to form a triamidoamine—called star-branched oligomer. The stepwise procedure is repeated to form succeeding generations of starburst dendrimers, each with twice as many terminal groups as its predecessor. (In this synthesis the Michael addition can be considered as the activation step.) If ammonia is used as the initiator core, the progression is 3, 6, 12, and so on, up to 1536 for the tenth generation. In contrast, if EDA is used as the initiator core, the progression is 4, 8, 16, and so on. With increasing size, the molecules tend to assume a spherical shape.

Generally, a large excess of EDA is added in the amidation step to hinder the dendrimers from bridging to one another. However, it is important that all EDA is removed before the next acrylation step. Otherwise, residual EDA molecules will act as new initiator cores, thereby destroying the monodispersity of the product. The higher-generation starburst dendrimers are fairly solid on the outside surface, they are porous and somewhat hollow in the interior. They tend to trap smaller molecules inside, and it becomes difficult to get all EDA out. However, by using ultrafiltration membranes and large quantities of solvent, it is possible to wash out all EDA and isolate PAMAM in high purity, even at the higher generations.

Finally, there comes a point, determined by the sizes and shapes of the molecules involved, at which the surface becomes so congested that there is no longer room for all terminal groups to react. In the PAMAM synthesis described above, the surface-dense packing starts to show up in the tenth generation; the eleventh generation becomes incomplete.

The *convergent synthesis* of dendrimers was developed as a response to the weaknesses of the divergent approach. Convergent growth begins at what will end up being the surface of the dendrimer and works inward by gradually linking surface units together with more monomers. See, for example, the reaction scheme in Figure 2.1. When the growing wedges are large enough, several are attached to a suitable core to give a complete dendrimer. We will again present only one example of convergent synthesis—a synthesis leading to an aromatic polyester dendrimer.

The surface group in Figure 2.1 is the phenyl group of polyester. The key intermediate is 5-(*tert*-butyldimethylsiloxy) isophthaloyl dichloride **10**. In the first step, the surface groups are attached by reacting with some hydroxy compound, e.g., a phenol producing a diester (diphenyl ester in this case). In the next step, the protecting siloxy group is removed by hydrolysis (focal point activation) and the resulting phenolic group is reacted with another portion (in an exact stoichiometric ratio) of the dichloride **10**. The activation and esterification reactions are repeated until a *dendron* of the required size is produced. (Dendron is a term used for description of wedgelike segments as the one above.) Finally, the dendrons are reacted with an appropriate core unit (e.g., 1,3,5-benzenetricarbonyl trichloride **11**) to yield the dendrimer. The main

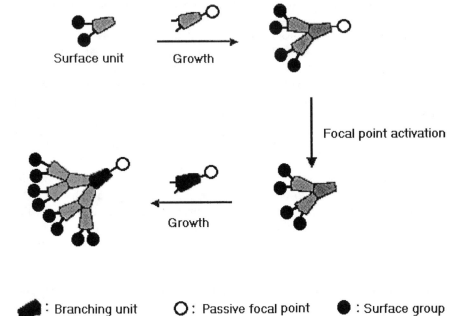

Figure 2.1. Convergent method of dendrimer synthesis.

advantage of the convergent growth is a relatively easy purification of the intermediate dendrons as well as of the final dendrimer.

$$OSi(CH_3)_2 Bu^t$$

10   **11**

Dendritic polymers have very different properties compared with their linear analogs. For instance, they have extremely high solubility in various organic solvents and low intrinsic viscosity in comparison to their linear cousins. Their intrinsic viscosity increases with size and then decreases after about fourth generation.

Many different types of dendrimers have been synthesized, and many different uses were developed or contemplated. For example, dendrimers with hydrophobic surfaces and hydrophilic cores are analogous to *inverse micelles*. The lower generations of such dendrimers resemble micelles; the higher generations, with their densely packed surface groups, are strongly reminiscent of the bilayers of liposomes, with the ability to enclose or exclude materials. Another dendrimer is a branched analog of poly(ethylene glycol) with a 2:1 carbon-to-oxygen ratio and multiple hydroxyl or protected hydroxyl chain ends. It is soluble in water and is found to be harmless (nontoxic) in animal studies. Many groups can be added to the core or to the periphery of dendrimers. Thus dendrimers are used as a polymeric matrix for catalysis, as a membrane for gas separation, etc.

The main disadvantage of dendrimers is their lengthy and costly synthesis. Thus much more easily accessible hyperbranched polymers that often can be synthesized in a single step may replace them for applications that do not require a precisely defined structure. However, monomers and procedures selected for their synthesis must satisfy certain requirements such as the absence of side reactions, equal reactivity of all B functionalities, and no internal cyclization reactions limiting the achievable molecular weight.

Hyperbranched polymers find applications as additives to conventional polymers, in blends, in thermosets, and in thermoplastics. Their addition enhances the mechanical properties and improves the processing characteristics because of their nonentangled nature. This feature is also valuable for formulation of coating materials. Optimum miscibility with other components of the blends can be achieved by modifications of their end groups. As an additive, a hyperbranched polyphenylene has been shown to reduce melt viscosity for polystyrene processing.

Synthetic procedures leading to dendrons, dendrimers, and hyperbranched polymers can be combined with other procedures known from organic and polymer chemistry. Thus dendrons may be attached to ends of linear polymer chains or used as multiple grafts on them. They can be converted to macroinitiators for various types of polymerization. At this moment, there seem to exist an almost infinite number of various achievable architectures and end uses, and researchers are busily exploring them.

2.1.2. Isocyanates

Isocyanates R—N=C=O are very reactive compounds that react under relatively mild conditions with a wide variety of other chemical groups. This gives a polymer chemist a broad range of ways to prepare various polymeric structures. Very often several different reactions of isocyanates are employed within the synthesis of a single polymeric material.

Generally, an isocyanate group reacts with all compounds that carry active hydrogen, such as water, alcohols, amines, and acids. Reaction rates are significantly enhanced by base catalysis; tertiary amines are very effective catalysts. For example, a catalyzed reaction with an alcohol can be depicted as

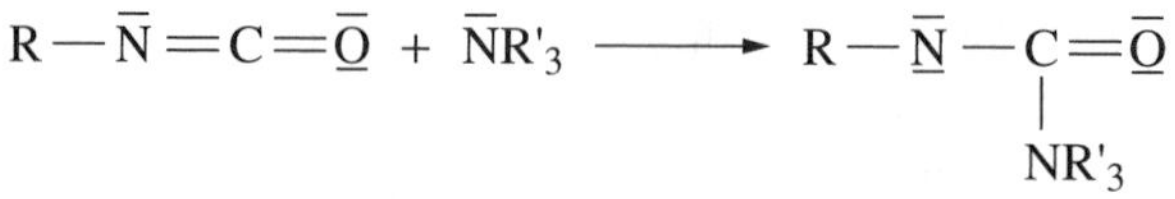

$$R-\bar{N}=C=\bar{O} + \bar{N}R'_3 \longrightarrow R-\underline{\bar{N}}-C=\bar{O} \atop \qquad\qquad\qquad NR'_3$$

$$(2.1.28)$$

12

Adduct **12** has excess electron density on nitrogen and readily accepts the active hydrogen from the alcohol. The reaction is completed by the elimination of the amine in a typical last step of an addition-elimination reaction on the carbonyl.

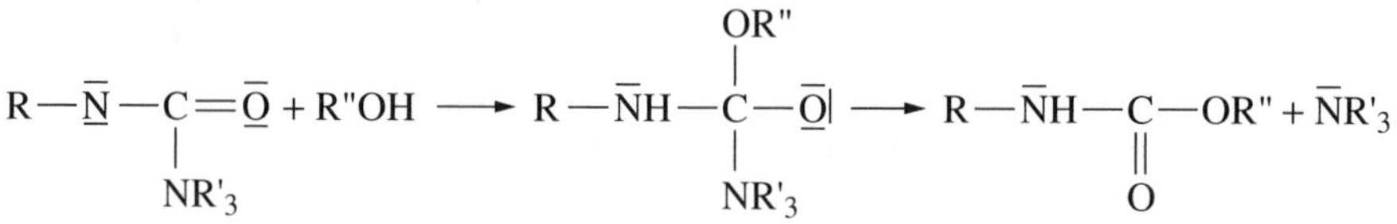

$$R-\underline{\bar{N}}-C=\bar{O} + R''OH \longrightarrow R-\bar{N}H-\underset{|}{\overset{OR''}{C}}-\bar{O}| \longrightarrow R-\bar{N}H-C-OR'' + \bar{N}R'_3$$

with NR'$_3$ under **12**, OR'' and NR'$_3$ under middle, O under **13**

$$(2.1.29)$$

Thus the product of the reaction of isocyanate with alcohol is a urethane **13**.

Some organometallic complexes MX_n are also very potent catalysts. They are believed to complex with the isocyanate oxygen, lowering the electron density on the carbon and facilitating the addition of alcohol:

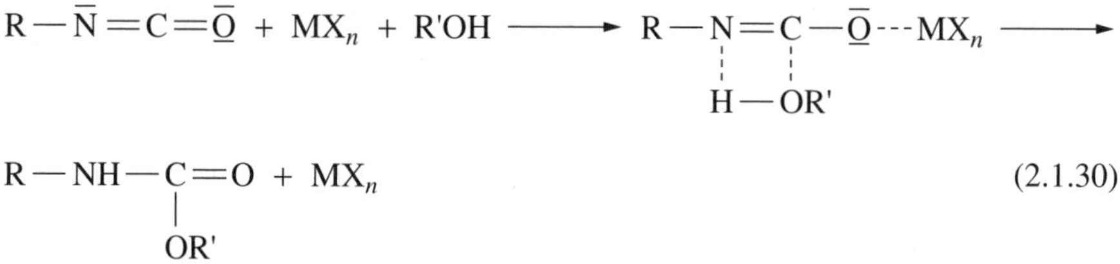

$$R-\bar{N}=C=\bar{O} + MX_n + R'OH \longrightarrow R-N=C-\bar{O}\cdots MX_n \longrightarrow \atop \qquad\qquad H-OR'$$

$$R-NH-\underset{|}{C}=O + MX_n \atop \quad\quad OR'$$

$$(2.1.30)$$

A complexation of the organometallic compound with the alcohol may also play a role in this scheme. Tin compounds are especially effective for this catalysis; tributyltin acetate or dibutyltin diacetate enhance the reaction rate by more than four orders of magnitude.

Reaction of isocyanates with amines proceeds under similar conditions and leads to substituted ureas **14**:

$$R-N=C=O + R'NH_2 \longrightarrow R-NH-CO-NHR' \qquad (2.1.31)$$

14

Actually, this reaction also proceeds (with slightly lower rates) with acyl-substituted amino compounds. Thus urethanes may react with another isocyanate to form allophanates **15**:

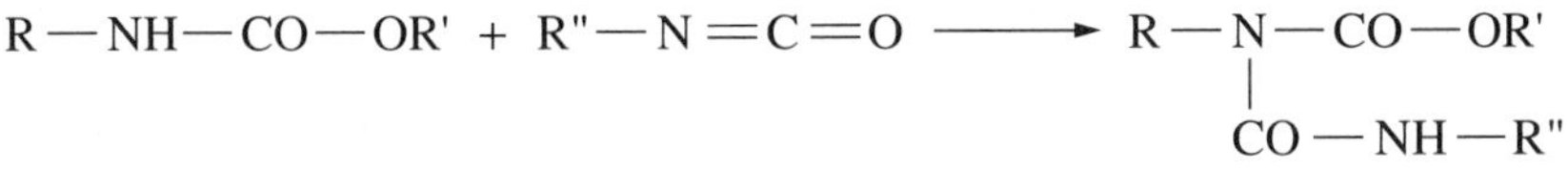

$$R-NH-CO-OR' + R''-N=C=O \longrightarrow R-N-CO-OR'$$

$$CO-NH-R''$$

15

(2.1.32)

Similarly, ureas may react with an additional molecule of isocyanate, yielding substituted biurets:

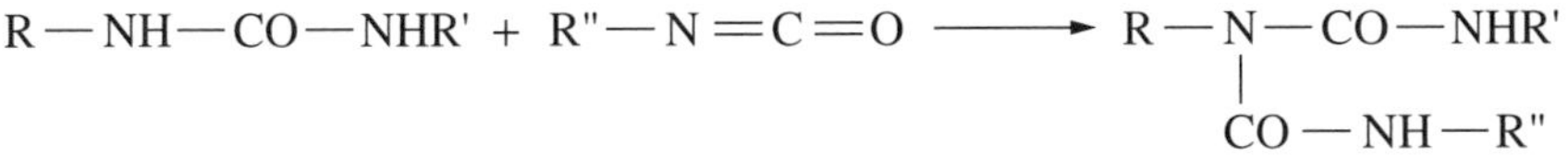

$$R-NH-CO-NHR' + R''-N=C=O \longrightarrow R-N-CO-NHR'$$

$$CO-NH-R''$$

(2.1.33)

Isocyanates also react vigorously with water, yielding an amine and carbon dioxide.

$$R-N=C=O + H_2O \longrightarrow RNH_2 + CO_2 \qquad (2.1.34)$$

When the isocyanate is in excess, the liberated amine reacts with another molecule of isocyanate to form urea according to scheme (2.1.31).

All the above reactions and their combinations are used for the preparation of different types of polymeric materials. Obviously, when high-molecular-weight materials are desired, bifunctional and multifunctional monomers must be used. Aliphatic diisocyanates [e.g., 1,6-hexanediisocyanate, $OCN(CH_2)_6NCO$] are used for the preparation of polyurethane and polyurea fibers. However, aromatic diisocyanates such as p-phenylene diisocyanate are used much more frequently.

For the preparation of linear polyurethanes or polyureas, the condensation of diisocyanate and glycol or diamine is best performed in solution, preferably in good solvents for the polymer (dimethyl sulfoxide, tetramethylene sulfone). Polycondensation in melt usually yields only low-molecular-weight products. The polymers and oligomers usually have high melting points (>200°C), and crystal formation during the early stages of polycondensation interferes with the completion of the reaction.

Diisocyanates are expensive. To reduce the cost of the final products, the other component of the polycondensation system is frequently chosen to have a rather high molecular weight—typically up to 5000. The other component is either a low-molecular-weight polymer of ethylene oxide or propylene oxide or their copolymer. These materials have two hydroxyls at the ends of the chain. Other low-molecular-weight polycondensation polymers with hydroxy end groups such as polyesters or polyamides can be also used. Eventually, these polymers will make a major portion of the final product and strongly influence its properties. When a three-dimensional network of a thermoset is to be prepared, a branched hydroxy oligomer may be used. Examples of such a material would be either glycerol or pentaerythritol, the hydroxy groups of which are extended by a polyethylene oxide chain. When a denser cross-linking is desired, a small excess of the diisocyanate is used. The excess isocyanate groups will react in the slow postcuring reaction with the urethane groups leading to allophanate cross-links.

Reaction with water adds an interesting twist to the chemistry of polyurethanes. A small amount of water will evolve some amount of gaseous carbon dioxide. It will also produce some urea linkages instead of urethane bonds; the cross-links formed in later stages of the reaction may be partially of the biuret instead of allophanate type. However, both these classes of bonds and cross-links lead to similar properties of the final product. What is more important is the action of the carbon dioxide. At first, it is dissolved in the reaction mixture; then gas bubbles are nucleated, and they convert the polymer to polyurethane foam. The relative timing of the events is all-important. If the bubbles are created too early, they escape from the mixture and the foam collapses. If they emerge during the process of gelling, they will be caught in the mixture but will grow quite freely. The walls between neighboring bubbles will get ever thinner; eventually, they will burst and rather open foam easily permeated by gases will result. If the gas bubbles emerge late in the process when the polymer is already very viscous and/or cross-linked, there will be many small bubbles isolated from each other. The foam will be tight and rather impermeable to gases. In either case, late-curing reactions will stabilize whatever morphology was created in the earlier stages.

The timing is influenced by two major factors—the relative rates of reaction of isocyanate with hydroxy groups and with water, and the rate of bubble nucleation. In an uncatalyzed system, the reaction with water is faster. However, tin compounds accelerate the reaction with hydroxy groups sufficiently to allow control of the sequence of events.

Nucleation also plays a very important role. Pure mixtures usually can be extensively oversaturated by gases. When the gases eventually evolve, a large number of bubbles will be nucleated simultaneously, and they will grow only to a small size. If the nucleation is stimulated early by the addition of appropriate components or of a volatile solvent, a smaller number of bubbles will occur sooner and they will grow to larger sizes.

Obviously, the density of the foam as well as the size and the openness of the gas cells can be controlled over a broad range by a judicious choice of the amount of reaction water and of volatile solvent, of the nature and amounts of the catalyst and the nucleation agents, and of the reaction temperature. On top of that, the nature of the hydroxy compounds strongly influences the properties of the foam skeleton.

There are several good reasons for the increasing popularity of polyurethanes. First, the reaction proceeds under rather mild conditions, at low temperatures and ambient pressure. Thus the manufacturing equipment can be quite simple. This led to the development of a new manufacturing process: reaction molding. The low-viscosity reaction mixture is injected into a mold, where it solidifies at moderate temperatures and pressures. If the same part were to be made by more traditional injection molding, the process would need high pressures and temperatures.

Second, the reaction yields no low-molecular-weight by-product, which would have to be removed. Reaction is easily forced to completion; the atmospheric moisture takes care of destroying all isocyanate groups that may be left unreacted.

Finally, the vast selection of hydroxylated components, the ease of controlling the cross-linking, and the possibility of forming foams of varying density and structure allow the manufacturer to match the properties of the polyurethane materials to an extremely broad range of properties required of the product.

2.1.3. Aldehydes

Carbonyl in aldehydes is generally more reactive than carbonyl in carboxylic acid derivatives. Thus the addition of compound $\overline{Y}$ carrying a free electron pair is even easier than its addition to a carboxylic derivative [compare equation (2.1.1)]

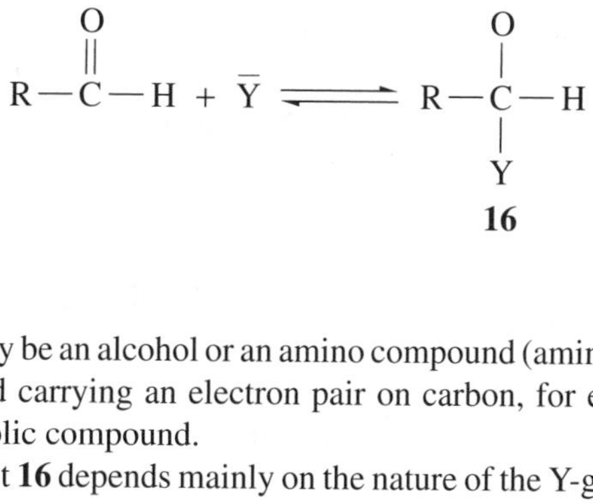

$$R-\overset{\overset{\displaystyle O}{\|}}{C}-H + \overline{Y} \rightleftharpoons R-\overset{\overset{\displaystyle O}{|}}{\underset{\underset{\displaystyle 16}{\overset{\displaystyle |}{Y}}}{C}}-H \qquad (2.1.35)$$

Compound $\overline{Y}$ may be an alcohol or an amino compound (amine or amide) but could also be a compound carrying an electron pair on carbon, for example, an enolized aldehyde or a phenolic compound.

The fate of adduct **16** depends mainly on the nature of the Y-group. Let us consider the main possibilities: When Y is an alcohol, the reaction requires acidic catalysis and yields a semiacetal.

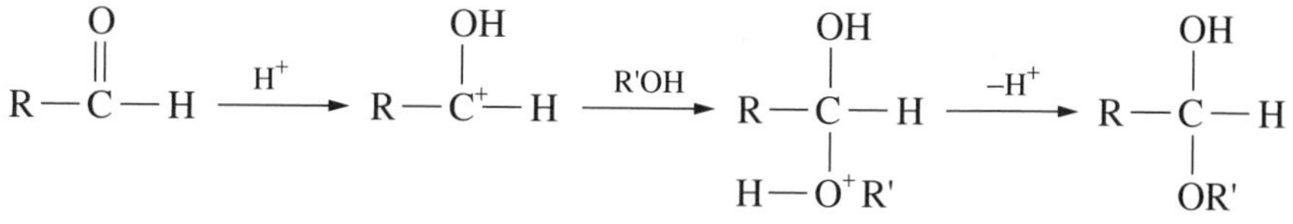

$$R-\overset{\overset{\displaystyle O}{\|}}{C}-H \xrightarrow{H^+} R-\overset{\displaystyle OH}{\underset{}{\overset{\displaystyle |}{C^+}}}-H \xrightarrow{R'OH} R-\overset{\displaystyle OH}{\underset{\underset{\displaystyle H-O^+R'}{\overset{\displaystyle |}{|}}}{C}}-H \xrightarrow{-H^+} R-\overset{\displaystyle OH}{\underset{\underset{\displaystyle OR'}{\overset{\displaystyle |}{|}}}{C}}-H$$

$$(2.1.36)$$

Semiacetals under conditions of acidic catalysis are very unstable. They split off a molecule of water; the resulting carbanion then adds on another molecule of alcohol to form an acetal:

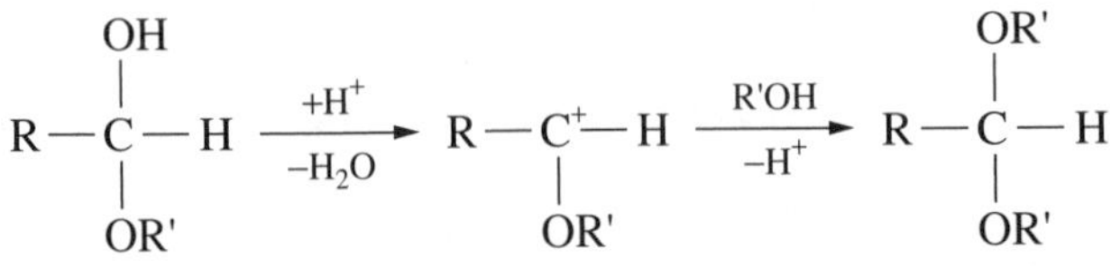

$$R-\underset{\underset{OR'}{|}}{\overset{\overset{OH}{|}}{C}}-H \xrightarrow[-H_2O]{+H^+} R-\underset{\underset{OR'}{|}}{\overset{\overset{}{+}}{C}}-H \xrightarrow[-H^+]{R'OH} R-\underset{\underset{OR'}{|}}{\overset{\overset{OR'}{|}}{C}}-H \qquad (2.1.37)$$

As reactions (2.1.36) and (2.1.37) are reversible, acetal interchange is an attractive route for preparation of more complex acetals, especially of polymeric ones. The reactions leading to acetals have high enough yields to be useful in the preparation of linear polymers with high molecular weights.

Useful polymers can be prepared by reaction of formaldehyde (or its dialkyl acetal) with glycols.

$$ROCH_2OR + HOR'OH \longrightarrow -OCH_2OR'- + ROH \qquad (2.1.38)$$

Glycols capable of yielding cyclic formals should, of course, be avoided in these polycondensations. Linear aliphatic glycols yield polymers with low melting points. High-melting cyclolinear polymers could be prepared from glycols such as 1,4-bis (hydroxymethyl) cyclohexane:

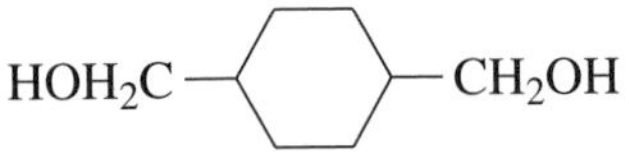

$$HOH_2C-\hspace{-0.5em}\left\langle\hspace{-0.3em}\bigcirc\hspace{-0.3em}\right\rangle\hspace{-0.5em}-CH_2OH$$

The expressed tendency of acetals to form six-member rings whenever possible can be utilized for polymer preparation in an interesting twist of the synthetic strategy. Pentaerythritol reacts with dialdehydes in a reaction where two six-member rings are closed in every molecule of the tetraol and a spiroacetal structure is formed:

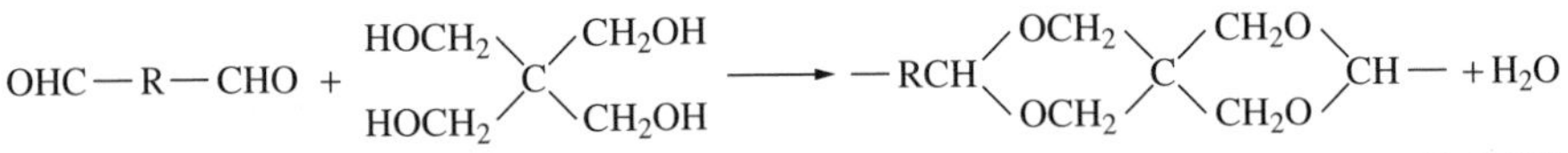

$$OHC-R-CHO + \underset{HOCH_2}{\overset{HOCH_2}{\diagdown}}C\underset{CH_2OH}{\overset{CH_2OH}{\diagup}} \longrightarrow -RCH\underset{\diagdown OCH_2}{\overset{\diagup OCH_2}{}}C\underset{\diagdown CH_2O}{\overset{CH_2O\diagdown}{}}CH- +H_2O$$

$$\qquad (2.1.39)$$

While the formation of acetals may lead to linear polymers, other reactions of aldehydes are usually not "clean" enough for the synthesis of linear chains. However, they are very useful for the preparation of cross-linked systems, which do not impose such high requirements on reaction mechanisms.

When compound $\overline{Y}$ in reaction (2.1.35) is an amine, the intermediate **17** is usually quite unstable; a molecule of water is expelled, and a Schiff base is formed:

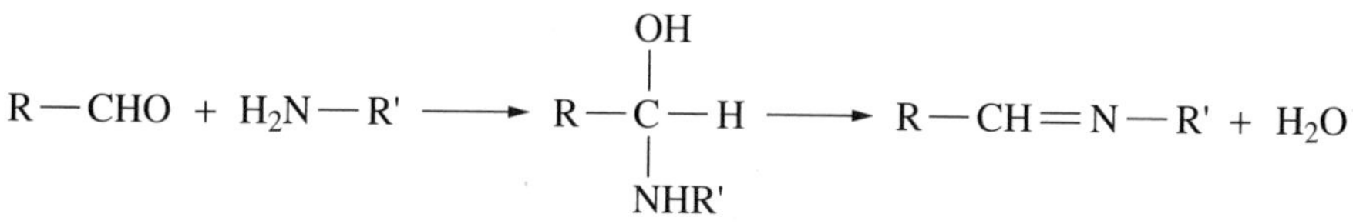

$$R-CHO + H_2N-R' \longrightarrow R-\underset{\underset{NHR'}{|}}{\overset{\overset{OH}{|}}{C}}-H \longrightarrow R-CH=N-R' + H_2O$$

17

(2.1.40)

The formation of Schiff bases is rather quantitative, and a polycondensation of diamines with dialdehydes or self-condensation of aminoaldehydes may lead to polymeric structures. However, Schiff bases easily undergo hydrolysis, and such polymers are not significant.

The intermediate **17** is more stable when the aldehyde used is formaldehyde. Thus polyamino compounds react with formaldehyde to yield polymethylol compounds. A typical example is melamine.

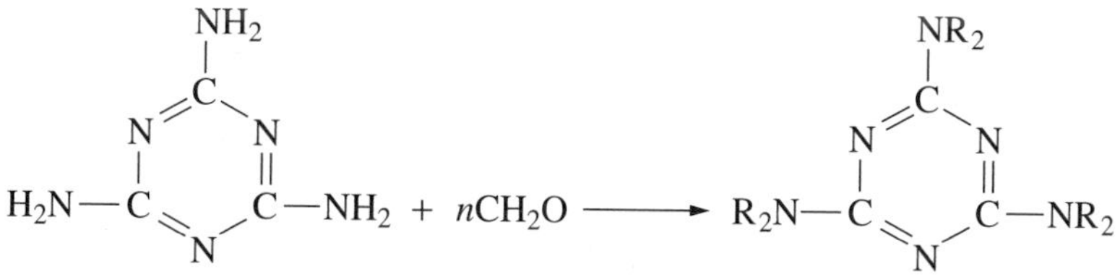

18

(2.1.41)

In compound **18**, R represents either hydrogen or a methylol group, $-CH_2OH$. Any number of methylol groups between zero and six may be present. The synthesis of melamine-formaldehyde resins proceeds in three phases. In the first phase, a mixture of compounds **18** is prepared under mild basic conditions. Then, at slightly acidic conditions, the methylol groups are alkylated (usually by butanol); these derivatives are quite stable.

$$\text{melamine}-N\underset{CH_2OH}{\overset{CH_2OH}{<}} + 2\,BuOH \longrightarrow \text{melamine}-N\underset{CH_2OBu}{\overset{CH_2OBu}{<}} + 2\,H_2O$$

19

(2.1.42)

During the same phase, some of the methylol groups react with some remaining amino hydrogens to form methylene bridges between two melamine moieties.

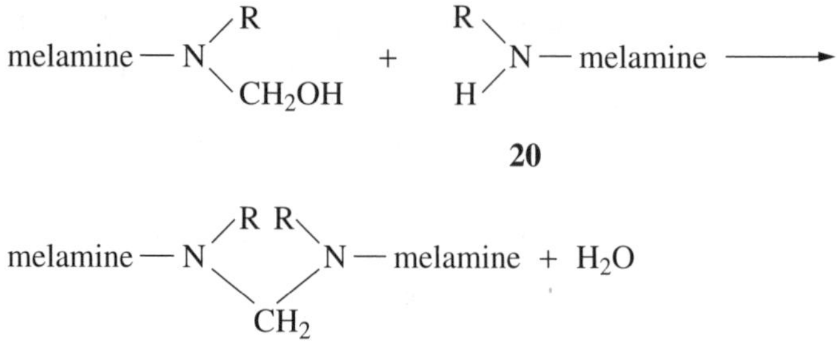

$$\text{melamine} - N\begin{array}{c}\nearrow R \\ \searrow CH_2OH\end{array} \quad + \quad \begin{array}{c}R \searrow \\ H \nearrow\end{array} N - \text{melamine} \longrightarrow$$

20

$$\text{melamine} - N\begin{array}{c}\nearrow R \quad R \searrow \\ \searrow CH_2 \nearrow\end{array} N - \text{melamine} \; + \; H_2O \qquad (2.1.43)$$

21

This second phase leads to prepolymers, which are stored until the final thermoset is to be produced. This is done by compression molding (Section 5.1.9) at high temperatures and pressures. During this step, reaction (2.1.43) proceeds further. Simultaneously, the butoxymethyl groups of **19** react with **20**, releasing butanol to form methylene bridges as in **21**, completing the thermosetting process.

Under basic conditions, formaldehyde also reacts with amino groups of amides, notably with urea. However, in this case, only one methylol group is attached to each NH_2 group:

$$NH_2CONH_2 \; + \; 2\,CH_2O \longrightarrow NH_2CONHCH_2OH \; + \; CH_2O$$

$$\longrightarrow HOCH_2NHCONHCH_2OH \qquad (2.1.44)$$

Under acidic conditions, the methylol groups are unstable and react with another amino group (possibly via intermediate formation of a Schiff base) and form methylene bridges between urea residues. Both branched structures and six-member rings **22** consisting of three urea residues are probably present in the final structures.

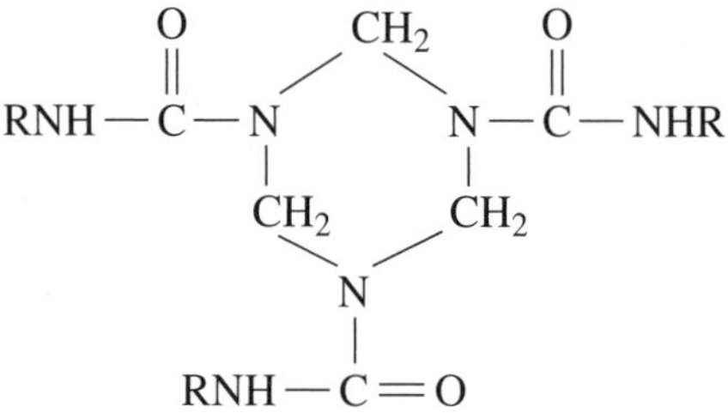

$$\begin{array}{ccccc}
 & O & CH_2 & O & \\
 & \| & \diagup \; \diagdown & \| & \\
RNH - C - N & & & N - C - NHR & \\
 & | & & | & \qquad R = CH_2OH \text{ or } H \\
 & CH_2 & & CH_2 & \\
 & \diagdown & & \diagup & \\
 & & N & & \\
 & & | & & \\
 & RNH - C & = O & &
\end{array}$$

22

As with melamine-formaldehyde resins, the final fabrication of urea-formaldehyde resins is done at elevated temperatures and pressures.

Formaldehyde also undergoes synthetic reactions leading to new carbon-carbon bonds. For a polymer chemist, the most important is the reaction with phenol leading to phenol-formaldehyde resins (Bakelite).

When the reaction is catalyzed by acids, a proton adds to formaldehyde, and the resulting carbocation (an electrophile) attacks the phenol in the ortho or para position:

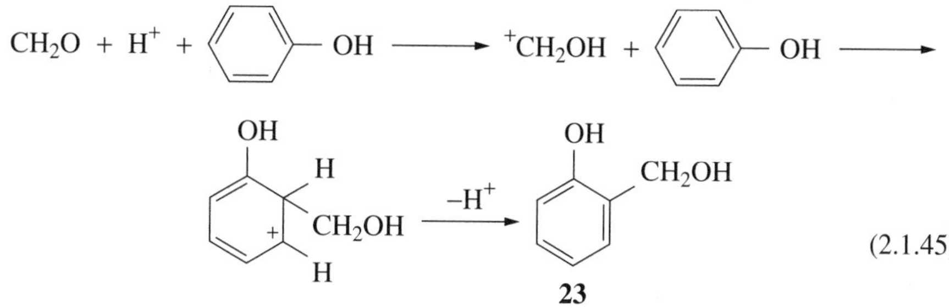

$$CH_2O + H^+ + \langle\text{benzene}\rangle - OH \longrightarrow {}^+CH_2OH + \langle\text{benzene}\rangle - OH \longrightarrow$$

$$(2.1.45)$$

23

However, in the presence of an acid, the benzylic hydroxyl in the hydroxybenzyl alcohol **23** is quite unstable. When it is protonated by the acid, it splits off water; the resulting carbocation attacks another molecule of phenol, forming a methylene bridge.

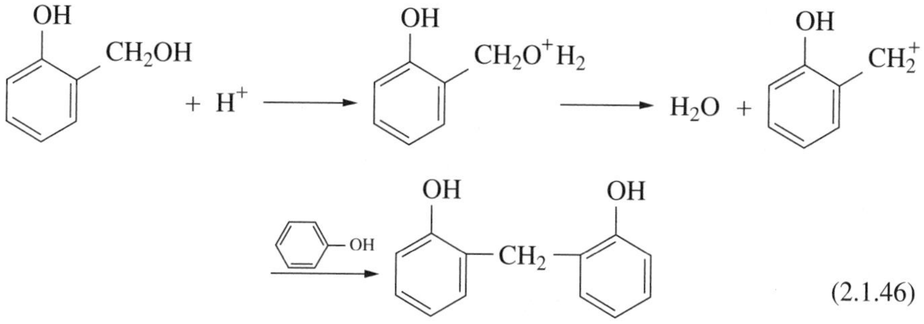

$$(2.1.46)$$

24

A phenol molecule has three positions capable of reacting with formaldehyde (two ortho and one para). Moreover, the substituted phenols of type **24** are even more reactive than phenol itself. Thus the reaction easily leads to large branched and eventually cross-linked polymers. To keep the reaction under control, a less than equimolar amount of formaldehyde is reacted with phenol. The product is a prepolymer with a molecular weight of the order of 1000. It is called *novolac*. During the final compression molding, novolac is reacted with additional formaldehyde [added as hexamethylenetetramine, $(CH_2)_6N_4$]. Besides the additional cross-linking by the methylene bridges, a number of other reactions are taking place: Two methylol groups may form benzyl ethers; ammonia from hexamethylenetetramine may be incorporated into the resin as a dibenzylamine arrangement; various dehydration and dehydrogenation reactions may also take place.

Phenol-formaldehyde resins of similar structure can be also prepared by a quite different mechanism using basic catalysis. The base deprotonates the phenol, and the resulting phenoxide anion attacks formaldehyde as a nucleophile. The attack may proceed through the phenolic oxygen. However, its result is a phenol semiformal **25**, a very unstable compound that easily decomposes back to the original reactants. When the attack proceeds through the carbon, ortho- or para-methylol phenol is the product [see reaction (2.1.47)].

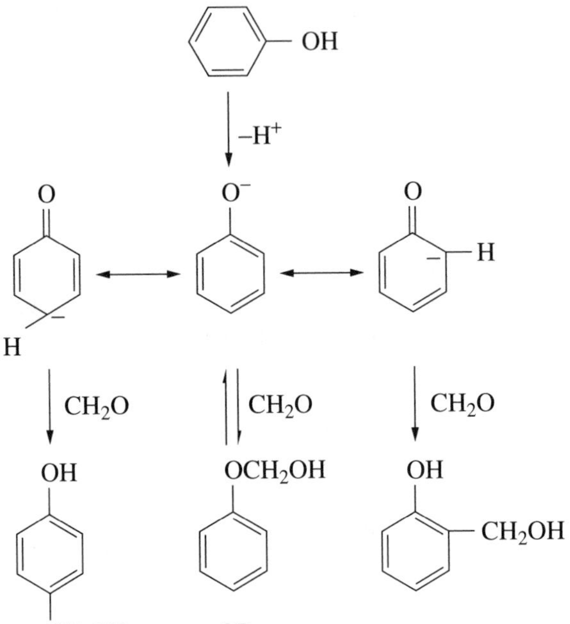

(2.1.47)

25

This reaction is usually carried on with an excess of formaldehyde; dimethylol and trimethylol phenols are also formed. The reaction between two molecules carrying methylol groups also occurs to a small extent, leading to methylene bridges; either water or formaldehyde is eliminated. Thus the reaction product, which is called *resole,* is composed of many mononuclear and paucinuclear species with varying numbers of methylol groups. In the second stage of the reaction, resole is slightly acidified; moderate heating then produces cross-linked but fusible structures called *resites*. After acidification, the reactions involved are very similar to those encountered in curing of novolacs. Finally, higher temperatures and pressures during the molding process give the final thermoset.

2.1.4. Condensations Forming Cycles

Cyclolinear polymers, especially those with many aromatic rings, have valuable physical properties and are often stable at elevated temperatures. Great synthetic effort has been directed toward the preparation of such structures, and, indeed, many rather complicated polymeric structures have been prepared. Some cycles are already present in the monomers; however, some of them are products of the condensation. The polycondensations of cycle-containing monomers belong to previous sections in this volume;

for example, condensation of terephthalic acid or its ester with glycols leads to commercially important polyesters; its condensation with diamines (either aliphatic or aromatic) yields excellent aromatic polyamides called *aramid fibers*.

In this section we will explore polycondensations during which new rings are formed in an essentially linear manner. Such reactions follow a common pattern. Synthesis of a new ring within the chain requires two condensation reactions for adding a single monomer to the growing chain. Thus each monomer unit must carry two bifunctional ends. Usually, one of the ring-forming reactions proceeds under mild conditions and yields a stable intermediate. Under harsher conditions, the second condensation closes the ring. There is always the possibility that the second reaction will be intermolecular instead of intramolecular, leading to cross-linked structures. However, in a skillfully designed monomer, the two groups that should participate in the second reaction are held close together by the already formed molecular skeleton. Thus the probability of ring closure is much higher than the probability of an intermolecular reaction, which would require a properly oriented encounter with a second molecule. Moreover, the rings to be closed are usually five- or six-membered and are frequently aromatic. They are therefore thermodynamically more stable than the competing branched structures. The thermodynamic stability is especially valuable when the condensation reactions are at least partially reversible.

Let us demonstrate the above principles on the synthesis of the best-known polymers of this group—polyimides. Cyclic imides are prepared by reaction of primary amines with diacids or (better) with cyclic anhydrides. Typically, phthalanhydride may react with aniline, giving phthalanilic acid **26** in the first step. The phthalanilic acid (or any corresponding *amic* acid) condenses upon heating to *N*-phenyl phthalimide **27**.

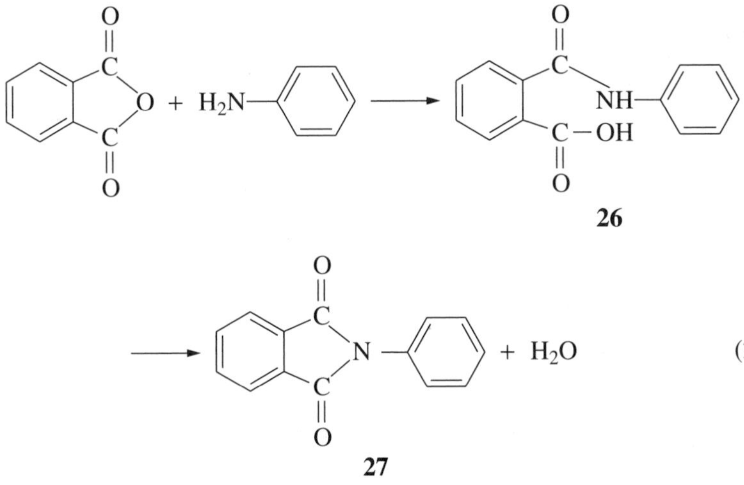

$$(2.1.48)$$

Polyamides are prepared by completely analogous reactions from aromatic dianhydrides, typically pyromellitic dianhydride **28**, and aromatic diamines as *m*- or *p*-phenylenediamine.

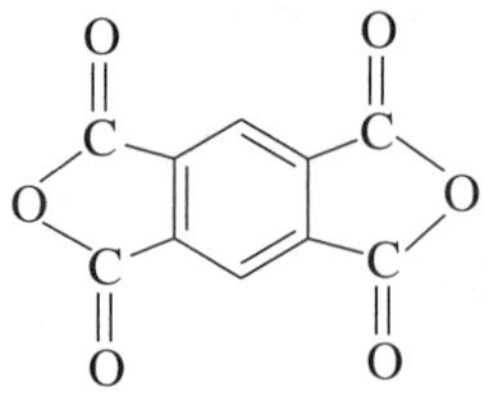

28

The role of the proximity of the two reactive groups in the polyamic acids can be seen from the fact that the cyclization reaction proceeds readily in the solid state. In fact, the methyl ester of phthalanilic acid self-condenses even in the crystalline state well below its melting point.

In the above reaction, two carboxylic groups reacted with one amino group to form an imide. Similarly, two amino groups may react with one carboxyl to form an amidine. When the two amino groups belong to an ortho aromatic diamine, the amidine is actually a heterocycle—an imidazole. In an actual synthesis, the carboxylic acids are replaced by their phenyl esters, which react more readily and are less prone to side reactions. A model reaction of this type is the one between phenyl benzoate and *o*-phenylenediamine. The diamine is usually used in the form of hydrochloride, which is more stable toward oxidation. The reaction again proceeds in two steps; the intermediate amino amide **29** is converted to a benzimidazol **30** at a temperature of 260–420°C.

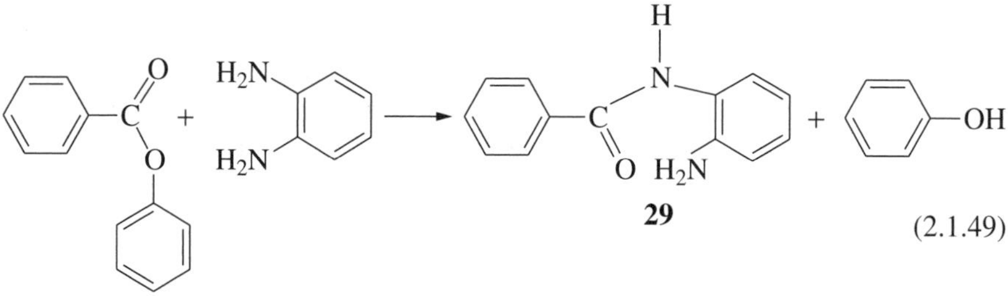

(2.1.49)

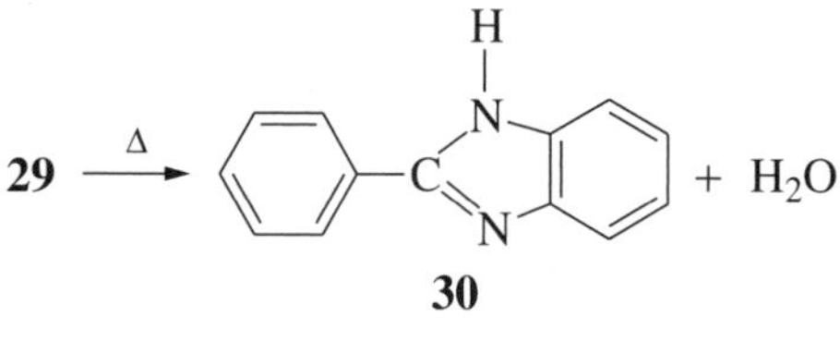

(2.1.50)

For the synthesis of high-temperature-stable polybenzimidazoles, phenyl esters of aromatic diacids, typically phenyl isophthalate or terephthalate, are used. The other component is a bis(orthodiamine), a tetramine; typically 3,3′- diaminobenzidine **31** in the form of a tetrahydrochloride.

$$H_2N-\text{[aromatic ring structure]}-NH_2$$

NH_2   NH_2

31

The commercial success of polyimides and polybenzimidazols led to synthesis of a number of other cyclolinear and ladder aromatic polymers based on the formation of other heterocycles: for example, hydantoin, oxadiazole, triazole, quinoxaline. However, the description of these synthetic procedures is beyond the scope of this textbook.

2.1.5. Siloxanes

The chemistry of polysiloxanes is also based on stepwise reactions. However, the typical reactions involving silicon are different from similar reactions of carbon compounds. The difference is related to the fact that silicon (similarly to its neighbors in the periodic table—aluminum and phosphorus) is a second-row element with vacant d orbitals; first-row elements do not have them. These d orbitals can participate in valence bonding, especially in some transition states when silicon may become penta- or hexacoordinated. (You may recall the pentacovalent compounds of phosphorus.) When silicon is bound to an electron-rich atom (typically oxygen), its free electron pairs can interact with the d orbitals of silicon to form a partial double bond. This effect is especially strong in the chain of silicon-oxygen bonds forming the backbone of polysiloxanes. As a result, the bonds are unexpectedly strong, and the O—Si—O angle is about 150° instead of the tetrahedral angle 109°.

Availability of d orbitals also influences the reactivity of silicon compounds. Thus, unlike the carbon-chlorine bond, the silicon-chlorine bond hydrolyses immediately on contact with water or any compound with active hydrogen. The silanol group, $\equiv$Si—O—H, is very unstable: two such groups form siloxane bonds, $\equiv$Si—O—Si$\equiv$, even in the presence of excess water. However, the reaction is sufficiently reversible to allow siloxane interchange [a reaction similar to transesterification, equation (2.1.2)] under both acidic and basic catalysis. Polysiloxanes are usually considered to be very stable toward hydrolysis. However, the stability is probably a result of their hydrophobicity; under the usual circumstances, water does not penetrate toward the siloxane bond. Siloxane compounds, which are water-soluble, are not hydrolytically very stable.

In the following, we will describe reactions involved in the synthesis of polydimethylsiloxanes. Other polysiloxanes are prepared by completely analogous procedures. The synthesis of polysiloxanes starts with hydrolysis of dimethyldichlorosilane $Cl_2Si(CH_3)_2$. The hydrolysis is a very vigorous reaction leading to disilanols, which immediately condense to siloxanes. Although some amount of linear chains is produced, the main product is the cyclic tetramer octamethyl cyclotetrasiloxane **32**. It is this compound that is used for the actual synthesis of polysiloxanes; either base or acid catalysis is employed.

The polymerization catalyzed by potassium hydroxide is believed to start with hydrolysis of the tetramer **32** leading to the dipotassium salt of the linear disilanol **33**. Alternatively, a similar salt can be used as an initiator.

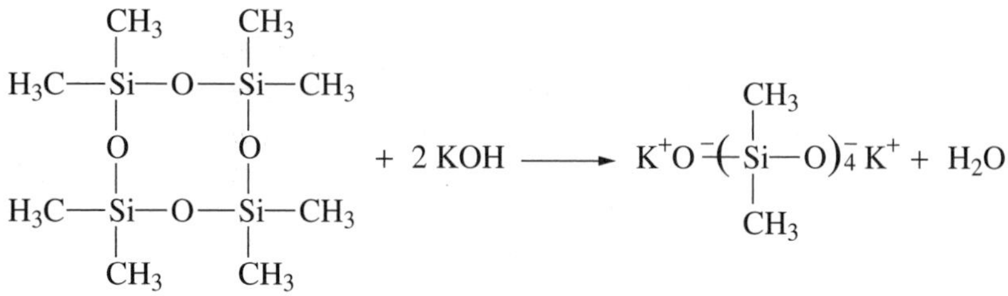

$$
\text{32} + 2\text{ KOH} \longrightarrow \text{K}^+\text{O}\!-\!\!\left(\!\overset{\displaystyle CH_3}{\underset{\displaystyle CH_3}{\overset{|}{\underset{|}{Si}}}}\!-\!O\right)_{\!4}^{\!-}\!\!\text{K}^+ + H_2O \tag{2.1.51}
$$

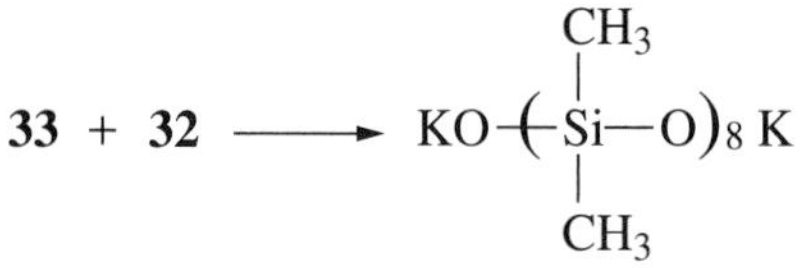

$$
\text{33} + \text{32} \longrightarrow \text{KO}\!-\!\!\left(\!\overset{\displaystyle CH_3}{\underset{\displaystyle CH_3}{\overset{|}{\underset{|}{Si}}}}\!-\!O\right)_{\!8}\!\!\text{K} \tag{2.1.52}
$$

The charge-carrying oxygen of **33** attacks another molecule of **32** (taking advantage of the d orbitals of the silicon), forms a siloxane bond, and cleaves the original siloxane bond on **32**. The result is a longer linear molecule capable of attacking another cyclic monomer and growing further.

So far the process has corresponded to a typical anionic chain polymerization (see Section 2.3). However, the active anions are attacking not only the cyclic molecules but also the linear chains, causing siloxane interchanges. This scrambling reaction produces linear and cyclic molecules of all possible degrees of polymerization. The distribution of material into linear and cyclic species of different sizes follows the rules outlined in Section 2.1.1.2. The average degree of polymerization is determined by the number of anionic ends. After the reaction mixture reaches equilibrium, the linear chains are capped at the ends by trimethyl chlorosilane.

In an alternate procedure, the hydrolysis products of trimethyl chlorosilane and of methyl trichlorosilane are added at the beginning of the polycondensation. The former compounds act as chain ends, the latter as branching points (when a rubberlike network is desired). In this case, the reaction is completed by neutralization of the alkali; the liberated silanols immediately condense to siloxanes.

For acidic catalysis, either sulfuric acid or trifluoromethylsulfonic acid (triflic acid) is employed. The reaction mixture consists of the cyclic tetramer and an endcapping compound, for example, hexamethyldisiloxane **34**. The catalyst protonates the oxygen of the siloxane, and in a siloxane interchange reaction the cyclic tetramer is incorporated into the growing linear chain. Thus the first step of the reaction proceeds as shown in reaction (2.1.53).

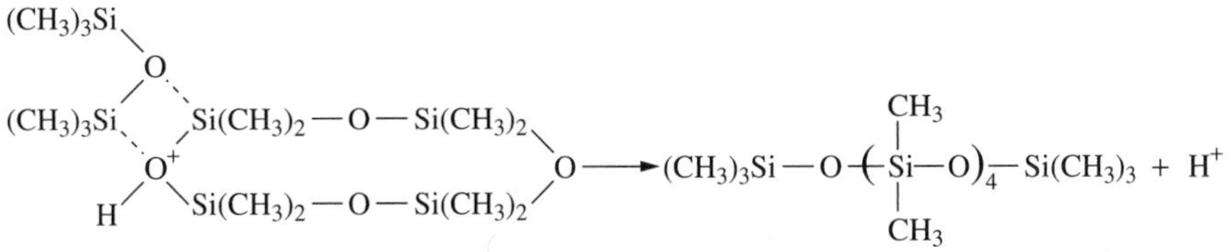

$$(CH_3)_3Si-O-\left(\!Si-O\!\right)_{\overline{4}}-Si(CH_3)_3 \;+\; H^+$$

34   (2.1.53)

The concerted reaction (2.1.53) is an *insertion reaction*. In the later stages of the reaction, the insertions of the cyclic molecules occur at any place in the growing linear chain. At the same time, the scrambling reactions with the same mechanism are gradually bringing the reaction mixture toward equilibrium encompassing the same mixture of linear and cyclic species as was prepared during the base catalysis.

At equilibrium, the amount of cyclic molecules in polydimethylsiloxanes is significant. At room temperature and in the absence of solvents, it amounts to about 8%. It increases with the dilution of the system and with increasing temperature. In fact, the degradation of polysiloxanes at temperatures above 250°C is caused mainly by decomposition into cyclic species, mainly tetramer (it is actually a re-equilibration reaction at high temperature).

The most stable and abundant cyclic species is the tetramer—an eight-member ring. Because of the preferred siloxane bond angle of 150°, the six-member cyclic trimer has a considerable ring strain and is almost absent at equilibrium. On the other hand, every other chain atom (oxygen) has no substituents; consequently, there is no overcrowding of hydrogens, which made the nine- to thirteen-member rings so strained in the hydrocarbon chemistry.

Whereas the preparation of other polysiloxanes from other monomers carrying two inert groups on each silicon (methyl, phenyl; diphenyl; methyl, cyanopropyl; etc.) proceeds along similar lines, hydrolysis of phenyl trichlorosilane, $PhSiCl_3$, follows a different path. Hydrolysis of this compound yields a three-functional silantriol, which we would expect to condense to a very heavily branched thermoset similar to a condensate of glycerol with phthalic acid. However, the phenyl groups are bulky. There would not be enough space available for all the branches. In other thermosets, this problem is alleviated by the fact that many reactive groups remain unreacted. In the present case, the reactivity of silanol groups is so high that unreacted silanols would be distinctly unhappy. The system tries to organize its structure in such a way that all silanol groups would form siloxane bonds. Such a structure is a highly regular ladder polymer, poly(phenyl sesquisiloxane) **35**.

$$
\begin{array}{ccc}
\text{Ph} & \text{Ph} & \text{Ph} \\
| & | & | \\
-\text{Si}-\text{O}-\text{Si}-\text{O}-\text{Si}-\text{O}- \\
| & | & | \\
\text{O} & \text{O} & \text{O} \\
| & | & | \\
-\text{Si}-\text{O}-\text{Si}-\text{O}-\text{Si}-\text{O}- \\
| & | & | \\
\text{Ph} & \text{Ph} & \text{Ph}
\end{array}
$$

35

The regular structure of the ladder polymer has lower entropy than a randomly cross-linked network. However, this loss of entropy is more than balanced by a decrease in enthalpy accompanying the increase in the number of siloxane bonds. There is no doubt that during the process of condensation of the monomer units, many branched structures are created. However, easy siloxane interchange and thermodynamic tendencies will eventually play their role in producing a soluble linear polymer with a perfect ladder structure.

2.1.6. Epoxy Resins

Epoxides react easily with nucleophiles and can thus yield polymeric structures by polycondensation mechanisms. The resulting thermosets are called *epoxy resins*. Their synthesis requires a number of steps and is based on special chemical properties of epichlorohydrin **36**. This compound is even more reactive than other epoxides because of the inductive effect of chlorine. Epichlorohydrin is reacted with dihydroxy compounds, preferably of phenolic nature (the attacking nucleophile—a phenoxy anion—is present at higher concentrations than its alkoxy equivalent because of the higher acidity of phenols).

$$
\text{R}-\text{O}^- + \text{CH}_2-\overset{\displaystyle}{\underset{\diagdown\diagup}{\text{CH}}}-\text{CH}_2\text{Cl} \longrightarrow \text{R}-\text{O}-\text{CH}_2-\overset{|}{\underset{\text{O}^-}{\text{CH}}}-\text{CH}_2\text{Cl}
$$

$$
\qquad\qquad\qquad\quad \text{O}
$$

36   **37**

$$
\longrightarrow \text{R}-\text{O}-\text{CH}_2-\text{CH}-\text{CH}_2 + \text{Cl}^-
$$

$$
\qquad\qquad\qquad\qquad\qquad \diagdown\!\diagup
$$

38   O

$$(2.1.54)$$

The uniqueness of epichlorohydrin is in the structure of the adduct **37**. With other epoxides, the reaction would be completed by neutralization of the charge by an active hydrogen and formation of a hydroxy group. Instead, the intermediate **37** eliminates a chloride anion and re-forms an epoxide **38**. This epoxide is again capable of reaction with another molecule of nucleophile (this time the reactivity is lower), leading to a polyether with pendant hydroxy groups **39**.

$$R-O-CH_2-CH\underset{\underset{O}{\diagdown\diagup}}{\;}CH_2 + ROH \longrightarrow R-O-CH_2-\underset{\underset{OH}{|}}{CH}-CH_2-O-R$$

38   **39**   (2.1.55)

Although a number of diphenols and glycols would react according to reactions (2.1.54) and (2.1.55), the monomer used most frequently is our old friend bisphenol **A3**. Novolacs (condensates of phenol and formaldehyde; Section 2.1.3) are also conveniently used as polyhydroxy compounds. The condensation is performed with an excess of epichlorohydrin; the product is a *prepolymer* (i.e., an oligomer with DP $\sim 3 - 10$) end-capped by unreacted epoxy groups as in **38**. The secondary hydroxyls (as in **39**) do not react appreciably with the epoxides at mild reaction conditions.

The prepolymers are then cured with various ingredients for the preparation of heavily cross-linked thermosets. The curing agents are compounds that react easily and preferably exothermally with the epoxide moieties. Amines and carboxylic acid anhydrides are used most frequently. Difunctional curing agents increase the length of linear chains. Cross-linking is achieved by using (1) polyfunctional curing agents or (2) prepolymers made from polyphenolic compounds (e.g., novolacs) and also by (3) taking advantage of the side reactions of the secondary aliphatic hydroxyl in the prepolymer. At moderately elevated temperatures, the hydroxyls react with the pendant epoxy groups, with the anhydrides and dianhydrides of the curing agent, and also with the carboxyls of the half-esters formed by reaction of anhydrides with pendant epoxides.

The popularity of epoxy resins is related to the broad variety of possible materials resulting from the use of different polyphenols and different sizes of prepolymers as well as different curing agents. Another advantage is in the ease of the curing reaction, which starts spontaneously after mixing the prepolymer and curing agent, which often does not even require a higher postcure temperature. Epoxy resins are very heavily cross-linked and mechanically very tough. When filled with inert materials such as sawdust, they retain most of their valuable properties at lower cost. However, the best results are achieved when the fillers reinforce the resin. Strong fibers, especially glass fibers and glass cloth, are used for this purpose. The resulting materials are known as *laminates* and are superior engineering materials. They are used for lightweight roofs, outdoor kiosks, boat hulls, and many parts of automobile bodies, among other things.

2.1.7. Miscellaneous Polycondensations

In the previous sections we have dealt with step-type reactions that have mechanisms uncomplicated by side reactions and have a high enough yield to be expected to lead to polymeric materials under most circumstances. Other types of reactions usually do not qualify for dependability, but occasionally they also produce polymers. This section is devoted to such polycondensations.

Ether bonds can be synthesized by the nucleophilic replacement of a halogen by an alkali salt of an alcohol or phenol. We will see later that there exist much easier and cheaper synthetic routes to polyethers. Consequently, this synthesis is used only for polyethers with special structures. Thus, m-bromophenolate **40** condenses (with copper as a catalyst) to poly(m-phenylene oxide) **41**.

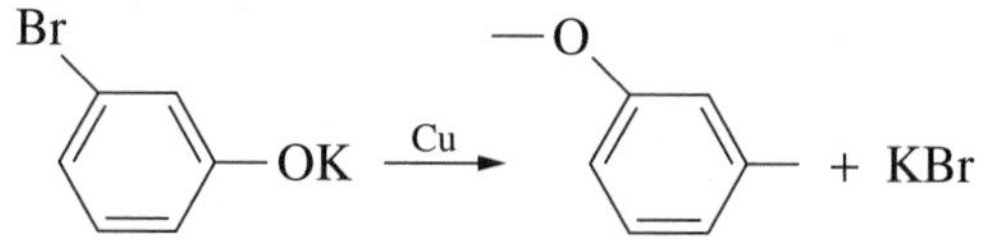

$$(2.1.56)$$

Polysulfones **42** are prepared by a similar reaction of bisphenol A with dichlorodiphenyl sulfone **43**.

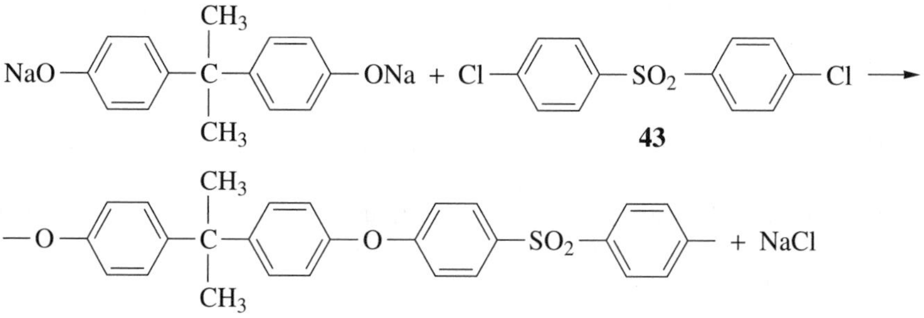

$$(2.1.57)$$

This reaction is carried on at 160°C in dimethyl sulfoxide as a solvent.

Sulfur analogs of glycols, dithiols, HSRSH, can be used in a similar scheme for the preparation of poly(alkylene sulfides) **44**.

$$NaS(CH_2)_x SNa + Cl(CH_2)_y Cl \longrightarrow -S(CH_2)_x S(CH_2)_y- \qquad (2.1.58)$$
$$44$$

This reaction yields only rather low-molecular-weight polymers. Alkaline polysulfides are much more reactive than sulfides. They can be reacted directly with dihalo compounds to yield high-molecular-weight poly(alkylene polysulfides) **45**

$$Na_2 S_x + Cl(CH_2)_y Cl \longrightarrow -(CH_2)_y-S_x- \qquad (2.1.59)$$
$$45$$

Electrophilic alkylation of aromatic nuclei (Friedel–Crafts reaction) does not have yields sufficient for formation of high-molecular-weight polymers. In a typical example, benzyl chloride self-condenses under the influence of most Friedel–Crafts catalysts. The resulting oligomeric polybenzyls are usually highly branched because the alkyl-substituted aromatic nucleus is more reactive toward electrophilic reagents than lower substituted nuclei. Typical polybenzyls have a structure like **46** and have no practical significance.

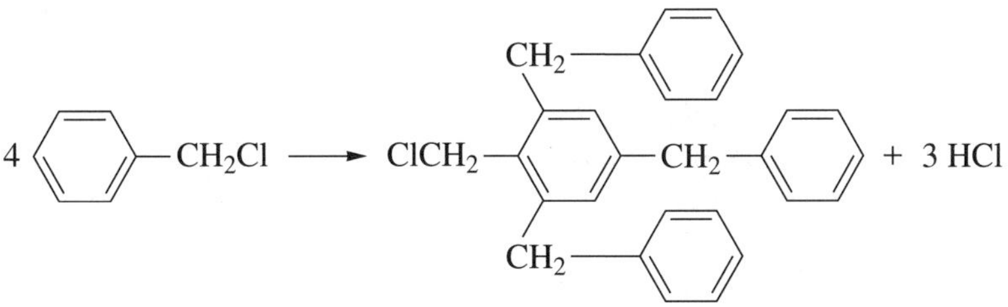

46
(2.1.60)

Diels-Alder condensation of dienes with olefins (dienophiles) is a reaction proceeding with good yields and at mild conditions. As such, it may qualify as a reaction leading to polymers. For example, 1,6-bis(cyclopentadienyl)hexane **47** reacts with quinone to form a low-molecular-weight polymer **48**.

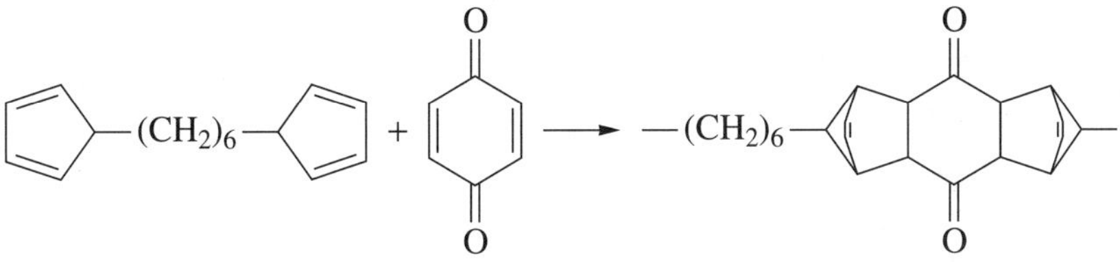

47   **48**   (2.1.61)

However, the Diels–Alder reaction is reversible at higher temperatures. Moreover, the cyclohexene double bond (which is synthesized by the reaction) is also a dienophile; its reaction with another diene group may lead to branched structures. Thus the theoretically attractive group of Diels–Alder polymers still remains a laboratory curiosity.

We will see in the next section that the introduction of radicals into susceptible monomers leads to their chain polymerization. It is one of the most important reactions leading to high-molecular-weight polymers. However, one of the typical reactions of radicals is their recombination, which can be used for synthesis of polymers in a stepwise reaction—radical coupling. In a typical example, radicals are produced by thermal decomposition of *tert*-butyl peroxide. These radicals can abstract hydrogen from susceptible groups, for example, from an isopropyl group on an aromatic ring; two resulting isopropyl radicals then recombine.

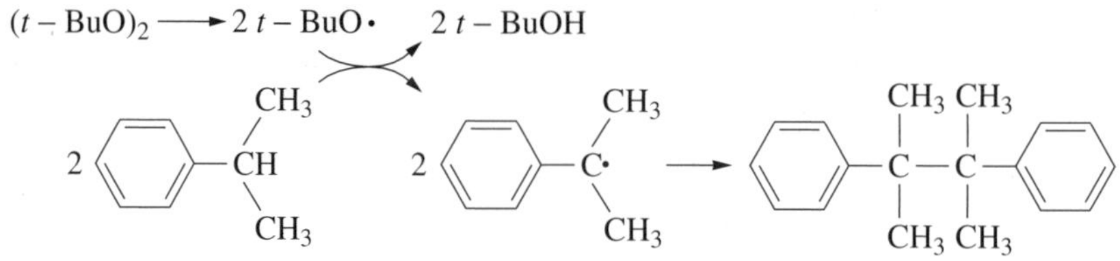

(t – BuO)2
2 t – BuO•
2 t – BuOH
2
CH3
CH
CH3
2
CH3
C•
CH3
CH3 CH3
C—C
CH3 CH3

(2.1.62)

If reaction (2.1.62) is performed using diisopropylbenzene instead of isopropylbenzene, a polymeric structure results.

Copper-catalyzed oxidations by molecular oxygen often proceed through radical intermediates. These radicals may recombine, leading again to polymers. Typical is the oxidation of *m*-diethynylbenzene **49**.

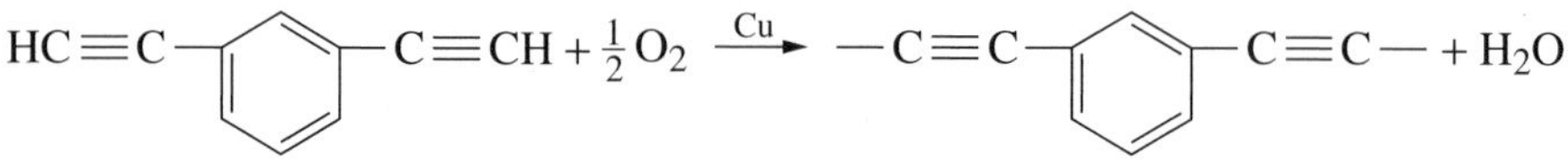

HC≡C
C≡CH + ½ O2
Cu
—C≡C
C≡C— + H2O

49

(2.1.63)

A similar oxidative coupling can be used to prepare poly(dimethylphenylene oxide) from 2,6-xylenol.

CH3
OH + ½ O2
CH3
CH3
O— + H2O
CH3

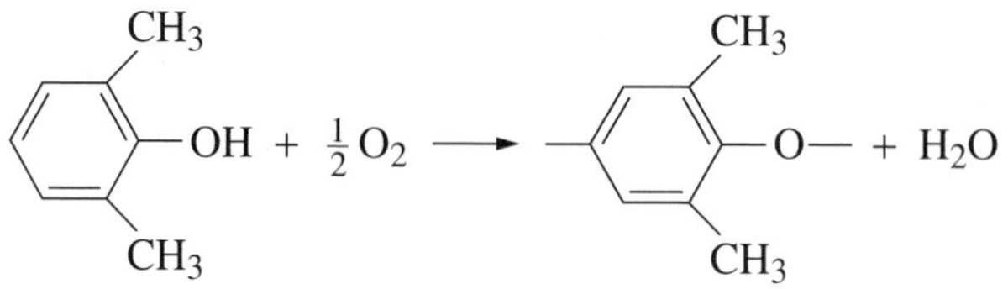

(2.1.64)

The catalyst in this case is a copper-pyridine complex, and the reaction proceeds at room temperature when oxygen is bubbled through the reaction mixture. The mechanism of the reaction is not quite clear, but it has been shown to be of radical character. Its stepwise nature is obvious from the gradual increase in molecular weight during the entire process of polymerization.

Coupling of alkyl halides by metals produces longer hydrocarbons. When applied to dihalides it would lead to polyethylene, but such a synthesis would not be economically attractive. However, an analogous reaction was recently employed for the preparation of polysilanes from dichlorodialkyl silanes, R_2SiCl_2, and metallic sodium. Unlike their hydrocarbon cousins, polysilanes are light-sensitive. This is a result of the presence of d orbitals on silicon atoms. In polysilanes, these orbitals may participate in bonding neighboring silicons, thus creating a polyconjugated structure that absorbs light in the near-ultraviolet range as polyconjugated hydrocarbons do.

2.2. RADICAL POLYMERIZATION

Among the chain polymerization methods, radical polymerization has been known for the longest time and is best understood. The method requires the introduction of some radical species into the monomeric system. The radical attacks a monomer molecule, transforming it into another radical. This process is called *initiation*. The monomer-converted-to-radical is an active species that adds other monomers while the growing chain preserves its radical character. This process is called *propagation*. A radical may react with a wide variety of compounds, but again the products have mostly a radical character. There is only one way a radical can be removed from the system—by reacting with another radical. This process is called *termination*. Initiation, propagation, and termination are necessary features of all radical polymerizations (in fact, of all chain processes), and their detailed interplay determines the outcome of the reaction and the properties of the resulting polymers. However, before studying these reactions, we need to review how radicals are born, how they behave, and how they perish.

2.2.1. Production of Radicals, Initiators

The term "radical" is used for those organic compounds that carry an unpaired electron. In most cases, radicals are very energetic species; they are therefore rather reactive and usually short-lived.

The most common way of generating radicals is by homolytic decomposition of covalent bonds. This can be achieved by imparting enough energy to the bond, either by increased temperature or by radiation. Indeed, any covalent bond will break when the temperature is raised enough (the process is called *pyrolysis*). However, for our polymerization purposes, we do not need indiscriminate breakage of bonds; we need to produce radicals in a controlled way under possibly mild conditions. The solution of the problem lies in selecting compounds with labile bonds, which can be broken by imparting to them only moderate amounts of energy, that is, by raising the temperature only moderately.

Such labile bonds are mainly single bonds between two like electronegative atoms such as nitrogen-nitrogen, oxygen-oxygen, fluorine-fluorine, bromine-bromine, and iodine-iodine bonds. The bond energy (i.e., dissociation energy) of all these bonds is about 38 kcal/mol; this should be compared with 83 kcal/mol for carbon-carbon and carbon-oxygen bonds. For simple molecules, the dissociation energy is equal to the activation energy for homolytic decomposition; the lower the energy, the lower the temperature sufficient for noticeable decomposition.

For more complex molecules, the dissociation of a bond may be accompanied by a simultaneous reorganization of the steric and electronic structure of the radical fragments. This reorganization usually lowers the dissociation energy and at the same time stabilizes the radicals. The effect is apparent even when alkyl radicals are compared with radicals made from single atoms (halogens, hydrogen). However, it becomes very important when the lone electron may become conjugated with olefinic and aromatic structures, and it can become the dominant feature when extensive aromatic structures are involved.

Substances that decompose to radicals at a convenient temperature are useful as initiators of radical polymerization. We will discuss just two classes of such initiators: peroxides and azo compounds.

The decomposition of peroxides is based on the low bond energy of the oxygen-oxygen bond. In solution, their decomposition follows first-order kinetics. A large number of peroxides are used for initiation purposes. The differences in dissociation energy make them useful at different temperatures. The aliphatic peroxides are convenient at 100–120°C; *tert*-butyl peroxide **1** is used quite often.

Aromatic peroxides—represented mainly by benzoyl peroxide **2**—are useful at 60–80°C. The difference is caused by the interaction of the radical free electron with the aromatic nucleus; electrons move rather freely throughout the conjugated system, and the unpaired electron may change its residency among several atoms; the energy is lowered by resonance effects.

(2.2.1)

Other favorite peroxides are peroxidicarbonates (for example, diisopropyl peroxidicarbonate **3**) and cumyl hydroperoxide **4**.

$$
\begin{array}{c}
\text{CH}_3 \\
\diagdown \\
\text{CH} \\
\diagup \\
\text{CH}_3
\end{array}
\text{O} \quad \text{O}
\quad
\begin{array}{c}
\text{CH}_3 \\
\diagup \\
\text{CH} \\
\diagdown \\
\text{CH}_3
\end{array}
\qquad
\begin{array}{c}
\text{CH}_3 \\
| \\
\text{C} - \text{O} - \text{OH} \\
| \\
\text{CH}_3
\end{array}
$$

$$\text{CH} - \text{O} - \overset{\text{O}}{\overset{||}{\text{C}}} - \text{O} - \text{O} - \overset{\text{O}}{\overset{||}{\text{C}}} - \text{O} - \text{CH}$$

3   **4**

In aqueous media, hydrogen peroxide, H_2O_2, and potassium persulfate, $K_2S_2O_8$, are also used frequently.

Molecular oxygen easily forms peroxides with several types of organic compounds, notably ethers and hydrogens on tertiary carbon. These peroxides may act as unwanted initiators in the presence of susceptible monomers.

The low dissociation energy of azo compounds is a result of a different phenomenon. They decompose into three fragments: two radicals and molecular nitrogen. Nitrogen molecules have unusually low energy; this leads to rather low overall dissociation energy. This argument applies mainly to aliphatic azo compounds. Aromatic azo groups bridge two aromatic systems of π electrons; the resonance effect lowers the overall energy. Dissociation of such compounds disrupts this extensive conjugation and is therefore energetically more demanding. Among aliphatic azo compounds, azobis(isobutyronitrile) **5** is the most frequently used initiator. It decomposes according to the following scheme:

$$
\underset{\substack{| \\ \text{CN}}}{(\text{CH}_3)_2\text{C}} - \text{N} = \text{N} - \underset{\substack{| \\ \text{CN}}}{\text{C}(\text{CH}_3)_2} \longrightarrow 2\,(\text{CH}_3)_2\overset{\bullet}{\text{C}} - \text{CN} + \text{N} \equiv \text{N}
$$

5   (2.2.2)

So far, we have been dealing with molecular decompositions yielding radicals in pairs. As both radicals were very energetic species, the amount of energy required was appreciable. It is advantageous to design decompositions that would yield only one radical for each decomposed molecule. Such reactions should display a concerted mechanism, where one of the molecular fragments is converted into a less energetic species. Some transition metals capable of existing in more than one valence state can donate or abstract an electron to or from a disintegrating molecule (this is a redox-type reaction). Typical is a reaction of ferrous ion with hydroperoxides, which yields one radical and a ferric ion. A mixture of hydrogen peroxide with ferrous salts is the well-known Fenton reagent used in organic chemistry for radical oxidations. It decomposes at room temperature according to the reaction.

$$\text{HO} - \text{OH} + \text{Fe}^{2+} \longrightarrow \text{HO} \bullet + \text{OH}^- + \text{Fe}^{3+} \qquad (2.2.3)$$

In the presence of susceptible monomers, the hydroxyl radical acts as an initiator. In addition to iron, many other metals (Mn, V, Cu, Co, for example) can decompose hydroperoxides; they are active in both the higher and lower oxidation states. Consequently, a small amount of the metal is sufficient for catalyzed decomposition. When using these initiators, it should be kept in mind that the metallic ions can also exchange electrons with the radicals that were just produced and deactivate them. Thus the admissible amount of the metals is very small.

The reaction of tertiary aromatic amines (*promoters*) with peroxides, although still not fully understood, proceeds with a similar mechanism.

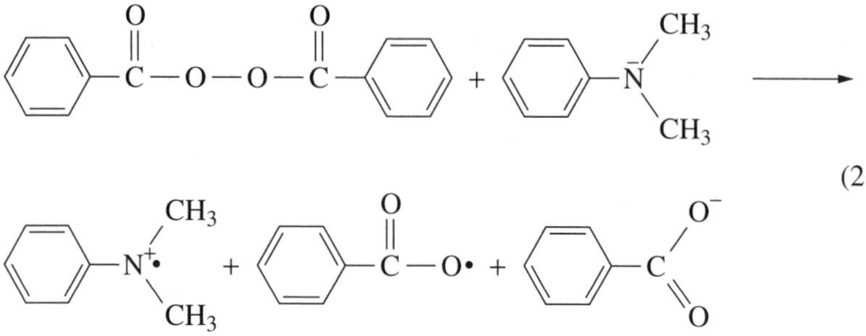

$$(2.2.4)$$

The energy-saving step is the large amount of resonance energy released by conjugation of the radical cation **6** with the aromatic ring. Thus promoters are used for lowering the decomposition temperature.

As we have already seen, homolytic decomposition of bonds can also be achieved by radiation. We will distinguish two cases: photochemical decomposition and high-energy irradiation. In the former case, susceptible compounds absorb radiation (usually in the near ultraviolet range); rearrangement of the orbitals in the excited molecule can then lead to spontaneous decomposition. Azobis(isobutyronitrile) **5** is a favorite initiator for photoinitiation. The photoinitiating process is essentially independent of temperature and can be used at temperatures much lower than those that would decompose **5** thermally. Another frequently used photoinitiator is benzoin methyl ether **7**, which decomposes as

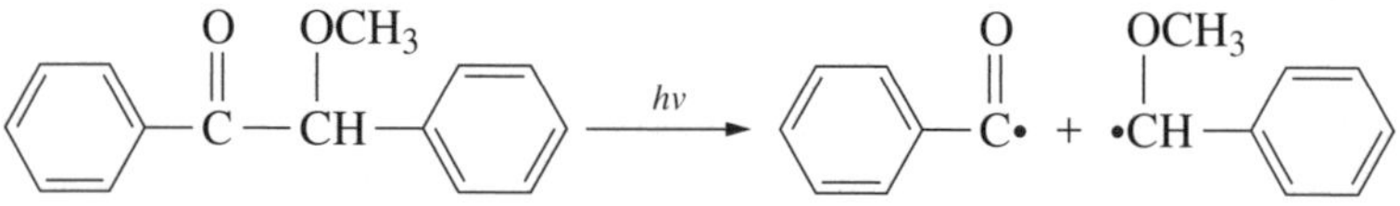

$$(2.2.5)$$

High-energy irradiation disrupts chemical bonds indiscriminately. Both radical and ionic fragments are produced. X-rays and γ rays, high-energy electrons, and other energetic particles all have very similar results. Radicals generated by this mechanism undergo all types of radical reactions including those useful in polymer chemistry. Random disruption of bonds is usually not desired. Thus, high-energy radiation is used only for some special purposes. It can generate radicals in locations that are inaccessible to chemical initiators, such as inside crystals, in glassy polymers, and in thermosets. It can also activate for radical reactions (cross-linking, grafting) materials that are rather unreactive otherwise such as hydrocarbons or fluorocarbons.

2.2.2. Reactions of Radicals

For a better understanding of the process of radical polymerization, we need to grasp the chemical personality of radicals: their tastes (types of reactions in which they are involved) and their temperament (reactivity). We will consider monomolecular reactions of radicals, their reactions with nonradical species, and reactions with other radicals.

In a monomolecular event, a radical may eliminate a small molecule with a very low energy. Typical is the elimination of carbon dioxide from peroxide radicals. The decomposition of benzoyloxy radicals (obtained by dissociation of benzoyl peroxide) is a relatively slow reaction competing with other radical reactions, but the decomposition of acetoxy radicals (from acetyl peroxide) is almost instantaneous.

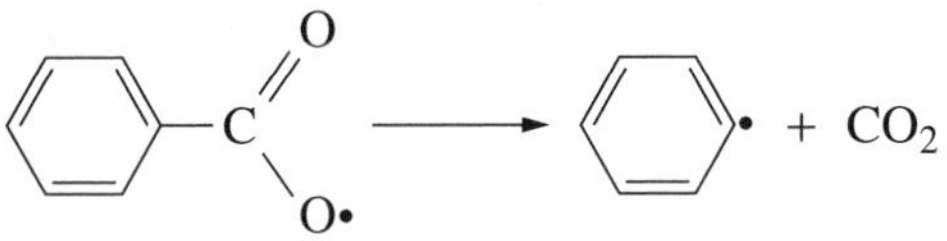

$$\text{(benzoyloxy radical)} \longrightarrow \text{(phenyl radical)} \cdot + CO_2 \qquad (2.2.6)$$

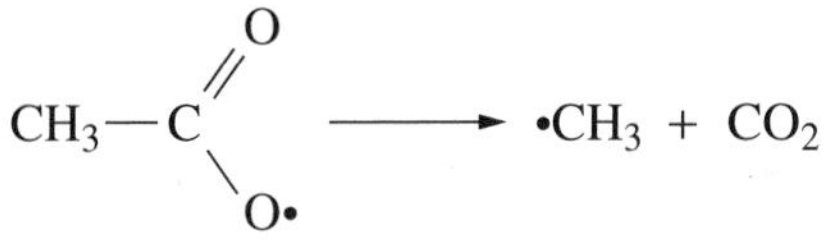

$$CH_3-C\overset{O}{\underset{O\cdot}{\big\langle}} \longrightarrow \cdot CH_3 + CO_2 \qquad (2.2.7)$$

Similarly, *tert*-butoxy radicals (from *tert*-butyl peroxide) and cumyloxy radicals (from cumyl hydroperoxide) can eliminate acetone and form methyl or phenyl radicals.

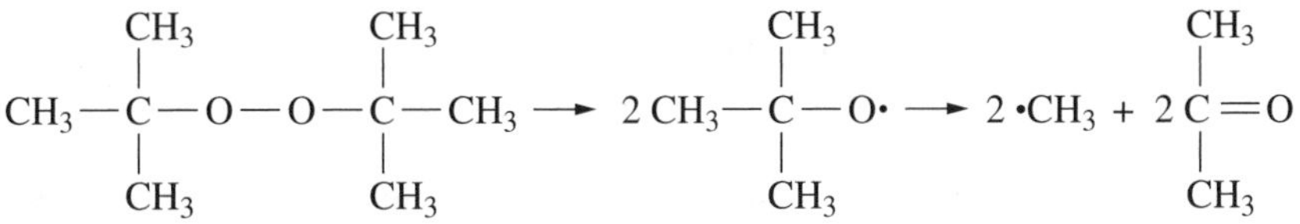

$$CH_3-\underset{\underset{CH_3}{|}}{\overset{\overset{CH_3}{|}}{C}}-O-O-\underset{\underset{CH_3}{|}}{\overset{\overset{CH_3}{|}}{C}}-CH_3 \longrightarrow 2\,CH_3-\underset{\underset{CH_3}{|}}{\overset{\overset{CH_3}{|}}{C}}-O\cdot \longrightarrow 2\,\cdot CH_3 + 2\,\underset{\underset{CH_3}{|}}{\overset{}{C}}=O$$

$$(2.2.8)$$

$$\text{Ph}-\underset{\underset{CH_3}{|}}{\overset{\overset{CH_3}{|}}{C}}-O-OH \longrightarrow \cdot OH + \text{Ph}-\underset{\underset{CH_3}{|}}{\overset{\overset{CH_3}{|}}{C}}-O\cdot \longrightarrow \text{Ph}\cdot + \underset{\underset{CH_3}{|}}{\overset{\overset{CH_3}{|}}{C}}=O$$

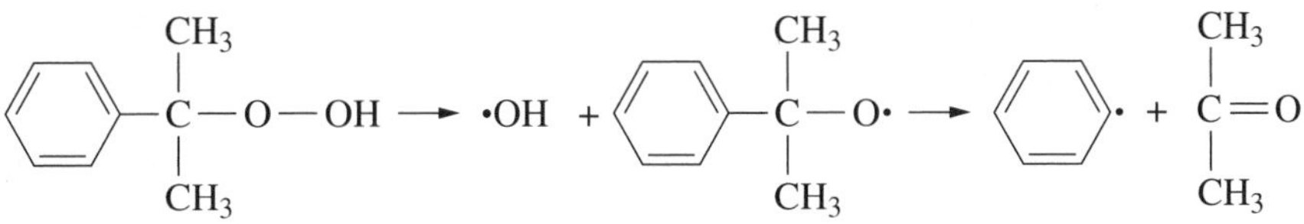

$$(2.2.9)$$

There are two important types of reactions of radicals with nonradical species: *additions* (usually on double bonds) and *abstractions* of single atoms. In both cases, the reaction does not change the number of radicals in the system; for each radical consumed in the reaction, a new radical is produced. The additions of radicals usually constitute the propagation step in the chain polymerization process, and we will treat them in much more detail in the following sections. On the other hand, abstraction reactions quite often represent unwanted side reactions and make the life of a polymer chemist more difficult. Of course, on a number of occasions, they may be utilized for process modification, cross-linking, and so on. In an abstraction reaction, the radical abstracts a hydrogen or halogen atom from a nonradical molecule, converting it to another radical.

$$R\cdot + X - R' \longrightarrow R - X + R'\cdot \qquad \text{where} \quad X = H, Cl, Br, I \qquad (2.2.10)$$

It is possible that a radical that was produced by abstraction abstracts an atom from another susceptible molecule. These two abstraction reactions may alternate in a chain reaction. A typical example is a reaction between hydrogen and bromine that is promoted by any radical initiation mechanism:

$$R\cdot + Br_2 \longrightarrow RBr + Br\cdot \qquad (2.2.11)$$

Bromine atoms produced by reaction (2.2.11) then start the chain reactions (2.2.12 and 2.2.13).

$$Br\cdot + H_2 \longrightarrow HBr + H\cdot \qquad (2.2.12)$$

$$H\cdot + Br_2 \longrightarrow HBr + Br\cdot \qquad (2.2.13)$$

Oxygen combustion of organic materials is also a chain radical process.

More important for a polymer chemist is the possibility that a growing chain (which is a radical) will attack some other molecule. This reaction leads to termination of the original polymer chain and formation of a new radical. If the new radical is sufficiently reactive, it may start another polymer chain; this process is called *chain transfer*. Radicals with low reactivity, which cannot start a new polymer chain, effectively stop the polymerization process; we are then talking about *inhibition*. Both chain transfer and inhibition will be subjects of later sections.

Although all hydrogen and halogen bonds are susceptible to radical abstraction reactions, the frequency of the reaction depends primarily on the energy of the radicals. The more energetic the attacking radical, the broader the range of molecules it may attack. Conversely, the less energetic the resulting radical, the more susceptible the parent molecule is to a radical attack.

As a general rule, the abstraction of hydrogen is about as difficult as abstraction of chlorine; abstraction of bromine is much easier, and abstraction of iodine is the easiest. Fluorocarbons are not subject to this reaction. Poly-halogenated compounds (chloroform or carbon tetrachloride, for example) are very powerful transfer agents.

Among hydrogens, the most easily abstracted are those of hydroperoxides **8** and thiols **9** (the arrows label the susceptible bonds).

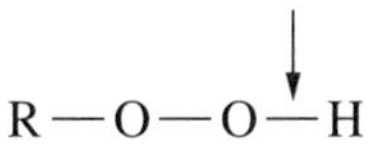

$$R-O-O\overset{\downarrow}{-}H \qquad R-S\overset{\downarrow}{-}H$$

$$\mathbf{8} \qquad\qquad \mathbf{9}$$

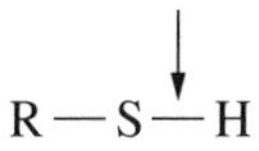

Among the hydrogens on carbon, aldehydic hydrogens **10** and α-hydrogens of alcohols **11** and amines **12** are most easily abstracted. In formulas **10–12**, R designates hydrogen or an alkyl or aryl group.

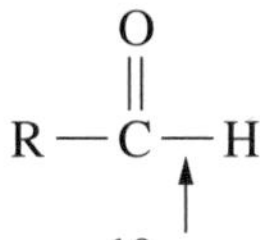

$$\underset{\mathbf{10}}{R-\overset{\overset{O}{\|}}{C}-H} \quad \underset{\mathbf{11}}{R_2-\overset{\overset{OR}{|}}{C}-H} \quad \underset{\mathbf{12}}{R'-\overset{\overset{NR_2}{|}}{C}-H}$$

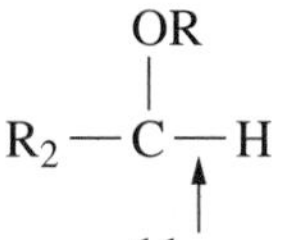

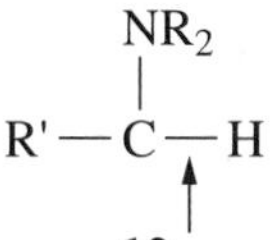

Allylic compounds **13** can react with radicals either by addition on the double bond or by an abstraction of hydrogen on a carbon adjacent to the double bond [reaction (2.2.14)]. The allylic radical **14** is one of the radicals well stabilized by resonance. Its formation often means a termination of the kinetic chain.

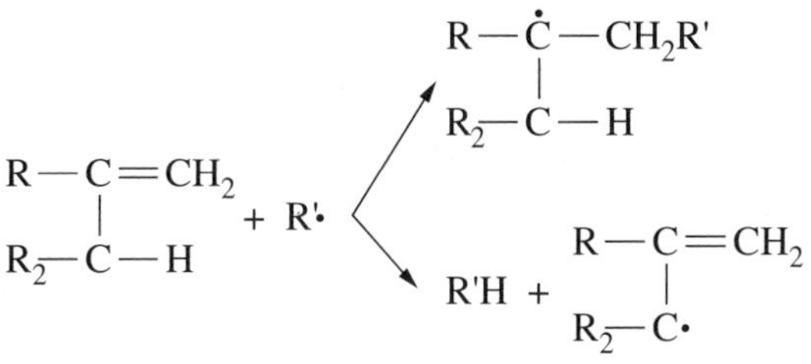

$$\underset{\mathbf{13}}{\overset{R-C=CH_2}{\underset{R_2-C-H}{|}}} + R\cdot \left\langle \begin{array}{l} R-\overset{\cdot}{C}-CH_2R' \\ \quad\;\; \overset{|}{R_2-C-H} \\[2mm] \qquad R-C=CH_2 \\ R'H + \quad \overset{|}{R_2-C\cdot} \end{array} \right.$$

$$\underset{\mathbf{14}}{} \qquad\qquad (2.2.14)$$

An encounter between two radicals in solution usually means their annihilation and effective termination of the kinetic chain of reactions. In the simplest case, the two radicals recombine and form a single molecule. Recombination of radicals releases a considerable amount of energy, and the reaction proceeds with a very low activation energy.

Besides the obvious significance of recombination for the termination of the chain reactions, recombinations play a number of other roles. Whenever low-energy radicals that are incapable of initiating a chain reaction are produced, they eventually decay by recombination. Radical recombination can also be utilized for the synthesis of polymers by a stepwise mechanism as we have seen in Section 2.1.7.

When an initiator, say, benzoyl peroxide, is thermally decomposed, the two benzoyloxy radicals are in a solvent cage. They need a finite time to escape from the cage. During that time, the two radicals may recombine again. The net result of the two reactions is, of course, no reaction at all.

We have seen, however, that some radicals can eliminate small molecules. If this happens while the radical is still in its original cage, the second-generation radical can recombine with its counterpart to form a molecule that is no longer susceptible to radical decomposition. In the case of benzoyl peroxide, the reaction proceeds as follows:

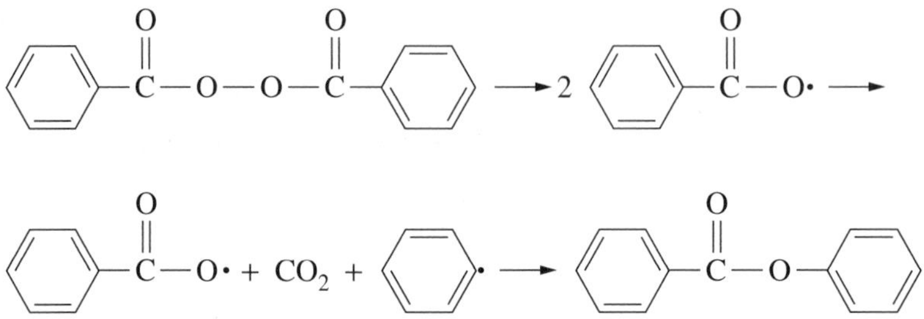

$$(2.2.15)$$

Reaction (2.2.15) represents a waste of the initiator. Not all radicals produced by the decomposition of the initiator are available for initiating the chain polymerization.

The encounter of two radicals may also proceed along different lines. One of the radicals may abstract a hydrogen atom from the other radical (from a carbon adjacent to the radical-carrying carbon); the ensuing diradical immediately isomerizes to an olefin.

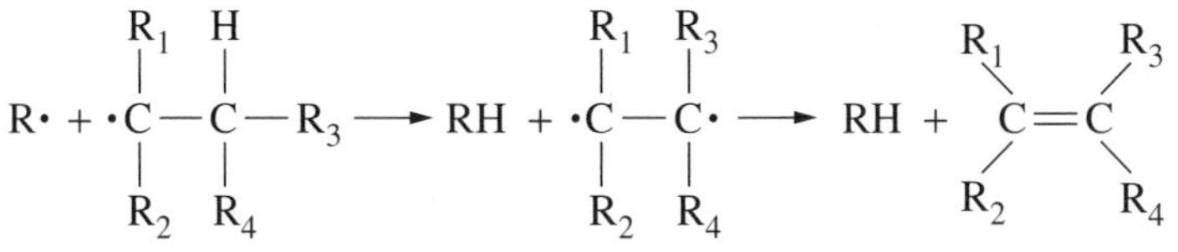

$$(2.2.16)$$

This *disproportionation* reaction competes with recombination for the termination of the radical reactions. Disproportionation is preferred when the substituents are bulky, because olefins are less crowded than the corresponding saturated compounds.

2.2.3. Reactivity of Radicals

Most radicals are much more reactive than nonradical species. Nevertheless, there is a very broad range of reactivities among different radicals. Some behave as perfectly stable molecules, whereas others attack virtually every molecule. Radical reactivity is closely related to the energy content of the radical: The higher the energy, the higher the reactivity. Knowledge of radical reactivities is especially important in the study of those reactions in which one radical produces another radical either by radical addition or by abstraction. Reactions for which the resulting radical is less reactive than the original one generally proceed rather easily (energy going downhill). When the resulting radical has slightly higher energy than the original one, the reaction is sluggish. When the energy difference is larger, the reaction does not occur at all.

When judging the reactivity of a radical, we must compare its energy in its radical form (having an unpaired electron) with that of the reacted form (the unpaired electron is now a part of a σ bond). The excess energy of the radical, which is due to the presence of the unpaired electron, is usually reduced by resonance and steric effects. The unpaired electron conjugates easily with all free and π electron pairs whether they sit on a single atom (halogen, oxygen, nitrogen, sulfur) or form a double bond (in olefins, carbonyls, nitriles, azo compounds) or are a part of a larger aromatic system (benzene, naphthalene, aromatic heterocycles, etc). Actually, unpaired electrons even conjugate with σ electrons of carbon-hydrogen bonds (hyperconjugation). The strongest resonance effect is observed when the unpaired electron mediates the conjugation of two or more aromatic systems, which are separated in the reacted form (e.g., diphenyl methyl **15**, triphenyl methyl **16**).

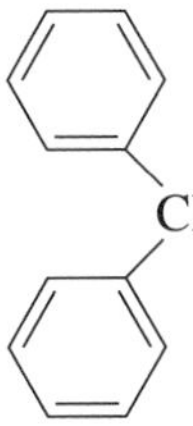

15 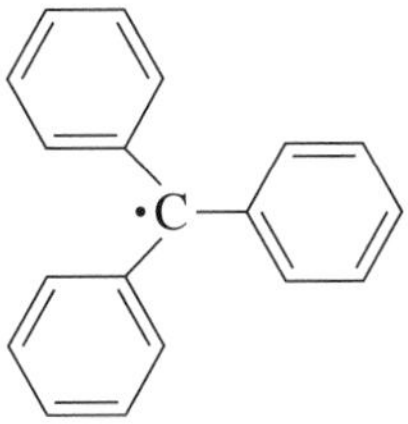   **16** 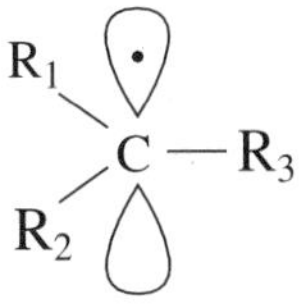   **17**

When the lone electron resides on a carbon atom, steric considerations may play a major role. The central carbon atom is believed to be in an sp^2 configuration **17**, in which the three bonded substituents are coplanar with the central carbon. The unpaired electron occupies a pure p orbital perpendicular to the sp^2 plane. The reacted form of the radical is usually in the tetrahedral sp^3 configuration. The transition from the planar sp^2 configuration to the sp^3 configuration causes crowding of the substituents. Thus the radical is sterically stabilized. This effect is the larger the bulkier the substituents are. We can see that steric effects and hyperconjugation quite often go hand in hand. Thus, among aliphatic radicals, the reactivity decreases as

$$H\cdot > \cdot CH_3 > CH_3\dot{C}H_2 > (CH_3)_2\dot{C}H > (CH_3)_3C\cdot \qquad (2.2.17)$$

Among unsaturated and aromatic radicals, reactivity decreases as follows:

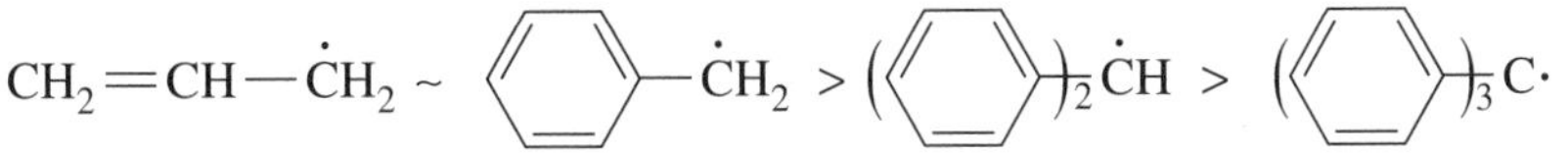

$$(2.2.18)$$

The triphenyl methyl radical is so stable that it does not dimerize quantitatively. At equilibrium, its dimer (hexaphenyl ethane) contains about 5% free radicals. This effect may be even stronger for compounds with still deeper conjugation. For example, diphenyl picryl hydrazyl **18** exists as a free radical that does not dimerize.

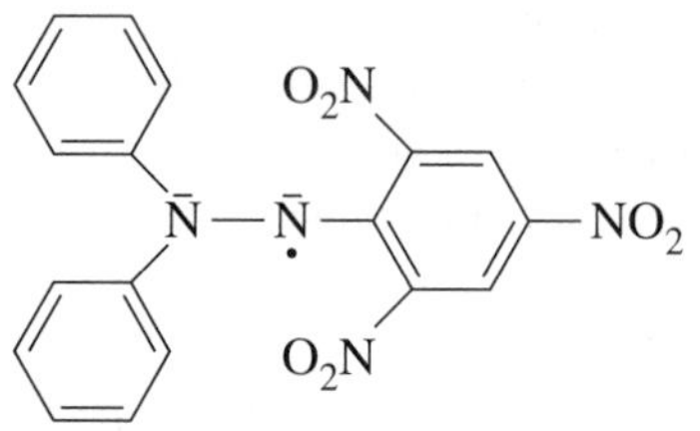

18

Diphenyl picryl hydrazyl is an extreme example of another group of unreactive radicals that are derived from hydroxy and amino derivatives of aromatic compounds. A typical case is hydroquinone, from which a hydrogen atom is abstracted very easily [reaction (2.2.19)]. The resulting radical can exist in many mesomeric forms and is quite unreactive. In fact, hydroquinone is one of the most popular radical inhibitors that are added to monomers to stop premature polymerization.

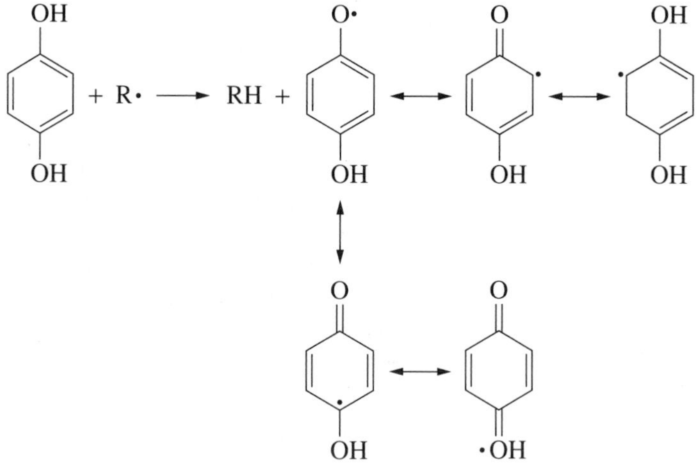

(2.2.19)

Up to now, we have described only those aspects influencing the reactivity of radicals that are inherent to the structure of the radical itself. However, the reactivity of a radical may be different with respect to different reaction partners (even after discounting for the difference in energies of these partners turned to next-generation radicals). Specifically, electron-rich radicals such as those derived from aromatic monomers (e.g., styrene) have a high affinity (read "reactivity") toward electron-poor unsaturated compounds (e.g., maleic anhydride). Of course, electron-poor radicals have a high affinity for electron-rich compounds. We will return to this topic in Section 2.2.5, which deals with radical copolymerization.

2.2.4. Kinetics of Radical Polymerization

The addition of radicals onto double bonds can be utilized for the formation of long polymer chains. In the simplest case, the radicals are introduced into the system in a process of initiation, they grow through propagation, and the whole process is completed by the termination of the polymer chain. The overall rate of polymerization, the molecular weight of the polymer and its distribution, as well as finer features of the polymer structure are determined by a delicate interplay of these three processes. In addition, radical abstraction reactions may terminate the physical polymer chain but may not disturb the kinetic chain. This process is called chain transfer. The presence of some materials in trace quantities (added deliberately or inadvertently) can inhibit or retard the reaction. In the following sections, we will study all these processes in some detail and then learn how to manipulate them and how to influence the structure of the polymer.

2.2.4.1. Initiation, Propagation, and Termination.   For the initiation of radical polymerization, thermally decomposing initiators are most popular; the thermal reaction is sometimes helped by a redox reaction [reaction (2.2.3)] or by the presence of a promoter [reaction (2.2.4)]. Photoinitiation has an advantage in the easy control of the radical production by manipulation of the intensity of the UV light. However, its disadvantage is in light absorption. As the light penetrates the reaction mixture and interacts with the photoinitiator, it gets absorbed, and its intensity decreases with depth. It may vanish completely within a relatively thin layer (a few millimeters). Hence, the production of radicals is very inhomogeneous and so is the structure of the polymer product. Consequently, photoinitiation is usually used only for polymerization of thin layers—for example, for coatings, films, and contact lenses. (Physical chemists use photoinitiation for the determination of the absolute kinetic constants from non-steady-state kinetics).

Thermal decomposition of initiators is a first-order reaction. Thus the concentration of the initiator and consequently the production of radicals decrease exponentially with time. When more or less constant radical production is required, the initiator and/or reaction temperature should be selected properly to make the half-time for decomposition of the initiator comparable to the desired overall reaction time. A true zeroth-order decomposition of the initiator can be achieved by using photoinitiation.

The radical fragments of the initiator are meant for initiating polymer chains. However, before they do so they must escape several dangers of being wasted. We have already seen that a molecule of initiator can decompose, eliminate a fragment while still in the original solvent cage, and recombine with its counterpart to form an unreactive molecule. Such molecules are wasted for initiation purposes. A radical that escapes from the cage is still not assured of reacting with a monomer molecule. It may first encounter a molecule of solvent, SH, and abstract hydrogen from it (we have treated such reactions in Section 2.2.2):

$$R \cdot + SH \longrightarrow RH + S \cdot \qquad (2.2.20)$$

Reaction (2.2.20) in itself is not a waste of a radical; a new radical S· is produced that can start a polymer chain. However, it may encounter another molecule of the initiator (e.g., benzoyl peroxide) and react with it.

$$
S\cdot + \underset{}{\text{(C}_6\text{H}_5)}\overset{\overset{\text{O}}{\|}}{\text{C}}-\text{O}-\text{O}-\overset{\overset{\text{O}}{\|}}{\text{C}}(\text{C}_6\text{H}_5) \longrightarrow
$$

$$
\text{(C}_6\text{H}_5)\overset{\overset{\text{O}}{\|}}{\text{C}}-\text{OS} + \text{(C}_6\text{H}_5)\overset{\overset{\text{O}}{\|}}{\text{C}}-\text{O}\cdot \qquad (2.2.21)
$$

Although reaction (2.2.21) produces another radical, it still represents the waste of one molecule of the initiator. This whole process represents induced decomposition of the initiator. The extent of induced decomposition depends on the interplay of properties of initiator, solvent, and monomer. Fast decomposition occurs when alcohols or ethers are used as solvents. The role of monomer is complex. Monomers forming low-energy radicals (styrene, methyl methacrylate) react easily with initiator fragments; they capture these radicals before they can start the sequence of induced decomposition. On the other hand, monomers forming such high-energy radicals (these are the less reactive monomers!) as vinyl acetate let the induced decomposition play its role fully. As a result, the kinetic constant for the decomposition of the initiator depends not only on temperature but also on the nature of the solvent and monomer. This dependence must, of course, be taken into account when designing the polymerization conditions. (Induced initiator decomposition is not observed with azobis(isobutyronitrile) 5 or with other azo initiators; this is why 5 is generally used in studies of polymerization kinetics.)

Radical fragments of the initiator that escape the dangers of wastage add onto the double bonds of monomer molecules and initiate the polymer chain. Although some initiator radicals may be stabilized by resonance, and their addition onto some monomers may be an uphill reaction energetically, common initiators can initiate most common monomers.

Some monomers, notably styrene and methyl methacrylate, polymerize even in the absence of any initiator. The rate of this self-initiation or *thermal* initiation increases very steeply with temperature. Thermal polymerization seems to proceed by a radical mechanism, but the exact mechanism of its initiation is still unknown.

The polymerization reaction proper is started by the addition of the initiating radical R· onto the double bond of a molecule of monomer:

$$
R\cdot + \underset{/}{\overset{\backslash}{\text{C}}}=\underset{\backslash}{\overset{/}{\text{C}}} \longrightarrow R-\overset{|}{\underset{|}{\text{C}}}-\overset{|}{\underset{|}{\text{C}}}\cdot
$$

19

The resulting radical **19** can then play the role of R• in the next addition onto the next molecule of monomer; this starts the chain propagation.

Not all compounds with carbon-carbon double bonds can participate in radical chain propagation. Successful propagation requires an interplay of favorable steric, polarization, and radical stabilization effects.

Sterically, the substituents on the radical slow down the reaction, but the effect is not overwhelming because the orbital of the free electron projects perpendicularly to the plane of the other substituents (formula **17**). It is thus relatively accessible in most situations. However, bulky substituents on the olefinic carbon shield it very effectively from the addition of a radical. In this context, all substituents are bulky, with the partial exception of fluorine. Consequently, most monomers that are polymerizable by radicals have structure **20**.

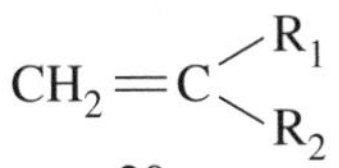

$$CH_2 = C \underset{R_2}{\overset{R_1}{<}}$$
20

The radical attacks **20** almost exclusively from the CH_2 end; this leads to almost pure head-to-tail addition. We will see that this effect is reinforced by the resonance of the radical.

Generally, the double bond should be polarized to make the monomer susceptible to radical polymerization. Both polarization by electron-rich groups—for example, by aromatic substituents as in styrene—and by electron-withdrawing groups—for example, by an ester group as in acrylates—are effective for this purpose. Alkyl groups have a very small polarizing effect. Consequently, olefins do not polymerize by radical mechanism. The only exceptions are ethylene and tetrafluoroethylene. However, steric factors for both of these monomers are very favorable. Nevertheless, they require rather drastic conditions for polymerization (180°C and oxygen, which produces hydroperoxides, as an initiator for ethylene).

At the other end of the polarity spectrum, strongly polar substituents may make even some 1,2-disubstituted ethylenes—for example, acenaphthylene **21**, maleic imides **22**, and vinylene carbonate **23**—susceptible to homopolymerization. These compounds have the double bond within a small cycle, and the substituents do not interfere severely with the approaching radical.

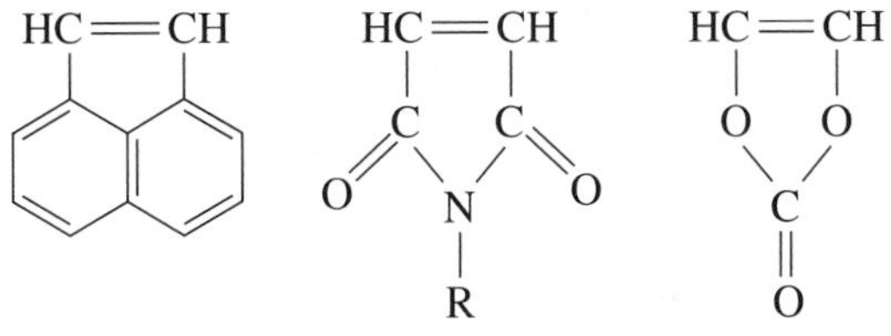

HC=CH   HC=CH   HC=CH

21   **22**   **23**

Polarization effects are especially important in copolymerization. Electron-rich radicals show preference for electron-poor monomers and vice versa. Of course, after adding an electron-rich monomer, the growing polymer chain has an electron-rich radical end that prefers to react with an electron-poor monomer. Thus a mixture of an electron-rich and an electron-poor monomer (e.g., styrene and methyl methacrylate) exhibits an enhanced rate of propagation and an alternating tendency. A similar effect makes some unsaturated compounds susceptible to copolymerization even if they do not homopolymerize. Typical examples are maleic anhydride **24** and crotonic acid, $CH_3CH{=}CHCOOH$. An extreme case of this situation is the copolymerization of maleic anhydride **24** with stilbene **25**. These two monomers copolymerize easily and yield a strictly alternating copolymer.

                24   **25**

We have seen in Section 2.2.3 that resonance within a radical lowers its energy and its reactivity. Resonance has two effects on chain propagation. A monomer that would convert to a well-stabilized radical has an enhanced reactivity. On the other hand, the resulting radical, being less energetic, is less reactive. The latter effect is stronger than the former. As a result, rates of propagation for monomers with well-stabilized growing radicals (e.g., styrene) are substantially smaller than for less-stabilized species (e.g., vinyl acetate). Radicals that are stabilized too well cannot continue in the chain reaction (this is a case of inhibition).

In the propagation step of a homopolymerization reaction, the resulting radical has the same structure, and hence the same energy, as the original radical. Thus the energy of the radical moiety does not contribute to the energy change accompanying the propagation step. The situation is different when the structure of the terminal radical changes upon addition. When the energy goes uphill, the rate is sluggish; for a downhill change, the rate is fast. Thus the addition of an initiator fragment to the first monomer may be easy or difficult depending on the species involved. Less reactive radicals cannot initiate monomers with very little stabilization, such as ethylene. This effect is more important in copolymerization. Whereas the more energetic radical may react with both monomers, the less energetic radical may refuse to react with the other monomer if the reaction requires too steep increase in energy. In this case, the more reactive monomer (which is transformed into the less energetic radical) may homopolymerize while the other monomer may stay unreacted. If the more reactive monomer is present only in small amounts, the whole process is effectively an inhibition. We will return to this topic when discussing copolymerization in Section 2.2.5.

In the absence of inhibitors and transfer agents, the propagation of a polymer chain will continue until it is interrupted by chain termination. Chain termination involves a reaction between two growing radicals leading to nonradical products. We have seen in Section 2.2.2 that two reactions are possible between two radicals: recombination and disproportionation. Recombination is favored at lower temperatures, whereas disproportionation prevails at higher temperatures. The reasons are both kinetic and thermodynamic. Kinetically, recombination does not go through an activated state (activation energy is low), whereas disproportionation, which is essentially a hydrogen abstraction [reaction (2.2.16)], has appreciable activation energy. In other words, the rate of termination increases with increasing temperature much faster for disproportionation than for recombination. Thermodynamically, recombination yields one molecule and disproportionation two. Hence, the product has higher entropy for disproportionation. As always, the entropy effect (which favors disproportionation) increases in significance with increasing temperature.

Steric effects are also important. Growing chains with crowded radical ends would recombine to rather strained structures. For example, methyl methacrylate chains react as in (2.2.23).

$$
\begin{array}{ccc}
& \text{CH}_3 & \text{CH}_3 & \text{CH}_3\ \text{CH}_3 \\
\text{~H}_2\text{C}-\text{C}\cdot & + & \cdot\text{C}-\text{CH}_2\text{~} & \longrightarrow & \text{~H}_2\text{C}-\text{C}-\text{C}-\text{CH}_2\text{~} \\
& \text{C} & \text{C} & \text{H}_3\text{CO}-\text{C}\ \ \text{C}-\text{OCH}_3 \\
& \text{O}\ \ \text{OCH}_3 & \text{O}\ \ \text{OCH}_3 & \text{O}\ \ \text{O}
\end{array}
$$

(2.2.23)

Less crowded chains consequently have a higher probability of recombining than more crowded ones.

In practice, typical polymers under typical reaction conditions terminate by both mechanisms simultaneously. The exact ratio of the appropriate kinetic constants depends on a number of factors and must be found experimentally if needed.

2.2.4.2. Rate of Polymerization and Molecular Weight.

In this section, we will analyze the reactions of initiation, propagation, and termination from the kinetic viewpoint. We will quantify the relations between the nature of the polymerizing system and the reaction conditions on one side and the rate of polymerization, the molecular weight of the polymer, and its distribution on the other side.

To simplify the analysis, we will again invoke the principle of equal reactivity of chemical groups without respect to the size of the molecule to which they are attached. In the present case, this principle implies the same reactivity of all growing radicals.

Even in the simplest case, we need to consider five reactions:

1. Decomposition of the initiator
2. Addition of initiator fragments on the monomer
3. Chain propagation
4. Termination by recombination
5. Termination by disproportionation

In the following sections, we will add to these reactions chain transfer and reactions with inhibitors.

We will use the symbol r for rates of reactions (in moles per liter per second) and k for kinetic constants. The subscripts d, i, p, tr, and td will refer to decomposition of the initiator, initiation, propagation, termination by recombination, and termination by disproportionation, respectively. The rate of termination r_t and kinetic constant of termination k_t are defined as

$$r_t = r_{tr} + r_{td} \qquad (2.2.24)$$

$$k_t = k_{tr} + k_{td} \qquad (2.2.25)$$

Symbols in square brackets will represent molar concentrations; I_2 is the initiator, $I\bullet$ the radical fragment of initiator, M the monomer, $P_i^\bullet$ a growing polymer radical with i monomers, and P_i an inactive polymer molecule of DP $= i$. Subscript zero refers to initial conditions. Thermal decomposition of the initiator will be assumed.

Decomposition of the initiator is described now as a first-order reaction:

$$r_d = -\frac{d[I_2]}{dt} = k_d[I_2] \qquad (2.2.26)$$

Thus, the concentration of the initiator decreases exponentially with time,

$$[I_2] = [I_2]_0 e^{-k_d t} \qquad (2.2.27)$$

If an approximately constant rate of initiator decomposition is desired during the total polymerization time t_{pol}, then reaction conditions must be selected such that $k_d t_{pol} < 1$. This selection is achieved by the proper choice of initiator; finer tuning requires manipulation of temperature. We have seen in Section 2.2.2 that k_d may depend on the reaction medium because of induced decomposition. This factor must be taken into account when selecting reaction conditions.

In an ideal case, the initiator fragments add onto monomer molecule in the actual initiation of the chain reaction. There are two radicals per molecule of initiator; consequently, the rate of initiation is twice the rate of initiator decomposition. In a real reaction, some radicals are lost because of recombination and induced decomposition. Thus, the *efficiency f* of the initiator is less than unity; usually it is between 0.5 and 0.7, but it may be much smaller if the induced decomposition is extensive or when redox initiators or promoters are used.

The initiation reaction and its rate are now as in (2.2.28) and (2.2.29).

$$I\bullet + M \longrightarrow P_1^\bullet \qquad (2.2.28)$$

and

$$r_i = 2fr_d = 2fk_d[I_2] \qquad (2.2.29)$$

The polymer radicals add monomers in a propagation reaction (2.2.30).

$$P_i^\bullet + M \longrightarrow P_{i+1}^\bullet; \qquad i = 1, 2, \dots n \tag{2.2.30}$$

The kinetic equation for this reaction reads

$$-\frac{d[P_i^\bullet]}{dt} = -\frac{d[M]}{dt} = \frac{d[P_{i+1}^\bullet]}{dt} = k_p[P_i^\bullet][M] \tag{2.2.31}$$

The same reactivity for all growing radicals implies that k_p has the same value for all sizes of molecules. It is convenient to consider all growing chains together and to define the total concentration of radicals as

$$[P^\bullet] = \sum_i [P_i^\bullet] \tag{2.2.32}$$

The total consumption of monomer (rate of propagation) can then be obtained by summing equation (2.2.31) for all radical species to obtain

$$r_p = -\frac{d[M]}{dt} = k_p[P^\bullet][M] \tag{2.2.33}$$

The two termination reactions can be written as

$$P_i^\bullet + P_j^\bullet \longrightarrow P_{i+j} \qquad \text{(recombination)} \tag{2.2.34}$$

$$P_i^\bullet + P_j^\bullet \longrightarrow P_i + P_j \qquad \text{(disproportionation)} \tag{2.2.35}$$

They are represented by the kinetic equations

$$\frac{d[P_i^\bullet]}{dt} = \frac{d[P_j^\bullet]}{dt} = -k_{tr}[P_i^\bullet][P_j^\bullet] \tag{2.2.36}$$

$$\frac{d[P_i^\bullet]}{dt} = \frac{d[P_j^\bullet]}{dt} = -k_{td}[P_i^\bullet][P_j^\bullet] \tag{2.2.37}$$

Alternatively, we can sum equations (2.2.36) and (2.2.37) for all species. We then obtain for the rate of termination [using also equations (2.2.24) and (2.2.25)]

$$r_t = -\frac{d[P^\bullet]}{dt} = 2k_t[P^\bullet]^2 \tag{2.2.38}$$

The factor of 2 in this relation reflects the fact that two radicals are destroyed in each termination event.

It should be noted that the derivatives in equations (2.2.31) through (2.2.38) represent the changes of concentrations only with respect to the reactions described by these equations. The overall change in the concentration of any species is obtained by summing the changes for all reactions in which this species participates.

In the first step of our analysis, we will evaluate the rate of polymerization, the molecular weight of the polymer, and its distribution during an interval of time short enough that the concentrations of initiator and monomer remain essentially constant. However, this period of time is much longer than the lifetime of a growing chain (which lasts for few seconds at most). With these assumptions (*steady-state assumption*), we expect the rate of polymerization to stay constant during the time interval. This implies a constant number of growing chains; this in turn, can be achieved only when the rate of radical production r_i is equal to the rate of radical destruction r_t. Thus equations (2.2.29) and (2.2.38) yield

$$2 f k_d[I_2] = 2k_t[P^\bullet]^2 \qquad (2.2.39)$$

Solving for $[P^\bullet]$ from equation (2.2.39) and substituting into equation (2.2.33), we obtain for the rate of polymerization

$$r_p = (k_p/k_t^{1/2})(f k_d[I_2])^{1/2}[M] \qquad (2.2.40)$$

Thus the rate of polymerization is predicted (not surprisingly) to be proportional to monomer concentration but (more surprisingly) only to the square root of initiator concentration. In writing equation (2.2.40), we have separated the kinetic factors into those that depend only on the nature of the monomer and temperature $(k_p/k_t^{1/2})$ and those that can be manipulated experimentally. We will see that the propagation and termination kinetic constants always come together as $(k_p/k_t^{1/2})$ when steady-state polymerization is studied. If separate values are wanted, nonsteady kinetic experiments must be utilized.

Let us first evaluate the average length of a kinetic chain, x_{av}, that is, the average number of monomer molecules incorporated into polymer between initiation and termination events. Obviously,

$$x_{av} = r_p/r_t \qquad (2.2.41)$$

Combining equation (2.2.41) with equations (2.2.38)–(2.2.40) yields

$$x_{av} = (1/2)(k_p/k_t^{1/2}) (f k_d[I_2])^{-1/2} [M] \qquad (2.2.42)$$

Thus the kinetic chain length is inversely proportional to the square root of the concentration of the initiator (or more generally to the square root of the initiation rate). At a given monomer concentration, the rate of polymerization and the average length of the kinetic chain are inversely proportional to each other. It follows that slow polymerization must be used when long chains are wanted. Fast polymerization produces short chains. We should note, however, that the nature of the monomer as reflected in the value $k_p/k_t^{1/2}$ plays a dominant role in expression (2.2.42). The same rate of initiation can lead to short chains with one monomer and long chains with another.

The relation between kinetic chain length and degree of polymerization depends on the chain transfer and the mode of chain termination. In the absence of chain transfer and when termination proceeds by disproportionation, the degree of polymerization (i.e., the physical length of the polymer chain) is equal to the kinetic length. When termination is by recombination, the number of chains is halved during termination; hence, the degree of polymerization is doubled. When termination occurs by both mechanisms simultaneously, the following relation holds for number-average degree of polymerization $\overline{DP}_n^o$:

$$\overline{DP}_n^o = x_{av} \frac{k_{td} + k_{tr}}{k_{td} + k_{tr}/2} \tag{2.2.43}$$

We will now consider the distribution of degrees of polymerization. During the growth process, after each monomer is added, there exists a probability p that another monomer will be added, that is, that the chain will not be terminated. This probability is independent of the length of the growing chain and is given by

$$p = \frac{r_p}{r_p + r_t} \tag{2.2.44}$$

We have dealt with the probability of a monomer unit being followed by another one in Section 1.6.2, and we have seen that systems to which this concept is applicable exhibit the most probable distribution of polymerization degrees. This distribution was described by equation (1.6.15), which is equivalent to equation (1.6.34) with the Schulz–Zimm parameter z equal to 1. In the present case, this distribution describes the growing chains just before they are terminated. If they are terminated by disproportionation, it also describes the resulting polymer, and the ratio $\overline{M}_w / \overline{M}_n$ is equal to 2. If the termination is by recombination, we are dealing with random selection and fusion of molecular pairs as described in Section 1.6.2, and the resulting distribution for the polymer is again the Schulz–Zimm distribution with $z = 2$ and $\overline{M}_w / \overline{M}_n = 1.5$. Finally, if both mechanisms contribute to the termination, the distribution is a superposition of both functions, and $\overline{M}_w / \overline{M}_n$ falls between 1.5 and 2.0.

At this moment we should stress that all the above considerations referred to molecular weight and its distribution for a polymer produced in a very short interval of time. For a real polymerization, they are applicable as reasonably good approximations during polymerization to low conversion (provided that the concentration of initiator does not decrease too much). Low conversion usually means up to 5–10%. At higher conversions, the polymerization rate and the molecular weight of the polymer being produced change continuously not only because of the changes in the concentrations of the initiator and the monomer but also because of other changes related to higher concentration of the polymer. (We will discuss these changes in Section 2.2.4.5.) The overall distribution function is a superposition of the distribution functions for all time intervals and becomes ever broader with increasing conversion.

We have seen in Section 2.1.1.4 that polycondensation reactions also produce polymers with a Schulz–Zimm distribution of molecular weights. However, there is an important distinction between these two situations. Polycondensation reactions are usually reversible reactions leading to equilibrium; for them the Schulz–Zimm distribution describes the final product.

Example. Selection of Polymerization Conditions

In this example, we will demonstrate how to select individual experimental parameters for the synthesis of a polymer with particular properties.

PROBLEM. Synthesize a sample of polystyrene by a radical mechanism with molecular weight $\bar{M}_w = 200{,}000$ and a possibly narrow distribution of molecular weight.

SOLUTION. The narrowest distribution of molecular weights in radical polymerization of a given polymer is achieved when the concentrations of the initiator and the monomer do not change appreciably during synthesis. This implies selecting the half-time of the initiator decomposition to be longer than the reaction time and keeping conversion to about 10%. Even in this case, the distribution will be somewhere between the Schulz–Zimm distributions with parameter z equal to 1 and 2. Such a distribution is rather broad. We will select polymerization of monomer in the absence of solvents; this reaction is the best understood. We will make the selections in several steps:

1. Selection of initiator and preliminary selection of temperature.
2. Selection of initiator concentration to achieve appropriate molecular weight.
3. Calculation of reaction time to achieve required conversion of the monomer.
4. Adjustment of temperature to assure sufficient lifetime of the initiator.
5. If necessary, iteration of steps 2–4.

Step 1. Benzoyl peroxide is a dependable initiator used frequently at 70°C. Its kinetic constant of decomposition in the absence of induced decomposition (which is very small in pure styrene) depends on temperature T as

$$k_d = 1.6 \times 10^{14} \exp(-30{,}000/RT)\text{s}^{-1} \tag{2.2.45}$$

Its value at 70°C is $7.8 \times 10^{-6}\text{s}^{-1}$, giving a comfortable half-life of 25 h; R is to be taken in cal/Kmol.

Step 2. First, we need to calculate the value of $\bar{M}_n$ corresponding to the required value of $\bar{M}_w$. The ratio $\bar{M}_w/\bar{M}_n$ depends on the mode of termination. For polystyrene, about 77% of the growing chains terminate by recombination and 23% by disproportionation. It is easy to show that the expected $\bar{M}_w/\bar{M}_n$ is equal 1.62 (If the mode of termination is not known, we would estimate $\bar{M}_w/\bar{M}_n = 1.75$). Thus the required $\bar{M}_n$ is

$$\bar{M}_n = 200{,}000/1.62 \sim 123{,}000 \qquad (2.2.46)$$

The molecular weight of styrene is 104; thus

$$\overline{DP}_n = 123{,}000/104 \sim 1187 \qquad (2.2.47)$$

The above ratio of termination constants together with equation (2.2.43) yields

$$x_{av} = 1187\left[\frac{0.23 + 0.77/2}{0.23 + 0.77}\right] \sim 730 \qquad (2.2.48)$$

The density of styrene is 0.858 g/mL at 70°C. From it we calculate the molarity of pure styrene as

$$[M] = 1000 \times 0.858/104 \sim 8.25 \text{ mol/L} \qquad (2.2.49)$$

The efficiency of benzoyl peroxide for the initiation of styrene is high: $f = 0.9$.
  The ratio of kinetic constants for styrene as a function of temperature is given as

$$k_p/k_t^{1/2} = 880\exp(-6600/RT)(\text{L mol}^{-1}\text{s}^{-1})^{1/2} \qquad (2.2.50)$$

The value at 70°C is $k_p/k_t^{1/2} = 0.055$ (L mol^{-1}s^{-1})$^{1/2}$; the value of R is again taken in cal/Kmol.
  Now we have all the values needed for calculating the concentration of initiator by rearranging equation (2.2.42) and substituting in the data:

$$[I_2] = \frac{(1/4)\left(k_p/k_t^{1/2}\right)^2[M]^2}{f\,k_d x_{av}^2} = \frac{0.25 \times 0.055^2 \times 8.25^2}{0.9 \times 7.8 \times 10^{-6} \times 730^2}$$

$$= 0.0138 \text{ mol/L} \sim 3.33 \text{ g/L} \qquad (2.2.51)$$

Step 3. We will assume that the rate of polymerization does not change during the early stages of reaction; we will use the initial rate for the calculation. Required 10% conversion implies that $\Delta[M]$, the decrease in the concentration of monomer, is equal to 0.1[M]. The $\Delta[M]$ is related to the required time of polymerization t_{pol} as

$$\Delta[M] = r_p t_{pol} \qquad (2.2.52)$$

where r_p is given by equation (2.2.40). Consequently,

$$t_{pol} = \frac{\Delta[M]/[M]}{\left(k_p k_t^{1/2}\right)(f\,k_d[I_2])^{1/2}} = \frac{0.1}{(0.055)(0.9 \times 7.8 \times 10^{-6} \times 0.0138)^{1/2}}$$

$$= 5842 \text{ s} \sim 1.62 \text{ h} \qquad (2.2.53)$$

Steps 4 and *5.* The reaction time is much shorter than the half-life of the initiator, and no adjustment of temperature is needed.

From the above example, you may get the impression that it is easy to determine the conditions for preparation of a polymer with any desired molecular weight. A few complicating circumstances exist, however. Apart from the fact that the values of kinetic constants and their temperature dependencies are known only for some polymers, polymerization kinetics is strongly influenced by chain transfer reactions, by degree of conversion, and by the presence of inhibitors. We will treat all these phenomena in the following sections. In addition, each monomer has a tendency to produce polymers in some particular range of molecular weight. These ranges are grossly different for different monomers. The reason for this behavior becomes apparent when equations (2.2.40) and (2.2.42) are combined.

$$r_p x_{av} = [M]^2 \left(k_p^2 / k_t \right) \qquad (2.2.54)$$

It is obvious that for a given monomer at a given temperature, the average length of the kinetic chain depends primarily on the value of k_p^2 / k_t, which depends only on the nature of the polymer. Table 2.1 shows the value of k_p^2 / k_t at 60°C for several polymers together with molecular weight estimated for a polymer prepared by bulk polymerization of monomer with the rate of polymerization adjusted (by proper selection of concentration of the initiator) to a convenient value of 1 mol L^{-1} h^{-1}. For this calculation, one kinetic chain was assumed to produce one polymer molecule. We will see in the next section that this assumption is quite unrealistic, especially for monomers like vinyl acetate, which exhibit appreciable chain transfer. Nevertheless, the table gives a fair description of the expected properties of various polymers. It is fairly easy to prepare polymers with lower molecular weight by either reducing the concentration of monomer or employing chain transfer agents. However, it is quite difficult to prepare a high-molecular-weight polymer from a monomer with a low value of k_p^2 / k_t. It is obvious that to achieve high molecular weights we have to choose slow polymerization. The technique of last resort is the use of thermal self-initiation at relatively low temperatures (e.g., ambient temperature) and reaction times up to several years (actually, polymers of the highest molecular weight for such monomers were prepared inadvertently when a bottle of monomer was forgotten on a shelf for a long time).

TABLE 2.1   Values of k_p^2 / k_t for Several Polymers at 60°C and Typical Values for their Molecular Weight

Monomer	k_p^2 / k_t (L mol^{-1} s^{-1})	M
Methyl acrylate	2.0	60,000,000
Vinyl acetate	0.2	6,000,000
Methyl methacrylate	0.03	900,000
Styrene	0.002	60,000

2.2.4.3. Chain Transfer.   In the previous sections we pretended that the growing radical had only two options: either to add a molecule of a monomer or to react with another polymer radical in a termination reaction. Actually, a growing radical can react with a number of other species. If such a reaction again produces a reactive radical, the propagation reaction may continue but the physical integrity of the chain may be disturbed. Such reactions are called *chain transfer* reactions. In this section, we will discuss several types of chain transfer reactions and study their effect on the structure of the resulting polymer.

Chain transfer reactions belong mainly to radical abstractions, which were described in Section 2.2.2. The moiety abstracted is either hydrogen or halogen or an initiator fragment (e.g., a benzoyloxy moiety).

The extent of the transfer reaction depends on the nature of the compounds present in the reacting system and, of course, on the reactivity of the growing polymer radical. Generally, the less reactive, less energetic radicals—for example, the well-conjugated radicals of growing polystyrene and poly(methyl methacrylate) chains—are *much less prone* to transfer reactions than the more reactive, more energetic, less conjugated radicals such as those in poly(vinyl acetate) chains.

Chain transfer may occur to any species present in the reacting system. All the following types of transfer reactions do occur in radical polymerization and have a significant influence on the properties of the resulting polymer: transfer to initiator, to monomer, to solvent or additive, and to polymer.

Molecules of initiators are rather labile (that's why they are used as initiators), and as such they are easily attacked by growing radicals. We have already met this reaction under the name of induced decomposition. The reaction proceeds as

$$P\bullet + I_2 \longrightarrow P\text{---}I + I\bullet \tag{2.2.55}$$

While describing the kinetics of transfer reactions, we will use double subscripts; the first subscript f will designate transfer, the second the species participating in transfer: I for initiator, S for solvent, M for monomer, and P for polymer. Thus the rate of transfer to initiator r_{fI} reads

$$r_{fI} = k_{fI}\,[P\bullet]\,[I_2] \tag{2.2.56}$$

The ratio of the rate of this transfer reaction to the rate of reaction of propagation [equation (2.2.33)] is

$$\frac{r_{fI}}{r_p} = \frac{k_{fI}}{k_p}\left(\frac{[I_2]}{[M]}\right) \equiv C_I\frac{[I_2]}{[M]} \tag{2.2.57}$$

where C_I, defined by the second part of equation (2.2.57), is the constant of transfer to initiator. For styrene and benzoyl peroxide at 60°C, $C_I = 0.055$; for vinyl acetate with the same initiator and temperature $C_I = 0.15$. Other peroxides exhibit similar values. However, transfer to initiators of the azo type is negligible.

The effect of transfer to initiator is twofold: It shortens the lifetime of the initiator, and it terminates the polymer chains (while influencing the kinetic chains only marginally). In the presence of transfer to initiator, equations (2.2.41) and (2.2.43) describing the degree of polymerization must be modified as

$$\overline{DP}_n = \frac{r_p}{r_{td} + r_{tr}/2 + r_{fI}} \tag{2.2.58}$$

Realizing that equations (2.2.41), (2.2.24), (2.2.25), and (2.2.43) can be combined to yield

$$\overline{DP}_n^o = \frac{r_p}{r_{td} + r_{tr}/2} \tag{2.2.59}$$

we can write the inverse of equation (2.2.58) as

$$\frac{1}{\overline{DP}_n} = \frac{1}{\overline{DP}_n^o} + C_I \frac{[I_2]}{[M]} \tag{2.2.60}$$

During the derivation of equation (2.2.60), the ratio r_{fI}/r_p was calculated using equation (2.2.57).

It is apparent from equation (2.2.60) that an increase in initiator concentration decreases the molecular weight of the polymer. (This is *not* an effect of increasing the initiation *rate* [cf. equation (2.2.42)]; the effect would persist even if the rate of initiation were lowered, for example, by lowering temperature.) Although the value of C_I is quite high, the ratio $[I_2]/[M]$ is typically of the order 10^{-3}. Thus the reduction in the degree of polymerization by transfer to initiator becomes significant only when $\overline{DP}_n^o$ (expected degree of polymerization in the absence of transfer) is of the order 10^4.

A growing radical can react with a monomer molecule not only by adding but also by abstracting hydrogen or another moiety from it. Like all transfer reactions, this transfer to monomer terminates a polymer chain and starts another one. The reaction can be written as

$$P^\bullet + M \longrightarrow P\text{---}H + M^\bullet \tag{2.2.61}$$

where $M^\bullet$ is a radical resulting from monomer after hydrogen abstraction. The rate of this transfer to monomer r_{fM} can be written as

$$r_{fM} = k_{fM}[P^\bullet][M] \tag{2.2.62}$$

Although the transfer constant k_{fM} is usually quite small, the reaction is significant because monomer is always present in appreciable concentrations. The ratio of reaction rates for transfer to monomer and for propagation is

$$r_{fM}/r_p = k_{fM}/k_p \equiv C_M \tag{2.2.63}$$

where C_M is the constant of transfer to monomer. The value of C_M increases with increasing reactivity of the growing radical and with the number and reactivity of the abstractable hydrogens in the monomer molecule. Thus, for the well-stabilized poly(methyl methacrylate) radical, $C_M = 0.7 \times 10^{-5}$; for polystyrene, $C_M = 6 \times 10^{-5}$; and for the rather reactive poly(vinyl acetate), $C_M = 21 \times 10^{-5}$ (all values at 60°C). Extremely high values of C_M are exhibited by allylic monomers of the general formula $CH_2{=}CHCH_2X$, where X is a halogen, ether, or ester group. Radical **26** derived from these monomers is not stabilized at all (and is therefore very reactive) and easily abstracts the labile α hydrogen of the monomer molecule.

$$\sim H_2C - \overset{\bullet}{C}H + CH_2{=}CHCH_2X \longrightarrow \sim H_2C - CH_2 + CH_2{=}CH - \overset{\bullet}{C}HX$$
$$\underset{CH_2X}{|} \qquad\qquad\qquad\qquad \underset{CH_2X}{|}$$

$$\textbf{26} \qquad\qquad\qquad\qquad\qquad\qquad \textbf{27} \qquad\qquad (2.2.64)$$

The resulting radical **27** is well stabilized because it facilitates conjugation between the double bond and the X substituent. As a result, C_M of allylic monomers is extremely high; its typical value is about 0.1. We will see that this value predestines these monomers to be polymerizable only to low-molecular-weight oligomers. On top of that, the well-stabilized allylic radical **27** has difficulty adding on any double bond, let alone attacking another molecule of allylic monomer and producing the very reactive radical **26**. Thus, in this case, transfer to monomer produces radicals that are incapable of (or at least very sluggish in) initiating new growing chains. They decay instead by other mechanisms, predominantly by recombination. Consequently, the transfer slows down the overall polymerization rate in contrast to other types of transfer. Such transfer is called *degradative*.

When the polymerization is carried on in solution or when some other compound is added to the system, transfer to the solvent or to the additive may occur. Again, transfer to solvent (or additive) can be described by the following relations (2.2.65–2.2.67);

$$P{\bullet} + SX \longrightarrow P{-}X + S{\bullet} \qquad\qquad (2.2.65)$$

$$r_{fS} = k_{fS}[P^{\bullet}][SX] \qquad\qquad (2.2.66)$$

$$\frac{r_{fS}}{r_p} = \frac{k_{fS}}{k_p}\left(\frac{[SX]}{[M]}\right) \equiv C_S \frac{[SX]}{[M]} \qquad\qquad (2.2.67)$$

Here X is abstractable hydrogen or halogen on the solvent molecule, and the transfer constant to the solvent is defined by the second half of equation (2.2.67).

Again, the extent of transfer depends on the reactivities of the growing radical and of the solvent or transfer agent. The more reactive growing radicals [e.g., poly(vinyl acetate)] have higher transfer constants than those that are less reactive (e.g., polystyrene). On the solvent side, aromatic hydrogens are much less reactive than aliphatic ones (there is a dramatic difference in C_s values for benzene and toluene).

TABLE 2.2   Chain Transfer Constants for Selected Solvents at 60°C with Respect to Polystyrene and Poly(Vinyl Acetate)

	$C_s \times 10^4$	
Solvent	Polystyrene	Poly(vinyl acetate)
Benzene	0.18	2.2
Toluene	1.25	21
Ethylbenzene	6.7	55
Isopropylbenzene	8.2	90
n-Heptane	0.42	17
Acetone	4.1(80°C)	11
Butanone	5.0(80°C)	65
Ethyl acetate	—	2.3
Ethyl butyrate	—	17 (50°C)
Phenol	—	220 (45°C)
Aniline	—	210
n-Butyl chloride	0.04	10
n-Butyl bromide	0.06	50
n-Butyl iodide	1.85	800
Methylene chloride	0.15	4
Chloroform	0.5	140
Carbon tetrachloride	92	960
Carbon tetrabromide	13,600	39,000

Reactivity increases from primary hydrogens or halogens to secondary to tertiary ones. Carbonyl in ketones activates the α hydrogens more strongly than carbonyl in esters. Phenols and aromatic amines are very effective transfer agents. The abstractability of halogens increases from chlorine to bromine to iodine (fluorine cannot be abstracted). Multiple halogens on a single carbon are especially easily abstracted. In Table 2.2 we have collected chain transfer constants for some representative solvents and additives with respect to polystyrene and poly(vinyl acetate).

We should now return to the relations predicting the degree of polymerization from kinetic considerations. In the presence of various types of transfer processes, equation (2.2.58) should be modified to

$$\overline{DP}_n = r_p/(r_{td} + r_{tr}/2 + r_{fI} + r_{fM} + r_{fS}) \qquad (2.2.68)$$

On appropriate substitutions, this relation may be converted to

$$\frac{1}{\overline{DP}_n} = \frac{1}{\overline{DP}_n^o} + C_I\frac{[I_2]}{[M]} + C_M + C_S\frac{[SX]}{[M]} \qquad (2.2.69)$$

It is obvious that the degree of polymerization will be severely limited if any of the last three terms in equation (2.2.69) has a significant value. The first term can be minimized by selecting an initiator with low C_I value and/or by keeping the initiator concentration low (and increasing temperature if a higher initiation rate is required). The last term can be minimized by avoiding solvents and impurities with significant values of C_s. However, the C_M term cannot be avoided: Monomers with high C_M cannot be polymerized to high molecular weights using radical polymerization techniques.

On the other hand, chain transfer can be utilized when reduction of degree of polymerization is required. This situation may occur with monomers having overly high value of k_p^2/k_t (cf. Table 2.1) or when low-molecular-weight polymers or oligomers are needed (e.g., as adhesive materials). Chain transfer agents are also used to control cross-linking in diene polymerization.

In the polymerizing mixture, there are two more species to which the chain transfer may occur: the polymeric product of the reaction and the growing chain itself. The mechanism of these transfer reactions is the same as before: The growing radical (especially if it is rather reactive) attacks an abstractable hydrogen either back on its own chain or on some other polymer molecule. The result of this attack is slightly different from other types of chain transfer. The new radical is now part of an already existing polymer molecule. Thus the number of polymer molecules is not changed; transfer to polymer does not change the degree of polymerization.

However, there is a change in the structure of the resulting polymer. Let us treat first the attack on the growing molecule itself. It would seem that such attacks are extremely rare—there are so many other molecules to attack compared with the single growing chain! However, parts of the growing polymer coil are always in the close vicinity of the radical end, and the possibility of their mutual encounter is high. This is especially true for those atomic groups that are situated back along the chain just far enough that their approach is very probable (remember the discussion of ring closures). Typical is the polymerization of ethylene by a radical mechanism. The attack can occur as in scheme (2.2.70).

$$(2.2.70)$$

The radical on the backbone continues growing to form the backbone of the polyethylene molecule while the original site of the radical is transformed into a butyl group (the formation of a shorter or longer side group is, of course, also possible). The process is known as *backbiting,* and it produces polymers with a number of short side chains. In the case of polyethylene, the result is low-density polyethylene with limited crystallizability.

When the growing radical attacks another polymer molecule, it creates a growth center in the middle of the molecule. The result is a long branch. Statistically, it is more probable that a big molecule will be attacked than that a small one will. Thus the degree of branching (average number of branches per molecule) will be larger for high-molecular-weight polymers and smaller for low-molecular-weight polymers. This effect has an unpleasant consequence for theoretical studies. The dependency on molecular weight of polymer properties that depend on the chain architecture can be distorted by the presence of a few long branches (which are difficult to detect).

The extent of long branching can be minimized, but not totally suppressed, by lowering the concentration of polymer in the polymerizing system. Thus, for research samples, the polymerization is not carried on beyond 10% conversion.

2.2.4.4. Inhibition and Retardation.   We have seen in the preceding section that the radical can be transferred to a molecule of low reactivity and that it may have difficulty restarting the kinetic chain. An extreme case of such a degradative transfer is inhibition of the polymerization. *Inhibitors* are compounds that (1) have very high transfer constants C_s and therefore allow the growth of only a very short polymer chain and (2) produce radicals of such a low reactivity that they cannot restart another polymer chain and must decay by recombination. Inhibitors effectively scavenge all reactive radicals in the system and inhibit the polymerization process until they are exhausted. After such an inhibition period, polymerization can proceed in the usual manner. (In inexperienced hands, this process can become quite dangerous. A chemist may try to start polymerization, but nothing happens because of the presence of an inhibitor. Imprudent addition of a major amount of an initiator may, after consumption of the inhibitor, start the reaction with such vehemence that overheating and explosion result.) Another class of compounds that has properties similar to those of inhibitors but lower transfer constants C_s allows some polymer formation. Also, the reactivity of the transferred radicals may occasionally be sufficient to restart the kinetic chain after some delay. Such compounds do not inhibit polymerization, they only retard it; they are therefore called *retardants*. There is no clear borderline between inhibitors and retardants; the difference is mainly in the extent of their action.

The inhibitors are either present in the system inadvertently or they are added on purpose. (Inhibitors are added to prevent premature polymerization during the storage of monomers; retardants are sometimes added to modify the rate of the reaction.) We will now describe the more important classes of inhibitors.

Industrially, the most popular inhibitors are chain transfer compounds from which hydrogen is abstracted: substituted phenols, aromatic dihydroxy compounds (e.g., hydroquinone), aromatic amines, and sulfhydro compounds. In the first step, a polymer radical abstracts an hydrogen from the OH, NH, or SH group. The resulting inhibitor radical usually recombines with another polymer radical. Thus one molecule of inhibitor may terminate two polymer chains. Because of the extensive conjugation of the inhibitor, the second polymer chain may react with a different part of the molecule of the inhibitor than the first one did. For example, it is believed that *p*-substituted phenols react with polymer radicals according to the scheme

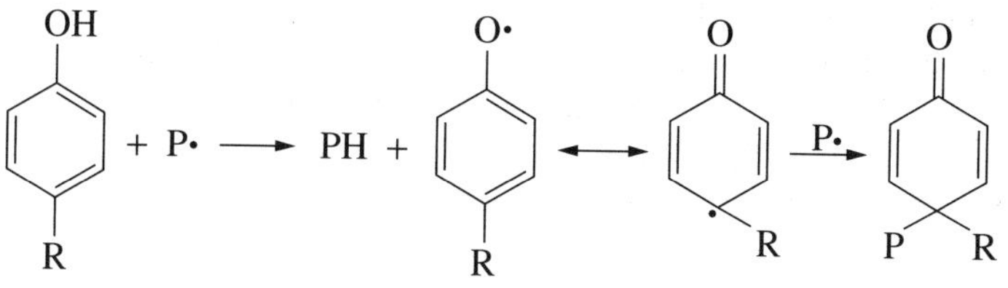

$$(2.2.71)$$

Another class of inhibitors are compounds to which the polymer radicals may add to form rather unreactive chain ends that will eventually recombine (or terminate by disproportionation) with a second polymer chain. Quinones and aromatic nitro compounds belong to this group. Reaction schemes (2.2.72) and (2.2.73) are typical. In polynitrocompounds (trinitrobenzene), each nitro group may stop two polymer chains.

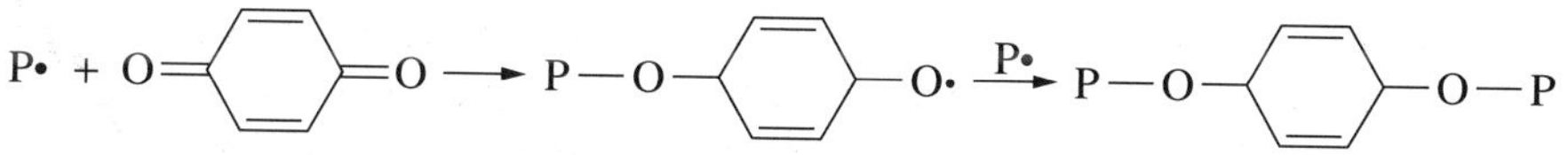

$$(2.2.72)$$

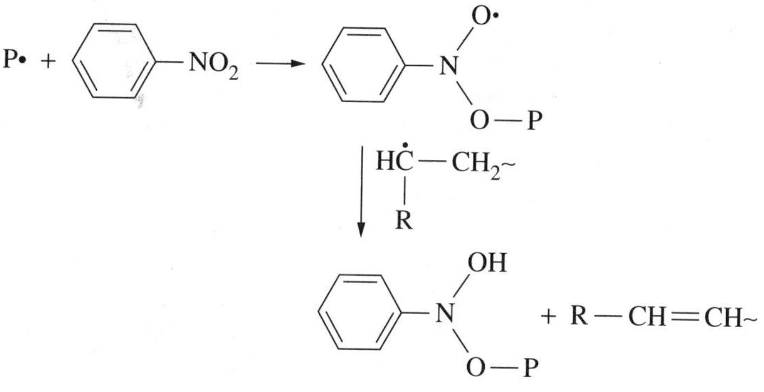

$$(2.2.73)$$

Rather active inhibitors are stable free radicals such as diphenyl picryl hydrazyl **17** or triphenyl methyl (which exists in equilibrium with its dimer, hexaphenyl ethane). These stable radicals recombine with growing chains to yield unreactive products.

In water solutions, transition metals in a higher valence state such as Fe^{3+} or Cu^{2+} may abstract an electron from a radical and convert it to a cation, which can combine with any counterion present in the solution:

$$P\bullet + Fe^{3+} + Cl^- \longrightarrow P-Cl + Fe^{2+}$$

$$(2.2.74)$$

Thus such metals are rather strong inhibitors. Recall that transition metals serve as agents facilitating the decomposition reaction of the initiator [equation (2.2.3)]. It follows that only minuscule concentrations of these metals are permissible for initiation without causing inhibition and excess induced decomposition of the initiator. In Section 2.2.8.3 we will see an ingenious way to circumvent this dilemma by placing the initiator and the growing chain into different phases in emulsion polymerization.

However, the most ubiquitous and always undesirable inhibitor is molecular oxygen. Strictly speaking, oxygen is a retarder; it reacts with radicals to form a peroxy radical.

$$P\bullet + O_2 \longrightarrow P\text{---}O\text{---}O\bullet \qquad (2.2.75)$$

The peroxy radicals have very low reactivity, but many of them will eventually add to the monomer and reinitiate the chain; the process is retardation. Besides the undesirable slowdown of the polymerization, the peroxy bonds built into the polymer (it is actually a copolymer with oxygen) have another unpleasant effect. They represent weak bonds in the polymer that will decompose when the polymer is employed at moderately elevated temperatures, and they may trigger a host of side reactions leading to degradation of the polymer.

Consequently, polymer chemists often go to appreciable lengths to get rid of dissolved oxygen from their reaction mixtures. Under ordinary circumstances, the system is just purged with nitrogen gas; the reaction is then performed either under nitrogen blanket or in a sealed reactor. However, for the preparation of research samples or when kinetics itself is studied, a special procedure must be applied. The reaction mixture is frozen in liquid nitrogen; most monomers and solvents crystallize at this temperature, and oxygen is excluded into the interstitial space between crystals. The container is then evacuated, and after that the mixture is allowed to thaw and oxygen bubbles out. It is purged with nitrogen, and the freezing-thawing process is repeated.

It should be mentioned that the added inhibitors are usually also removed from the system before polymerization. (Sometimes oxygen and other inhibitors are overrun by excess initiator.) The phenolic inhibitors can be extracted by alkaline solutions and the monomer distilled. However, a simple method is sometimes very convenient: The monomer is filtered over an active alumina column and the inhibitor is retained by polar adsorption on the column. Obviously, this method works only for some inhibitor-monomer combinations.

2.2.4.5. Kinetics at High Conversions—Gel Effect.

The kinetic schemes described in Section 2.2.4.2 are applicable only at low conversions or, more precisely, when the viscosity of the system is sufficiently low. With increasing conversion, the nature of the process is changing. We should now recall our discussion of kinetics of polycondensation in Section 2.1.1.3 and the postulate of independence of kinetic constants of the viscosity of the system. This postulate is based on a model in which two reacting groups collide repeatedly during an encounter but, because of extremely low reactivity of the groups, the probability of a reaction during an encounter is still quite low. Most probably, each reactive group would go through many encounters before reacting.

In a radical polymerization, the kinetic constants are much larger than in polycondensation, especially the termination constant. Thus the probability that a reaction will occur during an encounter is relatively high and increases with increasing duration of an encounter. When the system reaches sufficiently high viscosity and the mobility of the molecules is highly suppressed, the reaction rate becomes proportional to the number of encounters. The encounters are results of the diffusion movement, and the reaction becomes diffusion-controlled. The kinetic constants decrease sharply under such conditions.

However, the polymerization process is interplay of several elementary processes that are influenced by viscosity to different degrees. The diffusivity of small molecules (initiator, monomer, transfer agents) is proportional to the viscosity of the system and remains appreciable even at high conversions. Thus processes in which these molecules participate (initiation, propagation, transfer) are suppressed only moderately. On the other hand, the mobility of the radicals on the ends of the growing chain is severely restricted: The polymer molecules are extensively entangled, and movement of the chain end even over moderate distances would require appreciable disentanglement. The restricted mobility of the chain ends does not influence the propagation reaction too severely. The monomer molecules may diffuse to the waiting radicals and react. However, the termination reaction requires an encounter of two growing radicals, which becomes increasingly improbable: Two chains meet only when they grow toward each other!

Thus the termination reaction is severely suppressed, while initiation and propagation proceed almost at the same rate as before. The number of growing chains is increasing, and the overall polymerization rate accelerates enormously. The polymer chains are terminated only infrequently, and the molecular weight increases considerably. (Unlike in polycondensation, there is no re-equilibration of the degree of polymerization, and the distribution of molecular weights becomes bimodal.)

Although the kinetic constants start to decrease relatively early, the change in kinetics becomes very obvious at the moment when the polymerizing system gels, that is, when it loses macroscopic fluidity. Hence the term *gel effect*.

Sometimes the gel effect becomes quite dramatic. The reaction accelerates to such a point that the heat of polymerization cannot dissipate fast enough and temperature starts to increase. An increase in the temperature, of course, accelerates the process even more. The dependence of the temperature of the polymerizing system on time is called an *exotherm;* it peaks more or less sharply and then slowly returns to its steady-state value as polymerization approaches its end. Very sharp exotherms are undesirable; the properties of the polymer change in an uncontrolled way. Occasionally, the reaction escapes control altogether and the polymerization ends up as an explosion.

Under normal circumstances, after passing through the exotherm, the reaction effectively stops when the system goes through the glass transition. Usually, several percent of the monomer are immobilized in the glass and remain unreacted. Some unreacted radicals may also be frozen and be deactivated much later.

For controlling the gel effect and the exotherm, retarders are used to advantage. Nevertheless, because of the gel effect, bulk polymerizations are not favored by industry and are usually used only when the monomer is directly converted to the final article, for example, when Plexiglass articles are manufactured. In these cases, very slow polymerization is preferred to avoid possible mechanical and optical defects caused by the exotherm and the gel effect. For manufacturing other polymers, alternative procedures have been developed. We will treat them in Section 2.2.8.

2.2.5. Radical Copolymerization

The radical polymerization mechanism is frequently employed for preparation of copolymers. The overall reaction scheme for radical copolymerization is very similar to one for homopolymerization: Initiation, propagation, termination, and transfer reactions are again involved. However, the chemical nature of both monomers plays a role in a number of ways. It is not possible just to mix two monomers, add an initiator, and expect a copolymer to form. Although copolymerization may closely follow the kinetics of homopolymerization of the component monomers when the monomers are very similar to each other (e.g., ethyl methacrylate and butyl methacrylate), it may proceed much faster (e.g., for styrene and methyl methacrylate). Sometimes two monomers that would not homopolymerize can form a copolymer (e.g., stilbene and maleic anhydride). On the other hand, a small amount of styrene totally inhibits the polymerization of vinyl acetate. These phenomena are not unexpected; they follow from the principles governing the reactivity of radicals discussed in Sections 2.2.2 and 2.2.3.

A major difference between homopolymerization and copolymerization is in the propagation step. There are two types of growing radicals that can add two different monomers each. If the two monomers are designated M_1 and M_2, and the growing chains $P_1^\bullet$ and $P_2^\bullet$ according to the nature of the ultimate unit carrying the free radical, four reactions are possible that are characterized by four kinetic constants of propagation.

$$P_1^\bullet + M_1 \xrightarrow{k_{11}} P_1^\bullet \qquad (2.2.76)$$

$$P_1^\bullet + M_2 \xrightarrow{k_{12}} P_2^\bullet \qquad (2.2.77)$$

$$P_2^\bullet + M_2 \xrightarrow{k_{22}} P_2^\bullet \qquad (2.2.78)$$

$$P_2^\bullet + M_1 \xrightarrow{k_{21}} P_1^\bullet \qquad (2.2.79)$$

The constants for homogeneous addition k_{11} and k_{22} are obviously equal to the propagation constants k_p for the monomers in question. However, the constants for heterogeneous addition depend intimately on the nature of both monomers, on the reactivity of the radicals and monomers, and on the polar interactions between the two species.

We have seen that the reactivity of a radical depends on its energy content. When the energy of the free electron is lowered by conjugation (resonance-stabilized radicals), the radicals are less reactive than unstabilized highly energetic radicals. In homopolymerization, the lower reactivity of stabilized radicals (e.g., styrene) is partially (but far from fully) compensated for by the higher reactivity of their parent monomers (which are usually also highly conjugated and therefore prone to addition of radicals). Thus the rates of propagation for such monomers are much smaller than the rates for highly energetic radicals such as those derived from vinyl acetate. However, more reactive radicals are also more reactive toward other reactions, notably termination; this effect slows the overall rate of polymerization. The above phenomena together explain why monomers with widely different reactivities can be induced to homopolymerize with similar results.

In copolymerization, these compensating effects are absent. Radicals will preferably react with the more reactive monomers and be converted into less energetic radicals. These less energetic radicals may have not enough energy to react with the less reactive monomer; they will react only with their own monomer, and a homopolymer will result. This will happen in the system styrene-vinyl acetate, which will produce homopolystyrene. If the amount of styrene is small, all growing radicals will be converted to styrene radicals, which have no partners to react with. Styrene acts as an inhibitor for vinyl acetate.

It follows from the above discussion that two polymers may copolymerize only when the reactivities of the two radicals are comparable (the reactivity of monomers is less important). Monomers whose radicals have grossly different reactivities will not copolymerize.

Not all copolymerization phenomena can be explained by the reactivities of participating radicals. Two pairs of monomers having similar relations between their reactivities may copolymerize in a completely different manner. The relative richness in electrons is responsible for the differences. An electron-rich radical will react with an electron-poor monomer in preference to its parent monomer, which is also electron-rich; the product will be an electron-poor radical, which will prefer to react with the electron-rich monomer and not with its parent. Thus two monomers with different availability of electrons display an alternating tendency in their copolymerization.

In the following, we will first classify monomers according to their polarity and reactivity. We will then formalize the differences by introducing reactivity ratios $r_1 \equiv k_{11}/k_{12}$ and $r_2 \equiv k_{22}/k_{21}$, and we will present a review of copolymerization kinetics. Finally, the so-called Q-e scheme, an empirical recipe for predicting reactivity ratios, will be described.

Electron-rich monomers are those in which the olefin is conjugated with an aromatic group or another double bond; styrene and its derivatives and butadiene are the prime examples. Conjugation with oxygen (as in vinyl acetate) produces a much smaller effect. The family of electron-poor monomers is much larger because of the greater variety of electron-withdrawing groups. The largest effect is exhibited by cyano groups (acrylonitrile) followed by carbonyls (acrylates and methacrylates). The electron-withdrawing effect of halogens is modest (e.g., vinyl chloride); it increases with the number of halogen atoms (e.g., vinylidene chloride).

In Section 2.2.4.2 we saw that the full kinetic analysis of homopolymerization is quite involved, yet we were dealing with only one propagation constant and one pair of termination constants (disproportionation and recombination). In copolymerization of two monomers, we have four propagation constants and three pairs of termination constants (two pairs for termination of like radicals and one pair for unlike radicals). Obviously, the full kinetic analysis is quite complicated, even if we disregard the experimental difficulty in obtaining all the necessary parameters. Thus we will concentrate on only a few aspects of the overall picture—the overall composition of the copolymer and the sequencing of monomeric units in the copolymer.

The key quantities for the analysis are the reactivity ratios r_1 and r_2, which, for the two types of growing radicals, give the ratios of kinetic constants for the homoaddition and heteroaddition. Reactivity ratios depend on both the relative reactivity of the two types of radicals and their relative polarity. We recall that high reactivity of a monomer is usually accompanied by low reactivity of the corresponding radical and vice versa. We will discuss three experimentally important situations and one hypothetical case.

1. The reactivities and polarities of the two polymers are very similar as, for example, in the copolymerization of two acrylates or two methacrylates. Both reactivity ratios are close to unity in this case, and the two monomers are incorporated into the polymer in the same ratio as exists in the polymerizing mixture of monomers.

2. The reactivity of one monomer (let us identify it as M_1) is much higher than that of the other (M_2). Thus both types of radicals would prefer to react with it; on top of that the less reactive radical $P_1^{\bullet}$ may be not energetic enough to react with M_2. An extreme case is the mixture of styrene and vinyl acetate discussed previously. Obviously, $r_1 \equiv k_{11}/k_{12} \gg 1$ and $r_2 \equiv k_{22}/k_{21} \ll 1$. Less extreme examples of this group are characterized by $r_1 > 1$ and $r_2 < 1$.

3. The reactivities of both radicals are similar, but their polarities are of opposite sign. Each radical prefers to react with the unlike monomer, and alternating sequences are prevalent. In this case, both reactivity ratios are less than unity. An extreme case of this situation is a radical that cannot react with its own monomer for steric or other reasons but can still react with monomers of opposite polarity (maleic anhydride is a prime example). The reactivity ratio for such a monomer (with respect to any other monomer with which it can copolymerize) is zero; the reactivity ratio of the other monomer may, of course, still be finite.

4. Most textbooks also present hypothetical systems in which both radicals prefer to react with their own monomers. Both reactivity ratios are larger than unity; when they are very large, either block copolymers or two homopolymers result. However, as far as we know, no physical or chemical property can simultaneously make *both* radicals prefer their own kind. Hence, no example of such a system is known.

Let us calculate the composition of a copolymer. The ratio of the number of monomeric units of species 1 and 2 in a copolymer must be equal to the ratio of their rates of disappearance from the polymerizing system, $d[M_1]/dt$ and $d[M_2]dt$. For the latter quantities, the reaction scheme of equations (2.2.76) through (2.2.79) yields

$$-\frac{d[M_1]}{dt} = [P_1^\bullet][M_1]k_{11} + [P_2^\bullet][M_1]k_{21} \tag{2.2.80}$$

$$-\frac{d[M_2]}{dt} = [P_1^\bullet][M_2]k_{12} + [P_2^\bullet][M_2]k_{22} \tag{2.2.81}$$

Furthermore, the steady-state assumption introduced in Section 2.2.4.2 for the calculation of the total concentration of radicals must hold for each radical species separately. Although polymer radicals are created in the process of initiation and destroyed during termination, the main turnover of their concentration occurs during reactions (2.2.77) and (2.2.79), where a radical reacts with the unlike monomer and is converted into the radical of the other type. Hence, within the steady-state assumption we may write

$$\frac{d[P_1^\bullet]}{dt} = 0 = -[P_1^\bullet][M_2]k_{12} + [P_2^\bullet][M_1]k_{21} \tag{2.2.82}$$

Let us designate the molar fractions of the two monomers in the copolymer as x_1 and x_2. Then combining equations (2.2.80) through (2.2.82) together with the definitions of reactivity ratios gives the *copolymerization equation*

$$\frac{x_1}{x_2} = \frac{d[M_1]}{d[M_2]} = \frac{[M_1]}{[M_2]}\left(\frac{[M_1]r_1 + [M_2]}{[M_1] + [M_2]r_2}\right) \tag{2.2.83}$$

When either r_1 or r_2 or both differ from unity, the composition of the copolymer will be different from that of the monomer mixture (with one important exception to be treated below). In Figure 2.2 plots of copolymer composition as a function of composition of the monomer mixture are shown for few selected pairs of r_1, r_2 values. Some plots intersect the diagonal; these intersections represent monomer mixtures having the same composition as the resulting polymer. This situation corresponds closely to azeotropic mixtures in distillation equilibria. Accordingly, we are talking about *azeotropic copolymerization* and *azeotropic copolymers*.

Solving copolymerization equation (2.2.83) for the same composition of monomer feed and copolymer yields the azeotropic condition

$$\frac{[M_1]}{[M_2]} = \frac{1 - r_2}{1 - r_1} \tag{2.2.84}$$

Clearly, only such pairs of monomers form azeotropic copolymers for which the copolymerization parameters are either both larger or both smaller than unity. However, as we have already seen, the former case never occurs; the latter situation is the only significant one.

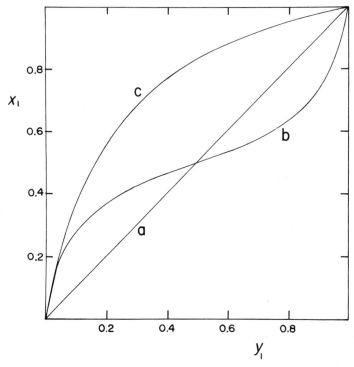

Figure 2.2. The dependence of copolymer composition on the composition of monomer mixture. (a) $r_1 = r_2 = 1$; (b) $r_1 = r_2 = 0.2$; (c) $r_1 = 5$; $r_2 = 0.2$. x_1 and y_1 are molar fractions of component 1 in the copolymer and in the monomer mixture, respectively.

Knowledge of the reactivity ratios is the key to predicting the composition of a copolymer. It is therefore very desirable to be able to predict these ratios to avoid painstaking measurement of them for each monomer pair. Among possible predicting schemes, the so-called *Q-e scheme* has gained wide acceptance. This empirical scheme is based on a postulated relationship between the propagation constants and characteristic properties of monomers.

$$k_{12} = P_1 Q_2 \exp(-e_1 e_2) \tag{2.2.85}$$

Here, P is the reactivity of the radical, Q is the reactivity of the monomer, and e is a parameter characterizing the electron density of the monomer. Electron-rich monomers have negative e values, whereas electron-poor monomers have positive values. The same e value applies to a monomer and its radical. Styrene was arbitrarily assigned values $e = -0.8$ and $Q = 1.0$. From equation (2.2.85) and the definition of r, we easily derive

$$r_1 = (Q_1/Q_2) \exp[e_1(e_2 - e_1)] \tag{2.2.86}$$

$$r_1 r_2 = \exp[-(e_1 - e_2)^2] \tag{2.2.87}$$

TABLE 2.3 Q and e Values for Selected Monomers

Monomer	e	Q
α-Methylstyrene	-1.27	0.98
Butadiene	-1.05	2.39
Styrene	-0.80	1.00
Vinyl acetate	-0.22	0.026
Ethylene	-0.20	0.015
Vinyl chloride	0.20	0.044
Methyl methacrylate	0.40	0.74
Butyl methacrylate	0.51	0.78
Methyl acrylate	0.60	0.42
Acrylonitrile	1.20	0.60

Q and e values for some monomers are collected in Table 2.3. Of course, it should be kept in mind that the Q-e scheme is a phenomenological concept with limited validity. (This is sometimes expressed by stating that Q and e values for a given monomer change according to the nature of the comonomer.) Nevertheless, the scheme is valuable in assessing the copolymerization behavior of monomers without having any actual information about it.

From the previous discussions it should be apparent that the composition of the copolymer is very rarely equal to the composition of the monomer mixture (i.e., only for closely related monomers or at azeotropic compositions). Generally, the two compositions will differ, often considerably. In a batch copolymerization, the monomer consumed faster will be depleted from the polymerizing mixture. The composition of the remaining monomers will shift, and the composition of the copolymer being produced at the moment will shift with it. Again, the situation is quite similar to batch distillation, in which compositions of the distillate and pot residue change continuously (unless it is an azeotropic system). However, the fractions of the distillate are easily separated, but copolymer molecules produced during various stages of copolymerization remain mixed together. The composition of the final copolymer may then be very heterogeneous. When a less heterogeneous product is required, either the conversion must be kept low or the batch process must be modified by adding monomers at the same rate as the polymerization process is consuming them.

Up to now, we have considered only the reaction ratios and their effect on the composition of the copolymer. We will now consider, at least qualitatively, the overall polymerization rate and the average degree of polymerization. These quantities depend on the rate of initiation and on the ratio of propagation and termination constants in a similar way as for homopolymerization, for which equations (2.2.40) and (2.2.42) were applicable. However, appropriate averages for the kinetic constants for propagation and termination should be used. Recall that the copolymerization of monomers with opposite polarity is characterized by reactivity ratios smaller than unity. This implies that the constants for the cross-reactions are larger than for the homoreaction and that the *average* value of k_p increases. This seems to indicate that both the overall rate of polymerization and the degree of polymerization must increase. However, instead of one pair of termination constants we have three pairs. Among them, the constant for recombination of radicals of different types is usually much larger than the others. The polar effects that enhanced the rate of reaction of radicals with monomers of opposite polarity are also operative when radicals of opposite polarity recombine. Thus the average termination constant increases also, and the overall rate of polymerization and degree of polymerization do not increase as much as we would expect otherwise.

2.2.6. Thermodynamics of Radical Polymerization

In the previous sections, we treated radical polymerization as if the polymerization reactions were completely irreversible. However, we have seen that radicals may eliminate nonradical molecules [reactions (2.2.6) through (2.2.9)]. If the radical is a growing chain and the eliminated molecule is a monomer, then this elimination reaction is the reverse of propagation reaction (2.2.30); it is sometimes called *depropagation*. The laws of thermodynamics require that two reversible reactions lead to equilibrium. Up to now, we have tacitly assumed that the polymerization equilibrium is completely shifted toward polymeric products. Such an assumption is frequently not warranted.

As in all equilibria, the equilibrium composition of a polymerizing system depends on the change of Gibbs energy ΔG_p^o for the reaction, that is, on the interplay of enthalpy and entropy effects (the subscript p is for polymerization). The reversible reaction to be considered reads

$$P_i^\bullet + M \rightleftharpoons P_{i+1}^\bullet \tag{2.2.88}$$

It can be described by an equilibrium constant K, where

$$K = \exp\left(-\frac{\Delta G_p^o}{RT}\right) = \exp\left(-\frac{\Delta H_p^o}{RT}\right) \exp\left(\frac{\Delta S_p^o}{R}\right) \tag{2.2.89}$$

It is important to realize that reaction (2.2.88) can occur and equation (2.2.89) will be applicable only if polymer radicals are present in the system, that is, either during the polymerization process itself or when a polymeric material acquires radical character. The latter situation occurs when the polymer is subjected to high temperature (breakage of bonds), to ionizing radiation, or to some radical reactant. As long as no radicals are present (unreacted monomer or "dead" polymer), the system cannot reach equilibrium for kinetic reasons.

Equilibrium constant K of equation (2.2.89) should be written in terms of the activities of the participating species:

$$K = \frac{a_{P_{i+1}^\bullet}}{a_{P_i^\bullet} a_M} \tag{2.2.90}$$

The ratio of activities of the two radical species is well approximated by the ratio of their concentrations, which in this case is very close to unity. As far as the activity of the monomer, a_M, is concerned, we will learn in Section 3.1 that it could be reasonably approximated by the volume fraction of the monomer in the mixture, ϕ_M. Thus the equilibrium value of ϕ_M can be expressed from equations (2.2.89) and (2.2.90) as

$$\phi_M = \exp\left(\frac{\Delta H_p^o}{RT}\right)\exp\left(-\frac{\Delta S_p^o}{R}\right) \qquad (2.2.91)$$

Let us discuss first the standard entropy change ΔS_p^o of reaction (2.2.88). It is dominated by the fact that during the polymerization reaction the number of molecules decreases; that is, the external degrees of freedom (translation and external rotation) are replaced by new internal degrees of freedom (internal rotation and vibration). The latter degrees of freedom get excited at a higher temperature than the former; hence they contribute less to the entropy. Thus most reactions where the number of molecules is decreasing (that includes all polymerizations) have negative values of ΔS_p^o. Moreover, these values are almost independent of the nature of the polymerizing system. For most olefinic monomers polymerized in a liquid phase, ΔS_p^o is about -26 ± 2 cal deg^{-1} mol^{-1}.

Thus, for a hypothetical athermal polymerization ($\Delta H_p^o = 0$), the expected equilibrium activity of monomer would be about $\exp(13) \approx 4 \times 10^5$, that is, far above the physical limits. (In the present case, unit activity corresponds to pure monomer.) In other words, the monomer would not polymerize. Appreciable negative values of ΔH_p^o are needed for a compound to be a polymerizable monomer.

In polymerizations of olefin compounds, the main source of polymerization enthalpy is the transformation of the carbon-carbon double bond to two single bonds. For the simplest olefin, ethylene, this transformation carries with it a standard change of enthalpy ΔH_p^o of about -21 kcal/mol. For other olefin compounds, this value is modified by two other phenomena: steric effects and resonance energies. In most cases, the negative polymerization enthalpy is lowered.

Among steric effects, the rearrangement of substitutents on the double bond plays a major role. The bonds of two geminal substituents on the double bond, which originally made a $120°$ angle, now make the tetrahedral angle of $109°$. This change brings the substituents much closer to each other; the repulsive forces increase the enthalpy of the polymer.

$$\begin{array}{c} \text{C==C} \\ 120° \end{array} \xrightarrow{\quad} \begin{array}{c} -\text{C}-\text{C} \\ 109° \end{array} \text{A} \qquad (2.2.92)$$

When one of the substituents is hydrogen, the effect is minor; for example, the polymerization enthalpy of propylene and other 1-alkenes is about -19 kcal/mol. However, two substituents lower the enthalpy appreciably; both isobutylene and methyl methacrylate have $\Delta H_p^o \approx -13$ kcal/mol.

Substitution by fluorine leads to the opposite effect: ΔH_{p}^{o} for tetrafluoroethylene is about -33 kcal/mol. The reason lies in the much higher carbon-carbon bond strength in fluorocarbons (124 kcal/mol in perfluoroethane) than in hydrocarbons (83 kcal/mol in ethane). The conversion of one double bond into two single bonds is thus more exothermic for fluorocarbons.

In many monomers the double bond is conjugated to another double or triple bond or to an aromatic system. Prime examples are butadiene, acrylonitrile, and styrene. In all these cases the conjugation is lost on polymerization and the polymerization enthalpy is lowered. For all these monomers, ΔH_{p}^{o} is about -17 kcal/mol.

The lowering of ΔH_{p}^{o} is most conspicuous when both the steric and resonance effects are involved simultaneously. Thus, α-methylstyrene has an ΔH_{p}^{o} of a mere -8 kcal/mol; it cannot be homopolymerized by the radical mechanism. On the other hand, the high affinity between electron-rich and electron-poor monomers, which led to the alternating tendency in their copolymers, manifests itself also in high copolymerization enthalpies. For example, the enthalpy of copolymerization of acrylonitrile and styrene is about -35 kcal/mol, whereas in homopolymerization both these monomers display only about half of this value.

According to equation (2.2.91), the volume fraction of monomer at equilibrium reaches unity when ΔG_{p}^{o} vanishes, that is, when

$$\Delta H_{p}^{o} = T_{c} \Delta S_{p}^{o} \tag{2.2.93}$$

This happens when temperature T reaches some specific value T_{c} [which is calculated from equation (2.2.93)] called the *ceiling temperature*. In an equilibrium state at or above ceiling temperature, no polymer can exist; that is, the monomer does not polymerize. At this point, it is appropriate to mention that the above treatment, which we applied to radical polymerization, is applicable to polymerization by any mechanism. Moreover, the polymerization entropies and enthalpies do not depend on the process by which the equilibrium is reached; that is, they do not depend on polymerization mechanism. Ceiling temperatures depend only on the nature of the monomer (and on pressure; they generally increase with increasing pressure).

A quick calculation using the values given above would yield results close to the actual values of ceiling temperatures, which are given in Table 2.4. (The discrepancies

TABLE 2.4   Ceiling Temperatures of Some Monomers

Monomer	Ceiling Temperatures ($^{\circ}$C)
Ethylene	610
Tetrafluoroethylene	1100
Isobutylene	175
Methyl methacrylate	220
Styrene	310
α-Methylstyrene	66

from the values obtained from data in the text are mainly due to the use of the same value of ΔS_p^o for all monomers.)

The high ceiling temperature of monomers like ethylene and tetrafluoroethylene should not be confused with thermal stability. The corresponding polymers are stable with respect to depropagation but not with respect to other degradation reactions.

Equation (2.2.91) implies that an appreciable amount of monomer should be present in an equilibrium mixture at temperatures well under the ceiling temperature. Combining equations (2.2.91) and (2.2.93) yields

$$\phi_M = \exp\left[\frac{\Delta S}{R}\left(\frac{T_c}{T} - 1\right)\right] \tag{2.2.94}$$

The application of equation (2.2.94) to methyl methacrylate reveals that the equilibrium concentration of monomer is 16% at 150°C; it is 4% even at 100°C. If the polymerization is performed in solution, the volume fraction of the unreacted monomer remains approximately the same. However, the conversion will be appreciably less because the sum of the volume fractions of monomer and polymer is now much less than unity.

2.2.7. Living Radical Polymerization

The radical polymerization has been the workhorse of polymer synthesis for many decades. Its mechanism is known in great detail, and it produces dependable results. However, its major limitation is the uncontrolled structure and broad polydispersity of the resulting polymers. This limitation is mainly due to the termination process that occurs from the very beginning of the polymerization. Thus the nature of the polymer molecules changes during the reaction. We will learn in Section 2.3 that some chain polymerizations may be performed in such a way that all molecules grow under identical conditions and are consequently quite monodisperse. These polymerizations are called *living*. For a long time, synthetic procedures were searched for a way that would allow also the radical polymerization to proceed as a living one while maintaining its versatility. Several procedures of this type were developed recently, and they are the subjects of this section.

In a perfect living polymerization all polymer chains would be initiated at the same time, grow at the same rate until all the monomer is exhausted, and then would stay alive (capable of resuming their growth) until deliberately terminated. This scenario implies the absence of chain termination during the growth process. In addition, the rate of polymerization must be kept slow enough so that the exothermic polymerization reaction does not get out of control. In classical radical polymerization the rate control is provided by the continuous chain termination that keeps the number of growing chains sufficiently small at all times. In a living polymerization the number of simultaneously growing chains is necessarily very large—it is equal to the final number of polymer molecules. Thus the rate of growth of individual chains has to be reduced. Even more importantly, the recombination of growing radicals (that are now present in much larger concentration than in classical polymerization) must be prevented or at least very substantially repressed.

Both the reduction of the growth rate and suppression of the termination are achieved by keeping the growing chain in a dormant state for most of the time. In the *atom transfer radical polymerization* (ATRP), a group that is easily removed by the catalyst present in the system temporarily terminates the growing chain. Another technique combines the growing radical with a stable radical that will form a very weak bond with it. This will keep the growing radical from any mischief except for occasional bond breakage that would allow for addition of a monomer molecule. (We will see in section 2.3.1 that during anionic chain polymerization the counterions play an analogous role.)

2.2.7.1. Atom Transfer Radical Polymerization (ATRP).

Transition metal catalysts in the lower-valence form (e.g., cuprous chloride in a complex with bipyridyl) catalyze the addition of alkyl halides to olefins. The reaction has a radical nature; the metal complex removes the halogen atom from the alkyl halide, leaving the alkyl radical that consequently adds to the olefin. The back transfer of the halogen completes the reaction. The metal complex itself goes first from the lower-valence form to higher valence and then back. When the concentration of the alkyl halide is very small compared with the concentration of the olefin, the result of the process becomes quite different. The product of the addition is again a halide that can undergo the reaction again and again: the reaction becomes a chain polymerization.

The above-described procedure was later improved in several ways. The initial halide that has become an initiator could be replaced by other initiators that add to olefins more easily, notably by phenyl sulfonyl chloride **28**. Besides copper, other transition metals are used: Fe, Ru, and Ni. The ligands complexing copper may be replaced by multidentate ones, for example, tris[2-(dimethylamino)ethyl]amine, $N(CH_2CH_2N(CH_3)_2)_3$. Triphenylphosphine is a useful ligand for iron.

28

The catalysts containing the metals in the low-valence form are often rather sensitive to oxygen, etc. An ingenious twist was developed by using a "classical" radical initiator as benzoylperoxide or azobis(isobutyronitrile) (AIBN) together with the higher-valence form of the catalyst—in this case $FeCl_3$ complexed with triphenylphosphine. Obviously, the molar ratio of the catalyst to the initiator-derived radicals must be larger than 1 to prevent chain termination in this system.

The ATRP method was successfully used for polymerization of monomers of the styrene type as well for acrylates and methacrylates. Under properly controlled conditions the polydispersities were quite narrow, with $\overline{M}_w / \overline{M}_n$ as low as 1.1.

2.2.7.2. Nitroxide-Mediated Radical Polymerization.   Nitroxides are stable radicals that do not dimerize and are not capable of initiating radical polymerization. However, they recombine with more active radicals in a reaction leading to alkoxyamines—compounds relatively stable at ambient and intermediate temperatures. Above 100°C they start dissociating to nitroxides and carbon-based radicals. This dissociation is fully reversible; it is the key to their use in radical polymerization.

Two different polymerization procedures are based on this chemistry. The older technique employed 2,2,6,6-tetramethylpiperidinyl-1-oxy (TEMPO) radical **29** and a classical radical initiator such as AIBN. The initiating radicals were trapped by TEMPO, and the polymerization could start only when the temperature was raised to about 120°C. It is interesting that it is sufficient to employ only a slight excess of nitroxides over the starting radicals to achieve a satisfactory control of the polymerization.

29

A more recent procedure uses the alkoxyamines both as initiators and the controlling agents. At about 120°C, the alkoxyamines dissociate and the carbon radicals initiate the polymerization. The nitroxy radicals exercise the control and eventually (on cooling) terminate the polymer chain. In this reaction there is no excess of the nitroxy radicals present in the system, yet there is almost no termination by recombination of two growing chains. Obviously, the nitroxy radicals must stay in close proximity to the growing chains at all times.

It is remarkable that the polymeric alkoxyamines are very stable compounds at ambient temperatures and may withstand many chemical procedures to which the polymer may be subjected. In fact, after purification, addition of another monomer, and raising the temperature the polymerization may resume leading to a block copolymer.

The nitroxy-mediated polymerization is a very robust reaction proceeding easily in bulk monomers with high yields and good control of polydispersity. Random copolymers are produced under rules similar to those in classical radical polymerization. (In anionic polymerizations this is often impossible to achieve.) Many alkoxyamines were synthesized that were superior to TEMPO-related ones. Alkoxyamine **30** is touted as a universal one. The only serious drawback of these polymerizations is the relatively high temperature that leads to self-initiation of some monomers (see

Section 2.2.4.1). The self-initiated chains start throughout the polymerization and may damage the narrowness of the molecular weight distribution.

30

Both fragments of the alkoxyamine remain as the end groups of the polymers. Because of the essential absence of the side reactions, almost all polymer molecules carry both groups. Moreover, the polymerization is usually so "clean" that it does not damage chemical groups that could be incorporated into both halves of the initiator. These groups can be then utilized for further reactions, opening a path to a large variety of copolymers as well as of molecular architectures.

2.2.8. Industrial Polymerizations

In the previous sections, we were mainly concerned with bulk and solution polymerization. These procedures are favorites of laboratory chemists; they are easier to understand and usually lead to better-defined polymers. However, from the viewpoint of an industrial chemist, who is usually interested in large-scale production and high yields, both these techniques have serious drawbacks. The main problem with large-scale bulk polymerization is that of temperature control. Toward the end of the polymerization, stirring of the highly viscous system is out of the question. Heat liberated by the polymerization (remember also the accelerated polymerization during the gel effect) cannot be dissipated easily. At best, this leads to structures with high internal stresses; at worst, the reaction gets out of hand completely. Thus bulk polymerization to high conversions is done mainly when it leads directly to the final articles. A prime example is the polymerization of methyl methacrylate to Plexiglas polymers of high optical quality. Such polymerization is usually performed extremely slowly to give the system enough time for dissipation of heat. Using a low reaction temperature lowers the residual amount of unreacted monomer as well, as discussed in the Section 2.2.6. However, the early onset of the glass transition may work in the opposite direction. Often a careful temperature regime is required. After polymerization of most of the monomer at moderate temperature, the temperature is raised above the glass transition point to complete "curing" and then again lowered carefully.

For preparation of polymers for other purposes, industrial chemists try to avoid the troublesome bulk polymerization. Solution polymerization offers better temperature control. However, most solvents are expensive and have high transfer constants that can make it difficult to achieve sufficiently high degrees of polymerization. Also, yields may be significantly lower because of the thermodynamic effects described previously. Most of these negative aspects do not apply, however, to water as a solvent; it is cheap and has low transfer constants. Thus water-soluble polymers (polyacrylamide, polyacrylic acid, etc.) are routinely prepared in water solutions. (Solution polymerization of acrylamide is sometimes also troublesome because of gelation. A newer process therefore utilizes droplets of acrylamide solution suspended in a hydrocarbon.)

Several techniques are used commercially that circumvent these problems and offer better control of the polymerization process. The resulting polymers, although more reproducible, often have lower purity and are usually more heterogeneous. In the following sections, we will discuss several such processes.

2.2.8.1. Suspension Polymerization.

In suspension polymerization, the system to be polymerized (monomer, initiator, and other possible ingredients) is added to water containing materials called *suspending agents*. These agents are either water-soluble organic polymers [gelatin, poly(acrylic acid), etc.] or inorganic water-insoluble salts in very fine form (silicates, aluminum hydroxide, etc.). Detergents are avoided; they would lead to micelles that would trigger a different process (see Section 2.2.8.3). Controlled stirring of the mixture creates droplets of more or less uniform size; suspending agents are adsorbed on their surface and prevent them from coalescing. It should be obvious that this procedure is applicable only to monomers that are only very sparingly soluble in water. For monomers that are slightly more soluble, electrolytes may be added to water to salt them out.

When the droplets have been formed, the temperature is raised to the polymerization temperature. The droplets are macroscopic, and within them the polymerization is the usual bulk polymerization. However, they are small enough to allow easy dissipation of the reaction heat to water under gentle stirring. The final product is in the form of pearls and is accordingly called *pearl polymer*. The suspending agents are usually washed off easily. The process is used extensively for the polymerization of styrene and many other monomers.

2.2.8.2. Precipitation Polymerization.

Some polymers, notably poly(vinyl chloride), poly(vinylidene chloride), and polyacrylonitrile, are insoluble in their monomers. This property leads to some new phenomena when they are polymerized in bulk. Insolubility of polymers implies quite different behavior from insolubility of crystalline substances. When an insoluble polymer is equilibrated with a solvent, it swells. The degree of swelling depends on the thermodynamic interaction coefficients; for polymers that are only marginally insoluble, the swelling may be appreciable.

When vinyl chloride or similar monomers are polymerized, the solubility of the polymer is soon exceeded and the polymer precipitates out of solution. However, the precipitate is in the form of swollen gel-like particles. These particles trap newly growing polymer chains, and the growing ends may be buried in the particle. Monomer molecules reach the growing ends by diffusion; this slows down the propagation rate somewhat but not too appreciably, because the swollen particles are still quite loose. The local environment of the growing end is quite similar to the situation in bulk polymerization after reaching the gel point. The termination rate is suppressed much more than the propagation rate. The result is an increase in the overall rate of polymerization, in the average molecular weight, and in the heterogeneity of molecular weight. At the same time, the macroscopic viscosity of the heterogeneous system is not excessive, allowing for reasonable control of temperature. Residual vinyl chloride is easily distilled off because of its low boiling point.

2.2.8.3. *Emulsion Polymerization.*

When the suspension polymerization system is modified by adding a detergent and by using a water-soluble initiator, the polymerization mechanism changes dramatically and acquires the name *emulsion polymerization*. The mixture contains several types of structure: (1) droplets of the monomer that are stabilized by a surface layer of the detergent; (2) a continuous water phase saturated by monomer and detergent and containing an initiator (usually hydrogen peroxide or potassium persulfate and a ferrous salt); (3) detergent micelles with a few molecules of monomer in their center.

For a better understanding of the mechanism of emulsion polymerization, we should invoke the basic thermodynamic rule for heterogeneous system: At equilibrium, the chemical potential of every component is the same throughout the system. When it is lowered by depletion in any part of the system, transport processes will bring it back to the equilibrium value.

Initiation and the first few propagation steps occur in the aqueous phase. The growing chain becomes increasingly hydrophobic and tries to enter a more suitable environment—monomer droplets or micelles. However, despite their much smaller total mass, the total surface of the tiny micelles is orders of magnitude larger than that of the droplets. Hence, virtually all growing chains enter the micelles. The chain continues to grow within the micelle, and the monomer inside it is depleted. This causes monomer molecules from the aqueous phase to diffuse into the micelle, increasing its size. In turn, the monomer depleted from the water is replenished by slow dissolution of the droplets. Thus, as the polymerization continues, monomer is gradually transferred from the droplets, which will eventually disappear, to the micelles, where it is polymerized, while the micelles slowly increase in size.

Simultaneously with this growth process, new polymer chains are initiated in the aqueous phase and enter the micelles. As the micelles are very small, the entrance of the second radical into the same micelle represents the creation of a very high local concentration of radicals; the radicals are terminated immediately. At this moment there is no growing chain in the micelle, and it idly waits for the entrance of the next radical to start the grow-stop cycle again.

This peculiar mechanism results in a peculiar kinetics of emulsion polymerization. At any moment, one half of the micelles contain one growing chain, while no polymerization occurs in the other half. Thus the overall polymerization rate is independent of such factors as overall concentration of monomer and concentration of initiator. It depends mainly on the number of micelles present, which in turn is given by the overall concentration of the detergent. The detergent is the main kinetic factor in this kind of polymerization. What about the degree of polymerization? The average value depends on the average time for which the chain within the micelle is allowed to grow undisturbed by the entrance of a hostile radical that is going to stop its growth. Hence, it is inversely proportional to the frequency of radical entries, which is proportional to the concentration of the initiator. Let us recall that in bulk polymerization the rate of polymerization is proportional and the degree of polymerization inversely proportional to the square root of initiator concentration.

It is apparent that emulsion polymerization is easily controlled. It can produce long polymer chains at a high reaction rate; heat dissipation poses no problem, either. The polymer produced is in the form of fine powder, which is handled easily. Thus, in industry, emulsion polymerization is frequently the polymerization method of choice. Its main disadvantage is in purification of the polymer from the detergent, which is always difficult and incomplete. Consequently, it is employed primarily when the presence of the detergent is not objectionable and the polymerization product can be used as such. This is frequently the case for polymers intended for compression or injection molding (Section 5.1.9).

2.2.8.4. Cross-Linked Structures.

Articles manufactured from three-dimensional polymer networks usually have superior shape stability compared with objects made from uncross-linked individual polymer molecules. Two strategies are generally employed for making such objects: (1) A prepolymer is prepared that is shaped into the required object and the three-dimensional structure is completed in the second reaction step (Sections 2.1.3 and 2.4.4); or (2) the desired structure is synthesized in a single step using multifunctional monomers. We already encountered this process in Section 2.1.1.5 when we studied three-dimensional networks produced by polycondensation reactions. The present section deals with networks produced by radical polymerization in the presence of divinyl monomers (cross-linking agents).

Monomers with two reactive vinyl groups are effectively tetrafunctional (each double bond is bifunctional because it can connect to two other monomers) and can therefore serve as cross-linking agents. As before, when the number of branch points reaches a critical value during the polymerization, the system gels. However, there is a basic difference between gelling in polycondensations and that in polymerizations. In the latter case, large molecules are present from the very beginning of the reaction. Let us outline a simplified analysis of the gelling process.

Let us designate conversion, the fraction of vinyl groups that have reacted, by θ. The average degree of polymerization of a polymer chain (discounting all cross-links) is x; the value of x depends on the interplay of kinetic constants of the polymerization process as described in Sections 2.2.4.2 and 2.2.4.3. The density of cross-links ρ is the ratio of the number of twice-reacted divinyl molecules, $N_0\theta^2\rho_0$, to the total number of reacted groups, $N_0\theta$:

$$\rho = N_0\theta^2\rho_0/N_0\theta = \theta\rho_0 \tag{2.2.95}$$

Here, N_0 is the original number of reactive groups and ρ_o is the original fraction of reactive groups belonging to the cross-linking agent. The system starts gelling when there is on average one cross-link per elementary chain, that is, when

$$\rho = 1/x \tag{2.2.96}$$

Thus, θ_c, the conversion at incipient gelling, is

$$\theta_c = 1/x\rho_0 \tag{2.2.97}$$

It is seen that a very small amount of the cross-linking agent is sufficient for early gelling (and later for extensive cross-linking) for polymerization conditions that yield long polymer chains. For example, a system with $x \equiv \overline{DP}_n = 1000$ and $\rho_0 = 1\%$ gels at about 10% conversion.

The above relations were derived assuming that all vinyl groups had the same or at least similar reactivity. [Thus, conjugated dienes, for example, butadiene, emphatically are *not* included in this scheme. The remaining double bond is much less reactive (or completely unreactive) once the first has reacted.] Recalling our discussion of copolymerization in Section 2.2.5, we realize that similar reactivity of the monomer and the cross-linking agent means that both reactivity ratios are approximately equal to unity. This can be accomplished only when the cross-linking agents are members of the same family as the monofunctional monomers. Otherwise, all the cross-linking agent may be incorporated into the polymer either very early in the reaction (when there is an alternating tendency) or not at all (when the cross-linking agent has low reactivity). Thus the proper cross-linking agent for monomers of the styrene family is divinylbenzene; for methacrylates, bis(methacroyl) esters of various glycols; similar acroyl esters for acrylates; and so on.

It has been observed that the actual degree of cross-linking is sometimes much less than the degree predicted by the above model. A telling symptom is the deformability of the cross-linked structure. A heavily cross-linked system should have elongation at break of about 1%, but much larger values are frequently observed. Two effects can cause this behavior: incomplete reaction and intramolecular cyclization.

In later stages of radical polymerization the mobility of polymer chains is severely restricted and diffusion of monomer molecules toward the immobile radicals becomes the main mode of chain propagation. This is the phenomenon that was the cause of the gel effect of Section 2.2.4.5. However, the unreacted vinyl group of the once-reacted cross-linking agent is firmly attached to the polymer chain and cannot diffuse toward the growing radical. Thus these effectively trapped vinyl groups contribute to the number of unreacted groups much more than simple statistics would predict.

We have already seen in Section 2.1.1.5 that intramolecular cyclization delays the onset of gelation even in polycondensation. However, in vinyl polymerizations kinetic effects enhance this statistical thermodynamic effect. The unreacted vinyl group of a molecule of cross-linking agent that was just incorporated into the growing chain dangles very close to the growing radical (and keeps dangling there through several subsequent propagation steps). Thus it has an unstatistically high probability of being incorporated immediately into the same polymer chain, causing a small loop but not contributing to the degree of cross-linking. This phenomenon manifests itself strikingly in the difference of cross-linking efficiency between cross-linking agents in which the two vinyl groups are connected either by a short and stiff chain or by a longer, flexible chain. Stiff chains keep the vinyl groups fairly far apart, preventing cyclization. Thus divinylbenzenes (meta and para isomers) are very effective cross-linking agents, as is 1,4-cyclohexanediolbismethacrylate **31**.

$$\text{H}_2\text{C}=\overset{\overset{\displaystyle \text{CH}_3}{|}}{\text{C}}-\text{CO}-\text{O}-\left\langle\bigcirc\right\rangle-\text{O}-\text{CO}-\overset{\overset{\displaystyle \text{CH}_3}{|}}{\text{C}}=\text{CH}_2$$

31

Ethylene glycol dimethacrylate **32** is also a rather effective cross-linking agent. Immediate incorporation of the second vinyl would lead in this case to a sterically unfavorable nine-member ring.

$$\text{H}_2\text{C}=\overset{\overset{\displaystyle \text{CH}_3}{|}}{\text{C}}-\text{CO}-\text{O}-\text{CH}_2-\text{CH}_2-\text{O}-\text{CO}-\overset{\overset{\displaystyle \text{CH}_3}{|}}{\text{C}}=\text{CH}_2$$

32

On the other hand, diethylene glycol dimethacrylate **33** and similar compounds cyclize easily and are not very effective.

$$\text{H}_2\text{C}=\overset{\overset{\displaystyle \text{CH}_3}{|}}{\text{C}}-\text{CO}-\text{O}-\text{CH}_2-\text{CH}_2-\text{O}-\text{CH}_2-\text{CH}_2-\text{O}-\text{CO}-\overset{\overset{\displaystyle \text{CH}_3}{|}}{\text{C}}=\text{CH}_2$$

33

Let us conclude this section with a few examples. Soft contact lenses are made by the bulk polymerization of hydroxyethyl methacrylate (HEMA) in the presence of a few tenths of a percent of either **32** or **33**. The resulting network is very loose, allowing extensive swelling of the material in water.

Acrylamide is polymerized in water in the presence of methylenebisacrylamide **34**.

$$\text{H}_2\text{C}=\text{CH}-\text{CO}-\text{NH}-\text{CH}_2-\text{NH}-\text{CO}-\text{CH}=\text{CH}_2$$

34

The resulting gel is used for gel electrophoresis, a technique for characterization of proteins (Sections 3.7.1.3–3.7.1.6).

Ion-exchange resins are based on small spheres (0.2–0.7 mm in diameter) of styrene-divinylbenzene copolymer. The reaction mixture contains about 8% mixed isomers of divinylbenzene in styrene; the technique of suspension polymerization is employed. Poly(vinyl pyrrolidone) of rather high molecular weight serves as the suspending agent, and the size of the particles is regulated by the speed of stirring.

The resulting particles of the copolymer are then derivatized, and the ion-exchanging groups are synthesized on the aromatic nuclei.

2.3. IONIC AND COORDINATION POLYMERIZATION

At first glance, chain polymerizations by radical and ionic mechanisms are very similar to each other. In both cases, an active center is introduced into the monomer by appropriate initiation and the center rapidly adds monomer units in a propagation reaction and is eventually terminated. However, two basic dissimilarities in the two processes make the actual procedures quite different.

1. Once the radicals initiate the polymerization, individual radicals are completely separated from each other. In other words, they do not mutually interfere during the chain propagation until the eventual termination. In most ionic processes, the counterions are always either quite close or at least in the very near vicinity of the active center and have a profound influence on all steps of the process. Thus ionic polymerizations depend strongly on the nature of the counterions and on the solvent.
2. Radical chains are terminated by reaction with another radical. Now, other radicals are always present during radical polymerizations, and the growing chains are usually terminated very quickly. Chains in ionic polymerizations are terminated by species that are not intrinsically necessary for the propagation process and can even be avoided. Thus ionic polymerizations often acquire a "living" character, because active centers may survive for quite a long time.

In anionic polymerizations, the initiating species is an anion that adds onto the monomer. The corresponding cation stands nearby—usually very close. Similarly, in cationic polymerizations, the initiating cation is closely associated with its anion. In many polymerizations of ever-increasing significance, a monomer molecule associates (coordinates) with some complex polar molecule (or a larger structure). It is not obvious whether the growing chain has cationic or anionic character or, indeed, whether it has an ionic character at all. Such polymerizations are called *coordination* polymerizations. In most instances, the catalytic center is part of a complex crystalline material and the polymerization has a heterogeneous character.

In the following sections we will treat all of the above types of polymerization processes.

2.3.1. Anionic Polymerization

In a typical anionic polymerization, some strongly basic anion A^- (often derived from some organometallic compound) reacts with an olefin.

$$A^- + \underset{/}{\overset{\backslash}{C}} = \underset{\backslash}{\overset{/}{C}} \longrightarrow A-\underset{|}{\overset{|}{C}}-\underset{|}{\overset{|}{C}}|^- \qquad (2.3.1)$$

The product of this addition is again an anion that can add further monomers in the process of propagation. Not all olefins are susceptible to anionic polymerization. We may imagine that in reaction (2.3.1) the anion adds to a double bond that has been totally polarized. To make it feasible, either the monomer must have a structure in which electrons shift with ease, or the double bond must be already partially polarized. Aromatic monomers such as styrene and its derivatives, vinylpyridines, vinyl naphthalenes, and dienes belong to the former class. The latter monomers possess a substituent on the double bond that has strong electron-withdrawing properties. Prime examples are acrylates, methacrylates, and acrylonitrile. An extreme example is ethyl cyanoacrylate, $CH_2\!\!=\!\!C(CN)COOEt$, which is polymerized by even the weakest bases with amazing speed.

Olefins are not the only monomers susceptible to anionic polymerization. Actually, many polar compounds—cyclic ethers, lactames, aldehydes, isocyanates, and others—are even more susceptible to anionic attack than olefins. Typical is the reaction of ethylene oxide,

$$A^- + CH_2\!\!-\!\!CH_2 \longrightarrow A\!\!-\!\!CH_2\!\!-\!\!CH_2\!\!-\!\!O^- \qquad (2.3.2)$$
$$\diagdown_O\diagup$$

that is initiated even by bases of intermediate strength.

However, before studying these processes, we must familiarize ourselves with the behavior of ions in nonaqueous media.

2.3.1.1. Ions in Nonpolar Media.   Electrostatic forces between ions of opposite sign are quite large, as is manifested by the high melting and boiling points of ionic substances. Consequently, oppositely charged ions will try to reside as close to each other as possible; this lowers the enthalpy of the system appreciably. In solutions, two effects counterbalance the large negative enthalpy of ion clustering:

1. High values of dielectric constant lower the electrostatic interaction, thus enhancing the solubility. Water, with its high dielectric constant ($\sim$80), is one of the best solvents for ionic substances.
2. Polar molecules may interact with ions, spreading the effective charge over a larger volume and thus reducing the electrostatic potentials. This interaction is known as *solvation of ions*. Again, water is one of the best solvating solvents.

For many anionic polymerizations, very strong bases are needed. Such anions would react with any hydrogen ion (active hydrogen in terms of organic chemistry) to form the associated (weak) acid, and their anionic character would be destroyed.

$$A^- + H^+ \longrightarrow A\!\!-\!\!H \qquad (2.3.3)$$

Such reagents can be dissolved only in solvents without any active hydrogen (aprotic solvents). Let us study the structure of a few typical solutions of ions in such solvents.

Every ionic solution must contain anions and cations. In the present case, the cations are usually alkali ions or ions of alkaline earth metals. These cations are completely insoluble in aliphatic hydrocarbons. Solubility can be achieved by complexing them either with aromatic hydrocarbons such as styrene, α-methylstyrene, naphthalene, and anthracene or with oxygen or nitrogen compounds such as ethers, amines, and phosphines. There is one major exception to this rule: lithium alkyls are soluble in hydrocarbons because of the atypical nature of lithium compounds. We will return to them later.

Solubilized ionic pairs do exist in several clearly distinguished forms. The most common form is a contact pair in which the two ions are in direct contact; the solvating molecules surround the whole pair. The tightness of the pairs (expressed as decreasing value of their dissociation constants) increases from lithium to cesium. In another form, the ions in the pair are separated by a solvent molecule "sandwiched" between them. Such pairs are called "loose" pairs and are generally much more reactive than the "tight" contact pairs. The relative concentration of the loose pairs (or better, the equilibrium constant for the interconversion of tight and loose pairs) depends on the nature of the solvent (it is much higher for the more polar dimethoxyethane than for the less polar tetrahydrofuran) but is generally small. The "loosening" process is usually exothermic ($\Delta H \approx -6$ kcal/mol) and is accompanied by a sizable decrease in entropy [$\Delta S \approx -(20\text{--}30)$ cal mol^{-1} K^{-1}]. Accordingly, the concentration of the more reactive loose pairs increases at *low* temperatures.

The third important form of the ions is "free" ions. Their concentration is always extremely small, but they are very reactive. Their formation from loose pairs is essentially athermal (the enthalpy-influencing step of solvation of the ions was already complete when the loose pair was formed), but it is accompanied by another decrease in entropy, $\Delta S \approx -(20\text{--}30)$ cal mol^{-1} K^{-1}.

The dissociation equilibrium between ion pairs and free ions obeys the usual rules for dissociation of weak electrolytes. Specifically, the dissociation can be suppressed and the free anions virtually eliminated by addition of a strong (i.e., easily dissociated) electrolyte having one ion (the counterion) common with the ion pair. A frequently employed electrolyte of this type is sodium tetraphenyl boride, Na$^+$ (BPh$_4$)$^{-1}$.

Other ionic aggregates have been postulated—for example, cations sandwiched between two anions or, conversely, anions sandwiched between two cations—but we will restrict our discussion to the three main types described above.

Organolithium compounds are characterized by the special bonding properties of the very small lithium atom, which can participate in multicenter electron-deficient bonds. For example, methyllithium forms very tight tetramers with lithium atoms at the vertices of a tetrahedron and methyl groups sitting above the tetrahedron faces (Fig. 2.3). Each carbon forms three bonds (containing a total of two electrons for the three electron-deficient bonds) with three adjacent lithium atoms; no lithium-lithium bonds exist. A common reagent, butyllithium, forms tightly bound hexamers. In this compound, the core of lithium atoms is completely surrounded by the hydrocarbon residues and the complex is soluble in hydrocarbons. Of course, the complexes do dissociate into smaller aggregates or even individual ion pairs, but the dissociation constants are extremely small.

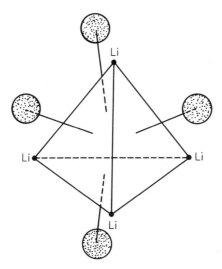

Figure 2.3. The arrangement of lithium atoms and methyl groups in a tetramer of methyllithium. (Reprinted from M. Szwarc, *Living Polymers and Mechanism of Anionic Polymerization.* Copyright © 1983 by Springer, Berlin. Used by permission of the publisher and the author.)

2.3.1.2. Anionic Initiators.   As in all types of chain polymerization, the first step in anionic polymerization involves conversion of the first molecule of monomer into a growing chain according to scheme (2.3.1) or (2.3.2). A prerequisite to this process is the introduction of a suitable anion, A^-, the initiator, into the system. There are several strategies for producing the anion.

It has been known for a long time that a reaction of alkali metals with various compounds produces strongly basic structures. The best-known reaction of this type is reaction with water or alcohols. The electrons of the alkali metal are used to reduce the hydrogen ions to molecular hydrogen; the strongly basic hydroxy or alkoxy anions are left in the solution (and they can be used as mild anionic initiators for susceptible polymers). However, in systems lacking active hydrogens, the fate of the electrons depends on the nature of the solvent.

The first interesting class of solvents possesses free electron pairs (on oxygen or nitrogen) capable of complexing atoms with vacant outer orbitals (in our case, alkali metals). A prime example of these solvents is liquid ammonia. (Interestingly enough, liquid ammonia does not seem to have active hydrogens; its reduction by sodium to sodium amide proceeds extremely slowly.) Other examples are hexamethyl phosphoric triamide, dimethoxyethane, and, to a much smaller extent, tetrahydrofuran and other ethers. Crown ethers (macrocycles with several ethereal oxygens) and kryptates (cages of a similar structure) belong also to this group.

These solvents dissolve alkali metals directly; the cations are complexed, and the electrons act as a separate species. The electrons reside in a sizable cavity surrounded by solvent molecules; they are often called *solvated electrons.* Their movement within the cavity is quantized (this is a classic example of a particle in a box), and they exhibit a broad absorption band in the red and infrared spectral regions; they are deeply blue. Of course, these solutions also contain other types of ions and ion pairs: alkali metal-electron pairs, free atoms, alkali metal anions with two electrons, and so on. When mixed with monomers, these solutions may transfer the solvated electrons to monomer, triggering the initiating sequence. However, this type of initiation leads to some unwanted side reactions and is not employed very often.

Many aromatic hydrocarbons (naphthalene is a prime example) react in a hetero-geneous reaction with an alkali metal.

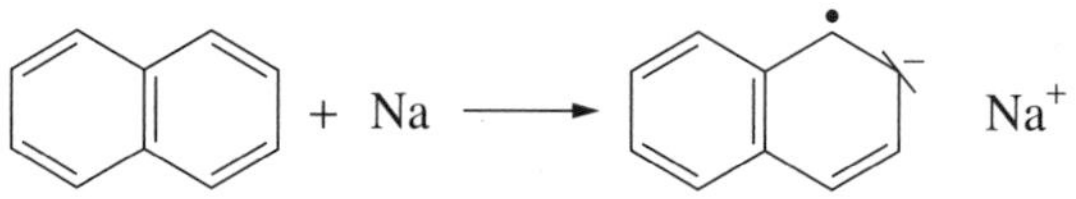

$$(2.3.4)$$

1

The hydrocarbon is converted into a *radical-anion,* which is associated with the alkali cation (sodium naphthalide **1** in this case). Sodium naphthalide solutions are characterized by their deep green color. The heterogeneous reaction (2.3.4) is reversible, which implies that at equilibrium the ratio of the concentrations of the hydrocarbon and its radical-anion is constant. The equilibrium composition depends on the reaction conditions or, more specifically, on the solvating power of the solvent with respect to the cations. For example, naphthalene at room temperature yields about 95% radical-anions when dissolved in tetrahydrofuran, but only about 1% in diethyl ether solutions. Sodium naphthalide and similar compounds are very stable in aprotic solvents; their solutions are routinely used as initiators for polymerizations of susceptible monomers.

Some aromatic monomers—typically α-methylstyrene—react with alkali metals in a similar manner. However, in this case, the preparation of the radical-anions is immediately accompanied by the reactions of initiation and propagation. We will return to the interplay of these reactions in the next section.

Another class of off-the-shelf initiators are alkali salts of very weak acids, such as alkoxides. In this context, we should recall that organometallic compounds are actually salts of extremely weak acids—hydrocarbons. However, they often form very tight ion pairs and are quite unreactive in initiation processes. More useful are salts of somewhat stronger acids—hydrocarbons with an active hydrogen. For example, fluorene **2** reacts with ethyllithium with evolution of ethane and formation of a salt **3** that is solvated more easily.

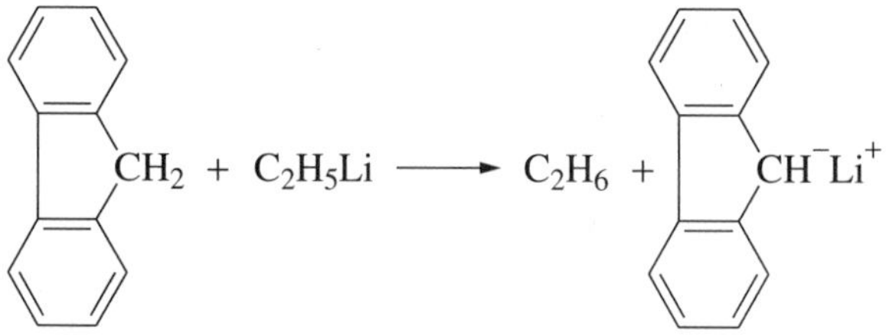

$$(2.3.5)$$

2   **3**

As we have already seen, alkyllithium compounds are soluble in aliphatic hydrocarbons and are used frequently as initiators. However, initiation by these compounds is a complex process to which we will return in the next section.

2.3.1.3. Initiation and Propagation.   When solutions of alkali metals or salts of the naphthalide type are added to susceptible monomers, they transfer the electron to the monomer (typically α-methylstyrene) and convert it to the corresponding radical-anion.

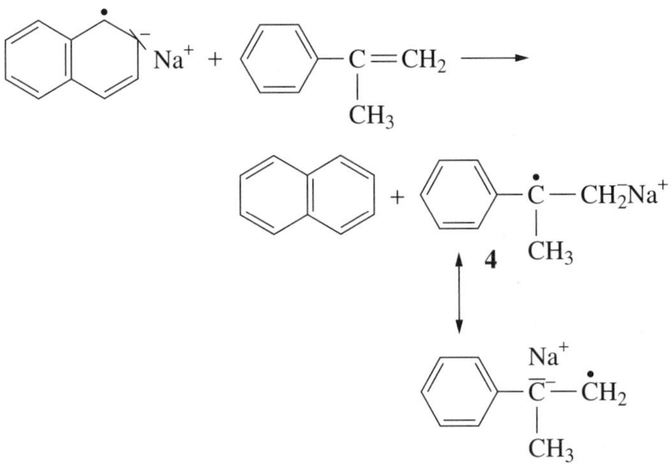

$$(2.3.6)$$

Between the two mesomeric forms **4** and **5**, form **4** is more stable. However, it cannot dimerize easily because of steric crowding, whereas form **5** dimerizes immediately to a dianion **6**.

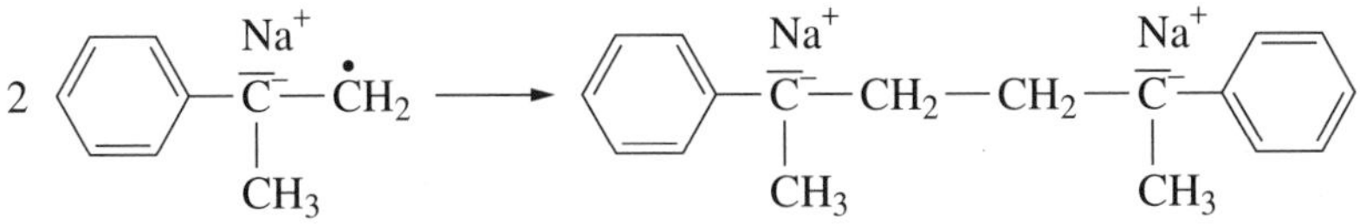

$$(2.3.7)$$

Reaction (2.3.7) is essentially irreversible, and sodium naphthalide converts α-methylstyrene immediately and quantitatively into the dianion **6**. The green color of naphthalide changes to the bright red color of **6**. Solutions of **6** are very stable and can even be stored before being used as second-generation initiators for actual polymerization of α-methylstyrene or any susceptible monomer.

The initiation by dianions follows the same mechanism as initiation by monoanions depicted by equation (2.3.1). At this moment, we should recall one fact from the process of radical polymerization: If the initiating radical is more reactive or only slightly less reactive than the growing polymer radical, the initiation is swift and easy; otherwise, rather unreactive radicals fail to initiate less reactive monomers (i.e., those producing rather reactive radicals). An exactly analogous situation exists in ionic polymerizations. The initiating ion must be energetic enough to produce the growing ion; if it is not, it will not initiate. Thus carbanions are generally required for polymerization of olefinic and diene monomers, whereas less reactive anions of the alkoxy type RO^- are usually sufficient for polymerization of aldehydes, cyclic ethers (ethylene oxide), lactons, and lactams.

We have seen in Section 2.3.1.1 that ionic compounds in media with low and intermediate polarity exist either in dissociated form or as ion pairs, tight or loose. Usually, most pairs exist in the tight form; the concentration of the loose pairs is much less than that of the tight pairs; the concentration of free ions is minuscule. However, the concentration of both loose pairs and free ions increases at low temperatures (remaining very low nevertheless). The three types of ionic species have grossly different reactivities in the propagation reaction.

Free anions add monomers according to scheme (2.3.1); the reaction has a very high kinetic constant. Thus, despite their extremely small concentration, free ions are responsible for the major part of the propagation reaction. This fact also leads to an unusual dependence of overall reaction rate on temperature. The concentration of free anions increases at low temperatures, and so does the reaction rate (the decrease in the kinetic constant due to the activation energy is not enough to compensate for the increased number of free ions). However, addition of fully dissociated electrolytes (e.g., sodium tetraphenyl boride) to the system suppresses virtually completely the presence of free ions and consequently slows down the polymerization considerably.

In the absence of free anions, the loose ion pairs are the most active species. The addition of an ion pair onto a double bond requires, among other things, severing the electrostatic "bond" between the ions of the pair and shifting the counterion to the newly added monomer. Even if the process is believed to go through the formation of an intermediate complex between the pair and the monomer molecule, the separation is easier with a loose pair than with a tight one.

$$-\overset{|}{\underset{|}{C}}|^-\,(M^+S_x) + \overset{\backslash}{\underset{/}{C}}{=}\overset{/}{\underset{\backslash}{C}} \longrightarrow -\overset{|}{\underset{\backslash C{=}C/}{C}}|^-\,(M^+S_x) \longrightarrow -\overset{|}{\underset{|}{C}}-\overset{|}{\underset{|}{C}}-\overset{|}{\underset{|}{C}}|^-\,(M^+S_x)$$

$$(2.3.8)$$

In reaction scheme (2.3.8), (M^+S_x) represents the solvated counterion. The intimate role played by the counterion explains the profound influence of the nature of the alkali metal as well as of the solvating agent (ethers, etc.) on the overall propagation process. First, they influence the equilibrium between loose and tight pairs and, through it, the overall polymerization rate. Second, steric and electronic interactions of the counterion with monomer determine the kinetic constant accompanying reaction (2.3.8). Third, the stereochemistry of the addition may be influenced as well. This is especially important in the polymerization of dienes, where the nature of the counterion strongly affects the ratio of 1,2 and 1,4 additions as well as the ratio of *cis* and *trans* isomers in 1,4 addition. For methacrylates the proportions of isotactic, syndiotactic, and heterotactic triads depend on the solvating agent. However, pure stereospecific forms cannot be prepared by this mechanism.

Equilibrium between loose and tight ion pairs depends on the nature and concentration of the complexing solvent S.

$$-\overset{|}{\underset{|}{C}}|^- \, M^+ + S \; \rightleftharpoons \; -\overset{|}{\underset{|}{C}}|^- \, SM^+ \qquad (2.3.9)$$

However, even in pure strongly complexing compounds like dimethoxyethane, the concentration of loose ion pairs is relatively low. Their concentration again increases with decreasing temperature, complicating the dependence of overall polymerization rate on temperature. In most cases, the propagation rate for tight pairs is much less than for loose pairs but is not negligible. The reactivity of tight ion pairs increases with increasing size of the alkali cation, that is, from Li^+ to Cs^+.

We will now consider a characteristic anionic polymerization. The initiator is added to a monomer and usually reacts quantitatively with it within a few seconds. A color change frequently accompanies this reaction. The growing chains are now distributed among tight ion pairs, loose ion pairs, and free ions, depending on the appropriate equilibrium constants, that is, on the nature of the components, temperature, and the concentrations of the components and added ionic compounds. Each type of active end group propagates according to its kinetic constant of propagation. However, the equilibria between various types of growing ends are established very quickly. Thus a given polymer chain changes the nature of its active end very frequently. It adds monomers fast when in the form of free ion, more slowly as a loose ion pair, and very slowly or not at all as a tight pair. Yet, because of the fast change from one form of the active group to the other, the fractions of time spent in the different forms are the same for all chains.

Polymerizations initiated by lithium alkyls in hydrocarbon solvents follow a slightly different scenario. The sodium and higher alkali-based initiators react quantitatively with the monomer in a very short time. Subsequently, the number of growing chains is constant and the rate of polymerization remains essentially constant. In the case of lithium initiators, the presence of lithium clusters is the dominant feature. It is believed that organolithium clusters do not react with carbon-carbon double bonds; to react, they must first dissociate to monomeric forms or even to free ions, and then they react very fast. However, the dissociation constant of the clusters is very low. Butyllithium is believed to exist in solutions as a hexamer, $(BuLi)_6$, and to dissociate as

$$(BuLi)_6 \rightleftharpoons 6\,BuLi \qquad (2.3.10)$$

Accordingly, the concentration of monomeric butyllithium at dissociation equilibrium is proportional to the sixth root of the total concentration of butyllithium, and so is the initiation rate. The initiation rate being quite slow, it takes a long time before all the initiator reacts with the monomer; lithium initiations are characterized by long induction periods. The dissociation processes can be speeded up by the addition of some complexing agent, typically of the ether type.

Growing polymer chains with lithium counterions form clusters, too. Polystyryl-lithium is believed to be dimerized, polyisoprenyllithium forms tetramers, and polybu-tadienyllithium may even form hexamers. The clusters are again unreactive with respect to propagation; only monomeric ion pairs propagate. Thus, because of the dissociation equilibria, the overall polymerization rate is proportional to the square root of the total concentration of the growing ends in the case of polystyrene and to the fourth root for polyisoprene. For polybutadiene, the power is even lower.

Another peculiarity of lithium chemistry is worth mentioning: Lithium of lithio-organic compounds has a tendency to form π complexes with double bonds. These complexes possibly serve as intermediates during the addition process. We will see similar mechanisms in Section 2.3.3, which deals with coordination polymerization. In the present case, lithium has an explicit tendency to form complexes with dienes, for example, **7**.

$$\begin{array}{c} \text{H}_2\text{C} \quad\quad \text{H} \\ \diagdown \quad\diagup \\ \text{C} \\ \vdots \quad\quad | \\ \text{R---Li} \quad\quad \text{C} \\ \vdots \quad \diagup \diagdown \\ \text{H}_2\text{C} \quad\quad \text{CH}_3 \\ \mathbf{7} \end{array}$$

This arrangement explains why lithium initiators in nonpolar hydrocarbon solvents polymerize isoprene, producing over 90% of the desirable *cis*-1,4 units. Polymerization of butadiene under similar conditions yields a much smaller fraction of *cis* units; nevertheless, it is substantially higher than the fraction obtained using sodium and potassium initiators.

In anionic polymerizations, the growing chains remain active even if the monomer is exhausted and polymerization cannot proceed further. Under such circumstances, depropagation may proceed easily, and the equilibrium between propagation and de-propagation must be considered. The thermodynamic situation is exactly the same as in the case of radical polymerization (which was treated in Section 2.2.6). Depropagation may be significant especially for monomers with low ceiling temperature, notably α-methylstyrene. With this monomer, combination of the propagation and depropagation reactions leads to an unexpected sequence of reactions, when an initiator is being prepared by reaction of metallic lithium with an equimolecular amount of monomer.

In the first stage of the reaction, lithium transfers its electron to α-methylstyrene and forms the radical anion **4↔5** of reaction (2.3.6). This radical anion immediately dimerizes to dianion **6** of reaction (2.3.7). The heterogeneous electron transfer reaction is slow compared to the rate of propagation of the resulting dianion. Thus the first few dianions propagate fast and consume essentially all monomer—producing chains with a relatively high degree of polymerization. However, the depropagation reaction provides enough monomer to keep the initial transfer reaction going. This reaction is effectively irreversible because the following dimerization step is irreversible. Consequently, the depropagation continues until all lithium metal is consumed; the monomer is converted into the dianionic dimer (or another low-molecular-weight *seed oligomer*), which is then used for the actual initiation of the same or a different monomer.

2.3.1.4. Living Polymers and Block Copolymers. The most characteristic feature of anionic polymerization is the lack of an inherent termination mechanism. The growing chains exist either in the more active (free ion or solvated ion pair) form or as less active tight ion pairs, but they preserve their ability to grow. In a typical experiment, the chains grow until all the monomer is exhausted (or until propagation-depropagation equilibrium is reached), and then they remain in the active form essentially indefinitely. When more monomer is added, the chains resume their growth. Such chains are appropriately called *living* polymers to distinguish them from *dead* polymers, which have been deactivated by some termination process. (During radical polymerization, most of the polymer molecules are dead.)

The living chains can be terminated by reactions that lead to anions that are not active enough to propagate the chain. For example, reaction with alcohols leads to alkoxides, which cannot initiate polymerization of carbon-carbon double bonds.

$$R|^-Na^+ + R'OH \rightarrow RH + R'O^-Na^+ \qquad (2.3.11)$$

Thus all impurities with active hydrogens (water, alcohols, amines, acids, etc.) interfere with anionic polymerization and must be removed from the polymerization system before the initiation by meticulous drying and purifying procedures.

Another possible terminating mechanism is the reaction of the growing ion pair (it is an organometallic compound) with some susceptible group on the monomer—either free or already incorporated into polymer. For example, organometallic compounds react with the ester group of methacrylates according to the scheme

$$R|^- Na^+ + R'-COOCH_3 \longrightarrow R'-\underset{\underset{R}{|}}{\overset{\overset{OCH_3}{|}}{C}}-O^- Na^+ \qquad (2.3.12)$$

However, the extent of this reaction is usually negligible compared with ordinary propagation and could be further suppressed by lowering the reaction temperature.

It is seen that it is possible to eliminate all the termination processes, and most synthetic procedures aim at doing so. The unique ability of living polymers to remain active is utilized in synthetic polymer chemistry for three different purposes: (1) preparation of polymers with a narrow distribution of molecular weights; (2) preparation of polymers derivatized at chain ends; and (3) synthesis of block copolymers. These procedures will now be described in some detail.

It is relatively easy to regulate molecular weight in anionic polymerizations. All initiator molecules produce polymeric chains, and reaction is usually allowed to proceed to complete conversion. Thus the number-average degree of polymerization, $\overline{DP}_n$, is equal to the molar ratio of monomer and initiator when alkyl metals are used as initiators. It is twice as large with initiators of the sodium naphthalide type, because two molecules of initiator produce one molecule of polymer with two growing ends.

The narrowness of the distribution of molecular weights is due to the fact that all molecules are allowed to grow for the same time under the same conditions. Let us inspect first how well these premises are fulfilled. The initiation is usually completed within a few seconds after addition of the initiator, which is much shorter than typical reaction times. (This is not true for polymerizations initiated by lithium alkyls; hence, these are not recommended for the present purpose.) However, care should be taken to accomplish the addition of the initiator and mixing within a very short time. Local reaction conditions are another concern. Like all polymerizations, the process is exothermic and may result in local overheating with a consequent local change in reaction kinetics. Thus appropriate thermostating and mixing are mandatory.

We will now explore a model for calculating the molecular weight distribution. Let us divide the total reaction time into short time intervals in such a way that during each interval only one monomer molecule could be added to a growing chain. Only a very small fraction of chains p will actually add a monomer during each time interval. Let us further specify the length of the intervals by requiring that the fraction p be the same for all intervals. (The length of the time interval increases as the reaction progresses, monomer is depleted, and the reaction rate slows down.) The number of intervals N is related to $\overline{DP}_n$ and p as

$$\overline{DP}_n = Np \tag{2.3.13}$$

The probability that a selected polymer molecule has a degree of polymerization x is equal to the probability that it added a monomer molecule at x intervals randomly selected from the total N intervals and did not add anything during the remaining intervals. According to the combinatory theorems of statistics, the latter probability $f_n(x)$ is given as

$$f_n(x) = \frac{p^x(1-p)^{N-x}N!}{x!(N-x)!} \tag{2.3.14}$$

where $N!$ is the factorial of N. It is easy to show that equation (2.3.14) is equivalent to the Poisson distribution function described in relations (1.6.37) through (1.6.39).

Under carefully controlled conditions it is routinely possible to achieve distribution so narrow as to have the ratio $\overline{M}_w/\overline{M}_n = 1.02$. The narrow distribution was obtained by utilizing the special *kinetic* factors of anionic polymerization. When a solution of a living polymer is allowed to rest long enough after all monomer has been consumed, the equilibrium propagation-depropagation reactions will eventually bring the system to full thermodynamic equilibrium, which is characterized by the most probable distribution described by equations (1.6.15) through (1.6.23).

Once the polymerization process is completed, the polymer is stabilized by "killing" the living polymer, that is, by destroying the active end groups. Adding alcohols that deactivate the chains according to reaction (2.3.11) routinely does this. The deactivation is nicely visible; the deeply colored solution of the living chains becomes colorless.

In the reaction with alcohols and similar reagents, the anion adds a proton and forms a saturated unreactive molecule. However, it is possible to kill the chain using more sophisticated means and to synthesize various end groups. For example, reaction with carbon dioxide leads to carboxylic salts, which hydrolyze to acids.

$$R|^- Na^+ + CO_2 \longrightarrow R'-COO^- Na^+ \longrightarrow RCOOH \quad (2.3.15)$$

Similarly, the addition of ethylene oxide leads to an alkoxide and an alcohol.

$$R|^- Na^+ + CH_2-CH_2 \longrightarrow R-CH_2-CH_2-O^- Na^+ \longrightarrow R-CH_2-CH_2OH$$
$$\underset{O}{\diagdown \diagup} \qquad\qquad (2.3.16)$$

In fact, most of the known reactions of organometallic reagents can be exploited in the termination reaction, leading to a great variety of custom-made chain ends.

In a number of cases, living polymer can initiate the anionic polymerization of another monomer. When the second monomer is added after full consumption of the first one, a block copolymer results. The procedure can be repeated, and polymers with several blocks can be prepared. (Note that addition of second monomer to living polymers with two active ends leads to a triblock copolymer in the first step.) The sequence of monomer additions manipulates the number of blocks; the amounts added control the length of the blocks. Obviously, the procedure is very flexible, and a polymer chemist can create a number of different structures at will.

From the industrial viewpoint, the formation of blocks is most important in copolymers of styrene with dienes (isoprene, butadiene). Especially important is styrene-butadiene-styrene triblock copolymer, which is the basis of reversibly cross-linked rubbers. (Polystyrene domains act as cross-links that may be reversibly separated by elevating temperature above the glass transition point of polystyrene.)

2.3.1.5. Ring-Opening Polymerizations.   As we have already mentioned, some saturated heterocycles can be polymerized using an anionic mechanism. These polymerizations are dominated by a thermodynamic phenomenon: free energy of polymerization. We have already studied it in Section 2.1.1.2 for a process that is the opposite of the present one—the formation of cyclic compounds. The thermodynamic principles described here are equally applicable to all polymerization procedures: polycondensations, anionic and cationic polymerizations, or whatever. Accordingly, only cyclic compounds having sufficiently high negative enthalpy of polymerization can be polymerized to high-molecular-weight products. The source of this enthalpy is the release of the strain in the ring on polymerization. Thus three- and

TABLE 2.5 Enthalpies of Polymerization for Some Cyclic Monomers

Monomer	ΔH_p, kcal/mol
Ethylene	-21
Cyclopropane	-27
Cyclobutane	-25
Cyclopentane	-2
Cyclohexane	$+1$
Cycloheptane	-5
Cyclooctane	-12
Ethylene oxide	-22
Trimethylene oxide	-19
Tetrahydrofuran	-3
Tetrahydropyran	$+1$
Caprolactam	-4

four-member rings have high polymerization enthalpies; five- and seven-member rings exhibit low values; for six-member rings, the value is essentially zero. In Table 2.5, polymerization enthalpies of several cyclic monomers are compared to ethylene. (No feasible polymerization mechanism is presently known for some of these monomers.)

In practice, only three types of cyclic monomers are polymerized anionically: epoxides, lactams, and N-carboxyanhydrides. In all three cases, the reaction mechanism is complex.

Ethylene oxide polymerization is usually initiated by sodium phenoxide.

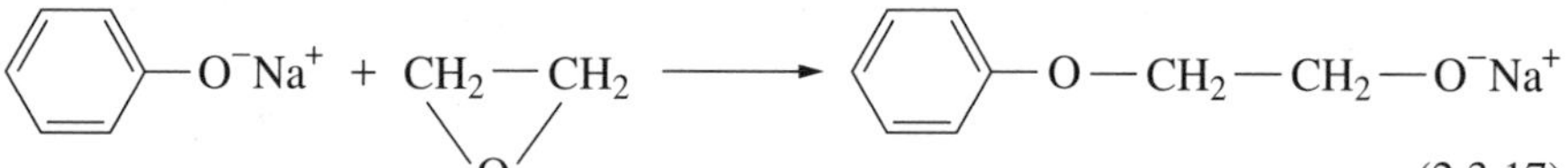

$$\text{(2.3.17)}$$

However, the reaction does not proceed if free phenol is not present in the reaction mixture. Apparently, it is needed for complexing the sodium cation to separate it sufficiently from the phenoxide anion so that the latter can proceed with its nucleophilic attack on the epoxide.

Lactams (mainly caprolactam) are anionically polymerized by alkali metals or any compounds that can convert the lactam into its salt **8**.

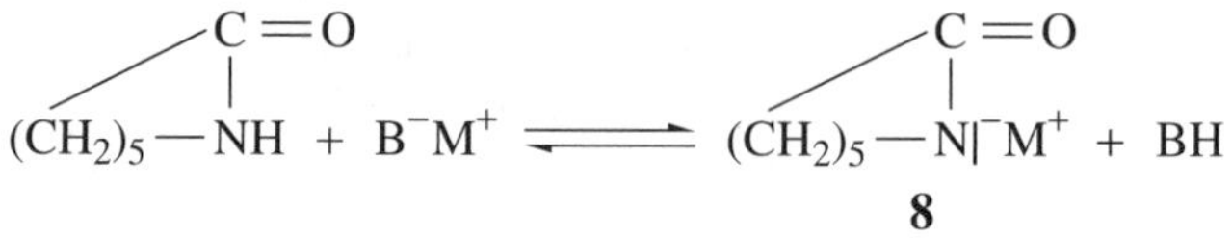

$$\text{(2.3.18)}$$

8

Here B$^-$ is a strong base; M$^+$ is a metal cation. The anion **8** attacks the carbonyl of the lactam ring in a nucleophilic reaction and opens it. However, this reaction is reasonably fast only when the lactam nitrogen carries another acyl group R–CO, that is, when it is an imide **9**. Then the reaction proceeds as in reaction (2.3.19).

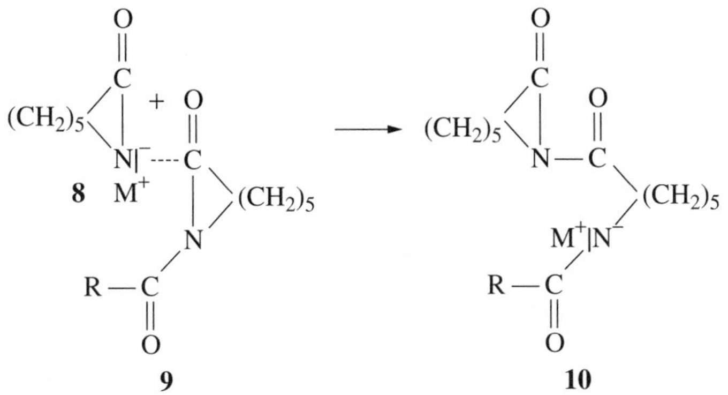

Obviously, the imide arrangement (which is the active species in this polymerization) is re-formed in the reaction product **10**. In the next step, **10** plays the role of B$^-$M$^+$ in reaction (2.3.18) and produces another molecule of salt **8**, which is ready for another attack on the imide arrangement **9**. An unusual feature of this chain polymerization is the existence of *two* active groups (the imide and the salt of lactam), which are both present in very small concentrations. Perhaps the two alternating steps of the mechanism [in reactions (2.3.18) and (2.3.19)] occur in a concerted fashion; the salt grouping in **10** is still in a close proximity to the imide. After it transfers its negative charge to an approaching molecule of lactam, the latter can then react with the imide immediately.

In practice, anionic polymerization of caprolactam displays a long induction period during which the imide is formed by a very slow addition of the lactam salt on another lactam molecule that is not activated by the second acyl group. When enough imidic groups have been formed, the reaction proceeds very fast. The induction period can be eliminated by adding either preformed N-acyllactams or acylating agents such as acyl chlorides or anhydrides to the reaction mixture. In the presence of such activators, even the very stable heterocycles such as pyrrolidone (butyrolactam) and piperidone (valerolactam) can be polymerized anionically.

Anionic polymerization of the N-carboxy anhydrides (NCA) of α-amino acids is the only synthetic procedure for the production of homopolypeptides. It is unique among chain polymerizations by virtue of the elimination of a by-product (carbon dioxide) during the process. It is believed that the basic initiator B$^-$(a tertiary amine or a methoxide ion) performs a nucleophilic attack on the carbonyl of the amino acid; carbon dioxide is then eliminated, and the charge shifts to the nitrogen atom.

The resulting nucleophile **11** then attacks other NCA molecules in a propagation reaction.

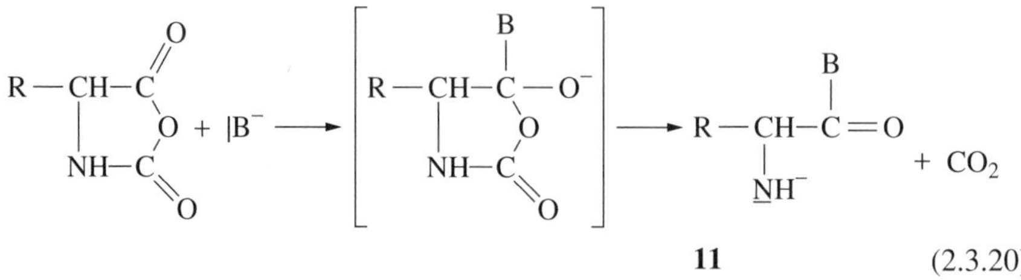

$$(2.3.20)$$

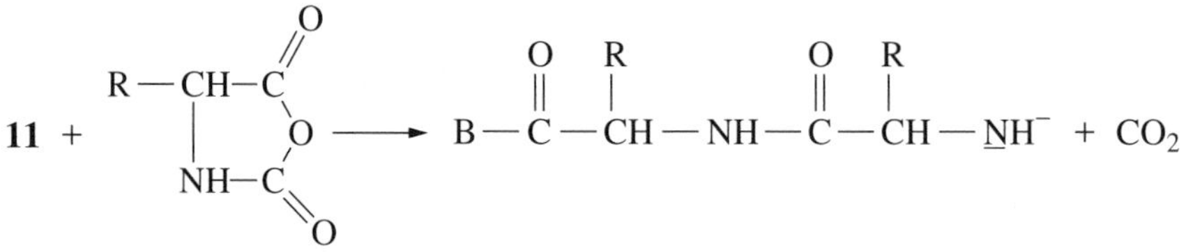

$$(2.3.21)$$

The reaction can yield polypeptides of a quite high molecular weight—up to 1,000,000. The actual course of the reaction may be more complicated, as is evidenced by some anomalies in molecular weight distribution.

2.3.2. Cationic Polymerization

Superficially, the mechanism of cationic polymerization is similar to the anionic mechanism. An initiating cation attaches itself to an olefinic, cyclic, or other monomer with the counteranion in close vicinity. The resulting cationic entity adds more monomers in a propagation reaction until the chain is terminated. However, the mechanisms of chain termination are quite different in these two types of polymerizations. We have seen that in anionic polymerization the counterion, although always in close proximity to the growing end, cannot react with it. At most, it forms a tight ionic pair, which later loosens again and keeps growing. Typically, the process ends with living ends of chains. Chain transfer is rare.

In cationic polymerizations, the counterion often reacts with the chain, terminating it. This usually regenerates the initiator, and a new chain is initiated; that is, the whole episode represents a chain transfer. This transfer process is actually the dominant feature of cationic polymerization. It must be carefully controlled. The control is very demanding and absorbs most of the attention of the experts of cationic polymerization. In particular, water and other impurities must be eliminated to a much higher degree than in other polymerizations. Very low temperatures ($-70°C$) must frequently be employed.

2.3.2.1. Cationic Initiators.   The most common initiating cation is the proton. A free proton cannot exist in condensed media; it is too energetic. It must be attached to some available electron pair. When the attachment is weak, the complex is an acid. The weaker the attachment, the stronger is the acid. The strongest acids known are complexes of compounds with acidic hydrogen (water, alcohols, hydrogen chloride), which are called *coinitiators,* with Lewis acids (*initiators*). Lewis acids are typically compounds with a central atom bound to a number of strongly electronegative atoms or groups (halogens, alkoxides) in such a way that the stable electronic arrangement around the central atom is incomplete. The best-known examples are boron trifluoride and aluminum trichloride with an incomplete sp^3 electronic shell. They eagerly accept into this vacancy any compound with an unbonded electron pair. For example, their reaction with alcohols proceeds as in (2.3.22):

$$
\begin{array}{ccc}
\overset{\displaystyle Cl}{\underset{\displaystyle Cl}{\overset{|}{\underset{|}{Cl-Al}}}} \;+\; \overset{\displaystyle H}{\underset{\displaystyle}{\overset{|}{|O-R}}} & \longrightarrow & \overset{\displaystyle Cl \quad H}{\underset{\displaystyle Cl}{\overset{|\quad|}{\underset{|}{Cl-Al-O-R}}}} \\
& & \mathbf{12}
\end{array}
\qquad (2.3.22)
$$

In **12**, the proton is bound to an oxygen that has three bonds (in an arrangement of bonds similar to that in an oxonium ion, H_3O^+). Moreover, the halogen atoms exert a strong pull on the electrons of the oxygen atom. As a result, the proton is bound extremely weakly; it is a powerful acid and a cationic initiator.

We should remember that the anion $AlCl_3(OR)^-$ will eventually serve as a counterion in our cationic polymerization. Its steric arrangement and proton affinity will strongly influence the polymerization process. Hence, polymer chemists have developed an entire palette of initiators in which other groups, typically alkoxides or alkyls, replace some of the halogens. Similarly, many different coinitiators are used.

Another group of Lewis acids is formed by halides of higher elements in the periodic table. They are capable of forming an sp^3d^2 electronic shell leading to an octahedral arrangement of six ligands around the central atom. If the shell is incomplete, the halides will react with the coinitiators in a way similar to that of scheme (2.3.22). These initiators include $SnCl_4, SbCl_5, TiCl_4, FeCl_3$, and $ZnCl_2$. (For the last ones, a tetrahedral arrangement is also conceivable.)

Some cationic polymerizations may he initiated by protonic acids: H_2SO_4, HCl, $HClO_4$. Recently, triflic acid, CF_3SO_3H, has been used to advantage. It is very strong and is soluble in many organic solvents.

The proton is not the only cation capable of initiating polymerization; alkyl and acyl cations can also do it. These cations are generated from acyl and alkyl perchlorates, tetrafluoroborates, etc. Tetrafluoroborates and similar complexes are formed by an addition of alkyl or acyl fluorides to BF_3:

$$
BF_3 \;+\; RF \;\longrightarrow\; (BF_4)^- R^+
\qquad (2.3.23)
$$

This is, of course, a reaction quite similar to reaction (2.3.22); in this sense, RF is a coinitiator.

2.3.2.2. Polymerization of Olefins.   As in all chain polymerizations, the first step in the process is an addition of the initiating cation R^+ to the olefinic monomer.

$$R^+A^- + \underset{R_2}{\overset{R_1}{\diagdown}}C{=}C\diagup \longrightarrow R{-}\underset{\underset{R_2}{|}}{\overset{\overset{R_1}{|}}{C}}{-}\overset{+}{C}A^- \qquad (2.3.24)$$

13

Then, in a familiar scenario, cation **13** assumes the role of the attacking entity and the chain propagates. Cation **13** is a very energetic species. Accordingly, reaction (2.3.24) is feasible only when the substituents R_1 and R_2 help to spread out the charge and lower the energy. Among such electron-donating groups, the alkoxy group is among the most effective. Thus alkyl vinyl ethers are rather prone to cationic polymerization. A similar effect is achieved by conjugation in dienes and styrene derivatives. (Conjugation serves as an electron reservoir; conjugated monomers are easily polymerized by all three basic mechanisms—radical, anionic, and cationic.) Hyperconjugation of methyl groups with the cationic site is much more effective than with radical sites. Thus α-methylstyrene is more reactive in cationic polymerization than styrene. For isobutylene, cationic polymerization is the only feasible one.

As we have already mentioned, the growing cations (similarly to radicals and unlike anions) are prone to termination and chain transfer. True termination occurs when the polymeric cation recombines either with any nucleophilic impurity or with its own counterion. It should be noted that recombination with the counterion immediately after addition of the first monomer is simply an addition of the initiator onto the double bond—a fairly common reaction in synthetic chemistry. Counterions derived from protonic acids ($HClO_4$, H_2SO_4) recombine directly, whereas the complexes of Lewis acids with coinitiators may eliminate the coinitiator residue, which will recombine with the growing chain. For example,

$$-CH_2{-}\underset{\underset{CH_3}{|}}{\overset{\overset{CH_3}{|}}{\overset{+}{C}}}\cdots[BF_3OH]^- \longrightarrow -CH_2{-}\underset{\underset{CH_3}{|}}{\overset{\overset{CH_3}{|}}{C}}{-}OH + BF_3 \qquad (2.3.25)$$

This type of termination consumes the coinitiator. However, the reaction with the counterion may proceed along a different line: The counterion may extract a β proton from the growing chain, regenerating the initiating species. For example,

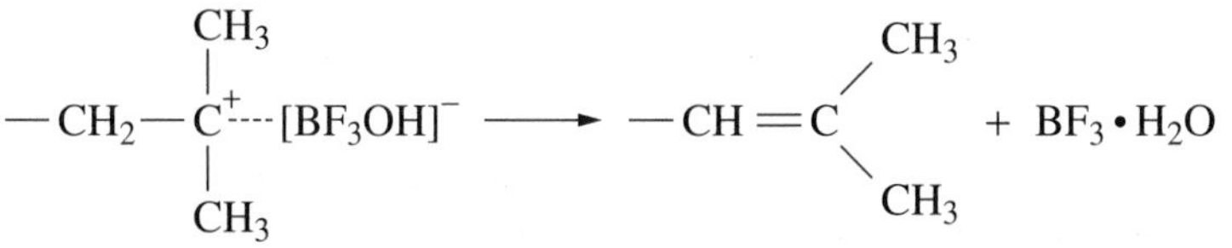

$$
-CH_2-\underset{\underset{CH_3}{|}}{\overset{\overset{CH_3}{|}}{C}}\overset{+}{\cdots}[BF_3OH]^- \longrightarrow -CH=C\overset{CH_3}{\underset{CH_3}{\diagup}} + BF_3 \cdot H_2O \quad (2.3.26)
$$

Carbenium ions are generally very reactive. Among their reactions, we need to mention an extraction of hydride ions, H^-. If the extraction occurs from another molecule, the process is a chain transfer, which is quite similar to the transfer in radical polymerization. Hydride ion can also be extracted from a neighboring location in the growing molecule. This process is actually an isomerization of the growing chain. Then the monomeric unit may not correspond to the parent monomer. A typical example is the cationic polymerization of 3-methyl-1-butene as in reactions (2.3.27).

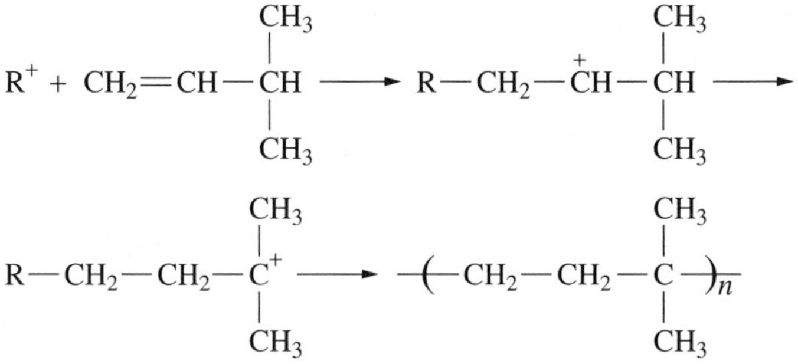

$$
R^+ + CH_2{=}CH-\underset{\underset{CH_3}{|}}{\overset{\overset{CH_3}{|}}{CH}} \longrightarrow R-CH_2-\overset{+}{CH}-\underset{\underset{CH_3}{|}}{\overset{\overset{CH_3}{|}}{CH}} \longrightarrow
$$

$$
R-CH_2-CH_2-\underset{\underset{CH_3}{|}}{\overset{\overset{CH_3}{|}}{\overset{+}{C}}} \longrightarrow \left(\!\!-CH_2-CH_2-\underset{\underset{CH_3}{|}}{\overset{\overset{CH_3}{|}}{C}}\!\!\right)_{\!\!n} \quad (2.3.27)
$$

Cationic polymerization frequently exhibits peculiar kinetics. Let us first analyze an oversimplified kinetic model in which the rate of initiation is slow and the termination is kinetically a first-order reaction. This scenario can be described as in reactions (2.3.28) through (2.3.30).

$$
H^+I^- + M \xrightarrow{k_i} P_1^+I^- \quad (2.3.28)
$$

$$
P_i^+I^- + M \xrightarrow{k_p} P_{i+1}^+I^- \quad (2.3.29)
$$

$$
P_i^+I^- \xrightarrow{k_t} P_i + H^+I^- \quad (2.3.30)
$$

Here, H^+I^- is the initiating complex, M is monomer, and $P_i^+I^-$ and P_i are living and dead polymers of polymerization degree i, respectively. Kinetic equations now read

$$
r_i = k_i [H^+I^-][M] \quad (2.3.31)
$$

$$r_p = k_p \sum_i [\text{P}_i^+\text{I}^-][\text{M}] \qquad (2.3.32)$$

$$r_t = k_t \sum_i [\text{P}_i^+\text{I}^-] \qquad (2.3.33)$$

As usual, r represents the rate of reaction, k the kinetic constants; subscripts i, p, t refer to initiation, propagation, and termination, respectively. With slow initiation, a steady-state kinetics is achieved, when $r_i = r_t$. The combination of equations (2.3.31) through (2.3.33) yields

$$r_p = (k_i k_p / k_t)[\text{H}^+\text{I}^-][\text{M}]^2 \qquad (2.3.34)$$

Hence the rate of polymerization increases with the concentration of the initiating complex and with the square of monomer concentration, in obvious contrast to radical polymerization. Quite often, the activation energy for termination E_t^* is larger than the sum of those for initiation and propagation ($E_i^* + E_p^*$). As a result, although individual chains grow more slowly at lower temperatures, a higher number of them are growing and the overall polymerization rate is increasing. This fact is often described by means of the so-called overall activation energy $E_a^* \equiv E_i^* + E_p^* - E_t^*$, which is negative in this case. (Negative activation energy is an oxymoron; it has no physical significance. Its apparent occurrence originates in an improper application of the concept of activation energy, which is defined for simple reactions only, to a more complex overall reaction.)

In practice, cationic polymerization of olefins (mainly isobutylene) calls for very low temperatures, typically $-80°\text{C}$ with $\text{BF}_3 \cdot \text{H}_2\text{O}$ as an initiator. The low temperatures not only increase the rate of polymerization but also suppress various deleterious effects such as chain transfer and termination by impurities, which would otherwise lower the molecular weight of the polymer substantially. Polymerization at ambient temperatures produces mainly oligomers. This is utilized in manufacturing oligomers of styrene that are used as components of pressure-sensitive adhesives.

The simple kinetic scheme of cationic polymerization described above is just one of several possible schemes. If the initiation reaction is fast, most of the kinetic chains will be started early. The polymerization will have a distinctly nonstationary course; it may even proceed explosively.

The nature of the counterion and of the solvent also play very important roles, as in anionic polymerization. The ion pairs may exist as tight pairs, loose pairs, and free ions. The prevalence of these types depends on the nature of the counterion and on the polarity of the solvent. The more polar the solvent, the greater the number of loose pairs and even free ions that are present. The kinetic constant of propagation increases enormously from tight pairs to loose pairs to free ions. Thus polar solvents are the ones used most often for cationic polymerization; they are the same ones as for the Friedel–Crafts reaction: halogenated hydrocarbons and nitrobenzene.

In another similarity to anionic polymerization, the close proximity of the counterion to the growing end may dictate the steric arrangement of the polymer. Thus cationically polymerized poly(vinyl ethers) are isotactic and crystalline. No doubt the low temperature of the polymerization also helps in producing a sterically pure polymer.

2.3.2.3. Living Cationic Polymerization. The successful development of anionic polymerization into a living process spurred an effort to find conditions that would work for cationic polymerization in a similar way. As in all living polymerizations, the key step was the reversible protection of the growing chain ends that would keep them in a dormant state.

The protection was found in the reaction with the counterion similar to the one that was described in reaction (2.3.25) that terminated the growing chain. However, to keep the chains alive, the terminating group must be removable by the catalyst/counterion still present in the solution. Chloride ions serve the purpose. In a typical example, a reactive alkyl chloride initiator combines with a Lewis acid, producing the initiating cation paired with its counterion as shown in the scheme below.

$$(2.3.35)$$

The growing chain is reversibly terminated by the chloride ions from the Lewis complex and is repeatedly reactivated by the transfer of the chloride back to the catalyst.

Different olefins are converted into cations with a varying ease. Isobutylene requires BF_3 as a catalyst; less active catalysts are needed for styrene and its derivatives; even weaker catalysts are sufficient for vinyl ethers. Accordingly, the strength of the Lewis acids has to be regulated to make the back-and-forth jumping of the chloride ions energetically appropriate. (In other words, the equilibrium between the active and dormant forms should be such as to keep the polymerization rate in the desirable range.) Thus a single isopropoxy ligand on titanium is advantageous for polymerization of styrene, whereas two such ligands are preferred for vinyl ethers.

It is advisable to prevent the growing cation from escaping the clutches of the counterions, that is, of the chloride ion complexed or covalently bonded. This is accomplished by an addition of some other electrolyte that supplies an excess of needed counterions, for example, tetrabutyl ammonium chloride $N(C_4H_9)_4Cl$. An analogous strategy was employed when the separation of the ion pairs in anionic polymerization was suppressed by addition of sodium tetraphenyl borate $NaBPh_4$.

The reader should note the similarity of the above cationic scenario with the ATRP mechanism in the radical chemistry. In the present case, the transition metal capable of binding additional ligands in its sp^3d^2 electronic shells is exchanging chloride anions with the active polymer cation. In the ATRP process a transition metal capable of changing its valence is participating in the shuffling of chloride atoms. In both cases appropriate ligands must be employed to keep the back-and-forth transfers under control.

Iodine chemistry was used for a development of an early initiator/catalyst system $HI/I_2/ZnI_2$ for living cationic polymerization. In this system, hydrogen iodide quantitatively adds to the double bond of the olefin creating the initiating species. However, the polymerization may start only after the addition of the other components of the catalyst. The known willingness of molecular iodine to act as a ligand to a variety of compounds undoubtedly helps the formation of the active center. The resulting catalyst is a mild one and is especially useful for polymerization of vinyl ethers.

Living cationic polymerization is technically rather attractive: it can be performed at ambient or near-ambient conditions under nitrogen atmosphere. Moreover, under selected conditions, many functional groups do not interfere with the polymerization. These groups include esters, ethers, silyl ethers, substituents on aromatic nuclei, electron-rich double bonds of the acrylic or methacrylic type, etc. Consequently, well-defined polymers carrying these groups (or groups into which they can be converted) are accessible by this living technique.

There exists still another path toward materials being developed by polymer engineers. The initiator may carry a reactive group (possibly protected) that after completion of the polymerization could be utilized for subsequent reactions. For example, it may serve for attachment of the polymer to other molecules or solid surfaces. Appropriate groups can be also converted into initiators for other chain-growing reactions (possibly anionic or cationic) and subsequently into block copolymers. The nucleophilic reagents used for the quenching of the original living polymerization may also be exploited as new reactive centers. It seems that living radical polymerization may become a preferred procedure for synthesis of specialty polymers.

2.3.2.4. Polymerization of Aldehydes and Cyclic Monomers.

The oxygen atom in carbonyl and in oxygen-containing heterocycles is a base in the Lewis sense. It can react with cations, and under suitable conditions this may lead to a polymerization.

Formaldehyde is easily polymerized by all types of cationic initiators. However, to obtain high-molecular-weight polyoxymethylene, it is necessary to carry out the reaction at low temperatures, preferably using Lewis acids as initiators. The reaction then proceeds as follows:

$$H_2C{=}O + H^+[BF_3OH]^- \rightarrow H_3C{-}O^+[BF_3OH]^- \qquad (2.3.36)$$

The ceiling temperature of polyoxymethylene is about $127°C$. Thus the polymer must be stabilized. Reacting the hydroxyl ends with acetanhydride and converting them to acetates does this.

Acetaldehyde and higher aldehydes and even acetone can be polymerized under similar conditions at very low temperatures. However, the resulting polymers have very low ceiling temperatures and are thus not stable enough to have any practical significance.

Cationic polymerization of cyclic ethers proceeds through the formation of a cyclic oxonium ion, which serves as the active species. The initiation and propagation steps for the simplest cyclic ether, ethylene oxide, can be described as follows:

$$
\begin{array}{c}
H_2C \\
| \quad \searrow O^+ - H \\
H_2C \quad [BF_3OH]^-
\end{array}
+
O\!\!\!<\!\!\!\begin{array}{c} CH_2 \\ | \\ CH_2 \end{array}
\longrightarrow
HO - CH_2 - CH_2 - O^+\!\!\!<\!\!\!\begin{array}{c} CH_2 \\ | \\ CH_2 \end{array}
\quad [BF_3OH]^-
\tag{2.3.37}
$$

$$
\left(\!\! O - CH_2 - CH_2 \!\!\right)_{\!\!n}\!\! O^+\!\!\!<\!\!\!\begin{array}{c} CH_2 \\ | \\ CH_2 \end{array}
\quad [BF_3OH]^-
+
O\!\!\!<\!\!\!\begin{array}{c} CH_2 \\ | \\ CH_2 \end{array}
\longrightarrow
\left(\!\! O - CH_2 - CH_2 \!\!\right)_{\!\!n+1}\!\! O^+\!\!\!<\!\!\!\begin{array}{c} CH_2 \\ | \\ CH_2 \end{array}
\quad [BF_3OH]^-
$$

$$\tag{2.3.38}$$

Ethylene oxide polymerized in this way forms mostly low-molecular-weight polymer. For industrial production, anionic polymerization is preferred. Under cationic catalysis propylene oxide and epichlorohydrin yield mainly cyclic oligomers (chiefly dimers). Useful polymers can be obtained from these monomers using initiators with less pronounced basic properties, such as Et_3Al-H_2O-acetyl acetone. Such initiators, however, belong more to the realm of coordination polymerization.

Cationic initiators are able to polymerize even monomers with polymerization enthalpy as low as that of tetrahydrofuran ($\sim$3 kcal/mol). This polymerization requires the most rigorous drying of the monomer (distillation over sodium mirror) and phosphorus pentafluoride as an initiator.

An industrially important cationic polymerization is that of trioxane (cyclic trimer of formaldehyde) to polyoxymethylene. The reaction can be initiated by almost any Lewis acid and yields highly crystalline polymers of high molecular weight. Interestingly, the reaction has an induction period that can be eliminated by adding a trace of formaldehyde.

Cyclic trimer and tetramer of dimethylsiloxane are also polymerized by cationic initiators, preferably by triflic acid, CF_3SO_3H. However, the chain transfer (read "transfer of the initiating entity from one molecule to the next") is so frequent that this reaction is usually considered to be an acid-catalyzed polycondensation (see Section 2.1.5).

2.3.3. Coordination Polymerization

In 1955 Ziegler discovered that triethylaluminum, $AlEt_3$, reacts with titanium tetrachloride, $TiCl_4$, in a complex reaction leading to a deeply colored heterogeneous system. This brew was a powerful initiator of the polymerization of ethylene, yielding a new highly crystalline type of polyethylene that was later shown to be strictly linear. Soon afterwards, Natta found that this initiator also polymerizes propylene (which had resisted all attempts at polymerization until then). Moreover, the resulting polymer was highly crystalline. A new class of isotactic polymers was discovered.

The following years were marked by an explosive growth of studies on and uses for the new initiators. The original *Ziegler–Natta initiators* are applicable for polymerization of vinyl compounds, mainly the less polar ones. Other types of initiators revolutionized the polymerization of epoxides and vinyl ethers. The common feature of these initiators is their intermediate standing between anionic and cationic initiators; the active complex is not ionized, not even to a tight ion pair. The monomer is first coordinated to the active complex; then in a concerted rearrangement reaction, it is incorporated into the growing chain.

2.3.3.1. Ziegler–Natta Initiators.

Ziegler–Natta initiators are usually prepared in situ by reacting a halogenide of a transition metal with an organometallic reagent. Among the former, titanium compounds, including $TiCl_4$, several crystalline modifications of $TiCl_3$, $TiBr_4$, and $TiCl_2Et_2$, are most popular. Useful halogenides of other transition metals include VCl_3, VCl_4, $VOCl_3$, $CrCl_3$, and $ZrCl_4$. The organometallic compounds are mainly aluminum trialkyls, AlR_3; aluminum dialkyl halides, AlR_2X (X = Cl, Br); and aluminum alkyl dihalides, $AlRX_2$. The alkyls are usually ethyls or isobutyls.

In no case is the reaction between the two components a simple one. Let us briefly consider the main features of the prototypical reaction, that of $TiCl_4$ with AlR_3. The initial step consists of alkylation of the transition metal:

$$AlR_3 + TiCl_4 \longrightarrow AlR_2Cl + RTiCl_3 \qquad (2.3.39)$$

Several alkyls can be bound to the titanium atom. Alkylated transition metals are unstable; they decompose spontaneously, with the evolution of various hydrocarbon fragments, to compounds with lower valence of the transition metal—typically trivalent titanium. The presence of this lower-valence compound is a prerequisite for a successful initiation. (In an alternative procedure, $TiCl_3$ may be used as the starting material.) Although $TiCl_4$ is a liquid easily miscible with hydrocarbon solvents, compounds of trivalent titanium are usually insoluble, intensely colored, crystalline materials. The catalyst is a semicrystalline mass consisting mainly of $TiCl_3$ with a fair amount of alkyl groups, especially on the surface of the microcrystals. Aluminum alkylated to various degrees is present in the mass, on its surface, and in the surrounding liquid.

In $TiCl_3$ crystals, titanium atoms are surrounded by six chlorine atoms in an octahedral arrangement. Obviously, the d orbitals of titanium participate in the bonding. The catalytically active site is on the surface of the crystal. Only four of the coordination sites of the active titanium atom are occupied by the lattice chlorines; one site binds an alkyl group, and the last one is vacant. During the polymerization, a molecule of alkene (ethylene) is coordinated to titanium at this vacant position. An overlap of the d orbital with the π electrons of the alkene is believed to contribute to this binding. Then in a rearrangement reaction, the π electrons form a true bond to titanium and the alkyl switches to the other end of the double bond.

$$\qquad (2.3.40)$$

The coordination site originally occupied by the alkyl is vacant after the rearrangement; it can coordinate the next molecule of alkene, and the chain can propagate. The whole process is called *coordination polymerization* because of the coordination step. Some researchers, however, prefer to stress that the newly attached monomer inserts into the metal-alkyl bond and call the process *insertion polymerization*.

There are two schools of thought about the detailed mechanism of the polymerization. The scenario just described includes only the transition metal atoms in the mechanism. Accordingly, we are talking about a *monometallic* mechanism. The other school assumes that the aluminum complex is always in the vicinity of the reaction site and helps in the coordination and rearrangement. Such a mechanism is called *bimetallic*.

It should be obvious that the reaction site is very crowded. Frequently, when a substituted ethylene approaches the site, it can do so in only one way. Stereospecific polymerization then results. Because of the heterogeneous nature of the initiator, several types of active sites are often present within the same initiator preparation. These types differ in the steric arrangement around the active titanium atom and produce chains with different stereoregularity. Thus the original $TiCl_4$-$AlEt_3$ initiator yields a mixture of atactic polypropylene that is soluble in cold heptane and isotactic polypropylene that is insoluble in hot heptane. (There is also a small fraction that is soluble in hot heptane but insoluble in the cold solvent; it is believed to be a stereoblock polymer.) It is rationalized that some less crowded catalytic sites produce the atactic (useless) form; the more crowded sites yield the valuable isotactic polymer. The less crowded sites are more reactive; it is possible to poison them selectively by adding a small amount of a compound such as water, pyridine, ether, or amine. These additives lower the polymerization rate but increase the yield of the isotactic polymer.

Chain transfer is a common reaction in coordination polymerization and can proceed by several mechanisms. We will mention two of them. Alkylaluminum compounds, which are present on the surface of the initiator as well as in the solution, are capable of exchanging their alkyls with the alkyls (actually growing chains) on the titanium atoms. The polymer chain, after being transferred to aluminum, stops growing and may even leave the surface of the initiator. Of course, the polymerization proceeds with the new alkyl, which was transferred to titanium. The transfer process may be repeated. The chain traveling with its aluminum complex can be reattached to another active polymerization center and continue growing. If this center happens to have a different stereospecificity than the original one, a molecule of a stereoblock polymer is formed.

Molecular hydrogen is a powerful transfer agent. In a reaction the mechanism of which is still not fully understood (does it coordinate first with the vacant orbitals of aluminum or titanium?), it terminates the growing chain and starts another one. The reaction is used for controlling the molecular weight of the polymer.

Besides α-olefins, Ziegler–Natta initiators are capable of polymerizing dienes (butadiene, isoprene) and cycloolefins (cyclobutene **14**, cyclopentene **15**, norbornene **16**).

$$
\begin{array}{ccc}
\underset{\underset{\displaystyle CH_2-CH_2}{\mid\quad\mid}}{CH=CH}
&
\underset{\underset{\displaystyle CH_2}{\diagdown\;\diagup}}{\underset{\displaystyle CH_2\qquad CH_2}{\diagup\qquad\diagdown}}{CH=CH}
&
\underset{\underset{\displaystyle CH_2-CH_2}{CH\diagdown\quad\diagup CH}}{\underset{CH_2}{\diagup\qquad\diagdown}}{CH=CH}
\end{array}
$$

$$
\textbf{14} \qquad\qquad \textbf{15} \qquad\qquad \textbf{16}
$$

The diene monomers can be incorporated into polymer chains either as 1,2 units or 1,4 units in the *trans* or *cis* configuration. The cycloolefins can be polymerized either by ring opening (cyclobutene yielding polybutadiene) or as polycyclobutene **17**.

$$
\underset{\underset{\displaystyle CH_2-CH_2}{\mid\quad\mid}}{CH=CH}
\begin{cases}
\nearrow & -CH_2-CH=CH-CH_2- \\
\\
\searrow & \underset{\underset{\displaystyle CH_2-CH_2}{\mid\quad\mid}}{-CH-CH-}
\end{cases}
\qquad (2.3.41)
$$

$$
\textbf{17}
$$

Initiator systems have been developed that give any of the above products in relatively high purity. The strategy consists of first finding a system capable of yielding the required polymer and then selectively suppressing the competing reactions by appropriate additives. The stereospecific initiators are mostly heterogeneous; this implies that their structure is often not known in much detail. As a consequence, preparation of the best initiating systems is both an art and a closely held secret of commercial manufacturers.

2.3.3.2. Metallocene Catalysts.

Introduction of the Ziegler–Natta catalysts opened an era of custom-made polyolefins. However, researchers were still not quite satisfied with the procedure: The heterogeneous nature of the process led to broad polydispersities and to products heterogeneous in their tacticity. Homogeneous soluble catalysts were expected to overcome these problems. The quest for such catalysts was long; finally, metallocenes filled the bill.

Metallocenes are compounds based on a relatively late entry into the family of aromatic molecules. Cyclopentadiene has a slightly acidic hydrogen on its only methylene group. Its dissociation yields a perfectly symmetrical five-member ring that possesses six π electrons. It differs from benzene—its six-member ring analog—by carrying a negative charge that is spread equally over the ring. This anion is perfectly willing to become a tightly bound ligand of cations of transition metals. In fact, two rings often sandwich the cation between them. Thus divalent cations form very stable neutral compounds—notably divalent iron forms the first known member of this family—ferrocene.

Multivalent cations bonded to two cyclopentadienyl rings—these are our *metallocenes*—still carry on the metal positive charges that have to be accompanied by counterions. Small counterions—typically chloride ions—actually form covalent bonds with the metal. This leads to neutral species that are soluble in hydrocarbons. Metallocenes of the fourth group—Ti, Zr, Hf—are the preferred catalysts for olefin polymerization, Zr being used most. Sterically, these compounds resemble an open jaw within which are the smaller ligands of the metal. Zirconocene dichloride **18** is a prime example.

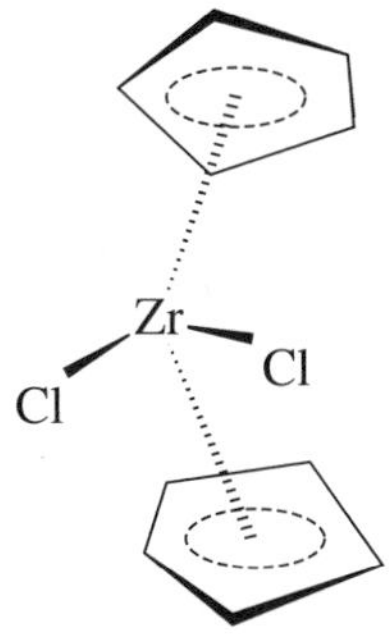

18

Reaction of zirconocene dichloride **18** with organometallic compounds leads to an exchange of chlorines with the organic group, typically with a methyl. One or both chlorines may be exchanged. The resulting compound is electron-rich thanks to the cyclopentadienyl rings and is willing to release one of the methyls (or the remaining chlorine) to compounds with an incomplete electronic shell. This produces an ion pair where zirconium acts as the cation and the other compound as the anion. In this configuration one of the coordination positions on the zirconium is empty. It was expected that this compound would coordinate olefins into this position and that the Ziegler–Natta polymerization would follow. However, the reaction was extremely sluggish when it occurred at all. It was found that the problem was in ion pairing: The counterions were so close and were loosened so rarely that the olefin had no chance of entering into play.

The situation could be remedied by using very large counterions or counterions extremely unwilling to share electrons even if only in ion pairing. Reaction of zirconocene dichloride **18** with MAO solved the problem. MAO (methyl aluminoxane) is a compound prepared by very careful partial hydrolysis of trimethyl aluminum. It is believed that its structure is mainly a linear or cyclic arrangement of $OAlCH_3$ groups, e.g., **19**. Larger clusters are probably also present.

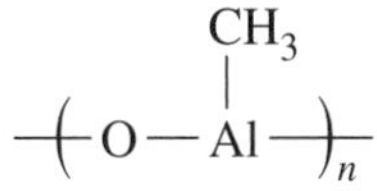

19

MAO serves multiple purposes. At the beginning, it is the organometallic compound exchanging chloride ions for methyl anions. Then, its coordinately unsaturated aluminum atoms are abstracting the methyl anions and creating the electronic vacancies on zirconium. Finally, in its very large anions the negative charge is so diluted that the zirconium-counterion attraction is relatively weak. However, the competition for the chloride or methyl anion between the zirconium and MAO is still stacked in favor of the metal. It takes a huge excess of MAO (the optimum atomic ratio Al:Zr is 1000 or higher) to produce a sufficient number of activated metallocenes for achieving an effective polymerization. But effective this polymerization is! One mole of the catalyst can produce hundreds of tons of the polymer. Each reaction center can add thousands of olefin molecules per second. This rate is comparable to very fast enzymatic reactions.

An example of a catalyst extremely unwilling to share electrons is trityl tetrakis (perfluorophenyl) borate, $[Ph_3C]^{(+)}[B(C_6F_5)_4]^{(-)}$. It abstracts a methyl anion from dimethylzirconocene, forming triphenylmethylmethane as a by-product. The perfluorated anion is then associated with the zirconium cation weakly enough to allow the polymerization to proceed.

The mechanism of the polymerization is the familiar one. The π electrons of the olefin fill the electron vacancy on zirconium. Then, in a coordinated electron shift accompanied by a small steric shift, the olefin forms a bond with zirconium and the other substituent on zirconium (first methyl, later the growing chain) attaches itself to the other end of the newly entering olefin. The electron vacancy is now shifted to the other zirconium orbital in the open cyclopentadienyl jaw. The reaction proceeds without any serious side products toward a high-molecular-weight, strictly linear polymer having much narrower polydispersity than analogous Ziegler–Natta products.

The above scenario is perfect for polymerization of ethylene as it produces highly crystalline well-defined polymers. However, polymerization of higher olefins—notably of propylene—yields atactic polymers that are generally undesirable. The quest for stereospecific catalysts had to continue. It was found that it was necessary to attach other structures to the cyclopentadienyl rings to allow the olefin to approach the zirconium only in some specific orientation. Very useful structure was discovered when the cyclopentadienyl anion in the zirconocene was replaced by the anion derived from phenylindene leading to a compound, **20**.

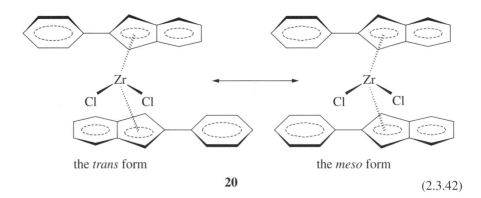

the *trans* form   the *meso* form

20

(2.3.42)

The complex **20** can exist in two conformations, *racemic* (*trans*) and *meso*. The transition between them is sterically hindered and proceeds with a low frequency. In our depiction, the zirconocene jaw is opened forward and the chlorine atoms point also forward. During the polymerization process, the chlorines are replaced by the growing chain and the coordinated olefin molecule as described above. As the individual olefin molecules are incorporated into the polymer, the reactive empty orbital switches in a pendulum-like fashion from left to right and back.

The polymerization proceeds with the two conformations of the catalyst in different ways. The indenyl groups exert the main steric influence. When the catalyst is in the racemic form, the propylene molecule approaches the catalytic center in such a way that the methyl group is pointing away from the indenyl group. Then the active center switches to the other side of the molecule, where the other indenyl group exerts the same effect. The result is an isotactic chain. With the catalyst in the *meso* form, on one side the two indenyl groups repel the methyl group of the propylene molecule equally. Thus it has a choice of orientation. On the other side, there are no indenyl groups and the free choice still prevails. The result is an atactic chain. However, as we have mentioned, the two conformations switch back and forth. The result is a stereoblock polymer. Manipulating the reaction conditions can influence the length of the blocks. This polymer is a valuable elastomer. The crystallites formed by the isotactic segments contribute to its toughness, whereas the atactic segments are responsible for the elasticity.

To obtain catalysts that would yield isotactic polypropylene without formation of the atactic blocks it was necessary to restrict the switching of the two forms of the catalyst. This was achieved by bridging the cyclopentadienyl moieties, thus preventing their mutual rotation when coordinated to the central metal of the metallocene. (The rotation is still possible when the bonding is loosened, for example, photochemically. This is exploited when the complexes are synthesized by reacting $ZrCl_4$ with the parent biscyclopentadiene compound. In this reaction the *meso* form is usually preferred kinetically; the irradiation converts it to the thermodynamically preferred racemic form.)

The cyclopentadienyl groups are tied together by a *handle* (either a dimethylsilyl group or, more frequently, an ethylene group). The bridged compounds are called *ansa*-metallocenes. (The Latin word *ansa* means *bent handle*.) A typical example is *rac*-ethylene-bis(1-indenyl)zirconium dichloride **21**.

21

Compound **21** catalyzes the polymerization of propylene in the same way as its *racemic* unbridged analog **20** did. However, no change toward the *meso* form is possible and the whole polymer molecule is isotactic except for a few improperly incorporated monomer units. Apparently, the control exerted by the six-member-ring part of the indene moiety is not perfect and occasionally allows an approach of an improperly oriented monomer molecule. The control may be improved by an introduction of some strategically placed group onto the cyclopentadienyl ring.

Once the principle of the stereochemical control was understood, *ansa*-metallocenes were designed that produced other stereotactic forms of the polymer. For example, isopropylidene(η^5-cyclopentadienyl-η^3-fluorenyl)zirconium dichloride **22** is a precatalyst leading to syndiotactic polymers.

22

The mechanism is easy to imagine, remembering that the methyl group of the propylene being incorporated is nudged to point away from the six-membered ring. When racemic forms of **20** or **21** were used as catalysts, the two relevant six-membered rings were on the opposite sides of the metallocene dissecting plane. This orientation, combined with the pendulum-like oscillation of the active orbital, led to the isotactic polymer. With **22** as a catalyst, the two rings are on the same side of the plane incorporating every other monomer in exactly the opposite way than the isotactic catalysts would. Of course, this implies a syndiotactic product. It is worthwhile to mention that this was the first practical procedure leading to syndiotactic polyolefins.

Although the active center keeps producing the polymer as long as the monomer is available, the polymerization is not a living one because chain transfer terminates individual polymer chains. This transfer is intimately related to the polymerization process itself. It is believed that the energy of the transition state in the process of incorporating the coordinated olefin molecule is lowered by the formation of the so-called *agostic* bond between the positively charged zirconium and α or β hydrogen of the growing chain. The agostic bond is a three-center two-electron bond between carbon, hydrogen, and some electron-deficient species (in this case zirconium). Occasionally, zirconium becomes electron-hungry and appropriates the hydrogen atom with both electrons (i.e., a hydride anion) for itself. The robbed polymer chain rearranges itself into an olefin and leaves the active center while the growth is transferred to the hydride ion.

The above-described mechanism of chain termination implies that the number-average degree of polymerization is equal to the ratio of rates of monomer incorporation and chain transfer. Moreover, after each monomer addition the probability of chain termination is equal to the inverse of this ratio. The situation is fully analogous to the situation described in Section 2.2.4.2, where molecular weight distribution was evaluated for polymers synthesized by radical polymerization (assuming termination by disproportionation). In both cases, the ratio $\overline{M}_w/\overline{M}_n$ is equal to two. (Of course, the above applies only to polymer produced in a short time interval. If the polymerization conditions are changing with progressing conversion, then molecular weight and its distribution are changing as well.)

When ethylene is polymerized, the rate of propagation is very high, the transfer occurs only after a very large number of monomer additions, and the molecular weight is quite high. It may be reduced by *hydrogenolysis,* that is, by addition of molecular hydrogen to the reaction mixture. We may imagine that the hydrogen molecule coordinates to the catalyst in a similar way as the olefin would. Its incorporation actually terminates the chain (producing a saturated chain end) and starts a new one from the hydride anion. A similar control by hydrogen is routinely applied in Ziegler–Natta polymerization.

Polymerization of propylene has a much smaller rate of propagation, and the transfer of β hydrogen has much better chance to play its role. Accordingly, the resulting molecular weight of the polymer is relatively low. However, polymerization at lower temperature produces polymers with higher molecular weight. It is somewhat surprising that the more sterically restricted *ansa*-metallocenes that yield isotactic polymers exhibit a smaller difference between polymerization rates of ethylene and propylene than do the simpler metallocenes that yield atactic polymers. Seemingly, the steric restrictions are suppressing the hydrogen transfer to some degree. As a result, the molecular weight of the polypropylene is higher with the more restrictive catalysts.

Metallocenes are well suited for copolymerization of olefins. Although the simple metallocenes incorporate ethylene more readily than propylene, the *ansa*-metallocenes produce essentially statistical copolymers having a composition very close to the composition of the feed. These copolymers are of considerable industrial interest. When the percentage of the propylene (or another higher olefin) is low, the product is *linear low-density polyethylene* (LLDPE) that has only short branches. It is more desirable than the *low-density polyethylene* (LDPE) prepared by high-temperature radical polymerization, the long branches of which, together with its large polydispersity, require more complicated technology of fabrication of the final products. Copolymers with the percentages of ethylene and propylene about equal are valuable elastomers.

Metallocenes also polymerize cyclic olefins, notably cyclopentene and norbornene. In these reactions, the double bond is opened but the ring structure remains intact. Polymerization of cyclopentene proceeds readily, but a complicated rearrangement at the active center after each addition leads to a 1,3-enchainment instead of the expected 1,2-enchainment (2.3.43). However, polynorbornene contains the expected 1,2-enchainment (2.3.44).

$$(2.3.43)$$

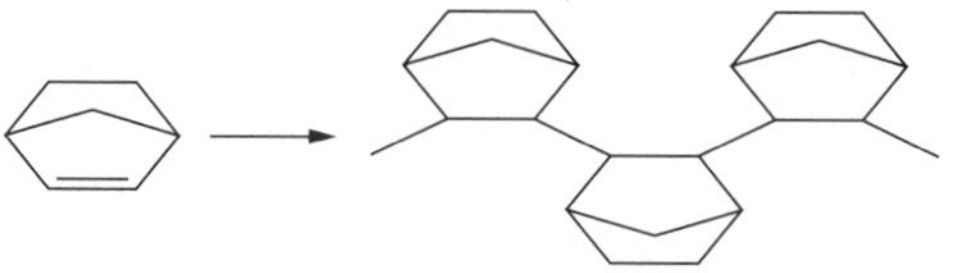

(2.3.44)

The homopolymers of cycloolefines have extremely high melting points (they usually melt with decomposition) and are therefore technically useless. However, the copolymers with ethylene that have a very high percentage of cyclic units (the cyclopentane ring is 1,2-enchained in this case) are easily prepared. They are amorphous and have very high glass transition temperature.

The chemistry of metallocene polymerization is undergoing a rapid development, and new catalysts with new properties and capabilities are emerging with high frequency.

2.3.3.3. *Metathesis Polymerization.* Many transition metals can coordinate π electrons of olefins and acetylenes in their coordination sphere, setting the stage for synthetic reactions of many types. The metathesis reaction has a very special mechanism and multiple uses. Several of them are employed for synthesis of polymers. The generic reaction may be described as follows.

$$\begin{array}{cc} R_1 & R_2 \\ | & | \\ CH= & CH \\ & \\ CH= & CH \\ | & | \\ R_3 & R_4 \end{array} \xrightarrow{\text{Catalyst}} \begin{array}{c} R_1 \\ | \\ CH \\ \| \\ CH \\ | \\ R_3 \end{array} + \begin{array}{c} R_2 \\ | \\ CH \\ \| \\ CH \\ | \\ R_4 \end{array} \qquad (2.3.45)$$

Two molecules of olefin are split across their strongest (double) bond and recombined in a different way. For example, two molecules of propylene may lead to a molecule of ethylene and a molecule of 2-butylene. It was soon found out that the reaction is more complicated. The key factor is the catalyst in which the transition metal (usually W or Mo, but also Ti, Ta, Nb, Rh) is double bonded to a carbene moiety =CHR and, at the same time, has an electron pair vacancy into which a molecule of olefin can coordinate.

The early metathesis catalysts were prepared in situ, usually by a reaction of a transition metal salt (e.g., WCl_6) with an alkylating agent such as $(RCH_2)_3Al$. This reaction is sometimes catalyzed by some Lewis acid; it may proceed slowly (meaning a long induction period for the methathesis reaction itself), and the exact nature of the catalyst is not known with any degree of certainty. More modern catalysts are prepared by sophisticated syntheses, carry various ligands on the central metal, have a well-defined structure, and lead to better-defined products.

The metathesis reaction starts by the coordination of the olefin molecule into the metal-carbene complex. The two double-bonded moieties then combine to form a metalacyclobutane. After this, the four-member ring separates again into two double-bonded molecules. This either restores the original situation or exchanges the carbene moiety for one-half of the olefin. (In the reaction scheme below, M represents the metal, Ln the ligands, and R_1, R_2, and R_3 alkyl groups.)

$$Ln_4M = CHR_1 + R_2CH = CHR_3 \longrightarrow \begin{matrix} Ln_4M - CHR_1 \\ | \quad\quad | \\ R_2CH - CHR_3 \end{matrix} \longrightarrow \begin{matrix} Ln_4M \\ || \end{matrix} \begin{matrix} CHR_1 \\ || \end{matrix} + \begin{matrix} CHR_1 \\ || \\ CHR_3 \end{matrix}$$

$$\begin{matrix} R_2CH \quad CHR_3 \end{matrix}$$

(2.3.46)

The reaction products are again olefins, and they participate in the metathesis again and again. When two different olefins are reacted together, the product consists of a mixture of all possible combinations of the original "half-molecules" connected by a double bond. Moreover, each olefin will be a mixture of *cis* and *trans* isomers. Eventually, an equilibrium mixture will ensue. The metathesis reaction among linear molecules changes neither the number of double bonds in the system nor the number of molecules. Hence, the change of Gibbs energy for all the reactions involved is small and the equilibrium mixture contains all the species in comparable amounts. This may necessitate extensive separations.

The basic metathesis reaction between two olefin molecules leads to another set of two molecules and does not yield any polymer. However, when the halves of the olefin are bonded together also by other means than the double bond, they will stay together even after the scission of the double bond. This may lead to a polymer. Cycloolefins and acetylenes fit the above pattern. Moreover, some nonconjugated dienes may also be polymerized by the metathesis mechanism. In the following, we will treat these three types of reactions.

Ring opening metathesis polymerization (ROMP) is used for polymerization of cyclic or multicyclic molecules with a double bond within the rings. Typical are polymerizations of cyclopentene (leading to polypentenamer) and norbornene [see reactions (2.3.47) and (2.3.48)]. Note that these polymers are completely different from polymers prepared from the same monomers using metallocene catalysts. Specifically, they contain isolated double bonds in the main chain. This feature predestines them as parts of rubber structures as well as intermediates for various functionalized polymers.

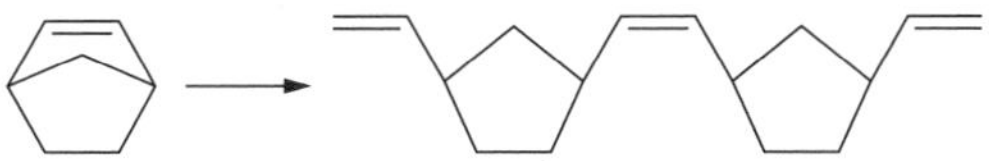

(2.3.47)

(2.3.48)

The mechanism of ROMP is as follows. The carbene-containing catalyst reacts with the double bond of the cycloolefin as described above. The metathesis attaches the original carbene group to one end of the now-split ring, thus forming the beginning of the polymer chain. The other end of the split ring is now converted to the metalacarbene, that is, back to the catalytic moiety that can cause the next addition. Further progress of the reaction is different for simple cycloolefins and the strained monomers of the norbornene type.

For cycloolefins the driving force of the polymerization, which did not exist for linear olefins, is the release of the strain of the cyclic molecule. Linear polymers are thermodynamically preferred to cyclic moieties. We have already seen (Sections 2.1.1.2 and 2.1.1.4) that 3-, 4-, and 8- to 14-member rings are at enthalpic disadvantage, whereas the disadvantage of the large rings has entropic origins. Consequently, polymers made from cycloolefins with the above numbers of ring members contain mainly linear molecules even at equilibrium.

However, the strain of five- and seven-member rings is only minor (six-member rings generally resist polymerization). Thus, with cyclopentene and cycloheptene, the reversible nature of the metathesis may play fully its role. Specifically, the catalyst may react with the inner double bond in the polymer, causing its scission. Polymer chain with an active end may react with another chain, resulting in an exchange of polymer segments of different lengths. Or the active end may react with a double bond of its own chain and split off a cyclic molecule. The reaction pattern is fully analogous to the one imparted to polycondensations by transesterification and transamidation reactions. At equilibrium, the distribution of molecular weights of the linear polymers is the most probable one, the amount of the cyclic monomer species is significant, and the larger cyclic molecules are also present.

The strain released by ring opening of norbornene is rather large. Consequently, the reverse reaction leading back to monomer does not play any role. Moreover, the reaction of the active moiety with the strained monomer is so much easier than its reaction with the less reactive inner double bonds of the polymer that it is possible to suppress the latter reaction completely by using specially designed, less reactive catalysts. The reaction may be even forced to become a living one. Catalysts were developed that led to various regular stereo-arrangements of the polymer backbone.

This development spurred the synthesis of advanced catalysts that allowed polymerization of norbornenes carrying substituents that would deactivate the catalysts of earlier generations. A fortuitous discovery that ruthenium metathesis catalysts were not poisoned by water but that their effectiveness was actually enhanced by its presence led to the development of water soluble catalysts, in which the water molecules were ligands of ruthenium. This widely expanded the applicability of the ROMP.

When the metathesis scission opens two bonds of the triple-bonded acetylenes, the ensuing fragments are still held together by the third bond. This opens the route to the synthesis of polyacetylenes. Acetylenes can be polymerized both by Ziegler–Natta catalysts and metathesis catalysts. It was found that the Ziegler–Natta catalysts produce polymers with a double bond located where the original triple bond used to be. They polymerize acetylene and its single substituted derivatives, preferably with less crowded substituents.

The metathesis catalysts convert the triple bond into a single one. Catalysts derived from WCl_6, $MoCl_5$, $W(CO)_6$, etc. combined with $Sn(CH_3)_4$ or similar compounds are used most often. They polymerize even the more crowded monomers. However, for the most crowded acetylenes, the best catalysts are based on niobium and tantalum.

Polyacetylenes have an extensive system of conjugated double bonds, which confer to them many interesting properties. They can be prepared in different forms depending on the proportion and organization of the *cis* and *trans* isomers. The *cis* configuration is usually preferred kinetically during the synthesis, whereas the *trans* form is more stable thermodynamically. The parent polyacetylene is a black powder unstable in air and insoluble in any solvent. However, it can be manufactured into freestanding films that after doping acquire semiconductive, sometimes even conductive properties. On the other hand, the substituted polyacetylenes are soluble in common solvents, are electrical insulators, and are quite stable; some of them, poly(diphenylacetylene), quite remarkably so. Their stiff main chain and bulky substituents provide them with a rigid structure combined with a substantial porosity, making them excellent materials for high gas permeability.

Acyclic diene metathesis polymerization (ADMET) produces polymers from linear α,ω dienes by a sequence of reactions that is formally identical with polycondensation.

$$H_2C=CH-(CH_2)_n-CH=CH_2 + H_2C=CH-(CH_2)_n-CH=CH_2 \longrightarrow$$
$$H_2C=CH-(CH_2)_n-CH=CH-(CH_2)_n-CH=CH_2 + H_2C=CH_2$$

$$(2.3.49)$$

Repetition of this reaction leads to a high polymer. The actual reaction is catalyzed by most ROMP catalysts and consists of a number of steps. During the reaction, the catalytic carbene centers carry either a methylene group or the carbene-bonded chain. The polymer molecules in the solution have two terminal methylene groups and several (later on, many) internal double bond. We need to consider reactions of both types of carbenes with both types of double bonds.

1. When the carbene carrying the chain reacts with the terminal double bond, the two chains combine (this is the growth process) and the catalytic center is converted into a methylene-carrying type.

2. When the methylene-carrying center reacts with the terminal double bond, the two methylene groups either exchange their place (no observable reaction) or they combine to form ethylene. The chain is transferred to the catalytic center. The reaction is usually performed under high vacuum that removes ethylene from the system. This is the process that drives the reaction to conclusion.

3. When the chain-carrying center reacts with an internal double bond, the reaction is unproductive. Its only effect is scrambling the chain lengths of the chains involved.

4. Finally, when the methylene center reacts with an internal double bond of a polymer, the chain is split into two parts; one of them produces a new terminal methylene, and the other one is attached to the catalyst. This reaction is also unproductive; it is actually counterproductive.
5. It is also possible that the active group reacts with an internal bond of the chain it is itself carrying. This represents an excision of a cyclic molecule.

As the reaction progresses with the continuous removal of ethylene, the number of terminal methylene groups decreases with the same (slow) kinetics that prevailed in the classical condensation reactions. The metathesis-mediated interchange of segments continues, leading to an equilibrium distribution of lengths of linear and cyclic molecules as is demanded by thermodynamics. This final equilibration is the same process that governed the final stages of the ROMP polymerization of cyclopentane.

The ADMET process has been applied to 1,5-hexadiene and to 1,9-decadiene, leading to polybutadiene and polyoctenamer, respectively. Copolymers of these two monomers were also prepared. However, the ADMET polymerization still awaits a major industrial application.

2.3.3.4. Group Transfer Polymerization (GTP).

Group transfer polymerization is based on special chemistry of ketene silyl acetals. Ketene acetals have a general formula **23**:

$$
\begin{array}{c}
R_1 \qquad\qquad O-R_3 \\
\diagdown\quad\qquad \diagup \\
C=C \\
\diagup\qquad\quad \diagdown \\
R_2 \qquad\qquad O-R_4
\end{array}
$$

23

In ketene silyl acetals the R_4 group is $-Si(R_5)_3$, usually $-Si(CH_3)_3$. Ketene acetals are actually derivatives of carboxylic acids, as is apparent from a hypothetical addition of water on the carbon-carbon bond. GTP is essentially the Michael addition of a nucleophilic reagent on a carbonyl conjugated with a double bond—in the present case onto methacrylic or acrylic esters. The nucleophilic reagent is our ketene silyl acetal that has been activated by a nucleophilic catalyst [e.g., tris(dimethylamino) sulfonium hydrofluoride, $(Me_2N)_3S^+HF_2^-$, or some other onium salt], which temporarily attaches itself to the silyl group and thus shifts the electrons toward the distant carbon of the ketene double bond. This nucleophilic carbon then attaches itself in the Michael reaction to the β carbon of the methacrylate. The transfer of the silyl group to the carbonyl of the methacrylate then completes the addition, transforming the ester group again into a ketene silyl acetal. The reaction then proceeds toward the polymer in a typical living manner.

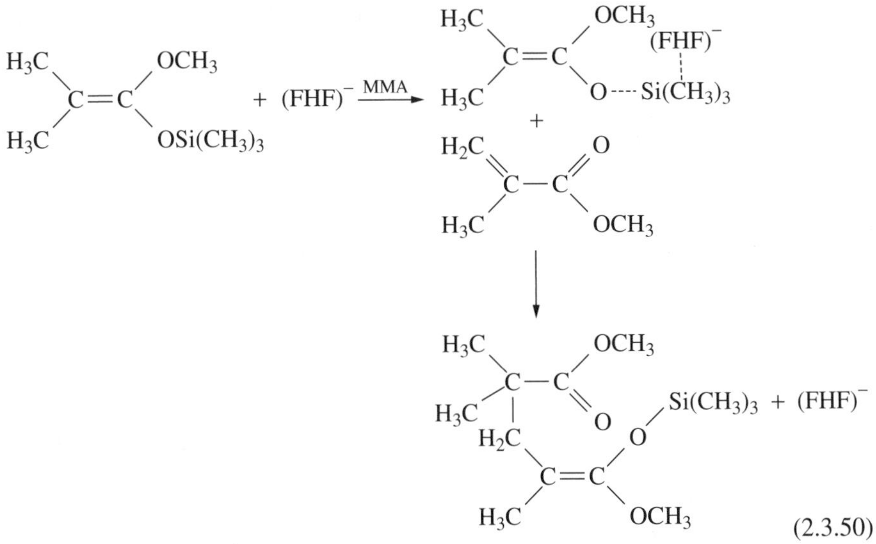

$$\text{(2.3.50)}$$

Interestingly, Lewis acids, electrophilic catalysts having a nature opposite to that of the nucleophilic ones, can perform the same feat. Weaker catalysts of this class are employed for best results: $ZnCl_2$, $SnCl_4$, R_2AlCl. With these catalysts, the reaction probably starts by the attachment of the catalyst to the carbonyl oxygen of the acrylate, pulling away electrons from its conjugated bonds and giving the β carbon a partially cationic character. Addition to the double bond of the ketene silyl acetal follows, and the —$Si(CH_3)_3$ group is kicked off as a cation (this converts the ketene acetal arrangement into an ester) and replaces the catalyst at the other end of the now joint molecule. Of course, this replacement recreates the ketene silyl acetal where the original ester group was. The freed catalysts than activate another molecule of the acrylate, and the polymerization proceeds.

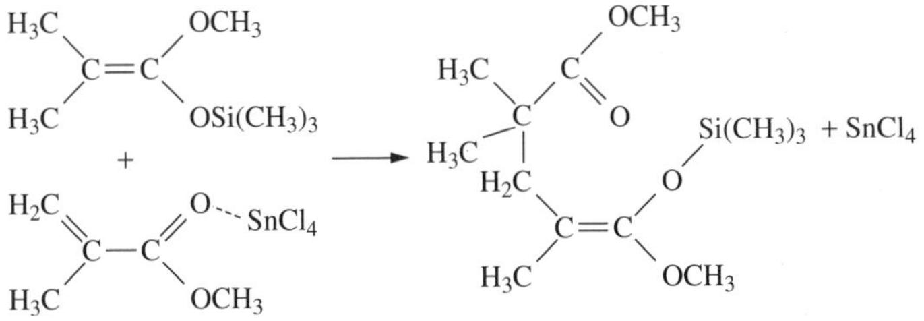

$$\text{(2.3.51)}$$

It should be stressed that in both the nucleophilic and electrophilic procedures the two four-atom reactants can approach each other in a way that allows for a transfer of the —$Si(CH_3)_3$ moiety over a relatively short distance. It is possible that the whole addition proceeds in a concerted manner and that the transferring group is never in an ionic form.

The GTP process is rather flexible, it proceeds under ambient conditions, has very few side reactions, and produces well-defined polymers with low polydispersity (not quite as good as for anionically prepared polymers). It is becoming one of the favored synthetic ways toward polymethacrylates and polyacrylates with a controlled structure. The nucleophilic catalysts are giving better results for polymethacrylates, whereas the Lewis acids work better for acrylic monomers.

2.3.3.5. Miscellaneous Coordination Initiators.

Many other coordination initiators have been developed. A complex of triisobutylaluminum with water and acetyl acetone soluble in hydrocarbons is an excellent initiator for epoxides, particularly for propylene oxide and epichlorohydrin. The mechanism of the reaction is not well understood; it is believed that aluminum becomes 6-coordinate in the propagation step. The reaction product is not stereospecific; the amorphous polymer of epichlorohydrin and especially its copolymer with ethylene oxide are excellent solvent-resistant rubbers.

Catalysts based on organozinc compounds are heterogeneous and often yield stereospecific polyepoxides. However, some soluble zinc complexes have been reported to yield stereospecific polymers as well.

In the industrial production of linear high-density polyethylene, more robust supported oxide catalysts have replaced the Ziegler–Natta initiators. The most common active metals of these catalysts are molybdenum and chromium; they are supported on alumina, silica, or a combination of the two. The support oxides are impregnated with solutions of salts of the active metals, for example, by ammonium molybdate or chromium nitrate, and calcined at 400–700°C. Alternatively, the oxides can be prepared by coprecipitating the hydroxides of the metals with aluminum hydroxide or silica and then calcining. To be active, molybdenum must be in its lower valence state, preferably Mo(V); the activation is achieved by reduction using hydrogen or activators such as CaH_2. During polymerization, the catalyst gradually loses activity because of poisoning by oxygen, water, etc. However, hydrogen or other activators can regenerate it. In contrast, chromium must be kept at a higher valence than the stable trivalent form. The active form is Cr(V) or Cr(VI).

Polymerization by supported oxides is performed at medium pressure (40–70 atm) at about 130°C in an inert solvent (xylene). The linearity of the resulting polyethylene is excellent. The regular adsorption pattern of ethylene molecules on the catalyst surface is believed to contribute to the polymerization mechanism. These catalysts also polymerize other olefins to polymers that are only slightly isotactic.

2.4. REACTIONS ON MACROMOLECULES

So far we have studied synthetic reactions that were designed to build macromolecules from smaller molecules. In this section, we will explore reactions that modify existing macromolecules.

We have already seen that the reactivity of chemical groups is essentially the same whether they are a part of a small or a large molecule. The differences between reactions of small molecules and macromolecules are related to the accessibility of the reactive groups and to the fate of the products of side reactions.

To react, reactants must come into contact. In homogeneous liquids (even rather viscous ones), thermal motion mediates these contacts. For this reason, polymer chemists prefer to modify their polymers after bringing them into solution. However, in many instances it is impossible or impractical to dissolve the polymer, and the reaction must be performed in a heterogeneous system. The penetration of the low-molecular-weight reactant into the polymer then plays a crucial role. The penetration is especially difficult when the polymer is either crystalline or glassy. The chemists are sometimes lucky; the progressing reaction may bring the polymer gradually into solution. Once dissolved, it will react as if it had been in solution from the beginning. If this does not happen, molecules that are close to the surface of the macromolecular particle will be modified extensively, whereas the inside molecules will react much less if at all. A heterogeneous product will result. Sometimes a nice-looking reaction turns sour: A dissolved polymer loses its solubility as the reaction progresses and precipitates from the solution, yielding again an incompletely reacted heterogeneous product.

Side reactions can have quite a detrimental effect when polymers are modified. In the chemistry of small molecules, side reactions are not too harmful. The desired product is usually purified without difficulty. When polymers are reacted, all the side products remain on the polymer molecule and can alter its nature profoundly.

2.4.1. Reactions of Side Groups

Polymers that cannot be prepared by direct polymerization can sometimes be obtained by conversion of the side groups of another available polymer. A classic example is the alcoholysis of poly(vinyl acetate), yielding poly(vinyl alcohol).

$$-\overset{\displaystyle |}{\underset{\displaystyle O-COCH_3}{CH}}-CH_2- \ + \ ROH \ \longrightarrow \ -\overset{\displaystyle |}{\underset{\displaystyle OH}{CH}}-CH_2- \ + \ CH_3COOR \quad (2.4.1)$$

Partial hydrolysis of poly(acryl amide) is the preferred method for manufacturing the copolymer of acrylamide and acrylic acid—a polyelectrolyte of multiple uses.

Another important reaction is the synthesis of polyphosphazenes from polydichlorophosphazene—the only polymer from this family that can be prepared by a direct polymerization (a thermal one).

$$-N{=}\overset{\displaystyle Cl}{\underset{\displaystyle Cl}{P}}- \ + \ 2\,ROH \ \longrightarrow \ -N{=}\overset{\displaystyle OR}{\underset{\displaystyle OR}{P}}- \ + \ 2HCl \quad (2.4.2)$$

$$\underset{\underset{\text{Cl}}{|}}{\overset{\overset{\text{Cl}}{|}}{-N=P-}} + 2\,NHR_2 \longrightarrow \underset{\underset{NR_2}{|}}{\overset{\overset{NR_2}{|}}{-N=P-}} + 2HCl \qquad (2.4.3)$$

The analogy between reactions of small and big molecules sometimes breaks down. The reasons may be steric. Crowded bulky substituents and unfavorable conformation of the chain can prevent the approach of some reagents. More important are electrostatic interactions. For example, when polymethacrylamide is hydrolyzed in an alkaline solution, the neutral amidic groups are converted into negatively charged carboxylate groups $-COO^-$. These groups prevent the hydrolyzing anion (HO^- or RO^-) from approaching the chain that carries the same charge, and the conversion is severely limited. On the other hand, it has also been observed that the neighboring carboxylate group helps to facilitate the hydrolysis of *p*-nitrophenyl methacrylate.

Reactions of side groups of polymers are also used to satisfy a growing demand for materials carrying some reactive and/or specific groups either on a macromolecule or on some solid support. Ion-exchange resins, immobilized catalysts and enzymes, and polymeric drug carriers are just a few examples of such materials. Although synthetic procedures for attaching the required groups to polymeric materials are limited only by the imagination of synthetic chemists, some stratagems are used repeatedly.

Polystyrene is commonly used as the supporting polymer. Its phenyl ring is easily sulfonated, chlorosulfonated, chloromethylated, or nitrated. Sulfonation of crosslinked beads of polystyrene starts at their surface but progresses readily through the bead, and a strongly acidic catex is produced. Chloromethyl substituents have the reactivity of benzyl chloride that can be utilized for many synthetic reactions. Chlorosulfonated polystyrenes have a similar reactivity. Nitrated polystyrene is easily reduced; the resulting amino groups offer a path to countless derivatives. Poly(*p*-lithiostyrene) is also a useful precursor of various special polymers because of its organometallic reactivity. It is prepared by reacting poly(*p*-bromostyrene) with butyllithium.

Among acrylic polymers, polyacrylhydrazide is a very versatile intermediate; it is made from poly(methyl acrylate) and hydrazine.

2.4.2. Cyclization Reactions

A profound modification of polymer properties is achieved when the flexibility of the polymer chains is reduced by cyclization reactions. We have already met such cyclizations when talking about polycondensations forming cycles in Section 2.1.4. The simplest reactions of this kind involve reactants that form a cycle by reacting with two neighboring side groups on a polymer chain. A typical example is the reaction of poly(vinyl alcohol) with aldehydes that leads to polyvinylacetals.

$$\begin{array}{c} -CH-CH_2-CH- \\ \;\;\;\;\;| \qquad\qquad\;\;\; | \\ \;\;\;OH \qquad\qquad OH \end{array} + RCHO \longrightarrow \begin{array}{c} -CH-CH_2-CH- \\ \;\;\;\;\;| \qquad\qquad\;\;\; | \\ \;\;\;O-CH-O \\ \qquad\;\;\;| \\ \qquad\;\;R \end{array} \qquad (2.4.4)$$

Another example is the dehalogenation of poly(vinyl chloride) by zinc. Removal of adjacent pairs of chlorine atoms forms cyclopropane rings.

$$\begin{array}{c} -CH_2-CH-CH_2-CH-CH_2- \\ \qquad\;\;| \qquad\qquad\quad | \\ \qquad\;Cl \qquad\qquad\;\; Cl \end{array} + Zn \longrightarrow \begin{array}{c} -CH_2-CH-CH-CH_2- \\ \qquad\qquad| \diagup \\ \qquad\quad CH_2 \end{array} + ZnCl_2$$

$$(2.4.5)$$

In another type of cyclization, two neighboring groups react together. For example, aldol condensation of poly(vinyl methyl ketone) proceeds as

$$\begin{array}{c} CH_2 \;\; CH_2 \;\; CH_2 \\ \diagup\;\diagdown\diagup\;\diagdown\diagup\;\diagdown \\ CH \;\;\;\; CH \\ |\qquad\quad | \\ C=O \; C=O \\ \diagup\;\;\;\;\;\;\;\diagdown \\ CH_3 \;\;\; CH_3 \end{array} \xrightarrow{-H_2O} \begin{array}{c} CH_2 \;\; CH_2 \;\; CH_2 \\ \diagup\;\diagdown\diagup\;\diagdown\diagup\;\diagdown \\ CH \;\;\;\; CH \\ |\qquad\quad | \\ C \;\;\;\;\; C=O \\ \diagup\;\diagdown\;\diagup \\ CH_3 \;\; CH \end{array} \qquad (2.4.6)$$

In the cyclizations described above, the pairs of neighboring residues reacting together are selected at random. When the reaction is irreversible, then, during later stages of the cyclization, groups must exist that are flanked on both sides by cyclized pairs. Such groups have no mate to react with and must remain single. Theory predicts that the maximum achievable yield of cyclization is about 83%; this was confirmed experimentally.

The above argument is not valid when the cyclization is reversible. Then continuous re-forming of the cycles leads to the most stable structure—a full ladder polymer. We met this type of reaction in Section 2.1.5 in the discussion of the synthesis of poly(phenyl sesquisiloxane). Chain polymerization of the side groups can also lead to a full ladder without leaving any unreacted groups. A typical example is the thermal polymerization of polyacrylonitrile at 175°C.

$$\begin{array}{c} CH_2 \;\; CH_2 \;\; CH_2 \;\; CH_2 \\ \diagup\;\diagdown\diagup\;\diagdown\diagup\;\diagdown\diagup\;\diagdown \\ CH \;\;\;\; CH \;\;\;\; CH \\ |\qquad\quad |\qquad\quad | \\ C \;\;\;\;\; C \;\;\;\;\; C \\ \|\qquad\;\; \|\qquad\;\; \| \\ N \;\;\;\;\; N \;\;\;\;\; N \end{array} \longrightarrow \begin{array}{c} CH_2 \;\; CH_2 \;\; CH_2 \;\; CH_2 \\ \diagup\;\diagdown\diagup\;\diagdown\diagup\;\diagdown\diagup\;\diagdown \\ CH \;\;\;\; CH \;\;\;\; CH \\ |\qquad\quad |\qquad\quad | \\ C \;\;\;\;\; C \;\;\;\;\; C \\ \diagdown\;\diagup\diagdown\;\diagup\diagdown\;\diagup \\ N \;\;\;\;\; N \;\;\;\;\; N \end{array} \qquad (2.4.7)$$

Oxidative pyrolysis of the resulting polymer then yields a condensed polyaromatic polymer.

Cyclization of natural rubber is still another type of reaction; it involves the polymer backbone. The reaction is initiated cationically (sulfuric acid or Lewis acids); its first step proceeds as in reaction (2.4.8).

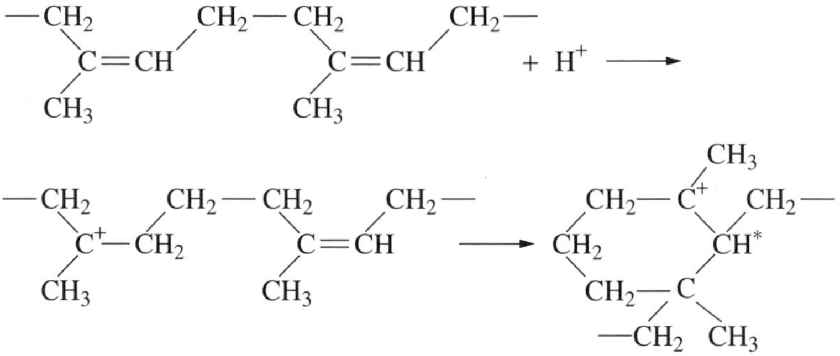

$$(2.4.8)$$

The polymeric cation **1** either transfers the proton designated by an asterisk to another monomeric unit, forming a double bond in the ring, or attacks the neighboring isoprene unit, forming a polycyclic structure. However, the number of fused rings in such structures is probably very low.

2.4.3. Modifications of Cellulose

Historically, polysaccharides offered chemists the first opportunity to work with polymeric materials, to modify them, and to convert them into new polymers with different, often quite valuable, properties. Presently, the material modified industrially in enormous quantities is cellulose.

For industrial purposes, cellulose is made either from wood or from cotton. Wood contains about 50% cellulose fibers; the remainder is mainly lignin and various hemicelluloses. Lignin is a polycyclic phenolic material; its main function is to glue the cellulose fibers together. Hemicelluloses are polysaccharides with less regular structure than cellulose; they have inferior properties from a technical viewpoint. To remove lignin and hemicelluloses, chips of wood are "cooked" with bisulfite-sulfur dioxide or with sodium sulfide-sodium carbonate liquor. In both cases, lignin is sulfonated and the lignin sulfonic acid dissolves. The hemicelluloses are partially hydrolyzed and dissolve as well. The remnants of lignin are removed by "bleaching," a reaction with chlorine that chlorinates and dissolves them. Cotton is much purer cellulose. The contaminants are mainly wax and pectins, which are removed by digestion in alkaline liquors followed by bleaching.

The resulting cellulose fibers have a complex architecture, but for our purposes they can be described as highly crystalline fibrils cemented together by more amorphous regions. The crystallinity is the major factor in the chemistry of cellulose; it severely restricts the accessibility of many reactants and makes cellulose fairly unreactive.

Chemically, cellulose is a polyhydroxy compound and exhibits all the characteristic reactions of hydroxyl groups. Of course, the reactivity of the hydroxyls can be exploited only after they are made accessible to reactants. Swelling performs this task. Strong acids and alkalis are effective swelling agents; first they penetrate the amorphous parts of the fiber, and then they swell the crystalline fibrils.

Sulfuric acid is an excellent swelling agent. It also reacts with cellulose to form esters (acid sulfates), but the reaction equilibrium is shifted toward reactants and the sulfate content is low. On the other hand, sulfuric acid works very effectively in tandem with nitric acid to convert cellulose into cellulose nitrate (routinely but incorrectly called nitrocellulose). Interestingly, the structure of the fiber is not destroyed by the swelling and the nitration; the nitrated fibers (or textiles) do not change their appearance. Nitration of cellulose is a reversible reaction that requires a strong acid catalysis. Consequently, when the unused acids are washed from the product, they can no longer catalyze its hydrolysis; cellulose nitrate is remarkably stable against hydrolysis at moderate values of pH.

Fully nitrated cellulose (a trinitrate) should contain 14.14% nitrogen. However, even the most exhaustive nitration yields products with at most 13.8% nitrogen. The discrepancy is ascribed to the side reaction: formation of cellulose sulfate. The sulfate groups are detrimental to the stability of cellulose nitrate. They can be removed by boiling the material with dilute alkalis; the nitrate groups are not affected, but the stability of the polymer is improved markedly.

Organic acids also swell cellulose to some degree, but the swelling does not lead to the formation of esters. Esterification is achieved when the acid-pretreated cellulose is reacted with an acid anhydride and a catalytic amount of sulfuric acid. The reaction proceeds from the surface of the fibers to their center. Cellulose acetate is by far the most important among these esters. Like the cellulose nitrate, it contains a small amount of bound sulfuric acid, which must be removed by a suitable hydrolytic procedure to stabilize the product. Fully acetylated cellulose (a cellulose triacetate) has poor compatibility when compounded with other plastics; it is soluble in chloroform but not in acetone. The partially hydrolyzed product is more convenient; it is soluble in acetone. It is the polymer of acetate textile fibers. Interestingly, it cannot be prepared by partial acetylation of cellulose; cellulose must be first fully acetylated and then partially hydrolyzed.

A mixed ester, cellulose acetate butyrate, has better compatibility with other polymers and overall better workability than cellulose triacetate. It is prepared in a similar esterification procedure by a mixture of acetanhydride and butyric acid.

At the other end of the acidity scale, sodium hydroxide is an excellent swelling agent for cellulose when used in concentrations of 10% and above. NaOH is believed to be quite firmly bound to cellulose. The swelling effectively disrupts the structure of the cellulose fiber. When the alkali is washed away, the crystallinity of the fibers is reduced and their structure is appreciably loosened; they interact better with dyes and have more luster, for example. This process, called *mercerization,* is applied either to cellulosic fibers or to finished textile articles.

The alkali cellulose in the presence of atmospheric oxygen undergoes rapid changes. It is degraded and highly oxidized. The process is called *aging.* Controlled aging is quite desirable for reducing the molecular weight of the cellulose; the molecular weight of native cellulose is too high for many industrial applications.

In the swollen alkali cellulose, most hydroxyl groups are accessible to reagents, and alkali-catalyzed reactions proceed with ease. Cellulose ethers are prepared by reacting alkali cellulose with alkyl halides; methyl chloride leads to methyl cellulose, ethyl chloride to ethyl cellulose. The reaction is heterogeneous; the alkyl halides penetrate the swollen cellulose by diffusion. The alkali is consumed in the etherification; obviously, a sufficient amount of it must be present. Moreover, the concentration of NaOH toward the end of the etherification should remain high to keep the reaction going even at higher degrees of substitution. Reaction with chloroacetic acid proceeds in a quite similar way; the product is carboxymethyl cellulose.

Reaction of alkali cellulose with ethylene oxide takes a different course. The alkali is not consumed in the process, but its presence is necessary for swelling the cellulose and for catalyzing the addition of ethylene oxide on the hydroxyls. The reaction is a typical anionic polymerization of ethylene oxide but is usually allowed to proceed to only short chains.

$$\text{cellulose}-\text{ONa} + n\,\underset{\displaystyle\diagdown\!\!O\!\!\diagup}{\text{CH}_2-\text{CH}_2} \longrightarrow \text{cellulose}\left(-\text{O}-\text{CH}_2-\text{CH}_2-\right)_{n}\text{ONa} \tag{2.4.9}$$

The industrially most important reaction of alkali cellulose is the one with carbon disulfide that converts the hydroxyls into xanthates:

$$\text{cellulose}-\text{ONa} + \text{CS}_2 \longrightarrow \text{cellulose}-\text{OCSSNa} \tag{2.4.10}$$

In a typical industrial process, one xanthate group is formed for approximately every two glucose units. This degree of substitution is sufficient to bring the cellulose xanthate into solution, which is called viscose. Viscose is not stable; it undergoes a number of reactions during which various new compounds of sulfur are formed and the degree of xanthation is decreasing. The alkaline conditions also lead to partial degradation of the cellulose chains. The process is called *ripening* and is very important for the properties of the final product. The valuable property of viscose is its instability in acidic media. When it is brought into contact with a dilute acid, it immediately decomposes back to cellulose. For the production of rayon fibers, the filtered ripened viscose is forced through very fine multiple holes in cup-shaped spinnerets immersed in a coagulating bath. The individual liquid jets are converted into filaments; all the filaments from one spinneret are combined into one fiber and are immediately drawn and spun. Cellophane sheets are produced similarly; in this case, viscose is forced through a slit. Annular slots produce sausage casings.

2.4.4. Cross-Linking and Vulcanization

Rubberlike materials consist of long flexible chains connected by a small number of cross-links into a loose infinite network. We have seen that such networks can be obtained when a few multifunctional monomers are added to a system undergoing polymerization or polycondensation. However, polymeric articles prepared in this fashion have the shape of the reaction vessel; that is not very convenient. Industry prefers materials composed from single polymeric molecules that can be easily shaped to the required form (tires, rubber bands, shoe soles) and then permanently cross-linked so that they retain their shape. This section will deal with some strategies employed in these processes.

To be cross-linked, the polymer molecules must have some reactive groups. Some polymers contain such groups because of their chemical nature. The most important examples are natural rubber and its synthetic cousin polybutadiene. Both polymers carry one unreacted double bond in each monomeric unit. Double bonds can be introduced into polymers not having them originally by copolymerization. For example, polyethylene copolymerized with a small amount of butadiene contains dangling vinyl groups. Another useful combination is polyisobutylene copolymerized with about 2% isoprene in the production of butyl rubber.

As long ago as 1839 Goodyear got an advise from Vulcan, the ancient Roman god of volcanoes, sulfur, and fire, to treat natural rubber with sulfur. Ever since, this cross-linking reaction has been called *vulcanization,* a term later transferred to all cross-linkings of elastomers. Chemically, vulcanization by sulfur is anything but simple. It requires temperatures in the range of 120–160°C, when the sulfur-sulfur bonds in the elemental S_8 sulfur rings start to break. The sulfur chain fragments attack the polymer chains. It is believed that the reaction is ionic, not radical. In any case, the primary sites of attack are hydrogens on the α carbon to the double bond, and structure **2** is the desirable structure produced. However, the reaction is quite inefficient; about 50 sulfur atoms are needed for every efficient cross-link. The average length of the polysulfide chain is about 12 atoms and decreases with the duration of the process to about 3. Besides the formation of sulfur bridges, numerous other reactions take place, and many of them involve the double bonds. Cyclic forms such as **3** are formed extensively.

The vulcanization by sulfur can be accelerated and formation of cyclic structures suppressed by organic *accelerators* (e.g., mercaptobenzothiazole **4**) usually combined with zinc oxide and long-chain fatty acids.

In modern vulcanization recipes, sulfur is replaced by other sulfur compounds, for example, by disulfur dichloride, S_2Cl_2, or by tetramethylthiaram disulfide **5**. When used together with accelerators, they lower the vulcanization temperature, increase the utilization of sulfur, and shorten the polysulfide bridges.

When the polymer that has to be cross-linked, has no convenient reactive groups, they can be created during the vulcanization itself. Cross-linking by organic peroxides is a technique of this type applicable for most elastomers. In this technique, the polymer is mixed with a peroxide (typically cumyl peroxide) and heated to about 150°C. The peroxide decomposes into radicals, which attack the polymer—usually by abstracting hydrogen atoms. With unsaturated polymers (polydienes), hydrogens on α carbons are abstracted preferentially, producing allylic radicals. However, all hydrogens are subjected to abstraction, including those of polyethylene, ethylene-propylene copolymer, and polysiloxanes. At temperatures below 200°C most radicals decay by recombination, thus forming the desired cross-links.

High-energy irradiation (electrons, γ radiation) also produces radicals and can serve for cross-linking purposes. However, scission of the polymer chains and other undesirable reactions sometimes complicate the process.

2.4.5. Polymer Grafting

We describe as grafted copolymers those macromolecules that possess a backbone from one homopolymer and one or more (or many) branches made from another homopolymer. Block copolymers are actually also grafted; the grafts are at the chain ends in this case. Both grafted and block copolymers have numerous valuable properties, especially when multicomponent or multiphase materials are desired. We will return to this topic in Chapter 4. For now we will consider several different methods for their synthesis.

The least sophisticated grafting method uses chain transfer in radical polymerization. The monomer to be grafted is radically polymerized in the presence of the backbone polymer. Either the initiating radicals or the growing chains abstract a hydrogen from the backbone polymer, creating a growth center in the middle of it. The monomer then polymerizes as a graft. There are numerous limitations. Peroxide initiators are active in this type of reaction, whereas azobis(isobutyronitrile) is not. The relatively unreactive growing chains of polystyrene generally do not abstract hydrogens from other molecules, whereas the highly reactive poly(vinyl acetate) chains do it with ease. The structure of the backbone polymer is also important. Tertiary hydrogens (as in polystyrene) and methyl hydrogens adjacent to carbonyl [as in poly(vinyl acetate)] are especially vulnerable. On the other hand, the radical produced may not be sufficiently active to initiate a graft of a nonreactive monomer such as vinyl acetate. Thus vinyl acetate is readily grafted onto poly(vinyl chloride) or onto polyacrylonitrile but not onto polystyrene. The transfer grafting always produces some homopolymers together with the grafted material; this is a serious drawback of the procedure.

Another group of methods converts the backbone polymer into an initiator for the other monomer. When radicals are introduced into the backbone polymer we speak of *activation grafting*. Irradiation using ultraviolet light or high-energy radiation can achieve it. Usually, the polymer is mixed with the monomer before irradiation. However, it is also possible to irradiate the polymer in the glassy state. The radicals are trapped and start polymerization after the polymer is swollen by the monomer. Another twist of the method irradiates the polymer in the presence of oxygen, which produces hydroperoxide groups that are later decomposed thermally.

$$-CH_2-CHR- \ + \ O_2 \longrightarrow \ -CH_2-\underset{\underset{OOR}{|}}{C}R- \qquad (2.4.11)$$

Some monomers are easily oxidized even without irradiation. In a copolymer of styrene and p-isopropylstyrene, the isopropyl group is autoxidized to a hydroperoxide. Initiation is achieved best in this case with redox decomposition by, for example, ferrous ions.

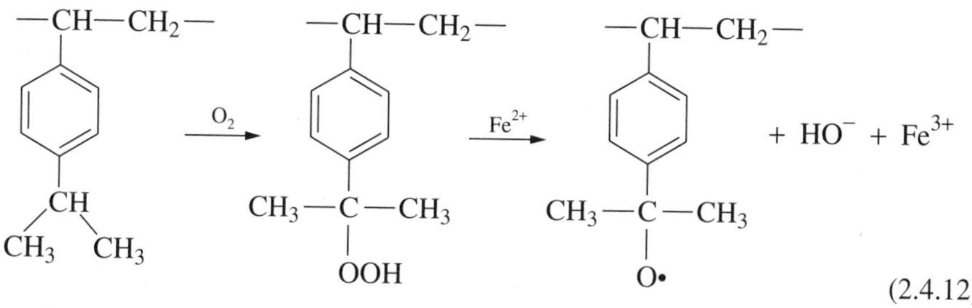

$$(2.4.12)$$

The discussion of derivatization of polymers in Section 2.4.1 mentioned the reaction of butyllithium with p-bromostyrene monomeric units leading to p-lithiostyrene units. These units can also serve as an anionic initiator for any monomer capable of anionic polymerization.

A third group of methods first synthesizes the grafts, converts them to monomers (sometimes called macromers), and then incorporates them into the backbone. Living polymers are very convenient intermediates for this purpose. For example, a growing dianion with two active ends can be reacted with ethylene oxide; after subsequent hydrolysis an α,ω-diol results, that can be used in any polycondensation (e.g., as a component of polyurethanes).

$$Na^+A^{-}{\sim\sim}\,A^-Na^+ + 2\,CH_2-CH_2 \longrightarrow Na^+O^-CH_2-CH_2-A\sim\sim A-CH_2-CH_2-O^-Na^+$$
$$\overset{\text{O}}{\diagup\!\!\!\diagdown}$$

$$\xrightarrow{\text{H}_2\text{O}} \ HO-CH_2-CH_2-A\sim\sim A-CH_2-CH_2-OH \qquad (2.4.13)$$
$$\mathbf{6}$$

Strictly speaking, this synthesis of a polyurethane is not grafting but the preparation of a block copolymer. A true grafting is achieved when the hydroxyls of **6** are converted into acrylic or methacrylic esters and then copolymerized with other acrylates or methacrylates (or any other monomer having suitable copolymerization parameters with acrylates). In the final synthesis, the macromer **6** would serve as a cross-link. However, when the whole synthesis is started with a living polymer having only one active end, a grafted copolymer would result.

Grafting via living polymers and macromers is a very flexible method capable of innumerable variations. Not only can the grafts and the backbones be selected from a large number of monomers, but it is also possible to regulate closely the length of the grafts (the living polymers are virtually monodisperse) and their average spacing (by choosing an appropriate stoichiometry of the monomers and macromers).

Finally, a reaction of living polystyrene with polyacrylates or polymethacrylates is an example of "true" grafting when pendant chains are covalently attached to an existing polymer chain.

$$(2.4.14)$$

2.4.6. Polymer Degradation and Stabilization

We have studied how macromolecules are born; now is the time to see how they perish. The death of a polymer (degradation is a more common expression) is a process in which the polymer loses its valuable properties and, eventually, its integrity. During this process, the macromolecular chain is usually broken into smaller pieces. We make many useful objects from polymers, and we want these objects to last. Hence we need to understand the degradation that leads to their destruction to be able to minimize it. However, there are some circumstances under which we want the polymer to decompose in a fast and clean reaction. One such case is that of packaging materials that are discarded when their contents are removed (soft drink containers, cereal boxes)—we want these materials to decompose and not to pollute our environment. Another example is *resists*—coating materials in the semiconductor industry. They are coated on silicon chips, and an image of the required circuit is projected on them using an energetic radiation. The resist should be either degraded to make it easily soluble (a positive resist) or cross-linked to make it quite insoluble (a negative resist).

There are three major degradation mechanisms—chemical, thermal, and irradiation—but they often work hand in hand. We would expect that the thermal stability of polymers (i.e., the temperature at which the breakage of bonds becomes significant) would be about the same as the stability of low-molecular-weight compounds with a similar structure. Actually, most polymers start to decompose at temperatures substantially lower—most of them above 200°C, many of them around 100°C. There are several reasons for this behavior.

1. When 1% of the backbone bonds undergo scission, the polymer is converted to an oligomer. The same amount of damage can go unnoticed with small molecules.
2. In considering the stability of small molecules, we need to consider only the stability of the constituent bonds. In the case of a polymer, there are always irregularities present, such as fragments of the initiator, branching points, and traces of oxygen built in as a comonomer. These irregularities frequently act as weak points where the degradation starts.
3. Many chain polymerizations are reversible. When an active group is created by some degradation reaction and the polymer is above its ceiling temperature, the chain may depropagate and unzip, that is, revert to the monomer.

We will now consider some more common scenarios of polymer degradation.

Atmospheric oxygen is the most ubiquitous agent damaging polymers. The reaction of oxygen with organic substances has a radical character. The ground state of molecular oxygen is a triplet state; hence oxygen behaves as a biradical. It recombines easily with any radical R• and forms a peroxy radical R—O—O•, which in turn can abstract an atom of hydrogen from susceptible molecules and start a chain oxidation:

$$R\bullet + O_2 \longrightarrow R\text{---}O\text{---}O\bullet \xrightarrow{R'\text{---}H} R\text{---}O\text{---}O\text{---}H + R'\bullet \qquad (2.4.15)$$

The resulting hydroperoxides, ROOH, are susceptible to homolytic breakage of oxygen-oxygen bonds, producing additional radicals. Traces of transition metals, which may be present as remnants of various initiators, greatly accelerate this decomposition as we have already seen in Section 2.2.1, reaction (2.2.3). Obviously, oxidation by molecular oxygen has an autocatalytic character, and it is frequently called *autoxidation*.

Allylic hydrogens are especially vulnerable to abstraction [compare equation (2.2.14)]. Polydienes have many such types of hydrogen. They oxidize easily, and rubbers based on them must be protected against oxidation. Tertiary hydrogen is also abstracted easily; that makes polypropylene sensitive to autoxidation as well. However, polypropylene is crystalline and oxygen cannot penetrate into the crystallites. Hence, the damage is confined to the amorphous regions in the polymer. This is enough to destroy it—when the amorphous parts of the polymer are extensively damaged, the polymer becomes brittle, and there is no more cohesion between crystallites.

Thermal degradation of polymers is triggered by thermal breakage of bonds. The initial breakage usually occurs at imperfections in the polymer structure: at chain ends, at branching points, or at hydroperoxy groups, for example. The next step in the degradation sequence depends on the nature of the polymer. One of the important mechanisms is *chain depolymerization,* which is also called *unzipping*. We have seen in Section 2.2.6 that polymers above their ceiling temperature become thermodynamically unstable with respect to their monomers and will revert to them when given an opportunity.

For some polymers—for example, poly(methyl methacrylate), poly(α-methylstyrene), and polytetrafluoroethylene—thermal depolymerization is a very clean reaction giving the monomer in a high yield. In fact, it is sometimes used for laboratory preparation of these monomers. For other polymers such as polyethylene and polypropylene, transfer degradation reactions prevail over unzipping, and these polymers decompose into fragments of varying sizes. Polystyrene and polyisobutylene are in the middle: They yield some monomer together with low-molecular-weight oligomers and other fragments.

In another group of polymers, high temperatures influence the side groups before they affect the main chain. Cyclizations of poly(vinyl methyl ketone) [reaction (2.4.6)] and of polyacrylonitrile [reaction (2.4.7)] belong here. The degradation of poly(vinyl chloride) is a free radical chain reaction involving hydrogen abstraction and chlorine atom elimination.

$$-CH_2-\underset{\underset{Cl}{|}}{CH}- + Cl\bullet \longrightarrow HCl + -\overset{\bullet}{C}H-\underset{\underset{Cl}{|}}{CH}- \longrightarrow -CH=CH- + Cl\bullet$$

$$\text{(2.4.16)}$$

The result of this dehydrochlorination is a highly unsaturated carbonaceous structure. Poly(vinyl acetate) eliminates acetic acid in a similar scenario.

Polymers with heteroatoms in the backbone, which may decompose into thermodynamically more stable heterocycles, will do so at higher temperatures. Thus polydimethylsiloxane will decompose into its cyclic tetramer, nylon 6 into caprolactam, polyoxymethylene into trioxane and formaldehyde, and polyorganophosphazenes into their cyclic trimers. These reactions are usually triggered by acidic or alkaline impurities.

Radiation is another cause of damage to polymers. We should distinguish the relatively low levels of radiation that polymeric materials receive during their lifetime (UV radiation from sunshine, encounters with electrons, X-rays, etc.) and more intense intentional irradiation. Photodegradation is important for polymers possessing a chromophore that absorbs radiation in the UV part of the spectrum (i.e., carbonyls, phenyls, etc.) together with an easily dissociable bond (tertiary hydrogens, etc.). Most common polymers qualify. Once a bond is broken within the polymer structure, many degradative reactions (which depend on the nature of the polymer) can follow: fragmentation or unzipping, cross-linking, discoloration. Low-energy radiation also assists oxidative and thermal degradations by supplying free radicals to them. The overall degradation is an interplay of all these processes.

High-energy radiation is used in the chemistry of resists. Partially cyclized poly-isoprene with incorporated azide groups is an example of a negative resist. The free radicals from the decomposition of the azide by UV light cross-link the polymer. A typical positive resist is poly(methyl methacrylate), which is virtually decomposed by intense electron beams.

We are not totally helpless against polymer degradation. An array of techniques exists that interferes with various steps in the degradation sequence. The first line of defense occurs during polymer synthesis: the elimination of prospective weak points. In this category we

1. Protect chain ends by converting them to nonreactive structures (e.g., poly-oxymethylene to its acetate).

2. Use radical initiators that produce unreactive chain ends (e.g., benzoyl peroxide).

3. Eliminate transfer agents that would create a reactive group.

4. Incorporate in the polymer chain another monomer that would interfere with unzipping [ethylene oxide in polyoxymethylene, carborane groups in poly(dimethyl siloxanes)].

5. Carefully remove traces of transition metals after using Ziegler–Natta initiators.

In the second line of defense, we try to prevent the usual first step of the degradation process: the photochemically facilitated reaction of oxygen with polymer. The simplest procedure consists in blocking the access of light to the polymer; compounding with carbon black does the trick for rubber. Alternatively, in *photostabilization* a compound is added to the polymer that captures the UV photons and is excited by them but then harmlessly dissipates the energy. Typically, alkyl- or alkoxy-substituted *o*-hydroxybenzophenones **7** are used for this purpose.

The third line of defense tries to destroy radicals that managed to be formed despite these precautions. Among radical inhibitors (see Section 2.2.4.4), which are called *antioxidants* in this context, sterically hindered phenols **8** are the most popular. In their formula R represents methyl or, more often, the *tert*-butyl group.

As already mentioned, sometimes we want manufactured materials to decompose at a predictable rate under the influence of naturally occurring enzymatic systems. Such a decomposition is called *biodegradation;* it is a desirable feature when used packaging materials are decomposed in dumps (soil bacteria and fungi supply the enzymes) or when polymers are used as drug delivery systems, as surgical sutures, as implants in arteries, and for similar purposes (the living body disposes of them after their mission is accomplished). Hydrolysis is the reaction used most frequently by living organisms for degradation of natural macromolecules. Consequently, for biodegradation of artificial materials, polymer chemists manufacture polymers with structures that can be mistaken by hydrolytic enzymes for natural structures. Esters and amides are usually selected as the target groups. For example, for implantation purposes, polycaprolacton and poly(lactic acid) are used to advantage. Among amides, copolymers of α amino acids (glycine, serine) with caprolactam are biodegraded quite easily.

Glycoside linkages are other structures that enzymes attack. Thus many cellulose derivatives (e.g., acetate, nitrate) are biodegradable if the derivatization is incomplete and some unsubstituted glucose unit (or even better, short sequences of them) still exists in the polymer.

2.A. SUGGESTIONS FOR FURTHER READING

Allport, D.C., and W.H. Janes, Eds., *Block Copolymers,* Wiley, New York, 1973.

Biesenberger, J.A., and D.H. Sebastian, *Principles of Polymerization Engineering,* Wiley, New York, 1983.

Bikales, N.M., and L. Segal, Eds., *Cellulose and Cellulose Derivatives,* 2 vols., Wiley, New York, 1971.

Boor, J., Jr., *Ziegler–Natta Catalysts and Polymerizations,* Academic, New York, 1979.

Brook, M.A., *Silicon in Organic, Organometallic and Polymer Chemistry,* Wiley, New York, 2000.

Campbell, I.M., *Introduction to Synthetic Polymers,* 2nd Ed., Oxford University Press, Cary, NC, 2000.

Cassidy, P.E., *Thermally Stable Polymers: Syntheses and Properties,* Dekker, New York, 1980.

Chung, T.-S., Ed., *Thermotropic Liquid Crystal Polymers: Thin-Film Polymerization, Characterization, Blends, and Applications,* Technomic., Lancaster, PA, 2001.

Clough, R.L., N.C. Billingham, and K.T. Gillen, Eds., *Polymer Durability, Degradation, Stabilization, and Lifetime Prediction,* Advances in Chemistry Series, No. 249 (ACS Publication), Oxford University Press, Cary, NC, 1996.

Grassie, N., and G. Scott, *Polymer Degradation and Stabilization,* Cambridge University Press, Cambridge, 1985.

Gupta, S.K., and A. Kumar, *Reaction Engineering of Step Growth Polymerization,* Plenum, New York, 1987.

Ham, G.E., *Copolymerization,* Interscience, New York, 1964.

Hsieh, H.L., and R.P. Quirk, *Anionic Polymerization: Principles and Practical Applications,* Dekker, New York, 1996.

Kennedy, J.P., *Cationic Polymerization of Olefins: A Critical Inventory,* Wiley, New York, 1975.

Kennedy, J.P., and B. Iván, *Designed Polymers by Carbocationic Macromolecular Engineering: Theory and Practice,* Hanser, Munich, 1992.

Lenz, R.W., *Organic Chemistry of Synthetic High Polymers,* Interscience, New York, 1967.

Lovell, P.A., and M.S. El-Aasser, Eds., *Emulsion Polymerization and Emulsion Polymers,* Wiley, New York, 1997.

Mishra, M.K., and S. Kobayashi, Eds., *Star and Hyperbranched Polymers,* Dekker, New York, 1999.

Moad, G., and D.H. Solomon, *The Chemistry of Free Radical Polymerization,* Pergamon, Oxford, 1995.

Morton, M., *Anionic Polymerization: Principles and Practice,* Academic, New York, 1983.

Noshay, A., and J.E. McGrath, *Block Copolymers: Overview and Critical Survey,* Academic, New York, 1977.

Odian, G., *Principles of Polymerization,* 3rd Ed., Wiley, New York, 1991.

Penczek, S., P. Kubisa, and K. Matyjaszewski, *Cationic Ring-Opening Polymerization of Heterocyclic Monomers: Mechanisms* (Advances in Polymer Science, Vol. 37), Springer, Berlin, 1980.

Penczek, S., P. Kubisa, and K. Matyjaszewski, *Cationic Ring-Opening Polymerization: 2. Synthetic Applications* (Advances in Polymer Science, Vol. 68/69), Springer, Berlin, 1985.

Ravve, A., *Organic Chemistry of Macromolecules: An Introductory Textbook,* Dekker, New York, 1967.

Reich, L., and A. Schindler, *Polymerization by Organometallic Compounds,* Interscience, New York, 1966.

Sandler, S.R., and W. Karo, *Polymer Syntheses (Organic Chemistry: A Series of Monographs,* Volume 29-I), Academic, New York, 1997.

Sawada, H., *Thermodynamics of Polymerization,* Dekker, New York, 1976.

Szwarc, M., *Living Polymers and Mechanisms of Anionic Polymerizaton,* Springer, Berlin, 1983.

Vogtle, F., K.N. Houk, H. Kessler, and J.-M. Lehn, Eds., Dendrimers II; *Architecture and Supramolecular Chemistry (Topics in Current Chemistry 210),* Springer, Berlin, 2000.

2.B. REVIEW QUESTIONS AND DERIVATIONS

1. Discuss the effect of mono- and trifunctional compounds on the structure of a polycondensation polymer.

2. Discuss the control of molecular weight in a polycondensation reaction.

3. Explain why: (a) polyacetylene is a conducting polymer when it is appropriately doped and (b) polysilane is a photodegradable polymer.

4. Why excess ethylenediamine is needed to obtain the starburst dendrimers of polyamidoamines?

5. Describe how the following quantities can be influenced in free radical polymerization: (a) molecular weight, (b) $\bar{M}_w / \bar{M}_n$, and (c) the rate of polymerization.

6. Describe a reasonable approach to prepare a lightly cross-linked polyphosphazene containing primarily the aliphatic acid groups ($-CH_2COOH$) along the backbone starting with the basic $-PCl_2N-$ polymer.

7. Explain why (a) polyesters based on aromatic diacids are much tougher than those based on aliphatic diacids and (b) polysilanes have grossly different properties than alkanes (polyolefins)

8. Which would you expect to have a higher chain transfer constant: (a) carbon tetrafluoride or (b) carbon tetrachloride?

9. Which of the following could serve as an initiator for an anionic chain polymerization (a) $AlCl_3.H_2O$, (b) $BF_3 \cdot H_2O$, (c) butyllithium, or (d) sodium metal?

10. What is the principal difference between propagation reactions with butyllithium and a Ziegler–Natta catalyst? Name the most widely used Ziegler–Natta catalyst for the production of HDPE.

11. Which of the two would be more reactive in free radical polymerization: (a) p-methylstyrene or (b) m-methylstyrene?

12. Name (a) a thermoplastic, (b) an elastomer, and (c) a fiber that is produced commercially by ionic chain polymerization.

13. In an anionic polymerization, stereoregularity of the polymer is strongly dependent on the closeness of the counterion to the growing end of the polymer. How can you influence the closeness of the counterion?

14. What types of transfer do you know in free radical polymerization and what is their effect on the polymer?

15. Describe the emulsion polymerization and its mechanism. How does it differ from the suspension polymerization?

16. Give a general discussion of different kinds of end groups present in polymers prepared by (a) free radical polymerization and (b) anionic polymerization.

17. Suppose that we want to produce a water-insoluble polymer that contains a pH indicator such that the polymer changes colors in response to the pH of water in contact with it. Describe the general approaches and considerations to produce such a polymer. Propose a specific approach with some common indicator and polymer combination.

18. Discuss the kinds of end groups that might be present in a polymerization initiated by the photodecomposition of diphenyl sulfide ($\phi_2S + SbF_6^-$) by assuming that polymerization can be initiated simultaneously by either a free radical or a cationic mechanism.

19. Use the general principles of anionic (living) polymerization to discuss different approaches to produce a tri-block polymer of the types: A-B-C and A-B-A. Discuss the considerations in the choice of initiator, monomer, and the order of addition of reagents.

20. Describe the role of methyl aluminoxane in metallocene polymerization reactions. Illustrate with an example.

21. What are the advantages of metallocene catalysts when compared with the Zeigler–Natta catalysts?

22. What conditions are favorable for the preparation of polymers by living cationic polymerization?

23. Discuss the mechanism of atom transfer radical polymerization using the transition metal catalysts.

24. What are the advantages of the nitroxide-mediated radical polymerization over that of free radical polymerization? Give examples.

25. List the characteristics distinguishing between the atom transfer radical polymerization and the group transfer polymerization with examples.

26. What is ring opening metathesis? Explain with at least two examples. What type of polymers can be obtained by this method?

27. Discuss how one might prepare a graft polymer of the type shown below that has a main chain (M = either organic or inorganic) very insoluble in water and in which some of the pendant polymer groups (A in the structure below) could be hydrolyzed off the main chain. [Note: there are many ways to accomplish this, and there is no single correct answer, but simply try to apply any of the different chemical schemes. In the structure, A does not have to be at each M site.]

$$-M-M-M-M-M-M-M-M-$$
$$\;\;\;\;|\;\;\;|\;\;\;\;\;\;\;\;|\;\;\;\;\;\;\;\;\;\;\;|$$
$$\;\;\;\;A\;\;A\;\;\;\;\;\;\;A\;\;\;\;\;\;\;\;\;\;A$$

28. Derive the most probable distribution of degree of polymerization for the polycondensation of a A-B type polymer.

29. Derive the reactivity ratios for polystyrene and methyl methacrylate from their Q-e values given in Table 2.3 in the text. What composition of the monomers would be required to produce a copolymer with a 1:1 mole ratio of these two monomer groups?

30. Derive the formula for the molecular weight distribution in case of condensation polymerization in which a small mole fraction of monofunctional material is present for A-B $\rightarrow$ –(A-B-A-B)–.

31. Discuss the physical and kinetic significance of the gel point.

32. Describe the two-step polymerization sequence in the formation of a polyether from an epoxy. What is the importance of curing agents in the formation of epoxy resins?

33. Describe the two main routes of forming network polyurethanes.

34. What is the difference between an inhibitor and a retarder in free radical vinyl polymerizations? Give examples of each.

35. Define chain transfer and the chain transfer constant. What practical use do chain transfer agents have?

36. Discuss the importance of poly(vinyl alcohol) in the suspension polymerization of vinyl chloride.

37. What is the importance of the presence and concentration of surface active agents (soaps) in an emulsion polymerization? Describe a micelle. What is the role of micelles in emulsion polymerizations?

38. In the emulsion polymerization, why does the degree of polymerization, but not the rate of polymerization, depend on the initiator concentration?

39. Give several examples of the types of catalysts that may be used for cationic polymerizations. Discuss the importance of a co-catalyst for each catalyst. For what solvent systems may each catalyst be used?

40. Describe the conditions for the synthesis of a graft copolymer of ethylene and styrene and a block copolymer of styrene and methyl methacrylate.

41. Polystyrene is soluble in benzene, whereas poly(vinyl alcohol) is soluble in water. How will a block copolymer of styrene and vinyl alcohol behave in benzene? In water?

42. The low-temperature cationic polymerization of vinyl ethers gives either isotactic or syndiotactic crystalline polymers depending on the polarity of the solvent. Explain the effects of the solvent for each stereoregular polymerization.

43. Oxidation is the major cause of polymer degradation with the resultant impairment of surface and quite possibly bulk properties. What are the functions of an antioxidant additive?

44. Biodegradation of plastics can be an effective method of alleviating the environmental pollution problems. Comment. (Do some literature search.)

45. In an anionic polymerization, the molecular weight distribution can be represented by a Poisson distribution function. Consider that in an anionic polymerization, a diblock polymer of the type, $-(A-)_n-(B)_m-$ is produced. Following the same logic as discussed in the text, derive the distribution function for the indices n, m, and $n + m$ (= degree of polymerization) as a function of the mole ratio of the two monomers, A and B, to the anionic initiator. Note that in general, one does not have to have the same number of moles of the two monomers.

2.C. NUMERICAL PROBLEMS

1. Polyester can be prepared by linear condensation polymerization. If the batch polymerization time to reach $\overline{M}_n$ of 60,000 is 3 h, how many hours are required to reach $\overline{M}_n = 100,000$? Assume that the overall reaction follows the second-order kinetics, the concentration of hydroxy and carboxy groups is the same, and the reaction by-product is being continuously removed from the system.

2. If the fractional conversion in an ester interchange polycondensation reaction is 0.999, what would be the $\overline{DP}$ of the polyester produced?

3. Assuming a value of 0.9999 for p in the Carothers equation, calculate the $\overline{DP}$ of a polyester prepared from a mixture of difunctional reactants to which 1.5 mol% of acetic acid was added. The mixture is equimolar in the reactive groups.

4. Styrene bulk polymerization is initiated at $60°C$ by adding 2% weight of benzoyl peroxide. Calculate the concentration of free radicals, $[P\bullet]$, in the reaction mixture in early stages of the polymerization using the data: $k_p = 176°$ L/mol·s, $k_t = 72 \times 10^6$ L/mol·s (assume that only recombination occurs), $f k_d = 1.92 \times 10^{-6}$ sec^{-1}, and density of styrene $= 0.908$ g/cm^3.

5. A polyamide is to be prepared from the reaction of adipic acid and hexamethylene diamine. If the molar ratio of the diacid to diamine is 0.99, what maximum $\overline{M}_n$ can be achieved?

6. A batch of poly(methyl methacrylate) with $\overline{M}_w = 300,000$ is to be prepared by free radical polymerization. Assume that benzoyl peroxide is used as an initiator at $60°C$ and the monomer is dissolved in benzene at a concentration of 1.0 M. Following the example in the text, calculate the concentration of the initiator that is required for polymerization. Find the necessary rate constants either from the example or from Table 2.1 of the text. If the solvent is toluene instead of benzene, the chain transfer by the solvent must be taken into account. What molecular weight, $\overline{M}_w$, would be obtained if $C_s = 1.5 \times 10^{-4}$?

7. Given the reactivity ratios derived from the Q-e scheme (Table 2.3 of the text), (a) calculate the composition curve for a copolymer of α–methylstyrene and methyl acrylate. In this plot, x-axis is the mole fraction of α-methylstyrene in the monomer and the y-axis is the mole fraction of α-methylstyrene in the final polymer. From the plot determine what monomer mix would be required to produce the final polymers with 0.2, 0.5, and 0.7 mole fraction of α-methylstyrene. For each of these copolymer compositions, compute the average sequence length of each monomer.

8. Two copolymers are considered. One has the reactivity ratios $r_A = 0.5$ and $r_B = 2$ and the other $r_A = 1$ and $r_B = 0.5$. Plot the dependence of copolymer composition on the composition of the monomer mixtures for both copolymers.

9. A copolymer of styrene (A) and butadiene (B) is made from a mixture containing 20 mole% of butadiene and 80 mole% of styrene monomers by free radical copolymerization. Calculate the copolymer composition at the initial stage. Take $r_A = 0.78$ and $r_B = 1.40$.

10. Styrene is added to a solution of sodium naphthalide in tetrahydrofuran so that the initial concentrations of styrene and sodium napthalide in the reaction mixture are 0.2 M and 1×10^{-3} M, respectively. After 5 s of the reaction at $25°C$, styrene concentration was determined to be 1.7×10^{-3} M. Calculate (a) the initial rate of polymerization and (b) $\overline{M}_n$ of the polymer.

11. At what extent of reaction will one reach the gel point of the following condensation reactions?
(a) Si $(OCH_3)_4 + R_2Si (OCH_3)_2$ (1:5 mole ratio of the two siloxane compounds).
(b) RSi $(OCH_3)_3 + R_2Si (OCH_3)_2$ (1:5 mole ratio of the two siloxane compounds). R is a nonreactive alkyl group.

12. A dendrimer was synthesized using a polyalcohol core and a branching unit shown below.

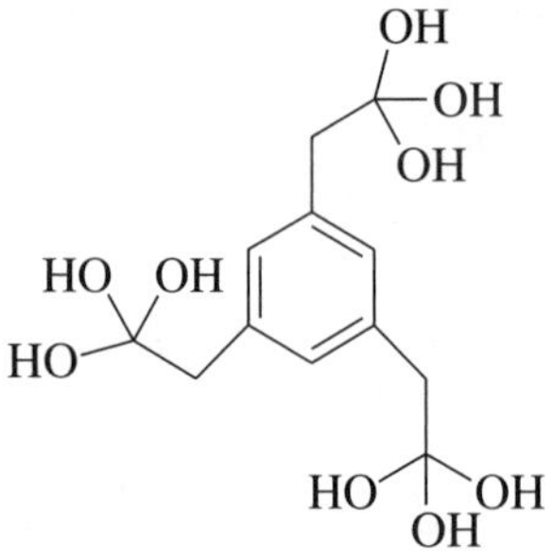

(Branching unit)

(Generation 0)

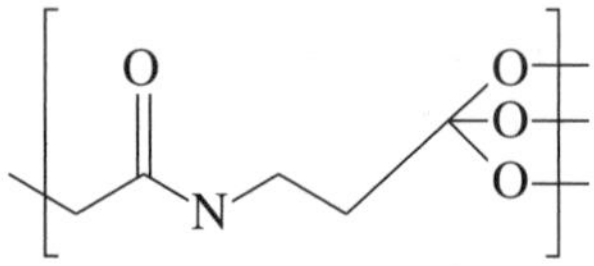

An analysis of generations from 1 to 9 indicated that the starburst limit is reached at the 7th generation. Assume that every terminal functional group is located on the surface of the spherical molecule and that the structure is stretched out to the full length of each arm. Use the contour length of the branching units along with the bond angles and bond lengths from the CRC Handbook to calculate: (a) the surface area required by each terminal functional group, (b) the molecular weight of every generation greater than 4, and (c) r_G of every generation up to the starburst limit.

3

MACROMOLECULES IN SOLUTION

In Chapter 1, we saw that macromolecules come in a great many structures. In Chapter 2, we learned some techniques for the preparation of these structures. The drive to prepare new types of macromolecules and macromolecular structures originates from an obvious fact: Different materials have different properties, which are useful for different purposes. Equally obvious is the necessity to correlate various properties of materials with their structure. Hence, we need methods for studying the structure of macromolecules. The important structural details of macromolecules, such as molecular weight, chain length, branching, and chain stiffness, are best studied when the individual molecules are separated from each other, that is, in solution.

Although the dissolution of a polymer permits the determination of its basic parameters, it also brings with it a host of new problems. For a correct interpretation of the behavior of macromolecular solutions we need to understand the thermodynamics of polymer-solvent interaction. Hence, there is tremendous interest in studying the thermodynamics of macromolecular solutions; a sizable fraction of the polymer literature is devoted to this subject. We will explore some of the basic underlying thermodynamic principles of polymer solutions in Section 3.1.

Thermodynamics itself forms a sufficient basis for description of many types of equilibrium. We will see that studies of various kinds of equilibrium provide a wealth of information about macromolecular systems. We will devote Section 3.2 to those methods of study of macromolecular solutions that deal with equilibrium and can be fully described by thermodynamic relations.

A large number of physicochemical techniques for the study of polymer solutions are based on forcing relative movement among parts of a solution. Relative movement in liquids always involves flow. Hence, we need to understand basic hydrodynamic principles and how they operate on the molecular level. We will study these principles in Section 3.3. Section 3.4 will deal with the hydrodynamic methods themselves.

When light or other kinds of radiation interact with matter, they are either absorbed or scattered. Scattered light contains a huge amount of information about the scattering system. We will study the scattering of light and other types of radiation by polymer solutions in Section 3.5. In Section 3.6, we will briefly review those spectral methods that are useful for the study of polymer solutions. Analytical separation techniques will be covered in Section 3.7. Finally, we will present in Section 3.8 some techniques that are used for the study of nucleic acids.

3.1. THERMODYNAMICS OF MACROMOLECULAR SOLUTIONS

Equilibrium properties of materials can be understood and/or predicted when their *Gibbs energy G* (frequently also called *free energy* or *free enthalpy*) is known as a function of the independent variables of the system, for example, of temperature, pressure, and composition. When the thermodynamics of mixtures is studied, the quantity of interest is the change in Gibbs function due to the process of mixing, ΔG_{mix}, defined as

$$\Delta G_{\mathrm{mix}} \equiv G - \sum_i n_i \, \overline{G}_i^{\mathrm{o}} \tag{3.1.1}$$

In this relation, n_i is the number of moles of component i; $\overline{G}_i^{\mathrm{o}}$ is the molar Gibbs function of pure component i at the pressure and temperature of the system; the sum includes all the components of the system. $\overline{G}_i^{\mathrm{o}}$ is equal to the chemical potential, μ_i^{o} of pure component i.

Recognizing that chemical potential μ_i is equal to the partial molar Gibbs function, that is, to the derivative $(\partial G/\partial n_i)_{P,T, n_{j\neq i}}$, it is easy to show that

$$\mu_i - \mu_i^{\mathrm{o}} = \left(\frac{\partial \Delta G_{\mathrm{mix}}}{\partial n_i} \right)_{P,T,n_{j\neq i}} \tag{3.1.2}$$

The last constancy subscript $n_{j\neq i}$ in these derivatives indicates that the numbers of moles of all components, except component i, are kept constant when evaluating the derivative. When the dependence of ΔG_{mix} on the independent variables is known (preferably in an algebraic form), all the basic thermodynamic properties of solutions are derived easily using standard thermodynamic manipulations.

Most theoretical procedures for predicting ΔG_{mix} start with the construction of a model of the mixture. The model is then analyzed by techniques of statistical thermodynamics. Different theories differ in the nature and sophistication of the underlying models, in the level of the statistical mechanical approach, and in the seriousness of the mathematical approximations that are invariably introduced into the calculation.

In the following sections several more frequently used theories and their underlying models will be introduced. Among the models we will find a few old friends, which were employed in Section 1.3.1.3 for the analysis of the effect of long-range interactions on the shape of a macromolecular coil. We will see that the problems encountered in the analysis of polymer coils and thermodynamic functions have many common points. This follows from the fact that both these phenomena depend strongly on interactions among polymer segments—on segments belonging to the same macromolecule in the case of coil shape analysis and on segments of different macromolecules in the thermodynamic case.

The Flory–Huggins theory, which was developed in the early 1940s, has a very prominent position among the many theories of polymer solutions. It is based on the pseudolattice model and a rather low-level statistical treatment with many approximations. The result is a theory that is amazingly simple, explains correctly (at least qualitatively) a large number of experimental observations, and serves as a starting point for many more sophisticated theories. The main drawback of the Flory–Huggins theory is its immense popularity. It has been repeatedly used and abused for analysis of situations for which it was clearly not suited because of the oversimplifications and approximations involved.

Another class of theories, known broadly as excluded-volume theories or perturbation theories, concentrates its detailed attention on the spatial and energetic interaction of polymer segments. The lattice model is used in some versions of the theory, whereas other versions do not need it. We have already seen these theories at work when talking about pseudo-ideal coils and the statistics of coil conformations.

Both the Flory–Huggins and perturbation theories build their thermodynamic functions on models that stress the spatial aspects of the mixture and the averaged interaction behavior but neglect other equally important aspects such as volume changes on mixing, compressibility, and the heterogeneous nature of molecular surfaces. Development of a theory that would take into account all these phenomena, although highly desirable, seems to be a formidable task. Two classes of theories (otherwise diametrically opposite)—equation-of-state theories and phenomenological theories—recognize the apparent impossibility of deriving the basic relations and postulate them instead.

Equation-of-state theories postulate a plausible, but nevertheless arbitrary, partition function with a few adjustable parameters (which have a well-defined physical significance) and derive all relevant quantities from it with a success that is fair at best. Similarly, the phenomenological studies assign some plausible form to some thermodynamic function (usually ΔG_{mix}) using empirical parameters. Usually, these parameters are assumed to be functions of all pertinent independent variables. In this sense, the relations are exact. However, the parameters sometimes have no clear physical significance. Such approaches have their value in describing and organizing experimental data and in preparing an experimental base for a future, more comprehensive, theory of mixtures.

In the following sections we will treat the above-mentioned theories in some detail.

3.1.1. Flory–Huggins Model

Flory–Huggins theory is based on splitting ΔG_{mix} into an enthalpy term and an entropy term according to equation (3.1.3) and evaluating these two terms separately.

$$\Delta G_{mix} = \Delta H_{mix} - T\Delta S_{mix} \qquad (3.1.3)$$

Physically, the pseudolattice model (see Section 1.3.1.3) is employed. The entropy of mixing, ΔS_{mix}, is computed from the number of possible arrangements of the molecules on the lattice. The enthalpy of mixing, ΔH_{mix}, is calculated as a change in interaction energies among molecular surfaces during the process of mixing.

The mathematical procedures employed in the Flory–Huggins theory, although straightforward, are somewhat lengthy. We find it convenient to develop first the theory for mixtures of low-molecular-weight solvents to get a better grasp of the principles involved; then we will return to macromolecular solutions.

3.1.1.1. Low-Molecular-Weight Mixtures. Let us consider a mixture of two substances, A and B, molecules of which have similar sizes and shapes. Let us model the mixture by a pseudocrystalline lattice; each lattice point is occupied by one molecule and has z neighbors. We will call z the *coordination number* of the lattice. Although the selection of the value of z is not obvious (physically, it should be between 4 and 12), the exact choice is not really important. We will see that its value will not appear in the more important results of the theory. Because of the similarity of both types of molecules, the same lattice may accommodate both types of molecules.

Let us first study the change in entropy of mixing, ΔS_{mix}. According to the rules of statistical thermodynamics, it is permissible to split the entropy of any system into several components: (1) entropy of external degrees of freedom, which for small molecules are essentially translational degrees of freedom; (2) entropy of internal degrees of freedom, for example, molecular rotations and vibrations; (3) entropy of intermolecular interactions; and (4) configurational entropy. The theory assumes that the entropy contributions of types 1–3 do not change upon mixing, that is, they do not contribute to ΔS_{mix}. This is a strong assumption, which is probably fulfilled only rarely; different neighbors may disturb the energy levels of a molecule in different ways and make the molecular entropy dependent on the nature of the molecular surroundings.

Nevertheless, the configurational entropy S_{conf} is considered to be the only source of ΔS_{mix}. The term S_{conf} is calculated from the well-known Boltzmann relation

$$S_{conf} = k \ln W \qquad (3.1.4)$$

where k is the Boltzmann constant and W is the number of different microscopic arrangements compatible with the same macroscopic state of the system. The use of relation (3.1.4) implies an assumption of total randomness of molecular arrangements; the preference for some lower-energy arrangements (usually accounted for by Boltzmann factors) is neglected. Solutions conforming to this assumption are sometimes called *regular solutions.*

For our model, the number of arrangements W is simply the number of different ways of selecting N_A locations for N_A molecules of substance A from $N_A + N_B$ locations needed for placement of all molecules A and B. According to the well-known combinatory formula, W can be written as

$$W = (N_A + N_B)!/(N_A! N_B!) \tag{3.1.5}$$

Substituting equation (3.1.5) into equation (3.1.4) and using the Stirling approximation for factorials,

$$\ln N! = N \ln N - N \tag{3.1.6}$$

we obtain, after some rearrangement,

$$S_{\text{conf}} = -k\left(N_A \ln \frac{N_A}{N_A + N_B} + N_B \ln \frac{N_B}{N_A + N_B} \right) \tag{3.1.7}$$

Realizing that the arguments of logarithms in equation (3.1.7) are molar fractions of the components, x_A and x_B, and substituting for numbers of molecules $N_i = N_{Av} n_i$, where n_i is the number of moles of component i and N_{Av} is Avogadro's number, we obtain the final form.

$$S_{\text{conf}} = -R(n_A \ln x_A + n_B \ln x_B) \tag{3.1.8}$$

In writing equation (3.1.8), we used the identity for the gas constant, $R \equiv N_{Av}k$.

By definition, ΔS_{mix} is the difference between the entropy of the system and the entropies of the components in separated (unmixed) form. According to our model, only configurational entropies need be considered. Moreover, for unmixed components, there is only one way to place N_i indistinguishable molecules of species i on a lattice with N_i lattice points. Consequently, the configurational entropy of unmixed components is zero; S_{conf} of the mixture is equal to ΔS_{mix},

$$\Delta S_{\text{mix}} = -R(n_A \ln x_A + n_B \ln x_B) \tag{3.1.9}$$

To calculate the enthalpy of mixing, ΔH_{mix}, the enthalpy is again separated into (1) internal and external degrees of freedom and (2) terms arising from intermolecular interactions. Again, the contribution of terms of type 1 is assumed to be independent of the state of mixing. Thus these terms do not contribute to ΔH_{mix}.

Intermolecular interactions result from so-called dispersion forces. Electrons moving within molecules produce a fluctuating dipole in all molecules, polar and nonpolar. Polar molecules also possess permanent dipoles. Both types of dipoles induce other dipoles (fittingly called *induced* dipoles) in neighboring molecules. The induced dipoles are attracted to the original dipoles, and this attraction lowers the energy of the system. The energy is lowest when the repulsive forces between the molecular cores and the dispersive forces are exactly balanced; the decrease in energy is called the *interaction energy*. This energy is larger for polar molecules than for nonpolar molecules. It is quite sizable; it is responsible for the cohesion of liquids and molecular crystals.

The interaction energy of pure substance i is expressed by means of *cohesive energy density* e_i, the energy needed to separate the molecules in 1 mL of liquid. This is experimentally achieved by vaporization. Thus e_i is related to the molar enthalpy of vaporization of substance i, $\Delta H_{vap,i}$, as

$$e_i = (\Delta H_{vap,i} - RT)/V_i \tag{3.1.10}$$

The term RT reflects the volume work during vaporization, and V_i is the molar volume of the liquid.

The dispersive forces decrease with the sixth power of the distance between the two dipoles. Thus they have significant values only when the two interacting entities (atoms, atomic groups, small molecules) are in direct contact. It is therefore reasonable to ascribe a characteristic value of interaction energy to each intermolecular contact (or better, to each contact of atoms or atomic groups that are not mutually bonded). The amount of interaction energy per contact obviously depends on the nature of the interacting groups.

In the pseudolattice model, each molecule has z contact points, z being the coordination number of the lattice defined above. In a mixture, there are three types of contacts: A–A, B–B, and A–B. They are assigned interaction energies ε_{AA}, ε_{BB}, and ε_{AB}, respectively. All these values must be negative. The model implies that the surface of each molecular species is homogeneous, that is, that all contact points of a molecule are equivalent. The enthalpy of mixing is calculated from the relation

$$\Delta H_{mix} = H - H_A - H_B \tag{3.1.11}$$

where the contributions of the contact energy to the total enthalpy, H, H_A, and H_B, are for the mixture, for pure component A, and for pure component B, respectively. H_A and H_B are calculated easily as

$$H_A = zN_A\varepsilon_{AA}/2 \tag{3.1.12}$$

$$H_B = zN_B\varepsilon_{BB}/2 \tag{3.1.13}$$

where the factor 2 takes into account the fact that two contact points are needed to form a single contact. In a mixture, all three types of contacts are present; thus the enthalpy of the mixture is a sum of three terms. Assuming that the number of heterogeneous contacts is N_{AB}, we can write

$$H = N_{AB}\varepsilon_{AB} + (zN_A - N_{AB})\varepsilon_{AA}/2 + (zN_B - N_{AB})\varepsilon_{BB}/2 \tag{3.1.14}$$

Substitution of the last three relations into equation (3.1.11) now yields

$$\Delta H_{mix} = N_{AB}[\varepsilon_{AB} - (\varepsilon_{AA} + \varepsilon_{BB})/2] \equiv N_{AB}\Delta\varepsilon_{AB} \tag{3.1.15}$$

where the excess contact energy of an AB contact, $\Delta\varepsilon_{AB}$, is defined by the second relation in equation (3.1.15).

It remains to estimate the number of heterogeneous contacts N_{AB}. It is equal to the number of contact points of type A multiplied by a probability p_B that a neighbor of molecule A is molecule B:

$$N_{AB} = zN_A p_B \tag{3.1.16}$$

The probability p_B is difficult to estimate, because the frequency of different types of contacts is governed by Boltzmann factors involving contact energies. It is therefore replaced by a probability that any molecule in the solution (i.e., not necessarily a neighbor of molecule A) is a molecule B. The latter probability is, of course, equal to the molar fraction of component B, $N_B/(N_A + N_B)$. The replacement is legitimate only when the arrangement of molecules on the lattice is completely random. This is the same assumption that was used in calculating ΔS_{mix}.

Combination of equations (3.1.15) and (3.1.16) now yields

$$\Delta H_{mix} = z\Delta\varepsilon_{AB}N_A N_B/(N_A + N_B) = n_A x_B N_{Av} z\Delta\varepsilon_{AB} \tag{3.1.17}$$

The product $N_{Av}z\Delta\varepsilon_{AB}$ has a well-defined physical meaning: it is an enthalpy change accompanying a transfer of 1 mol of A from pure A to an infinitely dilute solution of A in B. It is also the change in enthalpy for transfer of 1 mol of B from pure B to its infinitely dilute solution in A.

Combination of the expression for ΔS_{mix} from equation (3.1.9) with ΔH_{mix} from equation (3.1.17) according to equation (3.1.3) yields for ΔG_{mix}

$$\Delta G_{mix} = RT(n_A \ln x_A + n_B \ln x_B + n_A x_B N_{Av} z\Delta\varepsilon_{AB}/RT) \tag{3.1.18}$$

The last term in this relation is usually simplified by introducing the *interaction parameter* χ_{AB}

$$\chi_{AB} \equiv N_{Av}z\Delta\varepsilon_{AB}/RT \equiv z\Delta\varepsilon_{AB}/kT \tag{3.1.19}$$

It is seen that, according to this model, the interaction parameter χ_{AB} is inversely proportional to temperature whereas $\Delta\varepsilon_{AB}$ is assumed to be independent of temperature. Both these quantities can be either positive or negative.

Combining relations (3.1.18) and (3.1.19) yields

$$\Delta G_{mix} = RT(n_A \ln x_A + n_B \ln x_B + n_A x_B \chi_{AB}) \tag{3.1.20}$$

This equation describing the van Laar model of solvent mixtures is a predecessor of the Flory–Huggins equation. However, it is applicable only to mixtures of low molecular weight components with approximately the same molar volume.

The interaction parameter χ_{AB} is frequently associated with *Hildebrand's solubility parameter δ_i*, which is related to the cohesive energy density of relation (3.1.10) as

$$\delta_i = \sqrt{e_i} \qquad (3.1.21)$$

The concept of solubility parameters is based on a prediction that for a mixture of two types of spherical particles, the dispersion forces are such as to make the heterogeneous contact energy ε_{AB} a geometrical average of the homogeneous contact energies. (Such a relation would be valid if the forces were electrostatic.) Taking into account the negative sign of all ε_{ij} values, it is then possible to write

$$\varepsilon_{AB} = -\sqrt{\varepsilon_{AA}\varepsilon_{BB}} \qquad (3.1.22)$$

The cohesive energy of 1 mol of a pure component can be expressed either as H_i of equation (3.1.12) or as $e_i V_i$, where V_i is molar volume, that is,

$$e_i V_i = N_{Av} z \varepsilon_{ii} / 2 \qquad (3.1.23)$$

Combining the definition part of equation (3.1.15) with equations (3.1.21)–(3.1.23) and the definition of χ_{AB}, equation (3.1.19), gives the desired relation as

$$\chi_{AB} = V_m (\delta_A - \delta_B)^2 / RT \qquad (3.1.24)$$

where V_m is the molar volume, assumed to be the same for both components.

According to equation (3.1.24), the concept of solubility parameter implies that the χ_{AB} parameter must always be positive; the lowest possible value is zero, applicable to mixtures of components with exactly equal solubility parameters. Actually, negative values of χ_{AB} are possible when the contact energy of heterogeneous contacts is higher than that of homogeneous contacts. The discrepancy is due to the assumptions that only dispersion forces contribute to contact energies and that they can be described by equation (3.1.22). Whenever specific interactions (hydrogen bonds, donor-acceptor interactions, etc.) are involved, the solubility parameter concept must fail.

3.1.1.2. Macromolecular Solutions.

Let us now return to solutions of linear macromolecules. To place both the solvent molecules and macromolecules onto the same pseudolattice, we must subdivide the macromolecules into *segments* having the same volume as molecules of the solvent. If the ratio of molar volumes V_2/V_1 is equal to σ (from now on we will assign the subscript 1 to the solvent and the subscript 2 to the macromolecular solute), then each macromolecule occupies σ consecutive locations on the pseudolattice. For computation of configurational entropy, we must enumerate the number of distinguishable ways of placing N_1 molecules of solvent and N_2 macromolecules onto a lattice with $N_1 + \sigma N_2$ lattice points.

This enumeration is not at all as easy as it was for mixtures of small molecules. The computational technique calls for putting the polymer segments on the lattice one by one and counting the number of ways we can do so. Thus the first segment of the first molecule can be located on any of the $N_1 + \sigma N_2$ locations. For the second segment, we have z choices, and for the third, $z - 1$ choices (one of the neighbors of the second segment is already occupied by the first segment). However, with the fourth and all subsequent segments, we hit a snag. Some of the $z - 1$ neighboring lattice points may have already been occupied by previously placed segments of the same macromolecule or any macromolecule accommodated on the lattice before. It is impossible to calculate the number of *unoccupied* neighboring points by present mathematical means. We must again resort to probabilities. In the original derivation, it was assumed that the probability that a given neighboring point is already occupied is equal to the probability that any point is occupied; the latter probability is, of course, equal to the fraction of points occupied at the present stage of the placing process. This assumption works reasonably well at higher concentrations but fails miserably at low concentrations for the following reason.

Unlike in the case of a mixture of two solvents, when the placement of a molecule was completely independent of the placement of the previous one, we must place the segments of macromolecules in the immediate vicinity of the preceding ones. With long chains, the local density of segments may be quite high (meaning that the probability of a given neighboring point being occupied is not negligibly small), whereas the overall concentration of segments (assumed to represent the occupancy of the neighboring points) may be still vanishingly small. Thus the number of possible conformations is overestimated in the low-concentration region.

Despite this inconsistency, the above assumption gives quite useful results. We should also note that (unlike for low-molecular-weight mixtures) the number of configurations that a pure polymeric component may adopt is not 1. Individual molecules may still adopt huge numbers of conformations, restricted only by the condition that no segment may occupy a location that has already been claimed by another segment. The corresponding configurational entropy of pure polymers is sometimes called the *entropy of disorientation;* it is a major contribution to the entropy of fusion of polymer crystals—a process in which perfectly organized polymer segments in a crystal are converted into more or less random coils. In the present case, the entropy of disorientation must be subtracted from the configurational entropy of the macromolecular solution to obtain the change in entropy accompanying the mixing of solvent with an amorphous (i.e., disoriented) polymer.

Without further computational details, we will present the expression for ΔS_{mix} for a polymer solution represented by our lattice model and subject to the above-described approximations. It reads

$$\Delta S_{\mathrm{mix}} = -R(n_1 \ln \phi_1 + n_2 \ln \phi_2) \tag{3.1.25}$$

Here, the *volume fractions* ϕ_1 and ϕ_2 are the fractions of the total number of lattice points occupied by solvent molecules and polymer segments, respectively.

$$\phi_1 = n_1/(n_1 + n_2\sigma) \tag{3.1.26}$$

$$\phi_2 = n_2\sigma/(n_1 + n_2\sigma) \tag{3.1.27}$$

Calculation of ΔH_{mix} for macromolecular solutions follows almost exactly the outline depicted for low-molecular-weight solutions in equations (3.1.11)–(3.1.17). The only difference is in the probability p_2 [p_B of equation (3.1.16)] that a given lattice point is occupied by a polymer segment. In the present case, the probability must be approximated by the volume fraction ϕ_2, not a molar fraction. Thus the expression for ΔH_{mix} reads

$$\Delta H_{mix} = n_1 \phi_2 N_{Av} z \Delta \varepsilon_{12} = RT n_1 \phi_2 \chi_{12} \tag{3.1.28}$$

where χ_{12} is the Flory–Huggins interaction parameter defined by equation (3.1.19).

However, expression (3.1.28) is less plausible for macromolecular solutions than equation (3.1.17) was for the low-molecular-weight solutions. All of the z contact sites of a low-molecular-weight solute are available for contact with solvent molecules, whereas two of the contact sites of a polymer segment are permanently occupied by neighboring segments, and only $z - 2$ are available. Thus relation (3.1.28) is plausible only when the difference between z and $z - 2$ is negligible. It is therefore stated frequently that equation (3.1.28) is valid in the limit of z approaching infinity. This limit is, of course, physically meaningless. A more refined statistics taking into account finite z values reveals that χ_{12} should be concentration-dependent; the dependence is a slowly varying function; the absolute value of χ_{12} increases with increasing ϕ_2.

The influence of z on χ_{12} is a manifestation of a more general phenomenon: The molecular surface-to-volume ratio is usually less for polymers than for solvents. It is the surface of the molecule that is the locus of interactions. Hence, in many models the volume fractions in the interaction terms are replaced by surface fractions. However, we will not continue this line of argument.

Finally, after having discussed the assumptions of the model and approximations done during the derivation, we will combine the entropy term of equation (3.1.25) with the enthalpy term of equation (3.1.28) to obtain the expression for ΔG_{mix}:

$$\Delta G_{mix} = RT(n_1 \ln \phi_1 + n_2 \ln \phi_2 + n_1 \phi_2 \chi_{12}) \tag{3.1.29}$$

This equation is the famous Flory–Huggins relation, which has been the cornerstone of polymer thermodynamics for more than five decades.

3.1.1.3. Chemical Potentials.   Once the analytical expression for ΔG_{mix} is known, the calculation of chemical potentials and other thermodynamic functions (activities, activity coefficients, virial coefficients, etc.) is straightforward. Applying the recipe for the calculation of chemical potentials [equation (3.1.2)] to the low-molecular-weight mixtures described by equation (3.1.20), we get

$$\mu_A - \mu_A^o \equiv RT \ln a_A = RT \left(\ln x_A + \chi_{AB} x_B^2 \right) \tag{3.1.30}$$

$$\mu_B - \mu_B^o \equiv RT \ln a_B = RT \left(\ln x_B + \chi_{AB} x_A^2 \right) \tag{3.1.31}$$

In the limit of vanishing χ_{AB} (remember that for the pseudolattice model, this limit implies vanishing enthalpy of mixing or athermal mixing), equations (3.1.30) and (3.1.31) reduce to expressions valid for ideal solutions with activities a_i equal to molar fractions and activity coefficients γ_i equal to unity. For mixtures that are not athermal, activity and activity coefficients read

$$a_A \equiv x_A \gamma_A = x_A \exp(\chi_{AB} x_B^2) \qquad (3.1.32)$$

with similar expressions for component B. It is apparent that endothermic mixtures (mixtures absorbing heat during mixing), which have positive values of ΔH_{mix} and χ, exhibit positive deviations from Raoult's law. Exothermic mixtures, with negative values of ΔH_{mix} and χ, have negative deviations from Raoult's law.

For macromolecular solutions, we must apply equation (3.1.2) to Flory–Huggins equation (3.1.29), keeping in mind that volume fractions ϕ_i are functions of the numbers of moles. We find easily that the activity and chemical potential of the solvent are given as

$$\ln a_1 = \frac{\mu_1 - \mu_1^o}{RT} = \ln(1 - \phi_2) + \left(1 - \frac{V_1}{V_2}\right)\phi_2 + \chi_{12}\phi_2^2 \qquad (3.1.33)$$

Here, we have replaced the factor σ in the definitions of volume fractions [equations (3.1.26) and (3.1.27)] by the ratio of molar volumes V_2/V_1. It is left to the reader to prove that in the limit of extremely small volume fractions, the right-hand side of equation (3.1.33) is equal to $\ln x_1$, that is, that the chemical potential is equal to the chemical potential of the solvent in an ideal solution having the same molar fraction. This result is, of course, necessary; even macromolecular solutions must obey Raoult's law in extremely dilute solutions. However, at higher concentrations of macromolecules, the decrease in a_1 with increasing concentration of the polymer (expressed as its volume fraction) is much sharper for solutions obeying equation (3.1.33) (in the absence of enthalpy term) than for ideal solutions.

For many physicochemical calculations, especially when dealing with dilute solutions, it is convenient to express the solvent activity as a power series in terms of polymer concentration c_2 in mass/volume (g/mL) units. For the pseudolattice model (no change in volume in mixing), volume fractions and concentrations are related as

$$\phi_i = c_i v_i \qquad (3.1.34)$$

where the v_i values are specific volumes. Developing $\ln(1 - \phi_2)$ into a Taylor series and employing equation (3.1.34), we can transform equation (3.1.33) into

$$\ln a_1 = -\phi_2 - \frac{\phi_2^2}{2} - \frac{\phi_2^3}{3} - \cdots + \left(1 - \frac{V_1}{V_2}\right)\phi_2 + \chi_{12}\phi_2^2$$

$$= -\left[c_2 v_2 \frac{V_1}{V_2} + c_2^2 v_2^2 \left(\frac{1}{2} - \chi_{12}\right) + \frac{c_2^3 v_2^3}{3} + \cdots\right] \qquad (3.1.35)$$

Realizing that $V_2 = v_2 M_2$, we modify this relation further as

$$\ln a_1 = -c_2 V_1 \left[\frac{1}{M_2} + \left(\frac{1}{2} - \chi_{12} \right) \frac{c_2 v_2^2}{V_1} + \frac{c_2^2 v_2^3}{3 V_1} + \cdots \right]$$

$$= -c_2 V_1 \left[\frac{1}{M_2} + B c_2 + C c_2^2 + \cdots \right] \qquad (3.1.36)$$

where the *second virial coefficient B* and *third virial coefficient C* are defined as coefficients of the power series given by equation (3.1.36). Their interpretation in terms of Flory–Huggins theory is apparent from relation (3.1.36).

When the second virial coefficient becomes zero, relation (3.1.36) simplifies nicely, and many thermodynamic measurements are much easier to interpret. Such a situation deserves to be called ideal, but this adjective has already been reserved for systems whose activities are equal to the molar fractions. We will therefore call solutions with vanishing B *pseudo-ideal solutions*.

Inspection of equation (3.1.36) reveals that B vanishes when the parameter χ_{12} is equal to $\frac{1}{2}$. We should now recall that, according to its definition, equation (3.1.19), χ_{12} is inversely proportional to absolute temperature. If χ_{12} is positive (as it is for most polymer-solvent systems), then it acquires the value $\frac{1}{2}$ at some specific temperature.

It is seen that the Flory–Huggins relation also describes the temperature dependence of the thermodynamics of mixtures. However, it was soon realized that because of its conceptual limitations, the theory is not satisfactory for treating temperature dependences. Flory therefore modified the theory in the following way. The key factor for the calculation of B is $\frac{1}{2} - \chi_{12}$. The term $\frac{1}{2}$ originated from the Taylor expansion of $\ln(1 - \phi_2)$; we can trace it back to the entropy of mixing. The χ_{12} term has enthalpic origin. However, in real systems, molecular contacts may change the contribution of individual molecules to entropy. Hence, χ_{12} also has an entropic component. Similarly, in assigning the value $\frac{1}{2}$ to the entropy term, we have included only the configurational entropy of the lattice model and neglected possible contributions from changes in the volume in mixing and from contact interactions.

It is therefore expedient to abandon the particular values for both terms and to adopt more general unspecified values: ψ for the entropic term and κ for the enthalpic term. It is also convenient to recognize that the enthalpic term must be inversely proportional to temperature and replace κ by $\psi \theta / T$, where θ is a newly introduced parameter with dimension of temperature. The procedure could be described as

$$\tfrac{1}{2} - \chi_{12} = \psi - \kappa = \psi(1 - \theta / T) \qquad (3.1.37)$$

Using the new symbols, the expression for the second virial coefficient is written as

$$B = \psi(1 - \theta / T) v_2^2 / V_1 \qquad (3.1.38)$$

Thus, at some special temperature $T = \theta$, B vanishes and the solutions become pseudo-ideal. Sometimes, such solutions are also called theta solutions. At temperatures higher than θ, B is positive; at lower temperature B is negative.

We should, perhaps, inquire what happened to our model when we introduced relations (3.1.37). Formally, we have replaced one interaction parameter χ_{12} with two new parameters ψ and θ, adding flexibility to our relations. By doing so we have effectively abandoned the premises of the pseudolattice model; equation (3.1.38) belongs, strictly speaking, to phenomenological models. We will study the phenomenological approaches in more detail in Section 3.1.4.

Having discussed in detail the chemical potential of the solvent, we will now focus attention on the chemical potential of the macromolecular solute. It is again obtained from equations (3.1.2) and (3.1.29). The derivative is now taken with respect to n_2, and the result reads

$$\frac{\mu_2 - \mu_2^o}{RT} = \ln \phi_2 + \left(1 - \frac{V_2}{V_1}\right)(1 - \phi_2) + (1 - \phi_2)^2 \chi_{12} \frac{V_2}{V_1} \quad (3.1.39)$$

In discussing chemical potentials of solutes, Henry's law is frequently invoked, and the activity of the solute at vanishing concentration is set equal to dimensionless concentration c_2/c_0, where c_0 is unit concentration. This procedure can be formally described as the absorption of those terms in relation (3.1.39) that are not composition-dependent into the standard chemical potential, thus creating the *hypothetical standard state* having chemical potential μ_2^{oH}. After this maneuver, we again switch from volume fractions to concentrations and obtain [recalling the definition of virial coefficient B in equation (3.1.36)]

$$\frac{\mu_2 - \mu_2^{oH}}{RT} = \ln \frac{c_2}{c_0} + 2c_2 M_2 \left(B - \frac{v_2}{2M_2}\right) + c_2^2 v_2^3 \frac{M_2 \chi_{12}}{V_1} \quad (3.1.40)$$

In later sections of this chapter we will need an expression for the derivative of chemical potential of the solute with respect to its concentration. We obtain easily from equation (3.1.39)

$$\left(\frac{\partial \mu_2}{\partial \phi_2}\right)_{P,T} = RT \left[\frac{\phi_1}{\phi_2} + (1 - 2\chi_{12}\phi_1) \frac{V_2}{V_1}\right] \quad (3.1.41)$$

Alternatively, equation (3.1.40) yields the virial form

$$\left(\frac{\partial \mu_2}{\partial c_2}\right)_{P,T} = RT \left[\frac{1}{c_2} + 2M_2 \left(B - \frac{v_2}{2M_2}\right) + 2c_2 v_2^3 \frac{M_2 \chi_{12}}{V_1}\right] \quad (3.1.42)$$

Another useful form of this derivative is derived from the Gibbs–Duhem equation and equations (3.1.33) and (3.1.36) (second part).

$$\left(\frac{\partial \mu_2}{\partial c_2}\right)_{P,T} = -\frac{n_1}{n_2}\left(\frac{\partial \mu_1}{\partial c_2}\right)_{P,T} = -\frac{\phi_1 M_2}{c_2 V_1}\left(\frac{\partial \mu_1}{\partial c_2}\right)_{P,T}$$

$$= RT \frac{\phi_1}{c_2}\left(1 + 2BM_2 c_2 + 3CM_2 c_2^2 + \cdots\right) \quad (3.1.43)$$

Expressions (3.1.41)–(3.1.43) are, of course, fully equivalent.

3.1.2. Excluded-Volume Theories

During the derivation of the Flory–Huggins equation, it was stressed that it is not applicable at low concentrations of the polymer. Yet when we calculated chemical potentials and their virial expansion, we employed precisely the forbidden assumption of vanishing concentration of the polymer. The result had to lead to discrepancies. The most apparent of these was the dependence of the second virial coefficient on the molecular weight of the polymer. According to equation (3.1.36), B should be independent of molecular weight, but experiments show that it decreases significantly with increasing molecular weight. To overcome this discrepancy, it was necessary to develop new concepts for modeling macromolecular solutions. The excluded-volume model proved to be very fruitful.

The underlying idea of the excluded-volume model is the exploitation of simple thermodynamic behavior of ideally dilute solutions obeying Henry's law. In the ideally dilute limit, the molecules of the solute are completely separated from each other and are in contact only with solvent molecules. If more solvent is added to such a solution, the number of solute-solvent contacts does not change. Hence, no enthalpy change accompanies this process. However, when the solute concentration is increased (e.g., by removing some solvent), the solute molecules start to interact. In the first approximation, which is used mainly for solutes having large, compact molecules, the interaction is considered to be only steric in nature—solute molecules occupying some location *exclude* this location from the volume available to other molecules. The steric interference influences entropy but not enthalpy. Any enthalpic interactions are neglected.

In the next approximation, the enthalpic interactions (interpreted as net excess attractive or repulsive forces between polymer segments in contact) are included, but in excluded-volume language they are accounted for as changes in the excluded volume. The repulsive forces increase the excluded volume, but the attractive forces decrease it. Thus, within this concept, vanishing or negative excluded volumes are possible. This approximation is used mainly for polymer coils, where the main task becomes the evaluation of the number of segment-segment contacts when two polymer coils approach each other.

3.1.2.1. Compact Molecules.   Calculation of the entropy of mixing within the excluded-volume model is based on the concept that the contribution of a solute molecule to entropy depends on the number of ways in which we can place the molecule into solution. The latter quantity is proportional to the total volume of the system minus the volume that is not accessible because of the presence of solute molecules placed in solution previously.

Let us first consider a solution of spheres of radius R; the position of a sphere is fully described by coordinates of its center. It is apparent from Figure 3.1 that the center of one sphere cannot approach the center of another sphere closer than two radii. Hence, the volume excluded by one sphere equals eight times its actual

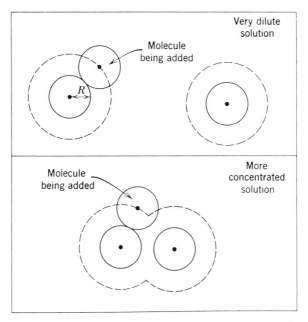

Figure 3.1. The excluded volume for solid spheres. An overlap of excluded volumes in more concentrated solutions. (Reprinted from C. Tanford, *Physical Chemistry of Macromolecules.* Copyright © 1961 by John Wiley & Sons, New York. Used by permission of the publisher.)

volume. For particles of different shapes, the multiple may be different. We will describe the excluded volume by a parameter u. It is also apparent from Figure 3.1 that at higher concentrations the volumes excluded by two or more particles may overlap: The total volume excluded by all particles is less than the sum of the volumes excluded by individual particles. Hence, the theory will run into complications at higher concentrations.

Let us now assume that the number of ways w_i of placing the ith molecule into liquid is proportional to the available volume. For the first molecule, the entire volume V is available; consequently $w_1 = AV$. Here, A is a proportionality constant that will be shown to be inconsequential. For the ith molecule, w_i is given as

$$w_i = A[V - (i - 1)u] \tag{3.1.44}$$

The total number of ways W is now

$$W = \left(\prod_{i=1}^{N_2} w_i \right) \Big/ N_2! = \left\{ \prod_{i=1}^{N_2} A[V - (i - 1)u] \right\} \Big/ N_2!$$

$$= (AV)^{N_2} \left\{ \prod_{i=1}^{N_2} [1 - (i - 1)u/V] \right\} \Big/ N_2! \tag{3.1.45}$$

The factorial $N_2!$ takes into account the number of ways we can permute N_2 indistinguishable molecules. Applying now the Boltzmann relation (3.1.4) to W, we obtain

$$S_{\text{conf}} = k\left[N_2 \ln AV + \sum_{i=1}^{N_2} \ln\left\{1 - (i-1)\frac{u}{V}\right\} - \ln N_2! \right] \qquad (3.1.46)$$

In dilute solutions, the total excluded volume (iu) is still much less than V; we can therefore develop the logarithms into Taylor series and keep only the first terms. We will also use the Stirling approximation [equation (3.1.6)] for the factorial $\ln N_2!$

$$S_{\text{conf}} = k\left[N_2 \ln AV - \sum_{i=1}^{N_2} (i-1)\frac{u}{V} - N_2 \ln N_2 + N_2 \right]$$

$$= kN_2[\ln AV - \ln N_2 - (N_2 - 1)u/2V + 1] \qquad (3.1.47)$$

Now $\mu_1 - \mu_1^{\circ}$ is most conveniently obtained from the well-known relation

$$\mu_1 - \mu_1^{\circ} = \left(\bar{H}_1 - \bar{H}_1^{\circ}\right) - T\left(\bar{S}_1 - \bar{S}_1^{\circ}\right)$$

$$= -T\bar{S}_1 = -T\left(\frac{\partial S_{\text{conf}}}{\partial n_1}\right)_{P,T,n_2}$$

$$= -RT c_2 V_1 \left(\frac{1}{M_2} + c_2 \frac{u N_{\text{Av}}}{2M_2^2}\right) \qquad (3.1.48)$$

Here we have employed the fact that for our model, $\bar{H}_1 - \bar{H}_1^{\circ} = 0$; $\bar{S}_1^{\circ} = 0$; and $V = n_1 V_1 + n_2 V_2$. Comparison of equations (3.1.36) and (3.1.48) yields for the second virial coefficient

$$B = u N_{\text{Av}}/2M_2^2 \qquad (3.1.49)$$

For compact particles of a given shape, the molar excluded volume, $u N_{\text{Av}}$, must be proportional to the molecular weight of the particles, M_2. Hence, the excluded volume model predicts that the second virial coefficient of particles of a given shape decreases with increasing molecular weight of the particles (it is inversely proportional to it).

3.1.2.2. Macromolecular Coils.

We will now explore how to adapt the concept of excluded volume to polymer coils. We have seen in Section 1.3.1 that molecular coils are very loose; the actual macromolecule occupies a few percent at most of the region of the coil. Thus the first impression is that two coils can easily interpenetrate and the excluded volume and its effects are very small. However, this impression is totally false. We must analyze the situation in more detail.

First, to get a feeling for the problems involved, let us consider a fictitious problem of two spherical particles, each composed of 1000 segments homogeneously distributed within its sphere. The segments occupy 1% of the sphere's volume. Let us place the centers of the two spheres on the same point; the spheres overlap fully. What is the probability that no segment of the first sphere occupies the same location as any segment of the second sphere? For the first segment, the probability is 0.99, because only 1% of locations are not available. For every other segment, the probability is the same. The probability that all segments find an unoccupied location simultaneously is therefore 0.99^{1000}, or about 4×10^{-5}, a very small number indeed. Consequently, for all practical purposes, the center of one sphere is excluded for centers of other spheres. As we remove the sphere further away from the reference sphere, the number of potential segment overlaps decreases as the overlap of the spheres decreases; it vanishes when the spheres no longer overlap.

The above example shows that the difference between compact and loose particles is mainly quantitative: For compact particles, a given location is either excluded or it is not; for loose particles, the degree of exclusion may have any value between zero and one. It is therefore possible to find the effective total excluded volume as an integral of the degree of exclusion over the region of coil.

The actual models are more sophisticated than this primitive model. The distribution of segments is modeled to conform more closely to actual coils. One possibility is to assume Gaussian distribution of densities in a particle of spherical symmetry (this represents fairly the actual distribution averaged over a long time). Other models take into account the continuity of the polymer chain and the distribution of intersegmental distances. These models usually require rather advanced mathematics for their treatment.

However, the above spatial problem represents only one half of the story. The other half is in intersegmental interactions. If the formation of a segment-segment contact lowers the overall energy (this corresponds to positive χ_{12} in the Flory–Huggins theory), then interpenetration of coils is enhanced. At some point, the physical exclusion and the intersegmental attraction balance each other exactly and the coils interpenetrate freely. We have met this situation already in Section 1.3.1.3 in discussing conformation of a polymer coil. These two situations are actually two manifestations of the same phenomenon; after all, when two segments meet in the solution, they do not "know" whether they are parts of two different molecules or belong to distant portions of the same macromolecule.

There are a large number of theories treating these two intertwined phenomena. They differ mainly in the way in which they model the spatial problem and in the rigor of mathematical treatment. However, they treat the problem of intersegmental interactions in the same way. They consider the situation when the exclusion effect and attractive forces balance each other as the perfect reference situation and calculate the perturbation of this perfect state that results if the balance is not complete. Accordingly, they are called *perturbation theories*. They are mathematically rather complicated; just considering their basic features and their main results will satisfy us.

Let us set the origin of our coordinate systems ($r = 0$) at the center of a reference segment. Centers of other segments are characterized by their radial coordinate r. We will describe the probability that some segment is located at a given point by a *radial distribution function* $g(r)$. This function is normalized to have a unit value for large intersegmental distances. We will consider first a hypothetical system of point particles. The radial distribution function of point particles depends on the interparticle interactions. It has a uniform unit value in the absence of interactions (e.g., in an ideal gas). However, any intermolecular potential $w(r)$ produces its characteristic $g(r)$ function. Conversely, any $g(r)$ function can be described by a properly tailored interaction potential $w(r)$. The two functions are related as

$$g(r) = \exp[-w(r)/kT] \tag{3.1.50}$$

Let us now compare the average number of segments around another segment with a similar average around a location without a segment. The difference (excess number of segments) is proportional to the so-called binary cluster integral β,

$$\beta = \int [1 - g(r)] dV \tag{3.1.51}$$

Here the integration is over all space. Obviously, β has the dimension of volume. [In the special case where $w(r)$ is infinite and $g(r)$ is zero within some region excluded to other segments, β is exactly the volume of such an excluded region.]

When real molecules are considered, $g(r)$ must vanish for intersegmental distances less than the combined radii of the segments. At larger distances, $g(r)$ oscillates; the first maximum corresponds to closest neighbors. The value of $g(r)$ at this maximum reflects the intermolecular interactions. If the segment-segment contacts are preferred (i.e., segment-solvent contacts are *not* preferred and χ is positive), the first maximum is higher than otherwise. In perturbation theories, equation (3.1.50) is applied even in this case. The appropriately chosen $w(r)$ function is now called the *potential of a mean force.* Both the segment exclusion and molecular interactions contribute to it.

Returning now to the analysis of coil expansion and mutual interaction of two coils, we construct ensembles of all possible coil conformations (or conformations of two coils when dealing with coil-coil interactions). In the partition function for these ensembles, the individual terms are modified by Boltzmann factors reflecting the potential of the mean force $w(r)$. The mathematical simplification arises from the fact that all quantities of interest are ensemble averages and are therefore calculated using integration. Thus the actual functional forms of $w(r)$ and $g(r)$ are not important for these calculations but only the value of the cluster integral, which figures in all results.

The next step is the difficult one: a detailed analysis of coil conformations with an evaluation of the frequency of segment-segment contacts. It is relatively easy to evaluate the frequency of single contacts; for multiple contacts the mathematical problems grow very fast. This is the weak point of perturbation theories; extensive literature covers the subject, and different simplifications lead to quite different results.

Nevertheless, it turns out that multiple contacts influence the computational results very little when the value of β is small. For such β values, the theoretical results are relatively simple. We will present some of the more important ones. In a solvent in which β equals zero, the conformational statistics of a coil is the statistics of an unperturbed Gaussian coil. Similarly, in the case of intermolecular interaction and $\beta = 0$, the presence of one coil does not influence the statistics of the other coil: the coils interpenetrate freely; the excluded volume and second virial coefficient are equal to zero.

Let us return to the nomenclature of Section 1.3.1.2, with N being the number of segments, A their length, and $\overline{h_0^2}$ the average end-to-end distance of unperturbed coils. Although N and A are somewhat arbitrary, the product NA^2 has direct physical significance. The value of β suffers from the same arbitrariness; however, all results of the perturbation calculations have β always in the combination $N^2\beta$; this product is then directly related to physically significant quantities. It turns out that NA^2 and $N^2\beta$ are the only sample-dependent quantities entering into the final results of many perturbation theories. Such theories are then often called *two-parameter theories*.

It is convenient to define a new parameter z as

$$z \equiv (3/2\pi NA^2)^{3/2} N^2\beta = \left(3/2\pi\overline{h_0^2}\right)^{3/2} N^2\beta = (3/2\pi A^2)^{3/2} \beta N^{1/2} \quad (3.1.52)$$

The introduction of the parameter z allows us to cast most of the results of perturbation calculations in a relatively simple form. The disadvantage lies in the fact that parameter z cannot be measured directly. It is evaluated from a large body of experimental data in a rather indirect way.

The expression for the second virial coefficient is

$$B = \left(N_{Av} N^2\beta / 2M_2^2\right) h(z) \quad (3.1.53)$$

The exact form of the function $h(z)$ has been the subject of extensive theoretical search that has resulted in many quite different relations. However, most researchers agree that $h(z)$ can be developed into a Taylor series; they even agree on the coefficient of the linear term:

$$h(z) = 1 - 2.865z + \cdots \quad (3.1.54)$$

This series converges rather slowly; consequently, the results are useful only for small values of z and β.

Let us study equation (3.1.53) in some detail. For small values of z, B is equal to the factor in parentheses. As N must be proportional to M_2, the virial coefficient is independent of molecular weight in the limit of small z. This is the same result as obtained from Flory–Huggins theory. Indeed, in this limit equations (3.1.38) and (3.1.53) are frequently combined to yield

$$N_{Av} N^2\beta / 2M_2^2 = \left(v_2^2 / V_1\right) \psi (1 - \theta/T) \quad (3.1.55)$$

This relation provides a useful link between $N^2\beta$ and Flory's constants ψ and θ. Combining equations (3.1.52) and (3.1.55) gives an alternative definition of the parameter z:

$$z = \left(\frac{3}{2\pi}\right)^{3/2} \left(\frac{\overline{h_0^2}}{M_2}\right)^{-3/2} \left(\frac{2v_2^2}{N_{Av}V_1}\right) \psi \left(1 - \frac{\theta}{T}\right) M_2^{1/2} \equiv B^* M_2^{1/2} \qquad (3.1.56)$$

In the second half of equation (3.1.56) we have also introduced a new thermodynamic parameter B^*, which is frequently employed in experimental work and is independent of molecular weight.

When the value of z is not very small, the function, $h(z)$ in equation (3.1.53) starts to differ significantly from zero. Combining equations (3.1.53)–(3.1.56) yields for the second virial coefficient

$$B = \left(v_2^2/V_1\right) \psi \left(1 - \theta/T\right) \left(1 - 2.865 B^* M_2^{1/2} + \cdots\right) \qquad (3.1.57)$$

Thus, for a given polymer-solvent system, B is decreasing with molecular weight in agreement with experiments. Unfortunately, because of the slow convergence of the series given in equation (3.1.54), the nice linear relationship (3.1.57) between B and $M_2^{1/2}$ is valid only within a limited range of z values.

It is also instructive to compare the perturbation second virial coefficient [equation (3.1.53)] with the one from excluded-volume theory for simple particles [equation (3.1.49)]. The apparent excluded volume of a coil is then

$$u = N^2 \beta h(z) \qquad (3.1.58)$$

As we have seen repeatedly, the same interaction among polymer segments operates in intracoil and intercoil interactions. It is therefore appropriate to return to the problem of coil expansion in good solvents, which we have treated in Section 1.3.1.3. Let us define the *expansion coefficient* α^2 as a ratio between the actual end-to-end distance, $\overline{h^2}$, and the unperturbed distance, $\overline{h_0^2}$,

$$\overline{h^2} \equiv \alpha^2 \overline{h_0^2} \qquad (3.1.59)$$

At the vanishing concentration of macromolecules, the perturbation theories relate α^2 to parameter z as another slowly converging series:

$$\alpha^2 = 1 + \tfrac{4}{3}z - \cdots \qquad (3.1.60)$$

Again, equation (3.1.60) is useful only for small values of z. And again, extensive literature is devoted to the search of an expression for α applicable to a broader range of circumstances. Among these results, the Flory expression is quoted quite often.

$$\alpha^5 - \alpha^3 = 2C_M \psi (1 - \theta/T) M^{1/2} \qquad (3.1.61)$$

$$C_M = \frac{1}{2} \left(\frac{9}{2\pi}\right)^{3/2} \left(\frac{\overline{h_0^2}}{M_2}\right)^{-3/2} \left(\frac{v_2^2}{N_{Av}} V_1\right) \qquad (3.1.62)$$

Comparison of equation (3.1.56) with equations (3.1.61) and (3.1.62) helps us to recast the Flory equation in terms of parameter z as

$$\alpha^5 - \alpha^3 = (3^{3/2}/2)z \qquad (3.1.63)$$

However, it should be noted that equation (3.1.63), when manipulated to yield a series for α^2, changes to

$$\alpha^2 = 1 + (3^{3/2}/2)z + \cdots = 1 + 2.598z + \cdots \qquad (3.1.64)$$

Thus the coefficient of the linear term is almost twice as large as that predicted by other forms of perturbation treatment [equation (3.1.60)].

Expansion of coils decreases with increasing concentration of the polymer. At higher concentrations, the coils slowly contract toward their unperturbed dimensions.

Monte Carlo calculation techniques, which helped us to gain insight into the factors influencing the coil conformation, have also been applied for evaluation of the interpenetration of polymer coils. It has been shown that the binary cluster integral (this time calculated for two whole, randomly coiled chains, not for two single segments) has a positive and a negative component in a poor solvent. The attraction between the chain segments dominates the interaction for a shallow interpenetration of the two coils, whereas for a deep interpenetration the dominant effect is the decrease in conformational entropy (the conformations of the two chains become interdependent). At the theta point, these two contributions balance each other. This is a quite different concept from that of freely interpenetrating coils in pseudo-ideal solutions that was originally implied by the perturbation treatment.

3.1.3. Equation-of-State Theories

Flory–Huggins theory and its refinements, excluded-volume theories, explain with various degrees of success a number of phenomena that are observed with polymer solutions: coil expansion, virial coefficients and their temperature dependence, and some phase equilibria. Nevertheless, some phenomena are beyond the reach of these theories. The pseudolattice model assumes that there is no change in volume during the mixing process and that the solutions are incompressible. Consequently, pressure does not figure in these theories, and there is no means to describe pressure-dependent phenomena. The theories also cannot explain why for some polymer-solvent mixtures the solvent quality worsens with increasing temperature and the solutions demix at some high temperature. (More recent research has shown that this behavior occurs for most polymer solutions if the temperature is raised enough.)

It was perceived that a more comprehensive theory must provide not only the entropy and enthalpy of mixtures but also the relationship between temperature, pressure, and volume of the system. The latter relation is, of course, the equation of state—hence the term *equation-of-state theories*. There exist a number of different approaches, but the concept originally developed by Prigogine and later elaborated by

Flory is the one employed most frequently in the literature. We will therefore restrict our attention to the Prigogine–Flory theory and some of its salient features.

The key quantity of statistical thermodynamics is the partition function Z. If it is known, all thermodynamic functions are derived easily. Specifically, the Helmholtz function A (which for liquids and solids at low pressures is virtually equal to the Gibbs function G) is related to Z as

$$G \cong A = -kT \ln Z \qquad (3.1.65)$$

The internal energy U (virtually equal to enthalpy for condensed systems), entropy, and pressure are related to A by the well-known expressions

$$H \cong U = \left(\frac{\partial(A/T)}{\partial(1/T)} \right)_V \qquad (3.1.66)$$

$$S = - \left(\frac{\partial A}{\partial T} \right)_V \qquad (3.1.67)$$

and

$$P = - \left(\frac{\partial A}{\partial V} \right)_T \qquad (3.1.68)$$

Like every theory, Prigogine–Flory theory starts by defining a model. This model is deliberately chosen to be very general; it may describe pure components as well as their mixtures. The ingenious idea is in *postulating* the mathematical form of the partition function. Because all relevant quantities are derived from the partition function, internal consistency of all results is automatically ensured. Moreover, judicious selection of the partition function leads to results that adequately describe experimental data while keeping the computations on a manageable level.

The construction of the model starts with an analysis of molecular degrees of freedom. The internal degrees (mainly vibrations and rotations of small groups) are not important for the process of mixing and are neglected. However, those degrees of freedom that can take a molecule or part of it from one place to the other are closely related to the entropy and must be accounted for. For small molecules, only translation represents external degrees of freedom; in this case, there are three. In large molecules, and especially in macromolecules, major parts of the molecule may move more or less independently. Such molecules have many external degrees of freedom; the exact number of them depends on details of molecular structure.

Another feature of the model is an assumption that molecules possess a hard core that cannot be compressed. The volume that a molecule occupies is always larger than its core volume. Ultimately, the model is intended to represent mixtures of molecules of grossly different sizes. It is therefore convenient to subdivide all molecules into segments of arbitrary but equal size. The core volume of a segment (the volume needed to accommodate a segment in the tightest possible packing) is designated v^*; the volume needed for a segment under actual conditions is v. There are r segments per molecule, and each segment has $3c$ external degrees of freedom. The segment size was selected arbitrarily, so the numbers r and c have no physical significance per se. However, the product rv^* is the core volume of the whole molecule; rv is molecular volume (a well-defined quantity); and $3rc$ is the total number of external degrees of freedom per molecule. The number of molecules in the system is N; thus, Nrv is the volume of the system, and $3Nrc$ is the total number of degrees of freedom.

Finally, for a pure liquid the partition function is postulated as

$$Z \approx Z_{comb} [g(v^{1/3} - v^{*1/3})]^{3Nrc} \exp(-E/kT) \tag{3.1.69}$$

$$E = -Nrs\eta/2v \tag{3.1.70}$$

Here, Z_{comb} is a factor representing the number of ways of arranging the molecules in the liquid. The Prigogine–Flory theory brings nothing new in this respect. It simply adopts the results of the pseudolattice model: $k \ln Z_{comb}$ is identified with S_{conf} of equation (3.1.4).

The expression in brackets in equation (3.1.69) is intended to describe the extent to which individual external degrees of freedom may express themselves. Although g is an inconsequential geometric factor, $v^{*1/3}$ is a linear characteristic of the segment; $v^{1/3}$ is the linear space in which the segment can move. Hence, the difference represents the *free volume* (in this case, it is actually free linear space) in which the segment can move. The exponent $3Nrc$ is the total number of degrees of freedom as defined above.

In the exponential term, E signifies the intermolecular energy of the liquid—that part of the energy that is due to intermolecular interactions. The concept of contact energies is again employed, and E is expressed through the number of contact sites per segment s and the energy parameter η characterizing a pair of sites in contact. The intermolecular energy per segment is $-s\eta/2v$; the individual values of s and η are not significant, only their product $s\eta$; η is positive in the present notation.

The intermolecular energy is described quite similarly to the lattice model of Section 3.1.1.1, except for the explicit introduction of molecular surfaces. However, the two models differ in the presence of the segmental volume v in the expression for intermolecular energy. This recognizes the expectation that the interaction energy should decrease when the molecules get farther apart from each other—because of changes in temperature and/or pressure. The functional dependence of E on volume is somewhat controversial. The inverse proportionality to volume simplifies the calculations and is claimed to be physically more plausible than other possible relationships. Nevertheless, its use may be the cause of some problems in the theory, as we will see below.

We can now apply equation (3.1.65) to combined equations (3.1.69) and (3.1.70) and obtain

$$A = -kT\{\ln Z_{comb} + 3Nrc \ln[g(v^{1/3} - v^{*1/3})]\} - Nrs\eta/2v \tag{3.1.71}$$

The volume of the system is $V = Nrv$; thus we can calculate the pressure by applying the recipe, equation (3.1.68), to our last relation:

$$P = ckTv^{-2/3}/(v^{1/3} - v^{*1/3}) - s\eta/2v^2 \tag{3.1.72}$$

The theory now calls for replacement of the parameters s, η, and c by new parameters: the characteristic temperature T^* and characteristic pressure P^*. These quantities are defined as

$$T^* \equiv s\eta/2v^*ck \tag{3.1.73}$$

$$P^* \equiv s\eta/2v^{*2} \tag{3.1.74}$$

Further, reduced thermodynamic variables $\tilde{v}$, $\tilde{P}$, $\tilde{T}$ are introduced as

$$\tilde{v} \equiv v/v^* \tag{3.1.75}$$

$$\tilde{P} \equiv P/P^* \tag{3.1.76}$$

$$\tilde{T} \equiv T/T^* \tag{3.1.77}$$

Combining equations (3.1.72)–(3.1.77) yields the well known Prigogine–Flory equation of state for liquids.

$$\frac{\tilde{P}\tilde{v}}{\tilde{T}} = \frac{\tilde{v}^{1/3}}{\tilde{v}^{1/3} - 1} - \frac{1}{\tilde{v}\tilde{T}} \tag{3.1.78}$$

It turns out that at ambient pressures the left-hand side of equation (3.1.78) is always negligibly small compared to the individual terms on the right-hand side. Hence, it can be safely dropped (at such pressures); the reduced volume and reduced temperature are then related as

$$\tilde{T} = (\tilde{v}^{1/3} - 1)/\tilde{v}^{4/3} \tag{3.1.79}$$

For further calculations, it is necessary to know the values of P^*, T^*, and $\tilde{v}$. They are obtained from experimental values of the thermal expansivity α and the thermal pressure coefficient γ. These quantities are defined as

$$\alpha \equiv \frac{1}{V}\left(\frac{\partial V}{\partial T}\right)_P \equiv \frac{1/V}{(\partial T/\partial V)_P} \equiv \frac{1/v}{(\partial T/\partial v)_P} \tag{3.1.80}$$

$$\gamma \equiv (\partial P/\partial T)_V \tag{3.1.81}$$

The more familiar isothermal compressibility β is related to the above quantities as

$$\beta \equiv -\frac{1}{V}\left(\frac{\partial V}{\partial P}\right)_T = \frac{\alpha}{\gamma} \tag{3.1.82}$$

Applying the recipe of equation (3.1.80) to equation (3.1.79), we obtain the desired relation after some manipulation as

$$\tilde{v}^{1/3} = 1 + \alpha T / 3(1 + \alpha T) \tag{3.1.83}$$

Knowing $\tilde{v}$, we calculate $\tilde{T}$ from equation (3.1.79); from $\tilde{T}$ and the experimental temperature, T^* is obtained easily.

Application of equation (3.1.81) to the equation of state, that is, equation (3.1.78) yields easily, in the limit of low pressures,

$$P^* = \gamma T \tilde{v}^2 \tag{3.1.84}$$

Agreement with experiments is only fair; the characteristic temperature T^* is found to increase significantly with the temperature at which it is measured. Thus, for temperature-dependent studies, this variation should be taken into account. Although significant results are still obtained, this inconsistency tarnishes the beauty of the whole treatment.

The main significance of the theory is in its application to mixtures. The whole theoretical machinery contained in equations (3.1.69) and (3.1.70) remains in place; only the number of external degrees of freedom $3Nrc$ should be replaced by an appropriate sum over components, that is, by $3(N_1 r_1 c_1 + N_2 r_2 c_2)$, where the indices 1 and 2 describe the components of the mixture. It is also necessary to express the intermolecular energy as a sum of terms related to three different contact energy parameters: η_{11} and η_{22} for homogeneous contacts and η_{12} for heterogeneous contacts. The appropriate relation reads

$$E = -(N_1 r_1 s_1 \theta_1 \eta_{11} + N_2 r_2 s_2 \theta_2 \eta_{22} + 2N_1 r_1 s_1 \theta_2 \eta_{12})/2v \tag{3.1.85}$$

Here, θ_i are the surface fractions defined as

$$\theta_1 \equiv 1 - \theta_2 \equiv N r_1 s_1 / (N_1 r_1 s_1 + N_2 r_2 s_2) \tag{3.1.86}$$

We also need to introduce volume fractions ϕ_i^* based on core volumes:

$$\phi_1^* \equiv 1 - \phi_2^* \equiv N_1 r_2 / (N_1 r_1 + N_2 r_2) \tag{3.1.87}$$

This definition is physically more pleasing than the older one, which was based on actual volumes of the components before mixing. The core volume fraction does not change with either temperature or pressure, nor is it influenced by changes of volume in mixing. However, everything has its price; before the core volume fractions can be calculated, the core volumes must be evaluated—this is not an easy task.

After the above-mentioned modifications are made, calculation proceeds along the same lines as before. The mixture is again described by equations of state [equations

(3.1.78) and (3.1.79)], where the reduction parameters for the mixture, P^* and T^*, can easily be shown to be related to their counterparts for pure components as

$$P^* = \phi_1^* P_1^* + \phi_2^* P_2^* - \phi_1^* \theta_2 X_{12} \tag{3.1.88}$$

$$T^* = \frac{\phi_1^* P_1^* + \phi_2^* P_2^* - \phi_1^* \theta_2 X_{12}}{\phi_1^* P_1^* / T_1^* + \phi_2^* P_2^* / T_2^*} \tag{3.1.89}$$

The newly introduced parameter X_{12} represents the differences in contact energies and is similar to $z\Delta\varepsilon_{12}$ of equation (3.1.28). Its definition reads

$$X_{12} \equiv (\eta_{11} + \eta_{22} - 2\eta_{12})(s_1/2v^{*2}) \tag{3.1.90}$$

We will now proceed to calculations of the thermodynamic functions of mixing. Routine calculations yield for the enthalpy of mixing

$$
\begin{aligned}
\Delta H_{\text{mix}} &= (N_1 r_1 + N_2 r_2) v^* \left[\phi_1^* \theta_2 X_{12} \tilde{v}^{-1} - \phi_1^* P_1^* \left(\tilde{v}^{-1} - \tilde{v}_1^{-1} \right) - \phi_2^* P_2^* \left(\tilde{v}^{-1} - \tilde{v}_2^{-1} \right) \right] \\
&= v^* \left[N_1 r_1 \theta_2 X_{12} \tilde{v}^{-1} - N_1 r_1 P_1^* \left(\tilde{v}^{-1} - \tilde{v}_1^{-1} \right) - N_2 r_2 P_2^* \left(\tilde{v}^{-1} - \tilde{v}_2^{-1} \right) \right] \\
&= n_1 V_1^* \left[\theta_2 X_{12} \tilde{v}^{-1} - P_1^* \left(\tilde{v}^{-1} - \tilde{v}_1^{-1} \right) - (\phi_2^*/\phi_1^*) P_2^* \left(\tilde{v}^{-1} - \tilde{v}_2^{-1} \right) \right]
\end{aligned}
\tag{3.1.91}
$$

The first term in brackets corresponds to the contact enthalpy term in the Flory–Huggins theory, where it was the only contribution to the enthalpy of mixing. In the present case, the two remaining terms in brackets in (3.1.91) contribute to enthalpy even if X_{12} vanishes. These terms are called *equation-of-state terms*.

The entropy of mixing is found to be

$$
\begin{aligned}
\Delta S_{\text{mix}} - \Delta S_{\text{comb}} &= 3(N_1 r_1 + N_2 r_2) v^* \left\{ \left(\frac{\phi_1^* P_1^*}{T_1^*} \right) \ln \left(\frac{\tilde{v}^{1/3} - 1}{\tilde{v}_1^{1/3} - 1} \right) \right. \\
&\qquad\qquad \left. + \left(\frac{\phi_2^* P_2^*}{T_2^*} \right) \ln \left(\frac{\tilde{v}^{1/3} - 1}{\tilde{v}_2^{1/3} - 1} \right) \right\} \\
&= 3 n_1 V_1^* \left[\frac{P_1^*}{T_1^*} \ln \left(\frac{\tilde{v}^{1/3} - 1}{\tilde{v}_1^{1/3} - 1} \right) + \frac{\phi_2^* P_2^*}{\phi_1^* T_2^*} \ln \left(\frac{\tilde{v}^{1/3} - 1}{\tilde{v}_2^{1/3} - 1} \right) \right]
\end{aligned}
\tag{3.1.92}
$$

Equation (3.1.92) is applicable both to mixtures of small molecules and to polymer solutions. In the former case, relation (3.1.9), which is valid for ideal mixtures, is used for ΔS_{comb}; the difference $\Delta S_{\text{mix}} - \Delta S_{\text{comb}}$ is then called the *excess entropy of mixing*, ΔS^E. For macromolecular solutions, the Flory–Huggins entropy of mixing, equation (3.1.25), is used for ΔS_{comb}. In this case, the difference $\Delta S_{\text{mix}} - \Delta S_{\text{comb}}$ is called the *residual entropy of mixing*, ΔS^R.

Unfortunately, the evaluation of the enthalpy of mixing and the residual entropy is not straightforward. Both expressions contain the reduced volume $\tilde{v}$, which must be calculated from T^* through $\tilde{T}$ using equation (3.1.79). T^* is found from equation (3.1.89); however, X_{12} is needed for its calculation. This latter parameter cannot be obtained directly, it is therefore considered to be an adjustable parameter that plays a role in all terms of ΔG_{mix}.

We will now construct an expression for ΔG_{mix}. We will cast it in a form very similar to the original Flory–Huggins equation (3.1.29):

$$\Delta G_{\text{mix}} = -T\,\Delta S_{\text{comb}} - T\,\Delta S^{\text{R}} + \Delta H_{\text{mix}}$$

$$= RT\left[\, n_1 \ln \phi_1^* + n_2 \ln \phi_2^* + n_1 \phi_2^* \left(\chi_{\text{H}}^* + \chi_{\text{S}}^*\right)\right] \tag{3.1.93}$$

$$\chi_{\text{H}}^* = \frac{V_1^*}{RT}\left(\frac{\theta_2}{\phi_2^*} X_{12} \tilde{v}^{-1} - \frac{P_1^*}{\phi_2^*}\left(\tilde{v}^{-1} - \tilde{v}_1^{-1}\right) - \frac{P_2^*}{\phi_1^*}\left(\tilde{v}^{-1} - \tilde{v}_2^{-1}\right)\right) \tag{3.1.94}$$

$$\chi_{\text{S}}^* = -\frac{3V_1^*}{R}\left[\frac{P_1^*}{T_1^*\phi_2^*}\ln\left(\frac{\tilde{v}^{1/3} - 1}{\tilde{v}_1^{1/3} - 1}\right) + \frac{P_2^*}{T_2^*\phi_1^*}\ln\left(\frac{\tilde{v}^{1/3} - 1}{\tilde{v}_2^{1/3} - 1}\right)\right] \tag{3.1.95}$$

We should stress that the volume fractions ϕ_i^* employed in the above equations are based on core volumes. They may differ significantly from the usual volume fractions, especially when the reduced volumes of the two components differ appreciably. Similarly, the enthalpy interaction coefficient χ_{H}^* and the entropy interaction coefficient χ_{S}^* may have different values from their usual counterparts, which are defined with respect to the usual volume fractions.

The reduced volumes of the mixture and of the pure components can be used for predicting the change of volume in mixing according to the simple relation

$$\Delta V_{\text{mix}} = V\left(\tilde{v} - \phi_1^* \tilde{v}_1 - \phi_2^* \tilde{v}_2\right) \tag{3.1.96}$$

where V is the combined volume of the mixture components before mixing.

Example

To get a better feeling of the factors involved, we will analyze a hypothetical polymer solution. The polymer and the solvent will have similar chemical structure and similar interactive properties. Thus we may quite realistically assume that the $s\eta$ values will be the same for both components and that X_{12} will be zero. Under these assumptions equations (3.1.74) and (3.1.88) yield $P_1^* = P_2^* = P^*$.

Large molecules have a smaller number of external degrees of freedom per segment than small molecules; that is, their c values are smaller: $c_2 < c_1$ (subscript 2 for polymer, 1 for solvent). Relation (3.1.73) interprets this inequality as $T_2^* > T_1^*$. Thus, at any temperature T, we have for the reduced temperatures $\tilde{T}_2 \equiv T/T_2^* < \tilde{T}_1$. We can now derive easily [cf. equation (3.1.89)] that for our example the reduced temperature of the mixture $\tilde{T}$ is a volume-weighted average of $\tilde{T}_i$:

$$\tilde{T} = \tilde{T}_1 \phi_1^* + \tilde{T}_2 \phi_2^* \tag{3.1.97}$$

For our example, we will select $P^* = 130$ cal/mL; $T_1^* = 5000$ K; $T_2^* = 8000$ K; $V_1^* = 100$ mL. These values are typical for polymer solutions. We will do the analysis for a concentrated solution, $\phi_1^* = \phi_2^* = 0.5$.

First, we will study the system at 300 K. At this temperature, $\tilde{T}_1 = 0.06000$; $\tilde{T}_2 = 0.03750$; and $\tilde{T} = 0.04875$. For these values, we find from equation (3.1.79) the values of reduced volumes [a numerical solution is needed for inversion of equation (3.1.79)]. The values are $\tilde{v}_1 = 1.267910$, $\tilde{v}_2 = 1.140057$, $\tilde{v} = 1.197820$. We have now all the values needed for the calculation of the quantities of interest from equations (3.1.94)–(3.1.96). The results are

$$\chi_H^* = -0.1680; \quad \chi_S^* = 0.6146; \quad \chi^* = 0.4466; \quad \Delta V_{mix}/V = -0.00616 \tag{3.1.98}$$

Let us repeat the calculation for 400 K. Now we have $\tilde{T}_1 = 0.08000$, $\tilde{T}_2 = 0.05000$, $\tilde{T} = 0.06500$, leading to $\tilde{v}_1 = 1.446023$, $\tilde{v}_2 = 1.204920$, $\tilde{v} = 1.304488$. The resulting values are

$$\chi_H^* = -0.3822; \quad \chi_S^* = 0.9017; \quad \chi^* = 0.5195; \quad \Delta V_{mix}/V = -0.02098 \tag{3.1.99}$$

The key feature of these results is the negative nature of the change in volume in mixing. Equation-of-state effects lead to contraction caused by mixing if the characteristic temperatures of the components are different. (This effect may, of course, be offset by the expansion effect due to unequal intermolecular interactions, which would be manifested in different P_i^* values and nonzero X_{12}.) This contraction naturally leads to a decrease in enthalpy [cf. equation (3.1.70)] and to negative values of χ_H^*. Thus the equation-of-state theories predict that mixing becomes more and more exothermic as temperature increases.

Even more spectacular is the effect of contraction on entropy. As the system contracts, the space in which molecules may roam decreases. The result is decreasing entropy and high positive values of χ_S^*. At some sufficiently high temperature, the negative contraction effect on entropy may dominate over the positive combinatory entropy; the entropy of mixing now becomes negative. The effect of this negative ΔS_{mix} on ΔG_{mix} is, of course, tempered by an ever-increasing negative value of ΔH_{mix}. This is an exact reversal of the roles that entropy and enthalpy played in the Flory–Huggins theory.

We can see from the values in relations (3.1.98) and (3.1.99) that the entropy effect dominates over the enthalpy effect. Our example yielded a sizable value of 0.4466 for χ^* even for a system with balanced intermolecular interactions at ambient temperature of 300 K. With increasing temperature, χ^* increases even more. We will see later that this behavior leads to the so-called lower critical temperature of demixing: The solution separates into two phases as the temperature is raised! This is, of course, in total contrast to the behavior of Flory–Huggins solutions.

It is worth mentioning that a similar reverse of the Flory-Huggins situation occurs also in some water solutions. Typical examples are solutions of poly(propylene oxide) and its copolymers (Pluronics). Paradoxically, when a hydrophobic molecule is added to water, water molecules in its vicinity get better organized, and enthalpy and Gibbs energy decrease. This improves the miscibility. With increasing temperature this effect decreases and the compounds revert to their (expected) immiscibility.

Although the Prigogine–Flory theory explained many phenomena satisfactorily from the qualitative viewpoint, its quantitative success was only moderate. Even if temperature- and pressure-dependent values of the supposedly constant quantities v^*, p^*, T^* were used for explaining the dependencies of thermodynamic functions on temperature and pressure, the agreement was still far from quantitative. The situation can be improved if the exchange enthalpic parameter X_{12} is accompanied by an additional term $Q_{12}T\tilde{v}_1$, where Q_{12} is a new adjustable parameter characterizing additional contributions to entropy; it has a major influence on the residual entropy and on the interaction parameter χ_S^*. Of course, such an improvement was bound to occur with the introduction of an additional adjustable parameter. This situation reminds us of the switch made within the Flory–Huggins theory when a single adjustable parameter $\left(\frac{1}{2} - \chi\right)$ was replaced by a pair of new parameters in the expression for $\psi(1 - \theta/T)$. The price for better agreement of the theory with experiment is again the same: The theoretical formulas for various quantities are no longer directly related to the underlying model. Thus the introduction of the parameter Q_{12} belongs in the realm of phenomenological manipulations.

3.1.4. Phenomenological Approach

All the theories of thermodynamic behavior of polymer solutions that we have described have some drawbacks. The Flory–Huggins theory is not applicable at low concentrations; the excluded-volume theories cannot be extended to higher concentrations. The Prigogine–Flory theory has only qualitative success. The most serious drawback is probably common to most theories of many phenomena (not restricted to the macromolecular field); the theory must necessarily work with a model that is simpler than the actual system. If the theory is good, it will describe the system qualitatively well, but there will be minor discrepancies. The discrepancies will become serious as the original assumptions become less and less satisfactory. In situations like that, it may be preferable to describe the phenomena using arbitrary mathematical relationships with adjustable parameters. Such relations are of great value when information is needed on a system that has been studied before. However, they may be worthless when predictions for other (even closely related) systems are needed.

Thus a compromise is being called for. The compromise usually starts with relations that describe some plausible theory. These relations are then modified by either adding new terms or altering the existing ones. Typically, an existing parameter is allowed to become a function of experimental variables. Such modifications characterize the *phenomenological approach*. We have seen one example of this approach when the expression $\left(\frac{1}{2} - \chi\right)$, characterizing polymer-solvent interactions, was replaced by $\psi(1 - \theta/T)$. Another example was the introduction of the parameter Q_{12}.

The advantage of this procedure consists in a reasonable expectation that similar systems will require similar modifications. There is an additional benefit: a skillfully designed modification may show the theoretician a way of modifying the original theory. If the attempt is successful, the phenomenological equation becomes a theoretical one.

When the phenomenological approach is applied to the thermodynamics of polymer solutions, the Flory–Huggins equation (3.1.29) is usually the one selected for modification. The interaction parameter χ_{12}, which was defined to be a constant at a given temperature, is allowed to be composition-dependent. It may also depend on pressure, temperature, and molecular weight of the polymer.

Before exploring the consequences of this maneuver, we need to address a formal point. Within the original Flory–Huggins theory, the parameter χ_{12} obviously has the same value in equations (3.1.29), (3.1.33), (3.1.39), and (3.1.41). As we will see below, this is not so for composition-dependent χ_{12}. Historically, the symbol χ_{12} is most closely associated with the meaning it has in equation (3.1.35) describing the chemical potential and activity of the solvent. We will therefore reserve this symbol for the activity expression and define a new function $g_{12}(\phi_2)$ to replace χ_{12} in equation (3.1.29).

$$\Delta G_{\text{mix}} \equiv RT[n_1 \ln \phi_1 + n_2 \ln \phi_2 + n_1 \phi_2 g_{12}(\phi_2)] \tag{3.1.100}$$

Equation (3.1.100) is a definition of g_{12}. It is therefore exact; any peculiar behavior of any binary mixture could be expressed by an appropriately behaving g_{12}.

We can now calculate the chemical potential of the solvent in the usual way by applying equation (3.1.2) to the phenomenological equation (3.1.100). The result reads

$$\ln a_1 \equiv \frac{\mu_1 - \mu_1^{\text{o}}}{RT} = \ln(1 - \phi_2) + \left(1 - \frac{V_1}{V_2}\right)\phi_2 + \chi_{12}\,\phi_2^2 \tag{3.1.101}$$

$$\chi_{12} \equiv g_{12} - \phi_1 \left(\frac{\partial g_{12}}{\partial \phi_2}\right)_{P,T} \tag{3.1.102}$$

Thus χ_{12} and g_{12} are equal to each other only if they are independent of composition. Otherwise, they are different functions of composition. Note that they are *not* equal even in the limit of vanishing ϕ_2.

In a similar manner we may calculate the chemical potential of the solute as

$$\frac{\mu_2 - \mu_2^{\text{o}}}{RT} = \ln \phi_2 + \left(1 - \frac{V_2}{V_1}\right)(1 - \phi_2) + (1 - \phi_2)^2\, X_{12}\frac{V_2}{V_1} \tag{3.1.103}$$

$$X_{12} \equiv g_{12} + \phi_2 \left(\frac{\partial g_{12}}{\partial \phi_2}\right)_{P,T} \tag{3.1.104}$$

Thus it was necessary to introduce another composition-dependent interaction function $X_{12}(\phi_2)$ differing from the previous ones. In this case X_{12} and g_{12} are equal in the limit of vanishing ϕ_2.

We need to calculate the derivative $(\partial\mu_2/\partial\phi_2)_{P,T}$ also. It reads

$$\frac{(\partial\mu_2/\partial\phi_2)_{P,T}}{RT} = \frac{\phi_1}{\phi_2} + (1 - 2F_{12}\phi_1)\frac{V_2}{V_1} \tag{3.1.105}$$

$$F_{12} \equiv \chi_{12} + \phi_2\frac{(\partial\chi_{12}/\partial\phi_2)_{P,T}}{2} \equiv X_{12} - \phi_1\frac{(\partial X_{12}/\partial\phi_2)_{P,T}}{2} \tag{3.1.106}$$

We will see later that the interaction function F_{12} defined by equation (3.1.106) is the experimental quantity measured by the methods of light scattering and sedimentation equilibrium. It is equal to χ_{12} in the limit of vanishing ϕ_2.

The concentration-dependent function χ_{12} can be calculated from experimental values of F_{12} as

$$\chi_{12} = \frac{2}{\phi_2^2}\int_0^{\phi_2} F_{12}\phi_2'd\phi_2' \tag{3.1.107}$$

Similarly, the function g_{12} can be evaluated by integration of equation (3.1.102) as

$$g_{12} = \frac{1}{\phi_1}\int_{\phi_2}^1 \chi_{12}d\phi_2' \tag{3.1.108}$$

Although for the calculation of χ_{12} it was sufficient to know the F_{12} function only in the interval $< 0, \phi_2 >$, for the evaluation of g_{12} we need to know χ_{12} in the remaining interval $< \phi_2, 1 >$. Thus, for this purpose, we need to have experimental data for all compositions of the mixture. Experimentally, it is very difficult to obtain such data. Hence, the g_{12} function is known for very few systems; even then its reliability is low.

The above treatment is valid both for polymer solutions and for mixtures of solvents. We will see that a detailed knowledge of the solvent-solvent interaction (preferably expressed as an appropriate function g_{12}) is needed for the proper interpretation of the behavior of polymers dissolved in mixed solvents.

3.1.5. Anisotropic Solutions and Liquid Crystals

In the preceding sections we tacitly assumed that the solutions are isotropic, that is, that the macromolecules (even if asymmetric) are oriented randomly. This appears plausible; any orientation of particles in solution (in the absence of outside orienting forces) seems to require a decrease in entropy. This is really the case of very dilute solutions. However, the situation is quite different in more concentrated solutions of rigid, rodlike particles. When the neighboring particles are packed in a more parallel fashion, the entropy increases. This should not seem paradoxical to anyone who has ever tried to accommodate a heap of matches in a small volume: Only packing them parallel in a box can do it. Similarly, long molecules in solution tend to align themselves in a parallel way—often in many layers. In this arrangement they still have a lot of freedom of movement along their axes; that is, they have considerable entropy. Haphazardly piled molecules cannot move in any direction easily; their entropy is lower.

A more detailed analysis would show that the orientational correlation of rods is negligible at high dilution and increases with increasing concentration. When the concentration reaches some critical value, the solution will separate into a dilute phase with randomly oriented molecules and a more concentrated phase in which the molecules are aligned to quite a high degree. The larger the length-to-diameter ratio of the rods, the lower the critical concentration. It should be stressed that this behavior is a result of steric-entropic effects and is operative even for athermal mixtures. Of course, for mixtures with positive enthalpy of mixing, the ordering phenomenon will be strengthened further.

Solutions with molecules packed in a parallel way are called *liquid crystals.* Their fluidity is usually quite large because of the ease of the axial movement of the molecules. In the absence of external forces, the molecules will form domains within which the molecules are aligned with each other. The domains themselves are oriented randomly. However, very small outside effects can lead to more or less complete orientation of all domains. Such outside effects include the proximity to the surface (or phase boundary), liquid flow, and electric or magnetic field.

The above-described effects can be observed with small molecules as well as with macromolecules. Small-molecule liquid crystals are mostly rigid linear molecules with two or more aromatic rings connected by other rigid groups. They usually have polar, strongly interacting groups. A typical example is *p*-phenylenebis(4-methoxybenzoate) **1.**

$$CH_3O-\langle\bigcirc\rangle-\overset{\overset{\textstyle O}{\|}}{C}-O-\langle\bigcirc\rangle-O-\overset{\overset{\textstyle O}{\|}}{C}-\langle\bigcirc\rangle-OCH_3$$

1

Such materials are usually used in neat form. Among macromolecules, liquid crystalline behavior in solutions is observed either with rigid particles of the tobacco mosaic virus type or with polymers that form stable helices in solution, most notably with poly(γ-benzyl glutamate). Other examples are polymers with rather rigid backbones, for example, solutions of aromatic polyamides in sulfuric acid.

Anisotropic solutions have many unusual properties: very strong optical activity (if any optical active center is present), strong optical anisotropy (birefringence), and rather unusual rheological behavior. Their viscosity increases sharply with concentration up to the critical transition concentration; then it decreases precipitously.

3.1.5.1. Types of Liquid Crystals.

The first observation of a liquid crystalline phase was made in 1888 by an Austrian botanist, Friedrich Reinitzer, who found that crystals of cholesteryl benzoate (thereby the name cholesteric) melted to form a cloudy liquid instead of the expected clear liquid. This cloudy liquid was stable over a 23°C range before finally melting to a clear liquid. It was soon realized that the cloudy liquid was a new phase occurring between solid and liquid states. Because the phase had some properties characteristic of crystals (it exhibited birefringence) and yet flowed like a liquid, it was named liquid crystal (LC). The liquid crystalline phase is also known as a *mesophase,* i.e., an intermediate phase. There are three different types of mesophase: *nematic, smectic,* and *cholesteric.*

In general, molecules making up the liquid crystalline phase have a mesogenic rigid core (which may be disc-, lath-, or cigar shaped) to which one or more flexible alkyl or alkoxy chains are attached, which usually carry dipolar groups like OCH_3, F, Cl, Br, NO_2, $N(CH_3)_2$, etc.

In liquid crystals the rodlike molecules are aligned with respect to the vector known as the *director,* **n**. In the absence of external perturbations the director is randomly distributed on a macroscopic scale. However, even small external fields can make the director uniform throughout the sample. This effect is amply utilized in many display devices.

In the nematic liquid crystalline phase molecules are aligned along the director, but there is no regular sidewise order among molecular neighbors. The molecular organization is illustrated in Figure 3.2(a).

Smectic phases possess varying degrees of sidewise ordering and molecular tilt in addition to the orientational order. Several smectic phases are known to exist. A cholesteric phase or chiral nematic phase may be formed if the constituent molecules are chiral.

The name of smectic liquid crystalline phases originates from the soapy feel given to them by their basic layer structure. Indeed, many soap and detergent molecules exhibit *lyotropic* (controlled by the solvent) smectic phases in solution. These structures result from the *amphiphilic* nature of the molecules that have both a hydrophobic and a hydrophilic end. In aqueous solutions, the molecules pack so that the hydrophobic ends come together—screening each other from water. A similar structure is found in lipid bilayers, which are fundamental constituents of cell walls.

There are several variants of smectics, but all are characterized by the *layered structure* caused by the segregation of the ends of the rod molecules onto common planes. The two most common variants are known as smectic A and smectic C. The smectic phase A [Figure 3.2c(i)] is characterized by a director that is normal to the layer, yet the molecules still possess a random positional order within the layer. At a lower temperature substances featuring the smectic A phase often exhibit the smectic C phase [Figure 3.2c(ii)]. In this phase, the molecules have the same random order within the layer but tilt relative to the layer normal. The tilt angle normally increases with decreasing temperature. Other smectic phases are even more crystalline in that they also feature some positional order within the layers. They may, for instance, exhibit hexagonal packing of molecules.

In *cholesteric* liquid crystals the molecules are also arranged in layers, in which the director is inclined with respect to the layer normal as in the smectic C phase. However, the orientation of the director slightly changes between layers. As a result, the molecules in successive layers are arranged in a *helical* way. The helix axis around which the director twists is parallel with the layer normal. The twist has a constant pitch (the distance needed for 360° helical turn) throughout the sample. Normally, it takes some hundred layers to complete one revolution in

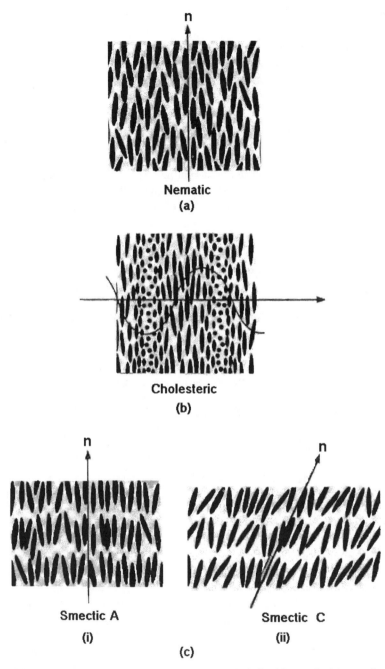

Figure 3.2. Representation of the molecular arrangements in liquid crystals: (a) nematic phase, (b) chiral nematic or cholesteric phase (molecules twist in a plane perpendicular to the plane of the paper), (c(i)) smectic phase A, (c(ii)) smectic phase C. (Reprinted from A.M. Donald and A.H. Windle, Eds., *Liquid Crystalline Polymers*. Copyright © 1992 by Cambridge University Press, New York. Used by permission of the publisher.)

the helix. Arrangement of molecules in cholesteric liquid crystals is depicted in Figure 3.2.

The periodic structure along the helix axis leads to spectacular optical phenomena. If we shine light with a wavelength equal to the pitch through the sample along the helix axis, the circularly polarized component with the same handedness as the helix is totally reflected. This gives the substance a beautiful intense color, easily observable by the naked eye. Because the pitch varies with temperature, the color of the substance will also vary. It will also vary with viewing angle because light entering at an angle sees a longer optical periodicity.

3.1.5.2. Liquid Crystalline Polymers.   Polymer chemists are busily trying to combine the useful properties of polymers and liquid crystals. This may be accomplished by incorporating the mesogenic groups into the polymer. The spontaneous alignment of the molecules leads to very strong structures. The macroscopic alignment is achieved easier than for the ordinary crystalline polymers, because the director is manipulated by outside fields relatively easily. The orientational order in liquid crystalline polymers (LCPs) is quite similar to that seen in small-molecule materials. Liquid crystalline polymers seem also likely to play an important role in nonlinear optical device manufacture.

The mesogenic groups may be a part of the backbone, or they may be attached as side groups. The *main-chain thermotropic* (controlled by temperature) *liquid-crystalline polymers* developed for molding applications are almost entirely polyesters or polyamides. They are polymerized using stepgrowth such as condensation reactions. DuPont's Kevlar® (polyamide made from terephthalic acid and *p*-phenylenediamine) is the best known example of these polymers. It forms high-strength fibers and is found in (among other things) tires, bullet-proof vests, cables, and sports equipment.

Liquid crystalline *side-chain* polymers are often made by radical initiated polymerization. Because of their ease of polymerization, preferred monomers are derivatives of acrylic or methacrylic acid, onto which the mesogenic groups are attached through a *flexible spacer*. The most common short flexible link is $-(CH_2)_2-$. Flexible spacers positioned between the polymer backbone and the mesogenic side groups progressively decouple the mesogenic part of the side group from the backbone and promote the formation of smectic phases in side-chain polymers. The nematic phase is normally observed in liquid crystalline side-chain polymers only when the length of the flexible spacer is between about two and six units. Many data are available to illustrate these trends, and some are presented in Table 3.1.

Many comblike polymers were designed and synthesized in the late 1970s. In these, the tail of the mesogenic unit was attached to the polymer backbone as shown for acrylate **2**, methacrylate **3**, and siloxane **4** main chains. In structures **2**, **3**, and **4**, R is a mesogenic head that is typically an aromatic (e.g., phenyl benzoate) or alicyclic group with an attached dipole group. An example of R is structure **5**, where X is CN, OCH_3, or NO_2 and Y is H or CH_3. Several hundred such comblike polymers have been prepared.

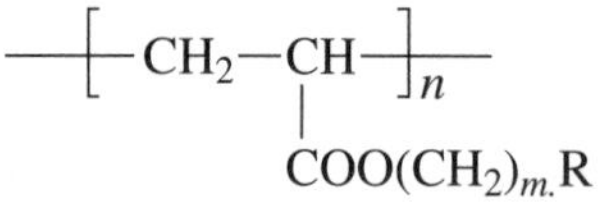

$$-\left[CH_2-CH \right]_n$$
$$\quad\quad\quad COO(CH_2)_{m.}R$$
2

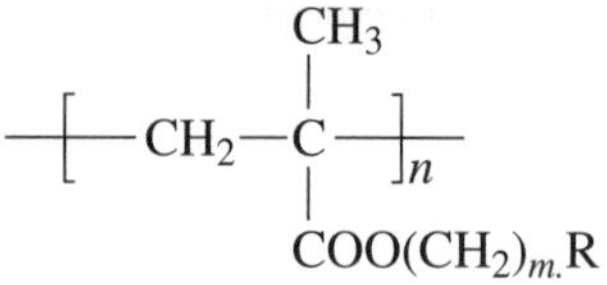

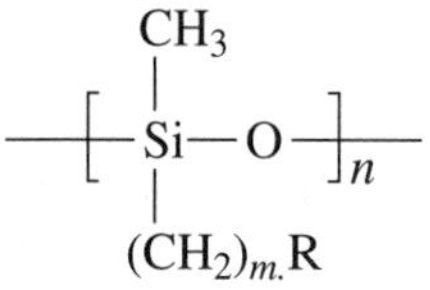

$$-\left[CH_2-\overset{\displaystyle CH_3}{\underset{\displaystyle COO(CH_2)_{m.}R}{C}} \right]_n$$
3

$$-\left[\overset{\displaystyle CH_3}{\underset{\displaystyle (CH_2)_{m.}R}{Si}}-O \right]_n$$
4

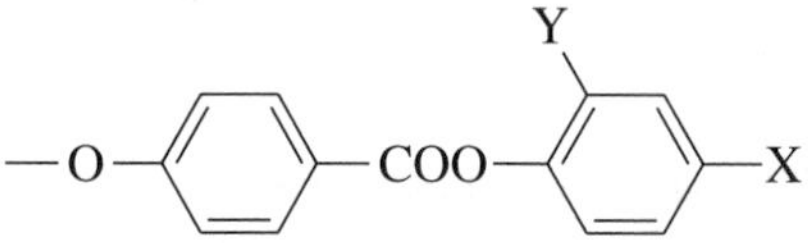

$$-O-\bigcirc-COO-\overset{Y}{\bigcirc}-X$$
5

Table 3.1 indicates the transition temperatures for glass (**g**) to nematic (**n**) to isotropic melt (**i**) for some representative polymers each having the same mesogenic head group R (which is the phenyl benzoate structure **5** with X = OCH$_3$ and Y = H). The transition from **g** to **n** occurs above room temperature for the stiff acrylate and methacrylate polymers and is below the room temperature for the more flexible

TABLE 3.1   Transition Temperatures for Glass (g) to Nematic (n) to Isotropic Melt (i) for Different Polymers

Structure	Transitions
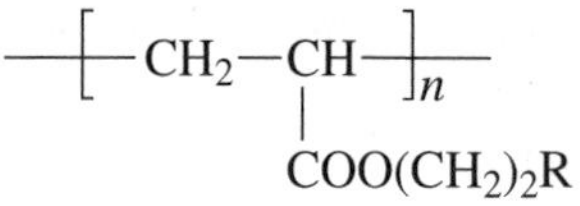 $-\left[CH_2-CH \right]_n$, $COO(CH_2)_2R$	$g \xrightarrow{47°C} n \xrightarrow{77°C} i$
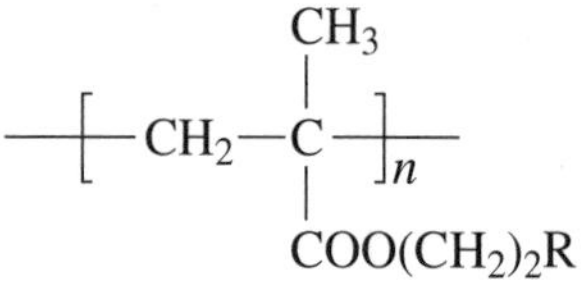 $-\left[CH_2-\underset{COO(CH_2)_2R}{\overset{CH_3}{C}} \right]_n$	$g \xrightarrow{96°C} n \xrightarrow{121°C} i$
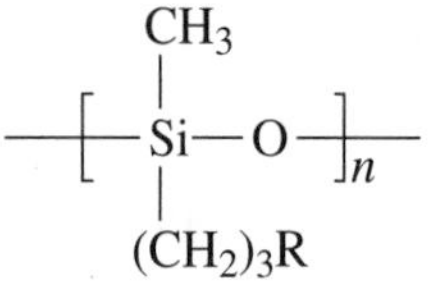 $-\left[\underset{(CH_2)_3R}{\overset{CH_3}{Si}}-O \right]_n$	$g \xrightarrow{15°C} n \xrightarrow{61°C} i$
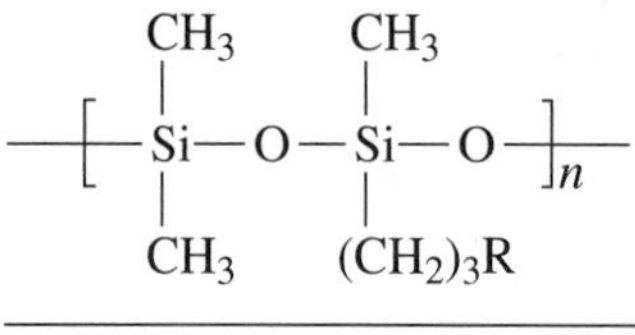 $-\left[\underset{CH_3}{\overset{CH_3}{Si}}-O-\underset{(CH_2)_3R}{Si}-O \right]_n$	$g \xrightarrow{3°C} n \xrightarrow{21°C} i$

siloxane polymers. The nematic LC range is sustained over several tens of degrees for each material before it becomes an isotropic melt. The LC to isotropic transition temperatures of most LC side-chain polymers prepared to date are well above the room temperature and are generally far higher than those found for low-molecular-weight compounds.

Mesogenic units may be incorporated into polymer chains in a number of ways. Polymerization of a single monomer (both in polycondensation and chain polymerization) produces polymers with the mesogenic unit at regular intervals along the chain. Copolymerization of a mesogenic monomer with a nonmesogenic one leads to mesogenic units at random positions. It is also possible to alternate sequences of nonmesogenic units with sequences of mesogenic units having similar length. LCPs in which mesogenic groups are included in both the side chains and the backbone have been synthesized, too.

3.2. EQUILIBRIUM METHODS FOR THE STUDY OF MACROMOLECULES IN SOLUTION

In the preceding section, we learned a few basic aspects of the thermodynamics of polymer solutions and familiarized ourselves with some of the more frequently encountered thermodynamic parameters. In the present section, we will study how thermodynamic properties manifest themselves in the actual behavior of polymer solutions. We will also explore the ways and means of measuring these properties. Thermodynamics is a science of equilibrium; its strength is in handling all kinds of equilibria. Not surprisingly, thermodynamic parameters are most directly obtained from the analysis of various types of equilibria. We will treat such equilibria in this section. Of course, thermodynamics pervades other fields of chemistry and physics as well. Thus thermodynamic properties can also be obtained from the study of some nonequilibrium phenomena that are heavily influenced by thermodynamics. We will therefore return to thermodynamics in Section 3.4 in dealing with hydrodynamic methods and in Section 3.5 in describing light scattering phenomena.

It is useful at this point to review the meaning of the equilibrium concept and the criteria for distinguishing equilibrium situations from nonequilibrium situations. For our purposes, the following definition of equilibrium should be satisfactory:

1. No property of the system changes during the duration of the experiment.
2. No matter or energy flows through the system. (This part of the definition distinguishes equilibrium and steady-state systems.)
3. Any small disturbance of the system caused by a change in outside constraints reverses itself when the outside change is reversed.

With this broad definition of equilibrium, there is only one completely general criterion of equilibrium: the sum of the entropy of the system and the entropy of its surroundings is at a maximum. However, this criterion is clumsy and difficult to employ. Consequently, other criteria are often used that lack sufficient generality. For example, the criterion of the Gibbs energy of the system G being at minimum applies only to systems kept at constant pressure and temperature. More dangerous is the often-quoted criterion of pressure being the same throughout the system. This applies only to systems that have no rigid walls and are not under the influence of inhomogeneous external potentials (e.g., gravitational potential). Similarly, the criterion of the chemical potentials of all components being homogeneous in the whole system is correct only for systems that have no partitions impermeable to any component; the external potentials must also be homogeneous in the whole system. In many situations we will deal with, the restrictions of the criteria are violated; hence the criteria are quite misleading.

For the sake of further discussion, we will separate physical systems into three categories.

1. *Homogeneous systems,* for which all intensive properties have the same values everywhere in the system.
2. *Heterogeneous systems* composed of a finite number of homogeneous *phases* with well-defined *boundaries*. The nature of the boundaries is very important— whether they are mechanically rigid (glass wall), thermally insulating, electrically insulating, permeable, or impermeable to individual components of the system.
3. *Continuous systems,* in which any or all intrinsic properties are continuous functions of spatial coordinates (any discontinuity represents a phase boundary).

There also exist, of course, heterogeneous systems, which have continuous phases. Such systems bring nothing new to our classification.

A sufficiently broad and simple-to-use set of equilibrium criteria can now be compiled as follows:

1. If the external potentials have the same value everywhere in the system, they are effectively absent. Then, for homogeneous systems, the pressure, temperature, and chemical potentials of all components will be homogeneous throughout the entire system at equilibrium. The same will apply to any *single* phase of a heterogeneous system.
2. In the absence of external potentials, the criteria for heterogeneous systems depend on the nature of the boundaries. The equilibrium pressure will be homogeneous throughout the system unless rigid boundaries fully separate some phases. The equilibrium temperature will be homogeneous unless some thermally insulating boundaries separate some parts of the system. The equilibrium chemical potential of any component will be the same throughout the system unless a boundary that is impermeable to this particular component exists in the system.
3. If the outside potentials change their values within the system, the whole system and/or its phases will have a continuous nature at equilibrium. However, in the

absence of rigid or thermally insulating or impermeable boundaries, we have a new broad criterion. The *total potential* of every component $\mu_{i,\text{tot}}$ remains constant throughout the system. The total potential is a sum of all relevant potentials for the component—its chemical potential μ_i defined in the usual way, its mechanical potential, electrical potential, etc. In our future analysis we will be concerned mainly with mechanical potentials; we will thus write our relations only for two representative potentials: mechanical, U_{mech}, and electrical, U_{el}. Thus

$$\mu_{i,\text{tot}} \equiv \mu_i + M_i U_{\text{mech}} + E_i U_{\text{el}} \tag{3.2.1}$$

Whereas chemical potential and total potential refer to 1 mol of the substance, mechanical potential refers to unit mass (1 g). Thus its multiplication by molecular weight M_i brings the second term to the same (molar) basis. Examples of mechanical potentials are gravitation potential, U_{gr}, and centrifugal potential, U_{ef}. Similarly, electrical potential is defined per unit charge; we must multiply it by molar charge E_i. For species without electrical charge, $E_i = 0$, of course.

We now have our thermodynamic tools ready, and we can proceed with the analysis of our equilibrium methods.

3.2.1. Osmometry

The phenomenon of osmosis and osmotic pressure is based on the existence of materials that are permeable to some substances and not to others. For example, metallic platinum is permeable to molecular hydrogen but not to other gases. A sieve may be "permeable" to sand but not to pebbles. These two examples show us that permeability is related to the chemical nature of the permeant and the barrier on one hand and to the size of the permeating particles and mechanical structure of the barrier on the other hand.

We are interested in systems somewhere between these extremes—in membranes that are permeable to solvents but impermeable to solutes. Again, such membranes exhibit a wide range of permeabilities. When a fritted glass disk is soaked from one side in a solution of ferrocyanide and from the other side in a solution of some copper salt, copper ferrocyanide will precipitate within the disk and form a membrane there. Such a membrane is permeable to water but impermeable to salts or any larger organic molecules. Indeed, such membranes were used in the nineteenth century in the early studies of osmotic phenomena.

However, for the study of the osmotic pressure of macromolecular solutions, membranes made from other macromolecular materials that are insoluble in the solvent employed are most commonly used. The most popular membranes are made from cellophane, which is permeable to water. If it is first treated with alkalis (which open its structure) and then transferred through water-alcohol mixtures to alcohol, then through alcohol-solvent mixtures to the solvent of choice, it becomes permeable to other solvents as well. Nitrocellulose membranes are also rather popular for water-soluble polymers.

Let us imagine two chambers separated by a semipermeable membrane with pure solvent on one side of the membrane and a polymer solution on the other side. The membrane is permeable to the solvent but impermeable to the polymer. The solvent on the solvent side has a higher chemical potential than the solvent on the solution side. Hence, it will diffuse across the membrane toward the solution side. If the flow is not counterbalanced by any mechanism, it will continue until all the solvent is transferred to the solution side.

In nature, such osmotic flow represents at least part of the mechanism by which plants and especially trees pump water from the ground to their topmost branches. Another phenomenon worth mentioning is a process in which a biological cell (which is enclosed by a semipermeable membrane) is placed in distilled water. Water will enter the cell, stretch it, and eventually burst it. This scenario may happen when distilled water is injected into the bloodstream. Red blood cells will burst, resulting in hemolysis—a fatal condition. On a more positive note, the same phenomenon is utilized by biochemists when they want to open bacterial cells in a gentle procedure avoiding grinding, sonication, etc. They place the cells first into a higher concentration of salt, building up the inside salt concentration. They then transfer them into distilled water, where the cells burst, releasing their contents essentially undamaged. The procedure is called *osmotic shock*.

In osmometric experiments, the flow of solvent across the membrane is utilized for building up pressure on the solution side. The increased inside pressure will drive the solvent in the opposite direction, until an equilibrium is reached and the solvent flow stops.

A pressure difference across the membrane at equilibrium may exist either when the membrane is rigid (the membranes themselves are very soft, but they may be supported by a rigid screen) or when an elastic membrane is stretched by the pressure and the elastic counterforce is the source of the pressure difference.

What equilibrium criteria are applicable in the osmotic case? Obviously, both phases (i.e., solvent and solution compartments) must be internally homogeneous. No thermally insulating boundary is present; hence the temperatures of both phases must be the same. However, the boundary between phases is rigid, so different pressures of the two phases are permissible. Similarly, the solute is not allowed to cross the membrane; thus its chemical potential on the two sides of the membrane may have different values. Consequently, only one useful criterion remains: the solvent may permeate freely across the membrane. Hence, at equilibrium its chemical potential must be the same on both sides of the membrane. We will utilize this condition for finding the relation between the difference in pressures and in the concentrations of the two phases.

It may be useful to review at this point a general procedure for calculating a difference in the values of any thermodynamic property of state between two systems. We consider the two systems as one system in two different states and follow the changes in the required property along a path connecting the two states. The difference does not depend on the path connecting the two states; hence we select a path that allows for the easiest calculation of the changes. This selection almost always consists of splitting the path into several segments in such a way that only one of the independent thermodynamic variables of the system changes in each segment.

In the case of osmometry, we change pressure in the first segment from the pressure on the solvent side, P_{ref}, to the pressure of the solution side, P', while keeping the concentration constant (zero in the present case). The change in the chemical potential of solvent accompanying this process is $\Delta\mu_1^I$. In the second segment we will change the concentration from zero to the solution side concentration c_2 at constant pressure P'. The change in μ_1 in the second segment is $\Delta\mu_1^{II}$. The value of $\Delta\mu_1^I$ is calculated as

$$\Delta\mu_1^I = \int_{P_{ref}}^{P'} \left(\frac{\partial\mu_1}{\partial P}\right)_{T,c_2} dP = \int_{P_{ref}}^{P'} \overline{V}_1 dP = \overline{V}_1(P' - P_{ref}) \equiv V_1\pi \quad (3.2.2)$$

In the last identity of equation (3.2.2) we have defined *osmotic pressure* π as the difference in pressures across the semipermeable membrane at equilibrium. We have also replaced the partial molar volume of the solvent, $\overline{V}_1$, in our dilute solution by the molar volume of pure solvent, V_1.

The expression for $\Delta\mu_1^{II}$ is calculated easily as

$$\Delta\mu_1^{II} = \int_0^{c_2} \left(\frac{\partial\mu_1}{\partial c_2}\right)_{P,T} dc_2 = \mu_1(P',c_2) - \mu_1(P',0) \quad (3.2.3)$$

where we have noted as arguments of μ_1 the applicable values of the variables P and c_2. In the range of osmotic pressures, the difference of chemical potentials [the rightmost side of equation (3.2.3)] is virtually independent of pressure.

At osmotic equilibrium, the two-segment path leads to a state with the same chemical potential as the starting state. Hence,

$$\Delta\mu_1 \equiv \Delta\mu_1^I + \Delta\mu_2^{II} = 0 \quad (3.2.4)$$

Equations (3.2.2)–(3.2.4) now yield the required relation between osmotic pressure and the (concentration dependent) chemical potential

$$\mu_1 - \mu_1^o = -V_1\pi \quad (3.2.5)$$

In highly dilute solutions, the difference in chemical potentials may be expressed by means of Raoult's law; equation (3.2.5) is then modified to

$$V_1\pi = -\left(\mu_1 - \mu_1^o\right) = -RT\ln x_1 = -RT\ln(1 - x_2) \approx RTx_2 \approx RTc_2V_1/M_2$$
$$(3.2.6)$$

Thus osmotic pressure is related to molecular weight of the solute, M_2, and its concentration c_2 as

$$\pi/RTc_2 = 1/M_2 + \cdots \quad (3.2.7)$$

where the ellipsis signifies neglect of terms with higher powers of concentration.

It should be stressed that Raoult's law is valid for all solutes at sufficiently low concentrations. Relation (3.2.7) is therefore valid for macromolecular as well as small-molecule solutes.

If the polymer is polydisperse, the first half of equation (3.2.6) is still valid. However, for x_1, we must now write

$$x_1 = 1 - \sum_i x_i = 1 - \sum_i \frac{c_i V_1}{M_i} \tag{3.2.8}$$

where the summation is over all polymer components. Equation (3.2.6) now changes to

$$\pi V_1 \approx RT \sum_i x_i \approx RT V_1 \sum_i \frac{c_i}{M_i} \tag{3.2.9}$$

We now divide equation (3.2.9) by $RT V_1 c$, with the total polymer concentration $c = \sum_i c_i$. The result is

$$\frac{\pi}{cRT} = \frac{\sum c_i / M_i}{\sum c_i} \equiv \frac{1}{\overline{M}_n} \tag{3.2.10}$$

Thus osmotic pressure yields for polydisperse polymers the number-average molecular weight $\overline{M}_n$, a quantity obtained otherwise only with difficulty.

Equations (3.2.6)–(3.2.10) are applicable only for ideally dilute solutions obeying Raoult's law. Macromolecular solutions obey this law only at concentrations so low that the measured values of osmotic pressure (and most other experimental quantities) disappear in experimental error. The measurements have to be performed at concentrations that are not ideally dilute. In such a case, we should relate the difference of chemical potentials in equation (3.2.6) not to $\ln x_1$ but to the logarithm of activity $\ln a_1$, for which we will use the power series given in equation (3.1.36). The result reads

$$\pi / c_2 RT = 1/M_2 + B c_2 + C c_2^2 + \cdots \tag{3.2.11}$$

Thus a single measurement of osmotic pressure is not sufficient for measurement of molecular weight. We need to measure $\pi / c_2 RT$ for several concentrations and extrapolate to the vanishing concentration of polymer c_2.

The reason that the higher virial terms are more important for large molecules than for small ones is not the high value of the virial coefficient B. After all, according to Flory–Huggins theory, B is independent of molecular weight; according to perturbation theories, it decreases with increasing molecular weight. It is the ratio of the value of the virial term to the leading term $1/M_2$ that is steeply increasing with increasing molecular weight, and so is the relative importance of B.

Equation (3.2.11) seems to suggest that experimental data be plotted as $\pi / c_2 RT$ versus c_2. The intercept of the plot is then $1/M_2$ (or $1/\overline{M}_n$ for polydisperse polymers), and the initial tangent is B. This plot is indeed useful for polymers in poor solvents,

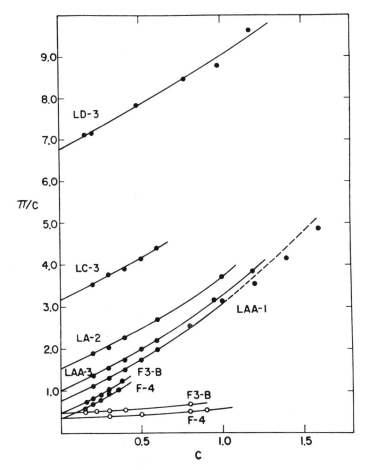

Figure 3.3. Plots of π/c against c for polyisobutylene fractions in a good solvent (cyclohexane, full points) and in a marginal solvent (benzene, open points) at 30°C. π in g/cm^2 and c in g/100 mL. (Reprinted from P.J. Flory, *Principles of Polymer Chemistry*. Copyright © 1953 by Cornell University. Used by permission of the publisher, Cornell University Press.)

which have low values of B. However, in good solvents, the third virial term Cc_2^2 becomes significant even at very low concentrations and the plot is curved (Fig. 3.3); the curvature makes the extrapolation somewhat complicated. The remedy was found in perturbation theories that claim that the third virial coefficient is closely related to the second, namely according to the relation

$$C = \gamma B^2 M_2 \qquad (3.2.12)$$

where γ is a slowly varying function of B, which for analyzing the osmometric data may be considered as $\gamma = 0.25$. This value of γ transforms the first three virial terms

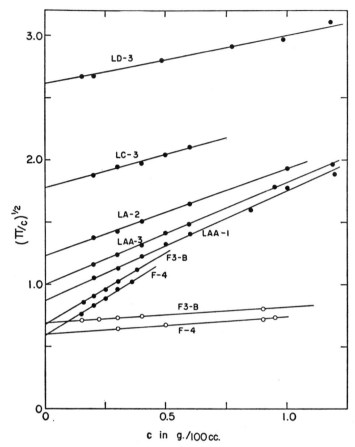

Figure 3.4. Plots of $(\pi/c)^{1/2}$ against c. The same data as in Figure 3.3. (Reprinted from P.J. Flory, *Principles of Polymer Chemistry.* Copyright © 1953 by Cornell University. Used by permission of the publisher, Cornell University Press.)

of equation (3.2.11) into a full square:

$$\pi/c_2 RT = \left(1/M_2^{1/2} + Bc_2 M_2^{1/2}/2\right)^2 + \cdots \tag{3.2.13}$$

This relation suggests that osmometric data should be plotted as $(\pi/c_2 RT)^{1/2}$ versus c_2. The intercept is, in this case, $M_2^{-1/2}$, and the product of intercept and slope is $B/2$. Indeed, experimental data plotted this way exhibited much better linearity (Fig. 3.4).

3.2.1.1. Macromolecular Solutions in Mixed Solvents. Equilibrium conditions for osmotic equilibria of macromolecules in multicomponent solvents require that the chemical potentials of all solvent components capable of permeating the membrane must be the same on both sides of the membrane. This requirement leads to an imbalance of solvent compositions in the two compartments.

We will consider only binary solvent mixtures and design one solvent component as a principal solvent with subscript 1; the other solvent component will have subscript 2, and the polymer, subscript 3. The independent variables are P, T, and the molalities m_2, m_3.

Let us consider the following experiment. After equilibrating in an osmotic apparatus a polymer solution in a mixed solvent, we want to add more polymer to the solution side (i.e., to increase its molality by dm_3) without disturbing the equilibrium. Obviously, the chemical potentials of both solvents are not allowed to change during the process. We can achieve this only if we also increase the pressure on the solution side by dP and add (or remove) some amount of solvent 2, changing its molality by dm_2. This experiment is described by two differential quotients: $(\partial P/\partial m_3)_\mu$ and $(\partial m_2/\partial m_3)_\mu$. The subscript μ recognizes the fact that the chemical potentials were kept constant.

The quotient $(\partial P/\partial m_3)_\mu$ describes the change in pressure on addition of the polymer; its integration yields the dependence of osmotic pressure on the molality of the polymer. The quotient $(\partial m_2/\partial m_3)_\mu$ characterizes the amount of solvent 2 that must accompany the polymer in our process. (This derivative becomes negative when solvent 2 is excluded from the vicinity of the polymer and solvent 1 is in excess.) The phenomenon is called *preferential adsorption*. This name is misleading: Preferential adsorption is not an attachment of solvent molecules to the polymer by some specific secondary bonds (of course, such bonds *may* play a role) but is a result of the interplay of interactions among all three components of the system. The excess molecules of the solvent are located somewhere in the polymer solution, not necessarily in contact with the polymer molecules. The effect is more conveniently described by means of the coefficient λ defined as

$$\lambda \equiv -v_2 \phi_1 \frac{M_2}{M_3} \left(\frac{\partial m_2}{\partial m_3} \right)_\mu \tag{3.2.14}$$

where v_i is the specific volume of the ith component. Coefficient λ uses the mass ratio instead of the molar ratio implied in $(\partial m_2/\partial m_3)_\mu$. How are the osmotic pressure and preferential adsorption related to the basic thermodynamic quantities that describe our three-component system? At a given constant temperature, differential changes in the chemical potentials of the two solvent components during any process are expressed through the changes in the independent variables during that process:

$$d\mu_1 = \left(\frac{\partial \mu_1}{\partial m_3} \right)_{m_2,P} dm_3 + \left(\frac{\partial \mu_1}{\partial m_2} \right)_{m_3,P} dm_2 + \left(\frac{\partial \mu_1}{\partial P} \right)_{m_2,m_3} dP \tag{3.2.15}$$

$$d\mu_2 = \left(\frac{\partial \mu_2}{\partial m_3} \right)_{m_2,P} dm_3 + \left(\frac{\partial \mu_2}{\partial m_2} \right)_{m_3,P} dm_2 + \left(\frac{\partial \mu_2}{\partial P} \right)_{m_2,m_3} dP \tag{3.2.16}$$

Realizing that for experiments at constant chemical potentials, $d\mu_1 = d\mu_2 = 0$, and dividing equations (3.2.15) and (3.2.16) by dm_3, we transform them to

$$a_{13} + a_{12}\left(\frac{\partial m_2}{\partial m_3}\right)_\mu + V_1\left(\frac{\partial P}{\partial m_3}\right)_\mu = 0 \tag{3.2.17}$$

$$a_{23} + a_{22}\left(\frac{\partial m_2}{\partial m_3}\right)_\mu + V_2\left(\frac{\partial P}{\partial m_3}\right)_\mu = 0 \tag{3.2.18}$$

Here we have replaced the derivatives of chemical potentials with respect to molalities by quantities a_{ij} defined as

$$a_{ij} \equiv \left(\frac{\partial \mu_i}{\partial m_j}\right)_{P,m_{k\neq j}} \tag{3.2.19}$$

Equations (3.1.17) and (3.1.18) can be solved for the two functions of interest, $(\partial m_2/\partial m_3)_\mu$ and $(\partial P/\partial m_3)_\mu$. From the latter derivative, osmotic pressure is calculated; it is again described by equation (3.2.11). The second virial coefficient now becomes a function of the derivatives a_{ij}; so does the coefficient λ, which is calculated from $(\partial m_2/\partial m_3)_\mu$. The derivatives a_{ij} can be found from ΔG_{mix} if it is a known function of composition variables. The Flory–Huggins equation generalized to three components is frequently used for this purpose.

$$\Delta G_{\text{mix}} = RT(n_1 \ln \phi_1 + n_2 \ln \phi_2 + n_3 \ln \phi_3 + n_1\phi_2\chi_{12} + n_1\phi_3\chi_{13} + n_2\phi_2\chi_{23}) \tag{3.2.20}$$

Here, n_i and ϕ_i are, as usual, numbers of moles and volume fractions; the χ_{ij} are binary interaction coefficients measurable in binary mixtures.

The results read

$$\lambda = v_3\phi_1\phi_2\frac{\chi_{12}(\phi_2 - \phi_1) + L\chi_{23} - \chi_{13} + 1 - L}{L\phi_1 + \phi_2 - 2\phi_1\phi_2\chi_{12}} \tag{3.2.21}$$

$$B = \frac{v_3^2}{2V_1}\frac{L - 2L\phi_1\chi_{13} - 2L\phi_2\chi_{23} + D\phi_1\phi_2}{L\phi_1 + \phi_2 - 2\phi_1\phi_2\chi_{12}} \tag{3.2.22}$$

$$D = 2\chi_{12}\chi_{13} + 2L\chi_{12}\chi_{23} + 2L\chi_{13}\chi_{23} - \chi_{12}^2 - \chi_{13}^2 - \chi_{23}^2 L^2 \tag{3.2.23}$$

$$L = V_1/V_2 \tag{3.2.24}$$

These intricate relations are best understood by noticing the dominant role of the solvent-solvent interaction parameter χ_{12} in the denominator of the expressions for both λ and B. For most solvent mixtures, χ_{12} is positive and frequently rather high. For such solvent mixtures both λ and B may have quite high values.

The detailed dependence of λ and B on solvent composition also reflects the interplay of the two polymer-solvent interaction parameters χ_{13} and χ_{23}. If these two parameters are approximately equal, then the minor solvent component is preferentially adsorbed; the adsorption changes sign in the middle of the concentration scale. In this case, B may go through a rather pronounced maximum as a function of solvent composition. A typical example of such behavior is afforded by solutions of polystyrene in mixtures of two marginal solvents, ethyl acetate and cyclohexane, which are quite dissimilar and have high value of χ_{12}. In an extreme case of this situation, a mixture of two nonsolvents for the polymer that do not mix well together may be a solvent for the polymer. An example of such a system is a solution of polystyrene in an acetone-cyclohexanol mixture. The phenomenon is called *cosolvency*.

If one of the mixed solvents is a much better solvent for the polymer (lower χ_{i3} value) than the other, it tends to be preferentially adsorbed onto polymer. However, even in this situation an inversion of the preferential adsorption may occur: In a solvent mixture of a small amount of cyclohexane (a marginal solvent) with an excess of benzene (a good solvent) it is cyclohexane that is preferentially adsorbed onto polystyrene. Of course, for a more balanced ratio of solvents, benzene is preferentially adsorbed. The virial coefficient B is predicted for such solvent mixtures to be higher in the good solvent than in the poorer one; however, it often goes through a maximum in the middle of the concentration scale.

All the above phenomena are easily understood from simple physical considerations. Two solvents with unfavorable enthalpy of mixing (high χ_{12} values) would trade their mutual contacts for any other contacts. Consequently, they prefer to form contacts with the polymer even if these contacts are not very favorable; even then they provide an opportunity to escape from contacting their hated cosolvent. Formation of these excess polymer-solvent contacts means, of course, disruption of polymer-polymer contacts; the solvent mixture becomes a better solvent with a high value of B. The same effect also influences the intracoil interactions. A decrease in the number of polymer-polymer contacts means in this case a sizable expansion of polymer coils.

Although the generalized Flory–Huggins theory explained qualitatively well the phenomena occurring in polymer solutions in mixed solvents, it failed from the quantitative viewpoint: The predicted increases in virial coefficients were many times larger than the increases actually observed. Again, the phenomenological approach came to the rescue. It has shown that a general description of three-component systems is not possible using only those parameters that are defined by the behavior of binary mixtures. Another function is needed that describes the difference between the real system and a hypothetical one constructed as a combination of binary systems. This function is usually called the *ternary interaction parameter* g_T, and the phenomenological expression for ΔG_{mix} of ternary mixtures is written as

$$\Delta G_{\text{mix}}/RT = n_1 \ln \phi_1 + n_2 \ln \phi_2 + n_3 \ln \phi_3 + n_1 \phi_2 g_{12} + n_1 \phi_3 g_{13}$$
$$+ n_2 \phi_3 g_{23} + n_1 \phi_2 \phi_3 g_T \tag{3.2.25}$$

Parameters g_{ij} and g_T are functions of composition. Equation (3.2.25) is exact; it is actually a definition of g_T. (The binary functions g_{ij} are also exact; they are defined by the same relation written for vanishing concentration of the component, which is absent in the binary.)

We will not present the relations for λ and B following from equation (3.2.25). They are even more complicated than relations (3.2.21)–(3.2.24), because they also contain derivatives of the g functions with respect to their arguments. Nevertheless, it is possible to use these relations for obtaining the function g_T and its derivative $\partial g_T / \partial \phi_3$ in the limit of vanishing ϕ_3. Thus equation (3.2.25) is a useful description of ternary systems. However, it suffers from the maladies of all phenomenological equations: The physical meaning of the function g_T is obscure, and it cannot be easily generalized even for similar ternary systems. At the present time g_T is known only for a very small number of systems, and it is waiting for a theoretical explanation.

3.2.1.2. Osmometry of Polyelectrolytes.

In the previous sections we have seen that osmotic pressure is directly related to the molar fraction of those components of the mixture that cannot permeate the osmotic membrane—at least for solutions that are ideally dilute with respect to such components. The molar fraction of the solute depends only on the number of solute molecules present in the system and not on their nature. It is often said that osmometry, together with other colligative methods (decrease of vapor pressure, elevation of boiling point, depression of melting point), counts the molecules in the system. It is reasonably easy to count molecules in nonionic solutions. However, ionic solutions present a quite different problem. Is one equivalent of NaCl one mole or two moles? Should one mole of neutralized poly(acrylic acid) count as one mole or several hundred?

The question was answered in the nineteenth century by Van't Hoff, who based his answer (you have guessed it) on the measurement of the osmotic pressure of electrolyte solutions using membranes impermeable to salts. It turned out that each ion should be counted as a separate molecule. This somewhat surprising finding follows from the basic rules governing combinatorial entropy and, through it, the chemical potentials. We have seen that combinatorial entropy depends primarily on the number of entities that can change their location more or less independently of other entities. Clearly, ions are such entities, which explains Van't Hoff's finding.

When we study the osmotic behavior of polyelectrolytes, we face a different problem. Our osmotic membrane is permeable not only to water but to small ions as well, but not to polyions. How should we calculate osmotic pressure in this case? Our model will contain polyanions with Z charges each. (Polycations would behave in an exactly analogous way). The counterions will be univalent cations. In addition to the polyelectrolyte, the solution will also contain a uni-univalent electrolyte with a cation identical to the counterion of the polysalt. We will count individual ions as molecules and assume that all components are sufficiently dilute to allow us to equate activities with molar fractions. Molar fractions of the polyion, the small cation and anion, and the solvent will be designated as x_p, x_c, x_a, and x_s, respectively. The side of the osmometer without polyions will be denoted by a prime, the other side by a double prime. Solutions on both sides of the membrane must be electroneutral, that is, they must contain equal numbers of positive and negative charges.

$$Zx_p'' + x_a'' = x_c'' \tag{3.2.26}$$

$$x_a' = x_c' \tag{3.2.27}$$

Furthermore, the activity of the electrolyte must be the same on both sides of the membrane. The activity of an electrolyte is known to be equal to product of the activities of its constituent ions. We have assumed that activity coefficients of small ions are equal to unity, hence,

$$x_a'' x_c'' = x_a' x_c' \tag{3.2.28}$$

Combining equations (3.2.26)–(3.2.28), we express the concentrations of small ions on the polyion side in terms of their concentration on the other side and of the polyion concentration as

$$x_a'' = x_a'\left[1 + \left(Zx_p''/2x_a'\right)^2\right]^{1/2} - Zx_p''/2 \tag{3.2.29}$$

$$x_c'' = x_a'\left[1 + \left(Zx_p''/2x_a'\right)^2\right]^{1/2} + Zx_p''/2 \tag{3.2.30}$$

It is apparent that the concentration of the cations on the polyion side is higher than on the other side; the concentration of anions is less. However, the *net* concentration of the low-molecular-weight electrolyte on the polyion side (x_a'' in the present case) is less than its concentration on the other side. This phenomenon is known as the *Donnan effect*.

We will now calculate the osmotic pressure π as the difference between the osmotic pressures π' and π'' that the two sides would exhibit against pure solvent (water).

$$\begin{aligned} \pi = \pi'' - \pi' &= -(RT/V_s)\left(\ln x_s'' - \ln x_s'\right) \\ &= -(RT/V_s)\left[\ln\left(1 - x_p'' - x_a'' - x_c''\right) - \ln\left(1 - x_a' - x_c'\right)\right] \\ &\doteq (RT/V_s)\left(x_p'' + x_a'' + x_c'' - x_a' - x_c'\right) + \cdots \end{aligned} \tag{3.2.31}$$

Substituting equations (3.2.27), (3.2.29), and (3.2.30) for x_c', x_a'', and x_c'' respectively, we may rearrange equation (3.2.31) to

$$\pi V_s/RT = x_p'' + 2x_a'\left\{\left[1 + \left(Zx_p''/2x_a'\right)^2\right]^{1/2} - 1\right\} + \cdots \tag{3.2.32}$$

At very low concentrations of the polyelectrolyte, Zx_p'' becomes much smaller than $2x_a'$. We may then expand the square root in equation (3.2.32) into a Taylor series and get

$$\pi V_s / RT = x_p'' + Z^2 x_p''^2 / 4 x_a' + \cdots \tag{3.2.33}$$

Switching from molar fraction of polyion to its concentration $c_p(x_p'' \approx c_p V_s / M_p)$ and from molar fraction of the small ions to their molality, $m_a'(x_a' \approx m_a' M_s / 1000)$, we transform the last equation to

$$\frac{\pi}{c_p RT} = \frac{1}{M_p} + 1000 \frac{v_s}{4 m_a'} \left(\frac{Z}{M_p} \right)^2 c_p + \cdots \tag{3.2.34}$$

It is seen that extrapolation to vanishing concentration of the polyelectrolyte still yields its molecular weight. However, the second virial coefficient may acquire quite high values if the charge-to-mass ratio of the polyions, Z/M_p, is high. For example, if there is one charge per 1000 units of molecular weight [this is a moderate charge, corresponding to 20 charges on a protein molecule with $M = 20,000$ or to poly(acrylic acid) with degree of ionization about 7%] and the molality of the salt is 0.1, while the specific volume of the solvent, v_s, is about 1 mL/g, then the second virial coefficient $B = 2.5 \times 10^{-3}$ mL/g. This value should be compared with B values for nonionized polymers in good solvents, which are rarely higher than 1×10^{-3} mL/g. When the molality of the salt is lowered further, the virial coefficients may rise dramatically.

These high values of B may make the extrapolation to vanishing concentration of the polyelectrolyte rather unreliable and are avoided in experimental studies. Virial coefficients can be reduced by reducing the charge on the macromolecules by adjusting the pH of the solution. For proteins, pH should be close to the isoelectric point of the protein. Employing higher concentrations of supporting electrolytes can also reduce the polyelectrolyte effect. Even then, the remaining effect is usually stronger than the nonideality of nonionized polymers.

What is the osmotic pressure at higher concentrations of polyelectrolytes when the concentration of the supporting electrolyte is very low; in other words, when the ratio $Z x_p'' / 2 x_a'$ is much larger than unity? Then equation (3.2.32) transforms into a limiting form,

$$\pi V_s / RT = x_p'' (1 + Z) \tag{3.2.35}$$

which represents the osmotic pressure exhibited by a Van't Hoff electrolyte using membranes that are impermeable to salt. The transition from the behavior described by equation (3.2.34) to that of equation (3.2.35) occurs at some concentration of polyelectrolyte that is lower with lower concentration of the supporting electrolyte (i.e., x_a'). In the complete absence of supporting electrolyte, equation (3.2.35) is applicable at all polyelectrolyte concentrations.

We need to consider another important question: The molecular weight of what entity is measured in osmotic experiments? Is it the molecular weight of the polyion itself, or does it include the counterions as well? If there are more types of counterions present, which one should be included in the molecular weight? Our derivation of equation (3.2.34) seems to suggest that we are measuring the molecular weight of the polyion itself. However, closer inspection reveals that the molecular weight calculated from equation (3.2.34) refers to the entity whose mass was used in calculating the polymer concentration c_p. If we weighed the sodium salt of a polyacid and used this weight for calculating concentration, then the molecular weight of the sodium salt is what we obtained!

The simple theory outlined above describes reasonably well the behavior of compact molecules of polyelectrolytes, for example, protein molecules. However, linear polyelectrolytes such as carboxymethyl cellulose or poly(acrylic acid) exhibit virial coefficients that, although still very large compared to nonionized polymers, are significantly lower than predicted by the above theory. A more sophisticated analysis revealed that in polyelectrolyte coils the counterions tend to be associated with the charged polyion and cannot fully participate in the Donnan equilibrium. At the same time, the repulsion of charges on the polyelectrolyte chain leads to a large expansion of the coils, which should be accompanied by a significant increase in the excluded volume effect and in the virial coefficient. Obviously, all the above effects are intertwined and contribute to the observable phenomena.

3.2.1.3. Technical Aspects of Osmometry.

In this section we will describe some of the major technical problems in osmotic measurement and their reflection in the construction of osmometers. There are two classes of problems: those associated with a slow approach to osmotic equilibrium and those related to imperfect membranes.

When the two solutions are introduced into the osmometer, the difference between the pressures on the two sides is generally different from the osmotic pressure. The imbalance between the actual and equilibrium pressure differences is the driving force of the osmosis—that is, of solvent transport across the membrane.

In most situations, this driving force is minuscule to begin with (a few millimeters of the solvent hydrostatic head) and asymptotically approaches zero as the system approaches equilibrium. Thus the transport of the solvent is extremely slow, yet it is used to generate the pressure difference characterizing the equilibrium. In older osmometers, both sides were attached to capillaries, and the solvent entering the solution side caused a rise in the level in the capillary, thus increasing the pressure. The positions of the menisci served as a measure of the pressure difference. Obviously, the narrower the capillaries, the faster the equilibrium could be reached. Unfortunately, capillary effects limit the choice of the diameter of the capillaries to values larger than about 1.0 mm. With such capillaries, the amount of transported solvent may be quite large and the equilibration time quite long. The amount of solvent passing through the membrane is proportional to the area of the membrane; thus larger membranes are advisable.

Full osmotic equilibrium requires that both compartments be homogeneous; the homogenization is achieved by diffusion. Thus the osmotic chambers should be very shallow with large membranes.

It is not necessary to let the solvent flow through the membrane be the sole factor that changes the pressures: We may give it a helping hand. In one design, the solution chamber is connected to the measuring capillary by another horizontal capillary into which a small bubble of air is introduced. This bubble is watched by a photocell, which registers the flow. Its signal is utilized through an appropriate servomechanism for changing the pressure in the other half-cell. The servocircuit may bring the system to equilibrium within minutes. This is a great achievement; the original capillary osmometers required one or two days for equilibration.

In another stratagem, rigid walls enclose the solution chamber. Any transport of solvent leads to compression of the liquid. Tiny amounts of solvent are sufficient for raising the pressure; hence the equilibration times may be quite short. The pressure inside the solution compartment is monitored by a very sensitive pressure gauge.

Another measuring concept applies various pressure differences to the two compartments and measures the flow rate, for example, by measuring the velocity of the meniscus rise. The dependence of velocity on applied pressure difference is interpolated to zero velocity; the pressure that produces no flow is the equilibrium pressure.

When mixed solvents are used, or when polyelectrolytes are measured in the presence of supporting electrolyte, the solvent compositions on both sides of the membrane are not the same. To achieve the proper mismatch of the solvent compositions, sometimes a large amount of solvent components must pass through the membrane and the equilibration time becomes very long. The methods for fast achievement of proper pressure balance do not allow for equilibration of solvent compositions. The remedy is often found in pre-equilibration—the solution and the solvent are extensively dialyzed against each other to achieve the appropriate compositions before the solutions are transferred to the osmometer. One of the disadvantages of this procedure is that extensive transfer of solvents may considerably change the concentration of the polymer, which was presumably known with good precision before the dialysis but must be remeasured after it. With some polymers, measurement of the concentration may be the least accurate part of the whole procedure. We should probably comment on the procedure of dialysis. It is essentially the same process as equilibration in an osmotic chamber, but measurement of the osmotic pressure is not the objective, and vigorous stirring is used to speed up the equilibration.

As if the above-mentioned problems were not enough, the real membranes are neither rigid nor strictly semipermeable. A nonrigid membrane bends under the influence of the osmotic pressure, accommodating the flow of solvent without registering any increase in pressure; this may lead to a false detection of an equilibrium and premature termination of the experiment. The effect is known as *bombing* of the membrane. Supporting the membrane by a rigid screen can minimize the bombing.

A more serious problem is imperfect semipermeability of the membrane. Membranes may be likened to sieves letting smaller particles through and stopping the larger ones. However, the pores (they are not really pores, they are voids in the network of polymer chains constituting the membrane) never have uniform size. Moreover, the flexible polymer coils in the solution may change their shape appreciably; a long molecule may creep through a narrow crevice. Thus the cutoff point for permeability is never very sharp. We can control the approximate value of the cutoff point by changes in the original conditioning of membrane. However, we face a dilemma: Very tight membranes with a low cutoff point will also slow down appreciably the flow of solvent, increasing the equilibration time even more. The problem is especially serious when polydisperse unfractionated polymers are measured. We recall from Section 1.6.2 that monomer is the most abundant species present in polymers that have most probable distribution of molecular weights. Experimentally, the penetration of oligomers through the membrane manifests itself in the form of the measured pressure dependence on time. It first rises as the pressure difference increases and approaches the osmotic pressure; then it decreases again as the oligomers slowly permeate the membrane, causing the osmotic pressure to decrease.

We must mention one more phenomenon. In most osmometers, the volume of the solution is the observed quantity (e.g., by measurement of the position of the meniscus). However, the volume of a liquid is very sensitive to changes in temperature (remember construction of mercury thermometers). Hence, the slightest change in temperature may cause false readings of osmotic pressure. Thus very good temperature control is necessary in osmometry.

A feeling for the construction of osmometers can be gained from Figure 3.5, which is a photograph of an osmometer used by Krigbaum and Flory. The membrane is vertical, and two identical half-cells contain a set of shallow channels that are connected by a circular channel close to the border of the cell. When the cell is assembled, the two sets of channels are perpendicular to each other, forming a system of "island" membrane supports. Each half-cell is equipped with a capillary (1 mm in diameter) and a needle injection valve. The whole assembly is immersed in a thermostat.

After considering all the problems plaguing experimental osmometry, it is fitting to assess its standing among the thermodynamic methods for the study of polymers. Despite its rather simple theory and very simple basic experimental arrangement, osmometry is experimentally very demanding, very slow, and prone to hosts of errors. In recent years, it has been used less and less. On the positive side, it is based on very firm theoretical foundations, and it provides a solid reference point for a number of methods requiring calibration. Most important, osmometry is the only practical method for an *absolute* measurement of number-average molecular weights. As such, it is bound to retain its significance in the future.

3.2.2. Equilibria in the Ultracentrifuge

As we have mentioned before, in the presence of outside potentials, chemical potentials may be nonuniform within the system at equilibrium; this nonuniformity is often accompanied by nonuniformity in concentrations, that is, by concentration gradients. Studies of such gradients provide valuable information about these systems. One outside potential that is always present on Earth is the gravitational potential. This potential indeed causes concentration gradients. However, the steepness of the gradients varies enormously with the size of the materials involved. Let us inspect an example: a sandy but murky pond in a salt marsh. The vertical concentration profile of the sand is rather steep indeed! It is best described as a step boundary forming the bottom of the pond. On quiet days the silt collects at the bottom but does not form a compact layer. The heavier particles hover at distances of up to a few centimeters from the bottom, and smaller particles may go to distances measured in feet. (Those

Figure 3.5. A two-chamber osmometer used by Krigbaum and Flory. (Reprinted from P.J. Flory, *Principles of Polymer Chemistry*. Copyright © 1953 by Cornell University. Used by permission of the publisher, Cornell University Press.)

of you who prefer suburban life may see the same phenomenon at the bottom of a swimming pool in the morning before the first swimmer stirs the water.) Finally, the salt in the water has virtually uniform concentration throughout the pond. Obviously, the smaller the particle, the less it is constrained by the gravitational forces. Distance may help; even molecules of air form a quite nonuniform concentration profile when studied between sea level and the high stratosphere.

It is obvious that we cannot learn much about the sand from the fact that it sits on the bottom; neither does the salt provide much information by being everywhere. However, the silt, having just the right size, has a lot of information hidden in its concentration profile. We can study even smaller particles (like air) if our apparatus is high enough—but this soon becomes impractical in the laboratory. Another possibility would be to increase the gravitational field, say by performing the experiments in the vicinity of Jupiter, but this concept has its flaws, too. Fortunately, we can replace the gravitational potential and forces by another mechanical potential: the centrifugal potential and its forces. The big advantage of the centrifuge is that its speed can be varied, and, through it, the mechanical potential can be tailored to the particles under study.

The behavior of macromolecular solutions in centrifuge cells gives us a true treasure of information. Equilibrium experiments yield two different averages of molecular weight (it is an absolute method!), virial coefficients, preferential adsorptions, and more. Velocity experiments provide another technique for measurement of molecular weight and molecular dimensions; they give us information, for example, about minor components in solutions of biopolymers. Indeed, it was the results of sedimentation analysis performed with the help of the ultracentrifuge that established once and for all the macromolecular nature of many materials previously described as colloidal.

The Svedberg in Uppsala developed the first ultracentrifuge in the late 1920s. In due time, the ultracentrifuge became the most treasured instrument in biochemical and polymer laboratories. However, the more sophisticated sedimentation techniques were rather demanding experimentally and mathematically. The ultracentrifuge was therefore replaced in most laboratories by faster techniques. Nevertheless, it is an instrument yielding experimental data with unsurpassed accuracy and reliability. It is a rather complex apparatus, and we will devote a short section to its description.

3.2.2.1. The Ultracentrifuge.

An ultracentrifuge is a centrifuge equipped with an optical system allowing for monitoring of the concentration profile in the cell during the whole experimental run. Centrifuges are designed to run at speeds of 2000–68,000 rpm. The lower limit is given by the requirement of mechanical stability of the rotor, the upper limit by engineering properties of the construction materials.

The heart of the ultracentrifuge is a rotor (Fig. 3.6). It is made from materials having the best strength-to-mass ratio: aluminum and, more recently, titanium. At a distance of about 6 cm from the axis, it has two holes for a cell and a counterbalance (recently, rotors with up to eight holes have become available, allowing multicell runs). The cell (Fig. 3.7) consists of a housing, a centerpiece that forms the body of

Figure 3.6. A two-cell rotor used in the Spinco model E ultracentrifuge. (Reprinted from *Mari-jo Zeller, Model E Analytical Ultracentrifuge: Instruction Manual.* Used by permission of Beckman Instruments, Inc.)

the solution compartment, and two thick (5 mm) windows that form the sides of the cell and allow for optical observation. The windows should not distort excessively under the action of centrifugal forces. Originally, they were made from fused quartz, but this material proved to be too soft. Sapphire windows are preferred for more demanding applications.

The centerpiece has either one or two compartments. In the latter case, one compartment is filled with the macromolecular solution and the other with reference solvent. The compartments are sector shaped with the apex at the rotor axis. This arrangement allows for undisturbed radial movement of molecules in the vicinity of the walls. The twin compartments can each hold up to 0.45 mL of liquid, forming a liquid column up to 10 mm long.

The rotor spins in a cylindrical chamber. Now, the centrifugal forces are huge (with accelerations equivalent up to 200,000 times gravity), and the kinetic energy of spinning rotor is enormous. The operator must be protected from the remote possibility of rotor explosion. Hence the chamber is made of 1/2-in.-thick steel armor, and another armored plate is on the front side of the instrument. This safety precaution adds heavily to the construction of the ultracentrifuge: It must contain heavy machinery for opening and closing the chamber (it is done by lifting it up and down). At the bottom and top of the chamber are lenses that perform an additional duty as windows for observation of the cells.

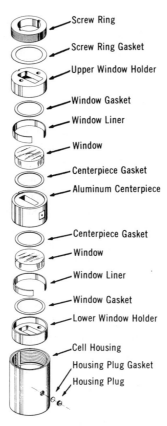

Screw Ring
Screw Ring Gasket
Upper Window Holder
Window Gasket
Window Liner
Window
Centerpiece Gasket
Aluminum Centerpiece
Centerpiece Gasket
Window
Window Liner
Window Gasket
Lower Window Holder
Cell Housing
Housing Plug Gasket
Housing Plug

Figure 3.7. Ultracentrifuge cell parts. (Reprinted from *Mari-jo Zeller, Model E Analytical Ultracentrifuge: Instruction Manual.* Used by permission of Beckman Instruments, Inc.)

Very good temperature control of the rotor is needed to prevent thermal convection of the liquid in the cell, which would ruin the measurement. (In fact, it frequently does; convection—caused either mechanically or thermally—is one of the worst enemies of ultracentrifuge users.) Two sources of heat constantly heat the rotor: thermal flow from the rotor drive and friction of the rotor in the air in the chamber. The latter heat source is very serious and must be eliminated by evacuating the chamber to about 10^{-3} torr. Thus the ultracentrifuge is equipped with a diffusion pump backed up by an oil pump.

The temperature detector is a thermistor located inside the rotor near its bottom. It is connected to the outside measuring and regulating system by a needle sticking out of the bottom of the rotor and spinning in a pool of mercury. The rotor is cooled and heated by radiation. The cooling coil is on the inside wall of the chamber and is controlled by a cooling unit, which is also a part of the ultracentrifuge. A wire heater intermittently radiates heat toward the rotor (and sometimes creates harmful temperature gradients in it).

Ultracentrifuges are frequently equipped with several optical systems: schlieren, interference, and UV absorption. The first two systems are designed to analyze the profile of refractive index along the radial direction in the cell, which closely parallels the concentration profile of the macromolecular solute. A high-pressure, high-intensity (1000 W) capillary mercury lamp is placed at the focus of the lens embedded in the bottom of the chamber. Thus the light beam is composed of strictly parallel rays traveling through the chamber parallel to the rotor axis. Once during every revolution, the cell becomes aligned with the light beam and optical measurements can be made. The light passing through the solution cell may be analyzed by means of a cylindrical lens and a schlieren bar (a dark hairline on a glass plate) inclined at an angle to the cylindrical lens. This optics produces *schlieren pictures,* which record the gradient of refractive index within the cell, dn/dr. Alternatively, the two half-beams, passing through the two cell compartments and a two-slit mask, produce a pattern of *interference fringes,* which depicts the difference of the refractive indices in the two compartments. Both interference and schlieren images are recorded on a photographic plate, which is later analyzed by means of a projector with a two-dimensional stage. Such projectors are now often used in conjunction with computers.

Ultraviolet optics can be extremely sensitive for the study of solutes with high absorbancy in nonabsorbing solvents. Solutions of nucleic acids are the prime examples. Originally, UV light source was photographed through the rotating cell and the concentration profile was deduced from the optical density of the photographic image. With the advent of fast electronics, the photographic film was replaced by a photomultiplier slowly moving in the imaging plane. The light intensities corresponding to the two cell compartments are electronically compared and recorded as a function of the radial position. With proper cuing, several cells can be monitored during the same run with a multicell rotor.

With so many massive components contained in the ultracentrifuge, it should not be surprising that this instrument, which is used for measurement of 0.10-mL samples of solutions with microgram quantities of solutes, is one of the most massive instruments in physicochemical laboratories, weighing about a ton (Fig. 3.8).

3.2.2.2. Sedimentation Equilibrium.

3.2.2.2. Sedimentation Equilibrium. When a macromolecular solution is centrifuged long enough, the macromolecules redistribute themselves in the cell and eventually reach sedimentation equilibrium, which is then recorded on the photographic plate. The sedimentation equilibrium presents one of the most dependable ways of measuring molecular weight and other thermodynamic quantities.

The starting point for our analysis is the expression for total potential, equation (3.2.1). For our present case, the electrical potentials are irrelevant; for the mechanical potential we substitute the well-known expression for the potential in the centrifugal field and obtain

$$\mu_{i,\text{tot}} = \mu_i - M_i\omega^2 r^2/2 \qquad (3.2.36)$$

Figure 3.8. Spinco model E ultracentrifuge.

where r is the radial distance from the rotor axis and the angular velocity ω is equal to $2\pi(\text{rpm})/60$. The homogeneity of the total potential at equilibrium implies that a derivative of it with respect to the coordinates must be equal to zero. Thus

$$\frac{d\mu_{i,\text{tot}}}{dr} = \frac{d\mu_i}{dr} - M_i\omega^2 r = 0 \tag{3.2.37}$$

Let us first study equation (3.2.37) for the simplest system—a pure solvent. Two independent variables are sufficient for a description of any one-component system; we will choose pressure and temperature. Then the total differential of the chemical potential μ_1 and the derivative $d\mu_1/dr$ will be

$$d\mu_1 = \bar{V}_1 dP - \bar{S}_1 dT \tag{3.2.38}$$

$$\frac{d\mu_1}{dr} = \bar{V}_1 \frac{dP}{dr} - \bar{S}_1 \frac{dT}{dr} \tag{3.2.39}$$

Here, $\bar{V}_1$ and $\bar{S}_1$ are the molar volume and molar entropy, as usual. In a thermostated system, dT/dr is zero everywhere and the second term in equation (3.2.39) may be omitted. Thus combining equations (3.2.37) and (3.2.39) gives

$$\bar{V}_1 \frac{dP}{dr} = M_1 \omega^2 r \tag{3.2.40}$$

We will now substitute for $\bar{V}_1$ the product of the molecular weight M_1 and specific volume $\bar{v}_1$, which for a pure substance is equal to its inverse density $1/\rho_1$. The result,

$$\frac{dP}{dr} = \omega^2 r \rho_1 \tag{3.2.41}$$

describes the dependence of pressure on the radial distance. This expression is, of course, well known from hydrostatics; however, we were able to derive it using purely thermodynamic arguments.

We are now ready to treat a two-component system. For this, equation (3.2.37) is applicable for each component. However, when calculating $d\mu_i/dr$, we must add one more independent variable to the total differential [equation (3.2.38)]; we choose for this the concentration of macromolecules c_2. The equivalent of equation (3.2.40) for the macromolecular component (subscript 2) now reads

$$\bar{V}_2 \frac{dP}{dr} + \left(\frac{\partial \mu_2}{\partial c_2}\right)_{P,T} \frac{dc_2}{dr} = M_2 \omega^2 r \tag{3.2.42}$$

Now substituting equation (3.2.41) for dP/dr (it is easily shown that this is valid even for mixtures; ρ is now the local density of the mixture) and switching again from molar volumes to specific volumes, we transform equation (3.2.42) to

$$\frac{dc_2}{dr} = \frac{M_2 \omega^2 r (1 - \bar{v}_2 \rho)}{(\partial \mu_2/\partial c_2)_{P,T}} \tag{3.2.43}$$

It turns out that for sedimentation analysis the square of the radial distance, r^2, is a more convenient variable than the radial distance itself. Its introduction modifies equation (3.2.43) to

$$\frac{dc_2}{d(r^2)} = \frac{M_2 (1 - \bar{v}_2 \rho) \omega^2}{2(\partial \mu_2/\partial c_2)_{P,T}} \tag{3.2.44}$$

The factor $(1 - \bar{v}_2\rho)$ plays a very important role in sedimentation processes; it is called the *buoyancy factor*. It is easy to show that, for solutions with negligible volume changes in mixing,

$$1 - \bar{v}_2\rho = \phi_1 (1 - \bar{v}_2\rho_1) \tag{3.2.45}$$

where ρ_1 is the density of pure solvent.

In the next step of our analysis, we substitute into relation (3.2.44) equation (3.2.45) and a relation for $(\partial\mu_2/\partial c_2)_{P,T}$. When we are interested in a measurement of molecular weight, relation (3.1.43) is most convenient for this purpose. The combined relations are then rearranged to give

$$\frac{c_2\omega^2(1 - \bar{v}_2\rho_1)}{2RT\,dc_2/d(r^2)} = \frac{1}{M_2} + 2Bc_2 + 3Cc_2^2 + \cdots \tag{3.2.46}$$

Experimentally, concentration is evaluated as a function of radial position at a large number of positions in the cell, and the left-hand side of equation (3.2.46) is evaluated for these positions. These values are then plotted against c_2; the intercept is equal to the inverse molecular weight, and the initial slope is equal to $2B$ (Fig. 3.9). Quite often, the plot will be curved because of the third virial term. In such a case, it is convenient to employ the same computational maneuver we did in Section 3.2.1 when analyzing osmotic data using a relationship between the second and third virial coefficients, equation (3.2.12). In the present situation, we would get an exact square from the first three virial terms if we selected the value of $\gamma = \frac{1}{3}$. Now $\frac{1}{3}$ is as reasonable a selection as $\frac{1}{4}$ was. Thus we may again use the square root plot, this time in the form

$$\left[\frac{c_2\omega^2(1 - \bar{v}_2\rho_1)}{2RT\,dc_2/d(r^2)}\right]^{1/2} \quad \text{vs.} \quad c_2$$

Again, the plot will exhibit much better linearity and the intercept will again be $M_2^{-1/2}$; the product of intercept and slope will be B. We should stress that for both types of the above analysis, the whole concentration dependence was obtained from a single equilibrium run. In contrast, both osmotic measurements and light-scattering measurements provide only one point per solution. This feature partially compensates for the greater time demands of the sedimentation technique.

In biochemical studies, the concentrations are frequently quite low and the nonidealities are small. Under these conditions, it is legitimate to neglect the higher virial terms in equation (3.2.46) and rearrange it to the following form:

$$\frac{d\ln c_2}{d(r^2)} = \frac{M_2(1 - \bar{v}_2\rho_1)\omega^2}{2RT} \tag{3.2.47}$$

According to this relation, for an ideally dilute solution (or for a pseudo-ideal theta

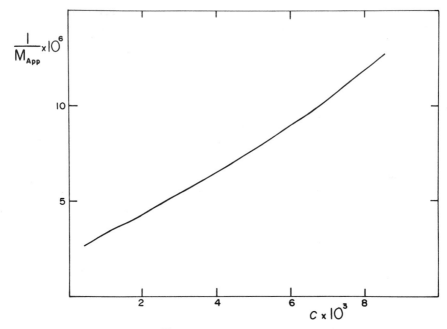

Figure 3.9. Dependence of $1/\overline{M}_{w,App} \equiv c_2\omega^2(1 - \bar{v}_2\rho_1)/2RT[dc_2/d(r^2)]$ on c_2 for polystyrene ($\overline{M}_w = 619,000$) in benzene at 20°C.

solution), $\ln c_2$ should be a linear function of r^2. From the plot of $\ln c_2$ versus r^2, the molecular weight is obtained easily.

Still another rearrangement of equation (3.2.47) is used:

$$\frac{dc_2}{d(r^2)} = \frac{M_2(1 - \bar{v}_2\rho_1)\omega^2}{2RT}c_2 \qquad (3.2.48)$$

This relation states that $dc_2/d(r^2) \equiv (dc_2/dr)/2r$ is a linear function of concentration. Again, the slope gives M_2. We should mention that all the above formulas for the calculation of molecular weight either contain concentration as a logarithm or employ a ratio of two concentrations. In either case, the units of concentration are not important. Indeed, arbitrary units, such as the number of fringes on the photographic plate or the shift in the fringe measured in micrometers, are used routinely. Of course, concentration in physically meaningful units must be known for the calculation of virial coefficients.

Throughout the above discussion, we have assumed that the macromolecular sample was monodisperse. This assumption is reasonable in biochemistry—hence the greater popularity of sedimentation analysis among biochemists than among polymer chemists. Polydispersity is more troublesome in sedimentation measurements than in other techniques. Each macromolecular species redistributes itself in the cell in a different way; consequently, the molecular weight distribution varies from point to point, considerably complicating the analysis. On the other hand, the redistribution provides an opportunity to evaluate the polydispersity. This task is reasonably straightforward for ideally dilute solutions, and we will restrict our attention to them. The general case of a polydisperse polymer in a nonideal solution has been studied many times, and a number of evaluation techniques have been offered, but they are all less than satisfactory.

We will now analyze a polydisperse polymer in sedimentation equilibrium in an ideally dilute solution; we will assume that the specific volume has an identical value $\bar{v}$ for each polymeric species. In an ideally dilute solution, the redistribution of one molecular species is not influenced by the presence of other species. Hence, we can write equation (3.2.48) for the ith species as

$$\frac{dc_i}{d(r^2)} = \frac{M_i c_i (1 - \bar{v}\rho_1)\omega^2}{2RT} \tag{3.2.49}$$

Summing both sides of equation (3.2.49) over all polymer species and defining $c \equiv \sum_i c_i$, we get

$$\frac{dc}{d(r^2)} = \sum_i \frac{M_i c_i (1 - \bar{v}\rho_1)\omega^2}{2RT} = \frac{\bar{M}_w c (1 - \bar{v}\rho_1)\omega^2}{2RT} \tag{3.2.50}$$

Here, the last equality was obtained employing the definition of $\bar{M}_w$. The last equation is very similar to equation (3.2.47). It tells us that the tangent to the $\ln c$ versus r^2 plot yields the weight-average molecular weight for polydisperse polymers. However, there is a major difference between these relations. For polydisperse polymers, $\bar{M}_w$ is changing from point to point within the cell and the above plot is curved. Although the dependence of $\bar{M}_w$ on the cell coordinates is interesting, the more important quantity is $\bar{M}_w$ of the original sample. The formula for its calculation is deduced in the following way. Equation (3.2.49) is integrated between the solution meniscus and the bottom of the cell to give

$$c_{i,\text{bot}} - c_{i,\text{men}} = \frac{M_i(1 - \bar{v}\rho_1)\omega^2}{2RT} \int_{r_{\text{men}}^2}^{r_{\text{bot}}^2} c_i d(r^2) = \frac{M_i(1 - \bar{v}\rho_1)\omega^2 c_i^0 (r_{\text{bot}}^2 - r_{\text{men}}^2)}{2RT} \tag{3.2.51}$$

Here, the subscripts bot and men denote bottom and meniscus, respectively. The second equality follows from the fact that the integral actually represents the total amount of species i in the sector-shaped cell; it must therefore be equal to the amount present before the start of the run, that is, to $c_i^0 (r_{\text{bot}}^2 - r_{\text{men}}^2)$. The superscript zero refers to the original, nonredistributed concentration in the cell. Summation of equations (3.2.51) for all macromolecular species, together with the definition of the weight-average molecular weight, yields

$$c_{\text{bot}} - c_{\text{men}} = \frac{\overline{M}_w^{\,0} c^0 \left(1 - \bar{v}\rho_1\right) \omega^2 \left(r_{\text{bot}}^2 - r_{\text{men}}^2\right)}{2RT} \tag{3.2.52}$$

from which the molecular weight $\overline{M}_w^{\,0}$ of the original sample is easily calculated. It should be noted that, for this calculation, only the concentrations at the cell ends are needed, not the entire concentration profile.

Similar, but a little more tedious, calculations would show that for polydisperse samples our other plot $dc/d(r^2)$ versus c is also curved upward. Its tangent at any point yields the centrifuge average of molecular weight $\overline{M}_z$, applicable to polymer at this particular location; $\overline{M}_z^{\,0}$ for the original sample is given by the formula

$$\overline{M}_z^{\,0} = \frac{dc/d(r^2)_{\text{bot}} - dc/d(r^2)_{\text{men}}}{c_{\text{bot}} - c_{\text{men}}} \left(\frac{(1 - \bar{v}\rho_1)\omega^2}{2RT}\right) \tag{3.2.53}$$

Clearly, the first fraction in this expression represents the slope of the line connecting the endpoints in the $dc/d(r^2)$ versus c plot. This type of evaluation is especially convenient when the concentration profile is recorded using schlieren optics. Then $(dc/dr)_{\text{bot}}$ and $(dc/dr)_{\text{men}}$ are simply the coordinates of the schlieren line at its endpoints; $c_{\text{bot}} - c_{\text{men}}$ is the area under the schlieren curve.

Sedimentation equilibrium experiments are time-consuming. Redistribution of the components within the cell can be viewed as a diffusion process that gradually reduces the deviations of the actual concentration profile from the equilibrium profile. The time required for this approach to equilibrium sharply increases with the length of the liquid column. Consequently, experimentalists try to shorten the column as much as possible. The most often employed length of the column is 3 mm; the time needed for its full equilibration is typically 2–3 days.

Sedimentation equilibrium is a valuable method for studying preferential adsorption in mixed solvents. We will present the final formula without a detailed derivation, which is somewhat tedious. (The subscript 3 refers to polymer in a mixed solvent.)

$$\frac{c_3\omega^2}{2RT\,dc_2/d(r^2)} = \frac{1}{M_3\left(1 - \bar{v}_3^*\rho^0\right)} + \frac{2Bc_3}{1 - \bar{v}_3^*\rho^0} + \cdots \tag{3.2.54}$$

The only difference between equation (3.2.54) and equation (3.2.46), which is valid for a simple solvent, is the replacement of the polymer buoyancy factor $(1 - \bar{v}_3\rho_1)$, by a new quantity, $(1 - \bar{v}_3^*\rho^0)$. Here ρ^0 is the density of the mixed solvent; $\bar{v}_3^*$ is a new quantity called the *specific volume of polymer at constant chemical potentials of solvents*. This quantity would be obtained if the specific volume were measured by comparing the density of the solution with the density of its dialysate (i.e., a solvent mixture in osmotic equilibrium with the solution). The ordinary specific volume $\bar{v}_3$ is measured by a similar comparison of the solution and the mixed solvent in which it was dissolved. The coefficient of preferential adsorption λ is related to these quantities as

$$\lambda = \frac{\left(\bar{v}_3 - \bar{v}_3^*\right)\rho^o}{d\rho^o/d\phi_1} \qquad (3.2.55)$$

The derivative in the denominator represents the change in the mixed solvent density with its composition.

Our sedimentation equilibrium experiment is evaluated by plotting the values of the left-hand side of equation (3.2.54) against polymer concentration. If M_3 is known (for example, from sedimentation equilibrium of the same polymer in a single solvent), then $\bar{v}_3^*$ can be evaluated from the intercept. The slope then provides the virial coefficient B. Thus sedimentation equilibrium gives us both quantities needed for evaluation of polymer-mixed solvent interaction: λ and B.

3.2.2.3. *Equilibrium in a Density Gradient.*   In the last paragraphs of the preceding section, concerning concentration distribution of the polymer in a mixed solvent, we have tacitly assumed that the density of the mixed solvent is constant throughout the cell. This assumption is reasonable when the densities of the two solvent components are close to each other. However, when they differ appreciably, the denser solvent will sediment toward the bottom and a density gradient will be established in the cell.

When a polymer is added to the solution as a minor third component, it will travel to a position in the cell where its buoyant density matches the local density and will form a band there (Fig. 3.10). If the polymer is composed of several components differing in density, each component will band at a different location in the cell, giving us a very sensitive tool for studying this kind of heterogeneity (Fig. 3.11).

Historically, this phenomenon played a crucial part in recognition of the role of DNA in cell reproduction. It showed that the two strands of the double helix of DNA preserve their identity during cell division. The study was based on the fact that DNA forms very narrow bands in a density gradient. The experiment was designed as follows. The bacteria were grown for several generations in a medium in which the only source of nitrogen was its isotope ^{15}N. From these cells DNA was isolated that had its heterocyclic bases made exclusively with this isotope. The density of such DNA was, of course, higher than that of normal DNA with the ^{14}N isotope. When the heavy DNA was subjected to density-gradient ultracentrifugation (in a concentrated solution of CsCl, and employing UV optics), it formed a band at a location with higher density than the normal DNA band. In the next experiment, the bacteria were washed and transferred into a growing medium with an ordinary ^{14}N source of nitrogen and were allowed to grow for just one generation. DNA prepared from these bacteria again showed a single band, which was located in the middle of the positions for the heavy and light DNA. Next, the growth was continued for another generation. This time, two equally large bands were formed, one at the middle density and the other at the light density. The next generations preserved this pattern except that the ratio of the band sizes favored the light DNA more and more. A continuous distribution of densities was never observed. Only a model in which two strands are

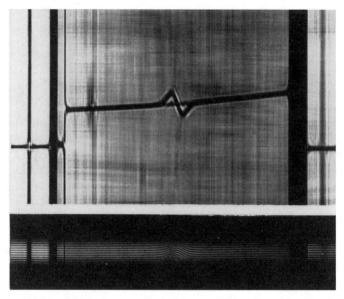

Figure 3.10. Sedimentation equilibrium of a polystyrene sample in a density gradient formed by an isorefractive mixture of cyclohexane and CBrClF. CBrF$_2$, at 44,000 rpm. Schlieren and interference photographs.

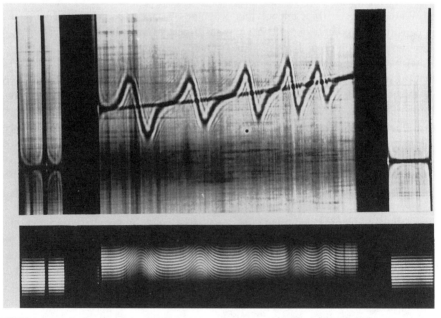

Figure 3.11. Sedimentation equilibrium in a density gradient. A mixture of polystyrene, poly(methyl methacrylate), and 25/75, 50/50, and 75/25 copolymers of styrene and methyl methacrylate.

separated during cell division and complementary chains are synthesized using them as templates but never compromising the integrity of the original strands could explain these phenomena.

Let us study the density gradient itself. We can estimate it from the distribution of the concentration of the denser solvent, which obeys the same rules that we met in the preceding section. It is convenient to start from equations (3.2.44) and (3.2.45), but we must take into account the thermodynamic behavior of the mixtures at high concentrations when we evaluate $(\partial \mu_2 / \partial c_2)_{P,T}$. Assuming the validity of the Flory–Huggins equation (3.1.29) for our mixture of solvents, we find the following relation for the density gradient:

$$\frac{d\rho}{d(r^2)} = \frac{\omega^2 (\rho_2 - \rho_1)^2}{2RT(1/V_1\phi_1 + 1/V_2\phi_2 - 2\chi_{12}/V_1)} \tag{3.2.56}$$

It is apparent that the steepness of the density gradient increases sharply with the rotor speed and with increasing density difference between the solvent components. Mixtures of solvents that do not mix well (high χ_{12} values) produce steeper gradients. It is customary to collect the thermodynamic properties of density gradient-forming mixtures into a single parameter β, which is defined as

$$\frac{d\rho}{d(r^2)} \equiv \frac{1}{2r}\frac{d\rho}{dr} \equiv \omega^2 \beta \tag{3.2.57}$$

It is obvious that β depends not only on the nature of the solvents but also on the composition of the mixture (which varies throughout the cell). However, for narrow bands it could be considered as a constant within the band.

We will now study the behavior of an ideally dilute monodisperse polymer in such a density gradient. We will neglect the variation in the preferential adsorption with the solvent composition within the band (this is actually a strong assumption, poorly satisfied in most systems). Under these assumptions, the equilibrium distribution of polymer concentrations is again described by equation (3.2.47). However, in the buoyancy term, $(1 - \bar{v}_3 \rho)$, we must use the specific volume at constant chemical potentials, $\bar{v}_3^*$, as in equation (3.2.54); it may differ quite appreciably from the inverse density of polymer. The solvent density is now a function of the radial position in the cell. We will express it as a Taylor series developed around a point with density $\rho_0 = 1/\bar{v}_3^*$; this point has radial coordinate r_0.

$$\rho = \rho_0 + \left(r^2 - r_0^2\right)\frac{d\rho}{d(r^2)} = \frac{1}{\bar{v}_3^*} + \left(r^2 - r_0^2\right)\omega^2 \beta \tag{3.2.58}$$

Equation (3.2.47) now reads

$$d\ln c_3 = -\frac{M_3 \bar{v}_3^* \omega^4 \beta}{2RT}\left(r^2 - r_0^2\right) d(r^2) \tag{3.2.59}$$

The integral form of this relation is

$$c_3 = c_3^o \exp\left[-(M_3 \bar{v}_3^* \omega^4 \beta / RT)\left(r^2 - r_0^2\right)^2\right] \tag{3.2.60}$$

Inspection of equation (3.2.60) reveals it is actually the Gaussian error function with r^2 as the independent variable and centered around r_0^2. [The error function is $A \exp(-x^2/2\sigma^2)$, where A is a normalization constant and σ is its standard deviation.] For our function (3.2.60) the standard deviation reads

$$\sigma^2 = 2RT / M_3 \bar{v}_3^* \omega^4 \beta = 2\rho_0 RT / M_3 \omega^4 \beta \tag{3.2.61}$$

Hence, M_3 is easily computed from σ obtained from the experimental curve. The method is especially advantageous for extremely high molecular weights, because ω can almost always be adjusted to produce an appropriately broad Gaussian curve.

For polydisperse polymers, the curves are not Gaussian any more. It can he shown that molecular weight averages of the whole samples can be calculated according to the approximate relations, which we present without proof, because it is somewhat tedious

$$\overline{M}_n^o = \frac{\rho_0 RT}{2\beta \omega^4 r_0^2} \frac{\int c \, dr}{\int (r - r_0)^2 \, c \, dr} \tag{3.2.62}$$

$$\overline{M}_w^o = -\frac{\rho_0 RT}{2\beta \omega^4 r_0^2} \frac{\int [(dc/dr)/(r - r_0)] \, dr}{\int c \, dr} \tag{3.2.63}$$

In these formulas, the integrals extend over the whole band.

The above expressions seem to offer a way to measure the number-average molecular weight of macromolecules with very high molecular weights—a feat no other measurement technique can achieve. However, there are a few difficulties. Correction for nonideality (which we have neglected) calls for tedious extrapolations to vanishing concentration of the polymer in the center of the band. Correction for the changes in preferential adsorption calls for the introduction of a multiplicative factor for molecular weight, which is difficult to evaluate. But the most difficult problem is a technical one: The integral in the denominator of equation (3.2.62) is actually a second moment of the Gaussian curve. The second moment weighs heavily the tails of the peak, that is, the parts of the curve that are measurable with the least precision.

Despite these problems, density gradient centrifugation is often one of the methods of last resort providing information no other method can yield: molecular weights in the extremely high range and a measure of heterogeneity in polymer density.

3.2.3. Phase Equilibria

When in a liquid mixture the interactions among molecules of different components become less and less favorable, the molecules will eventually cease to interact with the molecules of the other kind, and the system will separate into two phases. But thermodynamics will not allow complete separation of components: The two phases will still be mixtures of the same components, but they will have different compositions. An understanding of this phenomenon is important for itself, because demixing and limited solubility play a great role in many situations of practical interest. On top of that, a study of phase relationships is a powerful tool for studying the underlying thermodynamics, and it provides access to more precise values of interaction coefficients and their dependence on the experimental variables. We will therefore devote this section to the study of phase equilibria and related phenomena.

When studying phase relationships, it is customary to utilize the molar change in the Gibbs function of mixing, $\Delta G_{\text{mix}}^{\text{M}}$, for which it is implied that the total number of moles is one. Thus, although ΔG_{mix} was an extensive thermodynamic quantity, $\Delta G_{\text{mix}}^{\text{M}}$ is an intensive one; for its description in a two-component system one compositional variable is sufficient. This quantity is related to quantities in equation (3.1.1) as

$$\Delta G_{\text{mix}}^{\text{M}} \equiv \frac{\Delta G_{\text{mix}}}{n} = \sum_i x_i \left(\mu_i - \mu_i^{\circ} \right) \tag{3.2.64}$$

where n is the total number of moles and the x_i are molar fractions.

Alternatively, another quantity, $\Delta G_{\text{mix}}^{\text{V}}$, which refers to a volume unit of the system, is used—most frequently when studying polymer blends. Unfortunately, a volume unit of a material is not a very practical unit; the amount of matter contained in it changes with pressure and temperature as well as with volume changes in mixing. Thus the use of $\Delta G_{\text{mix}}^{\text{V}}$ is convenient only in conjunction with the assumption that no volume changes accompany mixing. In this case, the ratio of the number of moles of the ith component to the total volume, n_i / V, is equal to ϕ_i / V_i—the ratio of volume fraction to molar volume. These quantities are related as

$$\Delta G_{\text{mix}}^{\text{V}} \equiv \frac{\Delta G_{\text{mix}}}{V} = \sum_i \frac{n_i}{V} \left(\mu_i - \mu_i^{\circ} \right) = \sum_i \frac{\phi_i}{V_i} \left(\mu_i - \mu_i^{\circ} \right) \tag{3.2.65}$$

Our task is to find out whether a given mixture will form one phase or will separate into two or more *conjugate phases*. What are the criteria for equilibrium in the present case? There are no internal boundaries in the system, and no external forces are significant. Hence, at equilibrium the whole system should have uniform temperature and pressure, and its Gibbs function as well as ΔG_{mix} should be at minimum. Obviously, ΔG_{mix} for a multiphase system is the sum of the ΔG_{mix} values for all the phases. We can now rephrase our problem in the following way: Is it possible, for a given homogeneous phase, to find a set of two or more conjugate phases into which we may decompose it (observing the rules of stoichiometry) that would have a lower ΔG_{mix} than our original phase? If no such set of phases exists, the homogeneous phase is stable. If one such set exists, there are probably many more. The set exhibiting the lowest ΔG_{mix} is the one that is thermodynamically stable at the given conditions; our task is to find out which one it is.

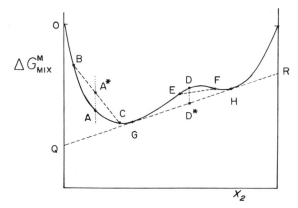

Figure 3.12. A plot of $\Delta G_{\text{mix}}^{\text{M}}$ versus x_2 for a partially miscible binary. See text for a detailed explanation.

We will demonstrate the computational procedures on a binary mixture. A plot of $\Delta G_{\text{mix}}^{\text{M}}$ versus x_2 (Fig. 3.12) is very helpful for this purpose. In accordance with equation (3.2.64), its slope is

$$\left(\frac{\partial \Delta G_{\text{mix}}^{\text{M}}}{\partial x_2}\right)_{P,T} = \left(\mu_2 - \mu_2^{\text{o}}\right) - \left(\mu_1 - \mu_1^{\text{o}}\right) \tag{3.2.66}$$

Consequently, the tangent to the curve at any point has an intercept on the axis $x_2 = 0$ that is equal to $(\mu_1 - \mu_1^{\text{o}})$; the intercept on the axis $x_2 = 1$ equals $(\mu_2 - \mu_2^{\text{o}})$. Points on the curve represent values of $\Delta G_{\text{mix}}^{\text{M}}$ for a single phase. For a system consisting of two phases, the additivity of extensive functions requires the representative point to sit on a straight line connecting the points that describe the two phases. Its x_2 coordinate refers to the overall composition of the system.

Is a mixture with composition at point A of Figure 3.11 stable with respect to the pair of phases at B and C? $\Delta G_{\text{mix}}^{\text{M}}$ for this pair will be depicted by the point A*, which has a higher value of Gibbs function than point A. Thus the single phase A is stable with respect to phases B and C (i.e., it does not demix into them). As a matter of fact, phase A is stable with respect to any pair of two phases. On the other hand, a sample with an overall composition represented by D is unstable with respect to the pair of phases E and F as well as to other possible pairs. From these, the pair G and H has the lowest value of the Gibbs function, namely, the value represented by point D*.

How do we find the composition of the conjugate phases at G and H? The equilibrium criteria of our system require that both components have the same chemical potentials in both phases. Hence, the tangents at points G and H must have the same intercept at both axes, that is, they must form a common double tangent. The chemical potentials of the two components in the conjugate phases are described by points Q and R. The composition of the conjugate phases can also be derived using the plot of $G_{\text{mix}}^{\text{V}}$ versus ϕ_2. This plot provides a much more convenient scale of volume

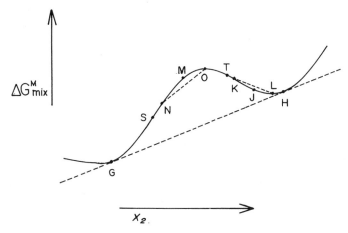

Figure 3.13. The unstable portion of $\Delta G_{\text{mix}}^{\text{M}}$ versus x^2 dependence. See text for a detailed explanation.

fractions for polymeric systems. Points connected by a double tangent again represent the conjugate phases; the intercepts on the axes are now $(\mu_1 - \mu_1^o)/V_1$ and $(\mu_2 - \mu_2^o)/V_2$.

It should be obvious that mixtures having compositions between G and H are unstable and will demix, whereas mixtures with compositions outside this range are stable. However, for full understanding of the phase behavior we must also consider the rate of demixing. Let us analyze the following experiment. A stable mixture is suddenly cooled, which brings it into an unstable state, and it starts to demix. What is the physical mechanism of demixing? Because of thermal motion, the concentration of the mixture fluctuates. For stable mixtures, the fluctuations reverse themselves and the mixtures remain homogeneous. In unstable mixtures, the fluctuations grow and end up in phase separation. In Figure 3.13 we have magnified the unstable portion of the $\Delta G_{\text{mix}}^{\text{M}}$ versus x_2 dependence. Let us consider a mixture with composition at point J. Thermal fluctuations will create regions with concentrations at K and L. However, the Gibbs energy for a two-phase system of K and L compositions is higher than for the J composition. Thus the J composition is stable to this kind of demixing, and such a fluctuation will reverse itself. If the demixing is to proceed, the fluctuation triggering the demixing must be large enough to bring point K beyond the region of the plot that is convex downward. The situation is in principle identical to the process of nucleation of colloidal particles, which we treated in Section 1.5.1. In fact, the nucleation and growth of colloidal particles is also phase separation. In the present case, the size of fluctuations depends primarily on the viscosity of the mixture. Mixtures with low viscosity usually separate readily; highly viscous mixtures (or glassy mixtures) may not separate at all during the duration of the experiment; such systems are said to be *metastable.*

The situation is quite different for mixtures that were brought to point M on our plot. The smallest fluctuation (e.g., fluctuation leading to compositions at points N and O) lowers the Gibbs function; this is an irreversible process that leads to a very fast phase separation. If the system is very viscous, the two phases will form very small domains, but they will be separated anyway. This type of process is called *spinodal decomposition*. The slower decomposition treated earlier, which led to two stable phases at G and H compositions, is called *binodal decomposition*. Inspection of Figure 3.13 reveals that spinodal decomposition may occur only between compositions at S and T, which mark the inflection points of the curve. The above considerations are quite general and are applicable irrespective of either the shape of the $\Delta G_{\text{mix}}^{\text{M}}$ versus x_2 dependence or its origin.

Having outlined the general principles governing the analysis of phase equilibria, we will now demonstrate them for the Flory–Huggins model of polymer solutions. We will use the plot of $V_1 \Delta G_{\text{mix}}^{\text{V}} / RT$ versus ϕ_2. (Multiplication by a constant V_1 / RT does not change the shape of the dependence.) For Flory–Huggins equation (3.1.29) this quantity is given as

$$V_1 \Delta G_{\text{mix}}^{\text{V}} / RT = \phi_1 \ln \phi_1 + (V_1/V_2)\phi_2 \ln \phi_2 + \phi_1 \phi_2 \chi \qquad (3.2.67)$$

The first two terms (the entropy contribution) are always negative. Moreover, the limiting tangents to these combined terms at both ends of the scale are vertical. This implies that $\Delta G_{\text{mix}}^{\text{V}}$ for mixtures extremely dilute in either component is always negative, no matter what the value of the enthalpy term $\phi_1 \phi_2 \chi$ is. It follows further that pure liquid components are never stable in the presence of the other component.

If the enthalpy term is positive and large enough, it will cause a hump on our dependence; this is the sign of demixing. In Figure 3.14 are depicted, for the simplest case $V_1/V_2 = 1$ (this is a good model for a mixture of two solvents), the entropy term (bottom line), the enthalpy terms for three different χ values (three upper lines), and their total—the Gibbs energy (three middle lines). It is apparent that for negative and small positive values of χ, the whole dependence of $\Delta G_{\text{mix}}^{\text{V}}$ versus ϕ_2 is convex downward; the components are miscible without restriction. What is the lowest value of χ for which the system will demix? To have a hump (i.e., a section of the curve concave downward), there must be two inflection points on the curve; we have already identified them as points of spinodal decomposition. An inflection point is a point where the second derivative is equal to zero; between two such points there must exist a point where the third derivative is equal to zero.

As the χ value decreases from high values, the inflection points get closer to each other; that is, the region of incomplete compatibility narrows. At some specific value of χ, the two inflection points merge—the two phases have identical composition. This value of χ separates the regions of complete and incomplete miscibility and defines the *critical point*. At the critical point both the second and third derivatives must be equal to zero simultaneously.

$$\left(\frac{\partial^2 \Delta G_{\text{mix}}^{\text{V}}}{\partial \phi_2^2} \right)_{P,T}^{\text{crit}} = \left(\frac{\partial^3 \Delta G_{\text{mix}}^{\text{V}}}{\partial \phi_2^3} \right)_{P,T}^{\text{crit}} = 0 \qquad (3.2.68)$$

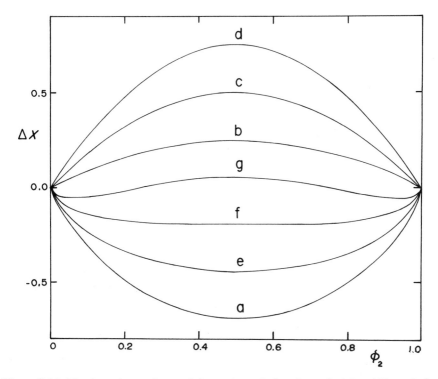

Figure 3.14. The dependence of several thermodynamic functions of mixing ΔX, on ϕ_2 for a mixture of components with $V_1/V_2 = 1$. (a) $V_1 \Delta S_{\text{mix}}^{\text{V}}/RT$; (b) $V_1 \Delta H_{\text{mix}}^{\text{V}}/RT, \chi = 1$; (c) $V_1 \Delta H_{\text{mix}}^{\text{V}}/RT, \chi = 2$; (d) $V_1 \Delta H_{\text{mix}}^{\text{V}}/RT, \chi = 3$; (e) $V_1 \Delta G_{\text{mix}}^{\text{V}}/RT, \chi = 1$; (f) $V_1 \Delta G_{\text{mix}}^{\text{V}}/RT, \chi = 2$; (g) $V_1 \Delta G_{\text{mix}}^{\text{V}}/RT, \chi = 3$.

These conditions specify both the critical value of χ and the composition at the critical point. For our present case of $V_1/V_2 = 1$ we obtain, by solving equation (3.2.68) together with equation (3.2.67), $\phi_2^{\text{crit}} = 0.5$ and $\chi_{\text{crit}} = 2.0$.

When the molar volumes of the two components are unequal (and that is emphatically the case for polymer solutions), the symmetry of $\Delta G_{\text{mix}}^{\text{V}}$ is lost and the critical point and compositions of *both* phases shift toward mixtures more dilute in polymer. Figure 3.15 shows the situation for a modest value of $V_1/V_2 = 0.1$; this is still an *oligomer* and not a polymer. The critical value of χ and the composition at the critical point are now

$$\chi_{\text{crit}} = 0.5 + \sqrt{V_1/V_2} + V_1/2V_2 \tag{3.2.69}$$

and

$$\phi_2^{\text{crit}} = \frac{\sqrt{V_1/V_2}}{1 + \sqrt{V_1/V_2}} \tag{3.2.70}$$

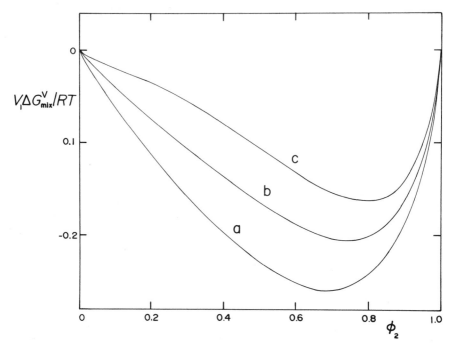

Figure 3.15. Dependence of $V_1 \Delta G_{\text{mix}}^V / RT$ on ϕ_2 for a mixture of components of unequal size: $V_1 / V_2 = 0.1$. (a) $\chi = 0.6$; (b) $\chi = \chi_{\text{crit}} = 0.861$; (c) $\chi = 1.1$.

In the limit of infinite molecular weight, χ_{crit} approaches 0.5; this is the value that characterized pseudo-ideal theta solutions. This coincidence leads to another definition of theta temperature: It is the temperature at which a polymer of infinite molecular weight (i.e., $V_1 / V_2 \to 0$) stops being completely miscible with the solvent. The critical composition of the mixture shifts with increasing molecular weight to vanishing concentrations of polymer.

In the above discussion we have implied that χ is changing with temperature. This is indeed the case for the Flory–Huggins model, where its temperature dependence is described by equation (3.1.19) or, for the newer version of the theory, by equation (3.1.37). With the temperature dependence of χ known, we may calculate the binodal and spinodal compositions for each temperature (numerical evaluation of the binodal is necessary) and use them for construction of a phase diagram, an example of which is presented in Figure 3.16. Although the Flory–Huggins model provides a good qualitative description of the phase diagram, it fails quantitatively. The main reason for this is the neglect of the dependence of parameter χ on composition. In fact, phase diagrams led to the discovery of this dependence.

When studying phase equilibria, we must make a clear distinction between critical point and point of *incipient precipitation*. The meaning of the latter is apparent from Figure 3.17, which describes the following experiment. A polymer solution

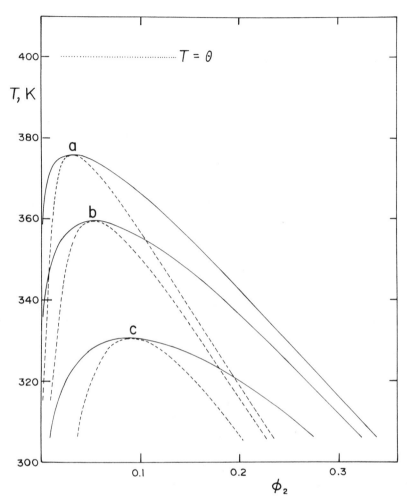

Figure 3.16. Phase diagrams for Flory–Huggins mixtures. Full lines are binodals; broken lines are spinodals; $\chi = 200/T$: (a) $V_1/V_2 = 0.001$; (b) $V_1/V_2 = 0.003$; (c) $V_1/V_2 = 0.01$.

at composition and temperature given by point B is cooled. When the decreasing temperature brings the solution into the unstable region, the solution separates into two phases. The temperature at which this happens (temperature of incipient precipitation) depends on the rate of cooling. At low rates and/or low viscosities, the demixing occurs when the temperature drops to point C on the binodal. At very high rates, or with very viscous systems, the demixing occurs only when temperature reaches point D on the spinodal. In real cases, the system precipitates somewhere in between. In any case, the temperature of incipient precipitation is always lower than the critical temperature (point A) unless the composition of the mixture happens to be exactly equal to critical composition.

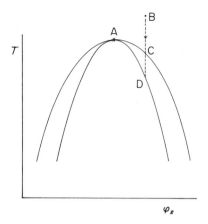

Figure 3.17. Critical point and point of incipient precipitation. See text for a detailed explanation.

In our example, which used the Flory–Huggins model, the full homogenization of the mixture was achieved when the temperature was raised high enough that the favorable entropy of mixing was able to overcome the unfavorable enthalpy of mixing for all compositions. Accordingly, the corresponding critical temperature is called the *upper critical solution temperature* (UCST). However, at higher temperatures, equation-of-state phenomena may start playing a dominant role, and the interplay of entropy of mixing and enthalpy of mixing may take an unexpected turn. We have seen in Section 3.1.3 that the contraction caused by mixing leads to a decrease of entropy, and this decrease may overpower the positive combinatory entropy at high-enough temperatures. Thus the entropy acts now against mixing. The equation-of-state contribution to enthalpy of mixing is always negative and contributes to mixing. Hence, at high enough temperatures, the roles of enthalpy and entropy are completely reversed. Again, with increasing temperature, the effect of entropy increases faster than the effect of enthalpy (see our calculations in Section 3.1.3), but this time it means demixing at high temperatures! However, we should stress that the basic rules of phase equilibria, and especially the analysis of $\Delta G_{\mathrm{mix}}^{\mathrm{M}}$ versus x_2 dependence as visualized in Figures 3.12 and 3.13, remain the same; what has changed is only the interpretation of the values of ΔG_{mix}. Thus the phase diagrams are almost mirror images of the diagrams we have seen before in Figure 3.16. The full homogenization of the mixture is reached when the temperature is lowered enough; the critical temperature is now called the *lower critical solution temperature* (LCST). For some polymer solutions using marginal solvents that are rather volatile, UCST and LCST are observed for the same system (Fig. 3.18). Note that in this case the lower critical temperature is always higher than the upper critical temperature. We have seen that the critical temperature depends on the molecular weight of the polymer; thus, for some polymer-solvent systems, there is a "window" of complete miscibility at a narrow temperature range when low-molecular-weight polymer is studied. For higher molecular weights, the

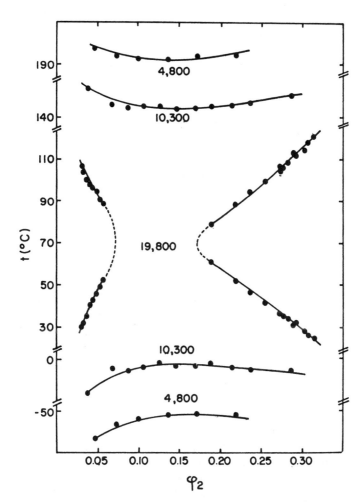

Figure 3.18. The phase diagrams for several fractions of polystyrene of indicated molecular weights in acetone. Low molecular weight fractions exhibit both UCST and LCST; the 19,800 fraction shows an hourglass diagram. Reprinted with permission from K.S. Siow, G. Delmas, and D. Patterson, *Macromolecules* **5**, 29 (1972). Copyright © 1972 by American Chemical Society. Used by the permission of the publisher.)

two two-phase regions merge and the phase diagram adopts the shape of an hourglass (Fig. 3.18).

Demixing of a polymer solution in the Flory–Huggins region can be achieved either by lowering the temperature or, especially when the solvent is a good one for the polymer, by addition of a nonsolvent. Of course, the polymer dissolved in a solvent-nonsolvent mixture is a three-component system; for example, polystyrene dissolved in benzene and cyclohexane. Such a mixture should be treated using the

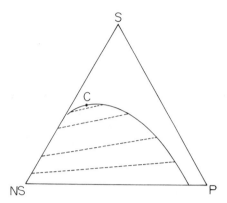

Figure 3.19. A triangular phase diagram. P denotes the polymer; S, solvent; NS, nonsolvent; C, critical point.

thermodynamics of three-component systems. Nevertheless, some phenomena are quite similar to those studied above. Thus, when a nonsolvent is added, a point of incipient precipitation is reached and the solution separates into two phases. These two phases have not only different concentrations of the polymer but also a different composition of the solvent mixture. The phase behavior is best depicted using triangular phase diagrams such as Figure 3.19. In this diagram, the full line separates the regions of full and partial miscibilities. The dashed lines are called *tie-lines;* they connect solutions that can exist in mutual equilibrium. (Obviously, if the system demixes into two phases connected by a particular tie-line, the overall composition of the system must be represented by a point on the same tie-line.) The critical point is the point for which the tie-line has zero length and is reduced to a point.

The exact form of the phase diagram for the polymer solutions in two-solvent systems is very sensitive to the detailed form of the expression for ΔG_{mix} as we have treated it in Section 3.2.1.1 and in equation (3.2.34). In principle, the interaction parameters can be obtained from the phase diagrams, but the actual theoretical analysis is very difficult.

3.2.3.1. Fractionation of Polymers.   The phase equilibria described in the previous section can be utilized for fractionation of polydisperse polymers. We have seen in Chapter 2 that most polymers possess a rather broad distribution of molecular weights—with the exception of some natural materials and some polymers prepared by living polymerization techniques. Fractionation of polydisperse materials serves two main purposes: (1) From the viewpoint of analysis, it allows for evaluation of the polydispersity of the sample that was fractionated; and (2) from the preparative viewpoint, it gives us access to polymer samples with a narrow molecular weight distribution, which may be difficult to obtain otherwise.

Polymer fractionation is based on an observation that different polymer homologs distribute themselves between two conjugate phases in such a way that species of higher molecular weight show much stronger preference for the concentrated phase than the smaller molecules do. We will now show that this behavior follows from the Flory–Huggins theory.

First, we will modify the Flory–Huggins equation for solutions of polydisperse polymers. A convenient form reads

$$\Delta G_{\text{mix}} = RT \left(n_1 \ln \phi_1 + \sum n_i \ln \phi_i + \chi n_1 \phi_2 \right) \qquad (3.2.71)$$

where the subscript 1 refers to the solvent, subscript i to individual polymer species, and subscript 2 to total polymer concentration, $\phi_2 = \sum \phi_i$; the summations are over the polymer species only. Equation (3.2.71) implies that interaction coefficients between the solvent and polymer are the same for all sizes of molecules and that there is no polymer-polymer (excess) interaction. Routine calculation now yields the chemical potential of each polymer component as

$$\mu_i - \mu_i^\circ = RT[\ln \phi_i - (x_i - 1) + \phi_2 x_i(1 - 1/\bar{x}_n) + \chi x_i(1 - \phi_2)^2] \qquad (3.2.72)$$

where $x_i = V_i / V_1$ and $\bar{x}_n$ is the number average of x_i. [Equation (3.2.72) should be compared to equation (3.1.39), which is valid for monodisperse polymers.]

When the temperature of the solution is lowered enough that χ achieves a value larger than χ_{crit}, the solution will separate into two phases. Calculation of χ_{crit} and ϕ_2^{crit} as well as of the total concentrations of the conjugate phases at subcritical temperatures is difficult, but we do not need to know these values. For our analysis it is sufficient to note that the chemical potential of each species must have the same value in the concentrated phase (all quantities referring to concentrated phase are designated by an asterisk) and in the dilute phase (without asterisk). This equality, in conjunction with equation (3.2.72) yields

$$\ln(\phi_i^*/\phi_i) = x_i \left[\phi_2^*(1 - 1/\bar{x}_n^*) - \phi_2(1 - 1/\bar{x}_n) + \chi(1 - \phi_2^*)^2 - \chi(1 - \phi_2)^2 \right]$$
$$(3.2.73)$$

The quantity in the brackets (which is quite difficult to calculate) is the same for all polymeric species and is always positive; we will designate it as σ. Consequently,

$$\phi_i^*/\phi_i = \exp(\sigma x_i) \qquad (3.2.74)$$

According to this relation, the ratio of concentrations ϕ_i^*/ϕ_i (the *partition coefficient*) is always larger than unity; that is, each species is more concentrated in the concentrated phase than in the dilute phase. The ratio of the concentrations increases sharply with increasing values of x_i, that is, with increasing molecular weight.

For fractionation, the ratio of masses in the two phases is more important than the ratio of concentrations; mass m_i equals concentration multiplied by volume. If we designate the ratio of the volume of the dilute phase to the volume of concentrated phase by r, then the ratio of masses is

$$m_i^*/m_i = \exp(\sigma x_i)/r \tag{3.2.75}$$

Thus, by selecting an appropriate value of r, we may get most of the high-molecular-weight polymer in the concentrated phase while most of the lower-molecular-weight components reside in the dilute phase. Of course, the transition is gradual, and both phases will contain polymer that is quite polydisperse. Nevertheless, the polymer in the concentrated phase will always have a higher average molecular weight than the polymer in the dilute phase.

The ratio of volumes is regulated mainly by the concentration of the original solution; the more dilute the solution, the larger the volume of the dilute phase and value of r. The values of σ and r depend not only on the original concentration and the final temperature (which determines the value of χ) but also on the distribution of molecular weights in the original sample, which is quite often unknown. Thus the proper selection of the conditions of fractionation still remains an art requiring experience.

Perhaps we should stress once more that in liquid-phase separations all components are present in both phases and the resulting fractions are quite polydisperse. A fraction with $\overline{M}_w/\overline{M}_n = 1.2$ is a good one; a fraction with this ratio equal to 1.1 is an excellent one. The notion that the high molecular weights precipitate first, then the middle ones, and so on, is totally false; fractionation cannot lead to pure components! This caveat should be kept in mind by biochemists, who frequently separate proteins by "precipitating" them with solutions of ammonium sulfate of increasing molarity. This process is again based on the formation of two liquid phases, and all proteins must be present in both phases. The process is an enrichment, not purification.

Experimentally, the temperature of a dilute polymer solution is lowered to a value designed to separate about 10% of the polymer in the concentrated phase. As the temperature is lowered, portions of the concentrated phase separate continuously; yielding a product that is not fully equilibrated. The equilibration is achieved by vigorously stirring the two-phase system for several hours at a closely controlled temperature. The concentrated phase (the precipitate) is then allowed to settle at the bottom, and the phases are separated. It is imperative to maintain good temperature control during the physical separation of the two phases; otherwise, the equilibrium may be disturbed. After the first precipitate has been separated, the temperature of the dilute phase is lowered further and the required number of fractions is successively separated.

In an alternative procedure, the polymer is separated first into two fractions of about equal size. These fractions are then refractionated in a cascade manner using schemes similar to those in fractional crystallization. This method is believed to yield final fractions of slightly better homogeneity.

When fractionating polymers with medium and lower molecular weights we may need to utilize a rather broad range of temperatures; this may become quite impractical. For this reason, polymer fractionation is more often performed using solvent-nonsolvent systems. In this procedure, we start with a polymer solution in a good solvent to which is gradually added a nonsolvent until an adequate amount of precipitate is formed. To achieve better equilibration, the resulting two-phase system is heated until it homogenizes again. It is then cooled slowly back to the temperature of fractionation, stirred, allowed to settle, and separated. After that, the addition of more nonsolvent will precipitate the next fraction.

In the solvent-nonsolvent fractionation, it is recommended to use moderately good solvents, not the very good ones, and, more importantly, moderate nonsolvents, not those completely incompatible with the polymer. For example, for the fractionation of polystyrene, it is better to use a butanone-hexane solvent system (a moderate solvent and nonsolvent) than a benzene-methanol system (a very good and a very poor solvent).

In this context, it is probably appropriate to mention another type of precipitation frequently used for the recovery of polymers from their solutions. In this case, fractionation of the polymer is undesirable. Moreover, polymer chemists prefer their samples in a form that is easily handled and dried—a powder or fiber, for example. In this case, we want to modify the precipitation procedure to obtain a dilute phase that contains only a negligible amount of polymer. The concentrated phase, on the other hand, should have only very small amounts of solvents to prevent coalescence of the droplets of the polymer-rich phase. The recommended procedure consists of dissolving the polymer in a good or moderate solvent to 1–3% concentration. This solution is then added dropwise to a large excess of a very poor solvent under stirring. The solvent diffuses fast out of the individual droplets of the polymer solution, leaving behind polymer particles with a very small amount of the solvent. Virtually all the polymer originally present in the droplet remains in the polymer particle. Of course, care must be taken to choose appropriately the volumes of the solvent and the precipitant. Otherwise, in the later stages of the dropwise addition, the system in the reactor will contain so much of the solvent that it will no longer act as a vigorous nonsolvent and the precipitate will not be homogeneous.

There is another way (which does not belong to this section) to prepare finely dispersed samples of polymers: Dissolve the polymer in a solvent having a freezing point in a convenient temperature range (water, benzene, cyclohexane, dioxane). Freeze the solution, and remove the solvent by freeze-drying. This procedure leaves very finely dispersed polymers that are solvent-free; the solvent is removed by this procedure more completely than by other methods of drying.

3.3. HYDRODYNAMICS OF MACROMOLECULAR SOLUTIONS

In equilibrium situations such as those described in Sections 3.1 and 3.2, the behavior of systems could be adequately described by the rules of thermodynamics, which are rather well established and do not allow any exceptions. This fact simplifies the analysis of equilibrium phenomena enormously. However, systems that are not at equilibrium also display a number of phenomena that could yield extensive information about the systems in question (in our case, about polymer solutions). Unfortunately, the rules governing irreversible (read nonequilibrium) processes are not so well understood and provide only very general guidance for the analysis of experiments. Having available little help from the basic principles, we must rely heavily on models during our studies.

In this section we will be interested in processes in which macromolecules are in relative motion with respect to their surroundings. Such processes are the subject of *molecular hydrodynamics*. A natural model for molecular hydrodynamics is macroscopic hydrodynamics, which has been perfected for centuries by engineers, sailors, and aviatic experts. In the following sections we will explore which rules of macroscopic hydrodynamics are applicable to molecular phenomena, what simplifications are possible, and where new approaches are needed.

Fortunately, extensive simplifications are justified. Macromolecular solutions under the experimental conditions we want to use are virtually incompressible. Furthermore, the velocities employed are very small, making the inertial effects negligible. These two facts allow us to remove the difficult terms from the basic equation of hydrodynamics (Navier–Stokes equation) and use instead much simpler relations for *viscometric flows* (i.e., for slow laminar flows of incompressible liquids).

3.3.1. Viscous Flow of Liquids

The concept of viscosity is usually introduced by means of a model of a fluid confined between two parallel plates: one stationary and the other moving with constant velocity (Fig. 3.20). The distance d between the plates is very small compared with their area A; thus the phenomena at the plate edges may be neglected. (This is often recognized by a statement that the plates are infinite.) The fluid is assumed to adhere to the plates: Infinitesimally thin layers of the fluid adjacent to the plates move with the same velocity as the plates. Under these circumstances, the velocity of the fluid has to change across the gap. In the present case, the velocity profile is linear and can be described by a constant velocity gradient. Such a flow is called *simple shear flow*.

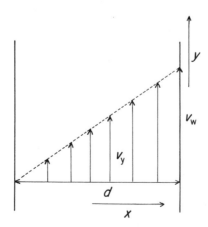

Figure 3.20. Velocity field in a shear flow between two parallel plates.

Before progressing further, we need to inspect the mathematical apparatus for studying flow phenomena. In our three-dimensional world, velocity must be treated as a vector with three components. Each component can be described by its derivatives with respect to three space coordinates. Hence, the velocity field is characterized by a set of nine velocity derivatives, which form a second-order tensor of velocity gradient **G**. Consequently, theoretical hydrodynamics is based on tensor calculus. The situation becomes further complicated because the geometric arrangement of our experiments often calls for the use of non-Cartesian coordinates: cylindrical or spherical in the less demanding problems. Some modern approaches use coordinates that are centered on the moving fluid element and deform with it; the coordinate system changes with time!

Recognizing that many readers are not conversant in tensor calculus, we will base our introduction to hydrodynamics on simplified approaches. Fortunately, it is possible to do so in many situations of practical interest. For example, in our model depicted in Figure 3.20, the velocity gradient tensor has only one nonzero component. Hence, with some caution, it can be treated as a scalar quantity; this will simplify the calculations tremendously. In other situations, the tensorial approach is mandatory; in such cases we will present the necessary relations but avoid lengthy proofs and detailed calculations. Whenever we use vectors and tensors, we will use the standard notation: boldface for vectors; bold sanserif symbols for tensors; centered dots for the scalar product of two vectors; a cross for vector product; two vectors side by side for diads; regular italics for absolute values (lengths) of vectors.

Returning now to our simple shear flow and to Figure 3.20, we will set our coordinate system in such a way that the x axis is perpendicular to the plates and the y axis is in the direction of plate movement. The xy plane is then called the plane of flow; the z axis is perpendicular to it. The only nonzero component of the velocity gradient tensor is G_{yx} (we will use for it simply the symbol G and call it the velocity gradient); for our model it is a constant.

$$G_{yx} \equiv \frac{dv_y}{dx} \tag{3.3.1}$$

In this relation v_i is velocity along the i axis. It is found experimentally that fluids between the plates hinder the movement of the plates, which will stop moving unless an outside force overcomes this resistance. Alternatively, we may say that the fluid exhibits a viscous force F that acts against the flow and must be counterbalanced by an outside force of the same magnitude and opposite direction to keep the plates moving. The force is related to parameters of our model as

$$F = \eta A G \tag{3.3.2}$$

where η is the coefficient of viscous resistance, or simply viscosity.

What is the origin of the viscous resistance? That depends on the nature of the fluid. In gases, the thermal motion moves molecules from a layer with a particular average velocity to another layer having a different velocity. A change in velocity

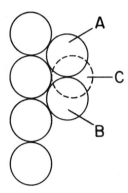

Figure 3.21. Molecular model of dissipation of energy in viscous flow.

requires force—in this case, the force driving our flow. Thus viscosity of gases is strictly a kinematic phenomenon not requiring any attractive forces among molecules for its explanation.

However, these attractive forces play a dominant role in the flow of low-molecular-weight liquids. Let us first consider the changes in energy accompanying the movement of a molecule in a liquid (Fig. 3.21). At some moment this molecule is located at a position A, where its distances to its neighbors correspond more or less to distances characterizing the minimum of the interaction energy. What happens to the energy of our molecule when it proceeds to position B? This position is another location with a minimum of the interaction energy; the energy content of the molecule has not changed. However, to get to B, our molecule has to pass through position C, where the intermolecular interactions are not maximized: Its energy is higher at C than at either A or B. This excess energy is again released when the molecule proceeds to position B. Where is this excess energy coming from, and where does it return? In ordinary liquids at rest, the thermal movement is able to supply the energy needed for this kind of transport. (If it is not, the liquid is in a glassy state.) However, the thermal energy is disorganized; it moves the molecules in all directions with equal probability, and no net flow results.

To achieve an organized motion of molecules (as in macroscopic flow of a liquid), the energy must be supplied by a force that is organized, that is, one that acts in a specific direction: It is our viscous force given by equation (3.3.2). (Some more refined models explain the flow again through the thermal motion, which is helped by the outside force when it moves the molecule in the direction of flow; the thermal motion is hindered by the outside force when the molecules move against the flow. The net result of both models is the same.) What happens to the energy that was supplied by the outside force to get our molecule to the energy-rich position C during its flow in the required direction? The energy is again converted to thermal energy when the molecule reaches its new energy minimum at position B.

We can summarize the above discussion in the following way. The disorganized thermal energy causes a random movement of molecules. Such a movement is part of liquid equilibrium. (Equilibrium does not prohibit molecular movement generally, it only prohibits macroscopic flows.) Macroscopic flow can be achieved only by an action of directed force. Work performed by a directed force has the character of free energy; it is converted into thermal energy by the molecular processes during the flow.

In concentrated polymer solutions and polymer melts there exists still another mechanism, leading to viscous resistance. The long polymer molecules are extensively entangled. In laminar flow, the entangled molecules are eventually separated. This separation requires massive disentanglement. The molecules resist this process in two ways:

1. Relative movement of two macromolecules or sections of them requires simultaneous displacement of many atomic groups from their positions of minimum interaction energy.
2. The disentanglement inevitably leads to the stretching of molecular chains, which imparts elastic energy to them. This energy must be supplied by the driving force but is eventually also dissipated as thermal energy.

Not surprisingly, the viscosity of polymer melts and concentrated solutions is very high.

Let us now calculate the power W expended by the outside force to keep the wall moving with constant velocity v_{wall}. This power is eventually dissipated within the liquid as thermal energy. Power expended by a force acting on a moving object is equal to a scalar product of two vectors: the force and the velocity of the moving object. In the present case, both vectors have the same direction, and their scalar product is simply equal to the product of their absolute values. For our plate model the velocity of the moving plate is related to the gap d between plates and to the velocity gradient as $v_{wall} = Gd$; the volume of the liquid between plates, V, is equal to the product Ad. Thus

$$W = \mathbf{F} \cdot \mathbf{v}_{wall} = F v_{wall} = \eta G^2 A d = \eta G^2 V \qquad (3.3.3)$$

This is a very important result that relates the power dissipated in a unit volume by viscous flow to the velocity gradient. Equation (3.3.3) was derived for a very special flow situation, for which G was constant throughout the liquid. For systems with a more complicated flow pattern, equation (3.3.3) is still applicable for differential volumes within which the velocity gradient may be considered constant. The total dissipated power is then given by an integral over the whole volume of liquid.

$$W = \eta \int_V G^2 dV \qquad (3.3.4)$$

Actually, equation (3.3.4) can be derived by the full tensorial apparatus of hydrodynamics for an arbitrary flow pattern. In this case, the quantity G^2 should be understood to mean the double scalar product of the tensor of velocity gradient with itself $G^2 \equiv \mathbf{G}:\mathbf{G}$. Equation (3.3.4) thus has a much broader range of applicability than equation (3.3.2). Either of these two equations is used as a definition of viscosity.

Throughout the foregoing discussion, we have assumed that viscosity is independent of the magnitude of the velocity gradient. Using this assumption, we followed Newton, who was the first to quantitatively describe viscous flow. Accordingly, liquids that conform to this assumption are called *Newtonian* liquids. Most low-molecular-weight liquids belong to this class. However, in the following sections we will meet many liquids whose viscosity depends on the velocity gradient. They are called *non-Newtonian* and often also exhibit other phenomena that are not observed with typical low-molecular-weight liquids.

3.3.2. Particles Moving Through a Liquid—Frictional Coefficients

In many situations of experimental interest, a macromolecule is moving through a liquid. We need to understand the phenomena accompanying this movement. Let us consider the action of an external force **F** on a particle suspended in a liquid (Fig. 3.22). The force will cause the particle to move. After some interval of time, the particle in our figure will move from the position marked by the solid line to the dashed line position. The liquid that was originally in the horizontally shaded area had to move elsewhere to make room for the particle. On the other hand, the vertically shaded space was vacated by the particle and some liquid had to move in to fill the void. We have seen that a moving liquid drags adjoining layers with it. Thus motion of one part of the liquid sets the whole liquid into motion. Although the velocity of this induced motion vanishes far from the source of disturbance, it has significant values up to quite appreciable distances. Now we have a liquid moving with different velocities at different locations, that is, exhibiting velocity gradients. This motion must dissipate power W according to equation (3.3.4). This power must be supplied by the force that set the particle and the liquid into motion: The supplied power is $W = Fv_p$, where v_p is the velocity of the particle.

Let us imagine that we change the velocity of the particle by some factor α. In Newtonian liquids, the distribution of liquid velocities will remain the same, except that all velocities and velocity gradients will be changed by the same factor α. In other words, the velocity gradient at every location in the liquid must be proportional

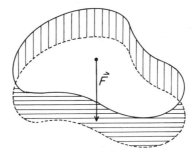

Figure 3.22. Displacement of a particle suspended in a liquid under the action of an external force.

to the particle velocity, that is, $G = \beta v_p$, where β is a function of the coordinates of this location with respect to the position of the particle, and of the size and shape of the particle, but is independent of the driving force and the liquid viscosity. We can now substitute this relation into the expression for the dissipated power and equate it with the power supplied:

$$W = \eta v_p^2 \int_V \beta^2 dV = F v_p \tag{3.3.5}$$

It is apparent that the integral in equation (3.3.5) depends only on the size and shape of the particle; it is a constant for any given particle. We can now express the force from the second equality in equation (3.3.5) as

$$F = \left(\eta \int_V \beta^2 dV \right) v_p \equiv f v_p \tag{3.3.6}$$

where f is the *frictional coefficient,* which depends on the shape and size of the particle and is proportional to the liquid viscosity.

An observant reader has probably noticed that in the above derivation we have neglected the vector nature of the force, velocity, and velocity gradient. Furthermore, equation (3.3.6) is written as proportionality between two vectors. Now, proportionality between vectors means that they are parallel to each other. Every sailor knows that the driving force is not always parallel to the particle motion—the boat may sail in quite a different direction than the wind is blowing!

Indeed, a less casual derivation would show that the frictional coefficient is actually a frictional tensor **f** and that equation (3.3.6) should be written as

$$\mathbf{F} = \mathbf{f} \cdot \mathbf{v}_p \tag{3.3.7}$$

The three principal axes of the frictional tensor have a definite orientation with respect to the particle in question; for symmetrical particles (ellipsoids, rods, disks, etc.), they coincide with their axes of symmetry. When the force is acting along the principal axis of the frictional tensor, the particle moves in the direction of the force; otherwise, it may move at an angle (this is the principle of sailing). The three axes are not equivalent: The frictional resistances for movements along them may have quite different values. (Indeed, a coin sliding through a liquid along its edge experiences less resistance than when it is dragged through perpendicularly to its face.) These three values are the three principal values of the frictional tensor, they are usually designated as f_1, f_2, f_3. Consequently, when analyzing the resistance of particles against movement, we must take into account their orientation. Fortunately, when we are studying frictional resistance on the molecular level, thermal motion simplifies the calculations for us. The orientation of the particle is unceasingly changing, and its average frictional resistance (which now has a scalar character) is given as

$$f = (f_1 + f_2 + f_3)/3 \tag{3.3.8}$$

For an actual calculation of the frictional resistance, the differential equation of flow (Navier–Stokes equation for incompressible liquids without inertial terms) must be solved for the following boundary conditions:

1. The liquid layer adhering to the particle moves with the same velocity as the particle.
2. The velocity vanishes at large distances from the particle.

The calculation is relatively simple for spheres: All three principal values of f are the same; the frictional coefficient f_0, (subscript zero for spheres) is related to the viscosity of the liquid η and to the radius of the sphere r as

$$f_0 = 6\eta\pi r \tag{3.3.9}$$

This relation is known as the *Stokes equation*. It is noticeable that the frictional resistance increases linearly with the radius, not with its cross section or volume.

The calculations rapidly become rather involved for particles with more complicated shapes. In fact, exact calculation was possible only for ellipsoids. It is customary to characterize the frictional behavior of nonspherical particles by the ratio f/f_0, where f is the frictional coefficient of the particle in question; f_0 is the frictional coefficient of a hypothetical sphere having the same volume as the actual particle and moving through the same liquid. In Figure 3.23, f/f_0, the frictional ratio of randomly oriented ellipsoids is plotted as a function of their axial ratio a/b. The ratio increases with increasing asymmetry of the particles, but for moderate asymmetries it does not deviate appreciably from unity. Nevertheless, proteins are frequently modeled by ellipsoids, and their asymmetries are estimated from f/f_0. (The frictional ratio is much less popular among polymer chemists.)

Experimentally, f is obtained by a suitable method (diffusion, sedimentation velocity), whereas f_0 is calculated from the molecular weight M of the particle and its specific volume $\bar{v}$ (these quantities must be known from other experiments).

$$f_0 = 6\pi\eta(3M\bar{v}/4\pi)^{1/3} \tag{3.3.10}$$

For particles that are neither spheres nor ellipsoids, and especially for particles of irregular shape, no simple relation exists between the frictional coefficient and their size (expressed by some characteristic dimensions). In such a case, the particles are assigned a *hydrodynamic radius* r_H (strictly speaking, a hydrodynamic radius for translational movement) defined by a relation analogous to equation (3.3.9):

$$f \equiv 6\pi\eta r_H \tag{3.3.11}$$

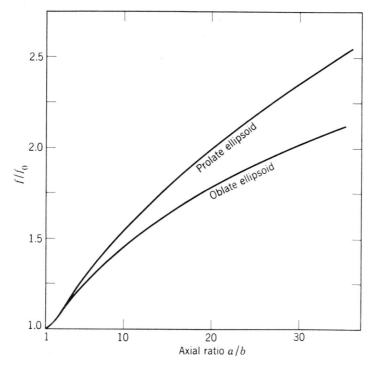

Figure 3.23. The frictional ratio f/f_0 as a function of the axial ratio of the ellipsoids a/b. (Reprinted from C. Tanford, *Physical Chemistry of Macromolecules.* Copyright © 1961 by John Wiley & Sons, New York. Used by permission of the publisher.)

Thus the hydrodynamic radius of a particle is equal to the radius of a sphere that has the same frictional coefficient in the same liquid as the actual particle. The concept of hydrodynamic radius is used extensively in connection with macromolecular coils. Their hydrodynamics is quite involved, and we will return to it in Section 3.3.5.

The above calculations were performed assuming that the liquid was continuous— that it did not possess molecular structure. Such an approximation is definitely appropriate for macroscopic particles. But what happens when the size of the particle becomes comparable to the size of liquid molecules? Somewhat surprisingly, the results of macroscopic hydrodynamics are virtually exact down to particles whose diameter is about 10 times larger than the diameter of liquid molecules. Even under this limit the deviations are relatively minor. When the particle size becomes comparable to that of the liquid molecule, the assumption that ceases to be valid first is the assumption of neighboring molecules adhering to the particle; a model of liquid slipping around the particle becomes more plausible. Of course, even with this model, the movement of the particle sets the liquid in motion and causes frictional resistance. For spherical particles in the "slipping" model, the frictional coefficient f_{sl} is only slightly less than the nonslipping f:

$$f_{s1} = 4\pi \eta r \tag{3.3.12}$$

This relation describes the real frictional coefficients reasonably well down to sizes equal to those of liquid molecules. Consequently, it is possible to use equations of macroscopic hydrodynamics for particles of all sizes; for particles comparable in size with molecules of the solvent, however, we should not adhere too strictly to the relation between the frictional coefficient and the size of the particle.

3.3.3. Particles Suspended in a Flowing Liquid—Viscosity Increase

The measurement of macroscopic viscosity of liquids is comparatively easy and can be carried out with very good precision. We will now show that a liquid with suspended particles has higher macroscopic viscosity than a liquid without these particles. Measurement of this excess viscosity yields a wealth of information about the nature of the suspended particles and even about the thermodynamic interaction of the particles with the liquid. Consequently, viscometry is one of the more popular techniques for studying the properties of polymer solutions. In the present section we will analyze the effects of suspended particles on the flow within the liquid. Viscometry itself will be the subject of Section 3.4.3.

Let us return to simple shear flow. In Figure 3.24 a hypothetical boundary A encloses a spherical element of the flowing liquid. The flow imparts different velocities to different parts of the spherical element. After some period of time, the liquid sphere will move to location B and later to C. Thus the flow has a translational component. The figure depicts the diameter of the sphere perpendicular to the flow lines. The flow brings this diameter closer and closer to the flow direction. Hence, the simple shear flow also has a rotational component. As the flow moves the volume element from A

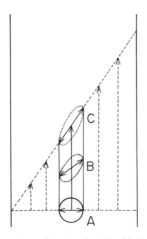

Figure 3.24. Deformation of a spherical liquid element in shear flow.

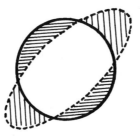

Figure 3.25. Additional flow of liquid caused by the inability of a rigid sphere to follow the deformation of a flowing liquid.

to B and then to C, the element loses its spherical shape and is deformed more and more. In fact, after a sufficiently long time the volume element is infinitely stretched and loses its coherence; however, we will restrict our attention to shorter times for which we can clearly observe the deformational component of the simple shear flow.

As a next step in our analysis, we replace the liquid within the hypothetical boundary by a real rigid sphere. Obviously, the flowing liquid around the sphere will try to force it into the same flow pattern that characterized the hypothetical liquid element. The real sphere can follow the translational motion—it also rotates with the liquid (its angular velocity ω is equal to $G/2$)—but cannot follow the deformation. Comparison of the space occupied by the real sphere and the hypothetical one is apparent from Figure 3.25. The situation is similar to the one we treated in the previous section and depicted in Figure 3.22. The liquid that is expelled from the horizontally shaded space by the sphere causes an additional flow of the liquid, which brings another portion of it into the void in the vertically shaded space. This additional flow causes additional dissipation of power, which must be supplied also by the external force maintaining the flow. (In the region occupied by the particle, there is no flow and no dissipation of power in the presence of the particle, but the power is dissipated there in the absence of the particle. Thus this region reduces the necessary input of power. However, this decrease is not sufficient to counterbalance the excess dissipation outside the space occupied by the particle.)

Before presenting the relations for its magnitude, we should first discuss what macroscopic viscosity is. It should be obvious that "true" (i.e., microscopic) viscosity cannot be changed by the presence of some particles at some distance from the liquid element under consideration. What is changed is the local character of flow! However, from the macroscopic viewpoint, we can still study the simple shear flow between two plates as in Figure 3.20, and we can still ask, What force is needed for maintaining the required velocity of the wall? The force is determined by the power dissipated, which is given by relation (3.3.4), where the velocity gradient is now a very complicated function of the position of all particles suspended in liquid; viscosity in relation (3.3.4) is, of course, still the original viscosity of the liquid. However, in the following discussion we will pretend that it is still legitimate to consider the flow to be simple shear flow, and we will describe it by equations (3.3.2)–(3.3.4). The apparent (homogeneous) velocity gradient is taken to be the ratio of the velocity of the wall to the gap width, $G = v_{\text{wall}}/d$. To force the validity of equations (3.3.2)–(3.3.4) we must now assign to the viscosity a new value η (macroscopic viscosity), which differs from the viscosity of the pure liquid η_0. This convention is very convenient, because all methods measuring viscosity (they actually measure either velocity with known force or force with known velocity) yield this quantity.

Even for a suspension of spheres, the calculation of the macroscopic viscosity is very involved, and we will present only the final formula. The calculation was first performed by Albert Einstein. [Besides his analysis of the viscosity of suspensions, Einstein in his early years made a number of important studies in the macromolecular field (theory of diffusion, role of fluctuations in light scattering) before engaging in less interesting studies in his later years.]

For a suspension of rigid spheres, the macroscopic viscosity can be expressed by means of a power series as

$$\eta = \eta_0(1 + \nu\phi \cdots) \tag{3.3.13}$$

where ϕ is the volume fraction of the suspension occupied by the spheres, and ν is a numerical factor characterizing the shape of the particles: for spheres, $\nu = 2.5$.

As is to be expected, calculation of the viscosity of suspension of ellipsoids is even more involved. Besides the hydrodynamic problems connected with the more complicated boundary conditions, another factor comes into play. The amount of additional power dissipated by an ellipsoid depends on the orientation of its rotary axis with respect to the flow plane. According to macroscopic hydrodynamics, the axis of each particle should follow a closed trajectory during the flow; different trajectories do not intersect. For macromolecular particles this highly organized pattern is disturbed by thermal motion. We will return to this question in Section 3.3.7.

When the disorienting effect of the thermal motion is accounted for properly, the macroscopic viscosity of the suspension of ellipsoids is described again by equation (3.3.13), in which the factor ν increases with increasing asymmetry of the ellipsoids. The factor ν is plotted in Figure 3.26 as a function of the axial ratio of the ellipsoids, a/b. Unlike the frictional ratio f/f_0, the factor ν changes quite appreciably with the axial ratio.

3.3.4. Hydrodynamic Interactions

We have seen that the force acting on a particle immersed in a liquid sets this particle in motion, which leads to a flow of the liquid as a whole. If there is another particle in the system, the flowing liquid will exert a force on it. Thus a force acting on one particle is transmitted through the liquid to other particles and sets them in motion, too. This sequence of events, known as *hydrodynamic interaction,* plays a crucial role in most flow-related phenomena.

Exact calculation of the flow pattern in a liquid in the presence of two particles and a force is a formidable problem that is quite complicated even for two spherical particles.

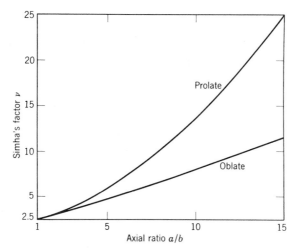

Figure 3.26. Simha's viscosity factor ν plotted as a function of the axial ratio of the ellipsoids a/b. (Reprinted from C. Tanford, *Physical Chemistry of Macromolecule.* Copyright © 1961 by John Wiley & Sons, New York. Used by permission of the publisher.)

To simplify the calculations we must adopt a model that leads to less demanding mathematics. The simplification is based on the fact that at larger distances from the particle the flow disturbance caused by the movement of the particle does not depend on the size and shape of the particle any more but only on the magnitude and direction of the force causing the motion. *Flow disturbance* denotes the difference (in a vector sense) between the velocities of the liquid at a particular point in the presence and in the absence of the force acting on the particle. Of course, the velocity of the liquid in the absence of the force (undisturbed velocity) may be nonzero, either because the fluid itself may exhibit a flow pattern (e.g., simple shear flow) or because other particles may be causing flow disturbances. (In the sense of our definition the latter disturbances are a part of the undisturbed velocity.)

Oseen has shown that at larger distances from the particle, the velocity disturbance $\mathbf{v}_d$ generally has a different direction than the force $\mathbf{F}$, to which it is related as

$$\mathbf{v}_d = \frac{1}{8\pi\eta}\left(\frac{\mathbf{F}}{r} + \frac{(\mathbf{F}\cdot\mathbf{r})\mathbf{r}}{r^3}\right) \tag{3.3.14}$$

This relation is often written in tensor notation as

$$\mathbf{v}_d = \mathbf{T}\cdot\mathbf{F} \equiv (1/8\pi\eta r)(\mathbf{U} + \mathbf{rr}/r^2)\cdot\mathbf{F} \tag{3.3.15}$$

In this notation $\mathbf{U}$ is the unit tensor, and the identity in equation (3.3.15) defines the *Oseen hydrodynamic interaction tensor* $\mathbf{T}$; $\mathbf{r}$ is a position vector from the point of action of the force to the position at which $\mathbf{v}_d$ is observed. Figure 3.27 schematically depicts the flow field described by equations (3.3.14) and (3.3.15).

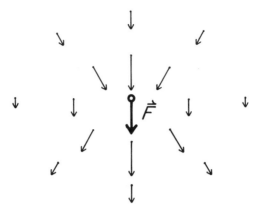

Figure 3.27. A schematic representation of the disturbance of the flow field generated by a force **F** acting on the liquid.

Our simplification consists of an assumption that equations (3.3.14) and (3.3.15) are valid not only at large distances from the central particle but at all distances. In fact, the simplified model disposes with the central particle altogether, and the force **F** is assumed to be a point force acting on the liquid directly.

Let us now consider a second particle situated at some distance **r** from the point of action of the force. The flowing liquid will carry it in the directions depicted in Figure 3.27. Note that for all directions of the position vector **r,** the flow has a major component in the direction of the force and only a minor component in the perpendicular direction. Thus the major part of the movement will be in the direction of the force **F**. This movement of the second particle does not require any outside force acting on it. However, if there is an outside force in the same direction (e.g., when a gravity force acts on both particles), the movement of the particle would be faster than in the absence of the central particle. Moreover, the second particle will accelerate the first one as well—the hydrodynamic interaction is a mutual phenomenon. In other words, in our example the particles cooperate in their movement: A cluster of particles (each of them acted on by an equal outside force) moves through the liquid faster than individual particles would move when alone.

What will happen if our cluster of particles is uniformly spread through the whole liquid body? According to our first impression, the foregoing argument is not changed for different spatial distributions of the particles; they should still show a cooperative motion. However, there is another factor in play. Inspecting Figure 3.27 once more, we see that the whole liquid moves on average in the direction of the force acting on the particle. More particles should cause more intensive movement. Such a movement is possible when the liquid is unbounded: Force acting on particles dispersed in a liquid will move the whole system in the direction of the force. However, in many systems of interest, the liquid is enclosed in a rigid container. Therefore, the net flow (i.e., the sum of the flows of all components of the system) is zero in such a case.

(Remember that our liquid is incompressible by definition.) Thus, if the particles move in the direction of the force, the liquid must move, on average, in the opposite direction. With increasing concentration of the particles, this reverse flow intensifies. To calculate the particle velocity under these circumstances we should write equation (3.3.9) for the relation between driving force and particle velocity in a more general way:

$$\mathbf{F} = \mathbf{f} \cdot (\mathbf{v}_p - \mathbf{v}_0) \tag{3.3.16}$$

Here $\mathbf{v}_0$ is the velocity that the liquid would have at the location of the particle if the particle were absent (all other particles remaining at their locations). In the simple situation treated in Section 3.3.2, $\mathbf{v}_0$ was equal to zero. In the present case, $\mathbf{v}_0$ has a direction opposite to that of $\mathbf{F}$; the velocity $\mathbf{v}_p$ (referred to the walls of the system) will decrease with increasing concentration in the system.

This example of a tight cluster on the one hand and uniformly distributed particles on the other hand manifests the complex role played by hydrodynamic interactions in macromolecular systems. In the following sections we will study the effect of hydrodynamic interaction on the hydrodynamics of polymer coils and on the behavior of moderately concentrated macromolecular solutions.

3.3.5. Hydrodynamics of Macromolecular Coils

While the flow of liquids in the vicinity of geometrically simple particles is complicated enough, the irregular shape of polymer coils, which varies from coil to coil, adds mightily to the complexity of the problem. Again, simplifications and modeling come to the rescue. Because the complexities of the calculation are great, so must be the simplifications; some models of the coil may have little resemblance to actual coils. In the present section, we will study the behavior of single coils subject to outside force (calculation of frictional coefficient) or subject to simple shear flow (contribution to viscosity). In both cases we will assume small forces and slow flow. In Section 3.3.6 we will address the hydrodynamics of moderately concentrated solutions. Section 3.3.7 will be devoted to the study of polymer solutions in stronger flow fields, which are capable of orienting and deforming polymer molecules.

In Section 1.3.1.2 we developed a model of a coil consisting of statistical segments. For hydrodynamic considerations, the spatial statistics of the segments is preserved, but the segments are given a new physical identity: The joints between linear sections of the model are now occupied by friction centers (usually but not always considered as spherical) connected by frictionless rods. Alternatively, when the model of Gaussian subchains is invoked, the rods are replaced by frictionless springs. Both these models are frequently called *pearl-string* or *necklace* models.

How does a pearl-string behave in a liquid when acted upon by an external force? Each friction center will move through the liquid, causing flow disturbances everywhere in the liquid, specifically at locations of other centers. How important are these disturbances? There are two limiting possibilities:

1. The friction centers are sufficiently far apart to make the hydrodynamic interaction among the segments negligible; each friction center moves as though other centers were absent. Such a coil is called a *hydrodynamically permeable coil* or *free-draining coil;* its frictional coefficient is equal to the sum of the frictional coefficients of all constituent segments.

2. In the other limiting case, the hydrodynamic interaction provides such a strong cooperative effect in the central parts of the coil that the liquid moves together with the segments. Closer to the surface (a better word is envelope) of the coil, the hydrodynamic interaction gets weaker and we can observe a relative velocity between the segments and the liquid. These outer reaches of the coil constitute a hydrodynamic transition layer. If the thickness of this transition layer is very small compared with the diameter of the coil, then the coil is called a *hydrodynamically impermeable coil*. If the transition layer has a significant thickness and/or the flow of the solvent is not completely hindered even at the coil center, then the coil is *partially permeable*.

In impermeable coils, most of the solvent inside the coil moves together with the macromolecule, forming its integral part from the hydrodynamic viewpoint. Thus hydrodynamically impermeable coils are similar to spheres and are characterized by a hydrodynamic radius r_H; this radius corresponds to the distance from some point within the transition layer to the center of the coil. The necklace model yielded for Gaussian coils in the limit of impermeability an approximate relation between r_H and the radius of gyration r_G.

$$r_H \sim \tfrac{2}{3} r_G \tag{3.3.17}$$

In the 1950s there was a protracted discussion in the literature as to whether real polymer coils were hydrodynamically impermeable or partially permeable. It turned out that, virtually in all circumstances, real coils are impermeable. The argument, which was used originally to support coil permeability, was shown to arise from a misunderstood effect of long-range thermodynamic interactions on the coil dimensions. Nevertheless, coil permeability is still occasionally invoked to explain some unexpected experimental findings. We should, of course, keep in mind that the concept of permeability/impermeability is physically meaningful only for those macromolecules that form coils. For example, low-molecular-weight fragments of DNA or helical molecules of poly(benzyl glutamate) behave hydrodynamically more like rods or bent rods than like coils. And, indeed, permeability is invoked mainly for macromolecules of this general type.

What is the frictional ratio f/f_0 of Section 3.3.2 for an impermeable coil? It must be equal to the ratio r_H/r_0, where r_0 is the radius of a coil after it has been compacted into a solid sphere. As we have already mentioned, in typical coils the actual polymer material occupies about 1% of the coil volume. Thus, $r_H/r_0 \approx 1/(0.01)^{1/3}$, or about 4 to 5. This frictional ratio is much higher than anything that could be achieved by particle asymmetry. Consequently, measurement of frictional coefficients helps to differentiate compact and coiled particles.

Another situation of interest is the behavior of our necklace model of a coil in simple shear flow. In principle, the effect of the polymer particle on the flow field (and vice versa) is the same as described for spheres in Section 3.3.3. The particle translates and rotates with the flowing liquid but cannot follow completely the deformation of the liquid (some deformation is, of course, possible for coils); this causes an increase in macroscopic viscosity as described before. Hydrodynamically impermeable coils are again modeled best as equivalent spheres that would produce the same viscosity increase as the coils. However, the value of the hydrodynamic radius for viscosity increase is larger than the value given by equation (3.3.17) (which was valid for frictional coefficients). For the present case it is

$$r_H \sim \tfrac{7}{8} r_G \tag{3.3.18}$$

The difference between these values should serve as a reminder of the limitations of the equivalent sphere model. Indeed, both relations for the equivalent hydrodynamic radius were obtained from a rigorous treatment of the behavior of the necklace model in a hydrodynamic field in the limit of coil impermeability.

We will now consider the behavior of the necklace model of coils in simple shear flow. At small flow rates, the flow forces that attempt to deform the coil are too weak to do so; it is therefore permissible to study the motion of the segments as though the coil were rigid. In this case, the integrity of the coil dictates it to rotate as a solid body. All segments rotate around the hydrodynamic center of the coil with the same angular velocity (to be determined). This uniform rotary motion does not allow the individual segments to follow faithfully the movement of the liquid, which rotates *and* deforms. The difference in velocity between the segment and the surrounding liquid has two major consequences: (1) Segment exerts a frictional force on the liquid, which causes velocity disturbances throughout the liquid. It also contributes to the stress tensor of the flowing liquid. (2) The relative motion of the segment and the liquid leads to additional dissipation of energy as described in Section 3.3.2.

Let us study the dynamics of the coil in some detail. The force $\mathbf{F}_i$ acting on the ith segment is the frictional force of equation (3.3.16):

$$\mathbf{F}_i = f(\mathbf{v}_{p,i} - \mathbf{v}_{0,i}) \tag{3.3.19}$$

Here, we are treating f as a scalar (spherical segments); $\mathbf{v}_{p,i}$ is the velocity of the ith segment and $\mathbf{v}_{0,i}$ is the velocity the liquid would have at the location of the ith segment if the segment were absent while all other segments were at their proper locations and moving with their actual velocities.

Velocity $\mathbf{v}_{0,i}$ may be written as

$$\mathbf{v}_{0,i} = \mathbf{u}_{0,i} + \sum_{j \neq i} \mathbf{v}_{d,j} \tag{3.3.20}$$

where $\mathbf{u}_{0,i}$ is the velocity the liquid would have at the location of the ith segment in the absence of the coil; $\mathbf{v}_{d,j}$ is the velocity disturbance at the location of the ith segment caused by another segment j; the summation refers to all segments except the ith one. The $v_{d, j}$ is given by the Oseen relation for hydrodynamic interaction, equations (3.3.14) and (3.3.15). Combining these relations, we obtain

$$\mathbf{F}_i = f\left(\mathbf{v}_{p, i} - \mathbf{u}_{0, i} - \sum_{j \neq i} \mathbf{T}_{ij} \cdot \mathbf{F}_j\right) \tag{3.3.21}$$

This relation represents N linear equations (N being the number of segments) among the forces $\mathbf{F}_i$ acting on individual segments. Relations (3.3.21) are very involved; the tensors $\mathbf{T}_{ij}$ depend on the distribution of intersegmental distances. (And the distribution is even non-Gaussian, if the polymer-solvent system is thermodynamically nonideal.) Kirkwood and Riseman were the first to solve this problem. Their solution included conversion of the matrix of velocity disturbances and related vectors as well as tensors to normal coordinates and extension of the set of differential equations to a very large number of segments. This allowed replacement of the differential equations by a single integral equation that could be solved. Once the forces $\mathbf{F}_i$ are known, the calculation of the macroscopic viscosity is straightforward. In the limit of a hydrodynamically impermeable coil, the result is the one we have already seen in equation (3.3.18); in this limit the necklace model behaves as an equivalent hydrodynamic sphere.

To demonstrate the calculation of viscosity we will study a fully permeable coil. For permeable coils, tensors $\mathbf{T}_{ij}$ are negligibly small and relation (3.3.21) simplifies considerably. The pertinent velocities and forces are depicted in Figure 3.28. In this figure, $\mathbf{F}$ is the force with which flowing fluid acts on the segment. It is convenient to decompose $\mathbf{F}$ into two components: tangential and radial. The tangential component $\mathbf{F}_t$ exerts a torque on the coil. The sum of the torques originating from all segments is the total torque acting on the coil. The angular velocity ω of the whole coil adjusts

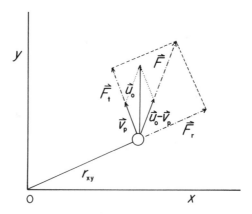

Figure 3.28. The force exerted by flowing liquid on a segment of a hydrodynamically permeable coil, and the velocity imparted by the force to the segment. See text for details.

itself in such a way that the total torque vanishes (this happens when $\omega = G/2$.) The radial component $\mathbf{F}_r$ tries to deform the coil. The actual deformation depends on the nature of the coil. At low shear rates it is negligible, at high flow rates the effect depends on the resistance of the coil to deformation. Soft coils will deform extensively, whereas rigid ones will change only marginally. We will return to this problem in Section 3.3.7.

Our present task is to calculate the excess dissipated power W_i due to our segment. This power is equal to the scalar product of the force and the velocity of the object on which it is acting. Thus

$$W_i = \mathbf{F}_i \cdot (\mathbf{v}_{\mathrm{p},i} - \mathbf{u}_{0,i}) = f(\mathbf{v}_{\mathrm{p},i} - \mathbf{u}_{0,i}) \cdot (\mathbf{v}_{\mathrm{p},i} - \mathbf{u}_{0,i}) \qquad (3.3.22)$$

where we used equation (3.3.21) to express $\mathbf{F}_i$ (neglecting all $\mathbf{T}_{ij}$). Recognizing that $\mathbf{u}_{0,i}$ is the unperturbed flow velocity characterized by the velocity gradient G and that $\mathbf{v}_{\mathrm{p},i}$ is determined by the angular velocity $\omega = G/2$, it is easy to show that the square of the vector $|\mathbf{u}_{0,i} - \mathbf{v}_{\mathrm{p},i}|$ is equal to the square of the segment velocity $v_{\mathrm{p},i}^2$. Hence,

$$W_i = f v_{\mathrm{p},i}^2 = f\omega^2 r_{xy}^2 = (fG^2/4)r_{xy}^2 \qquad (3.3.23)$$

where r_{xy} is the projection of the position vector of the segment into the plane of flow. To calculate the power W_{coil} dissipated by the whole coil, we need to sum up the contributions of all N segments and average over all possible conformations of the coil. Designating by n the number of molecules in the liquid, we then obtain the total excess dissipated energy W_{ex} as

$$W_{\mathrm{ex}} = n W_{\mathrm{coil}} = n N(fG^2/4)\overline{r_{xy}^2} \qquad (3.3.24)$$

For randomly oriented position vectors $\overline{r_{xy}^2}$ is equal to $\frac{2}{3}r_G^2$, where the radius of gyration r_G is determined by the statistics of the coil [compare equation (1.3.12)].

Recognizing now the relation between n and the polymer concentration c, as well as the relation between W_{ex} and viscosity, we can transform equation (3.3.24) into

$$\eta = \eta_0 \left[1 + cM \frac{f}{\eta_0} \left(\frac{A}{M_0} \right)^2 \frac{N_{\mathrm{Av}}}{36} \right] \qquad (3.3.25)$$

Here, η and η_0 designate the viscosity of the solvent and of the solution, respectively. We should note that the ratio of the frictional coefficient of the segment to the solvent viscosity, f/η_0, depends only on the nature of the polymer chain and is independent of the molecular weight M. The same applies to the ratio of the length of the statistical segment, A, and its molecular weight M_0. Thus according to equation (3.3.25), the increase in the viscosity of a solution of Gaussian fully permeable coils is proportional to the molecular weight of the polymer.

Let us compare this result with the viscosity increase caused by Gaussian hydro-dynamically impermeable coils. In this case, we may express the volume fraction ϕ in equation (3.3.13) through the concentration of the polymer c and the radius of gyration of its molecules. Transition to quantities characterizing coil statistics [equations (1.3.8) and (1.3.12)] then yields

$$\eta = \eta_0 \left[1 + bcM^{1/2} \left(\frac{A}{M_0} \right)^{3/2} N_{Av} \right] \qquad (3.3.26)$$

where b is a known numerical constant.

Thus, Gaussian impermeable coils lead to an increase in viscosity that is proportional to the square root of molecular weight. For non-Gaussian coils, the situation is more complicated; we will return to it in Section 3.3.4 in dealing with viscometry.

3.3.6. Concentration Effects in Macromolecular Hydrodynamics

We have already seen in Section 3.3.4 some effects of concentration on the hydro-dynamic behavior of macromolecular solutions. In this section we will study the concentration-related phenomena in more detail.

From the viewpoint of hydrodynamic interaction we can divide polymer solutions into four groups.

1. *Hydrodynamically ideal solutions,* in which the polymer molecules are so far apart that they do not influence the liquid flow in each other's vicinity. For such solutions, the hydrodynamic effect of a collection of molecules is simply a sum of the contributions of individual molecules. (We employed this model in the preceding section for calculating the contribution of an ensemble of polymer coils to the macroscopic viscosity.) However, as is apparent from equation (3.3.14), the interaction effects (i.e., the velocity disturbances) decay with the inverse first power of the mutual distance between molecules. This is an extremely slow decay. (The London interaction forces decay with the inverse sixth power of the distance; the electrostatic forces, which are considered to be long-range forces, decay with the inverse second power.) Consequently, molecules of a polymer must be extremely far from each other for the solution to qualify as hydrodynamically ideal. In fact, when solutions are studied that have sufficiently low concentration to be hydrodynamically ideal, the primary effect studied (e.g., the increase in viscosity) is usually so small that it is drawn into experimental error. It is, of course, an irony of nature that values measured for such solutions are the most valuable ones. In practice, they are obtained from measurements at higher concentration by a suitable extrapolation.

2. *Hydrodynamically dilute solutions* are solutions in which macromolecules mutually influence the liquid flow in each other's vicinity to a significant degree. Nevertheless, the qualitative character of the liquid flow is still the same as it was in the ideal solutions. For example, in the case of impermeable coils, the liquid flow is confined to the region in between individual coils and does not penetrate into the coil interior. Most hydrodynamic measurements are performed in dilute

solutions. The concentration effects in this region of concentrations are due to the same type of liquid velocity disturbances as we have studied in the preceding section, but this time hydrodynamic interactions play their role among different polymer coils in addition to intramolecular (intersegmental) interactions.

The theoretical calculations are extremely involved because hydrodynamic interactive forces decay with the inverse first power of the interparticle distance, as we have mentioned before. Consequently, divergent integrals result if an attempt is made to calculate the effect of hydrodynamic interactions on a test particle by summing the perturbances caused by all other particles in the solution. Various stratagems have been designed to overcome this difficulty; some of them are completely erroneous. For example, the so-called cell model for sedimentation of particles predicted falsely that the rate of sedimentation should decrease linearly with the cube root of concentration. Fortunately, more careful analyses have shown that most hydrodynamic properties can be expressed as power series in concentration. Extrapolation of the experimental data to the vanishing concentration of polymer can thus usually be done with confidence.

3. When the polymer concentration is increased further, at some *overlap concentration c^**, the molecular coils will fill all the available volume. In the region of *semidilute solutions* for which $c > c^*$, the individual coils overlap, form many contacts, and become entangled. The overlap concentration may be quite low; in a typical coil, the dry volume of polymer is about 1% of the coil volume, hence the overlap concentration is 1%. It is even lower for polymers with very high molecular weight.

In semidilute solutions, the liquid cannot flow around the coils; it must permeate them. The simple rules of hydrodynamic interaction used in our previous considerations are no longer sufficient and must be replaced by other concepts.

Let us consider the relative motion of polymer molecules and a solvent in a situation similar to one we treated in Section 3.3.2. In the present case, the solvent has to flow through the entangled network of all polymer coils. The situation is physically similar to the flow of solvent through a porous medium. The flow resistance of a porous substance depends on the porosity of the material (which corresponds to concentration of our polymer) and on the average cross section of the flow channels. For polymer solutions, the latter quantity is related to the homogeneity of the distribution of sections of polymer chains throughout the solution. In semidilute solutions, this distribution is still far from homogeneous. (As we would expect, solutions in thermodynamically good solvents are more homogeneous than those in poor ones.)

The structure and behavior of the entangled networks in the semidilute region are frequently described by *scaling laws,* which are based on the following model of a semidilute solution. The polymer network is characterized by a set of contact points between polymer chains. When the solvent and the polymer are in relative motion, the hydrodynamic interactions among polymer segments do not extend beyond the contact points. The average length of the polymer chains between two contact points is called a *screening length.* It decreases with increasing concentration of the solution; it is much shorter than the length of the whole chain and is independent of it.

The section of chain between two contact points is called a "blob"; it obeys the conformational rules of polymer chains (which depend, of course, on the thermodynamic interactions between the polymer and the solvent). The hydrodynamic interaction plays its full role within the blob, which moves as a more or less impermeable particle through the solvent. The hydrodynamic radius of the blob (for a given concentration of a given polymer sample) depends on the thermodynamic quality of the solvent. Scaling theory casts all the experimental dependencies in the form of power laws and predicts the applicable exponents; it does not predict the proportionality constants.

As an example, let us consider the sedimentation coefficient. We will see in Section 3.4.2 that in hydrodynamically ideal solutions it is proportional to the ratio of molecular weight and frictional coefficient, M/f; it is independent of concentration. According to scaling laws, in semidilute solutions the sedimentation coefficient is independent of the molecular weight of the polymer. It is inversely proportional to concentration in theta solvents and to the square root of concentration in good solvents. These predictions agree reasonably well with experiment.

The behavior of semidilute solutions in simple shear flow is complicated and is still not fully understood. It is in the nature of simple shear flow that particles that were close to each other are separated progressively. To separate, the entangled polymer chains must first disentangle. This process requires a large effort from the driving force, and the viscosity increases steeply with concentration. The disentanglement process is opposed by thermal motion, which keeps entangling the chains again. At low flow rates, the thermal motion prevails and the molecular structure of the flowing solution is essentially the same as for a solution at rest. However, at higher flow rates, the disentanglement prevails and the flowing solution shows an increasing tendency to separate into small domains containing one or a few molecular chains; these domains (possibly heavily entangled internally) easily pass each other in the flowing liquid, and the viscosity drops with increasing flow rate. Such solutions are called *non-Newtonian;* we will study them in more detail in Section 3.4.3.8.

4. *Concentrated solutions* make the fourth group of solutions. In these solutions, the individual polymer segments belonging to different chains and/or distant parts of the same chain are packed rather closely, and solvent molecules are distributed rather uniformly among them. This state of affairs is achieved only at rather high concentrations, typically above 30 volume %.

Mutual movement of molecules in concentrated solutions acquires a new character. The model of continuous solvent without molecular structure is no longer applicable. The relative movement of solvent and polymer is best treated as permeation or diffusion. A snakelike movement of molecules along their contours more and more dominates the mutual movement of polymer chains (i.e., flow). In all these respects, concentrated solutions behave very similarly to bulk polymers, and we will treat them together in Chapter 4.

3.3.7. Orientation and Deformation of Particles in a Flowing Liquid

Until now we have considered the hydrodynamic behavior of averaged polymer coils. An averaged coil has a spherical symmetry, whereas individual coils are asymmetric. Statistical analysis of random coils has shown that the largest dimension of a coil (i.e., the distance between two segments that are farthest apart) varies within a relatively narrow range. The largest distance between segments in a direction perpendicular to the direction of the largest elongation is about half that value. Consequently, a prolate ellipsoid with an axial ratio of 2 is a good model of a polymer coil. The hydrodynamic behavior of ellipsoids suspended in a liquid subject to simple shear flow has been studied theoretically. It has been shown that the flow tends to orient the particles; the extent of the orientation increases with increasing flow rate. The theoretical analysis is quite involved and will not be presented here. Instead, we will study the hydrodynamic behavior of another model of a macromolecule that, despite its extreme oversimplification of polymer coil characteristics, describes the hydrodynamic behavior of polymer solutions surprisingly well. This model is a *dumbbell:* two friction centers connected by a frictionless rod or spring. A dumbbell models not only the asymmetry of the coil but also its deformability; an elastic dumbbell models easily deformable polymer coils, whereas a rigid one represents nondeformable coils.

Let us consider first the kinematics of the dumbbell in simple shear flow. In Figure 3.28 several dumbbells (all of them oriented in the flow plane) are depicted together with arrows denoting the liquid velocity at the location of the friction center and its decomposition into tangential and radial components. We have already seen that the tangential velocities are connected with torques causing rotation of the dumbbell; the radial velocities and their related friction forces tend to deform the dumbbells. However, there is a major difference between the situations depicted in Figures 3.28 and 3.29. In the former case, the friction center was a part of a rigid structure rotating with a constant angular velocity $\omega = G/2$. In the present case, the rotary movement of the dumbbell follows faithfully the tangential component of the liquid flow. Consequently, the angular velocity varies with the orientation of the dumbbell; it is related to the velocity gradient as $\omega = G \sin^2 \varphi$, where φ is the angle between the direction of flow and the dumbbell axis.

To proceed with our analysis, we need to introduce a new concept: the distribution function of orientations, Φ. It is convenient to represent the function Φ in spherical coordinates φ and θ, where φ is the angle in the flow plane xy and θ is the angle between the position vector and the z axis. The distribution function Φ specifies how many molecules (dN) from the total number of molecules N_{tot} in unit volume are oriented in such a way that their axis points in a direction between φ and $\varphi + d\varphi$ and simultaneously between θ and $\theta + d\theta$:

$$dN/N_{\text{tot}} = \Phi(\varphi, \theta) \sin \theta \, d\theta \, d\varphi \qquad (3.3.27)$$

In solutions at rest, molecules are oriented at random and the function Φ has a constant value. Let us consider the effect of the rotation caused by simple shear flow

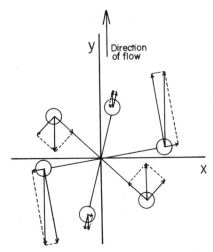

Figure 3.29. Liquid velocity at the locations of the friction centers, and its decomposition into tangential and radial components for three different orientations of a dumbbell in shear flow.

on the distribution function. As we have seen, every particle rotates fastest when it is oriented across the flow lines (i.e., along the x-axis) and slowest when oriented along flow lines (along the y-axis). The amount of time the particle spends in a region with a particular orientation is inversely proportional to its rotary velocity corresponding to this orientation. Hence, it spends the most time oriented along the flow lines and the least time oriented along the velocity gradient. Consequently, orientations corresponding to high rotary velocities lead to low values of the distribution function Φ, whereas low rotary velocities lead to high values of Φ. [Strictly speaking, our dumbbell rotates with zero velocity when oriented along the flow lines. Therefore it should stop when oriented in this direction; all molecules should be oriented along flow lines after a very short time. However, an ideal dumbbell is a poor model for a polymer coil. Even when the coil (or ellipsoid) is oriented along the flow lines, there exists a torque continuing the rotation. Even more important are the diffusion processes, which will be described in the next paragraph.]

One of the roles of thermal motion in nature is its tendency to reduce and eventually erase any inhomogeneities and deviations from randomness. Thus inhomogeneity in concentration is reduced by diffusion (see Section 3.4.1). Similarly, inhomogeneity in orientation is attacked by thermal motion through a process called *rotary diffusion*. There is a close analogy between diffusion (ordinary diffusion is actually a *translational diffusion*) and rotary diffusion. The former causes the flow of particles in ordinary space; the flow proceeds in the direction against the concentration gradient and is characterized by the (translational) diffusion coefficient D. The rotary diffusion causes flow in the space of orientations (independent variables φ and θ), proceeds against the direction of the gradient of the distribution function Φ, and is characterized

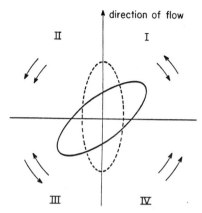

Figure 3.30. Effect of rotary diffusion on the distribution function of orientations of asymmetric particles in shear flow. Dashed line: distribution in absence of rotary diffusion. Full line: distribution as a result of combined kinematic orientation and of rotary diffusion flow.

by a *rotary diffusion coefficient* D_r. We will study the nature of D_r more closely in Sections 3.4.3 and 3.4.4. Presently, we will consider the effect of rotary diffusion on our dumbbells.

Our discussion of kinematic orientation led to a prediction that the distribution function of orientations would be rather asymmetric and independent of the flow rate and would have its maximum in the direction of flow. This function is schematically depicted in Figure 3.30 by a dashed line. The outer circle of arrows shows the direction of the rotary motion that generated this function. What will be the effect of rotary diffusion on this situation? The diffusion flow will drive the particles from the region of high concentration toward lower concentrations. The inner arrows in Figure 3.30 show the direction of this motion. It is apparent that in the second and fourth quadrants the directions of both types of motion coincide, whereas in the first and third quadrants they have opposite directions. The latter case obviously represents a slower combined motion than the former case. As we have already seen, a slower rotary motion means higher values of the function Φ. Thus the particles will be preferentially oriented in the first and third quadrants. At low values of velocity gradient, rotary diffusion is able to smooth out the deviation of the distribution function from homogeneity almost completely. The resulting distribution function is only slightly asymmetric; its maxima are oriented at a 45° angle with respect to the flow direction. The full line in Figure 3.30 schematically represents this function. With increasing velocity gradient, the kinematic orientation of particles starts to prevail over the rotary diffusion: The maxima of the distribution function shift gradually toward the direction of flow, and Φ becomes more and more asymmetric.

So far we have dealt only with the tangential component of the liquid flow that was imparted to our dumbbell. Let us now pay attention to the radial component. It tries to carry the friction center in its direction, but the cohesive forces of the coil prevent it from doing so. The resulting difference between the velocities of the friction center and the flowing liquid produces a radial force. As we have seen in Section 3.3.5, this force is responsible for excess viscous dissipation of applied power and leads to an increase in viscosity. However, the force also acts on the particle. If the particle is rigid, its rigidity cancels the force. Of course, if the particle is deformable, then the radial force will change its shape and size.

It is apparent from Figure 3.29 that radial forces stretch particles oriented in the first and third quadrants; they compress particles oriented in the second and fourth quadrants. However, the average deformation is not zero; the particle spends a longer time being oriented in the first and third quadrants than in the second and fourth. Thus on average the particle is stretched. For a detailed description of the particle deformation, it is important to consider the rate at which the particle can be deformed. If the particle can follow the periodic stretching and compressing forces, then its shape pulsates appreciably. If the particle resists fast changes in its shape (this property is called *internal viscosity*), then it adopts a deformed shape corresponding to deforming forces averaged for one rotation; after that it virtually does not change its shape any more during a single rotation.

A full hydrodynamic analysis has been performed for ellipsoids as well as for dumbbell and pearl necklace models in the limit of vanishing inner viscosity. Quantitative results of these calculations are important for understanding of non-Newtonian viscosity (Section 3.4.3.8) and streaming birefringence (Section 3.4.4). We will discuss them in more detail when we consider these phenomena.

3.4. HYDRODYNAMIC METHODS FOR THE STUDY OF MACROMOLECULES IN SOLUTION

After studying in the preceding section the basic phenomena related to mutual movements in liquids, we will now turn to situations of practical interest involving these hydrodynamic principles, and especially to the physicochemical methods that obtain information about macromolecular solutions from their hydrodynamic behavior. In Sections 3.4.1 and 3.4.2 we will outline the processes involving mutual movement of macromolecules and solvent: diffusion and sedimentation. Processes related to simple shear flow of macromolecular solutions and the methods based on them (viscometry, streaming birefringence) will be treated in Sections 3.4.3 and 3.4.4.

3.4.1. Diffusion

In Section 3.2 we introduced a very general criterion of equilibrium valid in the absence of rigid or impermeable boundaries: homogeneity of the total potential of every component. The rationale for this criterion was not presented because it is based mainly on nonequilibrium situations such as those to be described presently. In physics, the term *potential* (which is a function of coordinates) represents the potential energy an appropriate unit would have if it were present at the relevant location. In the case of the total potential, the appropriate unit is 1 mol of the substance. The negative gradient of potential energy is a force. In the absence of internal boundaries, the force must act on the substance and change its state of motion. This change of state of motion (i.e., flow) is incompatible with the definition of equilibrium. Hence, to have a system at equilibrium, any such force must be absent (or acting against an internal rigid or impermeable boundary). Of course, the force is absent only when the gradient of the total potential is zero, that is, when the total potential is homogeneous throughout the system. If it is not, the resulting force will lead to flow of that particular component.

In the present section, we are interested in the motion caused by such forces. In the simplest case, the external potentials (mechanical, electrical) are homogeneous and do not contribute to the forces; pressure and temperature are also homogeneous. Under these circumstances, the total and chemical potentials may be nonuniform only when the composition of the system is nonuniform. If it is, then components of the system will move with respect to each other. This process is called *diffusion*.

We will omit some finer points and will again study only the simplest situation: a binary solution enclosed within well-defined stationary boundaries, assuming that the volume of the solution does not change with the redistribution of the components. For such a solution, the net volume flow through any (hypothetical) boundary must be zero. The flows of the two components are not independent, and it is sufficient to consider only one of them.

In our simplest case, concentration changes occur along the x-axis only (one-dimensional diffusion). Diffusion flow I_i is defined as the amount of the ith component (mass or number of moles) passing through a boundary of unit area during a unit time. In the 1850s Fick described the diffusion flow by a phenomenological relation known as *Fick's first law:*

$$I_i = -D_i \frac{dc_i}{dx} \tag{3.4.1}$$

where c_i is the concentration of the ith component in the same units as were used in defining I_i. The flow proceeds against the concentration gradient. The *diffusion coefficient* D_i, which is defined by equation (3.4.1), generally depends on the nature of the system and on the concentration.

The law of mass conservation demands that the change in concentration in a volume element due to the flow must be equal to the net flow of the ith component into that volume element. In other words,

$$\left(\frac{\partial c_i}{\partial t}\right)_x = -\left(\frac{\partial I_i}{\partial x}\right)_t = \left\{\partial\left[D_i\left(\frac{\partial c_i}{\partial x}\right)_t\right]\Big/\partial x\right\}_t \tag{3.4.2}$$

where t is time. If the diffusion coefficient is independent of concentration, it is also independent of the position in the system (i.e., of x), and equation (3.4.2) adopts the more familiar form

$$\left(\frac{\partial c_i}{\partial t}\right)_x = D_i \left(\frac{\partial^2 c_i}{\partial x^2}\right)_t \tag{3.4.3}$$

To describe a three-dimensional distribution of concentrations, the Laplace operator ∇^2 must replace the second partial derivative in equation (3.4.3). For Cartesian coordinates, the result can be written as

$$\frac{\partial c_i}{\partial t} = D_i \left(\frac{\partial^2 c_i}{\partial x^2} + \frac{\partial^2 c_i}{\partial y^2} + \frac{\partial^2 c_i}{\partial z^2}\right) \tag{3.4.4}$$

Any of equations (3.4.2)–(3.4.4) represents *Fick's second law*. Solution of this equation together with appropriate initial and boundary conditions represents a complete description of the evolution of a diffusing system with time and is therefore of considerable practical interest. However, it is usually quite complicated except for a few simple situations.

A problem that is more important for us is the relation between the phenomenological quantity called the diffusion coefficient and the molecular properties of our solution. We will therefore follow the process of diffusion from the viewpoint of the motion of individual molecules. On the molecular level, diffusion transport is a result of thermal motion, as described in Section 3.3.1. As before, the important part of the molecular motion is that part that contributes to the macroscopic flow (it may represent only a minuscule portion of the overall thermal motion). In the ultimate analysis, this part of the flow is caused by the action of "organized" (i.e., nonrandom) forces acting on the molecules. In the present case, the force acting on 1 mol is the gradient of the chemical potential. To obtain the organized force F_2 acting on a single molecule of solute (subscript 2), we must divide by Avogadro's number. Hence,

$$F_2 = -\frac{d\mu_2/dx}{N_{Av}} \tag{3.4.5}$$

According to equation (3.3.8) the organized velocity of the solute molecule v_2 and the flow I_2 are

$$v_2 = \frac{F_2}{f_2} = -\frac{d\mu_2/dx}{f_2 N_{Av}} \tag{3.4.6}$$

and

$$I_2 = c_2 v_2 = -c_2 \frac{d\mu_2/dx}{f_2 N_{Av}} \tag{3.4.7}$$

Here, μ_2 is the chemical potential of the solute. Equation (3.4.7) is an expression for the diffusion flow that is based on a combination of relations from molecular hydrodynamics and thermodynamics. Its comparison with the phenomenological equation (3.4.1) yields the diffusion coefficient D_2:

$$D_2 = c_2 \frac{d\mu_2/dx}{f_2 N_{Av}(dc_2/dx)} = c_2 \frac{d\mu_2/dc_2}{f_2 N_{Av}} \tag{3.4.8}$$

Let us first evaluate the diffusion coefficient at vanishing concentration of solute (superscript zero). According to equation (3.1.42) or (3.1.43),

$$[c_2(d\mu_2/dc_2)]^o = RT \tag{3.4.9}$$

and

$$D_2^o = RT / f_2^o N_{Av} = kT / f_2^o \tag{3.4.10}$$

Equation (3.4.10) is of considerable interest. It relates the frictional coefficient (a molecular property) to a phenomenological quantity (diffusion coefficient) measurable by macroscopic techniques. Indeed, measurement of diffusion is a method of choice whenever we need to study frictional coefficients and to extract from them molecular properties of the solute (cf. Sections 3.3.2 and 3.3.4).

The diffusion coefficient is concentration-dependent. The simple relation (3.4.10) is valid only for the frictional coefficient and diffusion coefficient in the limit of vanishing concentration (superscript zero). To extrapolate experimental data to vanishing concentration we need to understand the concentration dependence of D_2. The required dependence is obtained from equation (3.4.8), into which we substitute for $d\mu_2/dc_2$ the expression from equation (3.1.43). The result reads

$$D_2 = D_2^o (1 - c_2 v_2) \frac{1 + 2BM_2c_2 + \cdots}{1 + k_f c_2 + \cdots} \tag{3.4.11}$$

The denominator and the coefficient k_f originated from the concentration dependence of the frictional coefficient (combined effects of hydrodynamic interaction and counterflow of the solvent). The concentration dependence is complicated: D_2 may, with increasing concentration, either increase, decrease or go through a minimum. Thus the measurements must be extended to the lowest concentrations attainable and the extrapolation must be done very judiciously when the limiting value is required.

3.4.1.1. Experimental Diffusimetry.

Although several methods exist for measuring diffusion coefficients in mixtures of low-molecular-weight compounds, essentially all measurements of diffusion in dilute macromolecular solutions use either the so-called *free diffusion* method or, more recently, quasi-elastic light scattering (Section 3.5.7). The former method uses a rectangular cell in which a sharp boundary is created between two solutions of different concentrations. The boundary is then allowed to diffuse freely. The cell is observed by means of an optical system (schlieren or interference) that registers the profile of refractive index, which mirrors the profile of concentration. The optical systems are similar to those used in the ultracentrifuge. The banes of diffusion measurements are convective disturbances, which may be caused by the slightest vibration or temperature variation during an experiment lasting many hours. Thus extreme stability of the instrument base and ultimate temperature control are the musts of diffusimetry.

In a successful experiment, the initial concentration profile at time $t = 0$ is virtually a step function; that is, $c = c_1$ for $x < 0$ and $c = c_2$ for $x > 0$, where $x = 0$ is the coordinate of the initial boundary. The derivative with respect to x is a Dirac function, $(\partial c/\partial x)_{t=0} = (c_2 - c_1)\delta(x)$. Here, c_1 and c_2 are the initial concentrations introduced into the half-cells. The cell is so long that measurable changes in concentration do not reach the ends of the cell during the time of the experiment.

If the solute is homogeneous (pure protein or monodisperse polymer) and its diffusion coefficient is independent of concentration, then the above boundary and initial conditions, together with Fick's equations, lead to the following dependencies of concentration on time and on the position in the cell.

$$c(x, t) = c_2 + \frac{c_1 - c_2}{2}\left[1 - \text{erf}\left(\frac{x}{2\sqrt{D_2\,t}}\right)\right] \tag{3.4.12}$$

$$\text{erf}(y) \equiv \frac{2}{\sqrt{\pi}}\int_0^y \exp(-z^2)dz \tag{3.4.13}$$

$$\left(\frac{\partial c}{\partial x}\right)_t = \left(\frac{c_1 - c_2}{2\sqrt{\pi D_2 t}}\right)\exp\left(-\frac{x^2}{4D_2 t}\right) \tag{3.4.14}$$

Thus the concentration profile is essentially the Gaussian error integral $\text{erf}(y)$; its gradient at time t is a Gaussian function with variance $\sigma^2 = 2D_2 t$.

Experimentally, photographs of the cell are taken at several times and evaluated in terms of the dependence of $(\partial c/\partial x)_t$ on x. The diffusion coefficient can be obtained from these dependencies using the area-to-height ratio. It is obvious from equations (3.4.12)–(3.4.14) that the height H of the curve is equal to the preexponential factor of equation (3.4.14). The area A under the curve is equal to $(c_1 - c_2)$. It is left to the reader to prove himself that

$$(A/H)^2 = 4\pi D_2 t \tag{3.4.15}$$

Hence, D_2 is easily obtained from the plot $(A/H)^2$ versus t.

Another method for evaluating D_2 is based on the relation among D_2, the variance σ^2 of the Gaussian function (3.4.14), and its width at half-height, $w_{1/2}$ (an easily measurable quantity):

$$4D_2 t = 2\sigma^2 = \frac{w_{1/2}^2}{4\ln 2} \tag{3.4.16}$$

Again, D_2 is obtained from a plot of $w_{1/2}^2$ versus t.

When the polymer is polydisperse, the diffusing boundary is a superposition of independent diffusing boundaries of all polymer species—a superposition of many Gaussian curves. The resulting boundary is still symmetrical, but it is not Gaussian any more. When the above evaluation techniques are applied to such a boundary, they yield well-defined but complicated averages of diffusion coefficients: Each method yields a different average. Hence it is possible to compare these averages and obtain an estimate of the polydispersity of the polymer. Unfortunately, the method is not very sensitive and requires extremely precise measurement of the concentration profile.

If the diffusion coefficient changes noticeably with polymer concentration, then it also changes with the position within the diffusion cell. Specifically, it has different values on the two sides of the boundary; they are spreading with different rates, and the boundary becomes asymmetric. Attempts to utilize such boundaries for the measurement of diffusion coefficients lead to ill-defined averages, which are sometimes called *integral diffusion coefficients*. In such circumstances, it is necessary to select the two original concentrations in the half-cells close enough to make the diffusion coefficient virtually constant within the cell. Several such experiments are performed, and the dependence of the diffusion coefficient on concentration is constructed and extrapolated to vanishing concentration.

3.4.1.2. Interpretation of Diffusion Coefficients.   Let us summarize the information that can be extracted from the knowledge of diffusion coefficients.

The diffusion coefficient at vanishing concentration of the polymer gives us the most direct access to the frictional coefficient [equation (3.4.10)] and, through it, to all properties obtainable from the latter: to the frictional ratio and its message about the shape and nature (compact or coiled) of the macromolecule; the hydrodynamic radius of the macromolecule and, through it, to the end-to-end distance of the coil and to its segment length, which represents the short-range structure of the polymer; etc.

The relation of the diffusion and frictional coefficients to the end-to-end distance of the polymer coil is frequently described by a quantity P:

$$P \equiv f_2^o / \eta_0 \sqrt{h^2} = kT / D_2^o \eta_0 \sqrt{h^2} \qquad (3.4.17)$$

Hydrodynamics predicts $P = 5.11$ for Gaussian impermeable coils; this value changes very little with non-Gaussian character of the coil.

Diffusion coefficients are also used for the calculation of molecular weights. An experimental relationship is established between diffusion coefficients of a series of polymer-homolog samples in a particular solvent and their molecular weight measured by some other, preferably absolute, method. For the method to be valid, unknown samples must belong to the same family of samples used for calibration; specifically, the degree of branching and broadness of the molecular weight distribution should be similar.

This method is one of a number of important methods (sedimentation coefficients, intrinsic viscosities) that use calibration. They all are based on the relation between molecular weight and coil size, relation (1.3.10).

$$\overline{h^2} \sim M^{1+2\varepsilon} \tag{3.4.18}$$

Combining relations (3.4.17) and (3.4.18) yields

$$D_2^{\text{o}} = K_{\text{D}} M_2^{-a_{\text{D}}} \tag{3.4.19}$$

where the exponent a_{D}, is equal to ($\frac{1}{2} + \varepsilon$). In practice, the constants K_{D} and a_{D} are obtained experimentally using the plot of log D_2^{o} versus log M_2.

3.4.2. Sedimentation Velocity

In Section 3.2.2 we studied the distribution of molecules in an ultracentrifuge cell at equilibrium. To reach the equilibrium, the molecules must first travel to their equilibrium locations. This flow process can provide valuable information about the system in question. When the cell is filled with a homogeneous macromolecular solution and brought to speed, the macromolecules are subject to a centrifugal force and move with more or less uniform velocity toward the bottom of the cell. The measurement of this velocity together with studies of related phenomena is known as a method of sedimentation velocity; it will be the subject of this section.

3.4.2.1. Homogeneous Solutes. Let us first calculate the force F_2 acting on a single molecule of polymer. It is again equal to the gradient of the total potential of the polymeric component divided by Avogadro's number. However, in the presence of a centrifugal field, we must use equation (3.2.36) to express the total potential. The result reads

$$F_2 = -\frac{d\mu_{2,\text{tot}}/dr}{N_{\text{Av}}} = \frac{M_2\omega^2 r - d\mu_2/dr}{N_{\text{Av}}} \tag{3.4.20}$$

Here r is again the radial distance from the rotor axis and ω is angular velocity. Further derivation follows the same lines as we used in Section 3.2.2.2 in dealing with sedimentation equilibrium. The gradient of chemical potential is expressed as

$$\frac{d\mu_2}{dr} = \overline{V}_2\frac{dP}{dr} - \overline{S}_2\frac{dT}{dr} + \left(\frac{\partial\mu_2}{\partial c_2}\right)_{P,T}\frac{dc_2}{dr} \tag{3.4.21}$$

For sedimentation experiments performed at constant temperature, the second term in equation (3.4.21) can be dropped. Molar volume of the polymer $\overline{V}_2$ is equal to

$M_2\bar{v}_2$ as usual. Equation (3.2.41) is still applicable for dP/dr. We can now substitute equation (3.4.21) together with these modifications into equation (3.4.20) and obtain

$$F_2 = [M_2\omega^2 r - \bar{V}_2\,(dP/dr) - (\partial\mu_2/\partial c_2)_{P,T}\,(dc_2/dr)]/N_{Av}$$
$$= [M_2\,(1 - \bar{v}_2\rho)\,\omega^2 r - (\partial\mu_2/\partial c_2)_{P,T}\,(dc_2/dr)]/N_{Av} \tag{3.4.22}$$

Let us return for a moment to sedimentation equilibrium. At equilibrium, the force F_2 must vanish. When this happens, equation (3.4.22) becomes identical with equation (3.2.43) describing sedimentation equilibrium.

Force F_2 is acting on a molecule suspended in a liquid. According to equation (3.3.5), the molecule must move with a velocity $v_2 = F_2/f_2$. Recognizing that velocity v_2 is a change of particle position with time, we obtain

$$v_2 \equiv \frac{dr}{dt} = \frac{M_2\,(1 - \bar{v}_2\rho)\,\omega^2 r - (\partial\mu_2/\partial c_2)_{P,T}\,(dc_2/dr)}{f_2 N_{Av}} \tag{3.4.23}$$

Let us calculate the velocity of a particle assuming that the concentration of the polymer is homogeneous in its vicinity. In actual experiments, this situation prevails in a major part of the sedimentation cell just after the cell filled with a homogeneous solution has been brought up to speed. Under this condition, $dc_2/dr = 0$ in the affected part of the cell (which is called the *plateau region*), and the second term in the numerator of equation (3.4.23) vanishes. The equation simplifies to

$$v_2 \equiv \frac{dr}{dt} = \frac{M_2\,(1 - \bar{v}_2\rho)}{f_2 N_{Av}}\omega^2 r \equiv s_2\omega^2 r \tag{3.4.24}$$

The collection of constants in the fraction is given a new symbol s_2 and a new name: *sedimentation coefficient.*

How does the position of a particle within the plateau region change with time? Integration of equation (3.4.24) yields

$$\ln r = s_2\omega^2 t + \text{constant} \tag{3.4.25}$$

where t is the time elapsed since the start of the experiment. The sedimentation coefficient is closely related to important macromolecular characteristics—the molecular weight and the frictional coefficient. Thus one of the goals of sedimentation velocity experiments is its measurement.

The position of a particular particle within a plateau region is not marked by any special feature and consequently cannot be measured. We need to find some label that would allow us to follow the movement of a particular particle. Molecules that were originally close to the liquid meniscus in the cell could be considered labeled in this sense. As long as all particles move with the same velocity, the concentration profile in the cell just shifts toward the bottom, and the step boundary, which existed at the meniscus, moves into the cell. The position of the boundary is easily detected by our optical techniques; its movement is interpreted as movement of the labeled particles and is used in calculating the sedimentation coefficient according to equation (3.4.25).

There seems to be a serious flaw in the above argument. The labeled particle was supposed to reside in the plateau region, yet we have selected one residing at the boundary where dc_2/dr has its maximum value. Consequently, that part of the flow that corresponds to the second term in the numerator of equation (3.4.23) (let us call it diffusion flow) should be strongly manifested precisely at our labeled position. This diffusion flow should move the molecules within the boundary toward the region of lower concentration—against the direction of the sedimentation flow. This flow should be quite analogous to the flow we analyzed when we studied the process of diffusion. In fact, this is exactly what is happening in the centrifuge cell. However, we should remember that in the free diffusion experiment, while the solute was moving toward the region of low concentration, the inflection point of the concentration profile remained at the position of the original sharp boundary. When schlieren optics was used to record the concentration profile, the peak of the schlieren line stayed at this position throughout the experiment. A similar situation exists in the sedimentation velocity experiment. The position of the schlieren peak is not influenced by the diffusion process but only by the sedimentation. Hence, the sedimentation coefficient can be calculated from its movement. Thus, during a sedimentation velocity experiment, the schlieren boundary separates from the meniscus and then gets broader and broader as it travels toward the cell bottom (Fig. 3.31).

We will now analyze the solute transport in the sedimentation cell more formally. We will start with the law of mass conservation, which in hydrodynamics is called the *equation of continuity*. We have met it already in equation (3.4.2); however, for centrifugal problems we need to cast it as a one-dimensional transport in cylindrical coordinates:

$$\frac{\partial c_2}{\partial t} = -\nabla \cdot \mathbf{I}_2 = -\frac{1}{r}\left[\frac{\partial (r I_2)}{\partial r}\right]_t \qquad (3.4.26)$$

Figure 3.31. Schlieren photographs of a sedimenting boundary of a sample of polystyrene. ($\overline{M}_w = 177,000$; narrow fraction; 0.004 g/mL) in ethyl acetate at 20°C as it progresses toward the bottom of the cell at 56,000 rpm.

The term $\nabla \cdot \mathbf{I}_2$ represents divergence of flow. Replacing I_2 by $c_2 v_2$, substituting equation (3.4.23) for v_2, and utilizing definitions of D_2 [equation (3.4.8)] and s_2 [equation (3.4.24)], we can transform equation (3.4.26) into

$$\left(\frac{\partial c_2}{\partial t}\right)_r = -\frac{1}{r}\left(\frac{\partial[c_2\omega^2 r^2 s_2 - D_2 r(\partial c_2/\partial r)_t]}{\partial r}\right)_t \qquad (3.4.27)$$

This equation is known as the *Lamm equation*. Together with appropriate initial and boundary conditions, it fully describes the development of a two-component system in a centrifugal field. It was originally derived for constant s_2 and D_2 (ideally dilute solutions) but is applicable also for nonideal solutions with concentration-dependent s_2, D_2.

Let us apply the Lamm equation to the plateau region. In this region $(dc_2/dr)_t = 0$ by definition, and the equation reduces to

$$\left(\frac{\partial c_2}{\partial t}\right)_r = -2c_2\omega^2 s_2 \qquad (3.4.28)$$

Its integral is

$$\ln\left(c_2/c_2^o\right) = -2\omega^2 s_2(t - t_0) \qquad (3.4.29)$$

where c_2^o and t_0 are the concentration in the plateau region and time at the beginning of the experiment, respectively. Combining equations (3.4.29) and (3.4.25) leads to

$$c_2/c_2^o = (r_0/r)^2 \qquad (3.4.30)$$

where r_0 and r are the initial and current positions, respectively, of some labeled particle in the plateau region (the boundary position is used in practice). Thus, although the concentration remains constant throughout the plateau region, it decreases as the boundary moves away from the meniscus. Equation (3.4.30) is called the *radial dilution law*.

Figure 3.32 serves as a more graphic explanation of this law. The amount of solute that occupied the region between the hypothetical surfaces A and B later fills the space between C and D. The latter volume is larger than the former on two counts: (1) cylindrical coordinates dictate that the area of C is larger than that of A; and (2) the distance between C and D is larger than the distance between A and B because particles at B (being farther from the rotor axis) are subject to a larger centrifugal force and therefore move faster than particles at A.

What part of the cell does the plateau region occupy, and how long does it last in the cell? At the beginning of the run, a plateau exists in the whole cell. Later development depends on the interplay of the sedimentation and diffusion terms in the Lamm equation. When the sedimentation term dominates the scene (large values of $\omega^2 s_2$), the boundary separates from the meniscus, leaving behind essentially pure solvent. As the boundary moves, it spreads because of diffusion. Simultaneously,

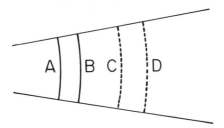

Figure 3.32. A graphical demonstration of the radial dilution law. A sedimenting solute originally confined between boundaries A and B is later found between C and D.

solute accumulates at the bottom. Both phenomena steadily compress the plateau region between the progressing and spreading boundary and the increasingly massive layer at the bottom. However, in well-designed experiments, the plateau region can often be observed until the boundary peak travels about three-fourths of its way to bottom. Eventually, the boundary will merge with the bottom layer, and a more or less exponential concentration profile will be established near the cell bottom corresponding to sedimentation equilibrium. Of course, runs that have values of $\omega^2 s_2$ large enough to produce a reasonably narrow boundary lead to an equilibrium distribution of concentration that is too steep to allow for a meaningful analysis.

When the sedimentation coefficient is too small or the rotor velocity was not selected large enough, the spreading of the boundary will be faster than its forward movement. The boundary may not be able to separate fully from the meniscus and the plateau region may disappear before the boundary gets very far. Even in this case, the solution of the Lamm equation allows for extraction of the sedimentation coefficient from the sedimentation velocity data. However, the simple procedure of following the boundary movement by measuring the position of the maximum of the gradient (schlieren) curve must be replaced by laboriously calculating the position of the first moment of this curve and following its movement toward the cell bottom.

We have seen in Section 3.3.4 that the hydrodynamic interaction combined with the backflow of the solvent slows down the movement of particles uniformly distributed in a closed container. Thus the sedimentation coefficient decreases with increasing concentration. This dependence is often expressed by a power series

$$s_2 = s_2^0 \left(1 - k_s c_2 + \cdots\right) \tag{3.4.31}$$

However, the inverse dependence shows better linearity:

$$1/s_2 = \left(1/s_2^0\right)\left(1 + k_s c_2 + \cdots\right) \tag{3.4.32}$$

The sedimentation coefficient at vanishing concentration, s_2^0, and the constant k_s have the same value in both these expressions. The latter constant is related to the ratio of hydrodynamic and dry volumes of the particle and consequently to the intrinsic viscosity (see next section).

Sedimentation coefficients are measured from the movement of the boundary that was selected as representative of the plateau region. Hence, the measured value corresponds to the concentration in the plateau region. However, this concentration decreases as sedimentation progresses. If the sedimentation coefficient depends on concentration, it increases with time, causing an upward curvature of the plot of ln r versus time.

This effect is partly counterbalanced by the pressure effect. As the pressure is increasing toward the bottom of the cell so is the viscosity of the solvent. Thus the frictional coefficient is increasing and the sedimentation coefficient is decreasing as the boundary moves toward the bottom. Consequently, when very precise values of sedimentation coefficients are required, the data must be extrapolated toward vanishing time of sedimentation to suppress the radial dilution effect and toward vanishing rotor speed to suppress the pressure viscosity effect.

Concentration also influences the shape of the boundary. As the boundary spreads because of the diffusion flow, a region of low concentration is left behind in the "tail" of the sedimenting boundary. Molecules within this region possess a larger sedimentation coefficient (corresponding to lower concentration) than the molecules in the plateau region. Therefore, they move faster and are "catching up" with the plateau region. Consequently, the boundary is narrower than it would be in an ideally dilute solution with the same values of sedimentation and diffusion coefficients. The effect is known as *self-sharpening* of the sedimenting boundary. When the concentration dependence of the sedimentation coefficient is large and the value of the diffusion coefficient is small (typical examples are solutions of nucleic acids), the self-sharpening effect totally dominates over diffusion and the boundary remains essentially a step boundary. This effect is called *hypersharpening* of the boundary.

What is the use of the sedimentation coefficient? The key relation is the second part of equation (3.4.24). If the molecular weight M_2 of the solute is known from an independent experiment, then it is easy to calculate the frictional coefficient f_2 and to obtain from it all types of information as we have seen in the previous section.

On the other hand, if we know the frictional coefficient from an independent measurement, we can calculate the molecular weight M_2 from equation (3.4.24). The frictional coefficient is usually obtained from the diffusion coefficient D_2^o by means of equation (3.4.10). Combining this relation with equation (3.4.24) yields the famous *Svedberg equation,* which served in the early days of macromolecular science for measurements of protein molecular weights and helped to prove their macromolecular nature.

$$s_2^o \big/ D_2^o = M_2(1 - \bar{v}_2 \rho)/RT \qquad (3.4.33)$$

Both the sedimentation and diffusion coefficients must, of course, be measured in the same solvent at the same temperature. They must be extrapolated toward vanishing concentration.

It is apparent that combined measurement of diffusion and sedimentation velocity is a very powerful method, which yields absolute values of molecular weights, frictional coefficients, and, through them, the sizes and shapes of macromolecules.

If the polymer belongs to a family of polymer homologs, the sedimentation coefficient itself can serve for evaluation of molecular weight. For such a polymer-homolog series, we may combine equations (3.4.19) and (3.4.33) and obtain

$$s_2^0 = K_D M_2^{1-a_D}(1 - \bar{v}_2\rho)/RT = K_s M_2^{a_s} \qquad (3.4.34)$$

Again, the constants K_s and a_s must be found by calibration, and all the measurements must be done using the same solvent and temperature. The exponent a_s should be equal to $(\frac{1}{2} - \varepsilon)$, where ε has values between zero and 0.1 and is the same quantity as was used in equation (3.4.18).

In biochemistry, the sedimentation coefficient is used for characterization and identification of proteins, RNA, ribosomes, etc. However, different researchers measure their samples using different buffers and different temperatures. These factors obviously influence the measured value of s_2^0. Changing the buffer density ρ changes the buoyancy term, $(1 - \bar{v}_2\rho)$; changing buffer viscosity changes the frictional coefficient. Normalization to some reference conditions is called for. Pure water at 20°C was selected as the reference solvent. The normalization procedure is based on an assumption that the partial specific volume $\bar{v}_2$ as well as the shape and size of the biochemical particle do not change with a change in buffer or temperature. Then the normalized sedimentation coefficient $s_{20,w}^0$ (subscripts refer to 20°C and water) is calculated as

$$s_{20,w}^0 = s_2^0 \frac{\eta}{\eta_{20,w}} \left(\frac{1 - \bar{v}_2\rho_{20,w}}{1 - \bar{v}_2\rho} \right) \qquad (3.4.35)$$

Here the values of viscosity η and density ρ without subscript refer to the buffer and temperature used for the velocity run; the subscripted values refer to water at 20°C.

3.4.2.2. Heterogeneous Solutes.

Although sedimentation velocity experiments provide a wealth of information when applied to solutions of monodisperse macromolecular samples, they are also extremely valuable for the study of heterogeneous samples. In the limit of ideally dilute solutions, each macromolecular component sediments independently of the others, and the sedimenting boundary is a superposition of the boundaries of all components. We need to distinguish two situations.

Heterogeneous biochemical samples are usually oligodisperse, that is, they consist of only a few components. When subjected to ultracentrifugation, they may form two or more separate boundaries, clearly revealing their heterogeneity. Even if the sedimentation coefficients of the components are close to each other (this prevents full separation of the schlieren peaks), the minor component may show up as a shoulder on the peak of the major component. Consequently, sedimentation velocity experiments are used as a favorite test of the homogeneity of biochemical samples.

However, this diagnostic test may fail under some circumstances. Mixtures of two proteins having identical or very close sedimentation coefficients will obviously be one of these cases. Other causes of failure are more insidious. Self-sharpening of the boundary and, especially, hypersharpening may suppress not only the spreading of the boundary under the influence of diffusion, but also the separation of multiple boundaries. Historically, one of these misdiagnosed tests considerably slowed down molecular genetics. Mammalian DNAs exhibited in sedimentation velocity experiments extremely narrow schlieren peaks, which were misinterpreted as an indication of a very narrow distribution of their sedimentation coefficients and hence their molecular weights. Actually, the sharpness of the peak was a result of hypersharpening. With the advent of UV absorption optics, which allowed lowering of the DNA concentration by orders of magnitude, it was firmly established that mammalian DNA is actually extremely polydisperse.

A technical problem may also mask sample heterogeneity. We have seen in Section 3.2.2.1 that thermal and mechanical disturbances may trigger convection in the sedimentation cell. Density gradients (which accompany concentration gradients of any component) tend to stabilize the run against convection. Hence, the regions in the cell where no gradients exist are most prone to convection: the plateau region and the region behind the boundary. The convective flow thoroughly mixes the affected regions; it may completely obliterate a small boundary belonging to an impurity. The effect is most dangerous with very dilute buffers and with single-component solvents (used mainly in polymer studies). These runs are not stabilized by gradients of the solvent components (e.g., salts). In any case, a perfectly flat concentration profile, either in front of or behind the boundary, should always raise the suspicion of convection.

Unlike biochemical materials, heterogeneous polymers possess a continuous distribution of molecular weights. Their sedimenting boundary reflects this distribution; it is, of course, strongly shaped by diffusion and concentration effects. Methods have been designed to overcome these effects. The shape of the boundary is first converted into the distribution of sedimentation coefficients, which is then extrapolated to vanishing concentration (to suppress the self-sharpening effect) and to infinite time. (The increase in the width of the boundary due to the heterogeneity of the sedimentation coefficients is proportional to time; the increase due to diffusion is proportional to the square root of time. Hence, at extremely long times, the effect of diffusion is negligible compared with the effect of sedimentation.) These extrapolation procedures are claimed to be the best method for characterization of samples with a very narrow distribution of molecular weights (the Poisson-type distribution). In less demanding applications, these techniques are useful when the effects of diffusion and concentration are both small, that is, for large, compact particles. Indeed, the sedimentation velocity method, appropriately modified, provides a very good description of the size distribution of latexes and suspensions.

3.4.2.3. Archibald Method. The *Archibald method* of sedimentation analysis is designed for obtaining the absolute molecular weight of macromolecular solutes from their sedimentation behavior using sedimentation times much shorter than those needed to achieve sedimentation equilibrium. The method is based on the obvious fact that there is no flow of the solute across the boundaries of the system, that is, across the meniscus and bottom. Clearly, the velocity v_2 must also vanish at these boundaries. Hence, the right-hand side of equation (3.4.23) is equal to zero and may be rearranged to

$$M_2(1 - \bar{v}_2\rho)\omega^2 r = \left(\frac{\partial \mu_2}{\partial c_2}\right)_{P,T} \frac{dc_2}{dr} \qquad (3.4.36)$$

This equation has exactly the same form as equation (3.2.43) describing sedimentation equilibrium. Of course, the present relation (3.4.36) is valid only at the meniscus and the bottom, whereas equation (3.2.43) is valid throughout the cell. Nevertheless, we may use the same computational approaches as before to obtain the molecular weight, for example, by employing equation (3.2.46). Of course, we need to evaluate dc_2/dr and c_2 at the boundaries; this must be achieved by extrapolating the data from the vicinity of these boundaries. This procedure is not very accurate; because of it, this theoretically very attractive method is used only sparingly. We should recall that for polydisperse polymers equation (3.2.46) yields the local values of $\overline{M}_W$, which in the present case change with time because of the redistribution of the polymer species within the cell. Thus another extrapolation toward zero time is needed to obtain a value pertaining to the original solution. This extrapolated value should be the same for both the meniscus and the bottom. Finally, the Archibald experiment must be repeated for several concentrations, and the apparent molecular weights must be extrapolated to vanishing concentration to eliminate the effect of nonideality.

Experimentally, the Archibald method requires rotor speeds somewhere between the high speeds needed for velocity runs and the low speeds appropriate for equilibrium runs.

3.4.3. Viscometry

Viscometry has a very special place among the methods for the study of macromolecules. Experimentally, it is much simpler than most other methods, yet it yields data that are rather precise and well-reproducible. In industry, measurement of viscosity serves as one of the fastest methods for monitoring molecular weights, conversions, and so on. On the theoretical side, viscometric data per se provide information that may be quite difficult to interpret. However, when they are combined with other, independently measured data (molecular weights, virial coefficients, frictional coefficients) they yield extensive, detailed information about thermodynamic properties of polymer solutions and the structure of polymers. After calibration by some absolute method, they are used for the fast and very precise measurement of molecular weights.

In the following sections, experimental techniques, evaluation of intrinsic viscosity (the key quantity in viscometry), interpretation of intrinsic viscosities, and phenomena related to the non-Newtonian character of some polymer solutions will be described.

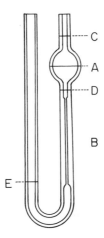

Figure 3.33. The Ostwald viscometer.

3.4.3.1. Viscometers. The simplest, most precise, and most popular instrument for measuring viscosity is a capillary viscometer. Its measuring principle is based on the Poiseuille equation relating the time t necessary for volume V of a liquid to flow through a capillary with radius r and length L to the viscosity of the liquid η and the pressure difference ΔP between the ends of the capillary that is causing the flow:

$$t = \frac{8LV}{\pi r^4 \Delta P \eta} \tag{3.4.37}$$

The *Ostwald viscometer* (Figure 3.33) has the simplest design. A well-defined amount of liquid is introduced into the viscometer and pushed into bulb A. Then, under the action of gravity, the liquid flows from the bulb through the capillary B. The flow time is measured between the moments when the liquid meniscus passes marks C and D. The average pressure difference ΔP is given by an average hydrostatic head h between liquid menisci in the bulb and in the other arm of the viscometer at E; h obviously depends on the amount of liquid in the viscometer. ΔP is then given as

$$\Delta P = hg\rho \tag{3.4.38}$$

where ρ is the density of the liquid and g is gravity acceleration. Equations (3.4.37) and (3.4.38) combine to yield

$$t = (8LV/\pi r^4 hg)(\eta/\rho) \equiv K(\eta/\rho) \tag{3.4.39}$$

Thus the time t is proportional to the quantity η/ρ known as *kinematic viscosity*. The instrument constant K follows from the dimensions of the viscometer; it is usually obtained by calibration.

In the above derivation we neglected the correction for kinetic energy frequently stressed in the older literature. This is related to the kinetic energy possessed by the jet of liquid flowing out of the capillary. It is completely negligible for properly designed viscometers.

For most viscometric work, the absolute value of viscosity is not important, only its relative value $\eta_r \equiv \eta/\eta_0$. Moreover, for dilute solutions, the densities of the solvent and of the solution are very close to each other, and their difference is routinely neglected. Then relative viscosity is evaluated using an extremely simple relation,

$$\eta_r \equiv \eta/\eta_0 = t/t_0 \tag{3.4.40}$$

where the subscript zero refers to pure solvent.

The amount of liquid in the Ostwald viscometer must always be the same. Hence, it is not possible to dilute the polymer solution inside the viscometer whenever concentration dependence of viscosity is needed. The *Ubbelohde viscometer* (Fig. 3.34) overcomes this inconvenience. Its capillary ends in a bulb B, which is equipped with side arm A. The side arm is closed at the top when the liquid is pushed up into the measuring bulb C; it is open during the measurement. The liquid outflowing from the capillary forms an inverted meniscus inside B. The hydrostatic head is the liquid column between bulb C and the outlet of the capillary; it is independent of the amount of liquid in the viscometer. Hence, it is possible to add more solvent after the first solution has been measured and to prepare and measure the whole concentration scale without cleaning the viscometer.

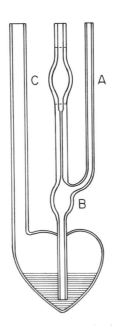

Figure 3.34. The Ubbelohde viscometer.

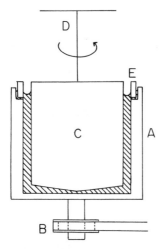

Figure 3.35. The Couette rotary viscometer.

The precision of capillary viscometry is remarkable. The typical efflux time of 100 s can be measured using a stopwatch with a routine precision of ±0.1 s. In modern viscometers the stopwatch is replaced by a photocell watching the capillary and connected to an electronic timer; this leads to a precision of about ±0.005 s. (Temperature must he controlled within ±0.01°C to suppress fluctuations of viscosity with temperature.)

The velocity gradient of a liquid varies inside the capillary from the highest value at the wall to zero in the center. Thus, whenever a measurement of the dependence of viscosity on the velocity gradient is required, rotary viscometers with much more homogeneous flow field are preferred.

For low- and medium-viscosity liquids, the *Couette rotary viscometer* with two concentric cylinders is the most frequently used design (Fig. 3.35). One of the cylinders (usually the outer one) is rotating, the other one is stationary, and the liquid is placed in the gap between them. When the gap is narrow, the velocity distribution within the liquid is very similar to that in simple shear flow. The flowing liquid exerts tangential stresses on the walls of the stationary cylinder; they add up to a torque, which tries to rotate it. The stationary cylinder is attached to a torsional fiber D that counterbalances the viscous torque and serves as a measuring element. The velocity gradient in rotary viscometers is calculated easily from the width of the gap and the rotor speed. The measured torque is proportional to the shearing stress on the surface of the cylinder, that is, to the product of the velocity gradient and the liquid viscosity.

For macromolecular samples with extremely high molecular weight (for example, DNA), viscosity varies with the velocity gradient down to very low values of the latter. At low velocities, the torque acting on the cylinder becomes extremely small and difficult to measure in the Couette arrangement.

The *Zimm–Crothers viscometer* (Fig. 3.36) is designed for the study of such solutions. Its stator S is a thermostated well filled with the measured liquid L. Rotor R,

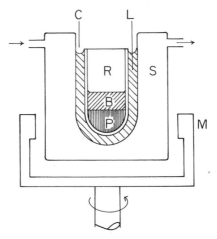

Figure 3.36. The Zimm-Crothers rotary viscometer.

which has embedded at its bottom a piece of paramagnetic material P (steel), is immersed in the liquid. Just enough balancing liquid B is added inside the rotor to make it float in the liquid. A properly installed rotor hangs on the meniscus and is centered by capillary forces. A magnet rotates fast around the cylinders, causing a constant torque on the paramagnetic rotor. The rotor will rotate with such a speed that the viscous torque will exactly balance the magnetic torque. The time needed for one rotation is proportional to the viscosity of the liquid; it is measured either by stopwatch or electronically.

3.4.3.2. Intrinsic Viscosity.   In Sections 3.3.3 and 3.3.5 we saw that an increase in the viscosity of macromolecular solutions is most easily interpreted for hydrodynamically ideal solutions. It is therefore desirable to separate the information about macromolecules from such relatively unimportant quantities as solvent viscosity and polymer concentration. This is accomplished by evaluation of *intrinsic viscosity*.

Let us first define the basic viscometric quantities: relative viscosity $\eta_r \equiv \eta/\eta_0$ [equation (3.4.40)], specific viscosity η_{sp}, reduced viscosity η_{red}, and intrinsic viscosity $[\eta]$.

$$\eta_{sp} \equiv \eta_r - 1 \equiv (\eta - \eta_0)/\eta_0 \tag{3.4.41}$$

$$\eta_{red} \equiv \eta_{sp}/c \equiv (\eta_r - 1)/c \equiv (\eta - \eta_0)/\eta_0 c \tag{3.4.42}$$

$$[\eta] \equiv \lim_{c \to 0} \eta_{red} \equiv \lim_{c \to 0} \frac{\eta_{sp}}{c} \equiv \lim_{c \to 0} \frac{\eta - \eta_0}{\eta_0 c} \tag{3.4.43}$$

Relative and specific viscosities are dimensionless, whereas the dimension of reduced and intrinsic viscosities is that of inverse concentration. In the older literature, they were evaluated in deciliters per gram because gram per deciliter was a favorite concentration unit often incorrectly called percent. Modern literature expresses $[\eta]$ in milliliters per gram.

Intrinsic viscosity is a measure of particle shape and of the ratio of its hydrodynamic volume to its dry volume. Biochemists, who deal mainly with compact particles, stress the shape aspect. The relevant relation is equation (3.3.13) relating viscosity increase to volume fraction ϕ of the solution occupied by macromolecules. A short calculation yields, for compact particles,

$$[\eta] = \nu\phi/c = \nu\bar{v} \tag{3.4.44}$$

where we have used the relation between volume fraction, concentration, and specific volume: $\phi = c\bar{v}$. The ν is the previously defined shape factor (see Fig. 3.26) and its value for spheres is 2.5 when the concentrations are expressed in grams per milliliter and 0.025 when units of grams per deciliter are used. Thus, for compact particles, measurement of intrinsic viscosity and specific volume yields ν and through it the asymmetry of the particle.

Polymer chemists prefer to think about macromolecules as hydrodynamically impermeable particles of low asymmetry. This approach is modeled by particles whose hydrodynamic volume V_h is larger than their dry volume V_d by a swelling factor F_{sw}. The hydrodynamically active volume fraction of the polymer in solution is then $\phi_h = \phi F_{sw}$, and intrinsic viscosity reads

$$[\eta] = \nu\phi_h/c = \nu(V_h/V_d)\phi/c = \nu F_{sw}\bar{v} \tag{3.4.45}$$

In this case, intrinsic viscosity yields the swelling factor F_{sw} if ν can be reasonably estimated. For coiled macromolecules, the swelling factor increases with increasing molecular weight. Thus high intrinsic viscosity usually means high molecular weight.

In the preceding paragraphs we have described most of the information that is obtainable from measurement of intrinsic viscosity alone. This information is certainly valuable, but it is meager. The situation changes radically when viscosity measurements are combined with other techniques.

3.4.3.3. Molecular Weight and Coil Dimensions.

The most common use of viscosity measurements is for the determination of the molecular weights of linear polymers. This, of course, requires calibration by an independent absolute method for measurement of the latter. The relation between intrinsic viscosity and molecular weight of polymer-homolog samples in a given solvent at a given temperature is described by the *Mark–Houwink–Sakurada* relation, which holds for almost all polymer-solvent systems with remarkable accuracy.

$$[\eta] = K M^a \tag{3.4.46}$$

We can derive this equation from equation (3.4.45), realizing that the hydrodynamic volume of an impermeable coil is proportional to the cube of its end-to-end distance and through it to $M^{3(1/2+\varepsilon)}$. The dry volume is equal to $M\bar{v}/N_{Av}$. Hence,

$$[\eta] = v\,(V_h/V_d)\,\bar{v} \sim M^{1/2+3\varepsilon} \tag{3.4.47}$$

Here, ε is the exponent from relation (1.3.10), which we have met repeatedly. Thus the viscometric exponent a should have the value 0.5 for pseudo-ideal theta solvents and 0.8 for very good solvents. Indeed, for most polymer-solvent systems, the experimental values lie between these limits. The only exceptions are solutions of cellulose derivatives and other polymers with a very stiff backbone; they often exhibit values of a up to 1.0 or even higher. This high value is usually rationalized by invoking partial hydrodynamic permeability of the coil and/or a too-small number of statistical segments, meaning that the limiting behavior of coils was not achieved, and/or high asymmetry of the hydrodynamically equivalent particle.

Intrinsic viscosity is frequently used for determination of characteristic dimensions of macromolecular coils. Let us return once more to equation (3.4.45) and again use the proportionality between V_h and the cube of the end-to-end distance $(\overline{h^2})^{3/2}$. We get

$$[\eta] = \Phi(\overline{h^2})^{3/2}/M \tag{3.4.48}$$

If it is assumed that all coils have the same value of the shape factor v and that the proportionality constant between V_h and $(\overline{h^2})^{3/2}$ is the same for all impermeable coils, then the *Flory constant* Φ should have a universal value for all linear polymers in all solvents. In reality, Φ decreases from its theoretical value 2.87×10^{23} valid for unperturbed Gaussian coils (2.87×10^{21} if $[\eta]$ is expressed in deciliters per gram) to about 2.1×10^{23} for polymers in good solvents. This decrease is caused by inhomogeneous expansion of polymer coils in good solvents. The periphery of the coil expands less than the more densely packed interior, and the ratio $V_h/(\overline{h^2})^{3/2}$ decreases slightly. Often an intermediate value 2.5×10^{23} ($[\eta]$ in milliliters per gram) is quoted as Φ for all polymer-solvent systems. Of course, the universality of Φ is expected to fail for permeable and/or rigid coils.

When the radius of gyration, r_G, is used for characterizing the size of the macromolecules, equation (3.4.48) is usually modified and another Flory's constant Φ' is defined as

$$[\eta] = (6^{3/2}\Phi)r_G^3/M \equiv \Phi' r_G^3/M \tag{3.4.49}$$

In Section 3.4.2.1 we derived the Svedberg equation (3.4.33) (a recipe for calculating absolute molecular weights) by combining diffusion and sedimentation coefficients. Similarly, we can calculate molecular weight from intrinsic viscosity and sedimentation coefficient. Combining equations (3.4.17) and (3.4.24) we obtain for the sedimentation coefficient s° a new expression,

$$s^\circ = M\,(1-\bar{v}\rho)/N_{Av}P\eta_0(\overline{h^2})^{1/2} \tag{3.4.50}$$

Eliminating $\overline{h^2}$ from equations (3.4.48) and (3.4.50) leads to

$$\beta \equiv s^0 \, [\eta]^{1/3} \, M^{-2/3} \eta_0 \, (1 - \bar{v}\rho)^{-1} \, N_{Av} = \Phi^{1/3} P^{-1} \qquad (3.4.51)$$

It turns out that the newly defined parameter β is a remarkably constant quantity. Its best value is about 1.27×10^7 (2.73×10^6 if $[\eta]$ is expressed in dL/g), and it is valid for linear polymers in good and poor solvents as well as for branched polymers and is even applicable for compact spherical and asymmetric particles. This insensitivity to the form of the particle is due to the fact that the hydrodynamic radii applicable for translational motion and for shear flow (rotary motion), although not equal to each other, depend on the form and/or expansion of the particle in a very similar way. Consequently, equation (3.4.51) can be used to estimate molecular weight from sedimentation and viscometric measurements with a high degree of confidence. Nevertheless, it is not an absolute method; the two hydrodynamic radii are not strictly proportional to each other.

3.4.3.4. Unperturbed Dimensions, Thermodynamic Parameters.

Measurement of intrinsic viscosity gives us access to an important characteristic of macromolecular chains—their unperturbed dimensions, which are usually expressed as h_0^2/M, where the subscript zero refers to Gaussian coils, that is, to theta solutions. In Section 1.3.1.1 we saw that this ratio is independent of molecular weight and that it is a measure of the coil stiffness. For example equation (1.3.7), which is valid for vinyl-type polymers, may be modified to

$$\frac{\overline{h_0^2}}{M} = \frac{N}{M} a^2 \frac{1 - \cos\theta}{1 + \cos\theta} \left(\frac{1 + \eta}{1 - \eta} \right) \qquad (3.4.52)$$

where N is the number of links in the polymer backbone, each of them with valence length a; θ is the valence angle; and η is the coefficient of steric hindrance. The ratio N/M is independent of molecular weight; it may be replaced by N_0/M_0 where N_0 is the number of bonds in a monomer unit (typically two) and M_0 is the molecular weight of the monomer unit. Thus steric hindrance around the bonds in the backbone can easily be evaluated from h_0^2/M; all other quantities in equation (3.4.52) follow from known principles of organic chemistry.

Unperturbed dimensions are obtained directly when intrinsic viscosity is measured in theta solvents. In this case, equation (3.4.46) reads

$$[\eta]_\theta = K_\theta M^{1/2} \qquad (3.4.53)$$

We should recall that equation (3.4.53) was derived assuming that the polymer coils are hydrodynamically impermeable. The fact that it is indeed satisfied for theta solvents (i.e., for solvents yielding a vanishing second virial coefficient) is a strong proof of the impermeability of polymer coils.

For theta solvents, equation (3.4.48) is modified to

$$[\eta]_\theta = \Phi_0 \left(\overline{h_0^2}/M \right)^{3/2} M^{1/2} \tag{3.4.54}$$

Combining relations (3.4.53) and (3.4.54) yields

$$\left(\overline{h_0^2}/M \right)^{3/2} = K_\theta/\Phi_0 \tag{3.4.55}$$

In these relations subscripts theta and zero refer to conditions of unperturbed coils. Once it has been established that a particular solvent is a theta solvent (for example, by finding that the second virial coefficient is equal to zero), K_θ and $\overline{h_0^2}/M$ can be obtained by measuring intrinsic viscosity $[\eta]_\theta$ and molecular weight on just one polymer sample.

For many polymers it is not possible to find a theta solvent. In such a case, an attempt is often made to extract the value of $\overline{h_0^2}/M$ from measurements of intrinsic viscosity in good solvents. Among a large number of methods designed for this purpose the Stockmayer–Fixman method is the most popular. It is based on the evaluation of the expansion coefficient α [see equation (3.1.59)] and on its expression through the parameter z [equation (3.1.56)]. Intrinsic viscosity is a measure of the hydrodynamic volume of the polymer coil. When it is measured for the same polymer sample in two different solvents, the ratio of viscosities is equal to the ratio of hydrodynamic volumes. If one of the solvents is a theta solvent, we have

$$[\eta]/[\eta]_\theta = \alpha^3 \tag{3.4.56}$$

Combining equations (3.4.53) and (3.4.56) then yields

$$[\eta] = K_\theta M^{1/2} \alpha^3 \tag{3.4.57}$$

The relation between α^3 and z is obtained from equation (3.1.60); it reads

$$\alpha^3 = 1 + 1.55z + \cdots \tag{3.4.58}$$

It turns out that the higher terms in this relation largely compensate for each other and that the linear term is sufficient for description of the coil expansion up to moderately high values of z. We can now combine equations (3.4.55)–(3.4.58) with equation (3.1.56). The result is

$$[\eta] = K_\theta M^{1/2} + 0.51\Phi_0 \left(2v^2 \psi \frac{1 - \theta/T}{V_1 N_{Av}} \right) M \tag{3.4.59}$$

This relation calls for plotting experimental data as $[\eta]/M^{1/2}$ versus $M^{1/2}$ (Fig. 3.37). The intercept of this plot is K_θ and from the slope, the thermodynamic

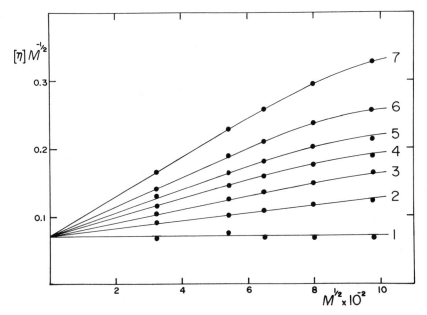

Figure 3.37. The Stockmayer-Fixman $[\eta]/M^{1/2}$ versus $M^{1/2}$ plot for polystyrene at 98.4°C in mixtures of a theta solvent, 3-methylcyclohexanol (a) and a good solvent tetralin (b). The volume fraction of (b) was: (1) 0.000; (2) 0.083; (3) 0.199; (4) 0.290; (5) 0.398; (6) 0.598; (7) 1.000. (Courtesy of Dr. A.-A. A. Abdel-Azim.)

interaction expression $\psi(1 - \theta/T)$ is easily obtained. If the measurement and the analysis are repeated at several temperatures, both ψ and θ can be evaluated.

Thus an analysis of viscometric data may provide both the values of polymer-solvent interaction parameters and the information about the structure of the polymer chain. Frequently, these data can be obtained without much help from other experimental methods. For many polymers, the Mark–Houwink–Sakurada constants K and are known for at least one solvent. It is then possible to measure the intrinsic viscosities of a set of polymer fractions in this solvent and calculate the molecular weights. After that, we measure the intrinsic viscosities of the same samples in other solvents and/or at different temperatures. Using these data we can derive the thermodynamic parameters of the polymer as a function of the temperature, the composition of a mixed solvent, etc. Consequently, measurement of intrinsic viscosity is probably the most popular method for obtaining thermodynamic data for polymer solutions. The main disadvantage of the method is the rather indirect evaluation of the pertinent quantities. On the theoretical side, the relation between α^3 and z is still subject to debate. Experimentally, the plot of $[\eta]/M^{1/2}$ against $M^{1/2}$ is often curved, complicating the analysis. A number of other evaluation procedures have been offered, but none of them is clearly superior of the others.

3.4.3.5. Branched Chains.   The above techniques for viscometric measurement of molecular weight, unperturbed dimensions, and thermodynamic interaction parameters are meaningful only when the branching and polydispersity of the polymer samples employed are carefully controlled. A branched macromolecule generally has a smaller radius of gyration and hydrodynamic radius than a linear molecule of the same molecular weight and composition. The relative decrease in these dimensions depends on the degree of branching (number of branches) as well as on the type of branching (short or long branches, random or regular, etc.). Moreover, the changes in radius of gyration and hydrodynamic radius are not proportional to each other; a branched coil exhibits a more uniform segment density than a linear one.

Branching may harm viscometric analysis in many ways. When the Mark–Houwink–Sakurada parameters are measured for a particular polymer-solvent system, it is imperative that branching be minimized. As we have seen in Section 2.2.4.3, branching is caused mainly by chain transfer by the polymer itself. In the case of radical polymerization, the transfer constant depends only on the nature of the polymer and cannot be modified. Thus limiting the conversion can reduce the branching, but it cannot suppress it totally. The molecules with a large molecular weight have a statistically higher probability of being branched; hence the intrinsic viscosity of high-molecular-weight samples or fractions is reduced proportionally more than for smaller molecules. Thus the dependence of $[\eta]$ on M may be distorted by branching, and lower values of the exponent a may be obtained.

Branching is harmful also when measurement of viscosity is employed for obtaining molecular weight. More compact branched molecules exhibit lower intrinsic viscosity, and their molecular weight will be underestimated. Similarly, the viscosity of branched molecules is less sensitive to changes in solvent and temperature. False results may be obtained when such polymers are employed for a study of thermodynamic parameters. On the other hand, when the molecular weight of a branched polymer is known from an independent experiment, the degree of branching can be estimated from its viscosity.

3.4.3.6. Polydisperse Polymers.   What kind of average of molecular weight is obtained from intrinsic viscosity? This question is more complex than it seems. The Mark–Houwink–Sakurada constants must be found by calibration using polymer samples that are always polydisperse; different absolute calibration methods yield, of course, different molecular weight averages. Thus, strictly speaking, the calibrating dependence must be established using samples that all have a similar molecular weight distribution (MWD). The method can be used only for unknown samples having the same type of MWD; it then yields the same type of average molecular weight as was used for calibrating samples.

This approach is too restrictive, and attempts are routinely made to circumvent it. Thus the polymers used for calibration are either carefully prepared narrow fractions, or, still better, they are polymers with a narrow MWD prepared using living polymerization techniques (Section 2.3.1.4). The resulting calibration is then assumed to be valid for monodisperse samples.

In ideally dilute solutions, the viscosity increases caused by all components of a polydisperse solute are additive. It follows that the measured $[\eta]$ is given as

$$[\eta] = \frac{\sum_i w_i [\eta]_i}{\sum_i w_i} \tag{3.4.60}$$

where w_i is the mass of the ith component of the solute. Substituting $K M_i^a$ for $[\eta]_i$ and $K \overline{M}_\eta^a$ for $[\eta]$, we obtain

$$\overline{M}_\eta = \left(\frac{\sum_i w_i M_i^a}{\sum_i w_i} \right)^{1/a} \tag{3.4.61}$$

where $\overline{M}_\eta$ is the *viscosity-average molecular weight*. For solvents with $a = 1$ (cellulose derivatives, for example), equation (3.4.61) becomes identical with equation (1.6.7) and $\overline{M}_\eta$ becomes equal to the weight-average molecular weight $\overline{M}_w$. For typical values of $a = 0.5$–0.8, the following inequality holds:

$$\overline{M}_n < \overline{M}_\eta < \overline{M}_w \tag{3.4.62}$$

However, in all cases $\overline{M}_\eta$ is much closer to $\overline{M}_w$ than to $\overline{M}_n$. Thus, for most practical purposes, calibration of the Mark–Houwink–Sakurada equation should be and usually is done using $\overline{M}_w$ values.

3.4.3.7. Concentration Dependence of Viscosity.

The viscosity of hydrodynamically ideal solutions increases linearly with concentration of the solute. However, this linearity exists only over a very narrow range of concentrations; the higher the intrinsic viscosity, the narrower the range. This behavior follows from equation (3.3.13), which we may write in terms of the hydrodynamically active volume fraction of the polymer, ϕ_h, as we did in equation (3.4.45)

$$\eta = \eta_0 \left(1 + \nu \phi_h + \nu_1 \phi_h^2 + \cdots \right) \tag{3.4.63}$$

The quadratic term arises from intermolecular hydrodynamic interactions, and its coefficient ν_1 should depend only on the shape of the hydrodynamically equivalent particle (as ν did). According to equation (3.4.45), $\phi_h = [\eta] c / \nu$. Substitution into equation (3.4.63) yields

$$\eta = \eta_0 \left\{ 1 + [\eta]c + (\nu_1/\nu^2)[\eta]^2 c^2 + \cdots \right. \tag{3.4.64}$$

and, further,

$$\eta_{sp}/c = [\eta] + k_H [\eta]^2 c + \cdots \tag{3.4.65}$$

In the last expression, we have replaced the combined constants ν_1/ν^2 by the *Huggins constant* k_H. According to the above derivation, k_H should have the same value for all polymers in all solvents (as long as the effective shape of the coil is the same); theoretical calculations yield for the Huggins constant a value of about 0.7.

In real experiments, theta solutions exhibit Huggins constants close to this theoretical value. However, good solvents usually yield values much lower: 0.2–0.4. For a given polymer-homolog series in a given solvent, k_H is either independent of molecular weight or decreases very slowly with increasing values of molecular weight. The reason is found in polymer thermodynamics. As the polymer concentration increases, individual polymer molecules experience surroundings containing more and more polymer segments belonging to other macromolecules. This means that the thermodynamic quality of this "solvent" is getting worse; the coil expansion is diminishing, and so is the hydrodynamically active volume. The net effect is a decrease of the Huggins constant.

The opposite effect is sometimes observed for polymers in thermodynamically poor nonpolar solvents. This effect is usually connected with the presence of strongly polar groups on the polymer (for example, from fragments of the initiator). These polar groups may aggregate in the nonpolar medium; the aggregated polymer clusters will then exhibit higher viscosity. The net effect is a high Huggins constant.

Equation (3.4.65) suggests plotting the experimental values of the reduced viscosity η_{sp}/c versus concentration. The intercept is $[\eta]$, and k_H is easily evaluated from the initial slope. Another evaluation technique employs the plot of $\ln \eta_r/c$ versus concentration. It can easily be shown that the intercept of this plot must also be equal to $[\eta]$; the initial slope is $\left(\frac{1}{2} - k_H\right)[\eta]^2$. Because both plots are slightly curved, it is useful to plot both η_{sp}/c and $\ln \eta_r/c$ versus concentration in the same graph and extrapolate them to a common intercept (Fig. 3.38). To avoid excessive curvature of the plots, the measurements should be done in such a range of concentrations that the highest value of η_r is less than 2.0 or, even better, less than 1.5 (if the precision of the measurement allows this).

3.4.3.8. Non-Newtonian Viscosity.

The viscosity of many polymer solutions depends on the velocity gradient G. This non-Newtonian behavior is more pronounced for more concentrated solutions. As the molecular weight of the polymer increases, non-Newtonian viscosity manifests itself at lower velocity gradients. In routine viscometry, it should be expected to play a role for polymers with molecular weights higher than about 1 or 2 million. To obtain the correct value of intrinsic viscosity for such polymers, we must extrapolate not only to vanishing concentration but also to vanishing G. This nuisance with extrapolation is compensated for by the additional information about the structure of the polymer that can be extracted from the non-Newtonian behavior. Moreover, knowledge of this information is very important in practice—in designing equipment for handling polymer solutions, for example.

Non-Newtonian behavior is caused by orientation and deformation of macromolecules in the flow field as described in Section 3.3.7. Orientation aligns particles along the flow lines. They interfere less with the flow of liquid, and the viscosity

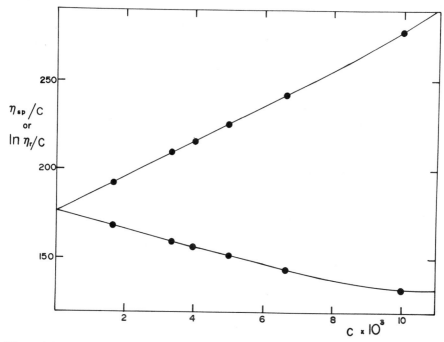

Figure 3.38. Concentration dependence for a sample of polystyrene ($\overline{M}_w = 619,000$) in benzene at 20°C. Upper curve, values of η_{sp}/c; and lower curve, values of $\ln \eta_r/c$.

decreases with increasing orientation. On the other hand, deformation—which, in the context of simple shear flow, is synonymous with elongation—causes the particles to interact more with the liquid, and viscosity increases. Thus orientation and deformation partially compensate for each other, but the orientation effect prevails. With increasing velocity gradient, the intrinsic viscosity decreases to about 50–80% of its initial value; at higher G, it levels off. The decrease is more pronounced for stiff-chain polymers, such as cellulose derivatives.

Interaction of molecular coils with a flow field is customarily expressed by means of the generalized velocity gradient parameter β_0, defined as

$$\beta_0 \equiv \eta_0[\eta]_0 GM/RT \tag{3.4.66}$$

where $[\eta]_0$ and M are intrinsic viscosity at vanishing G and molecular weight of the polymer, respectively. Physically, the dimensionless parameter β_0 is equal to the ratio of the energy dissipated by a single macromolecule during one rotation in the flow field to the thermal energy kT.

Theoretically, the dependence of intrinsic viscosity at finite G, $[\eta]_G$, on parameter β_0 is described as

$$[\eta]_G/[\eta]_0 = 1 - A\beta_0^2 + \cdots \tag{3.4.67}$$

Here, the value of A varies only slightly with the stiffness of the polymer chain. Thus the dependence of $[\eta]_G$ on G should have a horizontal initial tangent. In practice, the initial curvature is lost in experimental error and the dependence is linearly extrapolated toward vanishing β_0.

Rigid particles in a flow field become oriented but not deformed. They are usually modeled by ellipsoids, for which the theory has been developed fully. It turns out that the viscosity of suspensions of spheres is Newtonian, because spheres cannot be oriented. For ellipsoids, the decrease in viscosity with the velocity gradient becomes more pronounced as their asymmetry increases.

The non-Newtonian flow of more concentrated solutions of polymer coils is related to their entanglement. Entanglement of polymer chains is the main reason the viscosity of polymer solutions increases so steeply with concentration within the semidilute region of concentrations. At higher flow rates, the molecules disentangle progressively and form more compact units exhibiting less hydrodynamic interaction. Thus the decrease in viscosity with increasing G gets more intense as the polymer concentration increases. In concentrated solutions and melts, the viscosity may fall by several orders of magnitude when the velocity gradient is increased sufficiently.

3.4.4. Flow Birefringence

In a flowing liquid, suspended particles are kinematically oriented. A system in which particles are oriented is macroscopically anisotropic. This anisotropy manifests itself in a number of phenomena. Among them birefringence (anisotropy of refractive index) is the most sensitive to small degrees of orientation. Thus flow birefringence is the method of choice when orientation and deformation of molecules in dilute solutions is to be studied.

3.4.4.1. Optical Properties of Dielectrics.   External electric fields polarize dielectrics, that is, they induce macroscopic dipoles in them. There are two mechanisms of this polarization.

1. Molecules carrying a permanent dipole are oriented by the field and thus contribute to the macroscopic dipole. However, orientation of molecules is a relatively slow process that cannot follow the rapidly oscillating field of a light wave Thus permanent dipoles do not need to be considered in the context of optical phenomena.

2. In all molecules, there is a distribution of positive and negative electric charges (atomic nuclei and electrons). When an electric field $\mathbf{E}$ acts on the molecule, these charges are mutually slightly displaced; the molecules are polarized and acquire an induced dipole μ. This displacement of charges is a fast process, which is able to follow optical frequencies. The magnitude of the induced dipole is proportional to the polarizing field,

$$\mu = \alpha \mathbf{E} \qquad (3.4.68)$$

where the polarizability α describes the ease with which the electrons are displaced with respect to the molecular skeleton and depends on the chemical structure of the molecules.

Molecular polarizabilities are closely related to the *dielectric constant* ε (or better to the optical part of it) and to the refractive index $\tilde{n}$. For dilute gases the relation reads

$$\varepsilon = \tilde{n}^2 = 1 + 4\pi \sum_i \frac{\alpha_i}{V} \qquad (3.4.69)$$

For denser systems, the relation between ε and the polarizabilities becomes more complicated. The relation most frequently used in the literature (but not necessarily the most successful) is the one of Clausius and Mossotti:

$$\frac{\varepsilon - 1}{\varepsilon + 2} = \frac{4\pi}{3} \sum_i \frac{\alpha_i}{V} \qquad (3.4.70)$$

In these relations, the summations extend over all molecules in volume V.

So far, we have not considered the orientation of molecules with respect to the polarizing field. In most molecules, electrons are more easily displaced along some directions than along others. For example, in aromatic molecules, electrons shift more easily in the aromatic plane than perpendicularly to it. Consequently, the magnitude of a dipole induced by a given field depends on the orientation of the field with respect to the particle; the induced dipole generally has a different direction than the exciting field. In other words, molecular polarizability is a tensor of the second order; we will call it a *polarizability tensor*.

It is now obvious that the summation of molecular polarizabilities in equations (3.4.69) and (3.4.70) must be done in a tensorial manner. As long as the orientation of molecules in a liquid is random, the resulting polarizability tensor must be isotropic, even if individual molecules are anisotropic. However, once the molecules are preferentially oriented in some direction, the macroscopic polarizability tensor becomes anisotropic. The dielectric constant should be treated in this case as a *dielectric tensor* and the refractive index as a *refractive tensor*.

As we have already seen in Section 3.3.2, symmetrical tensors of the second order are conveniently described by their principal values. In the present case, these values refer to three mutually perpendicular directions, for which the direction of the excited dipole will be the same as the direction of the exciting field. Thus polarizability will have principal values $\alpha_1, \alpha_2, \alpha_3$; dielectric tensor $\varepsilon_1, \varepsilon_2, \varepsilon_3$; and refractive tensor $\tilde{n}_1, \tilde{n}_2, \tilde{n}_3$. The difference $\Delta n \equiv \tilde{n}_1 - \tilde{n}_2$ is measured in many birefringence experiments; it is called the *magnitude of birefringence*. It is also important to specify the directions of the principal axes of the refractive tensor with respect to instrumental coordinates.

The anisotropy of polarizability, dielectric and refractive tensors of bulk materials depends on the degree of orientation of the molecules of the system and on their molecular anisotropy. For example, in some crystals, all molecules are oriented in the same direction. Such crystals may exhibit birefringence of the order of $\Delta n = 0.1$–0.2. On the other hand, flow birefringence usually leads to values of $\Delta n = 10^{-8}$–10^{-5}.

3.4.4.2. Molecular Anisotropy.

The contribution of a single molecule to the anisotropy of polarizability depends on the anisotropy of its chemical structure and on the anisotropy of its surroundings. When an asymmetric but otherwise isotropic particle is suspended in a medium of different refractive index and subjected to an electric field, the induced polarization charges on its surface modify the external electric field. The net result is called *form anisotropy*. It is equal to the product of particle volume and an optical factor $(g_1 - g_2)$. The latter factor is zero for spheres; for asymmetric particles it increases with increasing asymmetry but approaches a saturation value. The optical factor is also proportional to $\left(\tilde{n}^2 - \tilde{n}_0^2\right)^2$, where $\tilde{n}$ and $\tilde{n}_0$ are the refractive indices of the particle and the solvent, respectively. Thus form anisotropy is always positive; that is, polarizability along the longer axis of the particle is always larger than along the shorter axis. It vanishes when $\tilde{n} = \tilde{n}_0$.

Macromolecular coils also exhibit form anisotropy. The coils are always asymmetric, and the average refractive index of the space occupied by the coil differs from $\tilde{n}_0$. When the coil is deformed, its form anisotropy increases until it reaches a saturation value.

Chemical structure of particles manifests itself in *intrinsic anisotropy*. Molecular polarizability is a tensor sum of polarizabilities of all electrons in the molecule. Valence shell electrons are responsible for the lion's share of the overall polarizability. Carbon-hydrogen bonds are believed to possess almost isotropic polarizability. Most other single bonds have greater polarizability along the bond than across it. Multiple bonds have greater anisotropy than single bonds. Conjugation substantially increases both polarizability and its anisotropy. Aromatic structures are very anisotropic; their greatest polarizability is in the aromatic plane. Anisotropy of rigid molecules is easily predicted from their structure. However, anisotropy of molecules with internal freedom of rotation must be calculated as a weighted average over possible conformations (the anisotropies of which may be grossly different).

Let us now focus attention on the intrinsic anisotropy of macromolecular particles. Anisotropy of rigid particles varies in the broadest limits. Some sols are composed of submicroscopic crystals; an example is a sol of vanadium pentoxide. The intrinsic anisotropy of such sols is very high. Suspensions of some organic dyes behave similarly.

In true macromolecules, high anisotropy is achieved when some anisotropic groups (usually aromatic) are arranged within the molecule in a very regular manner—most often in a helix. Typical examples are shorter fragments of the DNA double helix as well as the α helix of γ-benzyl glutamate in some solvents. On the other hand, in most globular proteins, the molecular chain changes direction so often that the molecules are virtually isotropic. For such molecules the molecular anisotropy is almost pure form anisotropy.

What is the intrinsic anisotropy of a coiled macromolecule? It is usually expressed in terms of the anisotropy of its statistical segments. The anisotropy of these segments, which are defined as strictly linear sections of the chain, depends again on the structure and size of the segments. It is characterized by the difference $\alpha_1 - \alpha_2$, that is, by the difference in the segment polarizabilities along and across the backbone. The anisotropy is positive for most chains with saturated substituents and negative for polystyrene and similar polymers with an aromatic group oriented perpendicularly to the backbone; it is very small for poly(methyl methacrylate), in which the anisotropy of side groups almost exactly balances the anisotropy of the carbon backbone.

Segmental anisotropy is large for rigid polymers; more monomer units contribute to it. High-molecular-weight DNA (which is coiled in solutions despite its backbone rigidity) is an extreme example. Its statistical segments are huge, and its heterocyclic bases are arranged perpendicularly to the helical axis. Thus its negative segmental anisotropy is extremely large.

How is the intrinsic anisotropy of the whole coil related to the anisotropy of its segments? Although the directions of individual segments are mutually independent, their orientation along the longer axis of the equivalent particle must be slightly preferred; this leads to nonzero overall anisotropy. Statistical analysis gives for the anisotropy of an average statistical coil

$$(\alpha_1 - \alpha_2)_{\text{coil}} = \tfrac{3}{5}(\alpha_1 - \alpha_2) \tag{3.4.71}$$

The result is somewhat surprising: The anisotropy of a coil is smaller than the anisotropy of any of its segments; it is the same for all chain lengths.

If the coils are deformed (e.g., by a strong flow field), their intrinsic anisotropy increases as

$$(\alpha_1 - \alpha_2)_{\text{coil}} = \tfrac{3}{5}(\alpha_1 - \alpha_2)\overline{h^2_{\text{def}}}/\overline{h^2_{\text{un}}} \tag{3.4.72}$$

This increase may be quite appreciable. The subscripts 'def' and 'un' refer to deformed and undeformed coils, respectively.

3.4.4.3. Birefringence of Systems Oriented by Flow.

A collection of anisotropic molecules, orientation of which is completely random, is perfectly isotropic at the macroscopic level. The system becomes anisotropic (birefringent) whenever the distribution function of molecular orientations Φ is even slightly anisotropic. Orientation of molecules of liquids may be achieved by electric field (Kerr effect), by magnetic field (Cotton–Mouton effect), or by the field of acoustic waves (Lucas effect). However, we will be interested only in orientation caused by flow of liquids (Maxwell effect).

Flow birefringent behavior of suspensions of ellipsoids was solved exactly. We will present only the main results. The distribution function Φ is a function of two parameters, reduced velocity gradient σ and the asymmetry factor b, which are defined as

$$\sigma \equiv G/D_r \tag{3.4.73}$$

and

$$b \equiv (p^2 - 1)/(p^2 + 1) \tag{3.4.74}$$

G being the velocity gradient and p the axial ratio of ellipsoids. The *rotary diffusion coefficient* D_r is related to the *rotary frictional coefficient* f_r in the same way as the translational quantities were. The latter coefficient is the proportionality coefficient (tensor) relating angular velocity of the particle ω with the torque $\mathbf{L}$ needed to sustain its angular velocity against the friction forces.

$$D_r = kT/f_r \tag{3.4.75}$$

$$\mathbf{L} = f_r \omega \tag{3.4.76}$$

For spheres, the frictional coefficient is given as

$$f_r = 8\pi \eta_0 r^3 = 6\eta_0 V \tag{3.4.77}$$

where r and V are the radius and volume, respectively, of the sphere. For ellipsoids, the relevant principal value of the frictional tensor is

$$f_r = 8\pi \eta_0 a^3 y(p) \tag{3.4.78}$$

where a is the longer semiaxis of the ellipsoid and $y(p)$ is a slowly varying function of the axial ratio.

We have qualitatively described function Φ in Section 3.3.7. It is symmetrical with respect to the flow plane; its maximum is in this plane. The angle between the flow direction and the direction of the maximum is the *orientation angle* χ. At small values of σ, $\chi = 45°$; it decreases toward zero with increasing σ. The asymmetry of the function Φ is expressed by means of the *orientation factor* $f(\sigma, b)$, which represents the fraction of the total number of particles that would produce the same anisotropy if oriented fully. The quantities $\chi(\sigma, b)$ and $f(\sigma, b)$ have been expressed as series. We will present the leading terms only.

$$\cot 2\chi = \sigma/6 - \cdots \tag{3.4.79}$$

$$f(\sigma, b) = \sigma \, b/15 - \cdots \tag{3.4.80}$$

At high values of σ, $f(\sigma, b)$ approaches a saturation value that is always smaller than unity. The orientation angle χ and orientation factor f are plotted as functions of σ for two values of p in Figures 3.39 and 3.40.

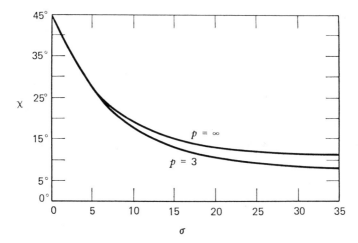

Figure 3.39. The orientation angle χ for prolate ellipsoids as a function of the reduced velocity gradient σ for two values of p. (Reprinted from A. Weissberger and B. W. Rossiter, Eds., *Physical Methods of Chemistry*, Part IIIc, *Polarimetry*. Copyright © 1972 by John Wiley & Sons, New York. Used by permission of the publisher.)

What is the optical behavior of our flowing system? The optical tensor of the liquid has the same symmetry as the function Φ. Thus, two of the principal directions are in the flow plane—one along the direction χ and the other perpendicular to it. The third principal direction is perpendicular to the flow plane; it possesses the intermediate value of refractive index.

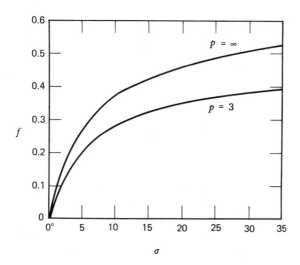

Figure 3.40. The orientation factor $f(\sigma, b)$ for prolate ellipsoids as a function of the reduced velocity gradient σ for two values of p. (Reprinted from A. Weissberger and B. W. Rossiter, Eds., *Physical Methods of Chemistry*, Part IIIc, *Polarimetry*. Copyright © 1972 by John Wiley & Sons, New York. Used by permission of the publisher.)

The magnitude of the flow birefringence Δn depends on the concentration of oriented particles c, on their anisotropy, $(g_1 - g_2)$, and on the orientation factor.

$$\Delta n = c\bar{v}(2\pi/\tilde{n}_0)(g_1 - g_2)\, f(\sigma, b) \tag{3.4.81}$$

Here, $\bar{v}$ is the specific volume of the particles and $\tilde{n}_0$ is the refractive index of the solvent.

It is customary to define two experimentally measurable characteristic values of the solution: *intrinsic orientation*, $[\omega]_0$, and *intrinsic birefringence*, $[\Delta n]_0$. Their definitions and interpretations for ellipsoids are

$$[\omega]_0 \equiv \lim_{\substack{c \to 0 \\ G \to 0}} \frac{\cot 2\chi}{G} = \frac{1}{6D_r} \tag{3.4.82}$$

and

$$[\Delta n]_0 \equiv \lim_{\substack{c \to 0 \\ G \to 0}} \frac{\Delta n}{\eta_0 c G} = \frac{2\pi}{15} \frac{(g_1 - g_2)\, b\bar{v}}{\tilde{n}_0 \eta_0 D_r} \tag{3.4.83}$$

From these two values D_r and $(g_1 - g_2)b$ are easily obtained; from D_r we can estimate the size of the particles. The most characteristic feature of flow birefringence of rigid particles is the saturation of Δn that occurs at high values of σ.

The evaluation of flow birefringence of coiled macromolecules proceeds along the same lines as for rigid particles. The main difference is the deformability of molecular coils; the main difficulty is the choice of a suitable model. Fortunately, some features remain the same for all models: deformable or rigid dumbbells and Gaussian subchains with or without internal viscosity.

The customary variable describing the interaction of the flow field with the coils is the parameter β_0, which we used when we studied non-Newtonian viscosity and defined by equation (3.4.66). Realizing that the hydrodynamically equivalent volume of a coil is proportional to $M[\eta]/N_{Av}$, the reader will easily find that the parameters σ and β_0 are very similar.

For coils, the orientation angle χ is related to β_0 as

$$\cot 2\chi = B\beta_0 - \cdots \tag{3.4.84}$$

The value of B predicted by theory depends on the model of the coil. The simplistic dumbbell model predicts $B = 1$, whereas the more sophisticated model of Gaussian subchains predicts $B = 0.2$. Both values refer to soft, easily deformed coils. For coils that resist fast deformation (i.e., have high *internal viscosity*), the theory predicts $B = 1.8$ in low-viscosity solvents; this value should decrease toward $B = 0.2$ with increasing solvent viscosity. Hence, at least in principle, experimental value of B and its dependence on solvent viscosity can serve for an estimate of the dynamics of the polymer coil in solution.

Two components contribute to the magnitude of birefringence: intrinsic and form anisotropy. Both components increase with increasing β_0 because of increasing orientation. However, the coil deformation influences them differently. The intrinsic anisotropy increases steeply at high values of β_0, whereas the form anisotropy approaches a saturation value. Hence, intrinsic anisotropy will dominate at high β_0. (If the intrinsic anisotropy is negative and, at low values of β_0, smaller than the always positive form anisotropy, then the birefringence Δn may, with increasing G, change its sign from positive to negative.)

From the viewpoint of polymer characterization, intrinsic anisotropy is a very valuable quantity whereas form anisotropy is more of a nuisance. Hence, we frequently need to separate the two. The most convenient way is to measure the polymer dissolved in a solvent with a matching refractive index; form anisotropy then vanishes. Another approach consists of measuring the birefringence for several samples with different molecular weights. Theory predicts that the form part of $[\Delta n]_0/[\eta]_0$ is proportional to $M/[\eta]_0$, whereas the intrinsic part is independent of molecular weight and can be obtained by extrapolation of $[\Delta n]_0/[\eta]_0$ to vanishing $M/[\eta]_0$. An example of this procedure is presented in Figure 3.41 for polystyrene in three solvents with different refractive indices.

Intrinsic anisotropy of a polymer segment ($\alpha_1 - \alpha_2$) is related to the intrinsic part of $[\Delta n]_0/[\eta]_0$ as

$$\frac{[\Delta n]_0}{[\eta]_0} = \frac{4\pi \left(\tilde{n}_0^2 + 2\right)^2}{45\tilde{n}_0 kT} (\alpha_1 - \alpha_2) \equiv 2K \qquad (3.4.85)$$

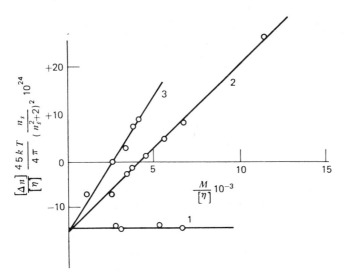

Figure 3.41. Separation of the intrinsic and form parts of the ratio $[\Delta n]/[\eta]$. Polystyrene (1) in bromoform; (2) dioxane; (3) butanone. (Reprinted from A. Weissberger and B.W. Rossiter, Eds., *Physical Methods of Chemistry,* Part IIIc, *Polarimetry.* Copyright © 1972 by John Wiley & Sons, New York. Used by permission of the publisher.)

In this relation, we have defined another very useful characteristic of a polymer, the *stress-optical coefficient K*.

Intrinsic anisotropy of polymer solutions at higher values of β_0 is related to the average squared end-to-end distance of a *deformed coil* [compare equation (3.4.72)]; this is a quantity that cannot be measured by any other method. The applicable relation reads

$$\overline{h_\beta^2}/\overline{h_0^2} = 1 + (2\beta/3)\,([\Delta n]_\beta/[\Delta n]_0)\cot 2\chi \qquad (3.4.86)$$

Here, the subscript β refers to finite values of β, whereas subscript zero indicates values extrapolated to vanishing β.

In many solutions, there are several components contributing to birefringence, for example, a solvent (molecules of which are oriented by flow as any other particle would be) and a polymer. The observed birefringence is a tensor sum of the component birefringences; the appropriate relations read

$$\Delta n = \left[\left(\sum_i \Delta n_i \sin 2\chi_i\right)^2 + \left(\sum_i \Delta n_i \cos 2\chi_i\right)^2\right]^{1/2} \qquad (3.4.87)$$

$$\cot 2\chi = \frac{\sum_i \Delta n_i \cos 2\chi_i}{\sum_i \Delta n_i \sin 2\chi_i} \qquad (3.4.88)$$

3.4.4.4. Birefringence and Stress.

For many substances there is a close relationship between optical and stress tensors, called the *stress-optical law*. This law states that anisotropies of both tensors are proportional to each other and that the principal axes of both tensors have the same orientation.

For polymer solutions, the law is valid only when form birefringence is either absent or negligible compared to intrinsic birefringence. The physical basis of the law is in the fact that polymer coils and their sections behave like springs. Both the tension of the spring and its anisotropy are proportional to h^2; hence, they are proportional to each other whatever the orientation and extension of the spring. The optical and stress tensors are both sums of the contributions of individual springs, and the stress-optical law follows. It is applicable for any system of springs or coils, be it a flowing polymer solution or a deformed swollen gel. In all cases, the axes of both tensors are parallel and their anisotropies are proportional to each other.

Of course, we must remember that the isotropic part of the stress tensor is simply hydrostatic pressure and that the isotropic part of the refractive tensor is just the mean refractive index. Neither of these quantities is influenced by the extension of the coils. Hence, we must exclude them from the analysis. This is done by considering the anisotropies only, that is, the differences between the principal values σ_i of the stress tensor and $\tilde{n}_i$ of the refractive tensor. Thus $(\sigma_1 - \sigma_2)$ is proportional to $(\tilde{n}_1 - \tilde{n}_2)$ and

$(\sigma_2 - \sigma_3)$ is proportional to $(\tilde{n}_2 - \tilde{n}_3)$, the proportionality constant being the stress-optical coefficient K defined by equation (3.4.85).

The situation is slightly complicated by the fact that birefringence is routinely reported as the difference of the principal values Δn and as χ, the orientation angle of the refractive tensor with respect to the flow direction (i.e., to instrument coordinates). On the other hand, the stress tensor is usually represented by its components expressed in instrument coordinates: the shear stress σ_{xy} and the difference of normal stresses, $(\sigma_{yy} - \sigma_{xx})$. The normal stresses are the components of the stress tensor acting perpendicularly on hypothetical planes that are normal to the direction of flow y (component σ_{yy}) or to the direction of velocity gradient x (component σ_{xx}). The rules of transformation of tensors from one coordinate system to another require that

$$2\sigma_{xy} = (\sigma_1 - \sigma_2)\sin 2\chi \tag{3.4.89}$$

$$\sigma_{yy} - \sigma_{xx} = (\sigma_1 - \sigma_2)\cos 2\chi \tag{3.4.90}$$

Thus the stress-optical law for simple shear flow is usually written as

$$\Delta n \sin 2\chi = 2K\sigma_{xy} = 2KG\eta \tag{3.4.91}$$

$$\Delta n \cos 2\chi = 2K(\sigma_{yy} - \sigma_{xx}) \tag{3.4.92}$$

For dilute solutions, solvent may contribute both to stress and to birefringence; only the shear components are influenced. (Pure solvents always exhibit $\chi = 45°$.) In such a case, subtraction of the solvent contribution from equation (3.4.91) yields for K (subscript zero refers to the solvent)

$$2K = \frac{\Delta n \sin 2\chi - \Delta n_0}{G(\eta - \eta_0)} \tag{3.4.93}$$

The stress-optical law is very valuable for studying the rheological behavior of polymer systems. Very sensitive birefringence techniques can be used for the measurement of small mechanical quantities (stresses), the direct mechanical measurement of which may be technically extremely difficult or impossible (this is true especially for the difference of normal stresses). Besides the obvious significance of these techniques for basic understanding of the mechanics of flow and the deformation of polymer systems, we will also mention one technical application. It is often necessary to know details of liquid flow around complicated objects. The flow pattern can be visualized when a liquid capable of flow birefringence is employed and the flow is observed in a system placed between crossed polaroid sheets; the streamlines are then clearly observed as a system of light and dark fringes.

3.4.4.5. Experimental Arrangements.

Most instruments for measuring flow birefringence are similar in design to Couette's viscometer: they have two concentric cylinders, one rotating and the other stationary. They are enclosed in a vessel with

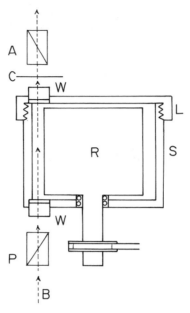

Figure 3.42. A flow birefringence instrument with an inner rotor.

windows W in the bottom and in the lid L, which allow a light beam parallel to the cylindrical axis to pass between the cylinders. A simple design with an inner rotor R and outer stator S is depicted in Figure 3.42. A parallel beam of light B passes through a polarizer P, the Couette cell, and an analyzer A.

The birefringent liquid in the cell acts optically as a birefringent plate. The polarizer and analyzer are crossed; in the absence of birefringence in the cell, the light beam is extinguished. The birefringent plate between crossed polarizers brightens the field of view unless its principal optical axes are parallel to the planes of polarization of the polarizers. To find this position, the coupled polarizers are rotated together until the brightened field is extinguished again. Then the orientation of the planes of polarization with respect to the direction of the flow of the liquid is characterized by an *extinction angle* (this is another name for the orientation angle; it is used almost exclusively in flow birefringent literature). The magnitude of birefringence is measured by a suitable compensator C located in front of the analyzer.

In some older flow birefringence instruments, the windows and polarizers cover the entire gap between the cylinders and the whole annulus can be observed. The extinction angle is the angle between the principal axes and the direction of flow. However, the direction of flow changes along the circumference of the cylinders, and so does the direction of the principal axes. For any given (permanent) orientation of the polarizers, there must exist four locations along the annulus where the principal axes are parallel to the polarizers and the light is extinguished. This leads to a characteristic dark cross in the view field. When the velocity gradient increases, the orientation angle changes and the dark cross shifts along the annulus. Hence the position of the cross can be used for measurement of the extinction angle.

The range of velocity gradients for which the flow birefringence can be measured depends on the nature of the liquid. The lower limit depends on the magnitude of birefringence: when $\Delta n < 10^{-8}$, the extinction angle usually cannot be measured with confidence. The detection limit for Δn is about 2×10^{-9}. For low-viscosity liquids, the upper limit of velocity gradients is given by the onset of turbulent flow. The laminar flow persists longer for liquids with higher viscosity, for narrower gaps between cylinders, and for outer rotors. The practical limit for velocity gradients is about 10,000 s^{-1} with an outer rotor and with a gap width of about 0.4 mm. For liquids of higher viscosity, the limiting factor is viscous heating. This heating coupled with removal of the heat by thermostated walls leads to thermal gradients that are accompanied by gradients of refractive index, which lead to a bending of the light beam, to reflections, and to changes in the polarization of the light, thus preventing the measurement. The practical limit is reached when the product ηG^2 is higher than about 0.1 W/mL.

3.4.4.6. *Interpretation of Flow Birefringence Data.*   We have seen that it is possible to extract the molecular weight of suspended particles from the values of extinction angle at low velocity gradients. However, this measurement is not very precise and cannot compete with either light-scattering or sedimentation analysis. Nevertheless, molecular weight and polydispersity influence the flow birefringence quite appreciably, and valuable information about polydispersity can be obtained from it. It is specifically advantageous for systems containing two grossly different particle types (e.g., molecular coils and microgels or microcrystals). It is often possible to separate the contributions to birefringence of these two types of particles and obtain information about both of them.

Optical anisotropy provides information about both macro- and microstructure of the macromolecule. Large segmental anisotropy $(\alpha_1 - \alpha_2)$ usually implies large size of the statistical segment. If the anisotropy of a monomer unit is known from independent measurements (e.g., for DNA), then the number of units in a segment can be estimated from the measured anisotropy of the segment. On the other hand, if the size of the segment is known independently (e.g., from light scattering or viscometric measurements), then it is possible to find from segmental anisotropy the anisotropy of a monomer unit; from this we may deduce the conformation of this unit—for example, the average orientation of the phenyl group with respect to the chain backbone in polystyrene.

Interaction of solvent with polymer chains may result in oriented adsorption of solvent molecules on the chains. In this case, the anisotropy of solvent molecules combines with the segment anisotropy and may change it considerably.

If the macromolecule (e.g., a copolymer) is composed of parts whose behavior is known, the higher structure of such molecules can be successfully studied. For example, statistical and grafted copolymers of styrene and methyl methacrylate exhibit quite different birefringent behavior. Similarly, coil-helix transition is manifested quite dramatically in flow birefringence.

The unique capability of flow birefringence to measure the deformability and deformation of single molecules in dilute solutions has already been mentioned; no other method can perform this task. Equally valuable is the capability to determine rheological properties of dilute and semidilute systems by employing the stress-optical law. In summary, flow birefringence is a very versatile method useful for solving problems of all sorts. It is especially informative if we can combine it with other techniques.

3.5. LIGHT SCATTERING

In order to learn the properties of macromolecules, we have handled them in many ways. We have subjected them to centrifugal and electric fields, let them diffuse, and forced them to flow through capillaries and even gels. However, there is a gentler way to study their intimate properties in great detail: simply shine light on them and look at them. Macromolecules do not like the light; they scatter it back and in the process reveal many of their secrets. Thus, the study of scattered light is one of the most powerful methods in the physical chemistry of macromolecules. It is also completely nondestructive and noninterfering; it is very suitable for kinetic studies.

In the following sections we will study why and how individual molecules scatter light; why larger molecules scatter much more light than small ones; why such large ensembles of molecules as liquids and solids scatter very little light; and what properties of the system we can learn from its light scattering. As a bonus, we will also find answers to such questions of general interest as why is the sky blue, why is the sunset red, why are clouds white, and what is the origin of the ring around the moon.

3.5.1. Scattering by a Single Small Isotropic Particle

The electric field of a light wave acting on a particle induces in it a dipole that oscillates with the same frequency as the incident light. An oscillating dipole produces a secondary oscillating field; it radiates electromagnetic energy. In other words, the particle scatters the incident light.

Let us follow this process more closely. Consider a single, optically isotropic particle with polarizability α. The particle is in vacuum and is much smaller than the wavelength λ_0 of the incident light. (In practice, it must be smaller than $\lambda_0/20$ to fulfill this condition.) Then, the electric field of the incident light, $\mathbf{E}$, is homogeneous within the particle; it induces in it a dipole $\mu = \alpha\mathbf{E}$. The electric field oscillates with frequency $\nu = \tilde{c}/\lambda_0$ and so does the induced dipole. Thus,

$$\mu = \alpha\mathbf{E} = \alpha\mathbf{E_0} \sin 2\pi \nu t \qquad (3.5.1)$$

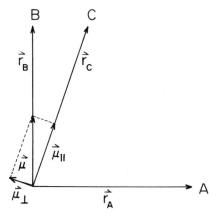

Figure 3.43. Electric field generated by an oscillating dipole and its dependence on mutual orientation of the dipole and the position vector of the detector. See text for a detailed explanation.

Here, E_0 is the amplitude of the electric vector, t represents time, and $\tilde{c}$ is the velocity of light. The magnitude of the electric field $\mathbf{E_s}$ generated by an oscillating dipole is found by solving Maxwell's equations. It depends on the vector $\mathbf{r}$ pointing from the dipole to the point of observation (Figure 3.43). At locations where $\mathbf{r}$ is perpendicular to $\boldsymbol{\mu}$ (point A in the figure), $\mathbf{E_s}$ is parallel to $\boldsymbol{\mu}$ and is found as

$$\mathbf{E_s} = \frac{d^2\mu/dt^2}{\tilde{c}^2 r} \tag{3.5.2}$$

At locations along the dipole ($\mathbf{r}$ parallel to $\boldsymbol{\mu}$, point B), $\mathbf{E_s}$ vanishes. For calculation of $\mathbf{E_s}$ at an arbitrary point C, we imagine that the dipole $\boldsymbol{\mu}$ is decomposed into dipole $\boldsymbol{\mu}_\parallel$ which is parallel to $\mathbf{r}$, and dipole $\boldsymbol{\mu}_\perp$, perpendicular to $\mathbf{r}$. Obviously, only the latter dipole contributes to the scattered electric field; the magnitude of $\mathbf{E_s}$ is then

$$E_s = \frac{d^2\mu_\perp/dt^2}{\tilde{c}^2 r} = \frac{d^2\mu}{dt^2}\frac{\sin\vartheta}{\tilde{c}^2 r} \tag{3.5.3}$$

$\mathbf{E_s}$ is parallel to $\boldsymbol{\mu}_\perp$; ϑ is the angle between vectors $\boldsymbol{\mu}$ and $\mathbf{r}$.
  Equation (3.5.1) substituted into (3.5.3) yields

$$E_s = E_0\alpha 4\pi^2 v^2 \frac{\sin\vartheta \sin 2\pi v t}{\tilde{c}^2 r} \tag{3.5.4}$$

In light-scattering experiments we are observing not the magnitude of the electric field E but the intensity of light I. Maxwell's theory relates I to E as

$$I = \frac{\tilde{c}\tilde{n}|E^2|}{4\pi} \tag{3.5.5}$$

where $\tilde{n}$ is the refractive index (equal to unity for vacuum in our problem) and $|E^2|$ is the time average of E^2. For a harmonic *field*, $|E^2| = E_0^2/2$. Using equation (3.5.5) to calculate the intensities of the primary light I_0 and the scattered light I_s, we find

$$R \equiv \frac{I_s r^2}{I_0} = 16\pi^4 \nu^4 \alpha^2 \frac{\sin^2 \vartheta}{\tilde{c}^4} = 16\pi^4 \alpha^2 \frac{\sin^2 \vartheta}{\lambda_0^4} \tag{3.5.6}$$

Here, R defined by equation (3.5.6) is the *Rayleigh ratio* named after the "father" of the theory of light scattering.

In order to appreciate fully the significance of equation (3.5.6), we need to understand the relation of angle ϑ to the experimental arrangement and to the polarization of the primary light. Let us choose our coordinate system in such a way that the primary light proceeds along the x axis and the scattered light is observed in the xy plane (which is often called the horizontal plane) from a direction inclined by an angle θ to the direction of the primary beam. When the primary light is polarized vertically (i.e., its electric field oscillates along the z axis), the position vector of the detector, $\mathbf{r}$, is always perpendicular to the induced dipole and $\sin \vartheta$ is equal to unity for all angles θ (Figure 3.44a). Thus, the Rayleigh ratio is

$$R = 16\pi^4 \alpha^2 / \lambda_0^4 \tag{3.5.7}$$

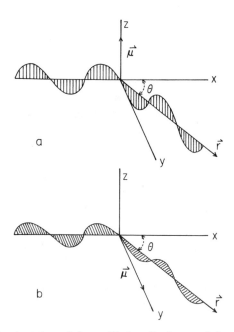

Figure 3.44. Mutual orientation of the oscillating dipole μ and the position vector of the detector, $\mathbf{r}$, for primary light polarized (a) vertically, (b) horizontally.

When the electric vector of the primary light oscillates in the observation plane (horizontally polarized light, Figure 3.44b), the angle ϑ is the complementary angle to the observation angle θ, that is, $\sin \vartheta = \cos \theta$, and the relation for scattered intensity reads

$$R = 16\pi^4\alpha^2(\cos^2 \theta)/\lambda_0^4 \qquad (3.5.8)$$

The scattered light is polarized horizontally in this case; no light is scattered at the angle $\theta = 90°$. Finally, when the primary light is not polarized (unpolarized light may be considered as a superposition of noncoherent vertically and horizontally polarized beams of equal intensity) the observed intensity is given as

$$R = 8\pi^4\alpha^2(1 + \cos^2 \theta)/\lambda_0^4 \qquad (3.5.9)$$

In this case, the light scattered at angle 90° is totally polarized vertically.

One of the important characteristics of scattered light is its angular distribution of intensities (scattering profile). Graphically, it is depicted as an envelope of vectors whose magnitude corresponds to the scattered intensity. The scattering profile for our small isotropic particle is shown in Figure 3.45 for vertically polarized, horizontally polarized, and unpolarized light.

What facts can we learn from relations (3.5.7)–(3.5.9) for the scattered intensity? First, the intensity is proportional to the square of particle polarizability. For particles made from a given isotropic material, polarizability is proportional to their volume and consequently to their molecular weight. Thus, the amount of light scattered by a single particle is proportional to the *square* of its molecular weight. We will see that under some circumstances the intensity of light scattered by many particles is additive. Then the amount of light scattered by a given amount of material (total mass of particles) will increase steeply with the coarseness of the division of the material. In fact, it will be proportional to the molecular weight of the particles, giving us an

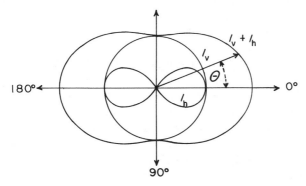

Figure 3.45. Scattering profile for a small isotropic particle for vertically polarized, horizontally polarized, and unpolarized primary light.

opportunity to measure the latter. We will return to this topic later. The potent scattering by large particles is best observed in a darkened room traversed by a sunbeam coming through a hole in draperies. The dust particles (which are still extremely small and invisible to the naked eye in diffuse illumination) appear as intense specks of light.

Another point of interest is the dependence of the scattered intensity on the wavelength of the light—the intensity increases with the inverse fourth power of the wavelength! In other words, the shorter wavelengths (the blue end of the spectrum) are scattered much more than the longer ones. We can observe the result of this any day (well, at least any sunny day). Molecules of air scatter the sunlight; the blue end of the spectrum is scattered much more than the red—we see the scattered light as blue sky. Now, why is the sunset red? When the sun is high in the sky, its rays penetrate the atmosphere more or less perpendicularly to its thickness. Scattering of the blue light modifies the spectral composition of the sunlight, but not very much. However, when the sun is on the horizon, its rays must travel tangentially to the earth's surface to reach the observer. The distance traveled through the air is much longer; most of the blue light is scattered in all directions, and only the less scattered red light reaches the observer. The effect is enhanced when the air contains other more powerful scatterers: dust particles or microscopic water droplets. This should explain the role that the colors of the evening sky play in weather forecasting. (We can still forecast weather from the appearance of the sky as well as official weathermen do.)

The third feature to notice is the complete polarization of light scattered at the angle 90° even when the primary light is unpolarized. This explains the fact that the light coming from the sky is polarized to a considerable degree.

3.5.2. Light Scattered by an Anisotropic Particle

Isotropic molecules are rare. Most real molecules are anisotropic and have an anisotropic polarizability tensor as we have seen in Section 3.4.4. When the electric field of a light wave acts on an anisotropic particle, the induced dipole generally has a different direction than the electric vector of the exciting light. It is the direction of this oscillating dipole that determines the polarization of the scattered light. Consequently, the scattered light will also have components that were absent in scattering by isotropic particles.

Calculations of the scattered intensities are straightforward but tedious. We will present only the results averaged for all orientations of the polarizability tensor of the particle with respect to the exciting field of the primary light. It is convenient to represent the Rayleigh ratio as a sum of intensities of light polarized vertically (symbol V) and horizontally (symbol H). The symbols V and H are further furnished with subscripts v, h, and u, describing the primary beam as polarized vertically, polarized horizontally, and unpolarized. The symbol R_t is used for total scattering by unpolarized primary light.

$$R_t = (V_u + H_u)/2 = (V_v + V_h + H_v + H_h)/2 \qquad (3.5.10)$$

Obviously, for isotropic particles and $\theta = 90°$, $V_h = H_v = H_h = 0$; V_v is equal to R of equation (3.5.7).

For anisotropic particles the following relations are found for light scattered at $\theta = 90°$:

$$V_v = \left(16\pi^4/\lambda_0^4\right)\left(\alpha^2 + 4\gamma^2/45\right) \tag{3.5.11}$$

$$V_h = H_v = H_h = \left(16\pi^4/\lambda_0^4\right)\left(3\gamma^2/45\right) \tag{3.5.12}$$

The average polarizability of the particle α and its anisotropy γ^2 are defined in terms of the three principal values of the polarizability tensor α_1, α_2, and α_3, as

$$\alpha \equiv (\alpha_1 + \alpha_2 + \alpha_3)/3 \tag{3.5.13}$$

and

$$\gamma^2 \equiv \alpha_1^2 + \alpha_2^2 + \alpha_3^2 - \alpha_1\alpha_2 - \alpha_1\alpha_3 - \alpha_2\alpha_3 \tag{3.5.14}$$

Thus, the total scattered intensity is

$$R_t = \left(8\pi^4/\lambda_0^4\right)\left(\alpha^2 + 13\gamma^2/45\right) \tag{3.5.15}$$

For isotropic particles, this equation reduces, of course, to equation (3.5.9) (with $\theta = 90°$).

The depolarization ratio or simply *depolarization* Δ is a quantity that can be measured experimentally with relative ease. It is defined as

$$\Delta = \frac{H}{V} = \frac{H_v + H_h}{V_v + V_h} = \frac{6\gamma^2/45}{\alpha^2 + 7\gamma^2/45} \tag{3.5.16}$$

For many purposes, it is convenient to split the total scattered intensity R_t into two parts: isotropic scattered intensity R_{is}, which is equal to the intensity of light scattered by a hypothetical isotropic particle having the same average polarizability α as the actual particle, and anisotropic scattered light R_{an}, which represents the excess scattering due to the particle anisotropy.

$$R_t = R_{is} + R_{an} \tag{3.5.17}$$

Eliminating α^2 and γ^2 from equations (3.5.9), (3.5.16), and (3.5.17) yields a relation between R_t and R_{is} as

$$R_{is} = R_t\,(6 - 7\Delta)\,/\,(6 + 6\Delta) \equiv R_tC_f \tag{3.5.18}$$

The *Cabannes factor* C_f is defined by the identity in this relation. The anisotropic scattering R_{an} is given as

$$R_{an} = R_t(1 - C_f) = (8\pi^4/\lambda_0^4) 13\gamma^2/45 \tag{3.5.19}$$

Thus, if we can measure the total scattering of a particle and its depolarization, we can calculate both the average polarizability and anisotropy of the particle.

3.5.3. Interference of Light Waves

When a light beam interacts with a system of particles, the waves scattered by individual particles interfere with each other. In fact, the intensity of the scattered light and its angular distribution depend primarily on this interference. In this section, we will review the principles governing interference of light waves. In the next sections, we will apply these principles to scattering by large molecules and to scattering by macroscopic collections of molecules (gases, liquids, solutions).

When several light waves arrive at the same location (e.g., at the detector) the electric field at this location is a superposition of the electric fields of all light waves. For our purposes, it is sufficient to consider only a special simple situation when all the waves have the same wavelength, are polarized in the same manner, and have the same amplitude, $\mathbf{E}_a$. However, they will arrive at the detector with different phases. Let φ_i be the phase difference of the ith wave with respect to some reference wave. We may thus write for the ith electric vector

$$\mathbf{E}_i = \mathbf{E}_a \sin(2\pi vt - \varphi_i) \tag{3.5.20}$$

The composite vector $\mathbf{E}_c$ for n waves is

$$\mathbf{E}_c = \sum_{i=1}^{n} \mathbf{E}_i = \mathbf{E}_a \sum_{i=1}^{n} \sin(2\pi vt - \varphi_i) \tag{3.5.21}$$

In Figure 3.46 are depicted several cases of interference of just two waves. When the waves are in phase, the electric vector doubles. When they are 180° out of phase, it vanishes. Other phase differences lead to intermediate results.

The intensity of the composite wave, I_c, is, according to equation (3.5.5),

$$I_c = \frac{\tilde{c}\tilde{n}}{4\pi} |E_c^2| = \frac{\tilde{c}\tilde{n}}{4\pi} v \int_0^{1/v} (\mathbf{E}_c \cdot \mathbf{E}_c) dt \tag{3.5.22}$$

According to equation (3.5.21), the scalar product $\mathbf{E}_c \cdot \mathbf{E}_c$ is the product of two sums. A little algebra using trigonometric identities will then show

$$I_c = \frac{\tilde{c}\tilde{n}}{8\pi} E_a^2 \sum_{i=1}^{n} \sum_{j=1}^{n} \cos(\varphi_i - \varphi_j) \tag{3.5.23}$$

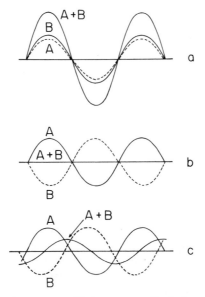

Figure 3.46. Interference of two waves with the same amplitude that are (a) in phase, (b) exactly out of phase, (c) 150° out of phase.

Let us consider a special case when all waves have the same phase φ. Then the double sum equals n^2. Thus, the intensity of n combined waves, which are in phase, is equal to the intensity of a single wave multiplied by n^2.

3.5.4. Scattering by Large Particles

When the scattering particles are larger than about $\lambda_0/20$, then the assumption of the electric field being homogeneous for the whole particle is no longer valid. For such particles, it is convenient to assume that they are composed of n scattering elements, all of them having the same polarizability α_a and all of them being much smaller than λ_0. Then the electric field, scattered by a *single* scattering element, is given by equation (3.5.4) ($E_s \rightarrow E_a$), and the Rayleigh ratio for the whole particle is

$$R = \frac{16\pi^4}{\lambda_0^4}\alpha_a^2 \sin^2 \vartheta \sum_{i=1}^{n}\sum_{j=1}^{n} \cos(\varphi_i - \varphi_j) \tag{3.5.24}$$

This equation was obtained by combining equations (3.5.4), (3.5.5), and (3.5.23).

Let us check equation (3.5.24) for a *small* particle with all φ_i having the same value and the double sum being equal to n^2. The polarizability of the whole particle is $\alpha = n\alpha_a$; hence in this case, equation (3.5.24) reduces to equation (3.5.6), as it should.

We will now evaluate a generic term of the double sum. The two scattering elements, i and j, are depicted in Figure 3.47. We need to calculate the paths

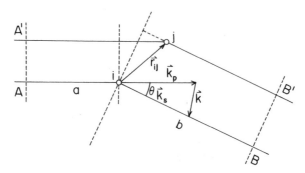

Figure 3.47. The path difference of light scattered by two elements. See text for details.

traveled by the light scattered by these two elements, Δ_i and Δ_j, and their difference $\Delta_{ij} \equiv \Delta_j - \Delta_i$. The distances should be measured between some reference plane perpendicular to the primary beam (AA′ in our figure) and another plane perpendicular to the scattered beam (BB′). Let us designate as $\mathbf{r}_{ij}$ the position vector from element i to element j; θ, the scattering angle; $\mathbf{k}_p$, a unit vector in the direction of the primary beam; $\mathbf{k}_s$, a unit vector along the scattered beam; and *scattering vector* $\mathbf{k}$, their difference, $\mathbf{k} \equiv \mathbf{k}_s - \mathbf{k}_p$. It follows from the definition of $\mathbf{k}$ that its length is $k = 2\sin(\theta/2)$. The distances of element i from the reference planes A and B are a and b, respectively. Light scattered by the ith element travels a distance $\Delta_i = a + b$. The distance traveled by light before it reaches element j is $a + \mathbf{k}_p \cdot \mathbf{r}_{ij}$; the distance from element j to plane B is $b - \mathbf{k}_s \cdot \mathbf{r}_{ij}$. Hence,

$$\Delta_{ij} \equiv \Delta_j - \Delta_i = (a + \mathbf{k}_p \cdot \mathbf{r}_{ij} + b - \mathbf{k}_s \cdot \mathbf{r}_{ij}) - (a + b) = \mathbf{k} \cdot \mathbf{r}_{ij} \quad (3.5.25)$$

The phase difference of the two scattered beams is related to Δ_{ij} as

$$\varphi_i - \varphi_j = 2\pi \Delta_{ij}/\lambda \quad (3.5.26)$$

It should be noted that the wavelength λ must be identified with the distance traveled by light during a time interval $1/\nu$. When the scattering is observed in media with refractive index $\tilde{n}$, then $\lambda = \lambda_0/\tilde{n}$.

We are primarily interested in particles whose orientation changes randomly with time. Thus, $\cos(\varphi_i - \varphi_j)$ must be averaged for all orientations of the position vector $\mathbf{r}_{ij}$ with respect to the scattering vector $\mathbf{k}$. Combining equations (3.5.24)–(3.5.26) followed by averaging over orientations (this calculation is somewhat tedious) yields

$$R = \frac{16\pi^4}{\lambda_0^4} \alpha_a^2 \sin^2 \vartheta \sum_{i=1}^{n} \sum_{j=1}^{n} \frac{\sin \mu r_{ij}}{\mu r_{ij}} \quad (3.5.27)$$

$$\mu \equiv 4\pi \sin(\theta/2)/\lambda \quad (3.5.28)$$

The Rayleigh ratio R depends on the scattering angle θ through $\sin^2 \vartheta$ and through μ. Equation (3.5.9) has shown us that for unpolarized incident light the factor $\sin^2 \vartheta$ should be replaced by $(1 + \cos^2 \theta)/2$. It is now convenient to introduce a new angle-dependent function R_θ:

$$R_\theta \equiv R/(1 + \cos^2 \theta) \tag{3.5.29}$$

This function depends on angle θ only through the factor μ. Let us evaluate this function for vanishingly small angle θ. In this limit, μr_{ij} is very small for all values of r_{ij}; $\sin \mu r_{ij} = \mu r_{ij}$; and the double sum in equation (3.5.27) equals n^2. Thus, R_θ extrapolated to $\theta = 0$ has the same value R_0 for all particles of the same polarizability irrespective of their size. This value is the same as for small particles, namely

$$R_0 = 8\pi^4 \alpha^2 / \lambda_0^4 \tag{3.5.30}$$

The size of the particle, however, plays an important role at $\theta \neq 0$. It is convenient to introduce the *angular scattering function* $P(\theta)$ defined as

$$P(\theta) \equiv \frac{R_\theta}{R_0} = \frac{1}{n^2} \sum_{i=1}^{n} \sum_{j=1}^{n} \frac{\sin \mu r_{ij}}{\mu r_{ij}} \tag{3.5.31}$$

Let us develop all the sine functions into a Taylor series ($\sin x = x - x^3/6 + \cdots$) and retain only the two leading terms. The result reads

$$P(\theta) = 1 - \frac{\mu^2}{6n^2} \sum_{i=1}^{n} \sum_{j=1}^{n} r_{ij}^2 + \cdots \tag{3.5.32}$$

We need now to recall from Section 1.3.1.2 the radius of gyration r_G; it is closely related to the double sum in equation (3.5.32). Assuming that the masses of all scattering elements are the same (we have already assumed that their polarizabilities are the same), we may derive easily for particles of any shape

$$\sum_{i=1}^{n} \sum_{j=1}^{n} r_{ij}^2 = 2n^2 r_G^2 \tag{3.5.33}$$

Combining the last two equations then yields

$$P(\theta) = 1 - \mu^2 r_G^2 / 3 + \cdots \tag{3.5.34}$$

Relation (3.5.34) alone would establish light scattering as one of the most important methods for the characterization of macromolecules. When $P(\theta)$ is measured as a function of μ^2 (i.e., of the scattering angle θ), the radius of gyration can be unequivo-cally calculated from the initial slope of the dependence. The only condition is that the particle must be homogeneous in the sense that it can be visualized as composed of identical scattering elements. (Thus, the method may fail for block copolymers.) When the character of the particle is known, its r_G^2 yields easily the characteristic dimension of the particle: radius in the case of spheres, length for rods, end-to-end distance for coils.

We have derived equation (3.5.34) for scattering by a single particle; however, it will become apparent that an equivalent measurement can be performed on solutions. The measurement of the radius of gyration by this method is absolute, requiring no calibration. In fact, it is the only absolute method for the measurement of the dimensions of macromolecules in solution. Its main limitation is of a technical nature: the parameter μ can have a maximum value (for $\theta = 180°$) equal to $4\pi/\lambda$. Hence, r_G must be sufficiently large to make the change in $P(\theta)$ within the experimental region large enough compared to experimental error. The practical limit is about $r_G = 0.05\lambda$; in this case, $P(\theta)$ and R_θ decrease by about 10% between $\theta = 0°$ and $\theta = 180°$.

So far, we have considered only the initial slope of the dependence of $P(\theta)$ on μ^2. At higher values of μ^2, the shape of the dependence is different for different types of particles. In Figure 3.48, $P(\theta)$ is plotted versus $\mu^2 r_G^2$ for spheres, thin rods, and Gaussian coils. These dependencies coincide more or less for all molecular shapes up to $\mu^2 r_G^2 \sim 1$. At higher values of $\mu^2 r_G^2$ they start to differ appreciably. It is noteworthy that $P(\theta)$ goes through zero for spheres at $\mu^2 r_G^2 \approx 12$; there is no light scattered for

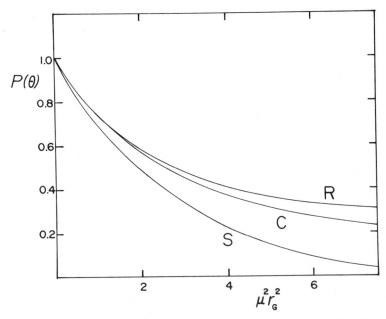

Figure 3.48. The angular scattering function $P(\theta)$ as a function of $\mu^2 r_G^2$ for thin rod (R), Gaussian coils (C), and spheres (S).

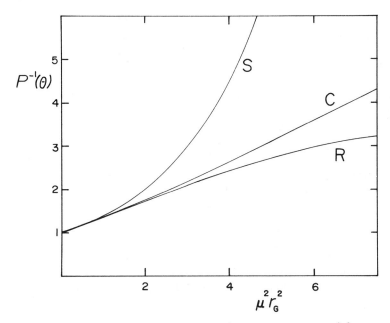

Figure 3.49. Inverse angular scattering function $P^{-1}(\theta)$ as a function of $\mu^2 r_G^2$ for thin rods (R), Gaussian coils (C), and spheres (S).

the corresponding angle. At still higher values the function goes through several minima and maxima. In the older literature, frequent attempts were made to estimate the particle shape from its angular scattering function. For this purpose, the inverse function $P^{-1}(\theta)$ is more suitable; it is plotted in Figure 3.49 for the three types of particles. Presently, it is believed that the shape of the $P^{-1}(\theta)$ function is unduly influenced by polydispersity and other experimental errors and that it does not give reliable information about molecular shape. However, the function $P^{-1}(\theta)$ turning sharply up or exhibiting maxima and minima provides a fair indication of a more or less spherical particle shape.

How good is the assumption that the scattering by a system of oscillating dipoles adequately represents the scattering by a real particle? It is good for small particles and also for large particles suspended in a medium of almost matching refractive index. For a description of the polarization of large particles suspended in a medium of substantially different refractive index, a system of dipoles is not sufficient; the particles should be represented by dipoles, quadrupoles, octupoles, and higher multipoles. As early as the beginning of the 20th century, Mie analyzed rigorously the scattering by spherical particles immersed in a continuous dielectric. His solution started with Maxwell's equations and required the use of an infinite series of electrical multipoles. It turned out that the angular scattering function was a very complicated function of the ratio r/λ (r being the radius of the sphere) and of relative index of refraction $m \equiv \tilde{n}/\tilde{n}_0$, where $\tilde{n}$ and $\tilde{n}_0$ are refractive indices of the sphere and outside continuous dielectric, respectively. The exact solution was presented in the form of an extremely slowly converging series of complicated functions. In the limit of either $r/\lambda \rightarrow 0$ or $m \rightarrow 1$, the solution simplified to the relations presented above. The actual evaluation of the scattering functions had to wait for the advent of modern computers. It turned out that the functions were very sensitive to minor changes in r/λ and m. However, in all cases the angular dependence exhibited a series of well-defined minima and maxima. From the positions of these extremes and from the knowledge of m, it is possible to calculate the sphere radius quite accurately.

The results of Mie's theory are very useful for meteorology; after all, water droplets in clouds are spherical and have a known refractive index. Even a lay person can observe the position of the first maximum on the angular dependence of the scattered moonlight: it forms the ring around the moon. When the water droplets are very uniform in size, a double ring can be observed. From the size of the ring, the size of the droplets can be calculated, and from it the imminence of the rain.

The Mie theory is exact and is applicable to spheres of any size. However, with increasing r, the evaluation of the functions soon shifts beyond the reach of even the fastest computers. It is therefore not surprising that no serious attempt has been made to extend the Mie theory to particles of other than spherical shape.

3.5.5. Scattering by Macroscopic Systems

Macroscopic systems are composed of a very large number of molecules, each of which will interact with the electromagnetic field and attempt to scatter it. It is obvious that the interference of scattered light will play a primary role in such systems.

Let us consider first a system that is perfectly homogeneous. Let us divide it mentally into identical small volume elements, each of them much smaller than the wavelength of the light but large enough to contain a large number of molecules. We have already seen that such a small volume element scatters the light as if all the scatterers were one single particle. The light waves scattered by individual volume elements will interfere. Let us consider two volume elements separated by a vector $\mathbf{r}$ and observed from a direction with scattering vector $\mathbf{k}$. The two scattered waves will display the phase difference $\varphi = (2\pi/\lambda)\,\mathbf{k}\cdot\mathbf{r}$ [compare equations (3.5.25) and (3.5.26)]. Given a particular volume element, a particular $\mathbf{k}$, and a particular λ, we can always select a conjugate volume element so as to make $\varphi = \pi$. Then the scattered waves will exactly cancel each other. It should be obvious that for any pair of elements with the same spatial relationship as the one above, the waves will cancel.

All volume elements can be paired in this way. This argument is applicable for all wavelengths (as long as they are much larger than molecules) and for all scattering vectors. Thus, no scattered light would be observed at any wavelength from any direction. There is only one exception: light waves propagated with scattering vector $\mathbf{k} = 0$. In this case, all the elements generate an electric field with the same phase and mutually reinforce. However, the resulting wave is propagated in the same direction as the primary light. It is considered to be a continuation of the primary beam, not scattered light.

In summary, no light is scattered by a perfectly homogeneous system. However, no real system is perfectly homogeneous. The properties of individual volume elements will fluctuate from the average properties. The light waves scattered by individual elements will not have identical amplitudes; hence, they will not be cancelled totally by interference.

Let us calculate first the intensity of light scattered by a single particle (or volume element) with volume V and dielectric constant ε embedded in a homogeneous medium with dielectric constant ε_0. According to the theory of polarization of dielectrics, an electric field $\mathbf{E}$ will induce in the particle a dipole μ,

$$\mu = \mathbf{E}V\frac{\varepsilon - 1}{4\pi} = \mathbf{E}V\frac{\varepsilon_0 - 1}{4\pi} + \mathbf{E}V\frac{\varepsilon - \varepsilon_0}{4\pi} \tag{3.5.35}$$

Here, we have split the dipole into two terms, the first term being equal to the dipole induced in the same volume of the outside dielectrics. Both dipoles oscillate and scatter light. The light wave scattered by the first term interferes with light scattered by the outside dielectrics and is totally canceled as explained above. Only the second term produces observable scattered light. In analogy with equation (3.5.6), we find its intensity to be

$$R = \pi^2 V^2 (\varepsilon - \varepsilon_0)^2 \, (\sin^2 \vartheta)/\lambda_0^4 \tag{3.5.36}$$

This relation is directly applicable to our fluctuating system; for it $(\varepsilon - \varepsilon_0)$ must be replaced by the fluctuation in dielectric constant, $\delta\varepsilon$.

The next task is the calculation of the light intensity scattered by all volume elements. A derivation along the lines that led to equation (3.5.24) yields

$$R = \frac{\pi^2}{\lambda_0^4}V^2 \sin^2 \vartheta \sum_{i=1}^{n}\sum_{j=1}^{n} \delta\varepsilon_i \delta\varepsilon_j \cos(\varphi_i - \varphi_j) \tag{3.5.37}$$

The fluctuations $\delta\varepsilon$ are not only different from volume element to volume element but also fluctuate with time. Thus, we need to average equation (3.5.37) over time. For this purpose, it is convenient to separate the terms in the double sum into terms for which $i = j$, and terms where $i \neq j$.

$$\sum_{i=1}^{n}\sum_{j=1}^{n} \overline{\delta\varepsilon_i \delta\varepsilon_j} \cos(\varphi_i - \varphi_j) = \sum_{i=1}^{n} \overline{(\delta\varepsilon_i)^2} + \sum_{i \neq j}\sum_{j} \overline{\delta\varepsilon_i \delta\varepsilon_j} \cos(\varphi_i - \varphi_j) \tag{3.5.38}$$

The fluctuations of $\delta\varepsilon_i$ and $\delta\varepsilon_j$ are completely independent of each other. The time average of their product must be equal to the product of their time averages. However, the time average of a fluctuation is equal to zero by definition; the second term in equation (3.5.38) must vanish. Moreover, the average of the squared fluctuation is the same for all volume elements of the same size. Thus, equation (3.5.37) simplifies to

$$R = n \left(\pi^2 / \lambda_0^4\right) V^2 \overline{(\delta\varepsilon)^2} \sin^2 \vartheta \qquad (3.5.39)$$

This relation is remarkable. Intensity of light scattered by a macroscopic system is equal to the sum of the intensities scattered by individual volume elements [the scattering power of each element being proportional to $V^2 (\delta\varepsilon)^2$]. This result is often stated in a more general way. When the scattering entities are completely unrelated to each other, then the total scattered intensity is equal to the sum of intensities scattered by individual entities. (Remember that for entities that are mutually related we should sum the electric fields, not the light intensities.)

Before proceeding further, we need to redefine the Rayleigh ratio R. In the preceding sections, we used it for describing the reduced intensity scattered by a single particle. In this section, it was used for the whole illuminated system. When observing macroscopic systems, it is customary to refer R to light scattered by a unit volume. Hence, we need to divide equation (3.5.39) by the total volume of the system, which is equal to nV. Further, we will assume that the primary light is unpolarized; then $\sin^2 \vartheta$ must be replaced by $(1 + \cos^2 \theta)/2$. The transformed equation (3.5.39) reads

$$R = \left(\pi^2 / \lambda_0^4\right) V \overline{(\delta\varepsilon)^2} (1 + \cos^2 \theta)/2 \qquad (3.5.40)$$

It remains to calculate the average fluctuation $\overline{(\delta\varepsilon)^2}$ in a volume element of size V. We will do that in the next section.

3.5.5.1. Theory of Fluctuations.

A macroscopic system under a given set of constraints (e.g., given temperature, volume, and overall composition) can exist in an extremely large number of different microscopic states. For example, either the molecules in a gas sample may be equally distributed among all volume elements or they may all reside in a single element. However, the latter situation is so extremely improbable that we do not need to consider it. The former situation is the most probable of all possible distributions. However, this does not mean that the system is always in the state with the highest probability. (In fact, it virtually never is.) It turns out that in the vicinity of the state with highest probability a large number of states always exist that also have high probability. These collective probabilities are such that the system can virtually never leave the vicinity. The deviation of the system from the most probable state is called a *fluctuation*.

How large are the allowed fluctuations? According to classical thermodynamics, the system in the state of the highest probability is characterized by the lowest value of the Helmholtz free energy A (at constant volume), and therefore it must always reside there. However, Einstein has shown that this rule is not strict: the system is allowed to spontaneously increase its free energy. Of course, the allowed increase (the fluctuation δA) is minuscule. Rather general statistical considerations show that the average fluctuation $\overline{\delta A}$ is given as

$$\overline{\delta A} = kT/2 \qquad (3.5.41)$$

Note that in this case the fluctuation is measured from the lowest value of A. It is always positive, and its average is nonzero. Equation (3.5.41) is equally valid for all systems irrespective of their size. Thus, large systems are allowed only minuscule relative fluctuations, while small systems may fluctuate appreciably.

Other properties of the system fluctuate as well. However, the extent of the fluctuation is always such as to produce $\overline{\delta A} = kT/2$ for every fluctuating independent variable. Free energy A as a function of any property of interest (we will use a generic symbol x for it) is always at a minimum for the most probable state, for which $x = x_0$. Let us develop this function into a Taylor series

$$A = A_0 + \left(\frac{dA}{dx}\right)_{x=x_0} (x - x_0) + \frac{1}{2}\left(\frac{d^2A}{dx^2}\right)_{x=x_0} (x - x_0)^2 + \cdots \quad (3.5.42)$$

The first derivative is zero by definition of minimum. Designating $\delta A \equiv A - A_0$ and $\delta x \equiv x - x_0$ and averaging over time, we get

$$\overline{\delta A} = \frac{1}{2}\left(\frac{d^2A}{dx^2}\right)_{x=x_0} \overline{(\delta x)^2} \quad (3.5.43)$$

Hence, from equation (3.5.41) and (3.5.43),

$$\overline{(\delta x)^2} = \frac{kT}{(d^2A/dx^2)_{x=x_0}} \quad (3.5.44)$$

3.5.5.2. Scattering by Gases and Liquids.

When dealing with one-component gases and liquids, it is customary to choose density ρ and temperature T as independent variables. For multicomponent systems, concentration variables must be added. Using the recipe of equation (3.5.44) and routine thermodynamic relationships, fluctuation of density is easily calculated as

$$\overline{(\delta\rho)^2} = kT\rho^2\beta/V \quad (3.5.45)$$

Here V is the volume of the fluctuating system and $\beta \equiv (1/V)(\partial V/\partial P)_T$ is its isothermal compressibility. Realizing that $\delta\varepsilon = (d\varepsilon/d\rho)\,\delta\rho$, we can combine equations (3.5.40) and (3.5.45) to obtain

$$R_d = \frac{\pi^2}{2\lambda_0^4}kT\beta\left(\rho\frac{d\varepsilon}{d\rho}\right)^2(1 + \cos^2\theta) = \frac{2\pi^2}{\lambda_0^4}kT\beta\left(\tilde{n}\rho\frac{d\tilde{n}}{d\rho}\right)^2(1 + \cos^2\theta) \quad (3.5.46)$$

The subscript d in R_d refers to the so-called *density scattering*. The second half of the relation was derived using the relationship between refractive index and dielectric constant, $\varepsilon = \tilde{n}^2$. We should note that the size of the volume element V in our scattering problem does not enter the final formula (3.5.46).

We have designated temperature as our other independent variable. Hence, we should consider its fluctuation also. However, it is generally believed that the dielectric constant does not change with temperature at constant density; that is, $(\partial \varepsilon / \partial T)_\rho = 0$. Thus, fluctuation of temperature does not cause fluctuation of the dielectric constant and does not lead to scattering of light.

Let us apply formula (3.5.46) to the scattering of dilute ideal gases whose dielectric constant is

$$\varepsilon = 1 + 4\pi \alpha N \qquad (3.5.47)$$

where α is molecular polarizability and N is the number of molecules in unit volume. From $\rho = MN/N_{Av}$ and equation (3.5.47) it follows that $\rho \delta \varepsilon / \delta \rho = 4\pi \alpha N$. The definition of β for ideal gases leads to $\beta = 1/NkT$. Substitution of these values into equation (3.5.46) gives

$$R_d = N \left(8\pi^4 / \lambda_0^4 \right) \alpha^2 (1 + \cos^2 \theta) \qquad (3.5.48)$$

Comparison of this relation with equation (3.5.9) shows that the light intensity scattered by a sample of an ideal gas is equal to the sum of the intensities scattered by individual molecules. This summing of intensities is an example of a situation mentioned before, when all individual scattering entities are totally independent of each other. This is obviously the case of ideal gases.

On the other hand, molecules of liquids are not independent of each other. Total light scattered by liquids is much less than the sum of light scattered by individual molecules. It is given by equation (3.5.46); it depends on temperature not only through kT but also through temperature-dependent compressibility β.

There is another source of fluctuation in liquids: molecular orientation. The distribution function of orientations is not perfectly homogeneous but fluctuates with time. We have seen in Section 3.5.2 that molecular anisotropy leads to additional light scattering. We have divided the total scattering into an isotropic and an anisotropic part. The isotropic scattering arises from the average molecular polarizability and results in density scattering as described above. The anisotropic scattering originates from fluctuations in molecular orientations. In most liquids, orientations of individual molecules are independent of each other. In such liquids, the anisotropic scattering is simply the sum of the scattered intensities of all molecules present. (If there is some degree of orientational correlation among neighboring molecules, it will be reflected in the intensity of anisotropic scattering. Hence, the latter can be utilized for the study of the former.)

It should be obvious that the interference of scattered light in liquids suppresses appreciably the fully polarized density scattering but not the strongly depolarized anisotropic scattering. Hence, the light scattered by liquids is depolarized much more than light scattered by gases; often the depolarization is quite large.

3.5.6. Light Scattering by Polymer Solutions

In the preceding section we omitted the fluctuation of liquid composition as a source of light scattering. This compositional scattering may yield valuable information for all types of mixtures, but we will discuss only the light scattering of polymer solutions.

Compositional scattering is again described by equation (3.5.40) in which we need to express the fluctuation of dielectric constant by means of the fluctuation of concentration c_2. These quantities are related as

$$\delta\varepsilon = \frac{d\varepsilon}{d\tilde{n}}\frac{d\tilde{n}}{dc}\delta c = 2\tilde{n}\frac{d\tilde{n}}{dc}\delta c \qquad (3.5.49)$$

Here we have used the known relation between the dielectric constant and refractive index. Fluctuation of concentration is evaluated according to equation (3.5.44). A routine thermodynamic calculation yields

$$\overline{(\delta c_2)^2} = \frac{kT M_2 \phi_1}{V(\partial\mu_2/\partial c_2)_{P,T}} \qquad (3.5.50)$$

Here, as usual, the subscript 2 refers to the polymer, μ is chemical potential, and ϕ is volume fraction. Substitution of equations (3.5.49) and (3.5.50) into equation (3.5.40) gives for the Rayleigh ratio

$$R = \frac{2\pi^2}{\lambda_0^4}(1 + \cos^2\theta)\tilde{n}^2 \left(\frac{d\tilde{n}}{dc_2}\right)^2 M_2\phi_1 \frac{kT}{(\partial\mu_2/\partial c_2)_{P,T}} \qquad (3.5.51)$$

Again, the size of the volume element V does not enter into the scattering formula.

3.5.6.1. Measurement of Molecular Weight and Size.

For solutions of macromolecules, equation (3.5.51) provides a direct means for measurement of molecular weight. Of course, we must first separate the compositional scattering from the anisotropic scattering and density scattering. It is routinely assumed that in dilute solutions it is only the solvent that contributes to the latter two forms of scattering. Thus, R of equation (3.5.51) should be understood as the difference between the scattering of the solution and that of the pure solvent. We will now simplify equation (3.5.51) by introducing the angular scattering function R_θ of relation (3.5.29) and the light-scattering function K defined as

$$K \equiv \frac{2\pi^2}{\lambda_0^4}\tilde{n}_0^2\frac{(d\tilde{n}/dc_2)^2}{N_{\mathrm{Av}}} \qquad (3.5.52)$$

$$R_\theta = K M_2\phi_1 \frac{kT}{(\partial\mu_2/\partial c_2)_{P,T}} \qquad (3.5.53)$$

Replacing further the derivative of chemical potential by its virial expansion, equation (3.1.43), we get after minor rearrangement

$$Kc_2/R_\theta = 1/M_2 + 2Bc_2 + 3Cc_2^2 + \cdots \tag{3.5.54}$$

Equation (3.5.54) calls for plotting experimental data as Kc_2/R_θ versus c_2. The intercept is equal to $1/M_2$, and the initial slope is $2B$. This plot and its underlying relation are analogous to equation (3.2.46) and Figure 3.9, which described sedimentation equilibrium of nonideal macromolecular solutions. Again, the plot may be curved when the nonideality is strong; a square root plot may again be preferable.

So far we have tacitly assumed that the macromolecules are small compared to the wavelength of the scattered light. If they are not, then the intensity of the scattered light is reduced by intramolecular interference as described in Section 3.5.4. In such a case, the relation between scattered intensity and molecular weight, equation (3.5.54), must be modified to

$$Kc_2/R_\theta = P^{-1}(\theta)/M_2 + 2Q^{-1}(\theta)Bc_2 + \cdots \tag{3.5.55}$$

where $P(\theta)$ is the angular scattering function of equations (3.5.31)–(3.5.34). $Q(\theta)$ is another function of θ; it usually does not differ much from unity [$Q(0) = 1$].

In order to evaluate the molecular weight of large particles, we must thus extrapolate values of Kc_2/R_θ not only to vanishing concentration c_2, but also to vanishing angle θ. Inspection of equations (3.5.34) and (3.5.28) shows us that the latter extrapolation is most conveniently performed by plotting experimental values of Kc_2/R_θ against $\sin^2(\theta/2)$. In practice, the data are almost always plotted in a way originally suggested by Zimm, which allows for simultaneous graphical extrapolation of concentration and scattering angle. In the Zimm plot, Kc_2/R_θ is plotted against the sum $\sin^2(\theta/2) + k'c_2$ (Figure 3.50). The arbitrary constant k' is selected to make the plot look nice. Two sets of lines are drawn through the experimental points: one for constant values of θ, the other at constant c_2. Both sets are extrapolated to yield lines at $\theta = 0°$ and $c_2 = 0$. The extrapolated lines must meet at the axis; their common intercept marks $1/M_2$. The line at $\theta = 0°$ represents the virial expansion of the chemical potential; from its initial slope the virial coefficient B is easily calculated. The line at $c_2 = 0$ reflects the function $P^{-1}(\theta)$, and its initial slope divided by the intercept yields the radius of gyration r_G^2 of the particles; from its further course we can guess the shape of the particle. This experiment yields a remarkable amount of information, indeed.

We would like to comment on Zimm's plotting technique, which is very general and is useful whenever we need to plot a function $z(x, y)$ of two independent variables x and y, such as viscosity as a function of concentration and velocity gradient, or of concentration and temperature, or of temperature and solvent composition. Zimm's plot is a projection of a three-dimensional diagram in coordinates x, y, z into the plane xz. The projection rays are parallel to the xy plane (thus, the y axis projects onto the x axis); the tangent of their angle with the x axis is our arbitrary constant k'.

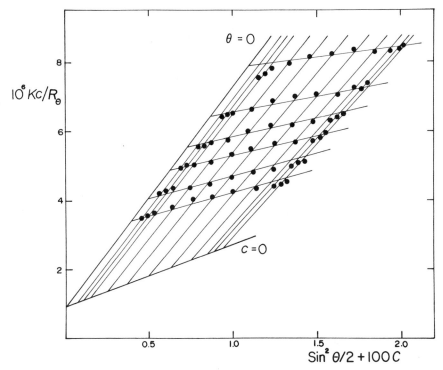

Figure 3.50. Zimm plot for a polystyrene sample ($\overline{M}_w = 980,000$) in tetralin at 98.4°C. (Courtesy of Dr. A.-A. A. Abdel-Azim.)

The two sets of lines are projections of intersections of planes $x =$ constant and $y =$ constant with the functional surface $z(x, y)$. The advantage of the plot is in correlation of experimental points; an erroneous point is more easily detected when it must be correlated with a large number of neighboring points.

Let us return to light scattering. Under some circumstances, it is not necessary to measure the whole angular dependence. When the molecules are known to be small (e.g., proteins), the measurement is performed at 90° only. The smallness of the particles can be checked by measuring *dissymmetry*, the ratio of intensities scattered at 45° and 135°, R_{45}/R_{135}. For small particles it is equal to unity. In some older studies, dissymmetry (if different from unity) was used for correcting the molecular weights measured at 90° and for estimates of the particle size. However, these procedures are not very dependable and are not used any more.

Another stratagem exploits the excellent parallelism of laser light. When it is used as the primary light, it is possible to observe the scattering at very small angles (2° or 4°). It is believed that no extrapolation to vanishing θ is necessary for such data. Laser light scattering also has the advantage that it is possible to observe very small scattering volumes. Dust particles enter such volumes only rarely and consequently disturb the measurement much less.

3.5.6.2. Effect of Polydispersity.   What kinds of averages of molecular weights and particle size do we measure by light scattering when the sample is polydisperse? In ideally dilute solutions, the intensities of light scattered by various components are additive. The constant K has the same value for all the polymer-homologes. Thus, in the limit of vanishing concentration, we can express the light intensity R_θ scattered by a polydisperse polymer as

$$R_\theta = \sum_i R_{\theta,i} = \sum_i K c_i M_i \left(1 - \frac{\mu^2 r_{G,i}^2}{3} + \cdots \right) \qquad (3.5.56)$$

Here, the expression we have substituted for Rayleigh's ratio $R_{\theta,i}$ of the ith component follows from rearranged equations (3.5.55) and (3.5.34). If we divide equation (3.5.56) by the total polymer concentration $c \equiv \sum_i c_i$, we get, after slight rearrangement,

$$\frac{R_\theta}{Kc} = \frac{\sum_i c_i M_i}{\sum_i c_i} \left[1 - \frac{\mu^2}{3} \frac{\sum_i c_i M_i r_{G,i}^2}{\sum_i c_i M_i} \right] + \cdots \qquad (3.5.57)$$

The first term on the right is, of course, the weight-average molecular weight $\overline{M}_w$. The coefficient of $\mu^2/3$ is, according to its definition, equation (1.6.11), the z average of the square of the radius of gyration $r_{G,z}^2$. Thus, these are the averages measured by light scattering.

   Polydispersity, besides affecting the intercept and initial slope of the plot of Kc/R_θ versus c, influences also its overall shape; it generally leads to curves that are flatter than the model dependencies of Figure 3.49. Sometimes an estimate of polydispersity can be obtained from the shape of this plot, but the results are not very reliable.

3.5.6.3. Polymers in Mixed Solvents.   Solutions of polymers in mixed (binary) solvents are three-component systems. The description of fluctuation of composition and of refractive index becomes much more complicated. Instead of presenting a rigorous argument, we will point to the relation between fluctuations and free energy. We need to understand the changes in the thermodynamic properties of the system upon introduction of the polymer component. This is exactly the same situation we analyzed in Section 3.2.1.1, in the context of osmometry, and in Section 3.2.2.2 for sedimentation equilibrium. We have learned that it is convenient to visualize the molecules of the polymer as carrying with themselves preferentially adsorbed solvent characterized by the coefficient λ of equation (3.2.14). In the present light-scattering experiments, it is again illustrative to consider the solvated complex as the entity whose concentration fluctuates in the system.

   The resulting relations are customarily cast in two different, but fully equivalent relations. The first modifies the constant K and yields the molecular weight of the unsolvated polymer.

$$K^* c_3 / R_\theta = 1/M_3 + 2Bc_3 + \cdots \qquad (3.5.58)$$

$$K^* \equiv \frac{2\pi^2}{\lambda_0^4} \tilde{n}_0^2 \frac{(\partial \tilde{n}/\partial c_3)_\mu^2}{N_{Av}} \qquad (3.5.59)$$

In these relations we designate the polymer by subscript 3, reserving subscripts 1 and 2 for the two solvent components. The derivative $(d\tilde{n}/dc_3)_\mu$ is a quotient obtained by comparing the refractive index of the polymer solution with its dialysate. The ordinary refractive index increment $(d\tilde{n}/dc_3)_{m_2}$ uses the solvent mixture in which the polymer was dissolved as a reference liquid. The situation is completely analogous to the case of two different density increments (or specific volumes of the polymer) that was studied by sedimentation equilibrium [cf. equation (3.2.54)]. Thus, measurement of refractive increment with respect to the dialysate leads to the molecular weight of the polymer without worrying about the preferential adsorption. This procedure is used routinely in biochemistry.

The other procedure utilizes the ordinary constant K (with $d\tilde{n}/dc_3$ at constant composition of the solvent mixture). In this case we obtain

$$Kc_3 / R_\theta = 1/M_3^* + 2B^* c_3 + \cdots \qquad (3.5.60)$$

$$M_3^* = M_3 \left[\frac{(\partial \tilde{n}/\partial c_3)_\mu}{(\partial \tilde{n}/\partial c_3)_{m_2}} \right]^2 = M_3 \left[1 + \lambda \frac{d\tilde{n}/d\phi_1}{(\partial \tilde{n}/\partial c_3)_{m_2}} \right]^2 \qquad (3.5.61)$$

$$B^* = B \left[1 + \lambda \frac{d\tilde{n}/d\phi_1}{(\partial \tilde{n}/\partial c_3)_{m_2}} \right]^{-2} \qquad (3.5.62)$$

In these equations we have used the relation

$$\lambda = \frac{(\partial \tilde{n}/\partial c_3)_\mu - (\partial \tilde{n}/\partial c_3)_{m_2}}{d\tilde{n}/d\phi_1} \qquad (3.5.63)$$

which is analogous to equation (3.2.55) describing the solution densities. The derivative $d\tilde{n}/d\phi_1$ refers to the dependency of the refractive index on the composition of solvent mixture in the absence of the polymer. If the true molecular weight of the polymer, M_3, is known (for example, from light scattered by the same polymer sample dissolved in a single solvent), then the coefficient λ can be evaluated from equation (3.5.61), and subsequently the second virial coefficient B is obtained from equation (3.5.62). Thus, light scattering produces both λ and B, from which the underlying thermodynamic parameters of the three-component system can be estimated. The sensitivity of the method increases with increasing value of $d\tilde{n}/d\phi_1$, that is, with increasing difference of refractive indices of the two solvents. We should recall that for sedimentation experiments the sensitivity increases for larger differences between the densities of the two solvents.

3.5.6.4. Turbidity. When a light beam traverses a polymer solution and the light is scattered, the intensity of the primary beam must be decreasing. Each layer of liquid scatters the same fraction of incident light, and hence the intensity I must be decreasing exponentially with the depth x.

$$I = I_0 e^{-\tau x} \qquad (3.5.64)$$

In this relation, τ is *turbidity;* it is a quantity quite similar to the extinction coefficient of light-absorbing liquids.

How is turbidity related to light scattering? In media that do not absorb light, we may state a law of light conservation: the flux of electromagnetic energy entering any volume element is equal to the flux leaving it. Thus, the intensity missing in the primary beam must be equal to the intensity of the scattered light integrated over all directions. As long as the scattering entities are small and the function $P(\theta)$ has a constant value of 1, the integration is easy and yields

$$\tau = (16\pi/3)R_{90} \qquad (3.5.65)$$

For such systems, the dependence of turbidity on concentration is obviously given by a relation very similar to equation (3.5.54), namely

$$Hc_2/\tau = 1/M_2 + 2Bc_2 + \cdots \qquad (3.5.66)$$

where the new turbidity constant H is given as $H \equiv (16\pi/3)K$. Thus, for small particles, measurement of turbidity is equivalent to the measurement of light scattering. In fact, in a sizable fraction of light-scattering literature, the results are reported as turbidity even if they were measured as light scattering—a slightly confusing practice.

When the scattering particles are larger than about $\lambda/20$, the intramolecular interference reduces the amount of light scattered at larger angles. The primary beam is reduced less and the turbidity is less than would be required by equation (3.5.66). In this case, molecular weight cannot be measured reliably.

The similarity of turbidity and extinction coefficient can be a source of error when light absorption is used for measurement of polymer concentration as is frequently done in biochemistry. We know that light scattering and turbidity increase with the inverse fourth power of the wavelength λ. Hence, they are quite large in the UV region, which is routinely used for spectrophotometric measurement of concentrations. Turbidity mimicking as absorption may lead to appreciable errors.

When discussing turbidity we should mention systems that are so large that the turbidity reduces the light intensity to a minuscule fraction of the original intensity. This may happen when the turbidity is large (emulsions, latexes, milk) or when the depth of the system is large (clouds). Under these circumstances, all the light is scattered in all directions; the scattered light is scattered again and again. Finally, the light is coming from all directions with the same intensity; all wavelengths are affected; that is, the light has the same spectral composition as the primary light (white in the case of sunlight). We have just described the reason milk and clouds are white. The above described *multiple scattering* has a strange result in astrophysics. The light changing its direction again and again does not move forward very fast. (In fact, this is another example of random walk, and diffusion equations apply again.) Thus, the light generated in the core of a star by nuclear reactions does not radiate immediately to space; it may take millennia to reach the star's surface.

3.5.7. Quasi-Elastic Light Scattering

Light scattering of macroscopic systems reflects fluctuations of their properties. It is in the nature of fluctuations that they change in time and space. The decay of every fluctuation is governed by the rules of dynamics (hydrodynamics) of the observed system. Hence, it is possible to derive hydrodynamic properties of solutions from their light scattering. The observed intensity of scattered light is characterized by irregular sudden onsets of fluctuations, which then decay with relaxation time τ_c. Thus, two intensity measurements taken within a very short time interval $\tau \ll \tau_c$ are usually quite close to each other (they may both differ from the average intensity), while two measurements taken at appreciably different times ($\tau \gg \tau_c$) will be unrelated. We want to extract the decay behavior from the fluctuating intensity. The most common technique for this purpose consists in evaluating the *autocorrelation function* $g(\tau)$ defined as

$$g(\tau) \equiv \lim_{T \to \infty} \frac{1}{T} \int_o^T I(t)I(t + \tau)dt \qquad (3.5.67)$$

Here, $I(t)$ is the scattered light intensity measured at time t, while $I(t + \tau)$ is the intensity measured after a (variable) delay time τ, and T is the overall experimental time. Presently, we are interested in the decay of fluctuations of concentration in macromolecular solutions. In this case, the function $g(\tau)$ converges when $T \approx 1$ min. Experimentally, the intensity is measured by counting photons (pulses detected by a photomultiplier) during a short interval Δt. The photon counts are fed into a special computer chip called an *autocorrelator*, which calculates the products $I(t) I(t + n\Delta t)$ for a large number of values of n (typically 128) and accumulates them in individual channels of the correlator.

It is possible (at least in some simple situations) to evaluate the relaxation time τ_c from the function $g(\tau)$. The τ_c is directly related to the diffusion constant of the solution, D, which can be found from it. The method is fast, precise, and absolute. In modern laboratories it has almost completely replaced the classical methods for measuring diffusion coefficients.

In order to understand the situation quantitatively, we need to return to the problem of scattering by a regularly arranged macroscopic system. This time we will assume that the density of scatterers along some vector **k** is changing in a harmonic way, while in the perpendicular directions it is not changing at all. Let us observe the light scattered by such a system. The interference of the scattered light will cancel it completely. The only exception would be the light scattered with a scattering vector **k** provided that the distance L between two successive maxima of the harmonic function is related to the wavelength λ_0 as

$$L = \frac{\lambda_0}{2\tilde{n}\sin(\theta/2)} \tag{3.5.68}$$

In this case the scatterers will reinforce each other. Equation (3.5.68) is, of course, an equivalent of the Bragg condition for diffraction of X-rays by crystals. The amplitude of the scattered wave will be proportional to the amplitude of the harmonic function describing the scatterers in our system.

How does this situation relate to a solution with fluctuating properties? Let us imagine that the density of scatterers (i.e., the fluctuations) is momentarily frozen. Then it is possible to express the spatial distribution of fluctuations by a three- dimensional Fourier series. Each term in the series scatters light, but only the term oriented along the vector **k** and having period L contributes to the observable light scattered under the angle θ. Previously, we represented the intensity of the scattered light by the time average of this particular scattered intensity.

We will now unfreeze the system. The fluctuations, which are causing an increase in the free energy, will tend to dissipate, and the system will approach the homogeneous state again. (Of course, thermal motion will generate new fluctuations that will be independent of the previous ones.) The return to homogeneity requires the molecules to move through the liquid; this movement is governed by the diffusion coefficient D and by Fick's equation (3.4.3).

We have seen that the only fluctuation that is active in light scattering has the form of a sine function with period L. The reader may easily satisfy himself that Fick's equation predicts that the concentration profile will retain its sinusoidal shape with increasing time, but the amplitude will exponentially decay. The relaxation time τ_c characterizing this decay is found as

$$\tau_c = \frac{L^2}{4\pi^2 D} = \frac{\lambda_0^2}{16\pi^2 \tilde{n}^2 D \sin^2(\theta/2)} \tag{3.5.69}$$

Further, if all the diffusing particles have the same diffusion coefficient, then the function $g(\tau)$ adopts a very simple form:

$$g(\tau) = I_0^2(1 + be^{-\tau/\tau_c}) \tag{3.5.70}$$

where I_0 is the average light intensity and b is a constant. The relaxation time τ_c is then found by an appropriate curve fitting, and D is calculated from it using equation (3.5.69).

The method runs into difficulties for polydisperse samples, for which the function $g(\tau)$ is a superposition of many exponential curves. The distribution of diffusion coefficients can be obtained from it in principle, but not in practice. When τ_c is evaluated from the initial portion of the autocorrelation function (small values of τ), the corresponding value of the diffusion coefficient is the z average, $\overline{D}_z$. Considerable effort has been devoted to attempts to derive more extensive information from the function $g(\tau)$, but the results have been meager.

The model of quasi-elastic light scattering, as presented above, is satisfactory only in the simplest circumstances. When the scattering particles are large, other modes of molecular motion—rotary diffusion and various types of particle deformation—also contribute to the autocorrelation function. However, these phenomena are beyond the scope of our treatment; they are also quite difficult to detect and interpret.

It should also be mentioned that fluctuations in the intensity of a light beam are inevitably accompanied by fluctuations in its frequency. The scattered light always has a finite spectral bandwidth even if the primary laser light is virtually monochromatic. This phenomenon gave rise to the term quasi-elastic: elastically scattered light has the same frequency; if the frequency is changed however little, the scattering is only quasi-elastic. The bandwidth broadening is similar to the Doppler effect known from acoustics. Light scattered by a particle moving toward the observer is perceived as having a higher frequency; particles moving away produce lower frequencies. *Self-beating spectroscopy* methods have been developed for the measurement of the bandwidth and band shape of the scattered light. Analysis of these bands provides the same information as the auto-correlation function. However, the two methods are somewhat complementary. The self-beating spectroscopy is preferable for fast fluctuations that produce broader bands, while the measurement of intensities and $g(\tau)$ is more convenient for slower fluctuations that correspond to narrow bands. The latter method is used almost exclusively in macromolecular studies.

3.5.8. Experimental Arrangements

The basic design of light-scattering photometers is simple. A cell containing the scattering sample is illuminated with a strong parallel beam of primary light and is observed with a photomultiplier. (In the early days of light scattering, the light intensities were measured visually with good success. The human eye adapted to full darkness is amazingly sensitive—it can detect signals of about 10 photons—and recognizes the equality of two light fields with a precision of about 2%.) The cell is rectangular when measurements are required only at 90°. Cylindrical cells are employed when the angular dependence of the scattered light intensity is needed.

In a typical experiment, the intensity of the scattered light is about five to six orders of magnitude smaller than the intensity of the primary light. Hence, great care must be devoted to the elimination of all sources of spurious light, notably light reflected on the glass-liquid and glass-air boundaries and, of course, on all parts of the hardware. The reflections on the air-glass boundary are advantageously suppressed when the measuring cell is immersed in a larger vat filled with a liquid (benzene, toluene, xylene) whose refractive index is close to that of glass.

The most dreaded enemy of light-scattering practitioners is dust. A single large dust particle may scatter more light than the measured liquid itself. Elimination of dust is not easy; it is accomplished by either centrifugation or filtration. Membranes

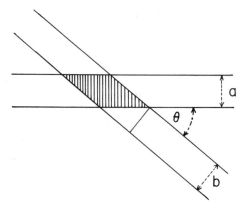

Figure 3.51. Illuminated volume in light-scattering cell that is observed by the photomultiplier. a is the width of the slit in the primary beam; b is the width of the slit in front of the photomultiplier.

with controlled porosity attached to a syringe have been used in more recent studies. Of course, the cell itself and all the filtering equipment must be dust-free to begin with. Moreover, both filtration and centrifugation can change the concentration of the macromolecular solution.

The intensity of the scattered light received by the photomultipler is proportional to the volume of the illuminated liquid observed by the photomultiplier. For a given instrument with a given arrangement of slits in the primary and scattered beams, this volume still depends on the angle of observation θ and on the refractive index of the observed liquid. The dependence on θ is apparent from Figure 3.51. The shaded area representing the illuminated volume is proportional to $ab/(\sin \theta)$, where a and b are the widths of the apertures restricting the primary and scattered beams, respectively. The effect of refractive index is less obvious. Light beams crossing the liquid-glass boundary are refracted; the degree of beam deflection depends on the refractive index of the liquid. Some light beams may thus miss the detector; they would hit it if not deflected. The needed correction is a complicated function of the apertures and refractive index. However, with a proper design (photomultiplier not seeing the edges of the primary beam), the observed volume is inversely proportional to $\tilde{n}^2$.

As the intensities of primary and scattered beams differ so much, it is impractical to compare them directly. In practice, the scattered intensities are compared to light scattered at 90° by some reference liquid; benzene is selected most frequently. [Usually, an intermediate standard is employed, typically a cell-shaped piece of poly(methyl methacrylate) or glass.] Putting all the factors together we get the formula for calculating the Rayleigh ratio R_θ (unpolarized primary light) as

$$R_\theta = R_{90,B} \frac{G}{G_B} \left(\frac{\tilde{n}}{\tilde{n}_B} \right)^2 \frac{\sin \theta}{1 + \cos^2 \theta} \qquad (3.5.71)$$

where G is the signal from the photomultiplier, and the subscript B refers to benzene. (The accepted value of $R_{90,B}$ is 1.67×10^{-3} m^{-1} for $\lambda_0 = 546$ nm.)

Rather intense light sources are needed to obtain sufficiently large scattered intensities. In older studies, mercury lamps were used, with filters isolating either the blue light ($\lambda_0 = 436$ nm) or the green light ($\lambda_0 = 546$ nm). The blue light has been more popular, because the scattered intensities are much higher. (Remember the dependence on λ_0^{-4}.) However, blue light may excite fluorescent radiation; even very small amounts of impurities may produce enough fluorescent light to vitiate the measurement. Green light is much safer in this respect.

The advent of lasers has changed the scenery of light scattering considerably. Lasers are a must in quasi-elastic measurements, but they have many advantages also in the classical scattering experiments. They produce perfectly parallel beams that are totally polarized and strictly monochromatic. Helium-neon lasers ($\lambda_0 = 632.8$ nm) are used most often. The use of lasers also allows for one shortcut. When the molecular weight of large particles is measured, extrapolation to zero scattering angle is necessary. With the traditional light sources, scattering angles between 25° and 150° are practical; for large particles, extrapolation may become difficult. When a laser light source is used, it is possible to place the detector directly against the light source, block the primary beam, and observe only a narrow annulus around it. The angle θ in this case is about 2°, and extrapolation is believed to be unnecessary. This design, however, does not permit the measurement of molecular sizes.

3.5.9. Small-Angle X-Ray Scattering (SAXS)

Both X-rays and visible light are electromagnetic waves. Hence, their scattering by matter must obey the same rules. The principal difference is the wavelength. While the wavelength of typical green mercury light is 546 nm, the wavelength of copper K_α radiation (the most frequently employed X-ray) is 0.154 nm. We have seen that the intensity of scattered light generally decreases with the increasing scattering angle—more precisely, with increasing value of the product μr, where r is the characteristic dimension of the scattering particle. [The X-ray scientists prefer to work with different symbols. Our parameter μ of equation (3.5.28) carries the symbol $h \equiv \mu$. More confusing is the definition of scattering angle θ; in X-ray work, it is only half of what we used to call θ in light-scattering studies. Hence, in X-ray studies $h \equiv 4\pi(\sin\theta)/\lambda$.] The dependence on hr implies that scattering of large structures is confined to small scattering angles while smaller structures scatter in a broader angular range. Small-angle X-ray scattering exploits this phenomenon thoroughly.

Before proceeding with the analysis of the X-ray scattering envelope, let us review the physical basis of X-ray scattering. The energy of X-rays is much greater than the dissociation energy of electrons in the sample (with the exception of the innermost electrons in the heaviest atoms). It follows that the electrons behave with respect to X-rays as free electrons. It follows further that every electron exhibits the same scattering behavior. Exact positions of electrons within molecules cannot be found; we have to work with electron density (number of electrons per unit volume). Inhomogeneities of the electron density (even on the atomic level) lead to X-ray scattering. On the higher level, the situation is fully analogous to light scattering. If the inhomogeneities (read "molecules") are regularly distributed, no scattered X-rays are observed; only fluctuations lead to scattering.

In SAXS studies, the experimental data are first evaluated as a dependence of the reduced scattered intensity $I(h)$ on the scattering parameter h. If necessary, the dependences are measured for several solutions and extrapolated to vanishing concentration of the macromolecules. The resulting curves can be studied at several levels. The scattered intensity extrapolated to zero scattering angle ($\theta = h = 0$) corresponds to the largest structures—to the whole macromolecules. From its magnitude, the molecular weight of the solute can be evaluated. This situation is completely analogous to light scattering; the difference in electron densities between the solute and the solvent plays the role of the difference of refractive indices. The scattering in the region of the smallest angles reflects the size of the largest structures present: the whole macromolecule and its radius of gyration, r_G. Thus, equation (3.5.34) for the angular scattering function applies again. However, in SAXS studies, equation (3.5.34) is usually replaced by the *Guinier approximation*,

$$I(h) = I(0) \exp\left(-h^2 r_G^2/3\right) \tag{3.5.72}$$

When the exponential function is expanded into a Taylor series, the first two terms are identical with those in equation (3.5.34). The Guinier expression usually works extremely well for most shapes of particles. It calls for plotting $\ln I(h)$ against h^2; the slope immediately yields the radius of gyration. The SAXS method is especially advantageous for the measurement of r_G of smaller particles, which possess an essentially flat scattering envelope when ordinary light scattering is measured.

However, the real strength of the SAXS method is manifested in scattering at angles beyond the Guinier region. At these angles, the interference of the scattered radiation reduces the intensity to very small values for all inhomogeneities of the size of the whole molecule. However, the inhomogeneities on a smaller (submolecular) scale still produce scattering of observable magnitude. Let us demonstrate the situation on scattering by thin, very long rods. Such rods are actually regular structures in one dimension; radiation scattered by one section of the rod must be exactly canceled by radiation from another section unless the rod is oriented perpendicularly (or almost perpendicularly) to the scattering vector **k**. In the latter case, the different sections mutually reinforce their scattering. The relevant inhomogeneity is now related to the cross section of the rod. Hence, from the angular scattering of the rods we can evaluate their cross section, or more precisely, the cross-sectional radius of gyration.

A more detailed analysis shows that different types of particles exhibit different limiting behavior at larger values of h; from this behavior we can determine the type of the particle and its characteristic dimensions. Without going into computational details, we will now present some of the more important relations in the theory of SAXS.

Intensity scattered by a single particle, $I_1(h)$, is related to its volume V and to the electron density difference between the particle and its surroundings $\Delta\rho$ as

$$Q_1 = \int_0^\infty h^2 I_1(h)dh = 2\pi^2 (\Delta\rho)^2 V \tag{3.5.73}$$

$$I_1(0) = (\Delta\rho)^2 V^2 \tag{3.5.74}$$

The invariant Q_1 depends only on the volume of the particle and on $\Delta\rho$ but not on the particle shape; its evaluation from experimental data is tedious but straightforward. Thus, the volume of the particle could be calculated from the intensity scattered at zero angle $I_1(0)$, and from Q_1. We should note that volume V is the actual (dry) volume of the macromolecule, *not* its hydrodynamic or solvated volume.

Scattering is caused by inhomogeneities. In solutions, the main locus of inhomogeneities is the surface of the particle. Hence, SAXS may yield the value of the molecular surface S. The appropriate relation reads

$$\frac{S}{V} = \pi \lim_{h\to\infty} \frac{I_1(h)h^4}{Q_1} \tag{3.5.75}$$

This relation implies that the limit exists; i.e., that $I(h) \sim h^{-4}$ at high values of h. Inspection of equation (3.5.73) reveals that the ratio $I_1(h)/Q_1$ must be independent of the units used for measurement of intensities; arbitrary units are also satisfactory. In other words, it is not necessary to measure the absolute values of intensities for the calculation of S. This will be true for most of our subsequent relations.

For long rods, $I_1(h)$ is the product of an *axial factor* $I_a(h)$, which at larger values of h is proportional to h^{-1}, and the *cross-sectional factor* $I_c(h)$. Hence,

$$I_c(h) \sim I_1(h)h \tag{3.5.76}$$

The dependence of the cross-sectional factor on h again has the form of the Guinier relation,

$$I_c(h) \sim \exp(-h^2 r_c^2/2) \tag{3.5.77}$$

Here, r_c is the cross-sectional radius of gyration, a quantity that is the two-dimensional equivalent of the usual radius of gyration r_G. (Note also the factor 2 in the denominator.) The r_c is easily evaluated from the slope of the dependence of $\ln I_c(h)$ vs h^2. Relation (3.5.77) is usually valid in an extended, but nevertheless limited, region. At very small values of h, the scattering by the whole rod becomes dominant and equation (3.5.72) applies. At very large angles, the Guinier approximation loses validity.

The assumption of a rigid rod with a straight axis was not necessary for the preceding consideration. It is valid also for slightly flexible threads provided that the "straight" portion of the thread is appreciably larger than its diameter. Then only the sections of the thread that are perpendicular to the vector **k** contribute to the scattering, which is still rodlike.

Similar relations hold for thin flat particles. In this case, the dependence of the thickness factor $I_t(h)$ on h yields the thickness radius of gyration r_t, from which the thickness can be evaluated.

$$I_t(h) \sim I_1(h)h^2 \sim \exp\left(-h^2 r_t^2\right) \tag{3.5.78}$$

SAXS is also valuable for studying molecular coils. We have seen that the radius of gyration r_G can be obtained from the scattering profile at small values of h. However, for Gaussian coils it can also be evaluated from the limiting behavior at large values of h, that is, from measurement in a more accessible region. In these calculations, the function $I_n(h) \equiv I_1(h)/I_1(0)$ is employed, which is identical with function $P(\theta)$ of equation (3.5.31). Its limiting behavior reads

$$\lim_{h \to \infty} h^2 I_n(h) = \frac{2}{r_G^2} \tag{3.5.79}$$

The larger the scattering angle, the smaller the scattering entities that determine the scattering profile. At some value of h, the segment model of a coil that led to equation (3.5.79) does not describe the scattering adequately; the wormlike model (Section 1.3.1.2) becomes more appropriate. This model behaves as a Gaussian coil when observed at small angles and as a rod at larger angles. Thus, the cross section of the chain can again be calculated from equation (3.5.77). More important, the persistence length of the chain, a, can be calculated from the transition in the scattering profile between the regions of segmentlike and wormlike scattering. Figure 3.52 is a schematic drawing of the scattering profile. The ordinate $I_1(2\theta)^2$ is proportional to $h^2 I_n(h)$, while the abscissa 2θ is proportional to h. The plot comprises an early ascending Guinier region, a flat central region corresponding to limiting behavior of a coil [equation (3.5.79)], and a late linear segment (pointing to the origin) characteristic of the rod behavior. The abscissa $2\theta^*$ of the point of intersection P is related to the persistence length as

$$a = 1.91\lambda/4\pi\theta^* \tag{3.5.80}$$

In real experiments, the central portion is usually not quite horizontal, but the transition is clearly observable.

In summary, SAXS under optimum conditions can provide a real treasure of structural information: molecular weight and dry volume of the particles, their radius of gyration, cross section of the molecular chain, its persistence length, and surface-to-volume ratio of the molecules.

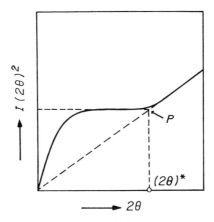

Figure 3.52. A schematic drawing of the scattering curve of a Gaussian coil. See text for details. From O. Glatter and O. Kratky, *Small Angle X-ray Scattering*, Academic Press, New York, 1982.

3.6. SPECTRAL METHODS

Many spectral properties are useful for gathering information about small and large molecules: ultraviolet and infrared spectra, dichroism, optical rotation and circular dichroism, fluorescence phenomena, nuclear magnetic resonance. Studies of these properties represent a major branch of physical and organic chemistry; an appropriate description of them would be well beyond the scope of this book. We will therefore assume that the reader is familiar with the general features of spectra, with their measurement and interpretation, and will not present their theoretical and experimental background. We will only demonstrate how various spectral properties contribute to the study of macromolecules.

3.6.1. Ultraviolet Spectrophotometry

Light absorption in the UV region is caused by *electronic excitations*. Only molecules with π electrons absorb in the experimentally accessible part of the UV region (190–400 nm). The absorption bands are few and very broad; very little structural information is obtainable from the spectra. On the other hand, both the shape of the spectra and the absorbances are measurable with very good precision. Measurement of absorbance serves as one of the better methods for finding concentrations of macromolecules, mainly of biochemical origin—proteins and nucleic acids.

The energy and probability of an electronic transition (i.e., position of the absorption peak and the extinction coefficient) depend strongly on the structural details of the chromophore. An extreme example is tyrosine, an amino acid present in most proteins. Its UV absorption is quite different in neutral solutions and in alkaline media where its phenolic hydrogen is dissociated. Thus, when protein solution is titrated by alkalies, a quite distinct change marks the titration of tyrosine. Now, in some proteins, part of the change occurs at pH values much higher than expected. In these proteins, some of the tyrosine residues are buried in the hydrophobic parts of the molecule and are not accessible to the titrating agent. They become accessible only at pH values at which the whole tertiary structure of the protein is disrupted.

The heterocyclic bases dominate the ultraviolet absorption of nucleic acids. When these bases are paired and stacked in the double helix, their electronic structures interact strongly. As a result, their extinction is about 40% less than extinction of the individual bases. The effect is called *hypochromicity*. When the double helix is disrupted, for example by thermal denaturation, the extinction increases substantially. Thus, UV spectroscopy is very useful for studying structural changes of nucleic acids. A similar effect is observed in optical rotation and circular dichroism and their dependence on wavelength in the UV region. These phenomena depend not only on the optical activity of the individual nucleotides, but also on their helical arrangement, which causes anomalies in the wavelength dependencies. Again, these anomalies disappear upon disruption of the helix. Similar phenomena are observed with other helix-forming polymers, such as poly(γ-benzyl glutamate). Optical activity is especially sensitive to conformational transitions.

The usefulness of UV spectrophotometry for synthetic polymers is much less. This is caused by, among other things, the paucity of solvents that are transparent throughout the UV region. Nevertheless, it was used for the study of the ionization of poly(vinyl pyridine) and for following the conformational transitions in polystyrene.

3.6.2. Fluorescence

Fluorescence occurs when some chromophores, after being electronically excited, reemit the energy. The whole process is quite complex. During the absorption of the UV photon, the chromophore is usually brought to a state that is excited both electronically and vibrationally. However, nonradiative relaxation very quickly brings the chromophore to its lowest vibronic level within the electronically excited state. When the remaining excitation energy is reemitted and the chromophore returns to the electronic ground state, the molecule usually retains some vibrational energy (which is subsequently also dissipated). Thus, the fluorescent radiation occurs at longer wavelengths than the exciting radiation; a whole fluorescence spectrum is observed because the molecules may end up in different vibronic states. The reradiation process is slow as electronic phenomena go; the lifetimes of the excited states are measured in nanoseconds, while most transitions among electronic states are completed in picoseconds or faster.

In macromolecular studies, the chromophore is either a part of the macromolecule itself [tryptophan residues in proteins, naphthyl groups in polyvinyl naphthalene)], or fluorescent probes are covalently attached to the macromolecule (e.g., on chain ends), or the probes are simply added to the solution, where they will interact with the macromolecule (e.g., by intercalating into the DNA double helix).

Many aspects of fluorescence yield information about macromolecular systems. The intensity and spectral distribution of the fluorescent radiation are very sensitive to the molecular environment of the probe. Typically, the radiation is much more intense when the probe is surrounded by a hydrophobic medium than when it is in water. This provides an important clue about the internal structure of more complex systems such as membranes.

Many things may happen to the excited chromophore during its long lifetime. It may collide with another molecule and transfer the excitation energy to it; nonradiative processes will eventually dissipate the energy. This phenomenon is called *quenching*; molecular oxygen, iodide ions, and many other molecules are efficient quenchers. This process substantially reduces the intensity of the fluorescence and its lifetime. Usually, each collision leads to quenching. Hence, the process is diffusion-controlled, and its observation provides information about the nature and rate of diffusion processes in the immediate vicinity of the probe.

In a similar process, the excited probe transfers its energy to another chromophore in the system that requires lower energy for its excitation. The observed fluorescence is then a combination of fluorescence spectra of both types of chromophores. When both chromophores are attached to the same macromolecule, study of this energy transfer yields clues about the dynamics of intramolecular motions. Sometimes the excitation energy is transferred repeatedly among chromophores of the same type (e.g., phenyls in polystyrene) before it reaches the chromophore of the other type (the trap). Fluorescence yields information about this migration and through it about the conformational details of the polymer chain. In another scenario, the excited chromophore forms a complex (an *excimer*) with its twin. This excimer yields a different fluorescence spectrum than the single excited chromophore (the excimer excitation can migrate too). Again, a deeper insight into the conformation and the dynamics of the coil is obtained.

The probability of absorption of a UV photon is at a maximum when the transition dipole of the chromophore is parallel to the electric vector of the radiation; it is vanishing when the vectors are perpendicular. The same applies for the fluorescent radiation. Let us imagine a system of randomly oriented chromophores, each of them rigidly fixed. Let us irradiate it with a polarized primary light. The properly oriented chromophores will be excited preferentially; the reradiated fluorescence will be depolarized only partially. Let us now return the freedom of motion to the chromophores. The excited ones will be able to change their spatial orientation during their long lifetime; the fluorescence will be increasingly depolarized. The degree of depolarization will depend on the interplay of the (independently measurable) lifetime of the excitation and the rate of reorientation of the chromophore. If the chromophore is attached to a rigid molecule (e.g., a protein), its rotational diffusion coefficient (see Section 3.3.7) is measurable. When the chromophore is part of a flexible coil, the intramolecular dynamics of the coil can be studied. In any case, the method of *fluorescence depolarization* provides valuable hydrodynamic information about the systems observed.

3.6.3. Infrared Spectra

In infrared spectroscopy, absorption of radiation is caused by *molecular vibrations*. Extensive study of model compounds has shown that a given chemical group (e.g., $-CH_3$, $-OH$, $-NH_2$, $>C=O$) is characterized by an absorption band in some region of the spectrum that is typical for this particular group. The exact position of the band within its region provides information about the molecular vicinity of the vibrating group. For example, aldehydic, ketonic, and ester carbonyls are clearly distinguished. This feature provides great diagnostic value for small molecules of natural origin in establishing their structure. It has similar value for analyzing the chemical structure of a polymer or copolymer of unknown origin. However, it is equally valuable for studies of the finer structural features of known polymers. For example, the characteristic bands of methyl groups have established their prevalence in low-density polyethylene and helped to solve the problem of short branches on polyethylene molecules.

In the production of polybutadiene, the monomers are incorporated either by 1,2 addition (leaving pendant vinyl groups on the polymer chain) or by 1,4 addition. In the latter case, the incorporated double bond is in the cis or trans configuration. The vibrational spectra of hydrogens attached to these different double bonds are clearly different. Hence, it is possible to quantify the proportions of the three forms of the monomeric unit; from these, we can infer information regarding the mechanism of polymerization.

The amide absorption of proteins and polypeptides occurs at different wavelengths for amides that are a part of the α-helix, of the β-sheet structure, or of a randomly coiled chain. This feature is particularly useful for identifying polypeptide conformations in solutions.

A particular vibration is active in the infrared absorption only when it is accompanied by a change (vibration) in the dipole moment of the molecule. Moreover, the dipole must vibrate in the same direction as the electric vector of the photon to be absorbed. In liquids, a certain fraction of molecules is always oriented in the appropriate way and the absorbance is independent of the polarization of the incident beam. However, if the vibrating groups are oriented along some direction, then absorption will depend on the mutual orientation of the dipoles and the polarization of light; we will observe *infrared dichroism*. This effect is mainly important when studying polymers in the solid state.

Infrared spectroscopy is also a useful technique for the measurement of the concentration of the end groups in a polymer. For linear polymers, the number average molecular weight may be determined from it. For branched polymers, this concentration together with information from other methods leads to the estimates of the degree of branching. It is a frequently used technique for the quantitative analysis of microstructure, stereoregularity, curing, branching, or crosslinking in polymers. The method is rapid, simple to operate and is inexpensive.

In recent years the advent of Fourier transform infrared (FT-IR) spectroscopy led to considerable improvement in both the quality and the interpretation of polymer IR spectra. Single scans through an empty sample compartment achieve a signal-to-noise ratio of 1000-to-1 or better at a resolution of 2–4 cm^{-1}. Such high ratio is critical for the isolation of small features of polymer spectra and for recording of the spectra with the full-scale sensitivity of one milliabsorbance unit. Thanks to the FT-IR the direct transmission sampling technique for recording of the IR spectra of polymers is not any more the only available procedure. Several new sampling techniques allow the polymer to be examined in its fabricated state; that is as a powder or fiber (diffuse reflection), film (reflectance absorbance), or bulk sample (photoacoustic).

Raman spectroscopy, a complimentary technique to IR, is also becoming a more common tool for the polymer analysis. Due to the differences in the nature of the selection rules, Raman spectroscopy yields vibrational information, which is not obtainable from IR.

3.6.4. Nuclear Magnetic Resonance (NMR)

Some atomic nuclei, notably protons and carbon-13, possess spin. This spin gives them a magnetic moment that may align with or against any external field felt at their location. These two alignments represent two slightly different magnetic energy states. If an oscillating magnetic field with energy corresponding to the difference between the energies of these two states is also felt at these nuclei, stimulated transitions between the two states will occur with a net absorption of energy. This phenomenon is called *nuclear magnetic resonance* (NMR). It is an invaluable tool in structure determination and much of our basic knowledge of polymer microstructures is based on the use of NMR.

The resonant frequency for a given type of nucleus is directly proportional to the magnetic field felt at the nucleus. This field, in turn, is the sum of the external magnetic field and the magnetic fields generated by the nearby atoms and bonds of the molecule in which the nucleus resides. The magnetic field generated by the moving electrons in nearby bonds is the principal reason for the different resonant frequencies (chemical shifts) of different atoms in a given molecule. The chemical shift of a given proton depends primarily on the types, distance, and angles of nearby bonds. Chemical shifts are thus highly characteristic for individual groups and serve as an excellent guide for identifying them. They are measured on the dimensionless δ scale (in ppm from the frequency of a reference compound).

What is the effect of other nuclei with a magnetic moment in the vicinity of the observed nucleus? The two moments may be parallel or antiparallel; these two arrangements will produce an opposite shift of the resonant frequency. The phenomenon is known as *coupling*. Two types of coupling play a major role in NMR studies.

Dipolar coupling results from the interaction between nuclei belonging to different molecules. Let us consider an interfering nucleus that is fixed in space with respect to the observed nucleus. The coupling of the nuclei will lead to splitting of the absorption band into a widely separated doublet; the separation will depend sensitively on the mutual distance and orientation of the nuclei. During an actual measurement, we do not observe a single nucleus (with its environment), but many of them. Each pair of interacting nuclei will produce a doublet with a different separation of lines. Obviously, we will observe only the superposition of all doublets (i.e., a very broad absorption band).

Let us now allow the molecules to participate in the normal thermal movement. The situation will change dramatically. The time needed for the absorption of energy during the transition is quite long (the radio frequencies employed are in megahertz).

During this time, the neighboring molecule will sample all possible mutual orientations and the coupling effect will average out; no splitting of lines will occur, and the lines will remain narrow. Consequently, when the molecules of interest are studied in solution, a spectrum of narrow lines is observed that could be evaluated in terms of chemical shifts, and so on, and utilized for determination of the molecular architecture. On the other hand, solid polymers yield broad NMR bands. When these bands are studied at different temperatures, they reveal the onset of thermal motion by a more or less sudden narrowing of the band. This effect is useful for studying glass transition and related phenomena.

When the magnetically interfering nuclei belong to the same molecule they may give a rise to a phenomenon of *nuclear spin coupling*. In this case the magnetic moments of the two atoms remain in the same mutual relation at all times. Their interaction is mediated by overlapping orbitals of the connecting bonds. As a result, the lines remain sharp and their splitting is clearly observable. The study of the multiplet structure of the absorption bands then provides detailed information about the molecular structure.

Let us briefly review the coupling and splitting rules of proton NMR of organic substances. Similar principles hold, of course, for carbon-13 NMR.

1. Hydrogens on the same carbon atom (geminal hydrogens) are strongly coupled and split mutually. However, when their environment is fully equivalent (this is always the case with a freely rotating methyl group), no splitting is observed.

2. Hydrogens on neighboring carbons are responsible for the multiplet structure; a single hydrogen produces a doublet, three hydrogens of a methyl group a quartet, etc.

3. Hydrogens on more distant carbons cause less prominent splitting, which is observable as a fine structure of the individual peaks of a multiplet.

Much of our basic knowledge of polymer microstructure (tacticity) was obtained from NMR studies. We will demonstrate the analysis of NMR spectra on a relatively simple case of poly(methyl methacrylate), $-CH_2-C(CH_3)(COOCH_3)-$. This polymer has three types of hydrogen atoms: three equivalent α-methyl hydrogens, three equivalent ester methyl hydrogens, and two β-methylene hydrogens. The environment of two methylene hydrogens is different in isotactic and syndiotactic polymers. In Figure 3.53, the two structures are depicted schematically; the filled circle represents the α-methyl group and the open circle the ester group. The syndiotactic arrangement (a) has a twofold axis of symmetry. The two geminal methylene protons are in equivalent environments; they have the same chemical shift and should appear in the spectrum as a singlet. In the isotactic arrangement (b) the two protons are in different environments. They should have different chemical shifts and should mutually split, producing two doublets in the spectrum.

Figure 3.54 shows the actual spectra (the tacticities are, of course, not pure). The β-methylene spectrum appears around $\delta = 2.0$; it is essentially a singlet for the

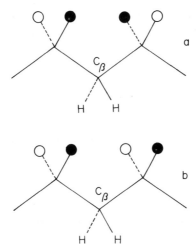

Figure 3.53. Environment of the two β-methylene hydrogens in poly(methyl methacrylate). It is equivalent for both in the syndiotactic arrangement (a) but not in the isotactic one (b).

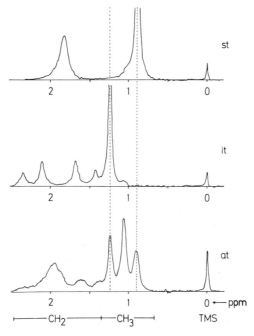

Figure 3.54. Sections from the proton resonance spectra of syndiotactic (st), isotactic (it), and atactic (at) poly(methyl methacrylate). The signals of the methyl ester protons are not shown. TMS is the reference signal of tetramethylsilane. (Reprinted from H.-G. Elias, *Macromolecules*, 2nd ed. Copyright © 1984 by Plenum Press, New York. Used by permission of the publisher.)

syndiotactic polymer; it has four bands for the isotactic one. The ester methyl hydrogens, being equivalent and having no close neighbors, form a singlet at $\delta = 3.6$; it contains no stereochemical information. We would expect similar behavior for the α-methyl hydrogens with their bands around $\delta = 1.2$. Actually, there are three bands in that region corresponding to three different chemical shifts for three possible monomer triads: syndiotactic, isotactic, and heterotactic. Thus, it is possible to deduce from the spectrum the statistics of both diads and triads in the polymer. The statistics can be extended even further when either higher-field proton NMR or carbon-13 NMR is used. Different tetrads will then be observed as a fine structure of the methylene spectrum; pentads will reveal themselves as a fine structure of the α-methyl spectrum.

Similar evaluation techniques can be used for studying the proportion and sequences of different monomeric units in copolymers. Obviously, NMR is the method of choice when the primary structure of polymers is being studied.

The proton spectra of polymers with severe multiplicity of line splitting show extensive overlap of resonances, making resonance-structure assignments difficult. Developments in Fourier transform techniques with pulsed NMR made it possible to obtain NMR spectra of the carbon backbone of the polymer chain. One could measure the carbon-C13 nucleus in spite of its low magneto-gyric ratio and low natural abundance. Even when the sensitivity of the carbon-13 NMR is much lower than of the proton NMR, the proton-decoupled carbon-13 spectra reveal more structural details than the proton spectra. The chemical shift range of carbon-13 is 250 ppm compared with the 10 ppm range of the proton and the proton-decoupled carbon-13 NMR spectra generally show only single resonances for chemically inequivalent carbons. Thus, ideal NMR analysis of polymers would utilize the high sensitivity of the proton NMR spectrum and the high specificity of the carbon-13 NMR spectrum.

Although high-resolution NMR is extremely important for characterization of polymers, it has one severe limitation: Polymer samples must be dissolved or melted to obtain the spectra. This limitation was removed with the advent of high-resolution solid state NMR spectroscopy. It is used for samples that are insoluble, such as cross-linked or intractable polymers.

In this technique, magic angle spinning overcomes the band broadening caused by dipolar coupling. The insufficient molecular motion in solids is compensated by rapid spinning of the solid sample along an axis inclined from the direction of the magnetic field by 57.4° (the *magic angle*). Thus, each molecule experiences a continuous change of orientation with respect to the external magnetic field resulting in averaging out the dipolar coupling. Narrow lines and enhanced sensitivity are achieved.

In addition to the information about the chemical structure of the sample, solid-state NMR spectroscopy provides information about the nature of the solid state, such as the conformation of the chains, crystallographic form, and the morphological character of the solid. The dynamic properties of solid polymers can also be studied by measuring the frequency of the molecular motion and the correlation times through relaxation times. The mechanical performance of polymers is determined almost entirely by their temperature-transition behavior and energy dissipation properties (brittle, tough, rubbery, etc.). It is therefore important to relate these dynamic properties to segmental motion in the polymer. Solid state NMR is useful in these studies.

3.6.5. Mass Spectrometry

Mass spectrometry (MS) is a useful technique for identifying the mass of an unknown polymer as well as for understanding the fragmentation pattern of a known polymer. In this technique, molecular ions are produced using a high electric field, they are separated, and the mass of the ions is measured in terms of their mass-to-charge ratios m/z. The MS spectrum records the relative abundance of ions as a function of m/z.

The unique features of classical MS are its high sensitivity at the ppm level, large dynamic range, linearity, and resolving power capable of distinguishing various isotopic forms for molecules with molecular masses up to ~ 4000 Da. The technique is fast and uses very little sample (usually less than 1 nmol). Recently, the technique was modified for studying much larger molecules including polymers. In the absence of excessive fragmentation of the polymer, it gives absolute molecular weights in the range from 100 to 400,000 Da and their distributions in a single measurement performed within one hour. Polymer additives, residual volatile chemicals, end groups and polymer degradation products can also be studied.

In ordinary MS of polymer samples, fragmentation of molecules during desorption and ionization cannot be avoided. To circumvent this problem, the 'soft' desorption-ionization methods were developed. In these methods, the sample is introduced in a condensed phase and subsequently vaporized or bombarded to emit useful intact ions. For rapid heating, either laser desorption (LD) or matrix-assisted laser desorption ionization (MALDI) are used. MALDI is very convenient for high molecular weight polymer characterization.

In MALDI, the sample dissolved in a matrix such as nicotinic acid, indole, or acrylic acid is subjected to laser desorption. The matrix has a dual function: It isolates the sample molecules from each other and absorbs most of the laser energy thus assisting in the formation of protonated or metalated ions. A dilute solution of the polymer (1×10^{-5}M) in an organic solvent (typically methanol, acetone or tetrahydrofuran) is mixed with the matrix (1 to 2×10^{-1} M) dissolved in the same solvent. Generally, the sample is mixed with the matrix material in ratios ranging from 500 : 1 to 10,000 : 1. For high molecular weight polymers, high ratios of the matrix-to-polymer concentrations are required. The matrix/polymer solution (0.5–2 μL) is placed on the probe to dry. After the drying, a laser pulse is directed onto the solid matrix to photoexcite the matrix material. This excitation causes the matrix to decompose explosively, resulting in the expulsion and soft ionization of the sample molecules. The ions are then subjected to the MS analysis.

To increase the sensitivity further, the MALDI experiment is coupled with the time-of-flight mass spectrometer (TOF-MS). In this method, the ions are released at one point and accelerated by an electric field. The masses of the ions are then determined by measuring the time required for them to traverse a fixed distance in a field-free drift tube. The TOF-MS operates on the principle that ions of different mass m, given the same kinetic energy E, achieve different velocities v according to $E = 1/2\, mv^2$.

Thus, an acceleration of a population of ions across a chosen potential drop U gives all of them the same kinetic energy eU (e is the charge of an electron) but different velocities. By measuring the time required for each mass to traverse a field-free drift region of known length, a mass spectrum is produced. The time-dependent output of a suitable ion detector placed at the end of the field-free region measures the relative abundance of ions having increasing values of m/z.

The spectra in a TOF-MS experiment are recorded by a sophisticated data-acquisition system based on fast digital oscilloscopes or flash analog-to-digital converter (ADC) computer boards that can distinguish pulses separated by a few nanoseconds. For each pulse TOF-MS detects all ions with mass-to-charge values within a mass range of interest and produces full mass spectrum in tens to hundreds of microseconds. When coupled to a pulsed ionization source, it generates thousands of complete mass spectra every second. Their integration over time provides rather reproducible molecular weight distribution curves.

The MALDI-TOF spectrum of a poly(methyl methacrylate) calibration standard with a nominal molecular weight of 10,900 g/mol using dithranol (1,8,9-anthracenetriol) as a matrix is presented in Figure 3.55. Well-resolved mass peaks in a mass range of 7000–12,000 g/mol are obtained. In addition, the spectrum indicates lower molecular weight peaks around 3000 g/mol. These peaks, however, are not well-resolved due to low peak intensity.

The molecular weight distribution curves obtained by TOF-MS provide more details than distribution curves obtained by any other means. However, some caution is needed. The distribution may be distorted by fragmentation of the polymer molecules that cannot be totally avoided. Another distortion may be due to occasional presence of ions carrying more than one charge. It is thus worthwhile to repeat the analysis under different experimental conditions, e.g., different matrix-to-polymer ratio or different intensity of the laser beam.

Mass spectroscopy is also extremely useful in studying the fragmentation of polymers under various conditions (typically after thermal decomposition). MS may analyze the degradation products either directly or after a preliminary separation by gas chromatography. Comparison of the elution times and molecular weights of the degradation products with known substances (expected to be among degradation products) can provide a qualitative and quantitative picture of the composition of the degradation products, from which the mechanism of the degradation may be deduced. When a similar analysis is performed on an unknown polymer, its nature may be found by comparison of the degradation products with those of known polymers.

3.7. SEPARATION TECHNIQUES

Most methods of studying macromolecules that we described so far gave us information that was averaged for whole samples. However, recent emphasis in characterization of macromolecular systems is on more detailed information about the distribution of various molecular species in the samples. Many separation techniques were developed for this purpose. The prominent ones are electrophoresis, gel permeation chromatography, field flow fractionation, and supercritical fluid chromatography.

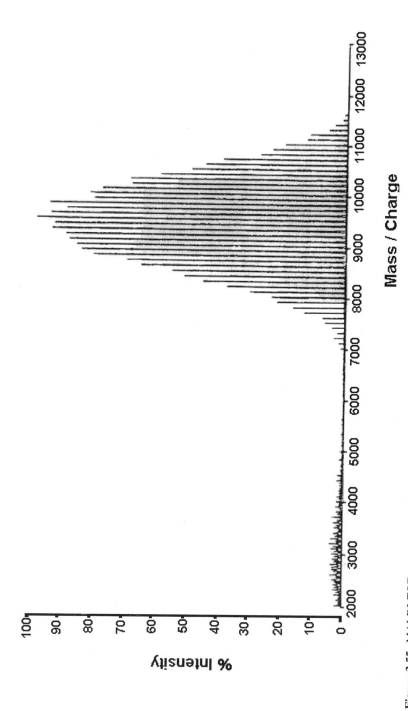

Figure 3.55. MALDI-TOF mass spectrum of poly(methyl methacrylate). (Reprinted from E. Esser, C. Keil, D. Braun, P. Montag and H. Pasch, *Polymer*, Vol. 41, p. 4040, 2000. Copyright © 2000 by Elsevier, UK. Used by permission of the publisher.)

3.7.1. Electrophoresis

Originally, electrophoresis was developed by biochemists for the separation of biopolymers. It has evolved as a sophisticated and powerful group of techniques, yet one that involves a great many hand operations.

In electrophoresis, charged particles move in an electric field. The particle velocity v is related to the strength of the electric field E [remember that E is the gradient of electric potential U_{el} of equation (3.2.1), i.e., $E = -dU_{el}/dx$] through electrophoretic mobility u:

$$v = uE \qquad (3.7.1)$$

We would expect mobility to be related to the charge of the particle, q and its frictional coefficient, f as

$$u = q/f = Ze/f \qquad (3.7.2)$$

where e is an elementary charge and Z is the number of elementary charges on the particle. Thus, measurement of mobility should allow us to calculate molecular charges provided that the friction coefficient can be obtained independently.

Actually, the situation is more complicated. Charged particles (polyions) require the presence of counterions. In solutions with a very low ionic strength, these counterions are distributed in the solution unequally, preferring to be close to the polyions. As the ionic strength increases, the effect diminishes but never fully disappears. The electric field pulls the counterions in an opposite direction than the polyions, effectively slowing the latter down. The amount of this retardation obviously depends on the ionic strength. The effect is most pronounced for coiled polyelectrolyte chains, for which many of the counterions are trapped inside the (hydrodynamically impermeable) coil and are moving together with the coil.

Analysis of the above phenomena is very difficult and, despite considerable theoretical effort, the relation between molecular charge and electrophoretic mobility of the particle is still not understood quantitatively and is no longer pursued in the literature. Of course, simple qualitative rules must always hold:

1. Everything else being the same, mobility increases with the net charge of the molecule.
2. Expansion of a molecule at a constant charge leads to decreased mobility.

Despite the above shortcomings, electrophoresis is still alive and well. In fact, for biochemists it is one of the most treasured methods for characterizing and separating proteins and other biological polymers. Proteins differ in their net charge and hence they have different mobilities; this makes it possible to distinguish them and/or separate them. Moreover, if two proteins happen to have the same or very similar mobilities at one set of conditions, it may be possible to clearly separate them at other conditions (change of pH, of the buffer, etc.).

Let us discuss the electrophoretic behavior of proteins in some detail. Most proteins in solution carry both positive and negative charges. The negative charges are located at dissociated carboxyls of aspartic and glutamic acids; at high pH even the phenolic hydroxyl of tyrosine and the sulfhydryl of cysteine may dissociate. Positive charges are carried by the basic amino acids arginine and lysine; at lower pH, histidine is also dissociated. Values of pK for the dissociating groups cover the region from about 4 to about 12. It follows that at any pH there exist particular groups that are only partially dissociated, that is, they are dissociated on some protein molecules but not on others. Clearly, different molecules have different charges and different mobilities. However, the dissociation and association of protons with the protein is a very fast process; the molecular charge fluctuates very rapidly, and the mobility of the protein corresponds to an *average charge*, which obviously does not need to be an integer. Average charge depends on pH; at some characteristic value of pH, it equals zero and the electrophoretic mobility vanishes. At these conditions the protein is said to be at its *isoelectric point*. The isoelectric point depends primarily on the nature of the protein; it is low for acidic proteins and high for the basic ones; it is a very sensitive characteristic of the protein. The isoelectric point depends also to some degree on ionic strength (which influences the dissociation constants) and on the nature of the buffer (some ions may be bound to protein more strongly than others).

In the original experimental arrangement for electrophoresis, a boundary is formed between the protein solution and the buffer in the same way as in the free diffusion experiment. An electric field is applied, and the movement and spreading of the boundary are observed using the same optics as in diffusion experiments. Accordingly, the method is called *free electrophoresis*; it is the only method capable of yielding electrophoretic mobilities. Its major disadvantage is the same as that of free diffusion: the experiment is quite prone to convective disturbances. As the interest in values of mobility was diminishing and separative powers of electrophoresis were becoming more important, the convections were restricted by performing the electrophoresis in a structured matrix preventing easy flow—on paper (hence the term *paper electrophoresis*). In a more recent development, the structure of the restricting matrix was made much finer by using gels with varying porosity. In this case, the particle experiences the resistance of the gel matrix that is a much steeper function of particle size than the hydrodynamic resistance in a liquid. Thus, in *gel electrophoresis*, proteins are separated employing two principles: differences in electrophoretic mobility and differences in molecular size. In the following sections we will treat these methods in some detail.

3.7.1.1. Free Electrophoresis.

A schematic diagram of a free electrophoresis cell is presented in Figure 3.56. The cell is U-shaped, and two boundaries are formed. Two electrodes are placed in the two buffer compartments, and an electric field is applied. In one arm of the cell, the boundary moves into the buffer and is called an *ascending boundary*; it is sharper and moves with a different velocity than the *descending boundary* in the other arm, which moves into the region occupied by protein solution.

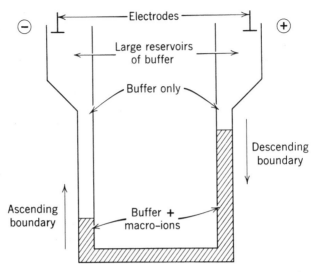

Figure 3.56. A schematic diagram of a free electrophoresis cell. The macro ions are assumed to have a positive charge. (Reprinted from C. Tanford, *Physical Chemistry of Macromolecules.* Copyright © 1961 by John Wiley & Sons, New York. Used by permission of the publisher.)

Why do the two boundaries behave differently? It is mainly the result of Donnan equilibria. The concentration of the supporting electrolyte is different in the protein region and in the buffer; so are the conductivity and (at given current density) the electric field. The original imbalance in concentration of the electrolyte leads to another artifact. As the protein boundary moves away from its original position, it leaves behind an "anomalous peak," which is formed by the electrolyte and is virtually stationary during electrophoresis. All the anomalies diminish with increasing ionic strength. However, the electrical heating increases with ionic strength causing convective disturbances. Thus, the highest practical concentration of salts is about 0.1 M.

Despite all the complications, free electrophoresis played an extremely important role in early biochemistry. For example, electrophoretic observations like the one depicted in Figure 3.57 led to the realization that blood plasma is a much more complex mixture than was believed previously. On the other hand, in free electrophoresis, the mobility of polyelectrolyte coils is essentially independent of chain length, and little information is obtained by studying it.

3.7.1.2. Paper Electrophoresis. The experimental arrangement for paper electrophoresis is quite similar to that of paper chromatography. A sheet of filter paper is fastened vertically between two buffer reservoirs equipped with electrodes. A drop of the protein mixture to be analyzed is placed on the paper, and a voltage difference is applied to the electrodes. Different proteins travel with different velocities and become separated. At this stage, nothing is visible on the paper. The proteins must be

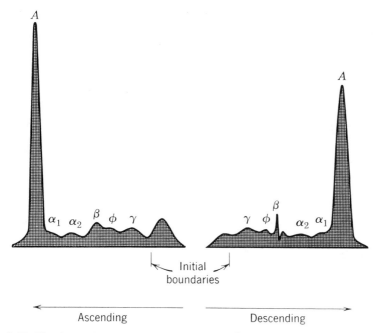

Figure 3.57. The electrophoretic schlieren pattern for normal human blood plasma. (Reprinted from C. *Tanford, Physical Chemistry of Macromolecules.* Copyright © 1961 by John Wiley & Sons, New York. Used by permission of the publisher.)

detected by staining. A solution of an organic dye (Coomassie brilliant blue is a good choice) is applied to the paper. It precipitates the protein and stains it. The excess dye is washed away, and the stained dots reveal the presence of the protein.

In actual experiments, the drop of the analyzed mixture is accompanied by a series of drops of solutions of known proteins, and all the samples are subjected simultaneously to electrophoresis and staining. Comparison of the positions of the stained dots then identifies individual proteins. This procedure avoids the measurement of mobilities.

The above basic technique is often adapted for other purposes. For example, the protein mixture may be applied not as a dot but as a line. After electrophoresis, the paper is dried and narrow strips are cut off the sides. These strips are used for detection by staining them in the usual way. After the positions of the protein bands have been determined, the main section of the paper is cut into horizontal strips, each strip containing just a single protein. The proteins are then eluted from the paper as purified specimens and may be used for other purposes.

Another variation is designed to overcome the possibility that several proteins may have the same mobility and may produce only a single dot. In this stratagem, after the electrophoresis, the paper is rotated by 90° (alternatively, a representative strip of the paper is sewn onto another paper), and the electrophoresis is performed a second time using buffer of quite different pH and/or composition. This procedure produces a two-dimensional map giving quite detailed information about the original sample. In another variation of the method, the second electrophoresis is replaced by paper chromatography using a suitable elution mixture. All these methods have been used successfully not only for the separation of protein mixtures, but also for separating and identifying peptides from the hydrolysis of proteins during studies aimed at finding the sequence of amino acids in proteins.

3.7.1.3. Gel Electrophoresis. During the last three decades paper electrophoresis was almost completely replaced by its younger cousin gel electrophoresis. In this type of experiment, the researcher has available several more adjustable features, which may be manipulated for optimum separation of any particular protein mixture. Among these the most important are the molecular sieving effect of the gel and the stacking effect that accompanies the use of discontinuous buffer systems.

Electrophoretic gels can be prepared from several materials of biological origin (agarose, starch), but recently the field has become dominated by polyacrylamide gels. The structure of the gel depends on the concentration of the monomer (acrylamide) in water (actually in a buffer) and on the concentration of the cross-linking agent (methylene bisacrylamide). We should visualize the gel as a liquid (water) that is spanned by long flexible chains of polyacrylamide. The length of the chains depends on the concentration of the cross-linker. A protein molecule moving under the influence of the electric field has to negotiate its way between the obstructing chains. Obviously, the mobility of the molecule is restricted the more the larger the protein, the more concentrated the obstructing chains, and the less stretchable they are (stretchability decreases with increasing cross-linking).

The gels are prepared using the same buffer to be used in the electrophoresis. Typical gels are made from 5–20% acrylamide; concentration of the cross-linker is optimally 5% of the total monomer. Radical polymerization at room temperature is initiated either with ammonium persulfate or with a photosensitive bioorganic dye, riboflavin. Tetramethylethylenediamine is used as an accelerator; degassing is required as in any radical polymerization.

The gels are prepared either as rods in glass tubes (4–6 mm in diameter, 10 cm long) or as slabs between glass plates (1.5 mm thick, 14 × 17 cm). The slabs are gaining in popularity; several samples can be applied to a slab, but only one to a rod. Readymade polyacrylamide slabs are now available commercially. Several rods or a slab with several samples are mounted between two buffer reservoirs and an electric field is applied in the usual way.

A useful variant of the method employs *concentration gradient gels* in which the concentration of acrylamide increases from the top toward the bottom; typically from 5% to 20%. These gels can accommodate proteins with a broader range of mobilities. A protein that traveled fast at the top of the gel is slowed down in the bottom part and does not leave the gel for a longer time, giving the slower proteins an opportunity to separate within the upper part of the gel.

Another technique allows for concentration of proteins into extremely narrow zones called *stacks* on the top of the column; they remain narrow during the subsequent electrophoretic separation. This very special technique is certainly beyond the scope of this book but we cannot resist the temptation to describe it—it utilizes so neatly some of the finer physicochemical principles.

Several different buffers and two gels are used—a short *stacking gel* at the top with low concentration of polyacrylamide, which does not restrict the movement of proteins, and a *resolving gel* tailored to the separation at hand. The reservoirs are filled with a Tris-glycine buffer of low ionic strength and relatively high pH (8.3). The gels are prepared using Tris-HCl buffer; the stacking gel has low ionic strength and low pH (6.7); the resolving gel has higher ionic strength and higher pH (8.9). The mixture of proteins in the same buffer as in the stacking gel is layered by injection above the stacking gel.

Let us now recall two basic principles of electrophoresis: (1) Electric current is the same at all cross sections of the cell. (2) Electric field is inversely proportional to the conductivity of the solution, that is, to its ionic strength and to the mobility of ions.

As the electric field is applied, the glycine-chloride boundary between the upper reservoir buffer and the protein solution moves into the latter, where the pH is lower. Glycine at pH 6.7 is only weakly dissociated, and its average mobility is low. Thus it has a tendency to move more slowly than chloride ions, and the boundary stays sharp. However, the continuity of the electric current forces both ions to move with the same velocity. The difference in mobility is compensated for by increased field strength in the glycine region.

In a properly designed experiment, the mobility of proteins in the stacking buffer is smaller than the mobility of chloride but larger than that of glycine. Thus, the chloride boundary overtakes the proteins, leaving them behind in the high electric field zone. Glycine stays even farther behind. When the chloride boundary reaches the resolving gel, all protein is concentrated into a very narrow zone (stack) between the chloride and glycine boundaries. The stack may be only few micrometers in thickness.

Within higher pH of the resolving gel, glycine dissociates almost fully; its mobility increases; it overtakes the proteins and follows immediately behind chloride. The stack of proteins is now surrounded from both sides by the same glycine buffer, and the stacking effect disappears. Proteins now travel with velocities corresponding to their mobilities in the gel and become unstacked, that is, separated.

The stacking procedure using several buffers and gels has been developed into a high-level art. Several thousand buffer systems have been designed, and the selection process has been computerized to allow optimization of the procedure for any desired separation problem.

3.7.1.4. Capillary Electrophoresis.

3.7.1.4. Capillary Electrophoresis.   Even though gel electrophoresis in its various forms is a common and highly effective method for separating charged molecules, it typically requires several hours for a run and is difficult to evaluate and automate. These disadvantages are largely overcome by the use of capillary electrophoresis (CE), a technique in which electrophoresis is carried out in very thin (1 to 10 μm inner diameter) capillary tubes made of quartz, glass, or plastic. Such narrow capillaries rapidly dissipate heat and hence permit the use of high electric fields, typically 100–300 V/cm, which is about 10 times higher than that of most other electrophoretic techniques; this reduces the separation time to a few minutes. This rapid separation, in turn, minimizes band broadening due to diffusion and leads to very sharp bands. Capillaries can be filled with buffer (as in free electrophoresis, but here the narrow bore eliminates the convective mixing), or with SDS-polyacrylamide gel (for separation according to molecular weight, Section 3.7.1.6), or with polyampholytes (for isoelectric focusing, Section 3.7.1.5).

The CE technique has very high resolution; it can be fully automated with automatic sample loading and on-line sample detection. However, it can only separate small amounts of material and hence is mainly used as an analytical technique. It has variously been referred to as high performance capillary electrophoresis (HPCE), capillary zone electrophoresis (CZE) and free solution capillary electrophoresis (FSCE); however, the term CE is now the most commonly used.

Capillary electrophoresis can be used to separate a wide spectrum of biological molecules including amino acids, peptides, proteins, DNA fragments (e.g., synthetic oligonucleotides) and nucleic acids as well as any number of small organic molecules such as drugs or even metal ions.

In CE, the migration time t for the solute depends on the tube length L, the electrophoretic mobility u of the solute and the applied voltage V as

$$t = \frac{L^2}{uV} \qquad (3.7.3)$$

Separation efficiency, in terms of the total number of theoretical plates, N is given by

$$N = \frac{uV}{2D} \qquad (3.7.4)$$

where D is diffusion coefficient of the solute.

Thus, the column length plays no role in separation efficiency, but it has an important influence on migration time and hence on time of analysis. High separation efficiency can be achieved by using high voltages (u and D are determined by the solute and are not easily manipulated). Consequently, high voltages and short capillaries should be employed. In practice, voltages of 10–50 kV with capillary lengths of 50–100 cm are commonly used.

3.7.1.5. Electrofocusing. The mobility of any given protein depends on the pH of the buffer. When the pH value corresponds to the isoelectric point pI of the protein, the latter does not migrate in the electric field any more. The method of electrofocusing is based on the preparation of electrophoretic gels (low concentration nonrestrictive gels in this case) that have a gradient in pH values. Now, it is possible to create such gradients by simply carrying out the electrophoretic experiment using a gel containing a mixture of *ampholytes* (low molecular weight compounds capable of carrying both positive and negative charges, e.g., amino acids). In a typical experiment, one of the reservoirs has a high pH, and the other a low pH. The ampholytes move in a direction given by their charge but do not penetrate into the reservoirs (pH barrier reverses their migration direction). The ampholytes with higher pI accumulate closer to the basic reservoir, and compounds with lower pI at the other end of the gel. Overlapping bands of ampholytes create the pH gradient. The formation of the gradient is completed in several hours; it may persist for several days. If a protein mixture is introduced onto the gel, each protein will migrate (with gradually decreasing velocity) toward a location where the pH matches its pI value; it will form a narrow band there. The technique is capable of separating proteins that differ by a few hundredths in their pI values.

3.7.1.6. SDS Electrophoresis. Detergents interact very strongly with proteins. They completely disrupt their tertiary and secondary structure and convert them into random coils. Sodium dodecyl sulfate (SDS) performs this conversion at a concentration of 0.1%. (When full disruption of the structure is required, the cystine sulfur-sulfur bonds must be also reduced; this is accomplished by adding a suitable sulfhydro reagent such as β-mercaptoethanol. Previous heat denaturation of the protein at about 100°C facilitates the procedure.) The detergent anions are strongly adsorbed onto the protein chains (about 1.4 g SDS/g protein) and impart a high negative charge to them completely overwhelming the original charge of the protein. All proteins now have an effectively identical structure and charge density; they differ only in their chain length, that is, in molecular weight.

When such solutions are subjected to electrophoresis on a nonrestrictive (low-concentration) gel, all proteins have essentially the same mobility. However, when higher-concentration, more restrictive, gels are used, the proteins are separated according to coil size, that is, according to their molecular weight. Typically, the molecular weight scale is provided by a number of known proteins that are run parallel to the unknown sample on the same slab or on a set of rods. Figure 3.58 is a photograph of a slab on which a mixture of calibrating proteins was applied to the central track. Protein samples from different parts of the same tissue were used in the remaining tracks. The distribution of various proteins throughout the tissue was thus visualized.

The mobility of proteins in SDS solutions (i.e., distance traveled in a given time) is a linear function of the logarithm of the molecular weight of the polypeptide chains and of the concentration of the gel (Figure 3.59). This makes the calculation of molecular weights by interpolation reasonably precise. In practice, more concentrated gels are used for smaller proteins; less restrictive gels with lower concentrations of polyacrylamide are employed for larger proteins.

SDS electrophoresis yields molecular weights that depend on proper calibration and are sometimes anomalous for proteins with very unusual amino acid composition. Nevertheless, it is one of the most sensitive and useful methods in biochemistry. Molecular weights are measured on microgram samples, and the proteins do not even need to be pure! Moreover, dissociation of complex proteins into their constitutive polypeptide chains helps to establish their subunit composition.

3.7.2. Gel Permeation Chromatography (GPC)

The separation method based on solute size is referred to as size exclusion or steric exclusion chromatography (SEC); the application in aqueous systems is traditionally

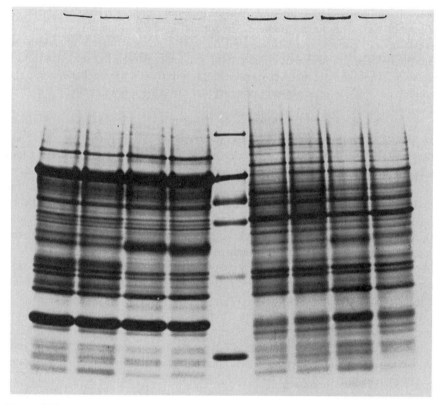

Figure 3.58. A photograph of a polyacrylamide slab used for SDS gel electrophoresis. A mixture of calibrating proteins was applied to the central track. Soluble proteins from different regions of rat epididymis were applied to the other tracks. (Reprinted from B.D. Hames and D. Rickwood, Eds., *Gel Electrophoresis of Proteins: A Practical Approach.* Copyright © 1981 by IRL Press, Oxford, England. Used by permission of the publisher.)

referred to as gel filtration (GF) and the application in nonaqueous systems is designated as gel permeation chromatography (GPC).

The heart of the GPC system is a porous gel, which actually effects the separation. In GPC, the porous material is divided into small particles (10–1000 μm) and packed into a chromatographic column. The particles are porous and the size of the pores is the main characteristic of the column.

Quite different materials are used. For aqueous solutions (used in biochemistry), the original gels were manufactured from cross-linked dextran (trade name Sephadex). For larger pores, low-concentration agarose gels are used. Synthetic polymers are represented by macroporous cross-linked polyacrylamide (Biogel). These gels, especially the less concentrated ones with large pores, are very soft. When the column is eluted, they are compressed and, consequently, the hydrodynamic resistance of the

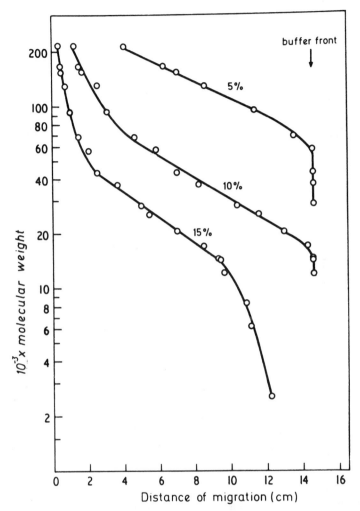

Figure 3.59. Calibration curve for SDS electrophoresis of polypeptides showing the relation between the distance traveled and the logarithm of molecular weight for three different concentrations of the gel. (Reprinted from B. D. Hames and D. Rickwood, Eds., *Gel Electrophoresis of Proteins: A Practical Approach.* Copyright © 1981 by IRL Press, Oxford, England. Used by permission of the publisher.)

column increases. Thus, they can be used only with moderate head pressures on the column. This is considered a serious drawback in modern chromatography, which emphasizes high-pressure, high-performance columns. More rigid porous particles are now gaining prominence. Porous glass beads can be used with both aqueous and organic solvents while the cross-linked macroporous beads from styrene-divinylbenzene copolymers (Styragel) are used for organic solvents.

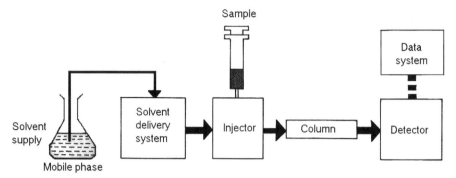

Figure 3.60. A typical GPC system.

A typical GPC system (Figure 3.60) consists of a pump, an injector (either manual or automated), column set, detector(s), and data-handling device. In addition, a degasser is used when using THF with a refractive index detector. The columns are almost always heated to elevated temperature (even for applications that can be handled at room temperature) to insure low pressure drop and uniform viscosities.

3.7.2.1. Principles of GPC.   In essence the GPC process is simple. A dilute polymer solution (0.1 to 1.0%) is injected into a stream of pure solvent, which is flowing through a set of columns. These columns are packed with porous gels having a range of pore sizes. As the sample passes along the columns the smallest molecules are following the solvent both into the flowing (mobile) phase and into the stationary solvent trapped within the gel pores. However, large solute molecules cannot enter the pores of the gel and are forced to stay in the outside mobile phase. Molecules of intermediate size enter the larger pores but not the smaller ones. The situation is schematically depicted in Figure 3.61. Distribution of the solute inside and outside the gel is a typical partition phenomenon.

In a GPC column, the volume of the solvent exterior to the gel particles (also called the interstitial volume) is V_0, and the internal volume of the pores is V_i. When a small sample of a solution is applied on the column and is eluted in the usual way, the newly introduced solvent together with very small molecules is eluted after volume, $V_T = V_0 + V_i$ has passed through the column. Very large particles cannot penetrate into any pores; they travel faster than the solvent front and elute with V_0 as their elution volume. Solute particles of the intermediate size penetrate larger pores but not the smaller ones and are eluted at volumes between V_0 and V_T. Samples thus elute within a relatively narrow, well-defined elution volume range that depends on the column geometry as well as on the nature of the packing material. The shape and position of the elution peak of the polymer depend upon distribution of the molecular weight or size of the polymer.

The accepted theory of GPC mechanism is based on the establishment of thermodynamic equilibrium between the solute in the interstitial volume and in the pore

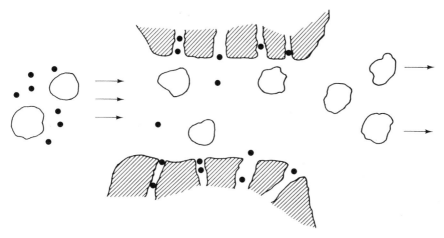

Figure 3.61. A schematic description of the partitioning effect of pores on the separation of particles of different sizes in GPC. (Reprinted from Harry R. Allcock and Frederick W. Lampe, *Contemporary Polymer Chemistry,* p. 395. Copyright © 1981 by Prentice Hall, Inc., Englewood Cliffs, New Jersey. Used by permission of the publisher.)

volume. Support for this theory is based on experimental findings that GPC results are independent of flow rate and that static equilibrium experiments agree with them well. At higher flow rates or for polymers with small diffusion coefficients, nonequilibrium effects, such as restricted diffusion, may play a role.

The retention mechanism in GPC is conveniently described using the distribution coefficient K_D, which is defined as the ratio of solute concentration within the pore volume of the packing, $C_i = x_i/V_i$, and solute concentration in the interstitial volume, $C_0 = x_0/V_0$, where x_i and x_0 are the amounts of solute in the respective phases. Thus

$$K_D = \frac{C_i}{C_0} = \frac{x_i V_0}{x_0 V_i} \qquad (3.7.5)$$

It is easy to show that for elution volume, V_e of the intermediate particles the following relation holds.

$$V_e = V_0 + K_D V_i \qquad (3.7.6)$$

Here, K_D can range from zero for a solute that is too large to diffuse into the pores of the packing ($C_i = 0$) to unity for a solute that can penetrate the total pore volume of the packing ($C_i = C_0$). K_D is related to thermodynamic quantities as

$$K_D = \exp(\Delta S/R) \qquad (3.7.7)$$

where R is gas constant and ΔS is entropy loss when the molecule enters the stationary pores of the packing. Here, it is assumed that enthalpic interactions of the solute with the packing are absent. For molecules of small size relative to the pores of the packing, $\Delta S = 0$ or $K_D = 1$. For larger solutes, ΔS will be negative and $K_D < 1$. According to equation (3.7.7), temperature does not play a major role in GPC. However, for nonrigid packings, temperature may influence the pore structure or V_i/V_0 of the packing and thus affect the separation. Depending on the nature of the solvent, polymer conformation may also be a function of temperature. Probably the greatest influence column temperature will have on GPC is on peak dispersion: at higher temperatures peak broadening will decrease because of the increased rate of mass transfer of polymer into and out of pores.

Classical GPC is used mainly for two purposes: for measurement of molecular weight of polymers with a narrow distribution of molecular weights and for finding molecular weight distribution of samples that have a broad one.

3.7.2.2. Molecular Weight and Universal Calibration Curve.   Gel permeation chromatography is routinely used for fast estimates of molecular weight of polymers. For a given polymer in a given solvent on a given column, the elution volume depends only on the effective hydrodynamic volume of the macromolecular coils, that is, on the molecular weight. First, the column is calibrated using well-defined narrow fractions of the polymer. Once the calibration plot is available, measurement of unknown samples becomes routine and fast. The calibration curve is constructed by plotting the logarithm of molecular weight, or some function of molecular size, versus V_e or K_D. For a well-behaved GPC system, the entire sample elutes within a K_D range of 0 to 1.

The calibration curve is necessarily S-shaped. Very small solutes elute with the solvent independently of their molecular weight. Similarly, very large particles elute at V_0, again independently of molecular weight. However, the intermediate part of the plot is often quite linear. Figure 3.62 shows calibration curves using standard polystyrene and a series of columns. The high linearity of the calibration curves on properly selected columns helps in calculating the molecular weight of polymers even with a wide molecular weight distribution.

The peak capacity of a typical GPC column (the number of peaks that can be clearly distinguished in a chromatogram) is usually less than 15. This makes GPC a less attractive tool for analysis of multicomponent samples consisting of monodisperse solutes. Nevertheless, three to five polymer fractions may be run together in a single calibration experiment.

When appropriate samples for calibration are not available or if there are branched or chemically heterogeneous polymers present, then the calibration curve should be based directly on the molecular size to obtain the correct molecular weight. This can be done by the use of a universal calibration method. Its theory is based on the fact that each elution volume will contain molecules having the same hydrodynamic volume. This volume is proportional to the cube of radius of gyration r_G^3 that is related to intrinsic viscosity $[\eta]$ [see the relation (3.4.49)] as

$$r_G^3 \ \alpha \ M[\eta] \tag{3.7.8}$$

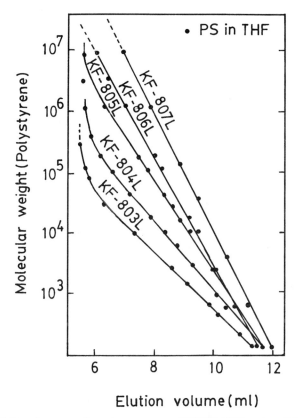

Figure 3.62. Calibration curves for polystyrene on a GPC KF-800L series of Shodex columns. Eluent THF. (Reprinted from Wu, C.-S. Ed., *Column Handbook For Size Exclusion Chromatography.* pp. 176. Copyright © 1999 by Academic Press, New York. Used by permission of the publisher.)

Consequently, secondary calibrants (fractions of other polymers with known molecular weight dissolved in the same solvent) can be used for calibration. For a given chromatographic system, the plot of hydrodynamic volume, $[\eta]M$, versus elution volume V_e, should fall on the same line for all types of polymers. This observation, first made by Henry Benoit, is amply confirmed by experiment as shown in Figure 3.63 for many polymers.

In practice, the elution volume of the sample under study is measured and its $[\eta]M$ value is read from the plot. Its molecular weight is then easily obtained if Mark-Houwink-Sakurada (MHS) viscosity coefficients [equation (3.4.46)] for this polymer are known. The MHS coefficients for many polymers in many solvents can be found in the literature. Of course, for these measurements all the polymers must be dissolved in the same solvent. A change of solvent would affect the swelling of the gel particles, their porosity, and the calibration curve.

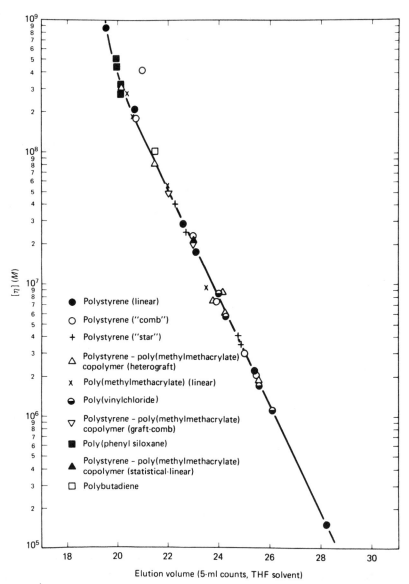

Figure 3.63. Universal calibration curve for GPC: a dependence of $[\eta]M$ on elution volume for a number of different polymers. (Reprinted with permission from Z. Grubisic, P. Rempp, and H. Benoit, *J. Polymer Sci., Polymer Letters*, Vol. 5, p.753. Copyright © 1967 by John Wiley & Sons, New York.)

In many publications that report GPC data so-called *polystyrene equivalent* molecular weights are presented. These are molecular weights that a hypothetical polystyrene sample would exhibit if eluted at the same elution volume. Needless to say, such values have no physical significance. They may be used for comparison of several samples of the same (presumably novel) polymer, for which the researcher cannot or does not want to measure intrinsic viscosity or any other meaningful molecular characteristic.

3.7.2.3. Molecular Weight Distribution.   The real strength of GPC is in studies of polydispersity. The principle of the method is simple: the elution band of a polydisperse polymer is transformed into the distribution function of molecular weights using the calibration curve. Of course, even monodisperse polymers produce elution bands of finite width due to diffusion of the polymer along the column, to slow equilibration of the solute between pores and outside solvent, etc. Thus the observed elution curve is broader than an "ideal" curve reflecting polymer heterogeneity only. This phenomenon is common to all types of chromatography but is rather prominent in GPC. Several methods have been designed to correct this band broadening, but none of them is quite satisfactory. Thus, GPC is useful mainly for the characterization of industrial samples with a broad and/or multimodal distribution of molecular weights. For this purpose, it has virtually superseded all other methods. It is much less successful for samples with a narrow distribution.

Raw data in GPC consist of a trace of detector response (proportional to the concentration of polymer in the solution) as a function of the elution volume. A typical GPC chromatogram follows the Gaussian curve. For evaluation of polydispersity, a baseline is drawn through the recorder trace, and incremental areas are measured for equal small increments of elution volume. The incremental area of the ith increment is m_i, the mass of polymer in the increment. It is converted into mass fraction w_i through division by the total area under the curve. Suitable calibration permits the translation of the elution volume axis into a logarithmic molecular weight scale and molecular weight M_i of the increment is read. A table of w_i and M_i values is constructed and the values of $\overline{M}_n$ and $\overline{M}_w$ are calculated using equations (1.6.5) and (1.6.7). From these the polydispersity, $\overline{M}_w/\overline{M}_n$ of the polymer can be obtained.

For obtaining molecular weight distribution curve, slight transformation of the data is needed in order to correct for the possible nonlinearity of the logarithmic molecular weight scale that was a result of possibly distorted calibration curve. Recently such calculations are routinely done (together with calculations treated above) by a computer that is a part of modern chromatography instruments.

3.7.2.4. Critical Point GPC.   In ideal GPC separations there is no contact interaction between the polymer sample and the material of the gel. The separation is achieved strictly on the basis of the steric exclusion that accelerates the movement of larger molecules along the column. In reality, there is always some interaction between these two materials. (In good GPC systems it is almost totally suppressed.) The adsorption of the polymer segments slows down the passage of the large molecules more than of the small ones. Judicious design of the interaction (achieved usually by the choice of the eluting solvent) may perfectly balance the steric (entropic) and adsorption (enthalpic) contribution to the polymer transport: all molecular weights are eluted in a single peak. (You may recall a similar balance of steric exclusion and contact interaction that determined the theta conditions and pseudoideal behavior of polymer solutions that was treated in section 3.1.1.3.)

The mode of chromatographic behavior that exploits this situation is called liquid chromatography at the critical point of adsorption. The polymeric nature of the sample (i.e., the repeating units) does not contribute to the retention of the species. Only defects (end groups, comonomers, branching points) contribute to the separation of the molecules. However, in order to shift the separation into such a chromatographic mode the experimental conditions must be carefully adjusted to the column that is used as well as to the problem at hand. In many cases adjusting the polarity of the mobile phase without buying new columns can do this.

Consider, for example, the separation of PMMA on a nonmodified silica column. PMMA elutes in size exclusion mode in a polar solvent like THF because the dipoles of MMA are masked by the dipoles of THF. However, when nonpolar toluene is used as the solvent on the same column, the separation is governed by adsorption because the dipoles of the carbonyl group in the PMMA interact with the dipoles on the surface of the stationary phase. Therefore, separation of PMMA in the critical point mode of adsorption can be achieved by selecting an appropriate THF/toluene mixture as the eluant. In this case, all the PMMA samples will elute at the same time regardless of their different molecular weights. However, samples with different end groups will be separated with high selectivity with no size separation effects.

3.7.2.5. Selection of Columns.   Column is the heart of the GPC machine.

The retention mechanism in GPC is mainly dependent on the dimensions of the solute and the physical characteristics of the packing. Recently, the old soft gels are virtually abandoned in favor of more rigid materials: porous glass beads or macroporous beads from styrene-divinylbenzene copolymers. These gels are capable of operating over a wide range of temperatures, although their lifetimes may be drastically reduced in highly polar solvents or at high temperatures. Under these conditions it is advantageous to use the porous inorganic type of gel. A range of different types of porous glass and silica gels are available; these are easily packed, can be recovered for cleaning and repacking and are indefinitely stable in most commonly used solvents and temperature conditions.

In order to cover a broader molecular weight range than a column with a single pore size may handle, many systems use extended range or mixed bed columns. These columns are blends of beads with different pore sizes. If blending of pores is done carefully, the column calibration curve will be linear over quite extended range of molecular sizes.

Everything has its price. The extended range of mixed bed columns is paid for by low resolution over a finite molecular range. Hence, use of several columns in a series is recommended to ensure a successful separation. The first consideration is to have enough pore volume in the column set to obtain the correct separation (i.e., the correct distribution profile of the polymer). The selection of column bed depends upon the molecular weight range of polymers under investigation. For example, for analyzing epoxy and phenolic resins in the molecular weight range of a few hundred to five thousand, a column set of 5 nm, 10 nm and 50 nm pore size would be ideal. For a polycarbonate having a weight average molecular weight of $\sim 28,000$ and a number average of $\sim 12,000$, a four column set of a 10 nm, 50 nm, 100 nm and 1000 nm would be best. Individual pore size columns are targeted at specific molecular weight range of polycarbonate.

On the other extreme of molecular weight, such as for ultrahigh molecular weight polyethylene with a weight average molecular weight of 4,000,000, number average of 300,000 and the z-average possibly higher than 10,000,000, a set of 10^4, 10^5, and 10^6 columns may be suitable. Thus, it is important to put three mixed bed columns together, or may be two mixed bed columns with a single 50 nm column. This will provide pore volume at low-molecular-weight end, where the low molecular weight tail of the sample may be unresolved from *impurity* peaks that are always seen at the end of the chromatogram. Alternatively, two mid-to-high molecular weight range columns in series with a single 50 nm size column would be a good *scouting* column set to achieve a good linear calibration range over several orders of magnitude in molecular weight. This column set would be appropriate for very broad polydisperse samples.

In biochemistry, GPC is used also for preparative purposes. Several components differing substantially in molecular size can be separated into individual zones. Another useful procedure, which is performed on gels with small pores, is exchange of buffers. The column is first equilibrated with the new buffer, then the sample in the original buffer is applied on the column. The macromolecules travel faster than the new solvent front, enter the new buffer, and are eluted.

Presently, GPC supports are available commercially that have pore sizes graduated from nanometers to micrometers. For broad size distributions, several columns with gels of different porosities are often connected serially. Such combined columns are used both in biochemical separations and in polymer distribution studies.

3.7.2.6. Detectors Used in GPC.   The most widely used detector for GPC analysis is the differential refractometer. It is a concentration sensitive detector that simply measures the difference in refractive index between the solvent in the reference side, and the stream of the eluate in the sample side. It is a "universal" detector (unlike an UV detector, for example) in that a response is obtained for any polymer that has a significant difference in refractive index when compared to solvent. If the difference in refractive index for the sample and the solvent is very small (polysiloxanes and THF, for example), and a poor signal results, another solvent should be selected that will dissolve the polymer and provide a significant refractive index difference.

The UV detector is another concentration sensitive detector that is often used for GPC. UV detection requires that in the solute a chromophore is present that will absorb light in the UV range. The UV detector is excellent for styrene type polymers, (polystyrene, styrene/isoprene, styrene/butadiene, ABS, etc.), epoxies, phenolics, polycarbonates, polyurethanes, and aromatic polyesters.

A photodiode array (PDA) detector, which is a step up from the UV, serves as a powerful, information-rich device. It looks at a wide range of wavelengths simultaneously. For example, one can set the PDA to look at a wavelength range from 190 to 800 nm, instead of looking at just one or two wavelengths as is usual for most UV detectors. Thus it is possible to observe the actual UV spectra of the polymer sample (or additives) and to determine the distribution of the chemical composition.

Recently, powerful molecular weight sensitive on-line detectors were developed. The use of low-angle laser light scattering and multiangle light scattering detection for water-soluble polymers has added a new dimension to characterization because it provides absolute molecular weight, size and conformation. In the light scattering detector a laser beam is focused into a cell (on-line in this case) that contains the sample solution and the intensity of the scattered light is measured. (The theory, instrumentation, and interpretation of light scattering is described in section 3.5.)

The viscosity detector exploits the Poiseuille law [equation (3.4.37)] stating that the pressure drop along a capillary, through which a liquid is flowing with a given constant flow rate, is proportional to the viscosity of this liquid. The pressure drop during a GPC run is monitored by sensitive pressure gauges and is compared with the drop in a reference capillary through which flows pure solvent.

MALDI (matrix-assisted laser desorption ionization) technique of mass spectroscopy (see section 3.6.5) was also employed for the analysis of the GPC eluate. It simultaneously provides concentration and molecular weight information of probes taken from the eluate. GPC and MALDI-TOF mass spectrometry can also be effectively coupled via a robotic interface, where chromatographic fractions are sprayed directly onto a moving MALDI-TOF target through a heated capillary nozzle. The target is pre-coated with an appropriate matrix, such as dithranol. For polydisperse synthetic polymers a continuous track of matter is deposited onto the matrix surface. After deposition of the sample fractions, the MALDI-TOF target is introduced into the spectrometer, and the spectra are taken from different positions of the polymer track. As a result, the well-resolved spectra are obtained which are characteristic for different molecular weight fractions of the sample. In the case of copolymers, information on molecular weight and copolymer composition is accessible.

For more demanding applications double detection is being used with increasing frequency. Two detectors following different aspects of the solution serially observe the eluate. For example, measurement of refractive index, which gives the total concentration, is combined with light absorption, which gives the concentration of some particular species. Some success has been achieved with block copolymers by using three detectors in series: first, a refractometer measures polymer concentration; second, the viscometer estimates molecular weight, and third, an ultraviolet spectrophotometer, sensitive only to one component of the polymer, gives concentration of this component. The comparison of the two concentrations shows the composition of the copolymer as a function of molecular weight.

Refractive index and light scattering detectors are a powerful combination that allows simultaneous measurement of concentration and molecular weight as a function of the retention volume. If the intensity of the scattered light is measured at various angles it is possible to monitor the radius gyration of the sample. It should be noted that with this combination of detectors the molecular weight of the eluted sample is measured directly and no calibration of the column is needed.

Combination of viscosity and refractive index detectors allows for measurement of intrinsic viscosity as a function of the elution volume. If the MHS coefficients of the polymer are known and the polymer molecules in the sample are linear, this measurement provides the complete molecular weight distribution function. Integration of the whole viscosity curve yields also the intrinsic viscosity of the whole sample. With modern instrumentation this procedure may be less demanding than the classical measurement of $[\eta]$.

The most extensive information about the sample is by using both viscometer and light scattering detectors in tandem with the refractive index detector. Besides the molecular weight distribution function, such a triple detector approach may provide also the MHS viscosity coefficients for the system under investigation as well as some information about the branching and branching distribution in the sample.

An experimentalist should be aware of several pitfalls when using multiple detectors. A serious error may be committed when evaluating those parts of the chromatogram where one signal is large while the other is on the threshold of observability. Such a situation may occur for the low-molecular-weight tail of the chromatogram where the light-scattering signal may be small or for the high-molecular-weight tail where the concentration may become too small for a meaningful measurement.

Another problem is in the fact that the detectors are in a series. The signals are shifted in time and the time difference depends not only on the geometry of the instrument but also on the fluctuation of the flow rate. Even minor misjudgment of the difference may be damaging in those parts of the chromatogram where the dependences are steep.

Still another problem is in the peak spreading. While the spreading may be insignificant in the capillaries connecting the detectors it may become important within the detectors themselves where the flow channel broadens and becomes irregular.

3.7.3. Field Flow Fractionation

When a liquid is flowing through a channel, different layers of the liquid flow with different velocities. If the composition of the liquid varies from layer to layer, fractionation can be achieved.

In field flow fractionation, the channel has a thin (fraction of a millimeter) ribbon-like form. The velocity profile of the liquid within the channel is essentially parabolic (Figure 3.64). When an outside field is applied to the liquid perpendicularly to the channel, particles may migrate toward one of the walls, setting the stage for the fractionation process. Various types of outside fields can be utilized for this purpose: a centrifugal field for colloidal particles, an electric field for charged macromolecules, a cross-flow field, or a thermal field. In the cross-flow field, the walls of the channel are made from semipermeable membranes and liquid is forced across the channel. As

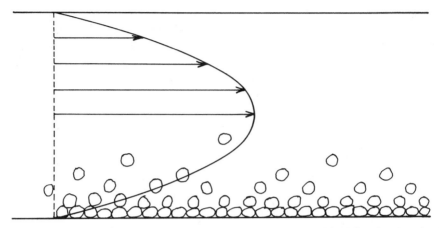

Figure 3.64. A schematic description of the partitioning principle of field flow fractionation.

the liquid is filtered out through the membrane, macromolecules accumulate near the wall. The thermal field makes use of the phenomenon of thermal diffusion—polymer molecules in temperature gradients tend to move toward the cold wall.

All the fields affect the distribution of the polymer molecules in a similar way. The field pushes the molecules toward the wall, and diffusion tries to drive them back. We met an example of this situation when we studied sedimentation equilibrium in Section 3.2.2.2. The result is a distribution profile of concentrations that has a more or less exponential form. (It is depicted schematically in Figure 3.64.) The steepness of the profile depends on the interplay of the field strength and the diffusion coefficient: the higher the molecular weight, the steeper the profile and the closer the molecules are packed toward the wall.

When the solvent is forced through the channel, various layers move with velocities decreasing toward the walls. The macromolecules are confined to layers near the wall; their average velocity is much less than the velocity of the eluting solvent: the higher the molecular weight, the smaller the velocity. Consequently, different species move along the column with different zonal velocities producing elution curves similar to those in GPC; also, the information extracted from them is similar.

Presently, the thermal field method is being exploited the most. Its basic design is simple: the separation channel is sandwiched between two copper blocks, one of which is heated by powerful heaters, the other cooled by running water. This arrangement allows for achievement of temperature differences larger than 100°C across the channel. The temperature gradient and the flow rate of the eluting solvent are optimized for the separation problem studied.

Generally, field flow fractionation yields the same information as GPC. Its main advantage is the simple geometry of the apparatus, leading to a better theoretical insight into the separation process. The same column can be used for widely different sizes of particles with proper manipulation of the elution rate and the applied field.

Also the hydrodynamic field in a smooth channel is less apt to shear degrade very large molecules than the very heterogeneous field within the packing of the GPC column. On the negative side, field flow fractionation is usually more time-consuming (many hours per experiment). The large amount of eluent that flows through the middle part of the channel substantially dilutes the sample, complicating detection. (This extra solvent can be diverted into a second outlet, but this may lead to loss of part of the sample.) The utilization of the method will probably require further technical improvements.

3.7.4. Supercritical Fluid Chromatography

In recent years polymer scientists are interested in the use of supercritical fluids as these offer alternative routes for polymer formation, modification, and purification. What are supercritical fluids? When a substance is subjected to pressures and temperatures above its critical point, a highly compressed gas known as a supercritical fluid is formed. Supercritical fluids exhibit properties somewhere between those of classical liquids and gases. Their densities are high enough to offer the solvating power of liquids, while their viscosities are low enough to permit the fairly rapid solute diffusion (characteristics of gases).

In recent years supercritical fluid chromatography (SFC) has been developed as an effective separation method that uses as mobile phases fluids beyond their critical points. Supercritical fluids possess some unique advantages over the ordinary liquid mobile phases. Their high volatility aids in coupling to mass spectrometers. Several chromatographically useful supercritical fluids such as carbon dioxide and xenon have extensive regions of transparency in the infrared spectral region, permitting an effective coupling with infrared spectrometers.

The solvating power of supercritical fluids has a major effect on the way analytes are separated. The solubility parameters of supercritical fluids are a strong function of fluid density and can be controlled precisely by varying the temperature or pressure of the system. Thus SFC allows separations that would be difficult or impossible to achieve using conventional methods.

3.8. TECHNIQUES FOR THE STUDY OF THE STRUCTURE OF NUCLEIC ACIDS

Recent fast progress of molecular genetics is based on the discovery of the details of the structure of nucleic acids—molecules of which determine every aspect of terrestrial life. Despite the complexity of these macromolecules, the structure of nucleic acids is studied by methods that are an extension of the known techniques of macromolecular science. This section will be devoted to the description of the key techniques used in the molecular genetics: Polymerase chain reaction (PCR), separation of DNA fragments of different length, sequencing of DNA, recognition of particular nucleotide sequences, and DNA profile analysis.

3.8.1. Polymerase Chain Reaction

Quite often there is not enough material available for the experimental procedures used for the study of a particular sample of DNA. This problem was solved by the *polymerase chain reaction* (PCR) technology in a way that reminds one of miracles. PCR is capable of copying a single DNA into a virtually limitless number of replicas.

The technique is based on an enzyme called DNA polymerase. This enzyme has been forced to repeat in a test-tube the replication process used by living cells during cell division. The appropriate reaction mixture consists of the enzyme, the four building blocks of nucleic acids (nucleotide triphosphates), single stranded DNA template to be copied, and a *primer*. The primer is a short (radioactively or fluorescently labeled) polynucleotide that is complementary to the end of the molecule studied. Under proper experimental conditions the primer attaches itself to the DNA molecule and the enzyme starts synthesizing the rest of the complementary chain. From the viewpoint of polymer chemistry, this is a sequential polycondensation reaction in which a bifunctional monomer is attached to a growing chain renewing the active end.

In the PCR technique two oligonucleotides are needed to serve as primers: they have nucleotide sequences complementary to the two 3'-ends of the desired (double-stranded) DNA segment to be amplified. The primers initiate the synthesis of the two complementary strands of the desired segment of the target DNA. Other ingredients of the reaction mixture are a heat stable DNA polymerase and four nucleotide triphosphates. The mixture is heated above the denaturing temperature of DNA and the double helix separates into its constituent single-stranded chains. Upon cooling, the primers hybridize at the complementary regions of these strands and the polymerase extends the hybrid into the whole segment in the 5' → 3' direction. (It should be noted that the part of the chain behind the primer is not copied. However, in this first cycle the new chain is synthesized beyond the intended end.)

From the two one-stranded chains we now have two double strands. The mixture is heated again, the double helices separate again. Upon cooling, the primers that are present in large excess hybridize again (in the middle of the newly synthesized chain) and the whole sequence of reactions is repeated. In this and subsequent cycles the reaction has to stop at the end of the chain, which is at the place where was the primer attached in the previous cycle. Consequently, only the desired fragment is copied. The process is repeated again and again in cycles as shown in Figure 3.65. The temperature cycles are controlled automatically by a computer and the researcher finds a sizable amount of his cloned DNA by the time he comes back from lunch.

In many cases, one cycle of reactions is completed within minutes. We leave it to the reader to calculate the number of cycles needed for cloning a single DNA segment. (The number of molecules doubles in every cycle!). It is not necessary to have a pure sample of the DNA (or its fragment) of interest. It is sufficient to know the beginning and end of the DNA chain section that is to be duplicated and to synthesize appropriate primers. The hybridization will occur irrespective whether the primer finds its partner

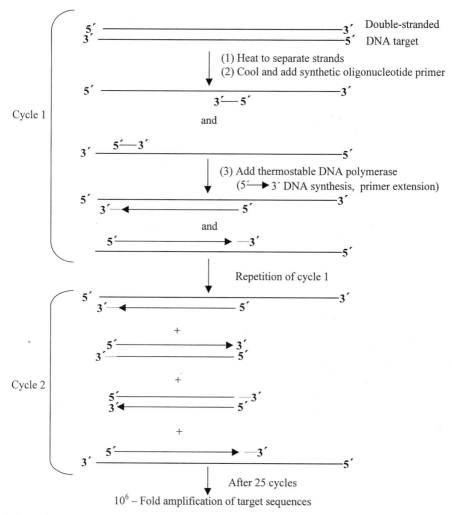

Figure 3.65. Illustration of the polymerase chain reaction for a specific segment of DNA. The dashed lines in front of the arrows indicate the possible extension of DNA synthesis. The dashed lines without arrows indicate the DNA strand generated in the previous cycles. The primers bind to the complementary sites on the target DNA (as described in the text).

at the end or in the middle of the DNA molecule. The molecules that do not carry the sequences that would hybridize with the primers are not duplicated and end up as an inconsequential minuscule impurity in the end product.

One of the early inconveniences of this method was low thermal stability of the polymerase. A special polymerase was identified and isolated from a thermophilic bacterium growing in hot springs. The high temperatures used during the thermal cycles do not degrade it.

3.8.2. Separation of DNA Fragments

In living organisms, DNA exists in (double-stranded) chains that contain millions of base pairs. Due to their size, it is virtually impossible to obtain detailed information about their structure in a single experimental step. Usually, these huge molecules are divided into smaller segments for subsequent characterization. The chain scission is performed by using *restriction enzymes (restriction endonucleases)*. These enzymes split the chain at precisely defined locations, namely at some specific sequence of bases. (It should be noted that even a relatively short sequence is not generally repeated in the chain very frequently. For example, any unique sequence of six bases occurs on the average only once within a chain of four thousand bases long.) The result of the splitting is a collection of a moderate number of fragments of convenient lengths that can be separated by conventional separation techniques such as gel electrophoresis (described in Section 3.7.1.3). For more demanding separations of long segments of DNA, a more powerful version of the method was developed—pulsed field gel electrophoresis.

3.8.2.1. Pulsed Field Gel Electrophoresis (PFGE). The conventional gel electrophoresis can readily separate small fragments of DNA up to 50 kilobases (kb) on a horizontal or vertical slab gel, but it often fails to clearly separate longer DNA fragments (in the million base pair range). The reason is that these DNA molecules are too large to move through the pores of the gel in their coiled form. However, the chain ends have more maneuvering capability than the centrally located parts of the molecule and they can worm their way through the maze of the gel in the direction of the electric field. They are pulling the rest of the molecule behind them thereby stretching it. More of the charged groups can now contribute to the forward movement of the molecule that starts travelling like a snake. Eventually, the whole molecule is more or less stretched out and it moves forward with a velocity nearly independent of length and no separation according to molecular weight is obtained.

However, the straightening of the chain needs longer time for longer molecules than for the shorter ones. If we can force the molecules to frequently repeat this alignment process then the longer molecules will travel slower than the shorter ones. The desired repeated realignment is achieved by changing the direction, in which the molecules are forced to travel, that is by periodically changing the direction of the electrical field.

Several experimental arrangements are employed for the manipulation of the electrical field. In the *orthogonal field gel electrophoresis* the gel slab is horizontal and the two alternating fields operate in the gel plane. The angle between the fields is typically between 100° and 150°. On the molecular level, the DNA molecules move in a zigzag manner. However, macroscopically their movement is straight in a direction bisecting the direction of the two fields. Many variants of this procedure are used. The duration of the pulses can be adjusted to the size of the DNA fragments (larger fragments require longer pulses). A sequence of pulses of varying length can be programmed—this improves the resolution of both the shorter and longer segments. Multiple pairs of the electrodes (each with a different direction with the respect to the direction of the macroscopic movement of DNA) may be employed and operated in a multitude of preprogrammed sequences.

In the *field inversion gel electrophoresis* the direction of the field alternates by 180 degrees in a normal gel electrophoresis apparatus. Each pulse in the primary direction lasts typically twice as long as the corresponding pulse in the reverse direction. It is not obvious why this technique improves the results, but it does. The main advantage of this technique is that it is not necessary to buy a second electrophoresis instrument.

The *transverse alternating field electrophoresis* uses electric fields both in the plane of the slab and perpendicular to it. The gel is positioned vertically within the chamber between two sets of linear electrodes: one in front and one behind the gel. The linear electrodes are parallel to individual DNA lanes. The main electrodes are pulling DNA along the lanes while the linear electrodes produce pulsating field across the gel. As a result, DNA zigzags through the thickness of the gel moving down at the same time. The orthogonal orientation of the fields allows an easy separation and resolution of very large DNA fragments.

The PFGE technique is routinely used for separation of higher molecular weight fragments of DNA. Like any other technique, it has its advantages and disadvantages. Long periods of electrophoresis are often required for good resolution, and migration of fragments is strongly dependent on experimental conditions. Therefore, it is difficult to compare gels even when they are run under identical conditions.

3.8.2.2. Southern Transfer Technique.

The Southern transfer technique developed by Ed Southern (known more colloquially as Southern blotting technique) is designed to identify fragments containing a specific base sequence. It utilizes a combination of fragmenting DNA by restriction enzymes with hybridization with specially developed probes that are called *markers*. The number of possible different sequences of fragments containing even a small number of nucleotides is huge. More than a thousand of different sequences exist in fragments with five bases, more than a million with ten bases, more than a billion with fifteen, and so on. Consequently, there exist many relatively short sequences that occur in the genome only once. Complementary sequences to these unique segments that are either radioactive or fluorescent are synthesized: they are our markers.

In the Southern technique, the total DNA is broken into fragments by restriction endonucleases and the DNA fragments are separated according to their size by agarose gel electrophoresis. The separated double-stranded DNA fragments are denatured by soaking the gel in 0.5 M NaOH solution to get single stranded DNA fragments. The gel is then overlaid by a sheet of nitrocellulose paper which, in turn, is covered by a thick layer of paper towels, and a heavy plate compresses the entire assembly. Nitrocellulose binds tenaciously single-stranded (but not double-stranded) DNA. The liquid in the gel is thereby forced (blotted) through the nitrocellulose so that the single-stranded DNA is transferred to the nitrocellulose where it binds at the same position it had in the gel. (An electrophoretic process named electroblotting can alternatively accomplish this transfer to nitrocellulose).

The nitrocellulose sheet is then moistened with a minimum quantity of a solution containing the radioactive marker that has the nucleotide sequence complementary to the specific DNA fragment in the DNA being analyzed. The moistened filter is held at a suitable renaturation temperature for several hours to permit the probe to hybridize to its target sequence. Then it is washed to remove the unbound radioactive probe, dried, and autoradiographed by placing it for a time over a sheet of X-ray film. Consequently, the specific DNA fragments to which the marker hybridizes can be identified by autoradiography.

3.8.2.3. DNA Profile Analysis.

In the last decade, a method called DNA profile analysis (or DNA finger printing or DNA typing) has revolutionized human genetics, diagnosis of inherited diseases, criminology, paternity suites, etc. To understand this, we need to take a brief look on the genome (that is the complete description of the genetic makeup of an organism). Living organisms may be classified as prokaryotes and eukaryotes.

Prokaryotes (bacteria) are simple unicellular organisms, genetic materials of which are situated typically on a single DNA molecule. During protein production, the individual genes (the coding sequences in DNA) are precisely transcribed into mRNA and sent to ribosomes to serve as a blueprint for protein synthesis.

In eukaryotes (higher organisms), DNA (which is segregated in the nucleus) has a much more complicated structure. Most genes are much longer than needed for coding their proteins. They contain coding sequences (called exons) that specify the amino acid sequence of the protein, and noncoding intervening sequences (introns), the purpose of which is presently unknown. The enzyme RNA polymerase transcribes the whole gene into pre-mRNA that includes the introns. The introns are excised by a series of splicing reactions that leads to the production of the mature mRNA that is then exported from the nucleus. The intron-related chain fragments are digested into individual nucleotides and recycled.

The sequences of the nucleotides in exons are conserved to a very high degree and vary little among individuals. Mutations (replacements of one nucleotide by another) in exons have a very important effect on the organism. Most of them destroy the information carried by the gene and are causing various genetic diseases. Some of the mutations may improve the gene and are a part of evolution.

However, introns that do not carry any known information as well as the nucleotide sequences outside the genes are highly variable. We will consider two types of variations: mutations and repetitive motifs. Mutations that are outside the exons have in most cases no effect on the organism and—once created—are replicated in the process of reproduction and spread in the population. Everybody has thousands of such mutations, but virtually no two individuals (except identical twins) have the same set of them. For most organisms, including humans, the noncoding parts of the genome contain long series of repetitive motifs. The number of motifs is highly variable among individuals. We are talking about the variable number of tandem repeats (VNTR).

Let us return to the profile analysis. The tested sample of DNA is digested with a mixture of several endonucleases that cut the whole molecule into segments at specific sites. Some of the mutations either create or destroy the specific sites for the endonuclease. As a result, the size of certain DNA fragments produced by digestion with a given set of restriction enzymes varies from individual to individual. These size differences are referred to as restriction fragment length polymorphism (RFLP). Other fragments contain the repetitive motifs. Again, their length varies among individuals.

In the next step of the analysis, the DNA fragments are separated using electrophoresis. Obviously, long sections of the genome will produce a huge number of fragments. We need to make visible only those containing the VNTRs and RFLPs. This is accomplished using DNA hybridization. A probe (or set of probes) is synthesized that is complementary to some known sequence in some highly variable part of the genome. Quite often it is the repetitive motif itself. The probe is either radioactive or it carries some fluorescent group. It is applied to the gel after electrophoresis and it recombines with its mate. The unbound probe is washed out, and the position of the fragments carrying the sequence of interest is appropriately visualized.

The resulting pattern of bands is virtually unique for every individual. Thus, when two samples are run side by side and the patterns are identical, it is highly probable that the samples belong to the same person (or to his/her identical twin). This obviously has enormous significance in forensic science or whenever unrecognizable mortal remains need to be identified. Moreover, the rules of heredity dictate that one half of the lines must also exist in father's DNA profile and the other half in mother's. Siblings (with the exception of identical twins) have on average also one half of common lines, while the half-siblings have common one quarter on average. Thus, the technique is also employed to resolve paternity disputes by comparing the DNA profiles of the child with those of the mother and the (alleged) father.

3.8.3. Sequencing of DNA Fragments

A detailed knowledge of the sequence of the nucleotides would mean a huge step toward the understanding of the chemical basis of life. The main problem in sequencing the human genome is the enormous number of nucleotides in it. The human genome contains about three billion nucleotide pairs. Special techniques of fast sequencing had to be developed. The sequencing starts with digesting the DNA sample by an endonuclease that will produce fragments with convenient lengths (in the order of a few hundred bases). The segments are separated by gel electrophoresis and each is sequenced separately. After that, the whole process is repeated using a different endonuclease. Comparison of the two sets of overlapping sequences then allows the construction of the overall sequence.

The sequencing technique is also based on DNA polymerase in a process very similar to the one described in Section 3.8.1. This time our primer is complementary to the end of the fragment to be sequenced. However, the amount of the template needs to be significant (we can prepare it by PCR).

Normally, the copying process would continue until the whole chain has been synthesized. However, our intention is to terminate the growth of some chains by adding to the reaction mixture a small amount of a *terminator*. The terminator is an analog of the appropriate nucleotide triphosphate (it is based on a sugar dideoxyribose) that is only monofunctional and consequently terminates the chain every time it is incorporated into it. The result is a mixture of fragments (starting with very short ones) each of which has the same base at its terminal nucleotide. The mixture is submitted to gel electrophoresis under conditions that allow determination of the number of nucleotides in each fragment. Thus, we know all the positions in the chain occupied by this particular base.

This procedure is repeated in parallel using four terminators corresponding to the four bases: adenine (A), thymine (T), guanine (G), and cytosine (C). It is advantageous to run the electrophoresis of the four mixtures side by side on a single gel so that all nucleotide lengths are represented in the four lanes of the gel. The sequence of the original chain can then be read directly as one "steps" up the bands on the gel one additional nucleotide at a time (ladder of bands).

The huge number of nucleotides even in the genome of the simplest organism called for development of automated sequencers that do all the steps starting from the PCR to the automated reading of the sequences. These machines can sequences up to 100,000 nucleotides per day.

Perhaps the most revolutionary approach (at least at the writing of this section) is the so-called "shotgun" technique. The whole genome is split into random fragments, the fragments are separated, cloned, and a very large number of them are sequenced. All the sequences are fed into a powerful computer with sophisticated software that finds the overlapping sequences and assembles the full map of the genome. During the time this section was written it was announced that the whole human genome was completely sequenced.

3.A. SUGGESTIONS FOR FURTHER READING

3.A.1. General Reading

Brostow, W., *Polymer Characterization*, Wiley, New York, 2000.

Campbell, D., J.R. White, and R.A. Pethrick, *Polymer Characterization*, Stanley Thomes Pub. Ltd., UK, 2000.

Cantor, C.R., and P.R. Schimmel, *Biophysical Chemistry*, Part II: *Techniques for the Study of Biological Structure and Function*, Freeman, San Francisco, 1980.

Cheremisinoff, N.P., *Polymer Characterization: Laboratory Techniques and Analysis*, Noyes Publications, New Jersey, 1996.

Elias, H.G. *An Introduction to Polymers*, Wiley, New York, 1997.

Eisenberg, H., *Biological Macromolecules and Polyelectrolytes in Solution*, Clarendon Press, Oxford, 1976.

Forsman, W.C., Ed., *Polymers in Solution. Theoretical Considerations and New Methods of Characterization*, Plenum, New York, 1986.

Gray, F.M., *Polymer Electrolytes,* Springer, Berlin, 1997.

Ke, B., *Newer Methods of Polymer Characterization,* Wiley, New York, 1964.

Morawetz, H., *Macromolecules in Solution,* 2nd ed., Wiley, New York, 1975.

Pethrick, R.A., and J.V. Dawkins, Eds., *Modern Techniques for Polymer Characterization,* Wiley, New York, 1999.

Richards, E.G., *An Introduction to the Physical Properties of Large Molecules in Solution,* Cambridge University Press, Cambridge, 1980.

Rudin, A., *The Elements of Polymer Science and Engineering: An Introductory Text and Reference for Engineers and Chemists,* Academic, New York, 1998.

Stuart, H.A., *Die Physik der Hochpolymeren,* 4 vols., Springer, Berlin, 1956.

Sun, S.F., *Physical Chemistry of Macromolecules, Basic Principles and Issues,* Wiley, New York, 1994.

Tanford, C., *Physical Chemistry of Macromolecules,* Wiley, New York, 1961.

Tsvetkov, V.N., V. Ye. Eskin, and S. Ya. Frenkel, *Structure of Macromolecules in Solution,* Butterworths, London, 1970.

Van Holde, K.E., *Physical Biochemistry,* Prentice-Hall, Englewood Cliffs, New Jersey, 1971.

3.A.2. Thermodynamics

Chung, T.-S., Ed., *Thermotropic Liquid Crystal Polymers: Thin-film Polymerization, Characterization, Blends, and Applications,* Technomic, Co., Inc., Lancaster, PA, 2001.

Ciferri, A., Ed., *Liquid Crystallinity in Polymers: Principles and Fundamental Properties,* VCH, Weinheim, 1991.

Cooper, A.R., Ed., *Determination of Molecular Weight,* Wiley, New York, 1989.

DeGennes, P.-G., *Scaling Concepts in Polymer Physics,* Cornell University Press, Ithaca, New York, 1979.

Donald, A.M., and A.H. Windle, *Liquid Crystalline Polymers,* Cambridge University Press, 1992.

Donth, E.-J., *Relaxation and Thermodynamics in Polymers: Glass Transition,* VCH, Weinheim, 1992.

Freed, K.F., *Renormalization Group Theory of Macromolecules,* Wiley, New York, 1987.

Hamley, I.W., *Introduction to Soft Matter: Polymers, Colloids, Amphiphiles and Liquid Crystals,* Wiley, New York, 2000.

Hildebrand, J.H., and R.L. Scott, *The Solubility of Nonelectrolytes,* Dover, New York, 1967.

Kamide, K., and T. Dobashi, *Physical Chemistry of Polymer Solutions-Theroretical Background,* Elsevier, New York, 2000.

Klenin, V.J. *Thermodynamics of Systems Containing Flexible-Chain Polymers,* Elsevier, New York, 1999.

Prigogine, I., *The Molecular Theory of Solutions,* North-Holland, Amsterdam, 1957.

Tampa, H., *Polymer Solutions,* Academic, New York, 1956.

Tai-Shung Chung, Ed., *Thermotropic Liquid Crystal Polymers: Thin Film Polymerization, Characterization of Blends, and Applications,* Technomic Publ. Co. Inc., Lancaster, PA, 2000.

Van Dijk, M.A., *Concepts of Polymer Thermodynamics,* Technomic, Publ. Co. Inc., Lancaster, PA, 1997.

Yamakawa, H., *Modern Theory of Polymer Solutions,* Harper and Row, New York, 1971.

Yamakawa, H., *Helical Wormlike Chains in Polymer Solutions,* Springer, Berlin, 1997.

3.A.3. Equilibrium Methods (Osmometry, Sedimentation Equilibrium, and Phase Equilibria)

Fujita, H., *Mathematical Theory of Sedimentation Analysis,* Academic, New York, 1962.

Fujita, H., *Foundations of Ultracentrifugal Analysis,* Wiley, New York, 1975.

Schachmann, H.K., *Ultracentrifugation in Biochemistry,* Academic, New York, 1959.

Tombs, M.P., and A.R. Peacock, *The Osmotic Pressure of Biological Macromolecules,* Clarendon Press, Oxford, 1975.

3.A.4. Hydrodynamics and Hydrodynamic Methods

Bohdanecký, M., and J. Kovář, *Viscosity of Polymer Solutions* (Polymer Science Library, Vol. 2), Elsevier, Amsterdam, 1980.

3.A.5. Light Scattering and Spectral Methods

Ando, I. and T. Asakura, Eds., *Solid State NMR of Polymers,* Elsevier, New York, 1998.

Berne, B.J., and R. Pecora, *Dynamic Light Scattering,* Wiley, New York, 1976.

Bovey, F.A., *High Resolution NMR of Macromolecules,* Academic, New York, 1972.

Burchard, W., and G.D. Patterson, *Light Scattering from Polymers* (Advances in Polymer Science, Vol. 48), Springer, Berlin, 1983.

Cebe, P., B.S. Hsiao, and D.J. Lohse, Eds., *Scattering from Polymers: Characterization by X-rays, Neutrons, and Light,* ACS Symposium Series, 739, Boston, Mass, 1999.

Chu B., *Laser Light Scattering,* Academic, New York, 1974.

Fabelinskii, I.L., *Molecular Scattering of Light,* Plenum, New York, 1968.

Glatter, O., and O. Kratky, *Small Angle X-Ray Scattering,* Academic, New York, 1982.

Huglin, M.B., Ed., *Light Scattering from Polymer Solutions,* Academic, New York, 1972.

Kerker, M., *The Scattering of Light and Other Electromagnetic Radiation,* Academic, New York, 1969.

Koenig, J.L., *Spectroscopy of Polymers,* 2nd ed., Elsevier, New York, 1999.

Kratochvil, P., *Classical Light Scattering from Polymer Solutions* (Polymer Science Library, Vol. 5), Elsevier, Amsterdam, 1987.

Partington, J.R., *An Advance Treatise on Physical Chemistry,* Vol. 4, *Physico-Chemical Optics,* Longmans, Green, London, 1953.

Roe, R.-J. *Methods of X-Ray and Neutron Scattering in Polymer Science* (Topics in Polymer Science), Oxford University Press, 1999.

Stacey, K.A., *Light-Scattering in Physical Chemistry,* Academic, New York, 1956.

Tonelli A.E., *NMR Spectroscopy and Polymer Microstructure: The Conformational Connection* (Methods in Stereochemical Analysis Series), VCH, Weinheim, 1989.

3.A.6. Separation Techniques

Glöckner, G., *Polymer Characterization by Liquid Chromatography* (J. Chromatography Library, Vol. 34), Elsevier, Amsterdam, 1987.

Hames, B.D., and D. Rickwood, Eds., *Gel Electrophoresis of Proteins: A Practical Approach,* IRL Press, Oxford, 1981.

Jančа, J., *Field-Flow Fractionation: Analysis of Macromolecules and Particles* (Chromatography Science Ser., Vol. 39), Dekker, New York, 1987.

Taylor, L., *Supercritical Fluid Extraction,* Wiley, New York, 1996.

Wilson, K., and J. Walker, *Practical Biochemistry—Principles and Techniques,* 4th ed., Cambridge University Press, Cambridge, 1994.

Wu, C.-S., Ed., *Column Handbook for Size Exclusion Chromatography,* Academic, New York, 1999.

3.A.7. Techniques to Study the Structure of Nucleic Acids

Burley, J. Ed., *The Genetic Revolution and Human Rights,* The Oxford Amnesty Lectures, 1998, ACS Publication, Oxford University Press, Oxford, 1999.

Lehninger, A.L., D.L. Nelson, and M.M. Cox, *Principles of Biochemistry,* 2nd ed., Worth Publishers, New York, 1993.

Voet, D., and J.G. Voet, *Biochemistry,* 2nd ed., Wiley, New York, 1995.

3.B. THERMODYNAMICS

3.B.1. Review Questions

1. Enumerate the assumptions used in the derivation of the Flory–Huggins theory. What assumptions are the most questionable? How do they affect the theoretical predictions?
2. Describe the ideas that led to the development of the Flory–Prigogine theory and its main results.
3. Explain why the Flory-Huggins theory fails to predict that the second virial coefficient depends on molecular weight of the polymer. Why is the Flory–Huggins theory not adequate at low concentrations of the polymer?
4. Compare the Flory-Huggins theory for the second virial coefficient with the theory of excluded volume.
5. The original Flory-Huggins theory contained the parameter $(1/2 - \chi)$. In the newer Flory's version this parameter was replaced by $\psi(1 - \theta/T)$. Give the argument for this refinement.
6. Show the dependence of Flory's expansion factor on the molecular weight of the polymer.
7. How does the mean square end-to-end distance vary with the molecular weight for theta and good solvents?
8. Polymers can exist either as helices or random coils in solution. Discuss several techniques that can be used to distinguish between these two conformations.

9. What is cohesive energy density? How are the interaction parameter χ and the solubility parameter δ related?

10. Define theta temperature. What special properties are exhibited by polymer solutions at theta conditions? Explain in detail why the Gaussian statistics of a macromolecular coil is valid at theta conditions but not in good solvents.

11. Many physicochemical relationships become simplified under theta conditions. (a). Explain what is meant by these conditions, (b). Describe three ways how to find the theta conditions of a polymer-solvent system.

12. What is the justification of the random walk model for the evaluation of the conformation of polymer coils? What are the main drawbacks of this model? How can the drawbacks be overcome?

13. Explain why it is possible for some real polymers to adopt a distribution of end-to-end distance that closely approximates the end–to-end distance distribution of an ideal Gaussian coil.

14. Describe how the technique of Monte Carlo is useful in determining the configurational properties of polymer molecules.

3.B.2. Derivations and Numerical Problems

1. One of the theories of liquids treats pure liquid as a mixture of molecules and holes of the same size. The density of the liquid is related to the "mole fraction of holes", x as $\rho = \rho_0 (1 - x)$ where ρ is the density of the liquid with holes and ρ_0 is density of liquid free of holes. Find the relation between density and solubility parameter, δ of the liquid. Hint: Use the Flory-Huggins approach, find the relation between the solubility parameter and the liquid-liquid contact energy, assume zero contact energy between holes and molecules.

2. Derive the dependence of the chemical potential of a polymer on its concentration (in mass/volume units) from the dependence of chemical potential of the solvent on concentration. Express the latter dependence by virial expansion.

3. According to the Flory-Huggins solution theory, the chemical potential of a solvent in a polymer solution is given by Equation (3.1.33). For dilute solutions (small ϕ_2) it is convenient to use a virial equation that is similar to Equation (3.1.36) and is given as:

$$\mu_1 - \mu_1^0 = -\bar{V}_1 RT \left[A_1 c_2 + A_2 c_2^2 + A_3 c_2^3 + \cdots \right]$$

Evaluate the first three virial coefficients (A_1, A_2, and A_3) for the Flory–Huggins theory. Derive also an expression for the dependence of chemical potential of solute, μ_2 on its concentration. The solute concentration is expressed as mass in unit volume of the solution.

4. Flory–Huggins solution theory for the enthalpy change, ΔH_{mix} on mixing n_1 moles of solvent with n_2 moles of polymer is given by Equation (3.1.28) and

for the entropy of mixing, ΔS_{mix} by Equation (3.1.25). Use these results to calculate the chemical potential of the solute in the solution, μ_2, relative to pure solute, μ_2^o.

5. The Flory–Huggins interaction parameter, χ is defined by either Equation (3.1.29) or (3.1.33) in the text. If χ depends on concentration, i.e, on ϕ_1, then these two parameters will not be equal. Name the first as $\chi_\mu(\phi_1)$ and the latter as $\chi_g(\phi_1)$. What is the relation between these two parameters?

6. Assume the validity of the Flory–Huggins relation as given by equation (3.1.29) in the text. Plot $\Delta G_{mix}/RT$ vs. ϕ_2 for the following set of parameters: (i) $M_1 = 100$; $M_2 = 100$; $\chi_{12} = 1.5$ and (ii) $M_1 = 100$; $M_2 = 5000$; $\chi_{12} = 1.0$ assuming $n_1 + n_2 = 1$ and specific volumes $\bar{v}_1 = \bar{v}_2 = 1$.

7. For regular solutions in which the components have molecules of the same size, the Gibbs energy of mixing is given by:

$$\Delta G_{mix} = RT \left[n_1 \ln x_1 + n_2 \ln x_2 + (n_1 + n_2) x_1 x_2 g \right]$$

where n_i denotes the number of moles of component i, x_i is mole fraction and g is a parameter that measures the deviation from Raoult's law. Show that for such solutions, the chemical potentials are given by equations (3.1.30) and 3.1.31). Further show that the activity coefficients are given by:

$$\ln \gamma_1 = x_2^2 g \quad \text{and} \quad \ln \gamma_2 = x_1^2 g.$$

If benzene and carbon tetrachloride form regular solutions with $g = 324$ J mol^{-1} at 298 K, then for an equimolar mixture, calculate ΔH_{mix} and ΔS_{mix}. Also, calculate γ_1 and γ_2 over the range of mole fractions from $x_1 = 0$ to 1, and plot the results.

3.C. OSMOMETRY, SEDIMENTATION EQUILIBRIUM, AND PHASE EQUILIBRIA

3.C.1. Review Questions

1. Discuss the role of diffusion in osmometry.
2. What is membrane asymmetry in osmometry? What kind of membrane would you use for the osmometric measurements of polystyrene in benzene?
3. Describe the factors affecting the dependence of pressure difference across the semi-permeable membrane on time during an approach to osmotic equilibrium.
4. Describe the dynamic method of measurement of osmotic pressure.
5. Explain why the results of density gradient ultracentrifugation may be used to detect the chemical heterogeneity in a polymer sample.
6. Describe the interference optics and schlieren optics in an ultracentrifuge experiment. Mention the advantages and disadvantages of schlieren

technique for determining the concentration gradient in an equilibrium experiment.

7. Discuss the arguments that led to the derivation of the formula describing the concentration profile in a sedimentation cell at equilibrium.

8. What does "boundary" mean in thermodynamics and how does it affect the equilibrium?

9. What is the advantage of the high-speed sedimentation equilibrium (Yphantis method) over the low speed Richards-Schachmann method?

10. What information can be obtained from a study of sedimentation in a density gradient?

11. When a polymer coil moves through a liquid under a gravitational force, it moves faster than the individual segments would move if separated. On the other hand, if a large number of polymer coils is present in the sedimentation cell, all of them will sediment slower than an individual coil. Explain.

12. Describe the fractionation of a polymer by the method of fractional dissolution.

13. Explain why some systems that should be phase separated at equilibrium can survive for a long time as homogeneous systems.

14. Discuss the binodals and spinodals for the phase relationships in polymer solutions.

15. How is the critical temperature of a two-phase system defined?

16. How do you find the composition of two phases in equilibrium in a partially miscible two-component system from the appropriate free energy vs. composition diagram. Sketch the diagram.

17. Describe the technique of fractionation of polymers by precipitation. Comment on the advantages and disadvantages.

18. What are the basic differences between the batch and the column fractionation methods? Which one gives better separation?

3.C.2. Derivations and Numerical Problems

1. Derive the van't Hoff law for the osmotic pressure of an ideal solution assuming the validity of Raoult's law.

2. Show that the molecular weight obtained from equation (3.2.50) in a sedimentation equilibrium experiment is a weight average.

3. Derive the critical value χ_{crit} that would lead to the phase separation of a two component mixture provided that the molar volumes of the components are $V_1 = 100$ mL and $V_2 = 5000$ mL

4. Calculate the molecular weight and second virial coefficient of a polymer solution that has an osmotic pressure of 80 mm water at 23°C at the concentration 10 mg/cm^3 and 30 mm of water at the concentration of 5 mg/cm^3.

5. The osmotic pressure (π) data for poly(methyl methacrylate) dissolved in three different solvents are given in the table. Use these data to compute the number average molecular weight and the second virial coefficient of the polymer.

Solvent/density	$100\ c/\rho$ (weight fraction)	π (cm of solution)
Toluene	0.893	0.968
$\rho_s = 0.865,$	1.075	1.350
g/cm^3	1.348	1.960
	1.720	2.874
	1.998	3.830
	2.052	4.069
	3.179	9.400
	3.184	9.470
Acetonitrile	0.642	0.440
$\rho_s = 0.780,$	1.582	1.001
g/cm^3	2.286	1.561
	3.094	1.841
Acetone	0.861	0.840
$\rho_s = 0.785,$	1.236	1.399
g/cm^3	1.800	2.531
	2.753	5.137

Does $\overline{M}_n$ depend on the solvent used? Explain.

6. Calculate the osmotic pressures in cm of solvent that would be established by the weight % solutions of each of the following three polymers A, B and C. Use 0.9 g/cm^3 for solvent density and 27°C for the theta temperature. Indicate how accurately you could measure the three osmotic pressures.

Polymer	Molecular Weight	Weight %
A	4,800	2.0
B	97,200	0.65
C	1,800,000	1.5

7. A sedimentation equilibrium experiment was performed at the rotor speed of 60,000 rpm using a 3 mm cell (distance from the bottom to the meniscus) mounted with its bottom 7 cm from the rotor axis. Specific volume of the polymer is 0.750 cm^3/g and density of the solvent is 0.995 g/cm^3. The polymer concentration at the bottom of the cell is 4 times that at the meniscus. Calculate the molecular weight of the polymer. Assume that the solution is pseudoideal.

8. In a sedimentation equilibrium experiment at 20°C with a polystyrene solution, the following data were obtained. If the density of solution is 0.902 g/cm^3, partial specific volume is 0.913 cm^3/g, the solution is pseudoideal, and the rotor speed is 9,000 rpm, calculate the molecular weight of this sample.

x cm	6.9	6.95	7.0	7.05	7.1
c g/L	1.30	1.46	1.64	1.84	2.06

9. A pseudoideal solution of a polymer with an approximate molecular weight, 200,000 and partial specific volume, $\bar{v} = 0.8$ mL/g is subjected to an equilibrium ultracentrifugation in a solvent of density, $\rho = 0.88$ g/cm^3 at 25°C. For the radial position of the meniscus at 6.8 cm and of the bottom at 7.1 cm, calculate the convenient speed of the rotor in rpm assuming that the optimum ratio of concentrations at the cell bottom and meniscus is 5.

3.D. HYDRODYNAMICS AND HYDRODYNAMIC METHODS

3.D.1. Review Questions

1. Discuss the concentration dependence of viscosity of a solution of polyelectrolyte (a) in distilled water and (b) in a buffer with high ionic strength.
2. Why is the viscosity of a macromolecular solution higher than the viscosity of the solvent? Explain in detail.
3. Explain why the movement of a particle through a viscous liquid is opposed by frictional force.
4. What is the frictional ratio, f/f_0 and what can be deduced from it?
5. Explain the origin of the frictional force acting against the movement of a particle through a viscous liquid.
6. Explain the meaning of the term hydrodynamic radius of a particle.
7. Starting from Einstein's relation for viscosity of a suspension of spheres and from the influence of thermodynamic quality of the solvent on the macromolecular coil, explain why the MHS exponent a is usually between 0.5 and 0.8.
8. How will the intrinsic viscosity of a given sample of poly(sodium methacrylate) depend on the ionic strength of the solvent (the solvent is aqueous NaCl)?
9. Describe two methods for the evaluation of the viscometric Huggins constant.
10. What is the driving force of diffusion? Base your answer on thermodynamic considerations.
11. Describe one of the methods for the measurement of diffusion coefficients including the experimental setup and evaluation of data.
12. Explain the self-sharpening and hyper-sharpening of the boundary in sedimentation velocity experiments.
13. Explain why sedimentation coefficients of macromolecules decrease with increasing concentration.
14. Two proteins with identical molecular weights and specific volumes travel with the same velocity in a sedimentation velocity experiment performed in a dilute buffer. When CsCl is added to the buffer (density increases) one of the proteins travels faster than the other. This phenomenon is explained by the preferential binding of water to the protein. Which protein binds more water?
15. Consider sedimentation of a dilute solution of polymer coils in a centrifuge cell, each coil being composed from segments (friction centers). Discuss the effect of hydrodynamic interaction on the sedimentation velocity of the coils.

16. What is the relation between the diffusion constant and frictional coefficient? How does the frictional coefficient depend on the viscosity of the solvent?
17. Discuss the appropriateness of the Stokes-Einstein relation for the diffusion coefficient of a macromolecular coil.
18. What is the extinction angle in streaming birefringence?

3.D.2. Derivations and Numerical Problems

1. The Lamm equation for sedimentation velocity experiments is given by Equation (3.4.27) in the text. Explain the symbols and derive the radial dilution law from the equation. State clearly the assumptions involved and formulate the law in words in detail.
2. A spherical protein with molecular weight $M = 100,000$ dissolved in water at 20°C has a diffusion coefficient, $D = 8.0 \times 10^{-7}$. The viscosity of water at 20°C is 1.0 cP and the partial specific volume of the protein is $\bar{v} = 0.75 \text{ cm}^3/\text{g}$. Calculate the hydration of the protein, i.e., the weight of water bound to 1 g of protein. Assume that there is an adsorbed layer of water moving with the protein.
3. The diffusion coefficient of bovine serum albumin of molecular weight 68,000 at zero concentration in water at 25°C is $6.75 \times 10^{-7} \text{ cm}^2/\text{s}$, its specific volume $\bar{v}$ is 0.78 cm^3/g, and the viscosity of water is 1 cP. Compute the frictional coefficient, f and frictional ratio, f/f_0.
4. How much power is dissipated in cm^3 water at 20°C subjected to simple shearing flow with a velocity gradient of 500 sec^{-1}? Viscosity of water at 20°C is 1cP.
5. How long will it take for a spherical air bubble of 0.5 mm in diameter to rise 10 cm through water at 20°C ? (viscosity of water is 1cP).
6. Ovalbumin has a diffusion coefficient $D_{20.w} = 7.76 \times 10^{-7} \text{ cm}^2/\text{sec}$ at 20° C. If the viscosity of water is 1 cP, calculate its frictional coefficient considering the molecule to be a sphere. What is the radius of the spherical protein? Calculate its molecular weight if its partial specific volume is 0.74 cm^3/g.
7. A 50 mg of polystyrene was dissolved in toluene to make 25 mL of solution. The flow time of 342.25 sec for 5 mL of this solution was measured using an Ubbelohde viscometer. Four additional measurements were done by adding solvent to the viscometer, mixing well with the solution in the viscometer and then recording the flow time of the new solution. These data are given in the following table.

Solvent added (cm^3)	3	3	5	5
Flow time (sec)	257.65	224.07	198.47	185.68

If the flow time for pure toluene is 148.20 sec, calculate the intrinsic viscosity and the Huggins constant.

8. Approximate relations exist between the radius of gyration, molecular weight, and intrinsic viscosity. [See text, equations (3.4.46) and (3.4.49).] Use these

relations to estimate the value of radius of gyration, r_G for polystyrene in toluene for the molecular weights of 100,000 and 300,000, respectively. The MHS parameters for this system are: $K = 1.23 \times 10^{-2}$ and $a = 0.72$ (while the units of K are peculiar, the units of $[\eta]$ are cm^3/g).

9. Using the data for osmotic pressure and relative viscosity given in the table for cellulose acetate dissolved in acetone at $25°C$ estimate its number average molecular weight and intrinsic viscosity.

$c/ (g/100 \ cm^3)$	π (cm of solvent)	η_{rel}
0.088	2.511	1.034
0.132	3.806	1.052
0.220	6.373	1.087
0.308	8.962	1.124

Take density of acetone at $25°C$ as $0.792 \ g/cm^3$.

10. A polymer sample with a molecular weight of 150,000 has an intrinsic viscosity of 0.14 g/dL, another sample of the same polymer has an intrinsic viscosity of 0.20 g/dL. What is the molecular weight of the latter if the intrinsic viscosities were measured in a theta solvent? If the solvent is a thermodynamically good solvent rather than a theta solvent, would the molecular weight (at $[\eta] = 0.20$ g/dL) be higher or lower than the one calculated above?

11. Using the data given in the table determine the MHS parameters for polyisoprene fractions.

$M \times 10^{-3}$	5.8	23.0	46.5	89.0	278	649	1,000	1,870
$[\eta]$ dL/g	0.11	0.30	0.50	0.80	1.84	3.40	4.70	7.40

12. A sample of 0.04 g of polystyrene was dissolved in 10 mL of benzene. A 4 mL of this solution was placed in the ubbelohde viscometer and the flow time was 107.443 sec. Then 2 mL of benzene were introduced into the viscometer and the time was 92.94 sec. Later, 2, 4 and 12 mL of benzene were sequentially added to the viscometer giving flow times 86.031, 79.488, and 73.306 sec, respectively. From an earlier experiment, it was found that the flow time for pure benzene is 67.457 sec. At $20°C$, the MHS parameters for polystyrene in benzene are $K = 0.0123$ mL/g and $a = 0.72$. Determine the intrinsic viscosity and Huggins' constant: (a) from the plot of η_{sp}/c vs. c, (b) from the plot of $\ln \eta_{rel}/c$ vs. c, and (c) calculate the standard deviation for your η_{sp}/c vs. c plot. Comment on whether it is small or large. Also, calculate the molecular weight of polystyrene.

13. Assume that a polymer solution conforms to a pseudo-lattice model, where two segments are not allowed to occupy the same lattice point (nonintersecting coils) and where the non-zero excess contact energies influence the coil statistics. Explain why the conditions under which the second virial coefficient is equal to zero are identical to the conditions for which MHS exponent $a = 0.5$.

14. Intrinsic viscosity of a polymer in water at 20°C is 50 cm^3/g. We plan to measure it more precisely using solutions of concentrations between 2 and 10 g/L. With what degree of constancy must be controlled the temperature of the viscometer if we want the correct final result within 1%?

15. Use the data given in the table for the movement of sedimentation boundary in a velocity experiment for an enzyme. If the rotor speed is 48,500 rpm, compute the sedimentation coefficient of the enzyme.

t (min)	0	12	24	36	44	52	60	68
Distance from center (cm)	6.57	6.62	6.67	6.72	6.76	6.80	6.84	6.86

16. A biopolymer with the molecular weight of 100,000 and partial specific volume 0.7 cm^3/g dissolved in a buffer of density of 1.0 g/cm^3 and viscosity of 1 cP at 20°C has a sedimentation coefficient equal to 2 Svedbergs (1 Svedberg = 10^{-13} sec). Predict whether the biopolymer is compact or forms a molecular coil. Watch the units during calculations.

17. A sedimentation velocity experiment was performed with the subunits of a protein at 3°C at a rotor speed of 51,970 rpm. Data for the boundary position vs. time are given in table below (assume that the first photograph was taken at zero time).

r_b(cm)	6.2438	6.3039	6.4061	6.4920	6.5898	6.6777
$t - t_0$(sec)	0	3,360	7,200	11,040	14,840	18,720

Calculate the sedimentation coefficient of the polymer. Calculate $s_{20,w}$ assuming partial specific volume = 0.73 cm^3/mol at 3°C and 0.74 at 20°C. Look for other pertinent data in standard handbooks. Assume that the buffer has the same viscosity as water.

18. For a spherical protein the sedimentation coefficient is 8.0 Svedberg (Svedberg = 10^{-13} sec) in water at 20°C and the partial specific volume is 0.8 cm^3/g. Estimate the molecular weight of the protein, clearly indicating any approximations you have made in the calculations. Describe how the diffusion coefficient for the protein can be computed from the data.

3.E. LIGHT SCATTERING

3.E.1. Review Questions

1. Analyze the light scattering from a perfect crystal.
2. Define the polarizability of a particle.
3. Discuss the light scattering of pure liquids.
4. Explain how does the thermodynamic quality of a solvent with respect to a given polymer influence the light scattering of its solution?

5. Explain why the scattering envelope is different for horizontally and vertically polarized light.
6. What is the internal destructive interference of scattered light?
7. Explain why (a) sky is blue, (b) sunset is red, (c) milk is white, and (d) clouds are white?
8. Explain the origin of the circle or ring around the moon.
9. Why does an isolated particle scatter light?
10. Explain how does the intensity of light scattered by a solution of polymer at moderate concentration (ca 0.005 g/cm^3) in a solvent depend upon temperature?
11. Suppose a polymer sample is bimodal in its molecular weight distribution. Describe qualitatively the appearance of the scattered photon-correlation function. Would the "non-standard appearance" of the correlation function be accentuated in thermodynamically "good" or "poor" solvents? What would one expect to be the effect of changing the temperature of the solution? (Note: Do not carry out elaborate calculations, but analyze this situation qualitatively).
12. Sulfur sols having extremely narrow distribution of radii may be prepared in water solution. When the light scattered from a white beam of light is observed, the interference colors (mainly alternating green and red) are observed in the angular dependence. Explain the physical basis of this phenomenon.

3.E.2. Derivations and Numerical Problems

1. Reduced intensity of light scattered by a solution of a monodisperse polymer at very low concentration is given as:

$$\frac{Kc}{R_\theta} = \frac{1}{M\left(1 - \mu^2 r_G^2/3 + \cdots\right)}$$

Explain the symbols and derive what sort of average molecular weight and r_G^2 can be obtained from the above equation for a polydisperse polymer.

2. Use the Zimm plot given in Figure 3.50 of the text to calculate the values of molecular weight, M_w, second virial coefficient, B, and radius of gyration, r_G of the polymer. The required physical constants are: solvent refractive index, $n_o = 1.454$; wavelength of plane polarized light, $\lambda_0 = 633$ nm; refractive index increment, $dn/dc = 0.10$ cm^3/g; the units of c are in g/cm^3. Indicate the estimated values in your calculations.

3. For some polymer solution the light scattering intensities in cps (counts per second) are given in the following table; the $\sin\theta$ correction to the raw data was already applied.
   Table. Data for scattering intensity from the polymer solution in cps/10^3 at different polymer concentrations in g/L.

$\theta°$	5 g/L	2 g/L	1 g/L
45	8.30	5.58	4.17
60	7.86	5.58	4.09
75	7.75	5.38	4.09
90	7.56	5.38	4.00
105	7.29	5.28	3.92
120	6.96	5.11	3.78
135	6.81	4.95	3.65
150	6.74	4.8	3.54
RI of solution	1.56058	1.56025	1.56012

The scattering wavelength (in vacuum) is 633 nm, corrected scattering intensity from solvent (refractive index 1.56000) is 1.50×10^3 cps (assume that it is independent of angle). The scattering intensity from benzene standard is 0.95×10^3 cps and the Rayleigh ratio of benzene standard at 90° is 8.765×10^{-6} cm^{-1}). Temperature is 20°C. Compute the values of M_w, B, and $\overline{r_G^2}$ of the polymer. For the latter quantity, treat each concentration data set as independent and obtain the corresponding $\overline{r_G^2}$ value and then extrapolate $\overline{r_G^2}$ to zero concentration. Do not produce the Zimm plot. (Note that you need to compute K from the data.)

4. Using the light scattering data given in problem (3) above, calculate the corrected scattering intensity in cps of a 7.5 g/L solution of the polymer at $\theta = 25°$. Explain your steps carefully as to how you are carrying out the required extrapolations. What assumptions are being made?

5. The autocorrelation function obtained from a quasi-elastic light scattering experiment at 90° scattering angle with incident light polarized vertically for some polymer sample in water has the time-dependent form given as:

$$<I(t)/I(0)> = A + B \exp(-\Gamma t)$$

Take the molecular weight of the polymer as 200,000, $n_o = 1.33$, $dn/dc = 0.1$ dL/g and $\Gamma = 1575$ sec^{-1}. Calculate the hydrodynamic diameter (in nm). At 20°C, take $\eta = 1.0$ cP, $\lambda = 633$ nm and refractive index of water $= 1.33$. Equal the average end-to-end distance $\overline{h}/2$ to the hydrodynamic diameter obtained above. Using the approximate form for the inverse of the scattering function $P(\theta)$ for random coils, calculate the numerical values of $Kc/\Delta R(\theta)$ and plot them vs. $\sin^2(\theta/2)$ for θ in 20° steps from 20° to 160°. Assume the second virial coefficient to be zero. $\Delta R(\theta)$ is the (solution–solvent) scattering intensity difference. (Note: since the incident light is vertically polarized, you need not to use the $\cos^2 \theta$ term in the scattering function.) Plot this function.

6. For some light scattering experiment, the refractive index of the solvent ($n_0 = 1.5$), refractive index increment ($dn/dc = 0.15$ mL/g), incident wavelength

($\lambda = 633$ nm), concentration by weight of the polymer ($c = 20$ mg/cm^3), molecular weight of polymer ($M = 10^5$ g/mol), average end-to-end distance square (random coil) ($\sqrt{\overline{h^2}} = 40$ nm), and zero second virial coefficient are given. Using these data, calculate $R(\theta)$ for θ angles from $20°$ to $160°$ in steps of $10°$ for two cases: (a) ignoring the effect of $\overline{h^2}$ (better of the molecular size) and (b) considering $\overline{h^2}$. Plot the results in such a way that the difference between these two cases is clear.

7. Consider the scattered photon correlation function for this polymer at $\theta = 45°$, $90°$, and $135°$. Assume that it may be treated as a "Stokes-Einstein particle" with its hydrodynamic radius given by $r_H = 6\sqrt{\overline{h^2}}$, where $\overline{h^2}$ has the same value as the polymer coil in problem (6) given above. Taking viscosity of the solvent as 0.50 cP (1 cP = 0.1 kg m^{-1} s^{-1}) and $T = 20°$C, calculate the predicted correlation time. Discuss the accuracy of this method for estimating $\overline{h^2}$.

8. A very dilute solution of some homogeneous particles with density, $\rho = 1.25$ g/cm^3) in water at $25°$C (find all the necessary constants for water in the literature) was studied by quasielastic light scattering (QELS) and by sedimentation velocity: (a). In the QELS measurements, the wavelength of light used was 633 nm and the scattered light was observed at $90°$. The analysis of the autocorrelation function yielded a single characteristic decay time $\Gamma_c = 2.5$ ms. (b). The sedimentation velocity experiment was followed by schlieren optics; photographs were taken repeatedly and the distances d were measured (on the photographic plate) between the maximum of the schlieren boundary and the image of a reference point on the rotor. The rotor speed was 52,000 rpm, the reference point was 5.6 cm from the rotor axis and the schlieren optics magnified all the radial distances by a factor 2.1. The time dependence of d values is given below.

T (min)	11	16	21	26	34
d (mm)	10.200	11.736	13.233	14.647	17.156

Calculate the molecular weight, hydrodynamic radius, sedimentation coefficient, and diffusion coefficient of the particles.

3.F. SPECTRAL METHODS

3.F.1. Review Questions

1. Explain how infrared spectroscopy can be used to differentiate between the isomers formed in the polymerization of isoprene.
2. Show how ultraviolet spectroscopy may be used to measure the percent composition of any styrene-butadiene copolymer.
3. From an NMR analysis of some vinyl polymers it is found that it contains 75%

of isotactic triads, 24% of syndiotactic triads and 1% of heterotactic triads. How would you classify this polymer?

4. Discuss the basis and limitations of the use of NMR to determine the stereo-regularity of polymers.

5. High resolution NMR may be used for the (a) characterization of polymer tacticity, (b) identification of two stereoregular polymers as isotactic and syn-diotactic, and (c) study of the distribution of the monomer units in a copolymer. Give an example of each case.

6. Poly(methyl methacrylate) may exist in isotactic, syndiotactic, and atactic forms. Show how NMR can be used to differentiate between these forms.

7. Explain why the high resolution NMR spectrum of poly(methyl methacrylate) polymerized with butyllithium in toluene at $-78°C$ shows a quartet of peaks for the methylene protons.

8. What are the advantages of the MALDI-TOF technique over that of the classical mass spectrometry?

3.G. SEPARATION METHODS

3.G.1. Review Questions

1. Separation of several components with very different molecular sizes is planned by a GPC experiment. Which one of the following two columns (geometrically identical) will have better separation power? (a) gel particles with carefully selected homogeneous pore size and (b) mixture of several types of gel particles with different sizes. Explain.

2. Describe the different types of detectors used in a GPC experiment mentioning the conditions under which these can be used.

3. How does the universal calibration curve in a GPC experiment help the data analysis?

4. What are the criteria for the selection of column material in GPC?

5. What is meant by critical point GPC? Describe this phenomenon with an example.

6. What are the advantages and disadvantages of mixed bed columns in GPC? (Do some literature search.)

7. Explain the procedure to determine the molecular weight distribution of a polymer using the GPC.

8. Define isoelectric conditions. Does the isoelectric point depend on the composition of the buffer? How is it used in protein purification?

9. The molecular weight of proteins and nucleic acids is determined using the pulsed field gel electrophoresis (PFGE) and agarose electrophoresis. Explain why a single technique is insufficient for the above purpose?

10. Comment why capillary electrophoresis is considered to be a high performance technique when compared to the conventional techniques?

3.G.2. Derivations and Numerical Problems

1. To a first approximation the calibration curve in a GPC experiment may be written $\log \overline{M}_n = a + bV$ where V is the elution volume and $\overline{M}_n$ is the number average molecular weight. Derive a general expression for the polydispersity $(\overline{M}_w/\overline{M}_n)$ of a polymer sample based on $c(V)$, where $c(V)$ is the concentration in g/cm^3 of the polymer that elutes at volume, V.

2. In order to obtain the calibration curve for poly(methyl methacrylate), the following elution volumes were obtained in a GPC experiment at 30°C for a set of monodisperse polystyrene standards dissolved in chloroform.

$M(g/mol)10^{-3}$	1960	680	160	50	20	10
V_e (cm^3)	119	126	137	147	157	163

    The MHS constants for polystyrene in chloroform at 30°C are: $K = 5.8 \times 10^{-3}$ cm^3/g and $a = 0.75$. Assuming that the universal calibration is valid, construct a calibration curve for the molecular weight-elution volume of poly(methyl methacrylate) in chloroform at 30°C. The MHS constants for poly(methyl methacrylate) in chloroform at 30°C are $K = 4.8 \times 10^{-3}$ cm^3/g and $a = 0.80$.

3. A sample of poly(methyl methacrylate) in chloroform was injected into a GPC column equipped with a differential refractometer as a detector. The following data of refractive index difference Δn at elution volume, V_e were obtained.

$\Delta n \times 10^5$	0.3	1.7	6.2	7.5	6.0	2.05	0.5
V_e (cm^3)	160	156	152	149	145	141	137

    These data describe a continuous smooth curve with a single peak. Using the above data and the calibration curve from the previous problem calculate and show graphically the molecular weight distribution of this sample of poly(methyl methacrylate). Assume that Δn is proportional to the concentration and the proportionality constant is independent of molecular weight.

4. Based on the universal calibration curve for GPC given in the text (Figure 3.63) and the data given in the following table, calculate the elution curve for a polystyrene sample which has one fluorescent group at the end of each polymer for the case of: (a) detection by Δn (refractive index difference depends only on the concentration of monomer groups) and (b) detection by fluorescence (detects only the fluorescent end-group, one per molecule). While carrying out this calculation you must consider the relative detector sensitivity (in arbitrary units) as a function of elution volume. Therefore, scale everything to the maximum signal $= 1.0$. Present your data graphically. The intrinsic viscosity is given by the MHS relation, where for tetrahydrofuran and polystyrene system, $K = 3.82 \times 10^{-2}$ and $a = 0.58$. (Note: the units of $[\eta]$ in the MHS relation are dL/g and you need to convert the product $M[\eta]$ to cm^3/g.)

M (g/mol) $\times 10^{-3}$	Amount of Sample (g)	M (g/mol) $\times 10^{-3}$	Amount of Sample (g)
10	0.2	90	1.0
15	0.5	110	1.4
20	1.0	130	2.0
30	0.6	150	1.5
50	0.4	170	1.0
70	0.8		

5. Following data were obtained in a GPC experiments at 25°C for standard polystyrene samples dissolved in tetrahydrofuran (THF).

$\overline{M}_w \times 10^{-3}$ (g/mol)	860	411	173	99	50	20	10	5
$[\eta]$ (cm^3/g)	206	125	67	44	28	14	9	5
V_e (cm$^3 \times 1/5$)	30	31	35	37	40	44	47	51

MHS constants for a hypothetical polymer in THF at 25°C are: $K = 1.6 \times 10^{-2}$ cm^3/g and $a = 0.65$. Use these data to construct a GPC calibration curve for this polymer.

3.H. TECHNIQUES TO STUDY THE STRUCTURE OF NUCLEIC ACID

1. Why is pulsed field gel electrophoresis (PFGE) preferred over the conventional gel electrophoresis for the separation of DNA molecules of approximately 100,000 base pairs?
2. Discuss the different variations of the PFGE technique.
3. What are the advantages of Southern blotting technique in predicting the terminal base pairs?
4. How is the DNA profile analysis useful for identifying a diseased gene?
5. Describe the method of DNA sequencing. Give an example.
6. Compare the different techniques used to study the structure of nucleic acid.

4

BULK POLYMERS

In Chapters 1–3, we focused our attention on individual macromolecules. Most polymers, however, are used mainly in bulk form, and we should therefore familiarize ourselves with the properties of bulk polymers. They are quite different from the properties of bulk low-molecular-weight materials. In Section 4.1 we will be concerned with relations between the properties of bulk materials and the properties and structure of the molecules that comprise them. The following sections will then present a short overview of methods for studying bulk materials.

4.1. PROPERTIES OF BULK POLYMERS

The physical state of every material is determined by two factors: (1) an interplay of forces holding the material together (chemical bonds, ionic forces, intermolecular interactions) with the thermal motion that tries to pull them apart and (2) the regularity of molecular structure that does or does not allow for the formation of ordered structures (e.g., crystals).

Let us review first the classical states of low-molecular-weight structures: crystals, liquids, and gases. At low temperatures, these substances try to maximize the interaction energy. This is usually achieved in a crystalline lattice; in most cases, small molecules have enough symmetry to allow efficient packing in crystals. The price for the high (negative) value of the interaction energy is the loss of translational freedom: The molecules are confined to their lattice locations. The entropy of crystals is low. At higher temperatures, the thermal motion tries to increase the entropy by destroying the order of the crystalline lattice by melting it. The entropy gained in melting is paid for by the loss of part of the (negative) interaction energy in the form of enthalpy of melting. At still higher temperatures, the liquids vaporize, gaining more entropy but losing all the remaining interaction energy (enthalpy of vaporization).

Although their behavior is governed by the same factors, the physical states and properties of polymeric materials are quite different from those of small-molecule materials. The differences are due to their macromolecular nature. To understand these differences, we need to examine molecular motion in bulk materials.

The attractive forces among molecules (van der Waals forces, attractions between dipoles, hydrogen bonds) try to keep them close together; thermal motion moves them around. A molecule cannot move to a new position unless it finds enough available space (a hole) or unless some cooperative motion of several molecules is possible; even in the latter case, some space is needed to facilitate the cooperative shift. This space is usually called *free volume.* Sometimes it is believed to be divided into molecule-sized holes; in reality it is composed of holes varying significantly in size. Strong intermolecular forces try to shrink the free volume against the thermal forces expanding it. As the temperature decreases, the thermal forces lose the battle. The free volume decreases, and viscosity (its measure) increases. For most low-molecular-weight substances this process is interrupted by the intervention of another event, crystallization. However, some molecules have difficulty crystallizing (e.g., some sugars, especially impure ones), because fast cooling does not give them enough time to crystallize. In such a case, the attractive forces prevail, and the free volume is reduced so much that it cannot support the molecular movement at all—the substance is converted to a *glass.*

Chain macromolecules are forced by their long shape to respond differently to the interaction forces and thermal forces. They do not move as single units; free-volume holes of their size are not available. Movements of their segments mediate their movement. Most macromolecules have many bonds around which parts of the molecule may rotate. A rotation around a single bond in the middle of a long chain requires an extensive change of position of a large portion of the molecule and is therefore not feasible. However, a synchronous rotation around several neighboring bonds may result in displacement of a small section of a chain while more distant parts of the macromolecule remain unperturbed *(crankshaft motion).* A long sequence of such motions may eventually move the entire macromolecule. The ease of the segmental movement depends, of course, on the structure of the chain (sizable side groups need more space for participating in the segmental motion) and on the free volume as dictated by the attractive forces.

We should recall from Section 3.3.1 that the movement of molecules caused by thermal energy is aimless. Macroscopic flow or deformation occurs only when the thermal motion is supplemented by an organized outside force. Thus the ease of macroscopic deformation must decrease together with decreasing intensity of thermal motion. As the temperature decreases the segmental motion decreases as well; the segments are locked in place, and the deformability of the system is severely reduced. At still lower temperatures, even the motion of side groups becomes impossible. Bulk polymers whose segmental motion is frozen are classified as glasses. Other typical states of polymer systems are also characterized mainly by their way of segmental motion and by their mechanical properties determined by these motions. We will treat

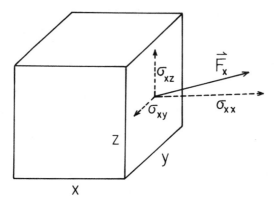

Figure 4.1. Cubic volume element of a continuous body and decomposition of the force $\mathbf{F}_x$ acting on one of its faces into three components of a stress tensor σ.

the typical states of polymer systems in the following sections. However, before doing that we need to introduce some basic concepts of *rheology,* a science concerned with the response of continuous bodies to outside forces.

4.1.1. Stress and Strain

Let us consider a cubic volume element of a continuous body; the edges of the element are oriented along the coordinate axes (Fig. 4.1). Generally, the body outside the element exerts forces on the faces of the element (the forces are proportional to the area of the face). Figure 4.1 depicts the force per unit area acting on the face perpendicular to the x axis as a vector $\mathbf{F}_x$; we may decompose it into three components along the edges of the volume element, which we designate σ_{xx}, σ_{xy}, and σ_{xz}. There are three pairs of faces in the volume element, and each of them has associated with it three force components. The three components perpendicular to the three faces are called *normal stresses;* the remaining six components are *shear stresses.* Together they form the stress tensor σ. The rules of mechanics require the stress tensor to be symmetrical.

Previous sections introduced tensors of the second order as entities relating two vectors. For example, in equation (3.3.7), the frictional tensor $\mathbf{f}$ related the frictional force $\mathbf{F}$ to the particle velocity $\mathbf{v}_P$. Stress tensor σ relates the stress force $\mathbf{F}$ acting on a plane with area A to the normal $\mathbf{n}$ of this plane:

$$\mathbf{F}/A = \sigma \cdot \mathbf{n} \tag{4.1.1}$$

We have already met the stress tensor twice. The viscous force F of equation (3.3.2) divided by the area A on which it is acting is actually the σ_{xy} component of the stress tensor. Our second encounter with the stress tensor was in Section 3.4.4.4, where we studied the relation between principal stresses and normal stresses as well as the relation between stress and optical tensors. We should also recall that a stress tensor inside a body that is subject to pressure as the only outside force is an isotropic tensor equal to negative pressure multiplied by the unit tensor $\sigma = -P\mathsf{U}$. Hence, whenever we may neglect the effect of pressure on the mechanical behavior of the system, we may add $P\mathsf{U}$ to the stress tensor.

In the following sections we will be concerned with only a few simple types of stress tensors. When a body is acted upon by a stress tensor in which only one normal component is nonzero (e.g., when stretching a string), we are talking about *tensile stress* or *tension*. When a sheet of plastic is stretched in two perpendicular directions, we are dealing with *biaxial tension*. Finally, a piece of material clamped between two plates, which are forced in opposite tangential directions, is subject to *simple shear.*

Different materials respond to stresses in different ways. *Elastic solids* deform instantaneously under stress. The term deformation implies that molecules or their parts change their mutual positions from their original equilibrium positions, which are determined by the interatomic and intermolecular forces, to new positions with higher free energy. The laws of thermodynamics compel the molecules to return to positions with the lowest free energy. This constitutes a returning *elastic force* that counterbalances the stress. When the forces are balanced, mechanics prohibits further macroscopic movement, that is, further deformation. When the outside stresses are removed, the elastic forces return the material to its original shape.

The deformation itself is described by another tensor—the *strain tensor.* Let us consider a vector **L** embedded in the stressed body. After deformation it changes to **L** + Δ**L**. The ratio Δ**L** $/L$ (L is the length of **L**) is a vector having three components. Three perpendicular vectors result in nine components that form a *displacement tensor.* The symmetrical part of this tensor is our strain tensor γ; the antisymmetrical part describes the rotation of the body and is not important when mechanical properties of elastic solids are considered.

Relations among the stress, strain, and rate of strain tensors (the latter tensor plays no role for ideal elastic bodies) are called *rheological equations of state* or *constitutive equations* of a body. For elastic bodies under small stresses, they are relatively simple. We have seen that every type of stress can be decomposed into a pressure part and a deviatory part. The pressure part causes uniform compression of the body, whereas the deviatory part causes deformation at constant volume. Accordingly, elastic materials are characterized by two independent parameters representing their resistance to bulk and shear deformation. In practice, more than two parameters are used; of course, only two of them are independent.

The *bulk modulus K* is defined as a proportionality factor between pressure P and the fractional change of volume, $\Delta V/V_0$:

$$P = -K \left(\Delta V/V_0 \right) \tag{4.1.2}$$

Bulk modulus is an inverse of isothermal compressibility as defined in equations (3.1.80)–(3.1.82) and used in Section 3.5.5.2.

The shear stress σ_{yx} and shear strain γ_{yx} in a simple shear experiment are related by means of a *shear modulus G* (do not confuse this symbol with the velocity gradient

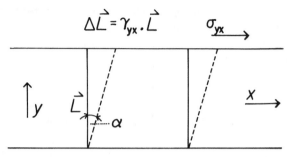

Figure 4.2. Components of stress and strain in a simple shear experiment.

of Section 3.3):

$$\sigma_{yx} = G\gamma_{yx} \qquad (4.1.3)$$

Shear modulus is also called *rigidity;* its reciprocal is the *shear compliance J.* A simple shear experiment is depicted schematically in Figure 4.2. The solid lines represent the shape before deformation; the dashed lines depict it after deformation. Closer inspection will reveal that the strain γ_{yx} is equal to tan α, as shown in the figure.

In the tensile experiment depicted in Figure 4.3, the length of the test piece increases from L to $L + \Delta L = L(1 + \gamma_{xx}) = L\lambda$, where λ is the elongation. If the material is incompressible, then the elongation along the x axis must be accompanied by contraction along the y and z axes. The condition of constant volume would yield (at small strains)

$$\gamma_{yy} = \gamma_{zz} \equiv \Delta d/d = -\gamma_{xx}/2 \equiv -\Delta L/2L \qquad (4.1.4)$$

It is, however, necessary to consider that the only nonzero component of the stress tensor σ_{xx}, also has a pressure part (negative in this case). The elongation must therefore be accompanied by volume expansion. As a consequence, the *Poisson ratio* μ defined as

$$\mu \equiv -\frac{\gamma_{yy}}{\gamma_{xx}} \equiv -\frac{\Delta d/d}{\Delta L/L} \qquad (4.1.5)$$

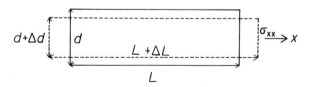

Figure 4.3. Elongation and lateral contraction in a tensile experiment.

is less than 1/2. (A value of 1/2 corresponds to incompressible bodies.) Thus the tensile experiment is described by two constants: the Poisson ratio and the *modulus of elasticity E* (often called *Young's modulus*) defined as

$$E \equiv \frac{\sigma_{xx}}{\gamma_{xx}} \equiv \frac{F/A_0}{\Delta L/L} \tag{4.1.6}$$

It should be noted that it is customary to calculate the tensile stress σ_{xx} as the ratio of the tensile force F and the cross-sectional area A_0 measured before deformation.

The three moduli and the Poisson ratio are not independent. A more careful analysis would show that they must be related as

$$E = 2G(1 + \mu) = 3K(1 - 2\mu) \tag{4.1.7}$$

The moduli are independent of the magnitude of the applied stresses as long as the stresses are small (Hookean behavior). At larger stresses, the relation between the stresses and strains may become nonlinear. In such a case, the moduli are described as functions of the elongation or the shear stress. When the applied stresses are increased beyond the capability of the material to balance them by elastic forces, the test specimen will break. In tensile experiments, the stress at the breaking point is called *tensile strength,* and the corresponding elongation is *elongation at break.* These two quantities together with the moduli are important characteristics of elastic solids.

Viscous liquids respond to stresses in a quite different way. Their deformation (strain) does not produce any elastic forces, and application of a shear stress leads to an unrestricted flow. However, the flow itself produces forces (we will call them viscous forces) that balance the applied stress and lead to a steady-state flow. The flow of liquids can also be described by constitutive equations. However, these equations use the rates of deformation (rates of displacement) instead of the deformations themselves. Rate of displacement is the time derivative of the position of a material point. The rate of displacement tensor is the same quantity as the tensor of velocity gradient $\mathbf{G}$ that we used extensively in Section 3.3. The symmetrical part of this tensor is called the *rate of strain tensor,* $\dot{\gamma}$ and is one of the fundamental quantities in rheology. The dot above the symbol represents the derivative with respect to time.

Many polymeric materials display behavior that is somewhere in between that of elastic solids and viscous liquids. Such materials are called *viscoelastic;* we will study their mechanical behavior in more detail in Section 4.1.5.

4.1.2. Glassy Polymers

In glassy polymers, segmental motion is virtually suppressed and the atoms are locked in certain mutual positions. How does such a body respond to an outside force trying to deform it? The distances between neighboring atoms must change. This implies that the valence bonds are stretched or compressed and the valence angles are deformed. The secondary bonds are deformed as well. All these modifications of the structure increase the potential energy of the molecules; this is the source of the elastic force. Deformation of the valence lengths and angles contributes most to it, but deformation of secondary bonds is still significant.

The magnitude of the modulus depends primarily on the number of bonds stretched and valence angles deformed per unit cross section of the test piece. Of course, it also depends on the force constants for the bonds and angles involved and on the force constants of the secondary bonds. For example, the shear modulus G for poly(methyl methacrylate) is about 1.5×10^{10} dyn/cm$^2 = 1.5 \times 10^9$ Pa $= 150$ kg/mm^2. Moduli of other organic polymers have similar magnitudes. Moduli of glasses formed by small molecules (e.g., glucose glass) depend more on deformation of the secondary bonds and are typically one-third as large. On the other hand, the number of bonds stretched is much higher in inorganic glasses of the silicate type. Consequently, their modulus is about one order of magnitude higher; the same applies to many metals such as steel and aluminum.

From the thermodynamic viewpoint, glasses are nonequilibrium systems. This fact manifests itself most clearly in the studies of the specific volume of glasses. In one-phase systems that are thermodynamically at equilibrium, specific volume at any given temperature is a material constant fully determined by the composition of the material. Specific volume of glasses depends on (1) the rate of cooling during the "freezing" of the melt to glass and (2) the time elapsed since this freezing. The more slowly the melt was cooled and the longer it rested, the smaller is the specific volume (i.e., the higher the density). Let us study this phenomenon in more detail; it will give us a better insight into the nature of the glassy state.

We have already noted that as the temperature decreases the thermal movement is less and less able to keep the molecules apart and the attractive forces tend to bring them together. Let us consider this tendency to be a stress and the shrinkage to be a strain. In the melt, the volumetric flow (shrinkage) succeeds in bringing the system to mechanical equilibrium very fast. However, below the glass transition temperature T_g, the slow relaxation phenomena take over (we will treat them in Section 4.1.5); the polymer remains in the nonequilibrium expanded state longer, and its approach to the equilibrium is slower the lower the temperature. It stays in the expanded state indefinitely at sufficiently low temperatures.

This process is frequently described in terms of *free volume*. The specific volume is imagined to be composed of *occupied volume* and free volume. The occupied volume is the volume needed to accommodate the molecular cores and their vibrations around their equilibrium positions. The intensity of these vibrations decreases with decreasing temperature; occupied volume slowly decreases with temperature. The free volume is associated with translational degrees of freedom and with the holes characterizing viscous liquids; it also decreases with decreasing temperature. At some temperature, the free volume fraction reaches a low value (given usually in the vicinity of 2.5%) that precludes the further volumetric flow that is needed for a further decrease in free volume; the material freezes. This process is, however, not completely abrupt; the lower the cooling rate, the longer the volumetric flow can follow the dictates of thermodynamics. Of course, once the system freezes, a further decrease in the specific volume reflects only the decrease in the *occupied* volume with temperature. (When

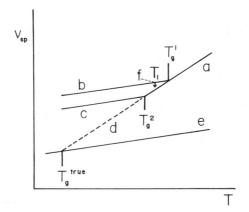

Figure 4.4. A schematic dependence of the specific volume of a polymer on temperature. (a) Melt; (b) polymer that was cooled fast; (c) slowly cooled polymer; (d) infinitely slow cooling; (e) hypothetical occupied volume; (f) annealing at temperature T_1. T_g^1 and T_g^2 are apparent glass transition temperatures; T_g^{true} is the temperature of a true second-order transition (hypothetical).

the system is kept at temperatures that are only slightly below the original freezing temperature—original T_g—the slow relaxation will allow for further shrinkage of the free volume as mentioned above. This process is called *annealing*.)

The dependence of specific volume on temperature is schematically depicted in Figure 4.4, where line a represents the volume of the melt and line b the volume of the polymer that was cooled fast; slow cooling corresponds to line c, and hypothetical infinitely slow cooling is depicted by line d. Line e represents the (hypothetical) occupied volume, and its intersection with d is the hypothetical true glass transition temperature. Annealing at temperature T_1 is shown by arrow f.

Glass transition temperature T_g is usually evaluated as the abscissa of the intersection of the dependences of specific volume on temperature for the melt and for the glass. Other thermodynamic properties such as enthalpy and compressibility show a similar picture. Such a behavior is characteristic of thermodynamic transitions of the second order, but glass transition is not one of them. For a true second-order transition, we cannot have two different values T_g^1 and T_g^2 depending on the cooling rate. Possibly, a true second-order transition exists at temperature T_g^{true}, but it cannot be observed experimentally.

4.1.3. Elastic Networks

Rubbers are materials that respond to stresses with a more or less instantaneous deformation that is fully reversible. In this respect they are similar to glasses. The similarity, however, soon ends. Rubbers can be reversibly stretched up to 1000% of their original length, whereas the limit for a typical glass is less than 1%. Also, the origin of the elastic force is quite different.

Rubberlike materials are lightly cross-linked networks with a rather large free-volume fraction that allows for fast rearrangement of polymer chains. At any given temperature, the free-volume fraction is larger for polymers with smaller intermolecular interactions than for more polar ones. Typical polymers of the former group are polybutadiene, polyisoprene, polyisobutylene, and polydimethylsiloxane; these are the best-known polymers used as rubbers. Of course, the free volume increases with increasing temperature; at higher temperatures even the more polar polymers become rubberlike (when cross-linked). On the other hand, at low temperatures rubbers lose their elastic properties. Remember that cold weather caused the sluggishness of the O-ring on the solid booster of *Challenger* that led to the destruction of the spacecraft.

Once a polymer network is in the rubberlike temperature region, its mechanical behavior is almost independent of the nature of the polymer chains and depends mainly on the architecture of the network. Let us develop a rather simple model of an elastic network and study its behavior under strain. It will be found that even the simplest model describes the behavior of real rubbers quite closely.

4.1.3.1. Theory of Rubber Elasticity.

4.1.3.1. Theory of Rubber Elasticity.   Our model will consist of a network of polymer chains connected at junction points; three or more chains originate from each junction point. Chemically, the junctions correspond to cross-links or branching points. The molecular chain between two junctions is an elementary chain. It is assumed that in the absence of strains the distribution of conformations of the elementary chains is the same as for an ensemble of identical free chains. In other words, the distribution of conformations is the most probable one, and the entropy of the system is at maximum. When the sample is deformed, the individual chains slip past each other and adopt new conformations. The new conformations still belong to the ensemble of all possible chain conformations, and their energy is the same as that of the original ones; the energy of the system does not change with deformation. Moreover, because of the large available free volume, the adjustment of the conformation is not opposed by intermolecular forces to any significant degree and is essentially instantaneous. On the other hand, the new set of conformations is not the most probable one any more, and its entropy is lower than before the deformation. The thermal motion that tries to bring the chains back to their original conformations represents the returning elastic force. Thus the elasticity of rubbers results from entropic factors, in contrast to the elasticity of glasses, where the deformation of bonds (i.e., an increase of energy) is the dominant factor.

At this moment we need to make a detour to some basics of thermodynamics. For reversible processes, heat entering the system is equal to TdS, and the first law of thermodynamics reads

$$dE = TdS + dW \tag{4.1.8}$$

We will consider isothermal processes in condensed systems only. Then we may write for the changes of Helmholtz energy A and Gibbs energy G

$$dG \approx dA = dE - TdS = dW \tag{4.1.9}$$

We are interested in phenomena related to the deformation of rubbers. We can therefore identify dW as the reversible work during the deformation of the sample. Moreover, we will introduce the stored-energy function W as an integral of dW from the relaxed sample to a deformed one. W is obviously a state variable. Mechanical properties of deformed systems are derived easily when W is known as a function of the strain. First we will evaluate W for a single polymer chain, and then we will extend the calculation to an elastic network.

First we need to find the entropy of a polymer chain with a particular end-to-end vector **r**. We will employ the model of a coil composed of statistical segments as we did in Section 1.3.1.2, where we used the symbol **h** for the end-to-end vector and calculated $\overline{h_0^2}$, the average of the squares of end-to-end distances for an ensemble of coils with all possible conformations. A more detailed analysis would have allowed us to calculate the distribution function of end-to-end vectors. The result of that calculation is

$$dN/N = \rho(r)\,dV = Be^{-3r^2/2\overline{h_0^2}}dV \tag{4.1.10}$$

Here dN/N is the fraction of the total number of chains N for which the vector **r** (its beginning is considered to be the coordinate origin) terminates in a volume element dV; B is a normalizing factor. Let us designate by Ω the number of possible conformations of a single chain with the vector **r** ending in dV; Ω must be proportional to dN/N; C is the proportionality constant. Then the Boltzmann relation, equation (3.1.4), yields for $S(r)$

$$S(r) = k \ln \Omega = k \left[\ln(BCdV) - 3r^2/2\overline{h_0^2} \right] \tag{4.1.11}$$

If we specify the chain with $r = 0$ as a reference and recognize that the energy of a coil does not depend on its conformation (i.e., $dE = 0$), then combining equations (4.1.9) and (4.1.11) yields

$$W(r) = 3kTr^2/2\overline{h_0^2} \tag{4.1.12}$$

Our next task is to relate $W(r)$ to the elastic force F acting on the end of the chain. For the extension of a chain, $dW = Fdr$ and

$$F = \frac{dW}{dr} = \frac{3kTr}{\overline{h_0^2}} \tag{4.1.13}$$

Hence, the force F needed for counterbalancing the retractive force of the coil is proportional to the end-to-end distance r. From the mechanical point of view, the coil behaves as a Hookean spring with zero unstretched length. We have already utilized this analogy in Section 3.4.4.4. (For a free chain, the retractive force is counterbalanced by thermal motion that may be described as diffusion of the chain ends. The equilibrium between the retraction of the spring and the diffusion is achieved when the end-to-end vectors acquire the most probable distribution.)

Let us now consider a network consisting of N chains of identical contour length. Let us subject this network to a uniform strain that deforms all embedded end-to-end vectors in a uniform way. It is convenient to select Cartesian coordinates along the axes of the strain tensor γ and to introduce extension ratios $\lambda_i \equiv 1 + \gamma_i$, where γ_i are the principal values of the strain tensor. Then the components x, y, z of the undeformed end-to-end vector and the components x', y', z' of the deformed vector are related as

$$x' = \lambda_1 x; \qquad y' = \lambda_2 y; \qquad z' = \lambda_3 z; \qquad (4.1.14)$$

Let us divide the N chains into groups in such a way that all chains in the pth group have the same end-to-end vector $\mathbf{r}_p$ with components x_p, y_p, z_p. Equation (4.1.11) transformed into Cartesian coordinates gives us the expressions for the entropy of a single chain before deformation $S_p(x, y, z)$ and after deformation $S_p(x', y', z')$ as

$$S_p(x, y, z) = k\left[\ln(BC dx_p dy_p dz_p) - 3\left(x_p^2 + y_p^2 + z_p^2\right)/2\overline{h_0^2}\right] \quad (4.1.15)$$

$$S_p(x', y', z') = k\left[\ln(BC\lambda_1\lambda_2\lambda_3 dx_p dy_p dz_p) - 3\left(\lambda_1^2 x_p^2 + \lambda_2^2 y_p^2 + \lambda_3^2 z_p^2\right)/2\overline{h_0^2}\right]$$
$$(4.1.16)$$

The change in entropy for all chains in the pth group, ΔS_p, then follows as

$$\Delta S_p \equiv N_p[S_p(x', y', z') - S_p(x, y, z)]$$

$$= N_p k\left\{\ln(\lambda_1\lambda_2\lambda_3) - \left[(\lambda_1^2 - 1)x_p^2 + (\lambda_2^2 - 1)y_p^2 + (\lambda_3^2 - 1)z_p^2\right](3/2\overline{h_0^2})\right\} \quad (4.1.17)$$

The change in entropy for the whole network, ΔS, is equal to the sum of ΔS_p for all groups:

$$\Delta S = \sum_p \Delta S_p = k\left\{\sum_p N_p \ln(\lambda_1\lambda_2\lambda_3)\right.$$

$$\left. - \left[(\lambda_1^2 - 1)\sum_p N_p x_p^2 + (\lambda_2^2 - 1)\sum_p N_p y_p^2 + (\lambda_3^2 - 1)\sum_p N_p z_p^2\right](3/2\overline{h_0^2})\right\} \quad (4.1.18)$$

The coordinates in equation (4.1.18) belong to undeformed molecules, therefore

$$\sum_p N_p x_p^2 = \sum_p N_p y_p^2 = \sum_p N_p z_p^2 = \frac{1}{3}\sum_p N_p r_p^2 = \frac{N}{3}\overline{h_0^2} \quad (4.1.19)$$

Substitution of equation (4.1.19) simplifies equation (4.1.18) enormously:

$$\Delta S = kN \left[\ln(\lambda_1\lambda_2\lambda_3) - \left(\lambda_1^2 + \lambda_2^2 + \lambda_3^2 - 3\right)/2\right] \qquad (4.1.20)$$

The result is surprising. The change in entropy depends only on the number of chains and not on their molecular weight, length, or any other specific parameter. It is therefore obvious that equation (4.1.20), which was derived for a network consisting of chains of identical length, must be valid also for any distribution of chain lengths. [It should be noted that our derivation neglected some finer points related to the connectivity of the network. Consideration of these points requires the term $\ln(\lambda_1\lambda_2\lambda_3)$ in equation (4.1.20) to be multiplied by 1/2.]

Recognizing again that the energy of the network does not change during an isothermal deformation, we find for the stored energy function W

$$W = (kTN/2)\left[\left(\lambda_1^2 + \lambda_2^2 + \lambda_3^2 - 3\right) - a\ln(\lambda_1\lambda_2\lambda_3)\right] \qquad (4.1.21)$$

where the constant a equals 2 for our derivation; $a = 1$ for the more sophisticated model mentioned above.

What are the moduli of rubberlike materials? Bulk modulus represents compressibility; it is of the same order of magnitude for liquids and rubbers. (The modulus of glasses is higher, but only moderately so.) On the other hand, we will see that the shear and Young's moduli are much smaller for rubbers than for glasses (they vanish for liquids). Consequently, even sizable shear or tensile strains result in relatively small stresses. Specifically, the negative pressure component of the stress tensor accompanying large tensile strains is far too small to change the volume of the sample. As a result, rubbers are effectively incompressible in shear and tensile experiments; the Poisson ratio is 1/2, and the product $\lambda_1\lambda_2\lambda_3$ (which represents the volume of the deformed unit cube) is equal to unity for all deformations.

Let us write equation (4.1.21) for rubber under tension. The second term in the brackets vanishes. The tensile deformation λ is equal to λ_1; $\lambda_2 = \lambda_3 = \lambda^{-1/2}$. Equation (4.1.21) then reads

$$W(\lambda) = (kTN/2)(\lambda^2 + 2/\lambda - 3) \qquad (4.1.22)$$

The retractive force F acting on a deformed unit cube under tensile strain is

$$F = \frac{dW}{d\lambda} = kTN_V\left(\lambda - \frac{1}{\lambda^2}\right) \qquad (4.1.23)$$

where N_v is now the number of elementary chains in a unit cube.

The modulus of elasticity E at smaller strains is calculated from equation (4.1.6); for a unit cube, both A_0 and L are equal to unity, and $\Delta L = \lambda - 1$.

The result is

$$E = 3kTN_v \tag{4.1.24}$$

$$G = kTN_v \tag{4.1.25}$$

Equation (4.1.25) for shear modulus G was obtained from equation (4.1.7) and from the Poisson ratio $\mu = 1/2$.

Equation (4.1.25) is remarkable; the shear modulus of a rubber is a direct measure of the number of elementary chains in a volume unit. Frequently, the number of chains is expressed through the number-average molecular weight of the elementary chain M_c, density of the rubber ρ, and Avogadro's number. The result reads

$$G = RT\rho/M_c \tag{4.1.26}$$

The molecular weight of the elementary chain, that is, of the segment between two cross-links, can be estimated from knowledge of the vulcanization procedure. The experiments yielded good agreement with equation (4.1.26), validating the concept of rubber elasticity as an entropy phenomenon. Typical values of G for vulcanized rubbers are $3 \times 10^5 - 3 \times 10^6$ Pa, that is, three to four orders of magnitude smaller than moduli of polymer glasses.

No theory is perfect. A number of deviations from the simple behavior described by equations (4.1.23) and (4.1.26) have been observed. It is instructive to study some of these deviations. It was found that rubbers with a low degree of cross-linking exhibited higher moduli than predicted by equation (4.1.26). In fact, high-molecular-weight polymers of the rubber family are rubbery even in the absence of cross-linking. In such rubbers individual chains become knotted and cannot slide past each other. Such knots are called *entanglements* and act as additional cross-links thereby increasing the modulus. On the other hand, in heavily cross-linked rubbers, some cross-links join two segments of the same elementary chain, forming a closed loop. Such loops are not elastically active, and the modulus is found to be too low.

Other discrepancies were found for the dependence of tension on elongation. Relation (4.1.23) predicts that the dependence is curved and the slope is gradually decreasing to one-third of its initial value (Fig. 4.5, curve a). Experimentally the decrease in the region of moderate elongations is more pronounced; at high elongations, the tension increases substantially (curve b). The behavior at low and moderate extensions is often described by the Mooney equation,

$$F = 2(\lambda - 1/\lambda^2)(C_1 + C_2/\lambda) \tag{4.1.27}$$

Here, the constant C_1 is an equivalent of $kTN/2$ of equation (4.1.23). The constant C_2 typically has a value of about $0.1C_1$; the existence of the second term is predicted by some general considerations in phenomenological rheology, but its physical origin is still obscure.

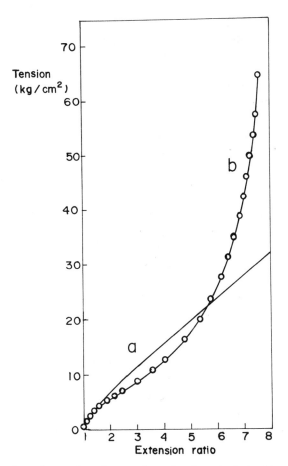

Figure 4.5. The dependence of tension on elongation for a typical rubber. (a) Theoretical dependence [equation (4.1.23)]; (b) experiment. (Redrawn according to L.R.G. Treloar, *The Physics of Rubber Elasticity.* Copyright © 1958 by Oxford University Press, Oxford, UK. Used by permission of the publisher.)

The steep increase in tension at high elongations is caused by the failure of the Gaussian model of a coil or, more specifically, of equation (4.1.10). No coil can be stretched beyond its contour length. When the end-to-end distance of a coil is larger than about one-half of the contour length, the retractive force becomes substantially larger than what equation (4.1.13) predicts. It would become infinite for coils that are fully stretched; however, the sample would break long before such an extension is reached.

Crystallinity is still another phenomenon influencing the properties of some rubbers. All rubbers are amorphous in the unstretched state. However, some of them (typically natural rubber) crystallize promptly and reversibly when stretched. The stretching aligns the polymer chains, thus reducing the entropy loss in crystallization.

The minimum necessary elongation is about fourfold. Let us recall that rubberlike elastic behavior is essentially athermal; no heat change accompanies the deformation. Crystallization is, however, an exothermal event; crystallizing rubber heats up during extension and cools down upon retraction. The crystalline regions act as additional cross-links that substantially increase the modulus at high elongations. This effect increases at lower temperatures, when tendency toward crystallization increases.

4.1.3.2. Swelling of Gels.

When a cross-linked polymer is brought into contact with a solvent that would dissolve its uncross-linked parent, it will swell and form a gel. If the degree of swelling is sufficiently large, the mobility of the polymer chains becomes extensive enough to give the gels rubberlike properties, even if the unswollen polymer is glassy. The driving force for swelling is the decrease in Gibbs energy during the mixing of solvent with the polymer particle; the elastic forces caused by the extension of the particle act against the swelling. At equilibrium, these two forces are balanced; this happens when the volume fraction of the polymer in the gel reaches a characteristic value ϕ_{sw}. Our task is to find how ϕ_{sw} is related to the basic properties of the gel.

Let us evaluate first the function W for swelling. The network swells uniformly in all directions; hence, $\lambda_1 = \lambda_2 = \lambda_3 \equiv \lambda = \phi_{sw}^{-1/3}$. Substituting these values into equation (4.1.21) we get

$$W = \frac{kTN}{2}[3(\lambda^2 - 1) - a \ln \lambda^3] = \frac{kTN}{2}\left[3\left(\phi_{sw}^{-2/3} - 1\right) + a \ln \phi_{sw}\right] \qquad (4.1.28)$$

It is convenient to express ϕ_{sw} as the ratio V_0/V, where V_0 and V are the volumes of dry and swollen gel, respectively. Obviously, $N = N_v V_0$, where N_v is number of elementary chains in the unit cube of the dry gel. Equation (4.1.28) changes to

$$W = \frac{kTN_v}{2}V_0\left[3\left(\frac{V}{V_0}\right)^{2/3} - 3 - a \ln\left(\frac{V}{V_0}\right)\right] \qquad (4.1.29)$$

The elastic forces that try to shrink the swollen network are equivalent to pressure; inside the gel there is an excess pressure π. Let us consider the transfer of a small volume dV of solvent from outside to inside the gel. The work needed for this transfer is πdV; it is also equal to dW. Hence,

$$\pi = \frac{dW}{dV} = \frac{kTN_v}{2}\left[2\left(\frac{V}{V_0}\right)^{-1/3} - a\frac{V_0}{V}\right] = kTN_v\left[\phi_{sw}^{1/3} - \frac{a\phi_{sw}}{2}\right] \qquad (4.1.30)$$

The present situation is quite analogous to an osmotic experiment. In both cases we have two phases with a pressure difference π, one of them pure solvent and the other a mixture of solvent and polymer. The solvent may freely permeate the phase boundary; this implies that chemical potential of the solvent is the same in both phases. Consequently, equation (3.2.5), $\mu_1 - \mu_1^o = -V_1\pi$, applies also for the swelling of gels. In this relation, the difference $\mu_1 - \mu_1^o$ refers as before to different concentrations but to the same pressure. If we choose the Flory–Huggins model of polymer thermodynamics, then equation (3.1.33) is the applicable relation for $\mu_1 - \mu_1^o$. (In this relation we must set the ratio of molar volumes, V_1/V_2, equal to zero because the molar volume of the solute V_2 approaches infinity.) Combining equations (3.2.5), (3.1.33), and (4.1.30) yields

$$RT\left[\ln(1 - \phi_{sw}) + \phi_{sw} + \chi_{12}\phi_{sw}^2\right] = -kTN_vV_1\left[\phi_{sw}^{1/3} - a\phi_{sw}/2\right] \quad (4.1.31)$$

Substituting again $N_v = N_{Av}\rho/M_c$, we obtain the condition for equilibrium swelling as

$$\ln(1 - \phi_{sw}) + \phi_{sw} + \chi_{12}\phi_{sw}^2 + (\rho V_1/M_c)\left(\phi_{sw}^{1/3} - a\phi_{sw}/2\right) = 0 \quad (4.1.32)$$

Thus, from an experimental determination of ϕ_{sw} and from a known value of χ_{12} for the given polymer-solvent system, we can calculate the average size of the elementary chain in the network. Alternatively, when M_c is known from the measurement of the modulus, χ_{12} can be evaluated from the swelling data. The modulus can be measured either on the dry polymer sample (if it is rubberlike) or directly on the swollen sample. The modulus of the swollen gel is also described by equation (4.1.25), which was presented for dry rubbers, but the number of elementary chains in a unit volume is now decreased by a factor of ϕ_{sw}. Thus we should replace equation (4.1.26) by

$$G = RT\rho/M_c\phi_{sw} \quad (4.1.33)$$

Here, ρ is still the density of the *dry* gel.

4.1.4. Polymer Melts

In the previous sections we discussed two idealized examples of bulk polymers: glasses, in which the segmental motion is totally frozen, and rubbers, in which the segmental motion is so free as to allow an essentially instantaneous change of chain conformation. The third idealized case is a polymer melt: a state of matter that would be described as liquid for small molecules. The main characteristic of melts is their ability to flow, to undergo an unlimited deformation. Although elasticity plays some role in the behavior of melts, it is not important in steady flow, that is, in situations when the velocity gradient tensor is not changing with time. In Sections 3.3 and 3.4 we dealt with steady flow of polymer solutions and we characterized it by means of steady-flow viscosity and its dependence on the velocity gradient. Now we will study these quantities for polymer melts. Knowledge of these quantities is extremely important when polymers are fabricated by extrusion, for example.

As for all liquids, the viscosity of melts depends on intermolecular forces. Under comparable circumstances viscosity increases from nonpolar to polar to hydrogen-bonded polymers. However, two other phenomena play even more important roles for macromolecules:

1. During the flow, molecular chains have to slip past each other. Molecules with sizable side groups or with branched chains have more difficulty passing one another and therefore higher viscosity.
2. Molecular weight of the polymer influences viscosity in a decisive way. Linear macromolecules adopt coiled conformations in melts. As long as the chain length is moderate, the movement of the flowing molecules follows the rules developed in Section 3.3.5. In the present situation, other polymer coils play the role of the solvent around the reference coil. During the flow, the reference coil cannot carry with it this polymeric solvent; instead, it has to extricate itself from its companions. This is, of course, an equivalent of hydrodynamic permeability. Consequently, viscosity increases with the first power of molecular weight [compare equation (3.3.25)].

The above scenario is valid as long as the chain length is not too long. If the chain length is long enough, the surrounding molecules can form loops and nooses around the reference molecule, effectively preventing it from rotary motion. Such melts still flow when shear stress is applied to them, but the character of the molecular motion is profoundly changed. It was mentioned in Section 3.3.1 that it is possible to view shear flow of liquids as a random thermal motion of molecules that is helped by the shear forces in one direction and hindered in the opposite direction. Let us therefore study the thermal motion of very long molecular chains in polymer melts. Other molecules (we will refer to them as a matrix) that restrict the lateral movement of the segments surround the individual macromolecule. Obviously, the restriction is far from complete; the neighboring chains may move in a concerted fashion, but this relative freedom does not extend very far.

We will model the above situation by imagining that the macromolecule is confined to a tube within the matrix. The diameter of the tube is much larger than the diameter of the chain; it has been estimated at about 35 Å for polyethylene and about 80 Å for polystyrene. The chain can move along the tube but not laterally; the motion is like that of a snake and is accordingly called *reptation*. The length of the tube is proportional to the molecular weight of the polymer, and so is the friction experienced by the chain moving through it. The movement is essentially diffusion and is characterized by the *disentanglement time* τ_d, needed for complete escape from the tube. As for all diffusion processes, this time is proportional to the square of the distance traveled (tube length) and to the friction coefficient. It is therefore proportional to the cube of molecular weight. We would expect that viscosity is proportional to the same power; in fact, most experiments are closer to the 3.4 power.

The two hydrodynamic regimes of polymer melts are clearly visible on the logarithmic plot of viscosity versus molecular weight (Fig. 4.6). The two linear segments intersect at the critical molecular weight, M_{cr}. The value of M_{cr} depends on the structure of the polymer; it varies from a low of about 4000 for polyethylene to a high of about 30,000 for polystyrene. The number of chain links in a molecule of critical size

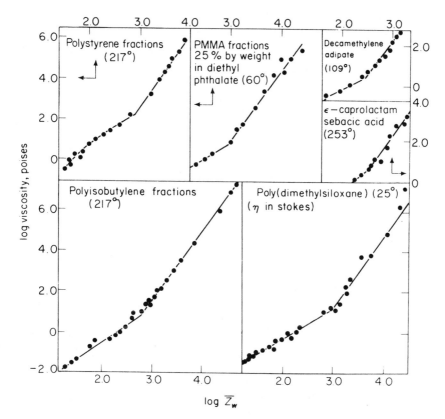

Figure 4.6. The dependence of melt viscosity on weight-average number of chain atoms Z_w for several polymers. (Reprinted from F. Rodriguez, *Principles of Polymer Systems*, 2nd ed. Copyright © 1982 by McGraw-Hill, New York.)

varies within even narrower limits: from about 300 for polyethylene to about 800 for polydimethylsiloxane.

The flow behavior of concentrated solutions of polymers is similar to the flow of melts, but the value of critical molecular weight $M_{cr,sol}$ is higher. It is given as

$$M_{cr,sol} = M_{cr}/\phi \tag{4.1.34}$$

where ϕ is the volume fraction of the polymer in the solution.

The viscosity of melts depends quite dramatically on temperature. Two phenomena contribute to this dependence. One of these phenomena is the effect of activation energy that speeds up all kinetic phenomena (including flow) at higher temperatures. The activation energy of flow is usually interpreted as the energy needed to create a hole large enough for a molecule or a polymer segment to jump into during flow. The needed holes and activation energies are larger for polymer segments than for small molecules. Although the concept of activation energy is sufficient for the description of melt viscosity at higher temperatures ($100°$ or more above T_g), it fails at temperatures closer to T_g, where viscosity increases by orders of magnitude over a temperature interval of several tens of degrees.

In the vicinity of T_g, the available free volume (or more properly, the lack of it) becomes the dominant factor in the play. We have already mentioned that the free-volume fraction is the same at T_g for all polymers. We may therefore expect that the viscosity-temperature relationship, although difficult to treat theoretically, will be similar for all polymers (theory of corresponding states) and all molecular weights. And, indeed, experimental data are described with surprising accuracy by the empirical Williams-Landel-Ferry (WLF) equation,

$$\log_{10} \frac{\eta_T}{\eta_{T_0}} = -\frac{a_1 (T - T_0)}{a_2 + T - T_0} \tag{4.1.35}$$

Here, η_T and η_{T_0} are viscosities at the experimental temperature T and the reference temperature T_0 respectively. If the glass transition temperature T_g is selected as T_0 then the same values of the empirical parameters (quoted as $a_1 = 17.44$ and $a_2 = 51.6$ K) are applicable for many polymers. Note that these values correspond to doubling of viscosity for a decrease of a single degree in temperature (in the vicinity of T_g). Because the viscosity at T_g is difficult to measure, T_0 is often selected as $T_0 = T_g + 50$ K. The WLF equation is still valid for this choice; the values of a_1 and a_2 are, of course, modified.

The above description of viscosity referred to the behavior of melts and concentrated polymer solutions at vanishing shear stress, that is, *zero-shear* viscosity. At larger shear stresses, the viscosity decreases steeply, often by several orders of magnitude. The experimental data are usually presented as flow curves, that is, as logarithmic plots of the rate of strain versus shear stress (Fig. 4.7). Typically, flow curves have three regions:

1. A Newtonian region at small stresses where the plot is linear with a unit slope.
2. A second Newtonian region at very large stresses (it may not be well developed).
3. An intermediate region where the flow curve is quite linear and has a larger slope. This slope is often called the *flow index*. It increases with increasing molecular weight and with polymer concentration, reaching values up to 2 or 3.

Technically important flows usually fall into the intermediate region.

4.1.5. Viscoelastic Materials

Real polymeric materials are neither purely elastic nor purely viscous. Sometimes the deviations from these limiting situations are minor, but some materials simultaneously exhibit both elastic and viscous behavior; they are said to be *viscoelastic*.

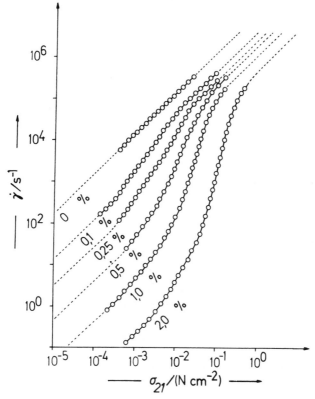

Figure 4.7. Flow curves of solutions of cellulose nitrate ($M = 294,000$) having different concentrations in butyl acetate at 20°C. Polymer melts exhibit a similar behavior. (Reprinted from H.-G. Elias, *Macromolecules,* 2nd ed. Copyright © 1984 by Plenum Press, New York. Used by permission of the publisher.)

4.1.5.1. Creep and Stress Relaxation.   The nature of viscoelastic behavior is perhaps most obvious in the *creep experiment,* in which a sample of a material is suddenly subjected to stress and the stress is kept constant afterwards. (We will consider shear stresses only, but other types of stresses evoke completely analogous responses.) The strain of a perfectly elastic material under these conditions would change instantaneously from zero to some finite value (determined by the compliance J of the material) and stay constant (Fig. 4.8, curve a). Strain of a perfectly viscous material would increase with time linearly and without limits (curve b). The slope of the line would be the fluidity (i.e., inverse viscosity) of the liquid. A viscoelastic material will respond to creep conditions by a more or less sudden increase in strain. At longer times, the strain will still be increasing but at a decreasing rate. For some materials, the rate of strain increase will eventually reach a constant value (Fig. 4.8, curve c). Thus the behavior of such materials at very long times is similar to that of liquids; accordingly,

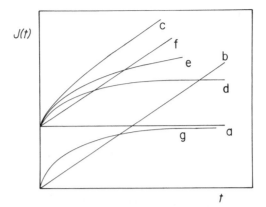

Figure 4.8. Time-dependent compliance for several model materials. (a) Perfectly elastic material; (b) perfectly viscous liquid; (c) viscoelastic liquid; (d) viscoelastic solid; (e) generic viscoelastic material; (f) Maxwell element; (g) Voigt element.

these materials are called *viscoelastic liquids.* For other viscoelastic materials, the strain itself will eventually become constant (Fig. 4.8, curve d). This behavior corresponds to that of elastic solids, and such materials are called *viscoelastic solids.* Of course, there also exist materials for which the rate of strain is ever decreasing but does not reach a constant or zero value within an experimentally accessible time (Fig. 4.8, curve e). However, the behavior of all these types of viscoelastic materials at intermediate times is similar. In Figure 4.8, the creep experiment is described in terms of time-dependent compliance, $J(t)$, the ratio of the strain to the (constant) stress.

What is the molecular basis of this behavior? At the instant the stress is applied, the interatomic distances are deformed as described earlier for elastic glasslike materials. The new arrangement does not correspond to the minimum of interaction energy any more, and a resulting elastic force tries to change the positions toward a state of lower energy. Unlike in glasses, in viscoelastic materials the mutual movement of neighboring atoms, atomic groups, polymer segments, and so on is allowed. Of course, the movement is slowed down by viscous forces as described for the flow of liquids. The movement of the smallest groups (e.g., side chains in polymethacrylates) is the easiest, and the elastic response related to their nonequilibrium arrangement is relieved the earliest. Gradually, the forces related to mutual displacement of larger and larger polymer segments are relieved as well.

In a creep experiment, a lesser and lesser portion of the outside stress is needed to maintain the original deformation. The excess stress causes additional flow; the sample creeps. There is a difference between molecular and cross-linked materials. In the former, mutual movement of the whole macromolecule is possible (even if it may be extremely slow). Hence, all possible elastic constraints (including entanglements) are eventually relieved, and the sample will deform indefinitely; these materials are viscoelastic liquids. (Steady-state flow of such liquids corresponds to the limiting

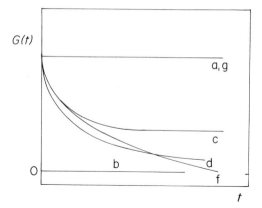

Figure 4.9. Time-dependent modulus for model materials. (a) Perfectly elastic material; (b) perfect liquid; (c) viscoelastic solid; (d) viscoelastic liquid; (f) Maxwell element; (g) Voigt element.

portion at long times of the creep curve c in Fig. 4.8. Its slope is the inverse viscosity of this material; it is the same quantity as the viscosity of a melt described in the previous section.)

In cross-linked materials, the excess stress is utilized for an ever-increasing deformation of the network (elastic response of molecular networks was studied in Section 4.1.3.1). Eventually, all the stress is needed for maintaining the deformation of the network, and the flow stops. Obviously, cross-linked materials behave as viscoelastic solids. The strain-to-stress ratio measured after the cessation of flow is the ultimate compliance J_c.

Viscoelasticity is clearly manifested also in a *stress relaxation experiment*. A constant strain is applied to the material, and the time-dependent stress needed for maintaining the deformation is recorded. The results are reported in terms of time-dependent modulus $G(t)$, the ratio of the stress to the (constant) strain (Fig. 4.9). $G(t)$ is time-independent for perfectly elastic materials (curve a); for perfect liquids it equals zero once the constant strain has been achieved (curve b). For viscoelastic solids the stress relaxes to some finite limiting value (curve c), while the limiting stress for viscoelastic liquids is zero (curve d).

The mechanical behavior of viscoelastic materials is frequently modeled by various combinations of springs and dashpots. In the following, we will use the subscript s for the quantities related to elastic behavior and modeled by springs, and the subscript v for viscous quantities modeled by dashpots. The relations between stress σ and strain γ will then read

$$\sigma_s = G_i \gamma_s = \gamma_s / J_i \tag{4.1.36}$$

$$\sigma_v = \eta_i \frac{d\gamma_v}{dt} \tag{4.1.37}$$

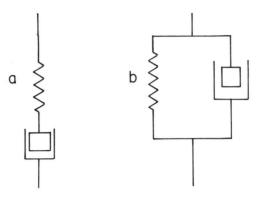

Figure 4.10. (a) The Maxwell element and (b) the Voigt element.

where G_i, J_i, and η_i are constants. The simplest combinations of the mechanical components are the *Maxwell element,* where the spring and the dashpot are in series, and the *Voigt* element, where they are parallel (Fig. 4.10).

The geometry of the Maxwell element calls for the relations

$$\sigma = \sigma_s = \sigma_v \qquad (4.1.38)$$

$$\gamma = \gamma_s + \gamma_v \qquad (4.1.39)$$

Combining equations (4.1.36)–(4.1.39) with the condition $\sigma = $ constant (for the creep experiment) or with $\gamma = $ constant (for stress relaxation) and a little additional calculus yield

$$J(t) = J_i + t/\eta_i \qquad (4.1.40)$$

$$G(t) = G_i e^{-t/\tau_i} \qquad (4.1.41)$$

$$\tau_i \equiv \eta_i/G_i \qquad (4.1.42)$$

According to equation (4.1.41), the modulus $G(t)$ decreases exponentially with time. The time constant for this decrease τ_i [defined by equation (4.1.42)] is the *relaxation time.* The functions $J(t)$ and $G(t)$ for the Maxwell element are depicted by curves f in Figure 4.8 and 4.9, respectively. It is seen that the Maxwell element is a good model for viscoelastic liquids and not as good for viscoelastic solids. The agreement with experiment is improved when several Maxwell elements are combined in a parallel arrangement (Fig. 4.11a). Then each element has its own relaxation time τ_i, and the expression for $G(t)$ reads

$$G(t) = \sum_i G_i e^{-t/\tau_i} \qquad (4.1.43)$$

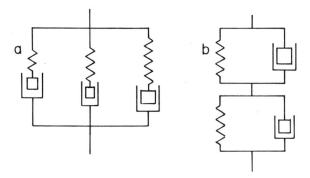

Figure 4.11. (a) Parallel arrangement of Maxwell elements and (b) serial arrangement of Voigt elements.

By a proper selection of values of G_i and η_i, any stress relaxation curve could be matched. It is even possible to model viscoelastic solids by assigning an infinite value of viscosity (and consequently an infinite value of relaxation time) to one of the Maxwell elements. If we describe the modulus of this particular element by $G_e \equiv 1/J_e$, we get the relation

$$G(t) = G_e + \sum_i G_i e^{-t/\tau_i} \qquad (4.1.44)$$

A model with several Maxwell elements adequately describes the creep experiment as well, but the expression for $J(t)$ is quite involved.

An alternative model, instead of using a set of arbitrarily selected Maxwell elements, utilizes a continuous spectrum of them, which is called the *relaxation spectrum* $H(\tau)$. Equation (4.1.44) is then changed to

$$G(t) = G_e + \int_{-\infty}^{+\infty} H(\tau) e^{-t/\tau} d\ln\tau \qquad (4.1.45)$$

Viscoelastic behavior can also be modeled by the Voigt element or by an array of such elements in a serial arrangement (Fig. 4.11b). The geometry of the Voigt element requires

$$\sigma = \sigma_s + \sigma_v \qquad (4.1.46)$$

$$\gamma = \gamma_s = \gamma_v \qquad (4.1.47)$$

and the time-dependent modulus and compliance (depicted by curve g in Figs. 4.8 and 4.9) are

$$J(t) = J_i(1 - e^{-t/\tau_i}) \qquad (4.1.48)$$

$$G(t) = G_i \qquad (4.1.49)$$

In this case, τ_i [again defined by equation (4.1.42)] is called the *retardation time*. It is obvious from the figures that the Voigt element is a good model for viscoelastic solids and less good for viscoelastic liquids. Again, agreement with experiment is better when several serial Voigt elements are used in the model and an element with zero modulus and viscosity η_0 is added to allow for the behavior of viscoelastic liquids. Then the relation for compliance becomes

$$J(t) = \frac{t}{\eta_0} + \sum_i J_i(1 - e^{-t/\tau_i}) \qquad (4.1.50)$$

Obviously, η_0 is identical with steady-state viscosity, which we studied in Section 4.1.4.

It is also possible to use an infinite spectrum of Voigt elements and characterize it by the *retardation spectrum* $L(\tau)$. The compliance is then expressed as

$$J(t) = \frac{t}{\eta_0} + \int_{-\infty}^{+\infty} L(\tau)(1 - e^{-t/\tau})d \ln \tau \qquad (4.1.51)$$

Expressions for modulus become very complicated when multiple Voigt elements are employed.

Relaxation and retardation spectra are both able to describe the behavior of viscoelastic materials, and therefore they must be interrelated. However, the relation is very complicated, and researchers prefer to present them both.

Before we leave the subject of stress relaxation and creep experiments, we need to ask what the response of the material is when the outside stresses are suddenly removed. Before the removal, one part of the outside stress balanced the internal elastic forces caused by the strain and the other part opposed the viscous forces while causing further deformation. On removal of the outside stress, the elastic forces acting as an internal stress (having opposite sign to that of the previous outside stress) try to bring the material to a state with the lowest entropy. For viscoelastic solids, this means back to the original shape. In viscoelastic liquids, the viscous flow caused permanent changes in the molecular arrangement; after removal of the stress, the material will try to reach a state without internal stresses. This new equilibrium state will, of course, be deformed with respect to the original one. In any case, this *elastic recovery* will be opposed by viscous forces that oppose *any* kind of internal motion and will require time before it brings the sample to its final shape. The phenomenon of elastic recovery is extremely important when articles are fabricated from polymeric materials by extrusion.

4.1.5.2. Dynamic Experiments.

Stress relaxation and creep experiments are capable of providing information about viscoelastic behavior at longer times, say from seconds up. At shorter times, these transient experiments must be replaced by periodic (or dynamic) experiments, in which the imposed stress or strain is a sinusoidal function of time. We will be interested only in those experiments in which the amplitudes of the stress and strain are small enough to make their ratio independent of the individual amplitudes. In other words, we will study materials in the region of *linear viscoelasticity* only. Under these conditions, a sinusoidal strain will be accompanied by a sinusoidal stress, but they will be out of phase; the stress will be led by a phase angle δ. The strain and stress can then be written as

$$\gamma = \gamma_0 \sin \omega t \qquad (4.1.52)$$

$$\sigma = \sigma_0 \sin(\omega t + \delta) = \sigma_0 \cos \delta \sin \omega t + \sigma_0 \sin \delta \cos \omega t \qquad (4.1.53)$$

Here, γ_0 and σ_0 are the amplitudes of the strain and stress, respectively; ω is the frequency in radians per second, and t is time. Alternatively, the expression for the stress is written as

$$\sigma = \gamma_0 [G'(\omega) \sin \omega t + G''(\omega) \cos \omega t] \qquad (4.1.54)$$

where $G'(\omega)$ is the *storage modulus* and $G''(\omega)$ is the *loss modulus*.

How much power per unit volume, W/V, is dissipated as heat during our dynamic experiment? In simple shear, the relation was given by the first part of equation (3.3.3); using the present symbols we may rewrite it as

$$W/V = \sigma \, d\gamma/dt \qquad (4.1.55)$$

Substituting relations (4.1.52) and (4.1.54) for γ and σ, respectively, and averaging over time, we obtain

$$W/V = \gamma_0^2 \, \omega \, G''/2 \qquad (4.1.56)$$

The power is thus dissipated (lost) only in relation to the G'' component of the modulus; hence the term loss modulus. The energy that is responsible for the in-phase oscillations is not dissipated; it is stored as kinetic/potential energy, and G' is called the storage modulus.

We should also recall the definition of viscosity given by the second part of equation (3.3.3) as

$$\eta \equiv \frac{W/V}{(d\gamma/dt)^2} = \frac{G''}{\omega} \qquad (4.1.57)$$

In calculating the second part of this relation, we found the time average of $(d\gamma/dt)^2$ to be equal to $\gamma_0^2 \omega^2/2$. It is noteworthy that the steady-state viscosity η_0 is related to the loss modulus as

$$\eta_0 = \lim_{\omega \to 0} \eta(\omega) = \lim_{\omega \to 0} \frac{G''(\omega)}{\omega} \qquad (4.1.58)$$

What is the behavior of Maxwell's and Voigt's elements in dynamic experiments? Equations (4.1.52)–(4.1.54) combined with equations (4.1.38) and (4.1.39) and with a little calculus yield, for the Maxwell element,

$$G'(\omega) = G_i \omega^2 \tau_i^2 / \left(1 + \omega^2 \tau_i^2\right) \tag{4.1.59}$$

$$G''(\omega) = G_i \omega \tau_i / \left(1 + \omega^2 \tau_i^2\right) \tag{4.1.60}$$

$$\tan \delta \equiv G''(\omega) / G'(\omega) = 1/\omega \tau_i \tag{4.1.61}$$

A similar calculation for the Voigt element yields

$$G'(\omega) = G_i \tag{4.1.62}$$

$$G''(\omega) = \omega \eta_i \tag{4.1.63}$$

$$\tan \delta \equiv G''(\omega)/G'(\omega) = \omega \eta_i / G_i \equiv \omega \tau_i \tag{4.1.64}$$

The Maxwell model predicts that the storage modulus will decrease with decreasing frequency, whereas the loss modulus goes through a maximum located at $\omega_{max} = 1/\tau_{i.}$. The loss angle will increase from $0°$ to $90°$. The Voigt model predicts that the loss modulus will decrease with decreasing frequency over the entire range of frequencies; $\tan \delta$ will decrease with decreasing frequency toward zero. Which of these two models is better? We have already seen when studying the creep and stress relaxation experiments that neither model is satisfactory and that multielement models are necessary for a more faithful description of actual materials. Nevertheless, inspection of Figures 4.8 and 4.9 reveals that the Maxwell model is to be preferred for short times. For long times, the Maxwell model is preferable for viscoelastic liquids, whereas the Voigt model is better for viscoelastic solids. At intermediate times, some combination of elements of both types is needed. We should realize that high frequencies correspond to short times and low frequencies to long times. As a result, the actual frequency dependence of the loss tangent (tan δ) frequently has a shape that is similar to the curves presented in Figure 4.12. The loss tangent increases from low values at high frequencies to a maximum in a transition region and falls off at lower frequencies. It stays low for viscoelastic solids but increases again at very low frequencies for viscoelastic liquids.

There is a similarity between the behavior of viscoelastic materials subjected to oscillatory stresses and the behavior of various electronic circuits; in both cases the relevant quantities are sinusoidal functions of time. Electrical engineers routinely utilize the following computational trick: They replace the sinusoidal functions with exponential functions of an imaginary argument, that is, with $\exp(i\omega t)$. The material constants become complex numbers as well. It is understood that only the real parts of the final complex expressions are physically significant. The computational techniques work well as long as only linear algebra is involved. The technique allows, for example, the lumping together of electrical resistance, inductance, and capacitance

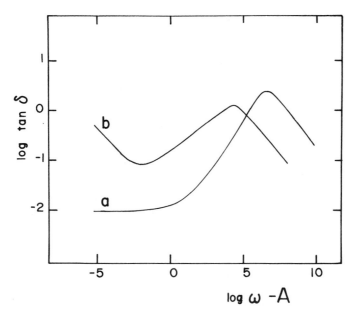

Figure 4.12. Characteristic shape of the dependence of the loss tangent on frequency for (a) viscoelastic solids and (b) viscoelastic liquids.

into electrical impedance; in optics, the refractive index and extinction coefficient are combined into the complex refractive index.

An analogous procedure combines the storage and loss moduli into a complex modulus $G^*(\omega)$ defined as

$$G^*(\omega) \equiv G_0(\omega) \exp(i\delta) \equiv G'(\omega) + iG''(\omega) \tag{4.1.65}$$

Similarly, complex compliance, $J^*(\omega)$, and complex viscosity, $\eta^*(\omega)$, are defined as

$$J^*(\omega) \equiv 1/G^*(\omega) \equiv J'(\omega) + iJ''(\omega) \tag{4.1.66}$$

$$\eta^*(\omega) \equiv \frac{G^*(\omega)}{i\omega} \equiv \frac{G''(\omega)}{\omega} - \frac{iG'(\omega)}{\omega} \equiv \eta'(\omega) - i\eta''(\omega) \tag{4.1.67}$$

The experimental data are usually reported either as $G_0(\omega)$ and $\tan\delta(\omega)$ or as $G'(\omega)$ and $G''(\omega)$. These descriptions are, of course, fully equivalent. Because of extensive ranges covered by these quantities, the plots are usually in log-log coordinates.

It is also possible to express the dynamic behavior of polymers by means of the relaxation and retardation spectra. However, in our treatment we will not engage in the rather special field of converting one viscoelastic function into another. Instead, we will ask what the origin is (on the molecular level) of the multitude of relaxation times that are so similar for all types of polymers. The structure of individual random coils is responsible for a major part of the relaxation spectrum. We have learned that a polymer chain is conveniently modeled as a string of segments; if the segments are large enough, then each of them behaves as a spring. Deformation of the segment-spring is opposed by the elastic forces of the spring and also by viscous forces opposing any relative movement of molecules and their parts. Hence, each segment corresponds to a Voigt element and is characterized by a retardation time τ. (Retardation and relaxation times are mutually related; it is customary to talk about relaxation times when analyzing molecular motions.) Independent segments all have the same value of τ. However, the spring-segments are not independent, they are connected into a chain; that is, they are mechanically coupled. Coupling removes the degeneracy in the characteristic times, and the system acquires a set (spectrum) of relaxation times.

This spectrum governs the relative movement of polymer segments. Recall that it was the loss of segmental movement that characterized the glass transition. At high frequencies, segmental motion is impossible and the polymer behaves as a glass. At lower frequencies (longer times), segmental motion becomes possible and we observe the glass transition. (It is usually reported as the frequency at which $\tan \delta$ exhibits a maximum.) This result is remarkable. In previous sections we characterized glass transition by *temperature,* not by *time* (frequency). Obviously, there is some time-temperature correlation governing viscoelastic behavior. Let us explore it.

The primary effect of temperature on polymeric materials is the change in the free volume. We have already seen that a given change in free volume affects viscosities of all polymeric systems in the same way irrespective of their chemical structure or molecular weight. This implies that the temperature-free volume effect operates on the segmental level by changing the frictional coefficients of individual segments. All frictional coefficients are changed by the same factor, which applies also to the viscous force that is coupled to our spring-segment and to its relaxation time. Actually, the whole spectrum of relaxation times is multiplied by the same factor. In other words, in a logarithmic plot, a change of temperature shifts the whole spectrum (together with all other viscoelastic quantities) by a uniform distance. If we shift it back, the plots corresponding to two temperatures are superimposed.

This feature allows for the evaluation of the dependence of viscoelastic quantities on frequency over many orders of magnitude (up to 15). The dependence of these properties on frequency (which can be measured typically over 2 orders of magnitude at most) is measured at a series of temperatures. All the dependences are then shifted along the log ω axis to form a master curve. This process is depicted in Figures 4.13 and 4.14, which show the storage compliance $J'(\omega)$ of poly(n-octyl methacrylate) as a set of curves measured at many temperatures and as a master curve, respectively. All the experimental curves are shifted to match a curve at a selected reference temperature; the necessary shift is the logarithm of the shift factor a_T (Fig. 4.15).

It should be obvious from the above discussion of the temperature dependences that the factor a_T is equal to the ratio of the viscosities at the experimental tempera-ture and at the reference temperature, η_T / η_{T_0}. Hence, the WLF equation (4.1.35) is applicable not only for the temperature dependence of steady-state viscosity but for the temperature dependence of other viscoelastic functions as well. Also, depen-dences on the product $a_T \omega$ (such as the one in Fig. 4.14) can be interpreted either as a

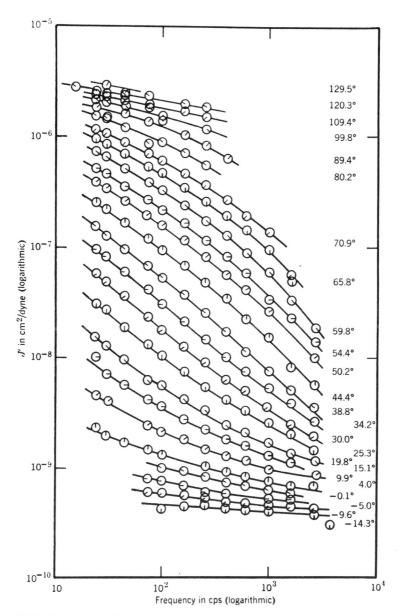

Figure 4.13. Storage compliance of poly (*n*-octylmethacrylate) plotted against frequency at temperatures indicated. (Reprinted from J.D. Ferry, *Viscoelastic Properties of Polymers,* 2nd ed. Copyright © 1970 by John Wiley & Sons, New York. Used by permission of the publisher.)

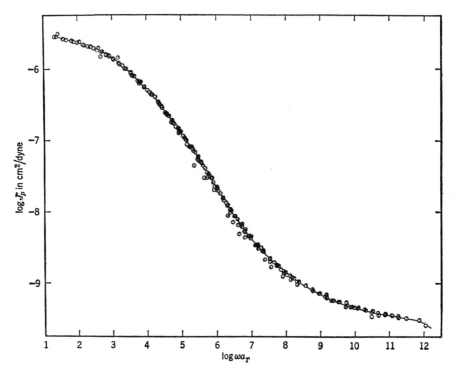

Figure 4.14. Master curve obtained by shifting the data from Figure 4.13 by a suitable factor log a_T along the log ω axis. Data reduced to 100°C. (Reprinted from J.D. Ferry, *Viscoelastic Properties of polymers.* 2nd ed. Copyright © 1970 by John Wiley & Sons, New York. Used by permission of the publisher.)

dependence on frequency at constant temperature or as a dependence on temperature at constant frequency (a_T in the latter case is related to temperature by means of the WLF equation).

When the moduli are plotted in either manner, they are seen to drop steeply in the glass transition region. The loss tangent has a maximum in the same region.

4.1.6. Crystalline Polymers

Many polymers can pack their chains into crystalline lattices. Crystalline polymers are among the most valuable ones: They usually have superior dimensional stability, do not swell easily, and often have fiber-forming properties.

As we have already mentioned, the driving force for crystallization is an increase in the (negative) interaction energy achieved by efficient molecular packing and by maximization of the number of strongly interacting groups that are in direct contact (e.g., forming hydrogen bonds). Once the molecules discover the most efficient arrangement, they repeat it endlessly in a crystalline lattice. The smallest repeating

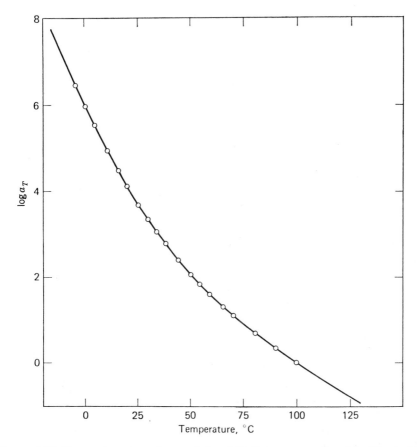

Figure 4.15. Temperature dependence of the shift factor a_T employed in Figure 4.14. (Reprinted from J.D. Ferry, *Viscoelastic Properties of Polymers,* 2nd ed. Copyright © 1970 by John Wiley & Sons, New York. Used by permission of the publisher.)

entity is the *unit cell;* it contains one or several identical building blocks. Individual molecules are the building blocks in crystalline low-molecular-weight substances. Because all molecules are identical, they may form perfect crystals whose size is limited only by the amount of available material.

Among macromolecular materials, only proteins consist of identical molecules. They form crystals just as small molecules do, but their unit cells are much bigger. (Protein molecules, with their intricate shapes, can only form crystals in which the interprotein spaces are filled with water molecules.) Most other macromolecules are composed of molecules of different sizes; moreover, their structure is seldom strictly regular. Consequently, they cannot form perfect crystals. However, monomeric units may play the role of building blocks within the crystals; a single polymer chain may run through many unit cells, and it does so. Of course, chain ends or any other kind of irregularity (branching points, etc.) cannot fit easily into the lattice. Thus the size of a crystallite is severely restricted. In fact, the structure of the chain must be regular to a very high degree to fit the chain into the lattice at all. A major group of important polymers that do not qualify are atactic vinyl polymers; they do not crystallize. [There are exceptions: atactic poly(vinyl alcohol) crystallizes, presumably because of the small difference between the sizes of the OH and H substituents and the strong hydrogen-bonding forces.] Random copolymers cannot crystallize either.

The tendency to crystallize is enhanced and the melting points are high whenever the polymer chains can interact strongly with each other. Typical examples are polyamides, aromatic polyesters, and cellulose. On the other hand, bulky side groups (especially when they have a highly irregular shape) diminish the crystallization tendency.

The detailed arrangement of monomeric units within the unit cell may be determined by X-ray diffraction techniques (Section 4.2.4). However, the organization of the unit cell is not sufficient for a full description of crystalline polymers. We also need to know the higher level of arrangement of the whole polymer chains in the material. That controls its morphology and its mechanical properties. We will study these topics in the following sections.

4.1.6.1. Morphology of Crystalline Polymers.

In the crystalline lattice, the polymer chains generally run parallel to each other. Individual chains adopt conformations with minimum energy. For polyethylene this means an all-*trans* zigzag conformation. This conformation is also adopted by other chains that possess larger number of unsubstituted methylene groups in their backbone: aliphatic polyamides and polyesters. Neighboring chains in these polymers are, of course, arranged in a way that maximizes the intermolecular interactions; in polyamides, all amide groups are involved in two hydrogen bonds.

If an isotactic chain is arranged in the zigzag conformation, the neighboring side groups are uncomfortably close to each other and the conformational energy is high. This repulsive interaction is completely relieved when the backbone adopts an alternating *trans-gauche* conformation. This arrangement is a helical one; the side groups form a helix around the winding backbone. The structure repeats itself after three monomer units, while the side groups make one turn around the helical axis. Accordingly, we are talking about a 3_1 helix. It is very popular among isotactic polymers: polypropylene, poly(butene-1), other poly(alkene-1)s, and polystyrene adopt it. Individual chains are packed in the lattice in such a way as to achieve maximum interlocking of the neighboring helices. The optimum arrangement obviously depends on the geometry and energetics of the side groups. Consequently, very similar helices may be packed into quite different crystalline lattices.

The comfort of the side groups sometimes prevails even over the requirement of the staggered conformation (*trans* or *gauche*) of the backbone. Helixes like 4_1, 7_2, etc. then result. A typical example is the 8_5 helix of polyisobutylene (this polymer crystallizes reversibly on extension). Polytetrafluoroethylene is a strange case. The fluorine atoms are slightly larger than hydrogens and marginally interfere with each

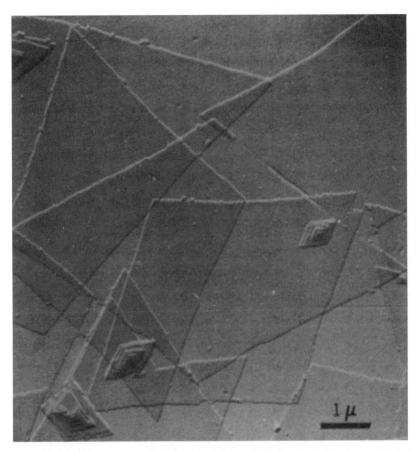

Figure 4.16. Electron micrograph of single crystals of linear polyethylene crystallized from xylene. (Reprinted with permission from V.F. Holland and P.H. Lindenmeyer, J. *Polymer Sci.*, **57,** 589. Copyright © 1962 by John Wiley & Sons, New York.)

other in the zigzag conformation. This interference is relieved by a small twist of the backbone: 13 CF_2 groups per turn of the helix have been found at temperatures below 19°C; at slightly higher temperatures, the number is 15.

It was as late as 1957 when the first single crystals of a linear polymer (polyethylene) were prepared using a very old technique: crystallization from a very dilute solution (less than 0.1%). The crystals were very small but large enough for electron microscopic studies. The crystals were lamellae or hollow pyramids about 100 Å thick (Fig. 4.16). Unexpectedly, the molecular chains were perpendicular to the lamellae. This implied that the chains were folding onto themselves in a very regular manner (Fig. 4.17). It was soon found that this folding was a very general feature of crystalline polymers. The lamella is thicker the higher the temperature during crystallization (i.e., the closer to the melting point), but the thicknesses remain generally between 100 and 200 Å .

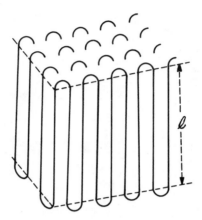

Figure 4.17. Model of chain folding in a polymer single crystal. (Reprinted from F.A. Bovey and F.H. Winslow, eds., *Macromolecules, An Introduction to Polymer Science.* Copyright © 1979 by Academic Press, New York.)

Folding of the polymer chains is in sharp contrast with the intuitive expectation that the molecular chains should be parallel to each other in a crystal. (Indeed, they are, but only over the short distance between two folds.) Before the discovery of chain folding, crystalline polymers were routinely described by a *fringed micelle model.* In this model, a bundle of chains forms a quite long crystalline fibril called a micelle. At the end of the fibril, individual chains leave it and enter an amorphous region. Most of them reenter another micelle. Thus most chains are participating in many micelles, ensuring the cohesion of the material.

The folds on the chain obviously do not benefit from the regularity of the crystalline lattice; they effectively increase the free energy of the crystal. Why then do the molecules not run parallel for the maximum possible distances as in the fringed micelle model? There are several reasons for this. First, the thermodynamic advantage of the micelle model is really not so large. The deviations from a regular arrangement at the end of the hypothetical micelle would be much more serious than those occurring at the folds. Second, a detailed analysis of the vibrational motion of polymer segments within the chains suggests that the free energy per monomer unit (the measure of thermodynamic stability) has two minima in its dependence on the length of the straight portion of the chain: one at lengths that correspond to actual lamellae thicknesses and another for infinitely long straight sections.

However, the decisive factor for the formation of lamellae is probably the kinetic one. Crystallization, like any other process during which two phases demix, requires nucleation. Given the high dilution and the coiled form of polymer molecules, it is much more probable that a single chain will fold onto itself to form a stable crystalline nucleus than that straight sections of several chains will meet to form a nucleus of a stable bundle.

Why is the thickness of lamellae so uniform, and why does it decrease with decreasing temperature? As in all nucleation phenomena, small nuclei are not stable;

they become stable only when their size is larger than some specific critical size. This critical size is the smaller the farther the nucleating system is from equilibrium conditions. In the present case, the distance from equilibrium is represented by the difference between the temperature of crystallization and the temperature of equilibrium melting. The larger the difference, the smaller the nuclei (the shorter the folds) that are stable enough to produce a crystal. The probability of forming a would-be nucleus decreases sharply with increasing size of the nucleus. Consequently, at any given temperature, the nuclei actually initiating the growth of crystals are very uniform in the fold length; shorter folds are not stable, and longer folds have a small probability of formation.

Once a stable nucleus has been formed, the rest of the chain will attach itself rather quickly to its regular arrangement; other chains will then follow, forming parallel folded layers. The molecule that is just folding itself into the crystal has no incentive to extend beyond the fold plane already formed; no stabilizing lattice forces exist there. Therefore, it will reenter the crystal in the shortest and most efficient way. In the case of polyethylene, about three methylene groups will form a loop with a non-*trans* (not quite *gauche,* either) conformation. As a result, the thickness of the lamella will not change during the crystal growth.

Do all polymers crystallize with chain folding? Chain folding is a result of thermodynamic and kinetic phenomena during a process in which the crystalline structure is formed from amorphous polymeric systems. When crystals are formed by different processes, linear structures without folds may result. Typical examples are crystalline structures formed by biological processes, such as cellulose fibers or β-keratins (silk). Another possibility is the drawing of crystalline polymers, during which the crystals are completely rearranged and linear structures result.

Polymer crystalline lattices are sometimes capable of incorporating a relatively large number of defects: chain ends, occasional wrong conformation, even an improper side group. For example, low-density polyethylene has a fair amount of randomly distributed side groups, yet it is still crystalline to quite a high degree. Similarly, copolymers of ethylene with a small amount of higher alkenes are also crystalline. Significantly, the unit cell dimensions of these materials are slightly larger than the dimensions of flawless polyethylene, indicating that the imperfections have been incorporated into the lattice and not left in the amorphous part of the polymer.

When the crystallization proceeds at higher polymer concentrations or in bulk polymers, the number of crystal imperfections increases. Specifically, quite often a chain folds back with a larger loop or fails to reenter the crystal at all. All these imperfections serve as convenient secondary nuclei; they lead to various types of epitaxial growth, and a number of polycrystalline structures of increasing complexity (dendrites, hedrites) are formed. However, we will restrict our attention to *spherulites:* crystalline structures prevailing in bulk polymers.

When a melt of a crystallizable polymer is cooled below its melting point, crystalline nuclei are formed and start growing. Again the prevailing mode of growth is by folding into lamellae. However, too many polymer chains want to participate in the growth. They attach themselves to the growing structure in any feasible way; the lamellae will branch and branch again. They will grow in all directions—the growing crystalline mass will have a spherical shape. The growth direction must be radial, but the direction of the chain folds must be perpendicular to it. Consequently, the chain segments are oriented tangentially to the surface of the spherulite. Eventually, the growing spherulites will come into contact; then they will grow in the only directions left. They will lose their spherical shape, become polyhedral, and fill most of the available space.

Spherulites are most conveniently observed when a section of the material is placed between crossed polarizers. Individual lamellae are anisotropic and therefore birefringent. Their principal optical axes are perpendicular and parallel, respectively, to the fold plane. However, individual lamellae are arranged within the spherulite with a spherical symmetry. Thin sections of them have a circular symmetry; one of the principal axes of the section is in the radial direction, and the other is tangential. From the optical viewpoint, these sections are quite analogous to birefringent liquids in the Couette cell, as described in Section 3.4.4.5. Thus we observe again between crossed polarizers two perpendicular dark bands, parallel to the principal direction of the polarizers. Each spherulite then displays the so-called *Maltese cross* (Fig. 4.18).

The overall morphology of crystalline polymers depends primarily on the temperature of crystallization. If the temperature is not at least 10°C under the equilibrium melting temperature, the polymer will not crystallize unless external nuclei are introduced. At slightly lower temperatures, a few nuclei will be formed and they will

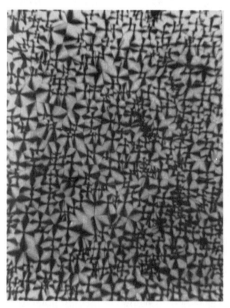

Figure 4.18. Spherulitic structure of a sample of low-density polyethylene photographed through a polarizing microscope. Two different magnifications. (Reprinted from H.R. Allcock and F.W. Lampe, *Contemporary Polymer Chemistry.* Copyright © 1981 by Prentice Hall, Englewood Cliffs, NJ. Used by permission of Dr. R. Stein.)

eventually grow to large spherulites. The growth will be slow, because the driving force for crystallization (the distance from equilibrium, i.e., the degree of undercooling) is small. At lower temperatures, the number of nuclei formed will be larger, leading to smaller spherulites; they will grow faster. However, at still lower temperatures, the polymer will approach the temperature of glass transition, and both the rate of nucleation and rate of growth will decrease. Thus there is an optimum temperature for fast crystallization. When a polymer melt is quenched under T_g, the polymer may stay amorphous indefinitely.

The degree of crystallinity depends on the regime of crystallization and on the regularity of the polymer structure. Although the imperfections and chain ends may be incorporated into the lattice, quite often they are left in the amorphous portion of the material. The chain loops in the folding plane also display amorphous characteristics. Finally, some chains may be left uncrystallized between the lamellae within the spherulites and between the spherulites. Thus the most perfectly crystalline specimens are about 90% crystalline. Usually the percentage is much lower.

4.1.6.2. Mechanical Properties of Crystalline Polymers.

Amorphous polymers in all their forms (melts, rubbers, viscoelastic materials, glasses) are homogeneous. Under their T_g they have a high modulus and usually are quite brittle. Above T_g, their modulus is much lower unless they are heavily cross-linked.

Crystalline polymers are heterogeneous; crystalline regions are interspersed with amorphous ones. They are characterized by glass transition temperature T_g and melting temperature T_m. The same interactive factors (chain mobility, secondary bonds, bulkiness of substituents) that determine the glass transition also play a crucial role in crystallization. Hence the two temperatures are for most polymers related as

$$0.5 < \frac{T_g}{T_m} < 0.8 \tag{4.1.68}$$

When the polymers are under their T_g, the rigid crystalline regions are embedded in a rigid glass matrix and the whole structure behaves essentially as a glass. Above their T_m, the polymers are homogeneous melts. The technologically important region is between these two temperatures. Within this region, the mechanical response of the material is mediated primarily by the amorphous domains; the crystallites act as numerous cross-links. Consequently, crystalline polymers are similar to very hard rubbers; their modulus is high, but they can support sizable deformation without breaking. They are simultaneously hard and tough, a very desirable combination of properties. Moreover, their modulus usually decreases only modestly with increasing temperature; of course, it falls precipitously at T_m.

Many crystalline polymers can be drawn into excellent fibers (nylons, aromatic polyesters, polypropylene). The drawing process is a remarkable reorganization of the crystalline structure that deserves special attention. We have already explained that moderate stresses cause deformation of crystalline polymers mainly in the amorphous regions. The molecular chains in these regions are stretched appreciably, whereas the overall deformation remains small. We should recall that individual polymer chains are alternatively folded in the crystalline lattice and coiled in the amorphous domains. Hence, the stretched coils are pulling on the folded portion of the chain. When the stress is sufficiently large, they succeed in pulling the chains out of the lattice and reorient them in a more or less parallel fashion along the drawing direction. The chains, which are now parallel to each other, re-form the crystalline lattice. The unit cell of the new crystal is essentially identical to the old one, even if the chain folding and overall morphology are completely changed. The new lattice is stable; when the stress is removed, the material does not return to its original shape but remains stretched.

At the beginning of the drawing, the reorganization of the structure and the accompanying elongation start at the weakest point (there is always a point that is weakest, even if by very little). The drawing decreases the cross section of the fiber at this specific location. The same force is now acting on a smaller cross-sectional area, the local stress is increasing, and the polymer continues to flow until it reaches a 4- to 10-fold extension. At this moment, the structure reorganization is complete and no more chain unfolding takes place. Further deformation would require changes in bond angles and valence lengths; this, of course, means a much higher modulus. Consequently, further drawing will occur at the next weakest point. However, this point now exists in the immediate vicinity of the drawn section where the cross section is already partially reduced. Macroscopically, the whole process is seen as a sudden formation of a neck on the sample that is extensively elongated. The length of the drawn portion then gradually increases until the whole specimen is drawn (Fig. 4.19). This process is known as *necking*.

The properties of drawn materials are markedly different from those of the parent (spherulitic) materials. The modulus is increased substantially (sometimes 10-fold or more); the tensile strength may increase even more. On the other hand, the extension at break is only a few percent, whereas for undrawn materials it could be several hundred percent (such extension is actually drawing).

4.1.7. Multicomponent and Multiphase Materials

In the preceding section, we saw that the introduction of heterogeneity into polymeric materials may have a very beneficial influence on their properties. Several types of heterogeneous materials are technologically important. Besides crystalline polymers, some other materials composed of more or less identical molecules can form heterogeneous systems. The best examples are block copolymers. However, many heterogeneous materials are prepared by mixing several components, which may or may not mix at the molecular level. In the following sections, we will study several types of such multicomponent and/or multiphase materials.

4.1.7.1. Plasticization of Polymers.   Properties of linear polymers depend primarily on the relation of the actual temperature to their glass transition temperature. Quite often, the type of behavior that occurs in the vicinity of T_g is most desirable: These polymers are pliable but still tough (uses include synthetic leather, protective clothing, and upholstery). However, the most convenient polymers [typically poly(vinyl

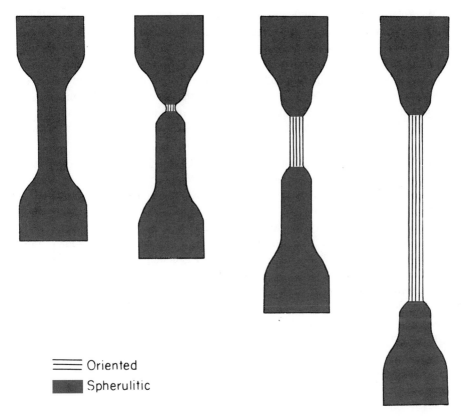

Figure 4.19. Schematic representation of necking on extension of a crystalline polymer. (Reprinted from F. Rodriguez, *Principles of Polymer Systems,* 2nd ed. Copyright © 1982 by McGraw-Hill, New York. Used by permission of Hemisphere Publishing Corporation.)

chloride) or its copolymers] may have a T_g too high for such purposes. The remedy is found in plasticization; the polymer is mixed with another material (usually consisting of smaller molecules) that has appreciable free volume, that is, is well above its T_g. The free volume of the mixture is approximately a weighted average of the free volumes of the components. In other words, T_g of the polymer is decreased to a more desirable value.

To make a good plasticizer, the additive must satisfy a number of conditions. It must be a solvent for the polymer, or at least it must swell it appreciably. On the other hand, it should not interact with the polymer too strongly; tight association of its molecules with the chain may even stiffen it. The plasticizing effect must last for a long time, often for many years. Hence, the volatility of the material and its diffusivity must be low to prevent it from escaping from the polymer. The most popular plasticizers are esters of multivalent acids such as dibutyl phthalate, dioctyl phthalate, or tricresyl phosphate; they are used for the more polar polymers. Nonpolar polymers used in rubbers are plasticized mainly by high-hydrocarbon oils of the naphthenic type.

Poly(vinyl chloride) is an interesting polymer. Its chains have a tendency to form parallel structures having a pseudocrystalline character. Presumably this is caused by the presence of syndiotactic sequences in this otherwise predominantly atactic polymer. These pseudocrystals serve as cross-links and affect mechanical properties. Poly(vinyl chloride) is dimensionally quite stable even in the plasticized state (i.e., above T_g), and, despite its relatively low molecular weight; this adds mightily to its industrial popularity. Other polymers under similar circumstances would form viscoelastic liquids incapable of maintaining their shape.

4.1.7.2. Polymer Blends.

As already mentioned in the discussion of copolymers in Section 1.2.4, it is often desirable to combine the properties of two polymers. It is sometimes possible to synthesize a copolymer, but it is much easier to take two existing polymers and blend them in any desired ratio. The difficulty of this approach is in the fact that two polymers are more often than not immiscible. In fact, only a few decades ago it was claimed that miscible polymer pairs are extremely rare. For blending purposes, even marginal miscibility is beneficial; when two polymers are mixed mechanically or otherwise, they will form domains. If they are totally immiscible (*incompatible* is the more appropriate term), they will not stick together, and the blend will have very poor mechanical properties. If they are sufficiently compatible to interpenetrate, at least in the vicinity of the phase boundary, they will form a heterogeneous structure that may have valuable properties.

Obviously, mastery of the process of blending requires a detailed understanding of the phenomenon of mixing, and we need to take a second look at it. We will start again with the Flory–Huggins expression for ΔG_{mix}, equation (3.1.29).

$$\Delta G_{mix} = RT \left[n_1 \ln \phi_1 + n_2 \ln \phi_2 + n_1 \phi_2 \chi_{12} \right] \tag{4.1.69}$$

In polymer blend studies, it is customary to use the change in Gibbs energy of mixing per unit volume, ΔG_{mix}^v, and the excess interaction energy per unit volume, B_{12}. These quantities are defined as

$$\Delta G_{mix}^v \equiv \Delta G_{mix}/V \equiv \Delta G_{mix}/(n_1 V_1 + n_2 V_2) \tag{4.1.70}$$

$$B_{12} \equiv RT \chi_{12}/V_1 \tag{4.1.71}$$

Here, as usual, V is the volume of the mixture; V_i is the molar volume of the ith component. Equation (4.1.69) then assumes the form

$$\Delta G_{mix}^v = RT \left[(\phi_1/V_1) \ln \phi_1 + (\phi_2/V_2) \ln \phi_2 \right] + B_{12} \phi_1 \phi_2 \tag{4.1.72}$$

For high-molecular-weight polymers, the ratios ϕ_1/V_1 and ϕ_2/V_2 are negligibly small, and the entire first term on the right-hand side of equation (4.1.72) can be dropped. We should recall that this first term is the always negative contribution of the configurational entropy of mixing that promotes the miscibility of small molecules.

For blends, the mixing relation reduces to

$$\Delta G_{mix}^{v} = B_{12}\phi_1\phi_2 \tag{4.1.73}$$

The interaction parameter B_{12} is a more convenient quantity than χ_{12}; it is essentially independent of molecular weight, whereas χ_{12} [as defined by equation (3.1.19)] is proportional to the number of contact points of molecules of species 1, z. If this species is a polymer, z and χ_{12} are proportional to its molecular weight and are very large.

According to equation (4.1.73), the whole burden of promoting miscibility must be carried by the excess interaction energy B_{12}; it must be negative for miscible blends. What values of B_{12} are predicted by theories of mixing? The Hildebrand theory, which we have presented in equations (3.1.21)–(3.1.24), was developed for interactions based primarily on dispersion forces that were expressed by solubility parameters. It predicted that χ_{12} and B_{12} must be always positive or equal to zero. A similar conclusion was arrived for mixtures with generalized nonspecific polar interactions. Thus the theory predicts that all the polymer pairs must be immiscible. However, some polymer pairs *are* miscible. Clearly, the theory is missing some important point.

The neglected factor is specific interaction between some pairs of chemical groups; this is mainly of the donor-acceptor type. The strongest interaction is the formation of a hydrogen bond. Also important are some more subtle interactions: between ester group and chlorine, between ether and phenyl, between phenyl and carbonyl, etc. Of course, these specific interactions may also be present in the homopolymers; their disruption in blending acts against miscibility.

Miscible polymer pairs must thus satisfy the following criteria:

1. Their solubility parameters must be close to each other to minimize the adverse Hildebrand interaction.
2. There must be a net increase in specific interaction on blending.

The larger the specific interaction, the larger the discrepancy in solubility parameters that can be tolerated. Polymer chemists are searching busily for compatible polymer pairs and for more quantitative rules for predicting compatibility. Let us introduce a few compatible polymer pairs: polystyrene-poly(vinyl methyl ether), poly(vinyl chloride)-polycaprolactone, poly(vinyl acetate)-polyepichlorohydrin, and polystyrene-polycarbonate.

Copolymers are often good candidates for blending; an adjustment of the monomer ratio may produce the desired solubility parameter. Moreover, copolymerization may force two monomers to be in contact that they would avoid otherwise. Blending reduces the number of undesirable contacts. This phenomenon is quite similar to cosolvency, when a mixture of two marginal nonsolvents is a good solvent for a polymer. Examples of this phenomenon are blends of polyesters with copolymers of styrene and acrylonitrile; polyesters are not miscible with either homopolymer.

Sometimes industrial chemists try to blend two polymers that are fully incompatible; no cohesion is expected between the two types of domains in such a case. The situation can be remedied by using compatibilizing agents. A successful agent is a block copolymer of the two monomers in question: Each block will dissolve in the domain of its kind, and the domains will be covalently linked. Even more promising are block copolymers, in which blocks of one type are compatible with one of the polymers to be blended, and blocks of the other type are compatible with the other polymer. The dissolution in the domains will be helped by the exothermic action of the negative coefficient B_{12}.

As explained earlier, the miscibility of blends depends fully on the negative value of B_{12}. This value is never very high, and the blends may demix under adverse conditions. Common causes of demixing are the equation-of-state effects affecting blends of polymers having different characteristic temperatures. The lower critical temperature may be observed for most blends; it shifts to lower temperatures as the coefficient B_{12} decreases. In another scenario, a compatible polymer pair may separate into two phases if cast from a solvent that interacts much more strongly with one of the polymers.

4.1.7.3. Heterophase Materials.   Materials with both rubbery domains and glassy or crystalline domains exhibit mechanical properties that are often different (and more desirable) than the properties of either part alone. Heterophase materials can be classified according to the size of the domains, the nature of the continuous phase (if any), the proportions of the phases, and the origin of the heterogeneity. Mechanical properties of the heterogeneous material depend primarily on the mechanical properties of the domains and on their proportions; the other factors play a relatively minor role.

When the size of the domains is macroscopic, we are talking about *composite materials*. A typical example is an epoxy resin combined with glass fibers or with fabrics made from glass fibers. Rubbery polymers are reinforced by *fillers,* that is, by particles with colloidal dimensions. A well-known example is a rubber compounded with carbon black. Glassy polymers are usually brittle; they are made *impact-resistant* by the inclusion of rubbery domains. Thus impact-resistant polystyrene is prepared by polymerizing a solution of polybutadiene (or of styrene-butadiene rubber, etc.) in styrene. (The resulting blend is immiscible and phase separates. Yet the phases have satisfactory cohesion because of grafting of polystyrene chains onto the rubber during polymerization.)

In the above examples, heterogeneity was caused by bringing into intimate contact two materials that are essentially immiscible, either for chemical reasons or simply because of their size and negligible diffusion. Heterogeneity can also be caused by demixing: A material that was originally homogeneous may tend to demix to lower its free energy. At the same time, total demixing may be impossible for structural reasons; the result is the formation of domains. Two classes of materials belong here: some block copolymers and *interpenetrating networks.* The most common way of preparing the latter materials consists of swelling a cross-linked polymer in a mixture of another monomer and a cross-linking agent and polymerizing again. The two networks may be completely independent or may be linked as a result of transfer reactions. As the polymerization progresses, the combinatory entropy of mixing decreases and the chains of the two networks separate and form domains.

Block copolymers made from incompatible blocks exhibit quite peculiar morphologies. The chains of each type of the block form separate domains. The size of the domains is governed by the size of the blocks; each block must be outside the domain of the other, and the surface of the domain must be just large enough to accommodate all chains crossing the boundary. In other words, a proper surface-to-volume ratio must be achieved together with an optimal packing geometry. When the volume fraction of one type of block is smaller than about 20%, it will form spheres of uniform size embedded in a very regular fashion in the continuum formed by the other block. When the volume fractions are about equal (30–70%), the most advantageous packing geometry is that of lamellae that again have very regular thickness. Finally, cylinders are formed when the fraction is intermediate between sphere-forming and lamella-forming values (Fig. 4.20).

An interesting and important type of these materials is a triblock copolymer ABA where the central block B is in the rubberlike region and the end blocks A are glassy. A typical example is styrene-butadiene-styrene copolymer, in which the volume fraction of the styrene blocks is 10–20%. Hence these blocks form spherical glassy domains embedded in a rubbery matrix. Many B chains span the distance between neighboring A domains. On the molecular level, A domains act as multichain cross-links, whereas the B elastic chains contribute to the elasticity. In addition, the heterogeneous structure enjoys all the benefits of a reinforced material. The main advantage of these rubbers is that they are thermoplastic. The glassy domains behave like cross-links below the glass transition temperature of the A blocks; above T_g, the polymer behaves as a melt and is easily formed into articles. Thus these rubbers can be reused, unlike the ordinary ones that are permanently cross-linked during vulcanization.

What is the molecular interpretation of the reinforcement of rubbers by the inclusion of hard domains? One effect is the increase in the degree of cross-linking, which increases the modulus in the familiar way. However, more important is the increase in tensile strength. Let us consider tensile failure of rubbery materials. At high elongations, some chains are fully extended and further elongation breaks them. At this moment, their load is transferred to a few neighboring chains, which may become overloaded, too; catastrophic breakage will occur soon thereafter. In particle-reinforced materials, many chains connect each pair of neighboring particles. If one chain breaks, its load is transferred to many others and the material remains macroscopically intact.

We also need to understand the mechanism of impact strengthening of glassy materials by rubbery inclusions. Tensile and impact failure of glassy polymers usually proceeds in two steps that follow each other in rapid succession. In the first step, a *craze* is formed; it is a narrow zone within the glass that presumably developed at some weaker point in the material. The walls of the craze are connected by a large number of fibrils (typical diameter 15 nm) that are formed by highly oriented bundles of polymer chains and are oriented in the direction of the applied stress. A large fraction of the craze volume consists of voids. However, the strength of the oriented fibrils is greater

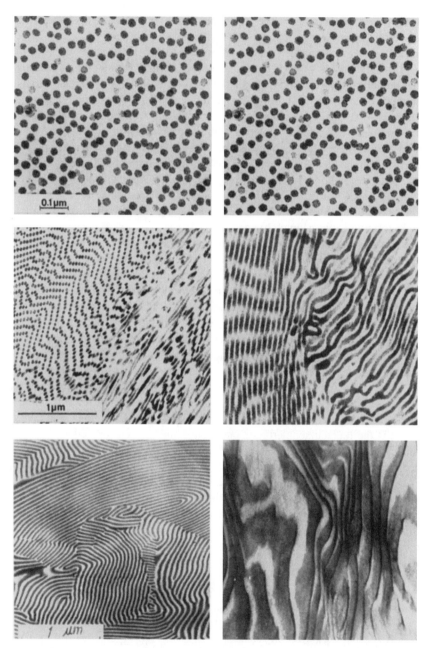

Figure 4.20. Electron micrographs of films of styrene (S) butadiene (B) block copolymers cut perpendicular (left) or parallel (right) to the film surface. Top row: SBS polymers with 20 mol% of B. Center: SB polymer with 40 mol% of B. Bottom: SBS polymer with 60 mol% of B. (Reprinted from H.-G. Elias, *Macromolecules,* 2nd ed. Copyright © 1984 by Plenum Press, New York. Used by permission of the publisher.)

than that of a bulk polymer; the crazed glass still has sufficient strength despite its battered appearance (the crazes scatter light and appear as white, hairy structures). Obviously, the strength of the crazes is limited: When the stress is too high, the fibrils break and a *crack* develops. In a crack, the walls are fully separated. Subsequent stressing will concentrate the stress at the ends of the crack; this will eventually lead to a catastrophic failure.

In glasses strengthened by rubbery inclusions, the impact will create crazes that will propagate until they reach the closest rubbery region. The rubber will deform easily and will dissipate the localized stress, which acts on the end of the craze, to the whole surface of the rubber. That will effectively stop the propagation of the craze and its development into a crack.

Finally, let us explore the strength of fiber-reinforced composites. The key to their properties is the high modulus of an oriented fiber compared with the modulus of the same bulk material. The modulus of a composite is a combination of the moduli of all components (with due attention to the orientation of fibers). Thus the fibers add high modulus to the composite, whereas the matrix (typically an epoxy resin) provides the bulkiness that is needed for dimensional stability. The reinforcing fibers may be of polymeric nature (cellulose, nylons, polyesters), but inorganic fibers are most popular because of their extreme modulus: glass, aluminum oxide, graphite. Carefully engineered composites with a high fiber content are often superior to metals.

4.1.8. Electrical and Optical Properties of Polymers

In the preceding sections, we studied bulk polymers primarily from the viewpoint of their mechanical properties. However, for many applications, their electrical and optical properties are important as well.

Among the optical properties, transparency is most important. We learned in Section 3.5 that macroscopic bodies let light pass through essentially unchanged unless it is reflected, absorbed, or scattered. Reflections occur at surfaces separating two phases, for example, on outer surfaces or on cracks. In this respect polymers behave exactly like all other materials. Most polymers do not carry chemical groups that absorb visible wavelengths; hence light absorption does not interest us, either. The transparency of bulk polymers is thus governed primarily by light scattering. Any inhomogeneity in the refractive index having dimensions comparable to the wavelength of the light scatters it. In liquids the inhomogeneities are caused by fluctuations, whereas in solid materials they usually exist at well-defined locations. However, the result is the same. They scatter light. When the inhomogeneities are large enough and numerous enough, we have a case of multiple scattering; the material is hazy, milky, or even opaque. Thus most crystalline polymers are opaque. The inhomogeneities are caused not only by differences in refractive index of crystalline and amorphous regions but also by differences of orientation of birefringent lamellae within the spherulites. On the other hand, extensively oriented crystalline fibers or sheets are rather homogeneous (even if anisotropic) and are sometimes quite transparent. Poly(vinyl chloride) is much less transparent than other, similar polymers because the presence of pseudocrystals (already mentioned in Section 4.1.7.1).

The transparency of other multiphase materials (composites, filled polymers, incompatible blends) is governed by the difference in the refractive indices of the two phases. These materials are usually opaque but may be quite transparent when the refractive indices are close to each other.

Many polymers are used in the electronics industry as insulators. Thus it is important to know their behavior in an electric field. We learned in Section 3.4.4.1 that molecules of dielectrics become polarized in an electric field. However, we considered only the polarization that was due to the shift of electrons within the molecules; such polarization is essentially instantaneous and can follow the changes in the electric field at all frequencies including the optical ones. An electric field, which changes more slowly, can induce additional polarization by another mechanism: the orientation of molecular dipoles. These dipoles try to lower their potential energy in the electric field by orienting themselves along the field. This orientation requires molecular movement and is opposed by friction forces. In low-viscosity liquids at low frequencies of the field, the molecules can follow the changes in the field without any delay. (The orientation is, of course, not complete. It is opposed by the Brownian movement, and a distribution function of orientation is established that is analogous to function Φ of Section 3.3.7). The polarization is thus rather high; this is manifested by a high dielectric constant of polar liquids. At very high frequencies, the molecular dipoles cannot follow the changes in the field, electronic shifts are solely responsible for the polarization, and the dielectric constant is equal to the square of the refractive index. At some intermediate frequencies, orientation of molecules lags behind the changes of the field; the polarization is out of phase with the exciting field.

Now, whenever the polarization is in phase with the field, the polarizing current is exactly out of phase. The dissipated power is the time average of the product of the field (voltage) and current; it is zero in this case (a capacitor made from such a dielectric has no ohmic resistance). However, when the polarization is out of phase with the voltage, there will exist a component of the polarizing current in phase with the voltage, and dissipation of power will result. The effect will be most pronounced when $\omega\tau$, the product of frequency ω and relaxation time τ, is equal to unity. The relaxation time is directly related to the rotary diffusion coefficient D_r, which is thus accessible from this experiment.

Let us now leave small molecules and consider the polarization of bulk polymers. If these polymers carry polar groups, the electric field will try to orient them. The magnitude of the opposing forces will depend on the physical state of the polymer and on the location of the polar group within the polymer molecule. Movement of side groups is obviously hindered much less than movement of the backbone. Hence, dipoles within side groups will display maximum out-of-phase polarization at higher frequencies than dipoles within the polymer backbone.

Reorientation of dipoles on a change in the electric field is known as *dielectric relaxation*. The strategy of its measurement and its formal description are completely analogous to mechanical relaxation. Thus, similarly to equation (4.1.65), a complex dielectric coefficient $\varepsilon^*(\omega)$ is defined as

$$\varepsilon^*(\omega) \equiv \varepsilon_0(\omega)\exp(i\delta) \equiv \varepsilon'(\omega) + i\varepsilon''(\omega) \qquad (4.1.74)$$

where the "real" part $\varepsilon'(\omega)$ describes the component when the field and the polarization are in phase; the "imaginary" part $\varepsilon''(\omega)$ reflects the out-of-phase component. The loss angle δ, defined in analogy to equation (4.1.64) as

$$\tan \delta \equiv \varepsilon''(\omega) / \varepsilon'(\omega) \tag{4.1.75}$$

is a measure of power dissipation in the dielectric.

The complex modulus and complex dielectric constant reflect the same phenomenon: mobility of molecular segments and groups. Both loss tangents usually display a maximum in the region of the main glass-rubber transition (T_g). Polymers with side groups capable of independent motion usually have a secondary maximum at low temperatures when this motion is frozen. However, there are differences between the mechanical and dielectric dependences. Any kind of movement contributes to mechanical relaxation, whereas dielectric relaxation requires the movement of groups carrying the dipoles. Thus polymers with nonpolar side chains display only a very weak if any secondary loss peak.

The loss component of the dielectric constant is very undesirable in practice if the polymer is to be used as an insulator. In this case, an insulator displaying a loss "window" in the frequency region of application should be selected. Most advantageous are, of course, nonpolar polymers of the polyethylene type, for which the dielectric losses are small at all frequencies.

Recently, growing attention has focused on ferroelectric polymers, electrets, as well as on conducting and semiconducting polymers. Some materials are permanently polarized when they are subjected to a very strong electric field. Such materials are called *ferroelectric* in analogy to ferromagnetic materials, which become permanently magnetically polarized in strong magnetic fields. The resulting polarized object is called an *electret*; it is the electrical analogy of a magnet. Electrets are strongly piezoelectric and pyroelectric and consequently very valuable technologically.

The most prominent ferroelectric polymer is poly(vinylidene fluoride). It is a crystalline polymer capable of crystallizing in several modifications. In one of these modifications, the polymer chains are in an extended zigzag conformation; all the fluorines are on the same side of the chain. This modification is important mainly for specimens oriented by stretching. We should recall that covalently bound fluorine has a van der Waals radius only slightly larger than that of hydrogen. Hence, very little resistance is encountered when the chain is rotated along its zigzag axis. It is this rotation by which the electric field accomplishes the alignment of many dipoles. The process is called *poling*. The chains in the poled polymer are still held in their mutual positions by crystalline forces, and the electret remains polarized until the polarization is reversed by an opposing field (the reorientation is met by a strong hysteresis) or until the temperature is raised above the Curie temperature, when the thermal movement destroys the alignment of the dipoles.

Nature has manufactured the best-known conducting polymer. It is graphite, and its conductivity is a result of its extended system of π bonds. Accordingly, synthetic conducting polymers are found among polymers having long chains of conjugated bonds. Presently, two types of chains qualify: polyacetylenes and polypyrroles. Polypyrroles are prepared by electrochemical oxidation of pyrrole; they are believed to have structure **1**.

1

Both polyacetylenes and polypyrroles are completely intractable, but it is possible to synthesize them in the form of self-supporting films. As such, they have rather low conductivity. The conductivity, however, increases dramatically with doping. They are usually doped with either I_2 or arsenic pentafluoride or sodium naphthalide. It seems that besides extensive conjugation and doping, conducting polymers must have a rigid planar structure. For example, even a slight deviation from planarity caused by copolymerization of acetylene with methylacetylene or of pyrrole with N-methyl pyrrole is sufficient for a drastic reduction in conductivity.

4.1.9. Transport Through Polymers

Unlike inorganic glasses and metals, both glassy and rubbery bulk polymers often allow a considerable transport of small molecules (gases, vapors and liquids) through them. This phenomenon is known as *permeability*. This property of the polymers is important in determining their usefulness for many applications. It is quite often undesirable (plastic soft drink bottles leaking CO_2, inner tubes and tennis balls losing pressure, vinyl chloride diffusing from PVC articles); however, sometimes it is very useful. The desalination of seawater and the separation of industrial gases or liquids are based on the differential permeation of different materials through a suitable polymer, which acts as a barrier material (often called membrane). Consequently, studies of permeability are technologically very important; in addition, they provide us with valuable information about the structure of polymers.

For convenience, we will first consider systems in which the concentration of the permeating species in the polymer is everywhere sufficiently low to not influence substantially the properties of the polymer. This condition is usually fulfilled well when permeation of gases (He, H_2, N_2, O_2, CH_4) is studied. It applies also to CO_2 at lower pressures.

In a typical experiment, a supported rubbery polymer membrane separates two chambers that contain the permeating gas at different pressures (often, the pressure is zero on one side), and the flow of the gas across the membrane is measured by suitable means. After a sufficiently long time, a steady-state flow is established; the concentration profile of the permeating gas in the polymer is linear. The gas in the infinitesimal surface layer is at equilibrium with the gas phase. Its concentration c

must, according to Henry's law, be proportional to the pressure p (in this section we designate pressure by p to reserve the symbol P for permeability).

$$c = Sp \qquad (4.1.76)$$

In permeability studies, it is customary to express the concentration c in units of cubic centimeters of the gas at its standard temperature and pressure (STP) dissolved in 1 cm^3 of the polymer; pressure is customarily expressed in centimeters of mercury; the units of *solubility* S follow. Equation (4.1.76) applies on both membrane boundaries, and the differences in concentration, Δc, and pressure, Δp, are related as $\Delta c = S \Delta p$. Within the membrane, the concentration gradient leads to a diffusion flow that follows Fick's first law, equation (3.4.1).

$$I = -D\left(\frac{\partial c}{\partial x}\right)_t = -D\frac{\Delta c}{L} = -DS\frac{\Delta p}{L} \equiv -P\frac{\Delta p}{L} \qquad (4.1.77)$$

In this relation, L is the thickness of the membrane and P is the permeability, which is defined by the last identity in equation (4.1.77). Its units are cubic centimeters (gas) STP per centimeter (thickness) per square centimeter (area) per second per centimeter of mercury. Thus permeability is equal to the product of solubility and diffusion coefficient. Both these quantities depend on the nature of the system and on temperature.

The diffusion coefficient reflects molecular friction; it is a steep function of molecular size. For instance, helium and hydrogen have much higher diffusivity than larger molecules. Solubility depends mainly on the thermodynamic interaction between the gas and the polymer. Most permanent gases are nonpolar and are much more soluble in nonpolar polymers than in polar ones. Polyacrylonitrile and polycarbonates are very poor solvents for gases and are therefore quite impermeable (they are good barrier polymers).

While comparing the numerical values of permeability of vapors we should keep in mind that permeability is referred to vapor pressure. This may lead to unexpected results. Let us consider a hypothetical case of a polymer that, when equilibrated with either of two different liquids, swells to the same degree. Let us also assume that the diffusivity of both liquids is the same. It follows that in an experiment in which the membrane is contacted from one side by the liquid with vacuum on the other side, the diffusion flow remains the same for both liquids. If the liquids are replaced with their saturated vapors (having the same chemical potential as the liquids), the situation remains unchanged. Now let us assume that the two liquids have very different volatilities. Say, one has saturated vapor pressure 10 times higher than the other one. Then the solubility and permeability of the *less* volatile liquid are found to be 10 times *higher* than those of the more volatile liquid!

Experimentally, permeability and diffusion coefficient are often obtained from a single transient experiment. The membrane is mounted in a dual chamber; both sides are carefully evacuated. Then the gas is admitted at the required pressure to one side (upstream side), and the accumulated amount on the other side (downstream side) is monitored as a function of time (Fig. 4.21). At the beginning, the permeation is

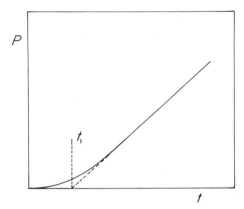

Figure 4.21. Schematic representation of the dependence of the amount of gas transported across the membrane (measured as pressure in the chamber on the downstream side of the permeation cell) on time in a typical permeation experiment. t_1 is lag time.

slow because no gas is dissolved in the polymer yet. At long times, the accumulated amount increases proportionally with time; the slope of the dependence yields the permeability. When the asymptotic tangent is extrapolated backwards, it intersects the time axis at time t_1, which is called *lag time*. Integration of Fick's second law [equation (3.4.3)] under proper initial and boundary conditions yields the relation between the lag time, the membrane thickness, L, and the diffusion coefficient as

$$t_1 = L^2/2D \qquad (4.1.78)$$

The diffusion coefficient is obtained from this relation.

How does temperature influence the molecular transport? The solubility of gases and vapors generally decreases with increasing temperature (for vapors, an increase in saturated vapor pressure is the main factor), whereas the diffusion coefficient generally increases. Thus the permeability may go either way; it is much less temperature-sensitive. It does not even change appreciably at the glass transition.

Permeability behavior of glasses differs in one respect from the behavior of rubbers. Glasses have a nonequilibrium free volume frozen in, and the diffusing molecules (diffusant) may fill the voids. The volume of the voids is limited; they eventually become saturated and display Langmuir-type nonlinear adsorption isotherms. Another portion of the diffusant is, of course, dissolved in the bulk polymer, obeying Henry's law in the same way as it did for the rubbery polymers. The Henry's law portion is most responsible for the diffusion transport; the Langmuir portion is much less mobile. As a result, the overall sorption is nonlinear, the permeability becomes pressure-dependent, and the time lag experiment is distorted, yielding wrong values of the diffusion coefficient.

Finally, we should consider transport of vapors and liquids that interact more strongly with the polymer. When the concentration of the diffusant is higher, the structure of the polymer may change. In particular, the free volume may increase and the internal friction may decrease with an accompanying increase in diffusivity. An extreme case of this situation can be observed when a glassy polymer is swollen in a good solvent. The solvent penetrates the glass slowly. However, once the structure is softened, further swelling proceeds fast. It is often possible to observe a glassy core that is surrounded by a fully swollen network. The swelling front then proceeds slowly into the core. Meanwhile, at the swelling front, the molecular chains are subject to large forces and some cracks may develop.

4.2. TECHNIQUES FOR THE STUDY OF BULK POLYMERS

In this section we will describe some more important methods for collecting information about the properties of bulk polymers. We will first be concerned with obtaining the mechanical properties, then the thermal and thermodynamic ones. Finally, a short description of methods for studying the structure and morphology of bulk materials will be presented.

4.2.1. Mechanical Methods

The experimental arrangements for measuring mechanical properties of polymeric materials depend on the nature of these materials. Different techniques are applied to viscoelastic liquids (which include melts), soft (rubbery) viscoelastic solids, and hard (glassy, crystalline) solids. The quantities of interest include the transient functions (creep, stress relaxation), viscosity and difference of normal stresses (for viscoelastic liquids only), and the dynamic quantities (storage and loss moduli, loss angle). This section presents only the broad outlines of the actual measuring techniques—too many different laboratory and commercial instruments have been designed.

The viscosity of bulk polymers and concentrated solutions is measured with viscometers based on the same principles as those described in Section 3.4.3.1. The higher viscosity, of course, demands some changes. Capillary viscometers are more robust, with wider and shorter steel capillaries and substantial outside pressure applied. The rotary Couette-type viscometers are also more robust and are operated at much lower speed. Two additional geometries are employed in rotary viscometers: cone and plate and parallel plate (Fig. 4.22). The former geometry provides an exceptionally homogeneous velocity gradient. The parallel-plate geometry yields a more complicated flow field, which is nevertheless still amenable to theoretical analysis. Its advantage is easy construction and an easier mode of operation. A major advantage of these two viscometers is their ability to measure not only the viscosity but also the difference in normal stresses, which is manifested by a thrust that the sheared liquid exerts on the cone (or upper plate) and is measured by means of a sensitive pressure transducer. Commercially, the Weissenberg rheogoniometer employs these measuring strategies.

For a proper evaluation of viscosity from the measurement, several precautions should be taken. For viscoelastic liquids, the term viscosity is meaningful only for those flows that have reached the steady state. In other words, viscosity is evaluated

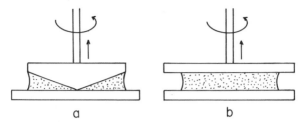

Figure 4.22. Schematic representation of (a) cone-and-plate, and (b) parallel-plate visco-meters. The arrows represent the torque and thrust that are measured.

from the asymptotic part of the creep curve. In rotary viscometers, the rates of rotation are very slow and the measurement is often taken during the first rotation or even during the first fraction of it. In such a case, the researchers should assure themselves that the asymptotic behavior was really achieved. Another artifact to be watched for is viscous heating, which may be appreciable for highly viscous materials. A rise in temperature, if unaccounted for, may vitiate the measurement. Finally, non-Newtonian behavior of these liquids (which is often quite pronounced) may accentuate strongly the nonhomogeneity of the flow field; its effect should be estimated for every measurement.

Instruments for the measurement of creep and stress relaxation are similar to each other and are conceptually very simple (Fig. 4.23). A flat specimen of the sample in the form of a "dog bone" is mounted between two clamps. The actual measurement is done on the narrow section of the dog bone; the wider ends serve to minimize the

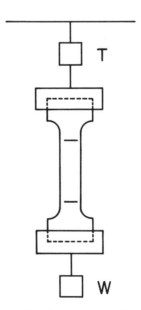

Figure 4.23. Schematic arrangement for the measurement of creep and stress relaxation. T is a stress transducer; W is a weight.

effect of clamping. In creep experiments, two marks are traced on the sample, and a constant stress is applied by means of a weight W. The distance between the marks is then monitored as a function of time. In a relaxation experiment, the sample is strained to the required elongation, and the stress is followed by means of a sensitive transducer T. Both these modes of operation are incorporated in a commercial instrument, Instron, which has a rugged frame to which the samples are attached. It is equipped with tensile or pressure transducers and a very sensitive control of the distance between the clamps. Instron can be programmed to keep stress or strain constant and to record the other quantity. However, the most common mode of operation is stretching of the sample at a constant rate of strain while both stress and strain are monitored and plotted against each other. The elastic modulus is evaluated from the initial slope of the dependence. The test is usually continued until the sample fails: Tensile strength and elongation at break are measured this way. The measured value of modulus is reasonably reproducible from sample to sample. However, the ultimate properties of the material depend strongly on the way the test specimen was prepared, and they exhibit a considerable scatter. It is recommended to measure at least 10 samples and evaluate the results statistically.

Many measuring principles and instrument geometries are used to determine the dynamic quantities of viscoelastic materials. For soft materials, the cone-plate and plate-plate geometries employed in the Weissenberg rheogoniometer can be adapted for dynamic measurements. One of the elements is set into oscillating rotary motion, and suitable gauges monitor the displacement and torque. They are generally out of phase, yielding simultaneously both the storage and loss moduli (from the measured ratio of amplitudes and from the phase angle). This technique is one version of the method of *forced oscillations.*

Forced oscillations are also used with hard solids, but the instrumental geometries are quite different. The sample is in the shape of a bar or rod and is deformed either by flexing or by torsion. The success of the methods depends on the availability of sensitive and precise stress and strain gauges. In the forced oscillation experiments, energy is continuously pumped into the specimen to keep the amplitude constant despite the losses. The lost energy is, of course, converted to heat that may be dissipated only slowly, and the temperature may rise, spoiling the measurement. The heating is reduced when small stresses are used, but then the strains may become very small and difficult to measure.

This problem is avoided in instruments using free *oscillation.* The sample is stressed and let loose. It will oscillate at a frequency that is the natural frequency of the oscillating system. This depends on the oscillating masses, geometry of the arrangement, and storage modulus of the tested material. Thus the modulus is calculated from the measured frequency and other known parameters. The loss modulus damps the oscillation; its amplitude decreases exponentially with time. From the rate of damping the loss tangent is easily evaluated.

The most popular geometry of this type is that of a *torsion pendulum* (Fig. 4.24). In this experiment, a moment arm with two masses at its sides acts as a horizontal pendulum. The sample acts as the restoring element as it is subjected to torsion during the pendulum motion. Adjusting the weights and their distance from the pendulum axis can change the frequency of the oscillation for a given sample.

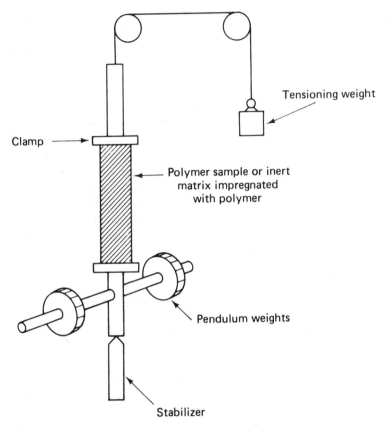

Figure 4.24. Scheme of a torsion pendulum. (Reprinted from Harry R. Allcock and Frederick W. Lampe, *Contemporary Polymer Chemistry,* p. 431. Copyright © 1981 by Prentice Hall, Inc., Englewood Cliffs, NJ. Used by permission of the publisher.)

The method of the torsion pendulum can also be used for soft polymers or even melts. In this case, the restoring element consists of a braided glass fiber or another matrix impregnated by the polymer. As long as the mechanical properties of the fiber or matrix are known, the properties of the polymer can be found from the measurement of this composite element.

Most of the instruments for the dynamic measurements can provide data only in a limited range of frequencies. When data in a broader region are required, the principle of frequency-temperature superposition (described in Section 4.1.5.2) must be utilized.

4.2.2. Differential Scanning Calorimetry (DSC)

All materials store energy in the form of thermal movement. This movement is different in different types of materials: simple translation of whole molecules in liquids and gases, rotation of molecules and vibration of molecular groups at sufficiently high temperatures, oscillation of crystalline lattices, motion of polymeric segments, etc. The amount of stored energy changes whenever the state of the system is changed. Every change must be accompanied by an input or output of energy—usually in the form of heat. In the simple process of heating, the amount of added energy is expressed in terms of heat capacity. The magnitude of heat capacity depends on the nature of the molecular movement in the particular material. It is large when many modes of movement are activated and small when they are frozen. Thus heat capacity is small for glassy and crystalline polymers and larger for viscoelastic materials; it changes more or less abruptly at the glass transition temperature. Besides heating, other processes are also accompanied by a flow of heat in or out of the system; these include melting and crystallization, vaporization, and chemical reactions such as polymerization and decomposition. It is thus obvious that knowledge of the amount of heat entering a sample provides a wealth of information about it.

Differential scanning calorimeters are designed to measure this heat. In a typical experiment, a sample of polymer (2–10 mg) is sealed in an aluminum pan and heated electrically. A temperature sensor is used for a feedback control of the heater to achieve a strictly constant rate of temperature increase. The electrical current is measured precisely, and it provides (after proper correction for the heating of the pan and other parts of the instrument) a record of the heat capacity of the sample as a function of temperature.

Usually, the same sample is observed twice. In the first scan, a very large number of phenomena are observed. Around $0°C$, water that is absorbed and adsorbed on the sample melts, and an endotherm results. The next phenomenon is usually glass transition. If the sample was originally quenched to low temperatures, then it possibly went through T_g before it had time to crystallize. In such a case, it will crystallize during the differential scanning calorimetry (DSC) experiment; an exotherm is observed. At $100°C$ or so, water leaves the sample, yielding another endotherm (much less distinct this time). The next observable endotherm signifies melting of the crystalline portion. Samples that were incompletely polymerized will complete the polymerization during the DSC experiment (an exotherm). Alternatively, the remaining monomer will vaporize. Finally, if the temperature is raised high enough, the polymer starts to decompose (this stage is avoided in the first run). Obviously, the amount of information obtained is very large but is often difficult to interpret; too many features may overlap.

A second scan is therefore performed, during which many previous phenomena are absent, namely, those related to water and unreacted monomer and possibly other volatile materials. The main features of this scan are shown in Figure 4.25. The region of T_g is characterized by three more or less linear sections. Section A at lower temperatures has the smallest slope and corresponds to glass; the middle section, B, has the largest slope and reflects the fast unfreezing of the segmental motion during the glass transition. The later section, C, has an intermediate slope characteristic of rubbers. The exotherm at D reveals the crystallization, whereas the endotherm at E shows melting. Finally, the terminal section at F reflects the decomposition of the polymer.

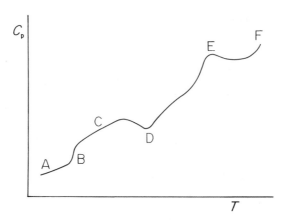

Figure 4.25. An idealized DSC scan. A, glassy region; B, glass-rubber transition; C, rubbery region; D, crystallization exotherm; E, melting endotherm; F, decomposition.

The DSC method is especially valuable for studying blends and phase-separated polymers. Two-phase systems display two separate T_g values; phases representing as little as 20% of the material can still be detected. On the other hand, compatible blends exhibit only one glass transition, which occurs between the T_g values of the two homopolymers; T_g of a compatible blend is a smooth function of its composition.

When the sample is placed into the DSC cell a purge gas is usually applied because in some experiments it is desirable to perform the study in an inert atmosphere such as nitrogen. In other cases the effects of oxidation may be studied by using an air or oxygen atmosphere.

The technique of DSC used to be limited to temperatures below 700°C. However, recent introduction of high-temperature differential scanning calorimeter has expanded the upper temperature limit to 1700°C. This increased capability allows the DSC analyses to be done on materials such as ceramics, metal alloys, silicates, and high-temperature composites. Most commercial instruments are also able to run at low temperatures, ranging to about –150°C, although in practice special care is required to obtain good data below about –100°C. The commercial instruments now use a computer interface to set up the temperature programs, control the DSC during the heating program, and collect data.

Differential thermal analysis is an older and simpler version of differential scanning calorimetry. In this technique, the polymer sample and a reference material (alumina) are placed side by side within a metal block that is heated at a uniform rate. Two thermistors or thermocouples are placed within the polymer and the reference, and the difference between the signals is measured. Generally, the sample with the larger heat capacity will lag behind in temperature. The difference will suddenly change whenever any transition takes place within the polymer (no transitions occur in the reference). Thus the information provided by DTA is qualitatively similar to that obtained from DSC; however, it is difficult to interpret quantitatively.

4.2.3. Inverse Gas Chromatography (IGC)

The only thing that is inverted in inverse gas chromatography is the interest of the researcher. He or she is not interested in the properties and composition of the material injected into the mobile phase but in the properties of the stationary phase. In a typical experiment, a polymer is coated onto a porous material that is packed into a chromatographic column. Various pure substances (*probes*) are then injected on the top of the column and their elution curves are recorded. Emphasis is on the precise measurement of the position of the peak maximum, peak width, peak asymmetry, moments of the curve, and so on. From these data it is possible to obtain information about the polymer-probe interaction coefficients and about the diffusion coefficient of the probe in the polymer. When the measurement is performed at different temperatures, the glass transition temperature can be determined as well as the degree of crystallinity of semicrystalline polymers. The measurements are quite fast, and extensive data can be accumulated within a reasonably short time. Technically, IGC is more demanding than ordinary gas chromatography. The flow rate of the carrier gas must be measured with high precision, and its fluctuation over an entire day must not exceed a few tenths of a percent. A very stable baseline, low electronic noise in the detector, stable temperature, and accurate measurement of time are also required.

Let us review the basic terminology of gas chromatography. The flow rate of the carrier gas (helium or nitrogen) is measured at the column outlet and is expressed either as F_0 in volume units (mL/min) or as u_0 in linear units (cm/s). Obviously,

$$u_0 = L F_0 / 60 \, V_0 \tag{4.2.1}$$

Here, V_0 is the *void volume* of the column, that is, of the gas space on the column; L is the column length. Chromatographic columns have appreciable hydrodynamic resistance; their inlet pressure P_i and outlet pressure P_0 are quite different. In a stationary gas flow, the mass flow along the column is constant, but the volume flow and linear velocity change along the column. The void volume is calculated from t_m, the time that a probe that does not interact with the column (a *marker*) needs to travel from the injector to the detector. The relation is

$$V_0 = t_m F_0 \left[\frac{3}{2} (P_r^2 - 1) / (P_r^3 - 1) \right]; \qquad P_r \equiv P_i / P_0 \tag{4.2.2}$$

The term in brackets accounts for gas expansion along the column.

When thermodynamic data are to be obtained by IGC, experimental conditions must be selected in such a way that the partition of the probe between the gaseous and polymer phases is virtually instantaneous. This implies that the diffusion of the probe within the polymer is fast enough, that is, that the viscosity of the polymer is low enough. This condition is usually, but not always, fulfilled when the experiment is performed at least 50°C above T_g. At phase equilibrium, the concentration of the probe in the polymer phase, c_p, is equal (at every location on the column) to its concentration in the gas phase, c_g, multiplied by the partition coefficient k. Another important quantity is the distribution coefficient K that relates the masses of probe in the two phases.

$$K = V_p c_p / V_0 c_g = k V_p / V_0 = k w_p v_p / V_0 \qquad (4.2.3)$$

Here, w_p, V_p, and v_p are the mass of the polymer on the column, its volume, and its specific volume, respectively. The probe in the gas phase travels along the column with the same velocity u as the carrier gas; the probe in the polymer phase does not travel at all. The mass-averaged velocity of the probe is thus $u/(1 + K)$, and the time t_{pr} needed for its elution is given by

$$t_{pr} = t_m (1 + K) \qquad (4.2.4)$$

In IGC experiments, the following quantities are routinely introduced: retention volume $V_R \equiv V_0 t_{pr}/t_m$; net retention volume $V_N \equiv V_R - V_0 \equiv V_0(t_{pr} - t_m)/t_m$; and reduced retention volume $V_g \equiv V_N/w_p$. All these quantities are experimentally accessible. By combining the definition of V_g with equations (4.2.3) and (4.2.4) we obtain

$$V_g = k v_p \qquad (4.2.5)$$

which relates the chromatographic quantity V_g with the partition coefficient k—a thermodynamic quantity. The partition coefficient expresses the solubility of the probe in the polymer. A more common description of solubility is through Henry's law,

$$P_1 = k_H c_g \qquad (4.2.6)$$

Henry's constant k_H is easily found to be given by

$$k_H = RT v_p / M_1 V_g \qquad (4.2.7)$$

where P_1 and M_1 are the partial pressure and molecular weight, respectively, of the probe. Another straightforward thermodynamic calculation links the temperature dependence of V_g to the molar enthalpy of sorption of the probe by the polymer, $\Delta \bar{H}_{sorp}$, according to the formula

$$\frac{d \ln(V_g/T)}{d(1/T)} = -\frac{\Delta \bar{H}_{sorp}}{R} \qquad (4.2.8)$$

The calculation of k_H and $\Delta \bar{H}_{sorp}$ is meaningful for all probes without regard to whether their critical temperature is above or below the column temperature. For probes that are under their critical temperature, IGC can be used for determination of the polymer-solvent interaction. When transforming the relations valid for gas sorption to relations valid for liquid sorption, we should refer the activities to a reference state applicable for liquids, namely to the pure liquid under its saturated vapor pressure, P_1^0. In this calculation, we must account for the nonideality of the probe vapor (up to its saturated vapor pressure). The virial coefficient of the probe vapor, B_{11}, will occur in the final formula, which is conveniently written as

$$\ln \Omega_1 \equiv \ln\left(\frac{a_1}{\omega_1}\right) = \ln\left(\frac{RT}{V_g P_1^0 M_1}\right) - P_1^\circ \frac{B_{11} - \overline{V}_1}{RT} \tag{4.2.9}$$

Here Ω_1 is the weight fraction activity coefficient of the probe (i.e., the ratio of its activity a_1 to its weight fraction ω_1). This expression is quite general and does not refer to any model of polymer solutions.

When the polymer-probe system is modeled by the modified Flory–Huggins relation, equation (3.1.100), the parameter χ_{12} of equation (3.1.102) is obtained by proper manipulation of equation (4.2.9) as

$$\chi_{12} = \ln\left(\frac{RT v_p}{V_g \overline{V}_1 P_1^\circ}\right) - P_1^\circ \frac{B_{11} - \overline{V}_1}{RT} - 1 + \frac{\overline{V}_1}{M_p v_p} \tag{4.2.10}$$

where $\overline{V}_1$ is the molar volume of the (liquid) probe and M_p is the molecular weight of the polymer. For high-molecular-weight polymers, the last term is negligible. The value of χ_{12} computed from equation (4.2.10) refers to vanishingly small concentration of the probe in the polymer. Inverse gas chromatography and equation (4.2.10) provide access to a large number of interaction coefficients; two dozen values for different probes can be obtained in a single day.

Inverse gas chromatography may be also used for thermodynamic studies of polymer blends. It can yield the polymer-polymer interaction parameter B_{23} [the same quantity we called B_{12} in equation (4.1.71)]. For this purpose, three columns are prepared and measured: two from homopolymers and a third one from a blend made using the same samples of the homopolymers. The three columns should be studied under conditions as identical as possible. If we use for the description of the blend-probe system the Flory–Huggins equation for three components [equation (3.2.20); subscript 1 for probe, subscripts 2 and 3 for homopolymers] and subsequently introduce the parameter B_{23} as in equation (4.1.71), we can derive

$$B_{23} = \frac{RT}{\overline{V}_1}\left[\frac{\ln[V_{g,\text{blend}}/(\omega_2 v_2 + \omega_3 v_3)] - \phi_2 \ln(V_{g,2}/v_2) - \phi_3 \ln(V_{g,3}/v_3)}{\phi_2 \phi_3}\right]$$

$$\tag{4.2.11}$$

Here the second subscript of V_g identifies the nature of the column; v_i and ω_i are the specific volume and weight fraction, respectively, of the ith component; ϕ_i is the volume fraction of the ith polymer in the blend. It is noteworthy that the parameters P_1°, B_{11}, $\overline{V}_1$, as well as molecular weights of the polymers, do not enter into equation (4.2.11).

It was found experimentally that the values B_{23} for a given blend depend on the nature of the probe. This should not be surprising because equation (3.2.20) is an oversimplified description of the thermodynamic behavior. The phenomenological relation (3.2.25) predicts that the quantity on the right-hand side of equation (4.2.11) must be dependent also on the ternary interaction parameter g_T and consequently on the nature of the probe. Nevertheless, a more sophisticated analysis using the measured values for many probes is capable of yielding the true B_{23} values.

Although the position of the elution band depends primarily on the thermodynamic factors, the width of the band and its asymmetry are governed mainly by diffusion phenomena—diffusion of the probe in the carrier gas along the length of the column (characterized by the probe-gas diffusion coefficient D_g) and diffusion of the probe into the polymer (described by the probe-polymer diffusion coefficient D_1 and the thickness of the polymer layer d). Both processes result in peak broadening. In gas chromatography, peak broadening is traditionally described by H, the height equivalent to one theoretical plate (the name is derived from analogy to separations using countercurrent extraction). H, which is a very useful quantity for description of the separating power of the column, is defined as

$$H = L \frac{(w_{1/2}/t_{\text{pr}})^2}{8 \ln 2} \tag{4.2.12}$$

where $w_{1/2}$ is the peak width at its half-height. When the penetration of the probe into the polymer is slowed down only moderately by diffusion, the probe peak elutes slightly sooner than in the case of instantaneous equilibrium (the differences are often only minor). The peak symmetry is also almost unchanged, but the peak width increases significantly. The analysis yields for H (after correction for some factors related to the nature of the inert support)

$$H = 2 \frac{D_g}{u} + 0.7 \frac{d^2}{D_1} \frac{u K}{(1 + K)^2} \tag{4.2.13}$$

Hence, the dependence of H on the flow rate u (the useful plot is Hu versus u^2) may yield the values of D_g and D_1/d^2.

When the penetration of the probe into the polymer is slower, the peak becomes very asymmetric, its maximum shifts to shorter times (eventually, it coincides with the marker), and the peak develops a long, low tail. This scenario is observed when the temperature of the IGC column is lowered through the glass transition region; IGC is useful for finding T_g of the polymer.

Inverse gas chromatography has also been used for studying crystalline polymers. This measurement is based on the observed fact that most probes cannot penetrate polymer crystallites, whereas they penetrate the amorphous phase (in the temperature region above T_g) more or less freely. Consequently, the dependence of V_g on $1/T$ (which is essentially linear in the rubbery region above the melting temperature T_m) exhibits a sharp break at T_m. Below T_m, the probe retention is smaller than the retention predicted by extrapolation from the region above T_m. From the difference, the degree of crystallinity can be estimated.

4.2.4. X-Ray Diffraction

X-ray diffraction is probably the most powerful tool for determination of the structure of crystalline and semicrystalline materials. X-ray diffraction is based on the same scattering principles as light scattering and SAXS. We studied these principles in Section 3.5, but we omitted the case of scattering by a perfectly regular array of identical scattering elements. (Unlike the homogeneous system described in Section 3.5.5, our present array is regular but not homogeneous. The properties of the material vary in the space between the array elements, and the distances between the elements are comparable to the wavelength, which is typically 0.154 nm for the K_α radiation of copper.)

Let us return to the interference of electromagnetic radiation scattered by two identical elements separated by a vector $\mathbf{r}$. According to equation (3.5.25), the path difference Δ of the two scattered beams is

$$\Delta = \mathbf{S} \cdot \mathbf{r} \qquad (4.2.14)$$

In this relation, we have replaced the symbol $\mathbf{k}$ used for the scattering vector in the light-scattering literature by the symbol $\mathbf{S}$ used in the X-ray literature. It follows from equation (4.2.14) that the observed scattered beam exhibits maximum intensity when the path difference is equal to an integer multiple of the wavelength λ. (Refractive index for X-rays is equal to unity; we do not need to differentiate between λ and λ_0 as we did in light scattering.) Thus for maximum intensity we have

$$\mathbf{S} \cdot \mathbf{r} = n\lambda \qquad (4.2.15)$$

Let us now consider a linear array of identical elements, the distance between any two neighbors being $\mathbf{r}$. When the distance between any pair of neighbors satisfies equation (4.2.15), all pairs (neighbors or not) satisfy it. The scattered intensity thus has a set of maxima for scattering vectors $\mathbf{S}$ defined by relation (4.2.15). What is the scattered intensity for vectors $\mathbf{S}$ not satisfying relation (4.2.15)? With two elements, the intensity is extinguished if the product $\mathbf{S} \cdot \mathbf{r}$ is equal to an odd number of half-wavelengths. For other values of $\mathbf{S}$, the intensity is finite. The situation is different for a large number of elements. For example, let us select a scattering vector $\mathbf{S}$ such that the phase difference of the scattered rays from the first and second element is very small, say $\lambda/2x$, where x is a large integer. Then the phase difference between the first and $(x + 1)$th elements is $\lambda/2$; they are exactly out of phase and cancel each other. We may then pair all the other elements in a similar way; no radiation is observed. This argument is valid, of course, only when the total number of elements is much larger than x; otherwise, some residual intensity will still be observed. Thus when the number of elements is large (as in crystals with millions of them) the X-rays are scattered in only a few well-defined directions.

How many diffraction maxima can we observe? Let us recall that the length of the vector $\mathbf{S}$ is $S = 2 \sin \theta$. (Remember that in X-ray literature—including this section—the scattering angle θ is one-half the angle between the directions of the primary and scattered beams.) If the angle between vectors $\mathbf{S}$ and $\mathbf{r}$ is α, then we can modify equation (4.2.15) to read

$$\mathbf{S} \cdot \mathbf{r} = 2r \sin\theta \cos\alpha = n\lambda \qquad (4.2.16)$$

Furthermore, we may write the inequality

$$1 \geq \sin\theta \geq \sin\theta \cos\alpha = n\,(\lambda/2r) \qquad (4.2.17)$$

Hence, r must be larger than $\lambda/2$ if any diffraction maximum is to be observed. (This condition is never fulfilled when r corresponds to crystalline lattice dimensions and the scattering of visible light is studied.) The number of observable maxima increases with the increasing distance r.

Our next problem is scattering by a three-dimensional array of identical elements, each of them sitting in a corner of a unit cell of a crystal. In this case, equation (4.2.15) must be satisfied for every pair of elements or, more specifically, for all three linear arrays that originate from an arbitrary element and proceed along the edges of the unit cell. A somewhat lengthy argument, which we are not going to reproduce, shows that the directions of the allowed scattered beams are the same as the directions of beams that would be reflected by imaginary mirror planes passing through arrays of identical lattice points. Obviously a large number of such planes exist; normals to these planes have a unique relation to the edges of the unit cells of the crystal. However, only those planes will give rise to diffraction maxima that satisfy the Bragg relation (4.2.18) between the scattering angle θ and the interplanar distance d_{hkl} (the integer subscripts hkl describe the inclination of the planes to the crystallographic axes).

$$\sin\theta = \lambda/d_{hkl} \qquad (4.2.18)$$

Thus, to produce an observable reflection from a given plane, the crystal must be properly oriented with respect to the primary X-ray beam; the interplanar distance is then easily calculated from the scattering angle. When $d < \lambda/2$, no reflection could be observed from that particular plane under any orientation of the crystal. It should also be noted that at any given orientation of the crystal only a very few or even none of the planes will give rise to diffraction maxima. To obtain a large number of reflections we need to explore many different mutual orientations of the crystal and the X-ray beam.

The most detailed structural information is obtained when well-developed crystals are available (they should be at least 0.2 mm in all dimensions). The crystal is mounted on a stage and rotated or oscillated around one of its axes. Individual planes come sequentially into proper orientation with the primary beam and produce a diffraction spot on the photographic plate. The pattern of the spots reflects the symmetry of the space group to which the crystal belongs (there are 230 possible space groups), and an experienced crystallographer can determine from it the space group and the dimensions of the unit cell. For many simple crystals with few atoms per unit cell, this provides sufficient structural information. It should be noted that such crystals usually have quite small unit cells and the Bragg relation allows for only a few reflections.

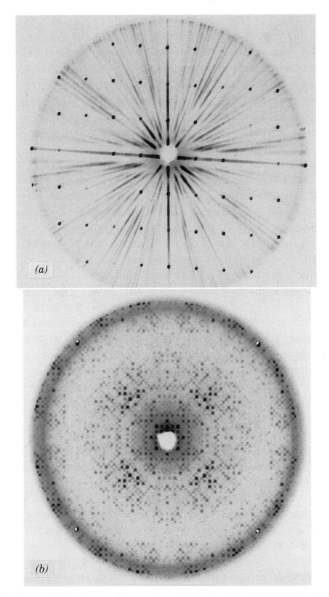

Figure 4.26. X-ray diffraction patterns of perfect crystals. (a) Methylamine tungstate; (b) histidine decarboxylase. (Courtesy of Dr. M. Hackert).

When the unit cells are larger and contain larger numbers of atoms, the number of measurable reflections increases dramatically. In Figure. 4.26 are presented two X-ray photographs: one for a simple inorganic salt, the other for a protein crystal. The photographs were taken with a precession camera that rotates the crystal and the film in a coupled manner designed to yield easily identifiable spots.

From the positions and the arrangement of the diffraction spots, it is possible to deduce the type of crystalline lattice and the dimensions of the unit cell. However, even more information is available from measurements of the relative intensity of the diffraction spots. In the case of small organic molecules crystallizing in a space group with a center of symmetry, the analysis of the diffraction pattern is reasonably straightforward. From the Fourier transforms of the properly phased spot intensities the three-dimensional distribution of electron densities can be deduced. (Remember, it is the electrons that scatter the X-rays.) The electron density function will have maxima at the locations of individual atoms. Hence, the positions of individual atoms in the cell, the interatomic distances, and the molecular architecture can be obtained from X-ray measurements with better precision than from any other method. When the crystals have lesser symmetry and/or when much larger molecules are involved, the analysis becomes lengthy and requires many iterative steps but is still feasible. Protein crystallographers are able to describe the conformation of proteins, often including the orientation of individual side chains of the amino acids, of bound inhibitors, etc.

The diffraction patterns of polycrystalline materials, such as crystalline powders or crystalline polymers (spherulites), look quite different. Individual crystallites are oriented at random. For each reflecting plane there exist a large number of crystals inclined by the proper angle θ to the primary beam to give a reflection. Many crystal orientations have the proper inclination; the diffracted beams form a conical surface. Its intersection with the photographic film produces a ring (or its section). Different reflecting planes yield a system of concentric rings (Fig. 4.27). Individual rings play the role of single diffraction spots observed in single-crystal diffractometry. From the analysis of their positions and intensities, structural information can be obtained as before: space group and size of the unit cell, the positions of the atomic groups within the unit cell, etc. This information is frequently sufficient for a full description of the crystal structure of a crystalline polymer.

X-ray diffraction is also invaluable for studying molecular orientation, such as that occurring in fibers. In a typical fiber made from a semicrystalline polymer, the individual crystallites and crystalline fibrils are oriented predominantly along the fiber axis. Of course, their lateral packing is random; a bundle of parallel, but laterally disoriented, crystallites is equivalent from the viewpoint of diffraction to a rotating single crystal. The resulting diffractogram has features that are between those of single crystals and crystalline powders. The diffraction rings no longer have uniform intensity; with increasing degree of orientation, they are broken more and more into individual diffraction spots arranged in rows (Fig. 4.28). From the distance between the rows of diffraction spots it is possible to evaluate the repeat distance of the polymer chain along the fiber axis. The repeat distance is a direct indication of the conformation of the chain. For polymers with zigzag conformation, it corresponds to single monomeric units; it is much longer for helical chains.

Another piece of information is contained in the width of the diffraction rings in the powder diagrams and in the size of the diffraction spots in the fiber diagrams. At the beginning of this section, it was claimed that a diffraction can be observed only when the Bragg relation (4.2.18) is satisfied exactly. This implied that diffraction spots are very sharp, their finite size being due to instrumental factors only. Such sharpness is predicted for crystals having infinite size. When the crystals are very

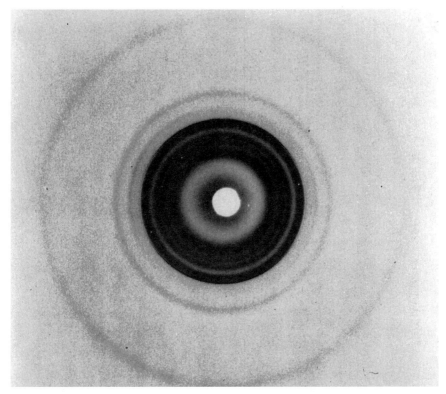

Figure 4.27. Powder X-ray diffraction pattern of isotactic polypropylene. (Reprinted from R.J. Samuels, *Structured Polymer Properties: The Identification, Interpretation, and Application of Crystalline Polymer Structure.* Copyright © 1974 by John Wiley & Sons, New York. Used by permission of the publisher.)

small, diffracted X-rays can be observed also at angles that differ from the angles prescribed by relation (4.2.18) by a small amount $\delta\theta$, such that the difference Δ of beams scattered by the most distant scattering elements of the crystal is still of the order of the wavelength λ. The destructive interference of the scattered radiation is then incomplete, and the diffraction spots get larger (or in powder diffractograms the width of the rings increases). From the size of the spots (after a correction for the instrumental factors), it is possible to evaluate the size of the individual crystallites in the polycrystalline material.

X-ray diffraction patterns can be produced also by monoaxially and biaxially oriented films. Their analysis yields results similar to those of fiber diagrams.

4.2.5. Neutron Scattering

Nuclear reactors can be operated in a mode that produces large fluxes of neutrons. These neutrons are valuable probes for investigating solid materials including

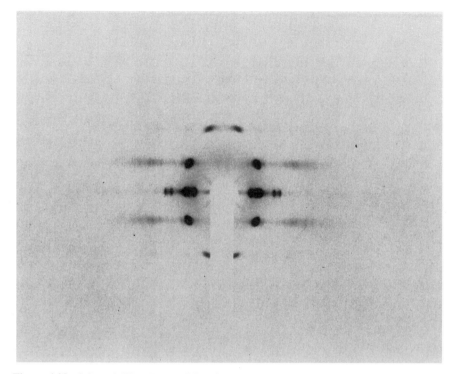

Figure 4.28. Oriented (fiber) X-ray diffraction pattern of isotactic polypropylene. (Reprinted from R.J. Samuels, *Structured Polymer Properties: The Identification, Interpretation, and Application of Crystalline Polymer Structure.* Copyright © 1974 by John Wiley & Sons, New York. Used by permission of the publisher.)

polymers. When a neutron approaches an atomic nucleus, it is repelled by the nuclear forces; the neutron beam is scattered. Modern physics has shown that the particle-wave duality is not restricted to photons; it applies to all particles. Hence, the scattered neutrons interfere as any other wave and can be used for structural determination in the same way as light or X-rays. The wavelength λ of any particle is related to its momentum mv by the de Broglie relationship

$$\lambda = h/mv \tag{4.2.19}$$

where m and v are the mass and velocity, respectively, of the particle; h is the Planck constant. It is possible to obtain a beam of neutrons having essentially uniform velocity; this means that they are monochromatic. Their wavelength is usually adjusted to 2–4 Å. Consequently, they can explore the same size structures as typical X-rays in the SAXS method. However, the two methods are not competitive; they are complementary.

The main similarity between scattering of neutrons and X-ray is the interference among the radiation scattered by individual atoms, which produces similar diffraction patterns. The main dissimilarity is in the nature of the scattering process itself. X-ray are scattered by electrons. Hence, their scattering provides information about the distribution of electron density. The electron cloud has the largest density around atomic nuclei. The higher the atomic number of the element, the higher the electron density and the more prominent will be its scattering. This provides for an easy identification of heavy atoms within the sample. On the other hand, light elements, especially hydrogen, are almost undetectable. Moreover, for hydrogen, the center of the electron cloud is not necessarily the position of the nucleus.

Neutrons are scattered by atomic nuclei; hence, they provide direct information about their position. The intensity of the scattering depends on the interaction of the neutron and its spin with the whole nuclear structure of the nucleus it is approaching. Different isotopes of the same element display very different scattering behavior. Isotopes with zero nuclear spin (^{16}O, ^{12}C) scatter neutrons coherently, whereas other isotopes produce a substantial portion of incoherent scattering that does not provide any information; it only adds to the background signal (no incoherent scattering is produced in light or X-ray diffraction). Neutrons scattered by protons are mostly incoherent; the effect is moderate for most other isotopes.

Intensity of the neutron scattering is usually expressed in terms of scattering cross section σ, and scattering length b. The scattering cross section is defined as the ratio of the total scattered neutron flux to the incident flux; it is a quantity very similar to the Rayleigh ratio of equation (3.5.6). It is useful to decompose the total scattering cross section into the coherent part σ_c and incoherent part σ_i

$$\sigma = \sigma_c + \sigma_i \qquad (4.2.20)$$

In light and X-ray scattering the scattered ray is always out of phase with the incident beam. This is also the case in neutron scattering for most isotopes but not for all of them; some of them scatter the neutrons in phase. The most prominent example of these isotopes is the proton. On the other hand, a deuteron scatters neutrons out of phase and much more coherently. To account for the phase behavior of scattered neutrons, the coherent scattering length b is introduced by the definition

$$\sigma_c = 4\pi b^2 \qquad (4.2.21)$$

It is assigned a positive value for isotopes scattering out of phase; it is negative for protons and other in-phase scatterers. The scattering length is related more closely to the amplitude of the scattered wave than to its intensity; its sign is very important when interference from larger structures is evaluated.

In macromolecular studies, the scattering properties of hydrogen isotopes play a major role. Their scattering lengths are comparable in magnitude to those of other elements (including the heaviest ones). Hence, in diffraction from crystals, the positions of hydrogen atoms can be deduced with the same precision as positions of other atoms, as we have already mentioned. For the studies of chain macromolecules, the major difference between scattering by protons and deuterons comes in very handy. A deuterated macromolecule embedded in a protonated matrix (or vice versa) provides a good scattering contrast that is utilized in a neutron equivalent of the SAXS method. Most of relations (3.5.72)–(3.5.80) are still applicable, allowing for the measurement of molecular characteristics of the particle. In a classic experiment a small amount of a narrow fraction of deuterated polystyrene was mixed with ordinary polystyrene and studied by neutron scattering. From the Guinier plot [cf. equation (3.5.72)], the radius of gyration of individual polymer chains in the bulk polymer was calculated. The experiment was repeated for several samples differing in molecular weight. The radius of gyration was found to be proportional to the square root of molecular weight. This result is expected for Gaussian unperturbed coils (see Sections 1.3.1 and 3.1.2.2). Hence, it is believed that in bulk polymers the molecular chains adopt the conformation of an unperturbed coil. This conclusion rests, of course, on an assumption (which may not be strictly correct) that the intermolecular interactions are not influenced by the replacement of protons by deuterons.

In another experiment, a deuterated polymer is dissolved in a different polymer, with which it forms a compatible blend. The scattering behavior of such a blend is similar to that of any other solution. It is thus possible to construct an appropriate Zimm plot (Section 3.5.6.1) from the scattering data and to evaluate from it the size of the coil and the virial coefficient. From these quantities, we may then proceed to the calculation of the polymer-polymer interaction coefficient. Thus neutron scattering is one of the very few methods giving experimental access to this elusive quantity.

4.3. TECHNIQUES FOR THE STUDY OF POLYMER SURFACES

Knowledge of the surface properties of polymeric materials is often very important. After all, polymers communicate with the outside world through their surfaces. Seeing is believing. That is the main attraction of the methods of direct visualization of macromolecular systems. Besides the classical techniques of microscopy and electron microscopy, surfaces can be visualized by more recent techniques: atomic force microscopy (AFM) and scanning tunneling microscopy (STM). Electron spectroscopy for chemical analysis (ESCA), also popularly known as X-ray photoelectron spectroscopy (XPS), is a powerful technique that maps the material surface together with the subsurface layers in terms of chemical composition.

4.3.1. Optical Microscopy

The informative power of optical microscopes is well recognized in biological sciences, metallurgy, and elsewhere. The main disadvantage of optical microscopy in macromolecular studies is its inadequate resolution. Two points can be adequately resolved only when they are separated by a distance of the order of the wavelength of the light, that is larger than about 2000–6000 Å. These distances are very large in the macromolecular world. Consequently, the use of the method is restricted to cases when the structures of interest are large enough. For example, spherulites are studied advantageously using a polarizing microscope, which reveals many details of the arrangement of individual crystalline fibrils. We have seen such a microphotograph in Figure 4.18.

Recently introduced technique of *near-field scanning optical microscopy* (NSOM) overcomes to some degree the inadequate resolution. If a strong enough light is passed through an aperture less than 100 Å in diameter placed close enough to the sample (within a distance significantly less than the aperture diameter) higher resolution could be obtained. The distance between the aperture and the sample is what is referred to as the near field. The use of this illumination technique allows one to construct optical images with resolution typically about 500 Å—well beyond the usual "diffraction limit."

4.3.2. Electron Microscopy

Electron microscopy utilizes the wave-particle duality of electrons. The wavelength of electrons is again determined by equation (4.2.19). The velocity of electrons emanating from a suitable cathode is easily manipulated by an accelerating electric field; wavelengths as short as 0.04 Å are routinely obtained. This provides an ample margin of resolution. Electron beams are used for diffraction experiments, and they provide the most precise information about the molecular architecture measured by any technique so far. (It is possible to measure interatomic distances with an accuracy of 0.001 Å.) However, electrons have only very weak penetrating power. Diffraction experiments of this kind are performed only on very dilute gases—a prospect not very appealing to a macromolecular scientist. For him or her, the electron waves are best utilized for direct visualization of objects according to principles governing ordinary microscopy, that is, in an electron microscope.

Scanning electron microscope (SEM) is useful mainly for studying surface features. In this version of the method, a narrow beam of electrons is focused on the object and scans its surface. The reflected (scattered) electrons provide a signal that modulates the beam of a cathode-ray tube. The two beams are synchronized as in a television camera and a TV set. Alternatively, the amplified signal is used for drawing the image on the photographic film. The main advantages of SEM are an impressive depth of field and a relatively simple sample preparation. The sample is placed onto a conductive support covered by conductive glue; it is sprayed with some heavy metal atoms (gold, platinum, palladium, tungsten) from a hot wire to increase the contrast of the surface features; and it is ready for observation. An example of a SEM photograph is shown in Figure 4.29. The disadvantage of SEM is its relatively low resolution (100 Å). Another problem (especially at higher magnifications) is a frequently observed distortion of the polymer sample under the intense beam of electrons.

Transmission electron microscope (TEM) has a much higher resolution of 2–4 Å. However, the sample preparation is much more demanding. The low penetrating power of electrons restricts the thickness of the samples to a few hundred Ångstrom units (this includes the thickness of the supporting membrane, if one is used).

Molecules and particles that are in a highly dilute solution or suspension may be placed or sprayed onto a support while the solvent vaporizes. The most satisfactory support is a thin film of carbon; evaporation of graphite from a carbon arc onto a glass slide prepares this. It is amorphous, tough, and quite transparent to electrons. The contrast of the deposited particles is enhanced by "shadowing": A thin layer of

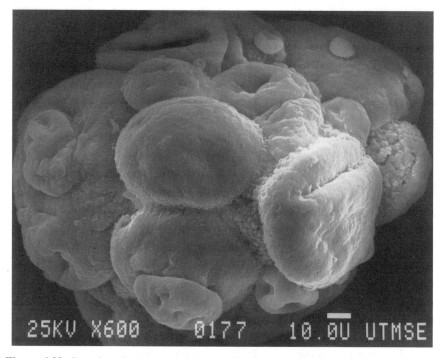

Figure 4.29. Scanning electron microphotograph of a particle of suspension polymerized poly(vinyl chloride). (M. Schmerling).

a heavy metal (platinum, for example) is deposited on the sample in such a way that the stream of metal atoms (from a hot wire) strikes the sample at an angle. The metal atoms pile up in front of the vertical features of the sample and leave a shadow behind them. This technique was found very useful for polymer monocrystals.

For studying surface features of polymer materials, surface replicas are first made by shadowing the material by a metal (chromium is used with advantage); they are then transferred (using sometimes quite elaborate procedures) to the grid in the microscope. The internal structure of polymeric materials can sometimes be studied by means of a similar technique. The sample is cooled to a very low temperature to make it brittle. It is then shattered, and replicas are taken of the tale-telling fractured surface.

Heterogeneous materials can be studied as thin sections. Specially designed micro-tomes can generate sections of a thickness of 200 Å or less. Such sections provide no contrast to the electrons; contrast must be developed by staining. In polymer studies, osmium tetroxide is a favorite staining agent. It selectively oxidizes some chemical groups, and the heavy osmium atoms are left at the site of oxidation. A classic case is the staining of styrene-butadiene block copolymers. Under properly selected con-ditions, the double bonds of butadiene are extensively oxidized while the monomeric units of styrene do not react to any appreciable degree. The electron micrographs presented in Figure 4.20 were produced using this technique.

Everyone who has seen some well-made electron micrographs understands the value of this technique. However, it also can lead to misinterpretations. The long manipulative sequence needed for bringing the structure of interest onto the grid of the microscope may lead to distortion of shapes. Biological structures may shrink on drying. Coiled macromolecules may collapse into spheres. Flow of the solvent on the supporting membrane (just before it evaporates) may orient sections of the molecules that were not oriented before. Equally dangerous is the subjective factor. An inexperienced operator may select for photographing the most prominent features observed, which may not represent the typical phenomenon at all. It is therefore recommended that several electron microscopic techniques be used simultaneously for each problem and that the results be checked against other independent methods whenever possible.

4.3.3. Atomic Force Microscopy (AFM) and Scanning Tunneling Microscopy (STM)

STM and AFM the recent surface imaging techniques with spatial resolution of 1 Å and 2–5 Å, respectively. These scanning probe microscopic methods form a class of extremely surface-sensitive tools known mostly for their ability to deliver atomic level resolution in real space. These methods use piezoelectric drivers in the tip scanners for compact constructions and low-voltage operations. AFM is usually done at ambient conditions mostly in air, and, more importantly, it does not require electrical surface conductivity. These two features allow a wide variety of inorganic, and polymeric materials to be studied with relatively simple and inexpensive AFM instruments. On the other hand, STM is applied to electrically conductive materials, such as metallic or highly doped semiconductor surfaces, or to over-layers, such as metallic or conductive inorganic surfaces. STM experiments are performed in an ultra-high vacuum environment with typical pressures less than 1×10^{-10} mbar.

AFM measures the interaction force between the sample and a probe tip. This force is essentially the force between the atom/atoms at the probe tip and atoms at the sample surface. It has an attractive van der Waals component typical for molecules in contact (recall the cohesive energy density) and a repulsive component that does not allow the molecules to overlap.

The AFM imaging is usually done at the atmospheric pressure. The probe tip is an insulator that is scanned in the x–y plane in contact with the specimen surface. The spatial scanning over the sample surface is performed by displacing the tip piezoelectrically in two planar (x and y) directions. Sample advancement and coarse positioning are done using a slit-stick (sawtooth) operation of the piezoelectric translation stages in two spatial directions. The piezoelectric transducers with a displacement of 5–50 Å/V allow an ultimate spatial resolution of 0.05 Å at 1-mV driving accuracy. A feedback mechanism keeps the tip at distance that leads to a constant interacting force. The tip deflections together with the x and y positions are stored in an image file on a workstation. Graphical routines are used to display the appropriate subsets of the data and for image processing. This arrangement provides information about the microstructure of a specimen at the atomic level.

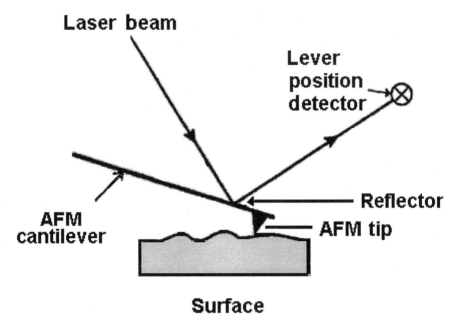

Figure 4.30. Schematic diagram of the basic AFM operation. (A sharp tip on a cantilever is moved in the proximity of the surface and the lever displacement is sensed through an optical readout system, mainly quadrant photo-detectors or interferometer designs.)

The basic concept of AFM is shown schematically in Figure 4.30. Its implementation in ambient-air conditions has proven remarkably efficient. A cantilever with an attached sharp tip structure is contacted to the surface with a small loading force of typically 10^{-10} to 10^{-7} newtons. Readout of the deflection signal is mostly done optically by bouncing an incident laser beam onto the cantilever toward a quadrant detector or into an interferometer.

AFM works in two modes of operation: (i) in contact with the surface employing the repulsive component of the interaction force and (ii) in a noncontact mode based on the attractive component. The contact mode essentially operates as a low-load, high-resolution surface profiler and extends the use of the conventional surface profilers into the nanometer range. Typical application of this mode includes the determination of surface flatness or roughness on coated or oxidized materials, surface profiling of patterns, and the assembly of polymer or Langmuir-Blodgett (LR) films. AFM is a well-established technique in the flatness analysis and in imaging small magnetic and replica structures as well as polymer surfaces.

The noncontact mode is more difficult to achieve experimentally; only a few truly atom-resolved results on semiconductor surfaces are available. This high-resolution mode has been studied under ultra-high vacuum conditions similar to those used in STM.

AFM is very handy for studying the surface structure and local properties of polymeric materials including biopolymers. It provides topographic information down to the angstrom level. Because it works at ambient and physiological conditions, it is particularly well suited for probing biological samples. When a special probe is employed, properties such as thermal and electrical conductivity, magnetic and electric field strength, and sample compliance can also be obtained. Many AFM applications require little or no sample preparation.

Scanning tunneling microscopy (STM) exploits the electron tunneling phenomenon. When two conductive bodies are in close vicinity, electrons may tunnel from one to the other. If there is a potential difference between the bodies, the tunneling in one direction will be enhanced. The magnitude of the tunneling current is a steep function of the distance between the bodies.

STM images are generated by the tunneling current between a biased metal tip and the substrate while rastering a cantilever with a tip in the x–y plane in a way that we already described for AFM. This lever can be made of micromachined silicon with the tip selectively grown and machined on the lever or the tip can be mounted on the lever, allowing a choice of tip materials, shapes, and lever responsiveness. The feedback mechanism keeps the tip at the distance corresponding to a constant tunneling current. STM uses a small tunneling current ($0.1-1$ nA) and a driving voltage typically in the range of 5 mV to 3 V.

The spatial resolution of STM is remarkable. Individual atoms may often be imaged, with best results obtained using ultra-high vacuum. With appropriate interactive regulation of the driving voltage and positioning of the tip it is even possible to move individual atoms on the surface. Some authors claim that by manipulating molecules they can even perform chemical reactions—one molecule at a time.

4.3.4. Electron Spectroscopy for Chemical Analysis (ESCA)

Electron spectroscopy for chemical analysis (ESCA) was largely developed in the 1970s and yields information about the composition and bonding within the first few atomic layers of the surface of almost any solid without damaging it, a capability unique among measurement techniques.

In an ESCA experiment, the sample surface is irradiated by a source of X-rays under ultra-high vacuum conditions. The photoionization takes place in the sample surface, and the resulting photoelectrons have a kinetic energy, E_k, given by the Einstein relation

$$E_k = h\upsilon - E_b - \phi \tag{4.3.1}$$

where $h\upsilon$ is the energy of incident beam of photons, E_b is the binding energy of the photoelectron and ϕ is the work function of the spectrometer. A schematic diagram of the physical ESCA process is given in Figure 4.31.

In ESCA highly monochromatic X-rays irradiate the sample and eject electrons, usually from the core levels of constituent atoms. ESCA measures the kinetic energy of the ejected electrons and produces a spectrum—a plot of number of emitted electrons vs. their kinetic energy. Because the energy of the X-ray photon is known, subtraction of the electron kinetic energy from it yields the electron binding energy. In commercial

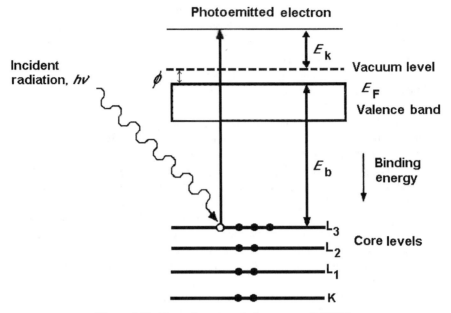

Figure 4.31. Photoelectron emission process in ESCA.

electron spectrometers, this subtraction is carried out electronically and the spectra of binding energy are produced directly.

The binding energy of core electrons is a characteristic of individual atoms present, and the ejected core electrons have different kinetic energies depending on their binding energies. Because the atomic structure of each element is unique, the elements of an unknown sample can be identified easily by recording the ESCA spectrum over a wide range (generally from 0 to 1000 eV). The binding energies of the intense lines present in the spectrum are then compared with the tabulated values for all elements.

The chemical environment of the atom and, specifically, the bonds by which it is bonded to its neighbors modify the binding energies of its core electrons. As a result, a shift in their positions in the spectrum is observed. This shift is called *chemical shift* and ranges from 0.1 to 10 eV. The shift arises from the variation of electrostatic screening experienced by core electrons as valence electrons are drawn toward or away from the atom of interest. The information about chemical states from the variations in binding energies of the photoelectron lines gives ESCA a major advantage over other techniques.

ESCA probes only the surface layers of the sample because the information-carrying photoelectrons must leave the sample unmolested. If they are scattered and lose energy in the solid before they escape into the spectrometer, they appear in the lower kinetic energy part of the spectrum and are useless in terms of chemical information. Electrons cannot travel great distances in solids without undergoing scattering and loss of energy. Only those that originate near the surface leave the sample with their full complement of energy and appear as a part of the main peak in the ESCA spectrum. The escape depth varies as a function of electron kinetic energy and of the type of material being observed. It is normally in the range 5–15 Å for metals, 15–25 Å for inorganic compounds, and 40–100 Å for organic and polymeric materials. It is the combination of high surface sensitivity and the ability to provide chemical information about species observed at surfaces that gives ESCA a unique position among surface analysis techniques.

In the photoelectron process, in addition to the emission of photoelectrons, Auger electrons are emitted as a result of relaxation of the energetic ions remaining after the photoemission. In the Auger process, an outer electron falls into the inner orbital vacancy and a second electron is emitted that carries the excess energy. The kinetic energy of these Auger electrons is the difference between the energy states of the singly charged initial ion and the doubly charged final ion and is independent of the mode of the initial ionization. These Auger electrons also constitute peaks in the ESCA spectrum.

The ESCA instrumentation operates at an ultra-high vacuum level adequate for electrons to leave the sample surface and to travel through the spectrometer to the detector without collision with gas molecules en route. This requires 10^{-5} to 10^{-6} torr. Typically, commercial instrumentation operates with the base pressures of 10^{-7} to 10^{-9} torr. Dry pumps are used in order not to contaminate the sample surface. The X-rays are normally produced using magnesium or aluminum X-ray anodes that emit an intense K_α emission. Its narrow line width allows high resolution to be readily achieved.

Precise measurement of binding energies of the core-level electrons led to the determination of chemical shifts. Chemical shift effects in organic polymers reveal information about the surface chemistry. Substitution by electronegative elements induces a shift to higher binding energy. This is illustrated by the spectrum shown in Figure 4.32 for poly(ethylene terephthalate), which is widely used in fibers and in packaging films. Here, carbon atoms that are bonded to oxygen appear at higher binding energies than those that are not. The presence of two oxygen atoms in the carboxyl group produces a larger chemical shift than does one oxygen atom bonded to ethylene glycol carbon.

ESCA is a very handy technique in surface analysis of polymers. The technique is useful for examination of polymers and other relatively fragile materials without causing any changes in surface composition. It is able to detect submonolayer surface concentrations. A typical polymeric material can be exposed to X-rays for much longer periods than required for an average ESCA experiment while no changes in its spectrum are observed.

ESCA is also used to determine the aging rate and chemical modification of a paint surface. Other aging characteristics of polymeric surfaces observable by ESCA include the appearance of titanium at the surface of TiO_2-pigmented paint formulations. If the TiO_2 pigment were exposed after brief weathering, one would expect that the paint would soon become chalky. ESCA is useful for studying metal surfaces to which paint will be bound to determine whether impurities are present that could cause the paint to chip or peel off. Of course, such tests require controlled

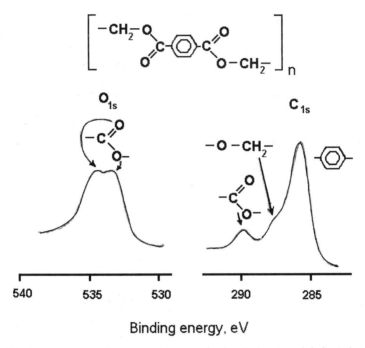

Figure 4.32. The 1s spectra of carbon and oxygen of poly(ethylene terephthalate) showing the chemical shift effect.

conditions and correlation of the ESCA results with those obtained previously by the more conventional techniques. Once these correlations are established, ESCA can give a quantitative, rapid, and convenient measure of surface modification.

Sometimes it is important to know not only what is present at the surface of a sample but also how thick the surface layer is or how the chemical composition varies as a function of depth. This added dimension of ESCA is provided with an accessory device known as the *ion gun*. This ion gun is used to direct a beam of ions, usually argon, to the sample surface. When the ions strike the surface they literally etch away the material. Thus, by sequentially observing the surface with ESCA and etching the surface with the ion gun, it is possible to obtain the composition of the sample as a function of depth. This capability is useful for studying corrosion. ESCA with ion etching can also be used for the solution of problems related to the formation of very thin polymeric protecting coatings on metals.

4.A. SUGGESTIONS FOR FURTHER READING

Alder, H.J., and K. Lunkewitz, Eds., *Polymer Sorption Phenomena,* Wiley, New York, 2000.

Alexander, L.E., *X-Ray Diffraction Methods in Polymer Science,* Krieger, Huntington, NY, 1979.

Bacon, G.E., *Neutron Diffraction,* Clarendon, Oxford, 1955.

Bailey, R.T., A.M. North, and R.A. Pethrick, *Molecular Motion in High Polymers,* Clarendon, Oxford, 1981.

Basset, D.C., *Principles of Polymer Morphology,* Cambridge University Press, Cambridge, 1981.

Bird, R.B., R.C. Armstrong, and O. Hassager, *Dynamics of Polymeric Liquids,* 2 vols., Wiley, New York, 1987.

Blumstein, A., *Liquid Crystalline Order in Polymers,* Academic, New York, 1978.

Blundell, T.L., and L.N. Johnson, *Protein Crystallography,* Academic, New York, 1976.

Bonnell, D.A., Ed., *Scanning Tunneling Microscopy: Theory, Techniques and Applications,* VCH, New York, 1993.

Briggs, D., *Surface Analysis of Polymers by XPS and Static SIMS,* Cambridge University Press, 1998.

Chapman. J.R., *Practical Organic Mass Spectroscopy,* 2nd ed., Wiley, New York, 1993.

Chen, C.J., *Introduction to Scanning Tunneling Microscopy,* Oxford University Press, 1993.

Cifferi, A., W.R. Krigbaum, and R.B. Meyer, Eds., *Polymer Liquid Crystals,* Academic, New York, 1982.

Crank, J., and G.S. Park, *Diffusion in Polymers,* Academic, London, 1968.

Cotter, R.J., Ed., *Time of Flight Mass Spectrometry,* ACS Symposium Series 549, American Chemical Society, Washington, DC, 1994.

Czanderna A.W., in S.P. Wolsky and A.W. Czanderna, Eds., *Methods of Surface Analysis,* Elsevier, Amsterdam, 1988.

DiNardo, N.J., *Nanoscale Characterization of Surfaces and Interfaces,* VCH, New York, 1994.

Donth, E.-J., *Relaxation and Thermodynamics in Polymers: Glass Transition,* VCH, New York, 1992.

Eirich, F.R., Ed., *Rheology,* 5 vols., Academic, New York, 1956–1967.

Ferry, J.D., *Viscoelastic Properties of Polymers,* 3rd ed., Wiley, New York, 1980.

Geil, P.H., *Polymer Single Crystals,* Wiley, New York, 1963.

Horton, O., and M. Amrein, Eds., *STM and SFM in Biology,* Academic, San Diego, CA, 1993.

Kakudo, M., and N. Kasai, *X-Ray Diffraction by Polymers,* Elsevier, Amsterdam, 1972.

Laub, R.J., and R.L. Pecsok, *Physicochemical Applications of Gas Chromatography,* Wiley, New York, 1978.

Mark, J.E., and B. Erman, *Rubberlike Elasticity, a Molecular Primer,* Wiley, New York, 1988.

Paul, D.R., and S. Newman, Eds., *Polymer Blends,* 2 vols., Academic, New York, 1978.

Ratner, B.D., and V.V. Tsukruk, Eds., *Scanning Probe Microscopy of Polymers,* ACS Symposium Series, No. 694, American Chemical Society, Washington, DC, 1998.

Rudin, A., *The Elements of Polymer Science and Engineering: An Introductory and Reference for Engineers and Chemists,* Academic, New York, 1998.

Samuels, R.J., *Structured Polymer Properties: The Identification, Interpretation, and Application of Crystalline Polymer Structure,* Wiley, New York, 1974.

Sawyer, L.C., and D.T. Grubb, *Polymer Microscopy,* Chapman and Hall, New York, 1987.

Sperling, L.H., *Interpenetrating Polymer Networks and Related Materials,* Plenum, New York, 1980.

Sperling, L.H., *Introduction to Physical Polymer Science,* Wiley, New York, 1986.

Shenoy, A.V., *Rheology of Filled Polymer Systems,* Kluwer Academic, 1999.

Settle F.A, Ed., *Handbook of Instrumental Techniques for Analytical Chemistry,* Prentice Hall, New Jersey, 1997.

Standing, K.G., and W. Ens, Eds., *Methods and Mechanisms for Producing Ions from Large Molecules,* Plenum, New York, 1994.

Stuart, H.A., *Die Physik der Hochpolymeren,* 4 vols., Springer, Berlin, 1956.

Stout, G.H., and L.H. Jensen, *X-Ray Structure Determination: A Practical Guide,* Macmillan, New York, 1968.

Treloar, L.R.G., *The Physics of Rubber Elasticity,* Oxford University Press, Oxford, 1958.

Vergnaud, J.M., *Liquid Transport Processes in Polymeric Materials: Modeling and Industrial Applications,* Prentice Hall, Englewood Cliffs, New Jersey, 1991.

Vinogradov, G.V., and A.Y. Malkin, *Rheology of Polymers,* Springer, Berlin, 1980.

Ward, I.M., and D.W. Hadley, *An Introduction to the Mechanical Properties of Solid Polymers,* Wiley, New York, 1999.

Wiesendanger, R., *Scanning Probe Microscopy and Spectroscopy, Methods and Applications,* University Press, New York, 1994.

4.B. STUDY QUESTIONS

1. What is the typical morphology of crystalline polymers?

2. Describe the crystalline structure of crystalline polymers. Mention the conditions necessary for a polymer to crystallize.

3. Which of the polymers has the higher T_g: isotactic polypropylene or atactic polypropylene?

4. Which is more likely to produce crystallites: HDPE or poly(butyl methacrylate)?

5. Which has the higher tendency to crystallize when stretched: unvulcanized rubber or ebonite? Which of these has the longer relaxation time (τ)?

6. Which is more likely to exhibit side chain crystallization: poly(methyl methacrylate) or poly(dodecyl methacrylate)?

7. What technique would you use to determine the crystallinity in a polymer?

8. Which will have the greater difference between T_m and T_g values: HDPE or LDPE?

9. What information can be obtained from an ESCA experiment?

10. Compare and contrast different types of microscopic techniques for the study of polymer surfaces.

11. Which of the following is viscoelastic: (a) steel, (b) polystyrene, (c) diamond, or (d) neoprene?

12. Would elastic modulus increase or decrease when (a) a plasticizer is added to the rigid PVC and (b) the amount of sulfur used in the vulcanization of rubber is increased?

13. In which model of a viscoelastic body (Maxwell or Voigt–Kelvin) is the elastic response of a polymer retarded by the viscous resistance?

14. What changes occur in a polymer under stress before and after the yield point?

15. How would you estimate the relative toughness of a polymer from the stress-strain curves?

16. Why are the specific heats of polymers higher than those of metals?

17. What thermal instrumental technique can be used to determine T_g?

18. What structural parameters influence the melting point, T_m, and the glass transition, T_g, of a polymer?

19. Mention the two methods used to measure the degree of crystallinity in a semicrystalline polymer. Explain how the percent crystallinity of an unknown polymer can be measured by DTA.

20. Explain how X-ray scattering results are useful in determining the size of the crystallites.

21. Discuss the phenomenon of chain folding in polymer crystals, giving the evidence in support of chain folding and the conditions under which it occurs.

22. Many crystalline polymers form spherulite structures when cooled from a melt or solution. Describe the growth and structure of such spherulites in a crystalline polymer. How are the molecules arranged in spherulites?

23. What is the structure of the crystallites of HDPE obtained by crystallization from a dilute solution and from a polymer melt?

24. Explain how the polarizing microscope is used to determine the melting point of a polymer.

25. Describe the differences in the behavior between a Hookean material and a linear viscoelastic material on phenomenological as well as a molecular basis.

26. What is Poisson's ratio? How does its value describe the mechanical properties of a polymer?

27. In many stress-strain curves for polymers, the region between the yield point and the break point is a plateau. Explain the significance of this region. What is the effect of temperature on the size of this plateau?

28. Describe the differences in stress-strain curves of a rubber, an amorphous polymer below its T_g, and a crystalline polymer above its T_g.

29. Describe a simple mechanical model that is used to understand the creep behavior of polymers. Derive the stress-deformation equations for this model. Sketch the stress-deformation curve and explain how this curve relates to the mechanical properties.

30. Discuss the effect of (a) molecular weight, (b) cross-linking, and (c) number of entanglements on a creep experiment:

31. Explain why the Boltzmann superposition principle is not applicable to crystalline polymers.

32. Describe an experimental setup of a stress relaxation experiment. Explain how

these data are useful in relating the mechanical properties to the molecular structure of a polymer.

33. The Williams–Landel–Ferry (WLF) equation relates viscosity of the polymer to the temperature. Explain the arguments underlying this equation. Why is it used to illustrate the principle of corresponding temperatures and how it can be used for constructing the standard flow curves?

34. Suppose that natural rubber is extended under a constant load for few minutes, and then the load is removed. Plot the elongation vs. time curves for experiments that are carried out at $-60°$, $20°$, and $120°C$. Explain the differences between these curves.

35. A plot of loss modulus vs. temperature for HDPE gives peaks at $140°$, $-40°$, and $-120°C$. Crystalline polypropylene gives peaks at $180°$, $130°$, $-20°$, and $-150°C$. Explain the meaning of each peak and the difference between the two sets of peaks.

36. Discuss the applications of piezoelectric materials.

37. One of the applications of polymers is as electrical insulation materials. Explain how polymers are used in this area.

38. Define mathematically the complex dielectric constant, the dielectric constant, and the dissipation factor.

39. Describe the effects of plasticizers and of rising temperature on the dielectric properties of a polymer.

40. Describe how you could obtain the length of the repeat period from the X-ray diffraction data.

41. How may X-ray scattering data be used to determine the size of the crystallite?

42. Draw a typical diagram obtained in a DSC experiment on a polymer. Explain what thermal and structural properties can be obtained from this diagram.

43. What is measured in DTA? What information can be obtained about a polymer?

44. Discuss the swelling of gels and describe how one can obtain the molecular weight between cross-links of polymeric gels using the swelling data.

45. Explain the merits and demerits of inverse gas chromatography in understanding the bulk properties of polymers.

46. Discuss and cite the evidence for the usefulness of the fringed-micelle theory of crystalline polymers.

4.C. NUMERICAL PROBLEMS

1. A natural rubber has a shear modulus of 10^7 dyn/cm^2 and a Poisson's ratio of 0.50 at room temperature. If a load of 5 kg is applied to a strip of this material which is 10 cm long, 0.5 cm wide, and 0.25 cm thick, calculate the % elongation of this material.

2. Calculate the tensile strength of a square sample (side 1.25 cm) of poly(methyl methacrylate) with a thickness of 0.35 cm, if failure occurs at 282 kg.

3. Calculate the compressive strength of a 10-cm-long plastic rod with a cross section of 1.30 cm × 1.30 cm, which fails under a load of 3500 kg.

4. If polypropylene with a length of 5 cm elongates to 12 cm, calculate its % elongation.

5. If tensile strength of a polystyrene is 705 kg . cm^{-2} and its elongation at break is 3%, calculate the value of modulus of elasticity.

6. At 200°C the Newtonian viscosities η_N (in poise, P) in the limit of zero shear rate of polystyrene samples of different molecular weights were measured and reported in the following table.

$\overline{M}_w \times 10^{-3}$(gmol^{-1})	η_N (P)
86	3.50×10^3
162	4.00×10^4
196	6.25×10^4
360	4.81×10^5
490	1.89×10^6
508	1.00×10^6
510	1.64×10^6
560	3.33×10^6
710	6.58×10^6

Using these data calculate the exponent value of molecular weight in its relationship with η_N.

7. Stress relaxation experiments were performed at 25°C and 45°C, respectively, on the samples of nylon yarn at a constant strain of 2%. The data for Young's modulus, E, are given below as a function of time.
Data at 25°C.

Log t (s)	0.6	1.0	1.3	1.6	2.0	2.3	2.7	3.0	3.3	3.8
$E \times 10^{-10}$ (dyn cm^{-2})	4.00	3.85	3.73	3.60	3.47	3.37	3.25	3.17	3.08	2.98

Data at 45°C

Log t (s)	0.6	1.0	1.5	1.75	2.15	2.65	3.15	3.40
$E \times 10^{-10}$ (dyn cm^{-2})	3.35	3.22	3.08	3.00	2.92	2.83	2.79	2.75

Plot E vs. log t (master curve) for the above set of data on the same graph. Find whether the time-temperature superposition is applicable for this experiment. If so, what is the shift factor?

8. In a dynamic experiment, $\gamma(t) = \gamma_0 \sin(\omega t)$, the power loss per cycle of oscillation is given by

$$\int_{\omega t=0}^{\omega t=v\pi} \sigma \, d\gamma$$

where v is the shear strain. Evaluate (a) power loss per cycle if the material is a Hookean solid: $\sigma = G\gamma$ and (b) power loss per cycle if the material is a Newtonian liquid: $\sigma = \eta(d\gamma/dt)$. Comment on the significance of these results.

9. The dynamic compliance data for poly(methyl acrylate) at several temperatures are measured. The curves measured at various temperatures were shifted to construct a master curve at 25°C. The following shift factors were obtained:

$T(°C)$	25.0	29.8	34.9	39.7	44.9	49.9	54.9	59.9	64.7	69.5	80.4	89.2
Log a_T	0	−0.98	−1.8	−2.4	−3.0	−3.5	−3.9	−4.3	−4.6	−4.9	−5.4	−5.7

Test whether these data obey WLF equation; if so, evaluate the constants of the WLF equation. Note that $T_o \neq T_g = 3°C$ in these data.

10. How much work is done on a rubber band that is slowly and reversibly stretched to $\alpha = 2.5$? The initial slope of the stress-strain curve is 2.5 MPa, and the volume of the rubber band is 4.0 cm^3.

11. Calculate the change in shear stress with time for a Maxwell model in a stress relaxation experiment in which Young's modulus at zero time, $E = 10^6$ dyn/cm^2, shear modulus $= 2.1 \times 10^5$ dyn/cm^2 and viscosity $\eta = 5 \times 10^8$ P.

5

TECHNOLOGY OF POLYMERIC MATERIALS

Innumerable useful articles are manufactured from polymeric materials. The manufacturing procedures are determined by the mechanical and thermal properties of the polymers. We have seen in Section 4.1 that these properties are quite different from the properties of other structural materials such as metals, glass, and ceramics. Hence, special manufacturing techniques have been designed for the production of various objects made from polymers. These techniques will be presented in Section 5.1.

In the production of polymeric materials and shaped objects, it is important to know not only the fundamental physicochemical properties of the polymers but also the more utilitarian properties that have direct bearing on the manufacturing process, on the behavior of the final products, and on their appearance. Polymer engineers have therefore designed a number of testing methods that are fast and relatively easy to perform to assess the range of application of any new polymer material. Section 5.2 is devoted to such tests.

High-performance polymers that exhibit good barrier properties are useful in a great variety of industrial separation processes. These properties depend on chain stiffness, order, crystallinity, orientation, presence of fillers, etc. The selective transport of liquids or gases through polymeric materials has important applications in the purification of air and water as well as in chemical and biotechnical industries. This will be discussed in Section 5.3.

5.1. FABRICATION OF POLYMERS

The material from which plastic objects are made is rarely a pure polymer. Usually, it contains a lot of additives: plasticizers for the control of T_g, antioxidants and UV

557

absorbers for extension of lifetime, reinforcing materials such as fibers or carbon black to give strength, fillers such as sawdust or kaolin for economy, vulcanizing agents for rubbers. Preparation of these mixtures is called *compounding;* we will treat it in the next section. The following sections will be devoted to the shaping of the raw plastics into the desired forms. The shaping techniques depend on the complexity and size of the shaped article and on the mechanical properties of the polymer to be shaped. Obviously, highly viscous as well as viscoelastic materials are more difficult to shape than moderately viscous liquids or powders; thus the latter are preferred whenever feasible.

5.1.1. Compounding and Mixing

Raw polymers are manufactured and delivered in several different forms. Powders are very popular. They are often the direct products of suspension or emulsion polymerizations and are quite porous. Other polymers are delivered as pellets or flakes. Rubberlike materials come in the form of bales. In some fabrication processes, notably in the production of thermosets, prepolymers are converted to the final polymers during fabrication itself. These prepolymers are marketed as powders (e.g., phenolformaldehyde resins) or as viscous liquids (e.g., epoxy resins).

Compounding of the raw material is essentially a process of mixing. Mixing of low-viscosity liquids or dry powders is easily achieved using stirrers, blenders, and similar equipment. Plasticization of powders is often performed in the dry state; the plasticizer enters the pores of the powder without causing any adhesion of the particles. Usually, at ambient temperature, the viscous plasticizer does not swell the polymer and is held in the pores by capillary forces; swelling and homogenization occur at elevated temperatures during the final fabrication process. Alternatively, when the powder is less porous, it may be dispersed in the plasticizer to form a paste. Again, no swelling occurs at ambient temperature.

Compounding of viscoelastic materials or glassy pellets (they must be melted first) is a more complicated process often requiring heavy machinery. The mixing is achieved by employing high local shearing rates. The basic instrument of this type is a *two-roll mill* consisting of two cylinders rotating in opposite directions with different speeds. The mass to be mixed is sheared between the rolls. The stresses involved are moderate; two-roll mills are used when more sensitive polymers require gentle handling. The *Banbury mixer* is much more robust. Its two rolls are equipped with short spiral blades that intensify the shearing in a rather complex flow pattern. Polymers are fed into the mixers under pressure; sometimes they are first heated if necessary for melting. However, in the Banbury mixer, the violent shearing and kneading often raise the temperature appreciably and the heat must be removed by cooling.

Instruments of this type are used also for treating some raw polymers, such as natural rubber. The process, called *mastication,* has a profound effect on the polymer, reducing its molecular weight by shearing and evoking a number of chemical reactions because of increased temperature. Consequently, the time and temperature of the masticating and/or mixing process should be controlled closely to avoid unwanted changes in the polymer.

5.1.2. Casting

Casting is conceptually the simplest technique for shaping polymer articles. It is applicable only when the starting material is a liquid with viscosity low enough for pouring. Two types of liquids qualify: polymer solutions and monomers (or prepolymers). Polymer solutions are used for casting films. The solution (typically, concentrations of 20% give adequate viscosity) is poured onto a flat surface that has been treated with a releasing (nonsticking) agent. After the solvent evaporates, the film is peeled off the support. The simplest casting surface for laboratory work is a sheet of plate glass. Other surfaces are also used, e.g., a Teflon sheet, which has a lower tendency to stick to polymer films, or a chromium-plated heated casting bench.

In the industrial version of the technique, the polymer solution is delivered through a slit die onto the surface of a slowly rotating metal drum. The solvent vaporizes during less than one rotation of the drum, and the dry film is continuously wound up on another receiving drum. Sheets of poly(vinyl chloride) and poly(vinyl alcohol) are often manufactured by this technique.

Monomers and prepolymers might be cast into molds of rather complicated shape and left there until the polymerization process (*curing*) is finished. Typical material for casting is methyl methacrylate. It usually has some poly(methyl methacrylate) dissolved in it that serves to increase the initial viscosity for easier handling. Epoxy resins are also frequently cast in a similar way (their prepolymers already have the right viscosity). Polystyrene, polyurethanes, and other polymers are also used for casting.

The actual polymerization during casting is usually achieved by heating in an oven. The heating should be carefully programmed to avoid runaway reaction in the beginning that may lead to evolution of gases and bubbles (note that N_2-producing azobisbutyronitrile is not an acceptable initiator for this process), to assure complete curing by raising the temperature above T_g toward the end of curing, and finally to cool the product slowly to minimize the internal stresses.

The density of most polymers is greater than that of their monomers. Hence, the polymer shrinks during curing, and the molds must be designed to compensate for this shrinkage. For example, when window-type sheets of polymeric glasses are cast, the two sides of the mold are held together by a spring clip; the gasket spacer between the sides is made of a soft rubber that can accommodate the shrinking. The shrinkage problem is less severe when prepolymers of higher molecular weight are employed or when a sizable part of the material is already in the form of polymer (e.g., in solution) before the casting.

5.1.3. Extrusion

The extrusion process is designed to continuously convert a thermoplastic material into a particular form. It is an extremely versatile process (often called melt extrusion), in which the polymer melt is forced through a die; it solidifies after leaving the die, forming an infinitely long object having a profile that is determined by the shape of the die. The final shapes include pipes, films or sheets, fibers, profiles, coatings for

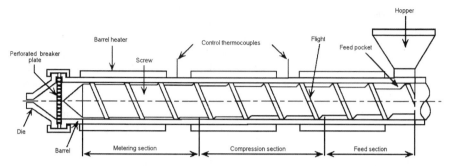

Figure 5.1. Schematics of a single-screw extruder. (Reprinted from N.G. McCrum, C.P. Buckley and C.B. Bucknall, *Principles of Polymer Engineering,* © Copyright, 1988 by Oxford University Press, Oxford. Used by permission of Oxford University Press.)

paper and other substrates, and coverings for wires and cables. Extruders are also used to feed the blow molding machines that produce bottles and other hollow articles.

The basic design of an extruder is shown in Figure 5.1. The most important part of the extruder is the screw (i.e., a cylinder with a rectangular ridge wrapped around it as a helix) that rotates inside a cylindrical barrel. The screw usually must be designed for every particular material. Screws are characterized by their length-to-diameter ratio and their compression ratio, namely the ratio of the volumes of one flight of the screw at the hopper end and at the die end. The screw is divided into three sections: feed, compression, and metering. The feed section conveys the material from under the hopper mouth to the compression section, where the gradually diminishing depth of thread causes volume compression of the melting granules and consequent removal of trapped air, which is forced back through the feed section. This ensures that the extrudate is free from porosity. The volume compression also leads to an increase in the shearing action on the melt, because of relative motion of the screw surfaces with respect to the barrel wall. This considerable shearing further homogenizes the material (and also possibly degrades it by mastication). Simultaneously, frictional heat is generated that leads to a more uniform temperature distribution in the molten extrudate. As the melt progresses toward the die, it is pressurized. This pressure then forces the melt through the die. The function of the final section of the screw is to meter the molten polymer through the die at a steady rate and to iron out pulsations. The screw is usually cored for steam heating or water-cooling. After the melt leaves the die, it must be solidified by cooling—air cooling for thin profiles, water cooling for thicker ones.

Different shapes of dies produce different final objects: circular die, a rod; annular die, a tube; slit die, a film. Many more intricate shapes can be produced by more sophisticated and/or multiple dies such as nets.

The cross section of the final article may be quite different from the shape of the die. One reason for that is *die swell*. In Section 3.4.4.4 we have seen that the stress tensor of a polymer solution flowing at a finite rate of shear also has normal components acting perpendicularly to the streamlines. Flowing polymer melts exhibit this phenomenon to an even higher degree. The flow of the melt in the die is essentially a shear flow. The normal component of the stress is acting against the wall of the die. Once the melt leaves the die, this stress component will cause the extruded material to expand in a radial direction—the die swell. This is a phenomenon very similar to the elastic recovery mentioned at the end of Section 4.1.5.1. The die swell is not too damaging when the profile of the extrudate has a circular symmetry. This symmetry is preserved; only the radii are changed. However, profiles with sharp edges may be severely distorted.

In many extrusion procedures, the profile of the extrudate is further modified after the material leaves the die. Specifically, the extruded films are often immediately stretched by being led through a pair of fast-turning rollers or over a fast-turning drum. Sometimes they are also stretched laterally. During this stretching, the thickness of the film decreases substantially (more than counterbalancing the die swell). More important, the polymer molecules are oriented with an appreciable increase in the film strength. A similar stretching procedure is also applied to fibers; this will be discussed in Section 5.1.7.

Nowadays, more plastic resins are processed by extrusion than by any other manufacturing technique. Modern extruders are sophisticated pieces of processing instruments that use microprocessor-based control systems to ensure consistency of the extruded product. Extruders range in size from tabletop experimental units to massive high-volume production machines that may be even longer than 50 feet.

5.1.4. Bubble Blown Film Extrusion

A typical setup for bubble blown film extrusion is shown in Figure 5.2. The molten polymer from the extruder head enters into a ring-shaped die from the side or through the bottom. The melt is then forced around a mandrel and emerges through the die opening in the form of a tube. While still in the molten state, the tube is expanded into a bubble of the required diameter (with consequent reduction in thickness) by blowing air through the center of the mandrel. The tubing is usually extruded upward, but it can also be extruded downward or even horizontally. The pinch rolls at one end and the die at the other end contain air in the bubble. It is important to maintain uniform air pressure to ensure uniform thickness and width of the extruded film.

Blowing air from a cooling ring placed around the die helps to cool the film bubble. After a few yards of free suspension, the bubble is flattened between two nip rolls and festooned through a series of rollers to the windup rolls at the end of the processing line. The frost ring refers to the ring-shaped zone where the bubble begins to change from a clear melt, which represents the heated amorphous molecular structure of the plastic, to the frosty appearance of the cooler melt, which represents the semicrystalline nature of the material. The resulting film is usually a tear-resistant and high-strength plastic that can be used to make heavy-duty trash bags or packaging materials. The film quality is affected considerably by such factors as density and molecular weight of the polymer (mostly polyethylene). High-molecular-weight polymers give tougher films than lower-molecular-weight polymers. High-density polymers will have increased stiffness and greater susceptibility to wrinkles during the windup.

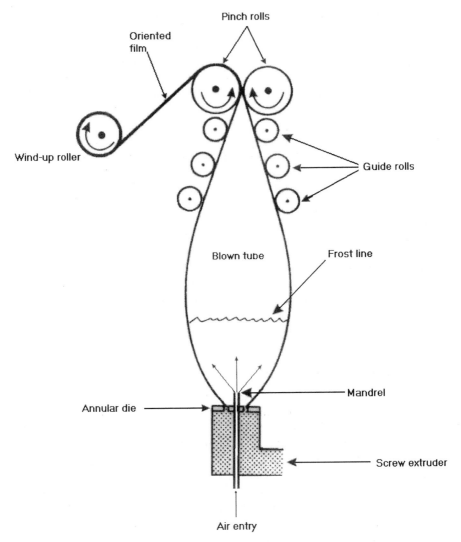

Figure 5.2. Bubble blown film extrusion.

The advantage of bubble blown film extrusion over the flat extrusion of films is the ability to produce films with a more uniform strength in both the machine and transverse directions. Also, the blown film extrusion is used in bag making, where only one seal across the bottom of the bag is needed, whereas with flat film extrusion, either one or two longitudinal seals are necessary. In flat film extrusion (particularly at high takeoff rates), there is a relatively higher orientation of the film in the machine direction than in the transverse direction. By using the bubble blown extrusion, blown diameters of 7 feet with the flat film widths up to 24 feet or even more can be produced.

Such large-width films of polyethylene find applications in agriculture, horticulture, and building.

5.1.5. Cast Film Extrusion

The cast film extrusion manufacturing of a film has a better tolerance control than is possible in the blown film process. The extruder used in this process is a standard flat-film die. The chilling and takeoff equipment is more sophisticated compared with other extrusion processes. By the cast film extrusion technique, it is possible to produce films having superior dimensional stability and better clarity and gloss than films produced by the conventional bubble blown film method.

In this process, the plastic melt, usually low-density polyethylene, is extruded through the die slit and cooled by the surface of two or more water-cooled chill or casting rolls. The hot polyethylene melt drops tangentially onto the first chill roll, so the alignment of this roll in relation to the falling film is critical. A gentle stream of air from an air knife is used to pin the molten plastic to the highly polished chrome-plated chill roll cylinders that have precise dimensions. The plastic is S-wrapped around the chill rolls and is taken across a series of takeoff rolls to the windup station. The cast film thickness is controlled by balancing the die opening, the chill roll size, spacing, and the takeoff speed.

5.1.6. Coating

Coating of other materials for protection, electrical insulation, better mechanical properties, or simply for a better appearance is one of the most important end uses of polymers. A number of different coating techniques are available for different purposes. Varnishes, lacquers, and paints are used to coat the wood, metals, and other materials by brush or by spraying. To achieve a uniform surface, the formulations should be carefully designed with a proper balance of viscosity and surface tension. Proper fluidity can be achieved either by using solvent-diluted systems or by employing systems that will solidify (cure) on drying and/or heating. Polymers that will cross-link under such conditions are particularly useful in these applications. Latex paints are easy to use in coating applications as they consist of droplets of liquid polymers or prepolymers emulsified in water. The emulsion can be easily applied because it is a fluid. When water vaporizes, the polymer droplets coalesce, and curing is done under the action of atmospheric oxygen.

Because of the high demand for covered wires and cables in the television and electrical industries, wire coating is becoming an important industrial process. In wire coating, a polymeric material is extruded over the metal wire in a continuous operation. The extruders used in wire coating are quite different from those used in other forms of extrusion. The basic difference is that the wire die is held in a crosshead that also holds the tapered guides mounted axially with the die. The extruder screw then forces the plastic melt down and over the guides through which the wire is drawn and towed through the die, in which coating is formed around it. Coating thickness ranging from 5 mil to more than 1/2 inch can be produced by this method. Wire coating processes use several different plastics and rubbers including polyethylene, cross-linked polyethylene, PVC, rubbers, nylon, and fluoropolymers.

Small metallic articles are coated by vinyl-type polymers using *dip-coating*. A hot article is dipped into a fluidized bed of polymer powder (the bed is kept at ambient temperature). The polymer particles hitting the hot object melt and stick to it, forming a layer that adheres well. If necessary, additional heat curing can be accomplished outside the fluidized bed. The same coating principle is used in charge spraying for coating larger metallic objects, such as car body parts. In a spray gun that is connected to high voltage, gas under pressure drives the charged polymer particles toward a hot-grounded part that is being coated. The electrostatic charge keeps the polymer particles separated and helps to focus the spray onto the coated object. The resulting coat is very uniform.

We mentioned the preparation of poly(p-xylylene) by oxidative pyrolysis of p-xylene via di-p-xylylene in Section 1.2.2. When this monomer is introduced at high temperature into a chamber containing cold objects to be coated, it dissociates into diradicals that then polymerize at the cold surface, producing a well-adhering and very tough layer of a polymer that forms an excellent insulator and a valuable dielectric for capacitors (no side groups; see Section 4.1.8).

5.1.7. Fiber Spinning

The production of synthetic fibers has completely revolutionized the textile industry. Fibers are fabricated by spinning, which is a modification of the extrusion process. There are two basic spinning techniques: *melt spinning* and *solution spinning*.

Conceptually, melt spinning is the simplest method and is used industrially whenever possible. Polymer melt is extruded through a single orifice, forming a *monofilament* or through a *spinneret* consisting of tens to thousands of holes when *yarn* is desired. After the extruded fiber leaves the spinneret, its temperature drops steeply and it solidifies. The solidification is a combination of crystallization and glass transition. A suitable windup device picks up the extruded fiber and applies a steady tensile stress that causes additional elongation in the solidifying fiber, greatly decreases its cross section, and increases its length. An industrial scale process for melt spinning of fibers is shown in Figure 5.3.

The pickup rate during spinning is limited by the hydrodynamic stability of the melt during the rapid elongational flow. Instabilities in this type of flow develop by two mechanisms. For melts and other highly viscous materials, *cohesive brittle fracture* is the limiting factor. In the elongational flow, the polymer molecules are deformed very quickly. When the rate of deformation exceeds the capability of the coil to relax (this capability is governed by the longest relaxation time in the mechanical spectrum), flow instabilities will develop, giving the fiber a distorted (*fractured*) surface. At still higher deformation rates, the jet will break. The longest relaxation time depends on the length of the polymer chain. A typical molecular weight for polyamides and polyesters used for production of fibers is about 20,000, that is, it is rather low. The relaxation times are short, and pickup rates up to 3000–4000 m/min are practicable. On the other hand, polymers of the polypropylene type need to have much higher

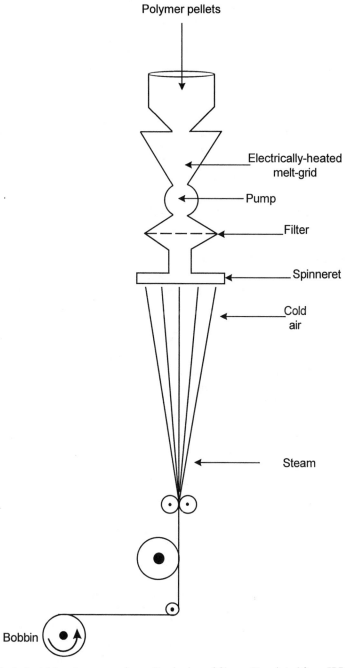

Figure 5.3. Industrial scale process for melt spinning of fibers. (Reprinted from H.R. Allcock and F.W. Lampe, *Contemporary Polymer Chemistry,* Copyright © 1981 by Prentice Hall, Englewood Cliffs, NJ. Used by permission of the publishers.)

molecular weight to make good fibers. This means longer relaxation times requiring much slower pickup rates (about 200 m/min).

For liquids of lower viscosity (polymer solutions) the limiting factor is *capillary waves* on the free surface of the liquid jet. Any fluctuation in the extrusion die will give rise to periodic oscillation of the fiber radius with increasing amplitude that will eventually break the fiber into droplets. High surface tension magnifies this effect, whereas higher viscosity suppresses it. Thus capillary waves do not play any significant role in melt spinning.

Only those polymers may be spun as melts that are thermally stable enough that they can be heated high enough above their T_g to lower the viscosity sufficiently for the spinning purpose. Many fiber-forming polymers decompose upon melting. Such polymers must be brought into solution before they are spun.

There are two versions of *solution spinning:* dry spinning and wet spinning. In *dry spinning,* the polymer is dissolved in a volatile solvent that vaporizes very fast after the solution leaves the spinneret. The elongation occurs while the fiber is drying. The process is more complicated than melt spinning because the solvent must be recovered. Moreover, the solvent evaporates first on the fiber surface, forming a solidified skin. The solvent from the fiber core evaporates a little bit later, causing a collapse of the outer skin and giving a slightly irregular appearance to the final fiber. Dry spinning is used for manufacturing polyacrylonitrile fibers from dimethylformamide solution and cellulose acetate fibers from acetone solutions.

In *wet spinning,* the solution is extruded through a spinneret into a coagulating bath. For example, polyacrylonitrile solution in dimethylformamide is spun into a water bath. Water is miscible with dimethylformamide; the solvent diffuses out from the solution, and the polymer precipitates. Again, the precipitation is accomplished first on the surface of the fiber; this may lead to inhomogeneities both along the fiber and in the radial direction. The coagulating bath may also react chemically with the polymer. Cellulose xanthate (viscose) is spun into dilute sulfuric acid; the xanthate groups are hydrolyzed, and the regenerated cellulose precipitates out in the fiber form.

Compared with melt spinning and dry spinning, the pickup rates in wet spinning must be much slower to give the coagulating fiber enough time in the bath (which is many meters long) to be sufficiently stabilized.

After spinning, the polymer in the fiber is still either amorphous or crystalline in a spherulitic form; the degree of its orientation is still very small. Consequently, fibers must be subjected to a process of *drawing* during which they are further stretched to elongations of 20–2000% depending upon the nature of the polymer. Quite often, the fibers are drawn directly as they are produced by spinning; it is a single continuous operation. However, when the spinning pickup rate is very high, engineering considerations require a two-step process.

The phenomena occurring during drawing have already been described in Section 4.1.6.2. Drawing must be performed at temperatures above T_g of the polymer but well below its T_m. It is accomplished by leading the fibers over a set of rollers whose surface velocity is increasing in a programmed sequence. It is during drawing that the polymer molecules become oriented, recrystallize, and acquire their strength.

5.1.8. Calendering

Calendering is a deceptively simple method for manufacturing wide polymer sheets. A preformed sheet, the extrudate from a slit die, or simply a polymer melt is fed between two rolls. The rolls are heated to bring the polymer into a plastic state; they form a thinner sheet by squeezing the feed. The exiting sheet enters the gap between one of the rolls and a third roll and is squeezed further. Typically, four rolls allow three passes of the sheet, which is then cooled, possibly stretched further by another (faster rotating) roll, and wound up.

Calendering is the preferred method of sheet fabrication from polymers whose melt has a high viscosity and from polymers that are heat sensitive. Typical materials include poly(vinyl chloride) and its copolymers. The thickness of calendered sheets varies from 100 μm to 2.5 mm (used for "vinyl" floor tiles). Calendering is used to produce shower curtains, vinyl upholstery materials, and vinyl floor tiles.

Although the physical concept of a calender is very simple, its design is quite demanding. It should produce thin, uniform sheets of rather large width. Eccentricity of the rolls must be controlled very tightly as well as their mutual adjustment. Consequently, calenders are the heavier and more expensive pieces of equipment in a polymer processing plant.

5.1.9. Molding

Molding is a technique for fabricating three-dimensional articles that is substantially faster than casting. *Injection molding* is a combination of extrusion and casting. The polymer is shaped in a preformed mold as in casting but is delivered to the mold under pressure, usually by a modified extruder. The heart of the injection molding machine (Fig. 5.4) is a screw rotating in a barrel as in the extruder. However, the screw is also operated axially and acts as a piston. The machine works in cycles. First, the feed is melted and compacted in front of the (in this phase rotating) screw. In the second phase, the melt is injected through a *gate* into a cold mold, where it solidifies. Then the molded article is ejected, and the cycle is repeated.

Polymers contract as they solidify and cool. To prevent formation of voids in the molded article during cooling, it is necessary to deliver some excess quantity of the melt into the mold. This is accomplished by injecting the melt in a compressed form using high pressure. The increase in volume during the eventual decompression will just compensate the contraction upon cooling. The first place where the melt solidifies is in the narrow gate. This serves an important purpose: it is now possible to withdraw the screw piston even before the whole polymer mass solidifies; the frozen gate prevents the melt in the mold from backflowing toward the receding piston.

Another phenomenon to watch for is the development of internal stresses in the molded article. During molding, the melt freezes while its molecules are severely deformed by the fast injection flow. Slow relaxation of the internal stresses may lead

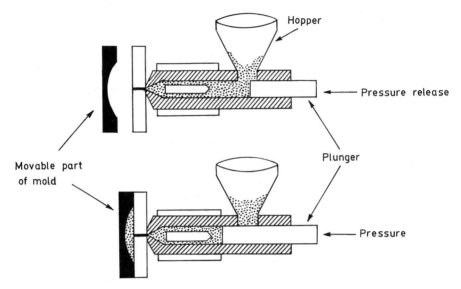

Figure 5.4. Cross-sectional view of the injection molding cycle. (Reprinted from H.R. Allcock and F.W. Lampe, *Contemporary Polymer Chemistry,* Copyright © 1981 by Prentice Hall, Englewood Cliffs, NJ. Used by permission of the publishers.)

to deformation of the article. The extent of the internal stresses can be reduced by a slower, programmed cooling of the mold that would allow some relaxation of the material.

Reaction injection molding(RIM) is a modification of injection molding that is very similar to casting into a mold. In both cases, the actual polymerization occurs in the mold, but the difference is in the rate. In casting, the monomer feed is mixed in a routine manner and easily poured into a mold; polymerization then proceeds over hours or days. In RIM, polymerization proceeds very quickly immediately after the reactants are mixed. Typically, two streams of low-viscosity reactants are mixed with a turbulent flow regime and are immediately injected into molds, where they are cured within minutes. The liquid that is injected has a low viscosity, and hence, the whole process requires only low pressures, which simplifies the instrumental design. Consequently, RIM is one of the most advantageous techniques of polymer fabrication. Unfortunately, the number of known systems that cure fast enough to be used in RIM is very small. The technique is used mainly with polyurethanes; the reaction of polyols with isocyanate reagents satisfies the rate demands. Rather large articles can be produced by RIM. In the automobile industry, it is used for manufacturing the whole body panels and bumpers.

Blow molding is used for manufacturing hollow articles such as toys and bottles. A piece of molten polymer tube called a *parison* (a term borrowed from the glassblowing industry) is transferred from an extruder to between two halves of a mold, which closes around it and pinches one end of the tube. Compressed gas is then injected through

Compressed
air

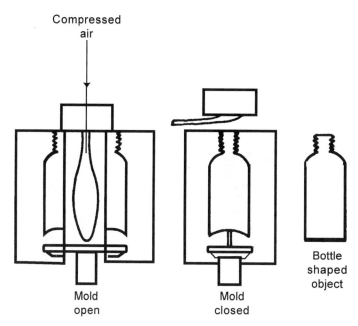

Mold
open

Mold
closed

Bottle
shaped
object

Figure 5.5. Cross-sectional view of the blow molding operation. (Reprinted from H.R. Allcock and F.W. Lampe, *Contemporary Polymer Chemistry,* Copyright © 1981 by Prentice Hall, Englewood Cliffs, NJ. Used by permission of the publishers.)

the other end of the parison and presses the melt against the mold, where it solidifies. Figure 5.5 shows the blow molding operation. For processing a more complicated article the extrusion can be modified in such a way that the parison is thickest in those parts that will be expanded most. This procedure improves the uniformity of the wall thickness of the article. The blow molding process is easy and inexpensive. It is used to make articles from polyethylene, polystyrene, polycarbonates, poly(vinyl chloride), and many other polymers.

Compression molding is a major fabrication technique. It is one of the oldest and prime methods of processing thermosets of phenol-formaldehyde and aminoplast-type resins. Dinnerware, ashtrays, wall sockets, and similar articles are produced by this method. It is also employed for processing thermoplastics and many types of rubber. In this technique, a known amount of molding powder is placed into an open two-part mold (Fig. 5.6). After the mold is closed, the pressure is applied normally in the range of 0.5–3 ton/in^2.

For complex components, another process has been developed that involves the application of pressure to molding powder before it is forced into the mold cavity. This is commonly called *transfer molding.* The actual process consists of charging a known amount of molding powder into a heated chamber outside the mold. After the powder has reached a sufficiently plastic state it is forced under high pressure through a suitable narrow opening (gate) into a closed mold.

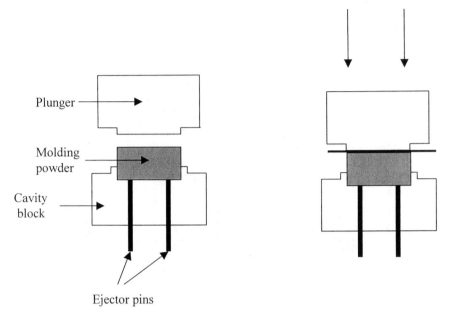

Ejector pins

Figure 5.6. Compression molding operation of thermosets. (Reprinted from H.R. Allcock and F.W. Lampe, *Contemporary Polymer Chemistry,* Copyright © 1981 by Prentice Hall, Englewood Cliffs, NJ. Used by permission of the publishers.)

5.1.10. Foam Fabrication

Polymer foams are very valuable materials. They are the basis of modern upholstery, they have excellent insulating properties, and they are used extensively as packing materials for fragile goods, all of this at rather low density and therefore quite economically.

Foams are manufactured both from cross-linked rubberlike polymers (polyurethane, rubber) and from glassy polymers like polystyrene. The production process must accomplish three tasks: formation of the foam, its shaping, and stabilization of the porous structure. In rubberlike foams the structure is stabilized by cross-linking that takes place during and after the shaping process.

In Section 2.1.2 we described in some detail the chemistry of foam formation in polyurethanes. Industrially, the diisocyanate is mixed quickly in an automatic mixer with the other ingredients and delivered on a wide moving belt. It rises and gels immediately, forming a continuous slab that is later cut into smaller pieces (e.g., mattresses).

In the production of polyurethanes, the foaming gas (carbon dioxide) is a product of the polymerization itself. For other polymers, the gas must be generated by some foaming agent. When foamed rubber is manufactured, rubber latex is mixed with a soap solution and with sodium fluorosilicate, Na_2SiF_6. The concoction is beaten like an egg white; during the aeration, the fluorosilicate hydrolyses to silica gel that stabilizes the froth long enough to allow for the vulcanization of the latex.

Other foaming techniques produce the gas within the polymer formulation either chemically or physically. Chemical *blowing agents* are substances that decompose with gas evolution during moderate heating (for example, ammonium bicarbonate or azobisbutyronitrile). Physical blowing agents are used mainly for the production of more rigid foams (*structural foams*) from polymers that are glassy at ambient temperatures, such as polystyrene. A low-boiling liquid (pentane or trichlorofluoromethane) is dissolved in the polymer before or during extrusion. The high pressure in the extruder prevents its vaporization. When the polymer leaves the die, the solvent vaporizes, simultaneously cooling the polymer to below its T_g and stabilizing the foam.

When formulating the feed materials for foam production, due care must be taken with the bubble nucleation process, which determines the size of the bubbles as well as the openness or closeness of the foam. This phenomenon was explained in Section 2.1.2 in the context of the chemistry of polyurethane foams.

5.2. TESTING OF POLYMERS

As mentioned earlier, for the development of new polymeric materials it is necessary to obtain useful technical data on individual polymeric materials to compare them with other known materials. This kind of testing is also important in production plants to test whether individual batches of polymeric materials have the expected properties. Testing is also necessary when raw polymers are bought. Nominally identical polymers can differ in many ways depending on the source of raw material, method of synthesis, fabrication process, etc. Thus their engineering properties may vary considerably.

5.2.1. Mechanical Testing

In Section 4.2.1, we presented a short review of techniques for studying the more fundamental mechanical properties, including creep and stress relaxation, relaxation and retardation spectra, and dynamic behavior. For an initial engineering assessment, the properties measured most often are modulus, toughness, impact strength, hardness, abrasion, scratch, and fatigue resistance, and T_g and T_m.

Modulus and tensile strength are usually measured with an Instron at a constant rate of strain as described in Section 4.2.1. However, tensile strength is not really a good measure of toughness; we consider as tough those polymers that can absorb an appreciable amount of energy before they break. The energy absorbed can be measured as an area under the stress-strain curve. In Figure 5.7 we can observe two stress-strain curves: curve a for a rigid, brittle polymer with high tensile strength, and curve b for a much softer polymer with lower tensile strength that yields at larger strains. It is obvious that the latter polymer can absorb much more energy before

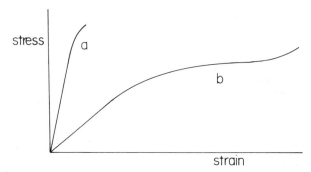

Figure 5.7. Typical stress-strain curves for (a) a rigid, brittle polymer and (b) a softer polymer that yields to stress.

breaking, that is, it is much tougher. However, this area method is also a poor method for measuring toughness. In the real world, a polymer is almost never strained slowly to the breaking point; it is supposed to sustain much more sudden attacks as in an impact collision.

Impact strength is usually measured by means of the *Izod notched impact test.* A bar with prescribed dimensions and with a notch of prescribed size is fastened at one end and attacked by a hammer pendulum (Fig. 5.8). At impact, the bar breaks and the movement of the pendulum is slowed down. The quantity measured is the difference between h_1, the height of the pendulum before its release, and h_2, the highest point of its trajectory after the impact. From this difference, and from the known mass of the pendulum, the loss of kinetic energy of the pendulum is evaluated; this equals the energy absorbed by the sample during the impact.

Hardness of materials is measured in terms of penetration of a test body (e.g., a small steel sphere) into the material at a given load. In the Brinell hardness test, the depth of indentation is measured after the load is removed; in the Rockwell test, it is measured while the sphere is still under load. *Abrasion* and *scratch resistance* are other important properties of polymer surfaces. Abrasion measures the wear of the material during long exposure of the surface to various forms of friction, whereas scratching involves the action of (specified) rough objects, such as sandpaper.

A polymer engineer wants to know how a product will perform not only as it leaves the production line but also after weeks or years of service. Mechanical properties of materials generally deteriorate with time, especially when they are under stress. This phenomenon is called *fatigue.* Fatigue is usually measured by flexing a bar of the polymer until it breaks. The process is easy to visualize on the molecular level. In every flexing cycle a few polymer chains are stretched too far and break. The damage is cumulative; eventually so many chains are broken that the macroscopic integrity of the sample is compromised. Accordingly, the number of flexing cycles the sample can withstand decreases with increasing severity of the flexing. With sufficiently mild loading, the sample may last indefinitely; the patience of the testing person may wear down first.

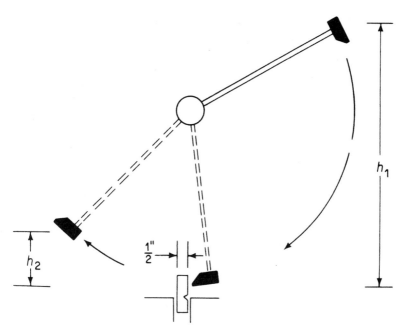

Figure 5.8. Scheme of notched Izod impact testing. (Reprinted from F. Rodriguez, *Principles of Polymer Systems,* 2nd ed. Copyright © 1982 by McGraw-Hill, New York. Used by permission of Hemisphere Publishing Corporation.)

5.2.2. Thermal Testing

In addition to differential scanning calorimetry and differential thermal analysis, which provide a wealth of information about thermal behavior of polymers and are gaining popularity even in testing laboratories, two more simple thermal methods are important in industrial testing.

Thermomechanical analysis (TMA) is conceptually similar to the measurement of hardness. A small spherical probe is pressed by a constant load against the polymer surface while the temperature of the polymer is raised slowly. The probe is connected to a sensitive position gauge. As the temperature increases through the glass transition region, the polymer becomes softer, the probe penetrates deeper, and T_g is registered. Another more or less sudden change of modulus/position/penetration may occur at the melting temperature T_m. TMA experiments can be run from about $-100°$C to $700°$C; however, the upper use temperature is limited by the thermal stability of the polymer.

Thermogravimetric analysis (TGA) measures the chemical stability of polymers. A polymer is heated within a microbalance, and its weight is recorded as a function of temperature. The operating temperatures for TGA typically range from about room temperature to $1000°$C. In a TGA experiment, a gas like nitrogen (inert) or air (oxidizing) is passed through the system to provide a suitable atmosphere for the measurement and to remove decomposition products from the sample chamber.

At lower temperatures, adsorbed water is lost first then the absorbed solvents and unreacted monomers. At higher temperatures, the polymer will start to decompose, often with a loss of volatile decomposition products. A polymer with a low ceiling temperature may depolymerize, converting back to monomer and leaving almost no residue. Other polymers may lose other small molecules. For example, poly(vinyl chloride) will split off hydrogen chloride.

Generally, the decomposition of polymers will occur at lower temperatures than the decomposition of model low-molecular-weight substances. This is caused by irregularities of the polymer structure; the decomposition often starts at the chain end, at a branching point, or at a group that was inadvertently oxidized. Incorporation of traces of oxygen into the polymer backbone during polymerization may severely compromise the thermal stability of the polymer. The thermogravimetric method may detect such substandard batches of materials.

The degradation products given off by the polymer sample in the TGA chamber may be analyzed by other methods such as Fourier transform infrared spectrophotometry, mass spectrometry, or gas chromatography coupled with mass spectrometry.

5.3. BARRIER PROPERTIES OF POLYMERS

Permeation of gases and liquids through polymeric membranes is attaining ever-increasing industrial importance. Today, separation membranes have become essential commodities not only in industries but also in daily human life. In this section we will review the principles governing permeation, discuss preparation and uses of various types of membranes, and describe their more important applications. The permeation (transport) of gases or liquids through a polymeric matrix was discussed in Section 4.1.9.

A membrane is a thin barrier that selectively transports some substances while retaining the others based on factors such as the molecular size of the transported substance or its interaction (solubility) with the barrier material. Depending upon their morphological setup, membranes are able to differentiate between the components of a mixture of gases, vapors, or liquids. This property has enormous impact in many industrial-scale separations, food packaging industries, drug delivery, etc. For instance, in food packaging applications, it is specifically desirable to impede oxygen permeation.

In most membrane processes, mass transport across a nonporous polymer membrane is caused by a chemical potential gradient or, in more general terms, by the difference in the free energy of the permeate between the membrane upstream (feed side) and downstream (product side) interfaces. Correspondingly, the mass transport through a membrane can be described by the Fick's first law [see equation (4.1.77)]. When the permeants are driven by more than just the concentration gradient as depicted by equation (4.1.77), the flux equation combines the concentration, electrical, pressure, temperature, and other gradients in the membrane. The effectiveness of the forces that drive the individual membrane processes depends to great extent on the type of the membrane employed.

Membranes can be made from a wide variety of organic (polymers) or inorganic (carbons, zeolites, etc) materials. The transport parameters of the membrane are primarily a function of the membrane chemical makeup, its structure, and its morphology. Such parameters as degree of polarity, interchain forces, ability to crystallize, and chain stiffness must be considered when selecting a particular membrane for a specific application.

The transport phenomenon in a membrane system is also highly influenced by the ongoing permeation process itself. Polymer-penetrant interactions play a major role. In addition, concentration polarization of the fluid mixture at the membrane surface is almost unavoidable; it is largely controlled by the permeability rate of the permeant and by the mode of fluid flow (laminar or turbulent) over the membrane. In extreme cases, at the membrane-fluid interface, an extensive boundary layer may develop that might determine the entire membrane transport process.

To be useful in an industrial separation process, a membrane must exhibit high flux and selectivity (or rejection) toward a particular component, and at the same time it should have a good mechanical strength, tolerance to feed components (fouling resistance), resistivity to temperature, and low manufacturing costs. Of these, flux and selectivity are important in deciding the separation abilities. The higher the flux of a membrane at a given driving force, the lower is the membrane surface area required for a given feed flow rate and therefore, be the capital costs of the membrane system. Selectivity determines the extent of separation. Membranes with higher selectivity are desirable because higher product purity can be achieved in a separation process.

The selective permeation of liquids or gases through the barrier polymers has important applications such as in the purification of air and water as well as in the chemical and biotechnological industries. Although the major uses of membranes are in the production of potable water by reverse osmosis and electrodialysis, they are used for many other important applications such as filtration of a particulate matter from liquid suspensions, air, and industrial flue gas, separation of liquid mixtures (organic-organic or organic-aqueous such as dehydration of ethanol azeotropes), drug delivery, etc.

5.3.1. Membrane Types

Membranes are available in several configurations depending on their end applications. There is no conceptual limitation for the form of a membrane, and they can be shaped in almost any desired configuration. From a practical viewpoint, membranes can be fabricated in a suitable device, generally referred to as *membrane module*. The membrane module design is closely related to the membrane process and its technical feasibility. To meet the flux requirements, membranes must have a very large surface area and a high area-to-volume ratio (packing density). For different applications, different types of modules are needed.

Flat sheet membranes are prepared in *plate* and *frame module*. The *spiral-wound modules* are prepared by sandwiching the alternate layers of flat sheet membranes, spacers, and the porous material around an inner porous permeate collection tube. *Hollow-fiber shapes* are prepared with the typical outer diameters ranging from 80 to 200 μm and a wall thickness of about 20 μm or more. These fibers are made by melt or wet spinning from cellulose triacetate, polyamide, polysulfone, etc. Modules containing fibers of larger diameters, called *capillary modules,* are also manufactured.

Even though hollow-fiber modules have the highest membrane packing density per module volume, spiral-wound and plate and frame modules are more commonly used in large-scale separations. In all cases, the process stream must be pretreated to remove the large-size particles that can plug the pores. When the feeds cannot be pretreated to remove the potential fouling contaminants, a *tubular membrane module* is sometimes used.

5.3.2. Membrane Preparations

Membrane preparation is both complex and fascinating. Different approaches are used to make membranes. Symmetric membranes have a uniform structure over the entire membrane thickness, whereas asymmetric membranes have a gradient in the structure. The dense region of asymmetric membranes controls the separation. Various types of membranes will be discussed in the next sections.

5.3.2.1. Symmetric Porous Membranes. Symmetric porous membranes can be pre-pared in cylindrical, sponge-, web-, or slitlike structures by a variety of methods such as irradiation, stretching of a melt-processed semicrystalline polymer film, vapor-induced phase separation, and temperature-induced phase separation.

Symmetric membranes with cylindrical porous structures are produced by an irradiation-etching method; they are also called *nucleation track* membranes. In their preparation, a dense polymer film (e.g., polycarbonate) is irradiated with charged particles that induce polymer chain scission along the nucleation tracks throughout the film thickness. The film is then passed through an etching medium, typically a sodium hydroxide solution. Uniform-size pores are formed during etching of the par-tially degraded polymer along the nucleation tracks. The porosity and pore size of the membranes can be controlled by both the irradiation and the etching time.

Membranes having symmetric slitlike porous structures are produced from semi-crystalline polymers like polyethylene and polypropylene with a melt extrusion/ stretching process (Section 5.1.3). A nucleated lamellar structure is first formed by melt extrusion of the semicrystalline polymer and recrystallization under high stress. The slitlike pores are then formed between stacked lamellae by stretching the mem-brane in the machine direction.

A vapor-precipitation/evaporation method is used to produce symmetric mem-branes with web- or spongelike pore structures. In this method, a polymer solution containing a solvent and a nonsolvent is cast onto a suitable substrate, which is then exposed to a water vapor-saturated air stream to induce phase separation. After phase separation, solvent and nonsolvent are removed by evaporation by blowing a stream of hot air across the membrane. In this method, the porosity and pore size of the membrane can be controlled by the polymer concentration in the casting solution as well as by the composition of the vapor atmosphere. Low polymer concentration, high humidity, and addition of solvent vapor to the casting atmosphere produce mem-branes with high porosity and large pore size. In this process, a constant polymer concentration profile is maintained throughout the entire membrane thickness at the onset of phase separation, thereby yielding a porous membrane with highly symmetric structure.

The formation of membranes by the phase separation of an initially stable polymer solution is called a *phase inversion* method. By this method, the porous membranes with a symmetric structure can be produced by bringing a thermodynamically stable polymer solution to an unstable state. A change in temperature, composition, or pressure that leads to a decrease in Gibbs free energy of mixing of the solution (Section 3.2.3) causes phase separation of the initially stable polymer solution into two phases having different compositions.

Thus symmetric porous membranes are produced by a thermally-induced phase inversion process. The phase diagram for a binary solution containing an amorphous polymer and a solvent displaying an upper critical solution temperature is shown in Figure 3.16. In this technique, a solution of the polymer in a mixed solvent, which is on the verge of precipitation, is brought to phase separation by a cooling step. The phase separation of the initially stable polymer solution occurs because of nucleation and growth or because of spinodal decomposition. The resulting membrane structure depends on the kinetics of phase separation and on the local distribution of the polymer-rich phase at the point of precipitation.

Dense symmetric membranes with thicknesses greater than $10\,\mu$m are prepared by solution casting and subsequent solvent evaporation or by melt extrusion. Because these membranes are relatively thick, they have very limited applications in gas separation studies. They are commonly used in electrodialysis and pervaporation.

5.3.2.2. Asymmetric Membranes.

Asymmetric (anisotropic) membranes have high permeation rates with excellent separation ability. Their efficiency is based on the fact that the transport rate through a (thin) dense membrane is inversely proportional to its thickness but its selectivity is independent of its thickness. The porous part of the membrane has no major effect on the performance but eliminates the mechanical integrity problems associated with the handling of ultrathin membranes.

Asymmetric membranes consist of a very thin $(0.1$–$0.5\,\mu$m) active layer, and an approximately 150- to 300-μm-thick porous support, which provides the mechanical strength while the active layer brings about separation. The active layer must be free from defects (i.e., micropores). Asymmetric membranes can be classified into three main categories: (a) integral-asymmetric membranes with a porous skin layer, (b) integral-asymmetric membranes with a dense skin layer, and (c) thin-film composite membranes. These types are shown in Figure 5.9. The porous integral-asymmetric membranes are prepared by the phase inversion technique. They find application in electrodialysis, ultrafiltration, and microfiltration, whereas the integral-asymmetric membranes with a dense skin layer are used in reverse osmosis and gas separation.

One of the first integrally skinned asymmetric membranes was that of cellulose acetate developed in the 1960s and used successfully in desalination studies. These membranes are typically prepared by an immersion precipitation method from solution of a polymer in a solvent. When the cast polymer solution is immersed into a nonsolvent for the polymer that is miscible with the solvent, an asymmetric structure results with either a porous or a nonporous skin layer. The structural gradients in

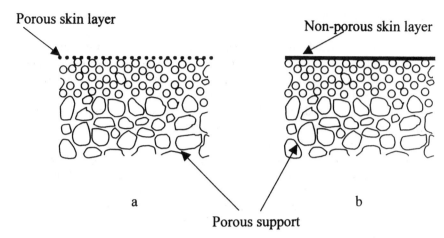

Figure 5.9. Schematic representation of asymmetric membranes: (a) integrally skinned (porous skin layer) and (b) integrally skinned (nonporous skin layer) and/or thin-film composite membranes.

these membranes are created from a very steep polymer concentration gradient in the nascent membrane at the onset of phase separation. In the immersion precipitation process, phase separation is induced by solvent/nonsolvent exchange during the immersion step.

The formation of membranes by the immersion precipitation method occurs in a very short time, typically in less than a few seconds. Different membrane structures can be obtained by carefully controlling the thermodynamic and kinetic variables involved in the immersion precipitation process. Most of the commercial membranes prepared by this method use multicomponent mixtures containing polymer, solvent, and nonsolvent or additives. The pore size and the skin layer thickness can be modified by adding to the casting solution nonsolvents (alcohols, carboxylic acids, surfactants, etc.), inorganic salts (lithium nitrate or lithium chloride), or polymers (polyethylene glycol, polyvinyl pyrrolidone, etc.). The presence of even small quantities of these compounds significantly affects the membrane structure and, hence, its separation performance.

The success of asymmetric membranes led to the development of thin-film composite membranes. Composite membranes consist of ultrathin semipermeable layers supported by a highly porous substrate providing a minimum resistance to the permeants. The properties of two or more different materials are combined to give the desired material. Different layers in the membrane are responsible for the overall separation and mechanical function of the membrane. The thickness of the deposited dense layers varies generally between 0.1 and 1.0 μm. The permeability of composite membranes depends on the porosity of the substrates and on the complex composition of the deposited selective layer. Composite hollow-fiber membranes consisting of a polysulfone porous support coated with polyethyleneimine cross-linked by toluene diisocyanate are widely used.

Thin-film composite membranes are an improvement over phase inversion membranes and offer several advantages: (a) an ability to control the thickness of the separating layer, (b) independent selection of the materials from which the separating layer and the porous support are formed, and (c) independent preparation of the skin layer and the porous support, making it possible to optimize each structural element.

The composite membranes can be prepared either as single-layer (a nonporous skin layer on the porous support) composites or multilayer composites. The multilayer composite has a porous support and several layers of different materials including a selective layer, each layer performing a specific function. The selective layer is usually applied by lamination, solution coating, interfacial polymerization, or plasma polymerization.

The solution coating method is by far the most widely used method for the commercial production. This method involves the deposition of a dilute polymer solution onto the surface of a porous membrane and subsequent drying of the thin liquid film. The simplicity of this process is very attractive for a large-scale production of these membranes. However, it is quite difficult to produce defect-free thin composite films with a thickness less than 1 μm using the solution coating method. Many methods have been suggested to overcome this problem; use of ultrahigh-molecular-weight polymers for the formation of the selective layer proved to be successful. Thin-film composite membranes are used in nanofiltration, reverse osmosis, pervaporation, and gas separation.

5.3.2.3. Ion-Exchange Membranes.

Ion-exchange membranes can be either homogeneous or heterogeneous. Heterogeneous membranes are prepared by incorporating the ion-exchange groups into the film-forming resins by (a) dry molding or calendering mixtures (Section 5.1) of the ion-exchange and film-forming materials, (b) dispersing the ion-exchange material in a solution of the film-forming polymer and then casting films from the solution and evaporating the solvent, and (c) dispersing the ion-exchange material in a partially polymerized film-forming polymer, casting films, and completing the polymerization. Heterogeneous ion-exchange membranes have several disadvantages over the homogeneous membranes; the most important ones being high electrical resistance and poor mechanical strength when swollen in dilute salt solutions.

Homogeneous ion-exchange membranes are better suited because the fixed ion charges are distributed homogeneously over the entire polymer matrix. The homogeneous membranes are prepared by (a) polymerization of mixtures of reactants (e.g., phenol, phenolsulfonic acid, and formaldehyde) that can undergo condensation polymerization (at least one of the reactants must contain a group that can be made anionic or cationic) or (b) chain polymerization of mixtures of reactants (e.g., styrene, vinylpyridine, and divinylbenzene) that can polymerize (at least one of the reactants must contain an anionic or cationic moiety).

Anionic or cationic groups can also be introduced into the polymer by techniques such as imbibing styrene into polyethylene films, polymerizing the imbibed monomer, and then sulfonating the styrene. Grafting techniques are also used. The monomers containing strongly acidic or basic ion-exchange groups are difficult to graft directly onto hydrophobic polymers. This is usually achieved by the radiation-induced graft polymerization technique using the selected monomers. The most commonly used anionic or cationic groups in ion exchange membranes are: $-SO_3^-$ or $-NH_3^+$. The other charged groups like $-COO^-$, $-PO_3^{2-}$, $-HPO_2^-$ as well as various tertiary and quaternary amines are also used.

To provide the structural support needed in electrodialysis, the membrane is prepared by applying a paste containing the cation- or anion-selective polymer onto a woven fabric generally made of poly(vinyl chloride). Membranes can be made in flat sheets and contain about 30–50% water. Each membrane has a network of molecular-size pores that are too small to allow significant water flow and that have electronegative (SO_3^-) or electropositive (NH_3^+) charges fixed to the membrane. To maintain electrical neutrality each of the fixed charges on the membrane is associated with an ion of the opposite charge. Thus the ion can easily move from one fixed charge to another, and the membrane can pass the electric current in the form of migrating ions. Because the fixed charged groups on the membrane repel the like-charged ions, the anions cannot enter the cation-selective membrane and vice versa. The high concentration of counterions in ion-exchange membranes is responsible for the low electrical resistance of the membrane. The resistance of ion-exchange membranes should be around 2–10 Ω cm^2 and the fixed charge density is about 1–2 mEq/g.

The widely used cation-exchange polystyrene has negatively charged sulfonate groups chemically bonded to most of the phenyl groups in polystyrene. The negative charges of the sulfonate groups are electrically balanced by the positively charged cations. Similarly, the anion-exchange membranes consist of cross-linked polystyrene having positively charged quaternary ammonium groups bonded to most of the phenyl groups; the counterions are negatively charged. In both cases the counterions carry the electric current. Nafion cation-exchange membranes have a polytetrafluoroethylene backbone with sulfonic acid groups attached at the end of short side chains of perfluoropropylene ether units. They exhibit excellent chemical stability and low electrical resistance. However, this membrane is more expensive than other membranes like polysulfones and polyethersulfones that also have good chemical resistance.

5.3.3. Membrane-Based Separation Processes

Separation by membranes depends on membrane selectivity and the driving force used. Membranes with higher selectivity are desirable because higher product purity can be achieved in a separation process. Reverse osmosis, ultrafiltration, and microfiltration use hydrostatic pressure gradient as the driving force. Electrodialysis utilizes an electrical potential gradient as a driving force and ion-exchange membranes as the discriminating barrier. Pervaporation involves a phase change. These processes will be discussed in the following sections.

5.3.3.1. Reverse Osmosis and Filtration Techniques.   Reverse osmosis and other filtration techniques are applied mainly to aqueous systems. In an osmotic experiment (see Section 3.2.1), a semipermeable membrane that is selective to certain components of the solution (usually the solvent) is used. The direction of solvent flow is determined by its chemical potential, which is a function of pressure, temperature, and concentration of the solutes. The osmotic flow from pure water to salt solution will occur across the membrane until the equilibrium of solvent chemical potential is restored. Thus osmotic equilibrium results when the pressure differential on two sides of the membrane becomes equal to the osmotic pressure. A further increase in pressure differential will raise the chemical potential of water in the solution and will cause a reversal of the osmotic flow toward water, now at a lower solvent chemical potential relative to the solution. This phenomenon is called *reverse osmosis* (RO), so named because of the reversal of flux. The external pressure applied to the salt solution must exceed the osmotic pressure of the solution for reverse osmosis to take place.

The mechanism by which reverse osmosis occurs across the semipermeable membrane is not fully understood. Nevertheless, it is certain that the chemical nature of the membrane must be such that it will exhibit a preference for water versus dissolved salts at the surface-solution interface and, consequently, reject the salt. This may occur by a weak chemical bonding of water to the surface. Pure water at the interface or within the membrane tends to repel the solute, and the transport of water across the membrane takes place through the pores by molecular diffusion.

Rejection R_i (also called retention or retention coefficient) is the term used to measure separation in an RO experiment and is defined as:

$$R_i = \frac{X_i - Y_i}{X_i} \times 100 \qquad (5.3.1)$$

where X_i is the feed side concentration of component i and Y_i is its permeate side concentration. This measure of membrane performance is dependent on the system in which it is measured and on the measurement conditions.

The technique of RO is used for obtaining potable water from brackish water or seawater. For a successful RO operation, the diameter of the pores should be no larger than 5–20 Å to retain all the dissolved microsolutes, such as salt ions, while allowing water to freely permeate through the membrane. Cellulose triacetate is the most frequently used RO membrane, although other polymers like polyamides, polysulfones, and thin-film composites are also employed. The presently used RO devices include tubular, spiral-wound, and hollow-fiber modules. For optimal performance, water feed to the RO device should be pretreated to remove the gross amounts of solids and to prevent membrane fouling by precipitation.

The pressure differences used in RO are generally up to 80 bars. If the membrane is expected to retain only macromolecules or particles with an insignificant osmotic pressure, the necessary operating pressure can be as low as 2–10 bars. In this case, the process is called *ultrafiltration* (UF). In ultrafiltration, the species are passed or rejected at the membrane surface on the basis of the size. The UF membranes have pore sizes ranging from 10 to 1000 Å and are suitable for filtering larger particles, including some bacteria and viruses as well as some moderately sized organic molecules like sugars. Because the transport of matter across a UF membrane involves viscous flow, physical structure of the membrane will control the flow rate as well as rejection.

As in RO, the pressure differential is the driving force for separation. In some cases, permeate is the desired product (e.g., when high-molecular-weight impurities must be removed from the feed solution), whereas in other cases, concentrated solution on the feed side is the valuable product.

There are three primary UF configurations: tubular, spiral wound, and hollow fiber. Typically, asymmetric membranes are used in UF applications. Membranes can be made from a variety of polymers such as cellulose acetate, PVC, polyacrylonitrile, polycarbonate, and polysulfone. Compared to RO, UF membranes have a larger variety of applications. These include treatment of industrial effluents and process water; concentrating, purifying, and separating macromolecular solutions in the chemical, food, and drug industries; clarifying and purifying biological solutions and beverages; and pretreating seawater in RO processes.

When membranes having pore sizes in the range of 0.01–10 μm are used, the process is termed *microfiltration* (MF). (There is no sharp distinction between UF and RO, nor is there any between UF and MF.) MF is used to separate very fine colloidal particles in the micrometer and submicrometer range, i.e., larger than 0.1 μm (100 nm) from liquids and gases. The process can be distinguished as *dead-end filtration* and *cross-flow filtration* depending upon the hydrodynamics of the feed flow. In the so-called dead-end filtration, the direction of feed flow is parallel to the permeate flow, whereas in cross-flow filtration it is perpendicular to the permeate flow. Dead-end filtration is suitable for suspensions with a very low solid content, and cross-flow filtration is used for much higher concentrations. MF membranes in tubular and capillary forms can be prepared from asymmetric membranes. Cross-flow filtration is particularly useful in concentrating and/or washing of various colloidal suspensions such as pigments, metal hydroxides, grinding effluents, etc., for separating emulsions (oil-polluted industrial effluents), and as a pretreatment for the RO plants.

Another growing process in membrane separation is the *nanofiltration* (NF); its performance falls between RO and UF. NF membranes have the pore sizes around 10 Å. The range of molecular weight cutoff (about 200) of NF membranes is the main distinguishing feature. Moreover, NF membranes are strikingly different from the RO and UF membranes because they have a charged layer as the separation zone: Their skin layer is made of polyelectrolytes like sulfonated polyether sulfone. NF membranes exhibit two separation mechanisms: molecular sieve effect and charge effect. Important parameters for solute separation include not only the size of membrane pores but also the charge polarities of the membranes and solutes.

The driving force in NF separation is the pressure difference, causing a volume flux through the membrane. In NF, the solutes having sizes larger than the pore size of the membranes cannot permeate and are consequently rejected, while the smaller solutes can permeate. This is the same mechanism as in UF. NF membranes retain sugars and some multivalent salts like $MgSO_4$ but permeate many monovalent salts like NaCl and undissociated acids. The separation of peptides and amino acids from protein solutions can be achieved by using sheet-type NF membranes. NF membranes have many advantages over RO because of their higher water flux and improved fouling resistance against hydrophobic colloids, proteins, and oils. They are also useful in demineralization of water, removal of heavy metals, etc.

5.3.3.2. Electrodialysis. Even though the technique of reverse osmosis has been widely used in desalination, it suffers from the practical problem of membrane *clogging* or *fouling*. Electrodialysis (ED)—an alternative technique to RO in desalination of ionic salt solutions—has been very successful because of recent successes in the development of ion-exchange membranes carrying fixed negative or positive charges. In ED, ions are removed from water by passing through semipermeable membranes that are impervious to water; direct electrical current transports the ions through the membranes.

In the basic ED design, the cation- and anion-selective membranes are placed alternatively in an electric field between the anode and the cathode (Fig. 5.10). The cation-selective membranes permit only the transport of cations, and anion-selective membranes permit only the transport of anions. The transport of ions across the membranes results in ion depletion in some cells and ion concentration in alternate cells. Water exiting from the ion-depleted cells is the desalted product, while the water leaving the ion-concentrated cells is the brine. During this process, hydrogen forms at the cathode and oxygen is produced at the anode. If the solution contains chloride ions, chlorine gas is formed at the anode. The gases formed at the electrodes are removed by an independent stream, which rinses the electrode cells. The base (OH^-) formed at the cathode compartment is neutralized by an acidified (pH = 2) cathode rinse stream. This prevents the precipitation of salts in the cathode compartment. The anode is usually a stainless steel, whereas the cathode is made of platinum-coated tantalum, niobium, or titanium.

Plate and frame modules are most commonly used in ED units. In a practical ED apparatus, about 100–600 cation- and anion-exchange membranes are stacked in parallel to form about 50–300 cell pairs. Spacer gaskets separate the membrane sheets from each other. The membranes should be as large as possible and the distance between them should be as small as possible, because these conditions are favorable to the economics of the process.

Electrodialysis is particularly more economical than RO for the desalination of brackish water and for the concentration of seawater for salt recovery. Similar to RO or UF, electrodialysis also requires the pretreatment step to achieve an optimum performance. If dense (nonporous) membranes are used, the electrically neutral substances remain quantitatively in the original solution, and thus the process can be used for both concentration of the dissolved ionic species and demineralization of neutral species. Because ED removes only ions, it is particularly suitable for separating nonionized from ionized substances.

5.3.3.3. Gas Separation. The phenomenon of transport of gases through polymers was discussed briefly in Section 4.1.9. The control of the gas transport through polymeric membranes is particularly important for their applications as protective coatings, membrane separation, and packaging for foods and beverages. Modern packaging materials control the exchange of components between the package contents and the external environment, acting as barriers for some gases while allowing others through. For example, protection from attack by oxygen, water vapor, or flavor losses are the most common functions in food packaging. Foods differ in their sensitivity to oxygen, and so the kind of oxygen barrier required to store them successfully

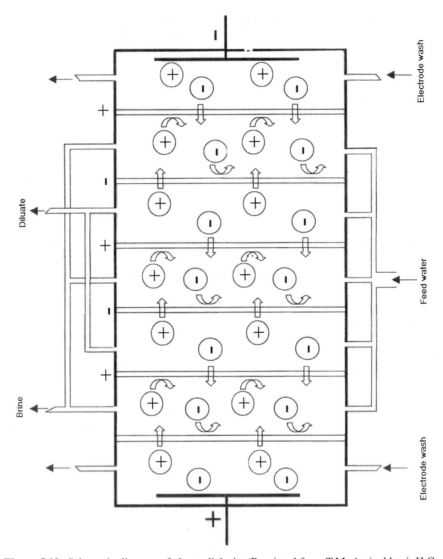

Figure 5.10. Schematic diagram of electrodialysis. (Reprinted from T.M. Aminabhavi, H.G. Naik, U.S. Toti, R.H. Balundgi, A.M. Dave and M.H. Mehta, *Polymer News,* Vol. 24, pp. 294–303 (1999). Copyright © 1999 by The Gordon and Breach Science Group, Malaysia. Used by the permission of the editor, K.S. Kirshenbaum, USA.)

differs accordingly. Other applications include the biogas separation. Biogas or landfill gas typically contains about 40–45 mol% CO_2, 60–55 mol% methane gas, less than 1 mol% water vapor and other compounds. If biogas is to be satisfactorily utilized as a methane source, CO_2 must be selectively removed from the collected gas stream by using membranes.

Gas mixtures can be effectively separated using both porous and dense membranes. A trans-membrane pressure difference of up to 70 bars or more leads to a gas transport through membranes. Separation through porous membranes is based on kinetic gas principles. Separation through dense membranes occurs as a result of the differences in sorption characteristics and diffusion rates of the gaseous components of a mixture. Gas-separating membranes utilize the difference of permeation rates at which the different gas molecules diffuse through the molecular chain gaps called *pores*. These pores arise from either the fluctuation of excess volume due to thermal segmental motion of the polymer chains or from the frozen excess free volume existing in the glassy polymer below its glass transition temperature.

In rubbery polymers above their glass transition temperature, thermal segmental motion is *not frozen* and the size distribution of molecular gaps depends on statistical probability. If the membranes contain pores in a more controlled order then certain gases will selectively permeate. The gas diffusion rate is more dependent on the size and shape of the gas molecule in a glassy polymer than in a rubbery polymer. Consequently, amorphous polymers like heat-resistant polymers with stiff aromatic chains (such as in polyimide) that have high glass transition temperature are promising candidates for high-performance gas separation.

The *hollow-fiber membrane* configuration is one of the more popular modules used in gas separation. These membranes can be dense or asymmetric with a thin surface layer as described in Section 5.3.2.2. *Thin-film composite* polysulfone hollow-fiber membranes coated with polydimethylsiloxane that plugs the aggregate pores are also widely used. Thin films can be formed either by laminating or coating the surface of a microporous membrane. Alternatively, thin separation layers are created on the membrane by interfacial polymerization, plasma treatment, or chemical reaction. Dry cellulose acetate membranes in *spiral-wound modules* are used in industrial gas separation applications.

Because gas permeation is more sensitive to membrane imperfections (micropores) than reverse osmosis, gas separation membranes should be pretreated (annealed) before use. Gas separation by membranes is commercially successful in separating hydrogen from synthesis loop gases and in the recovery of CO_2 from tertiary crude oil recovery, in separating methane gas from its mixture with carbon dioxide, in helium concentration from natural gas, in enrichment of oxygen, and in separating oxygen from its mixture with nitrogen.

Nitrogen production by membranes is a fast-growing industry. Nitrogen produced by separating air using hollow-fiber membranes costs 1/2 to 1/3 less than the nitrogen commonly purchased in cylinders. Silicone-coated asymmetric membranes are quite useful for this purpose. Ammonia cross-linked brominated PPO hollow fiber is also excellent for this separation.

5.3.3.4. Pervaporation.

Pervaporation (PV) is a preferential permeation followed by evaporation. It is a unique technique that is quite different from other membrane-based separation processes because the permeate undergoes a phase change from liquid to vapor, which is then condensed to give back the liquid. On one side of the membrane the preferentially sorbed fluid is in the liquid state, and it is withdrawn

Retentate

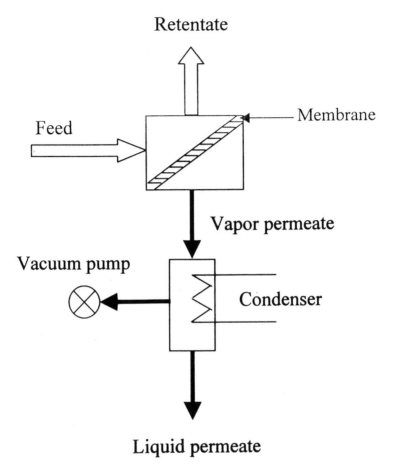

Figure 5.11. The operation of the pervaporation technique.

through the membrane as a vapor by applying a vacuum on the other side of the membrane. Thus the driving force for the permeating molecules is not achieved by elevating the pressures at the feed side (upstream) as in other pressure-driven processes but by a reduction in activity (reduction of partial pressure) on the permeate side (downstream). Because the partial pressure of the permeating components at the permeate side is below the corresponding saturation vapor pressure, the permeate will evaporate at the permeate side of the membrane; on condensation it will give the permeated liquid.

The operation of the PV process is displayed in Figure 5.11. The process is considered as an alternative method for the separation of liquid mixtures that are difficult or impossible to separate by using the conventional methods. It is used to achieve the separation of azeotropic mixtures and closely boiling mixtures, for concentration of fruit juices and recovery or removal of trace substances (aromas), as well as for extracting water from water-miscible organic mixtures.

Even though the mechanism of PV is not fully understood, it seems that the process is governed by the solution-diffusion principles. A good preferential sorption leads to good selectivity if the interaction between one of the components of the mixture and the polymer is strong. However, a membrane in contact with the mixture rich in the strongly interacting component could become very swollen. Excessive swelling dilates the polymer and results in a loss of mechanical strength and selective properties of the barrier. Conversely, insufficient swelling is detrimental to the permeation rate, which is a function of solubility (a thermodynamic property) and diffusivity (a kinetic property). Both solubility and diffusivity affect the selectivity of PV membranes. The ideal polymer for PV membranes would have such a balance between affinity and swelling capacity that would constitute a good selective barrier.

The *separation factor,* α_{AB} (more commonly called *selectivity*), of the component mixture AB is the term used to measure the membrane separation performance in a PV experiment. This term is defined by the equation:

$$\alpha_{AB} = \frac{Y_A/Y_B}{X_A/X_B} \qquad (5.3.2)$$

where X_i and Y_i refer to the feed side and permeate side composition of the i-th component, respectively. However, selectivity is more properly defined as the ratio of component permeabilities.

In a PV experiment, separation is governed by the chemical nature of the permeating species as well as that of the membrane material and by the morphology of the membrane itself together with the experimental conditions. Good PV performance can be achieved by incorporation of appropriately designed functional groups into the membrane and by using a *composite* or *asymmetric membrane* with a thin skin. Well-performing PV membranes are prepared by blending a polymer having high flux with another having high selectivity. PV membranes prepared from inorganic materials (*zeolites* or *ceramics*) have attracted much interest in view of their good thermal and mechanical stability, their structural integrity without the problem of swelling, and their chemical and microbiological resistance. Polyimides are also used because of their excellent thermal stability, chemical resistance, and mechanical strength.

Because of their high mechanical strength and chemical resistance, nonporous asymmetric membranes with a dense top layer and an open porous sublayer as well as composite membranes are used in PV separation. These membranes may be prepared either from glassy or rubbery polymers configured in a proper geometry, including flat sheet, hollow fiber, spiral wound, etc. Grafted membranes, blends, and composites are widely used. Rubbery polymers have higher permeation flux than glassy polymers because of their higher free volume space and chain flexibility, thus leading to both higher permeation flux and better selectivity.

Pervaporation is a more expensive process when compared with RO; thus the process is limited to cases where the traditional methods are costly, e.g., separation of isomers or dehydration of water-alcohol mixtures.

5.3.4. Polymeric Devices for Drug Delivery

An understanding of the *barrier properties* of polymers has become increasingly important in medical and pharmaceutical sciences, particularly because of the recent advances made in protein, peptide, and drug delivery to target sites. In these areas, polymers are used in various configurations such as *injectible particles, hydrogels, transdermal patches*, etc. The term *gel* (see Section 1.2.7.1) applies to swollen polymer network. Hydrogels are cross-linked polymeric matrices that can absorb large quantities of water. The properties important in *drug delivery* applications are their equilibrium swelling, sorption kinetics, solute permeability, and *in vivo* release performance. Polymers based on acrylic and methacrylic acids form the most versatile class of pharmaceutical hydrogels. In these systems, the drug is dissolved or dispersed uniformly throughout the device and the release is mainly governed by the diffusion mechanism, although other processes such as swelling and *erosion* may occur simultaneously, depending on the nature of the drug and the matrix material.

Other types of drug delivery systems include biodegradable polymeric *microspheres* that are nonaggregated particles less than 250 μm in diameter. Several names are given to these particles: *microparticles, microcapsules, nanocapsules, nanospheres*, and *nanoparticles*. To emphasize their small size, microspheres (microparticles) and microcapsules smaller than 1 μm are usually referred to as nanospheres (nanoparticles) and nanocapsules.

A microcapsule has its drug centrally located within the particles, where it is encapsulated within a unique polymeric membrane, whereas a microsphere has the drug dispersed throughout the particle, i.e., it is a homogeneous matrix of the drug and the polymer. Theoretically high-quality microcapsules will release drug at a constant rate (zero-order release), whereas microspheres typically give a first-order release of the drug. With proper formulation design and process manipulation, one can control the release of the drug at a specified rate.

Here the key factor is the choice of an appropriate polymer. These polymers must meet several requirements such as tissue compatibility, biodegradation, drug compatibility, drug permeability, and ease of processing. One class of polymers that is popular for this purpose includes poly(lactide)—a thermoplastic polyester—and a copolymer of lactide and glycolide refered to as poly(lactide-*co*-glycolide). Other polymers like albumin, poly(β-hydroxybutyrate), polyanhydrides, polycaprolactone, and polycyanoacrylate are also used.

After selecting the proper polymer, the next step is to choose an appropriate microencapsulation procedure. In micro/nanoparticles production, drug is entrapped, encapsulated, or attached to the polymer matrix by adopting different procedures.

Polycyanoacrylates can be prepared as nanoparticles of about 200-nm size by mechanically dispersing alkyl cyanoacrylate in an aqueous acidic medium in the presence of a surfactant. No initiator or irradiation is required. Polymerization is normally carried out in the liquid phase by bulk, suspension, emulsion, or micelle processes. A nonreactive, biologically active material may be incorporated into the polymer during polymerization.

The most commonly used processes for the preparations of microcapsules are *spray drying, solvent evaporation*, and *phase separation*. The technique one chooses depends on many factors: the proposed function of the microcapsules/microspheres, the desired microcapsule/microsphere size, the physicochemical properties of the drug and the polymer, compatibility of the process conditions with the drug and the polymer. Usually it is desirable to make a product with the highest ratio of drug to polymer to minimize the amount of mass administered to the patient without compromising the release kinetics.

Spray drying can be used as the primary means of *microencapsulation* for a wide range of drugs. In this process, the drug is dispersed in a solution of the coating material, which is then atomized, and the solvent is dried off using the heated air in a spray dryer.

In the so-called solvent evaporation method, the polymer is dissolved in an organic solvent like dichloromethane, chloroform, or ethyl acetate. Drug is dissolved or dispersed into the preformed polymer solution, and this mixture is emulsified into an aqueous solution to make an oil in water (O/W) emulsion by using a surfactant/emulsifying agent like gelatin, poly(vinyl alcohol), polysorbate, poloxamer, etc. After the formation of a stable emulsion, organic solvent is evaporated either by increasing the temperature, by placing under vacuum, or by continuous stirring.

Drugs that are moderately to very water soluble and are unsuitable for encapsulation by the methods discussed above can be encapsulated by using organic liquids in which the drug is insoluble but the coating polymer is soluble under certain conditions. Changes in temperature, addition of incompatible polymer, or changing the solvent may induce phase separation of the polymer. The coacervated polymer encloses the core materials to form the microcapsule wall. Usually, low polymer concentrations are required for encapsulation by the *coacervation* effect that involves separation into polymer-rich and polymer-poor regions. The phase separation must be gradual, enabling the concentrated polymer solution to deposit and cover uniformly the surface of the core material to form a satisfactory coating.

Recently, *supercritical fluid technology* (see Section 3.7.4) has been used to produce drug-loaded polymeric particles. The advantage of this method is that it avoids the use of toxic organic solvents. Two different approaches are employed. In the first method, the polymer and drug solution are sprayed into supercritical fluid (CO_2) through capillary tubes. The supercritical fluid disperses the polymer as well as the drug in addition to functioning as an *antisolvent* for both polymer and drug. The supercritical fluid then rapidly diffuses into the solution droplets, resulting in the precipitation of small drug particles. This method, known as the precipitation with *compressed antisolvent* (PCA), is used to produce relatively monodisperse submicrometer-size microparticles.

The second method is a spray drying method that uses a *rapid expansion of supercritical solutions* (RESS). This is accomplished by first dissolving the drug and polymer in the supercritical fluid, followed by a rapid expansion of the fluid to a state where the solubility is substantially lower. The resultant drop in the solution density causes a high degree of supersaturation and concomitant formation of small-size particles. Supercritical fluids have low solvating power; thus the throughput capacity of RESS is severely limited. This disadvantage has been overcome by adding modifiers to the supercritical fluid to enhance its polarity.

The most commonly used approach, and one of the easiest ways to deliver a drug to the patient is through skin permeation, called *transdermal delivery*. This technique involves the transport of drugs through the intact skin to attain systemic circulation in sufficient quantity to administer a therapeutic dose. During the last decade, it has received increasing attention because molecular diffusion-mediated transport across a flat sheet membrane is well understood. This makes transdermal delivery an effective and reliable means for the controlled release of drugs.

Devices for transdermal drug delivery are generally fabricated as multilayered polymeric laminate structures in which a drug reservoir or a drug/polymer matrix is sandwiched between two polymeric layers. The outer backing layer, comprising an impermeable polymer or a foil, is designed to prevent loss of the drug through the backing membrane surface. The other polymeric layer may function as an adhesive or a rate-controlling membrane. In the *monolithic* (or *matrix*) system, the drug is dissolved or dispersed in the polymer phase and its release from the drug/polymer matrix controls the overall rate of release from the device. In the *reservoir* (or membrane) system, the drug is present between the polymeric layers in an essentially pure form and the diffusional resistance across the polymeric membrane controls the overall rate of drug release. The selection of either a monolithic or a reservoir system depends on the desired rate of drug transport and delivery to the systemic circulation.

The types of polymeric materials used in the preparation of transdermal membranes (patches) are quite varied; these include olefinic polymers and copolymers, cellulose esters, polyamides, ethyl vinyl acetate copolymer, and poly(vinyl chloride).

5.A. SUGGESTIONS FOR FURTHER READING

Agassant, J.F., Ed., *Polymer Processing: Principles and Modeling,* Hanser Gardner Publications, 1991.

Baker, R.W., *Membrane Technology and Applications,* McGraw Hill, New York, 2000.

Hatakeyama, T., and F.X. Quinn, *Thermal Analysis: Fundamentals and Applications to Polymer Science,* Wiley, New York, 1999.

Matsuura, T., *Synthetic Membranes and Membrane Separation Processes,* CRC, Boca Raton, FL, 1993.

McCrum, N.G., C.P. Buckly, and C.B. Bucknall, *Principles of Polymer Engineering,* 2nd ed., Oxford University Press, New York, 1997.

Middleman, S., *Fundamentals of Polymer Processing,* McGraw-Hill, New York, 1977.

Miles, D.C., and J.H. Briston, *Polymer Technology,* 3rd ed., Chemical Publishing, 1996.

Mulder, M., *Basic Principles of Membrane Technology,* 2nd ed., Kluwer Academic, Boston, MA, 1996.

Ottenbrite, R.M., and S.W. Kim, Eds., *Polymeric Drugs and Drug Delivery Systems,* Technomic., Lancaster, PA, 2000.

Pinnau, I., and B.D. Freeman, *Membrane Formation and Modification,* ACS Symposium Series No. 744, Oxford University Press, Cary, NC, 1999.

Rautenbach, R., and R. Albrecht, *Membrane Processes,* Wiley, New York, 1989.

Stastna, J.,D. and De Kee, *Transport Properties in Polymers,* Technomic, Lancaster, PA, 1995.

Stevens, M.J., and J.A. Covas, *Extruder Principles and Operations,* 2nd ed., Chapman & Hall, New York, 1995.

Tadmor, Z., and C.G. Gogos, *Principles of Polymer Processing,* Wiley, New York, 1979.

Vieth, W.R., *Membrane Systems: Analysis and Design: Applications in Biotechnology, Biomedicine and Polymer Science,* Wiley, New York, 1993.

Wunderlich, B., *Thermal Analysis,* Academic, New York, 1990.

Ziabicki, A., *Fundamentals of Fibre Formation: The Science of Fibre Spinning and Drawing,* Wiley, New York, 1976.

5.B. REVIEW QUESTIONS

1. Describe in detail the process of extrusion and the type of polymers processed by this method.
2. What are the advantages of the bubble blown method as opposed to the flat extrusion method in polymer fabrication?
3. Discuss the applications of modern coating methods. What are the requirements of a wire coating method?
4. What type of polymers can be melt spun? How fibers are produced by the melt spinning method?
5. Discuss the difference between the wet spinning and the dry spinning of fibers.
6. Give the large-scale applications of the films produced by calendering.
7. Discuss and compare the different types of molding techniques used to make shaped objects.
8. What are the thermal analysis techniques used in polymer analysis? What kind of information can be obtained from each of these techniques?
9. What membrane qualities are required in electrodialysis? Explain your answer with examples. Refer to the literature and work out the economics of the ED process in comparison to reverse osmosis.
10. Give the characteristics distinguishing nanofiltration from other filtration methods such as ultra- and microfiltration. Mention the type of membranes used in each of these methods.
11. Describe different processes used to prepare asymmetric membranes and thin-film composite membranes with examples.
12. Explain the term "reverse osmosis." How is it different from osmosis? What types of membranes are used in reverse osmosis experiments?
13. What is membrane fouling? How can it be avoided?
14. Discuss in detail the mechanism of electrodialysis, giving an example of a grafted anionic membrane. Explain in what respect is electrodialysis better than reverse osmosis.

15. What are the advantages of ultrathin composite membranes over other asymmetric membranes? Explain the mechanism of transport through dense and ultrathin composite membranes.

16. Explain the mechanism of pervaporation separation based on the principles of sorption-diffusion. What types of membranes are effective in separating (a) water-ethanol mixtures, (b) DMSO-water mixtures, and (c) the isomers of xylenes?

17. Discuss the commercial applications of gas permeation through membranes.

18. Explain the role played by the biodegradable polymeric membranes (fabricated as microspheres or nanoparticles) in drug delivery studies. What methods are normally used to prepare such membranes? Explain the mechanism of release kinetics through such matrices based on diffusion principles.

19. Discuss the advantages of the supercritical fluid technology in producing drug-loaded polymeric particles. In what way is the method superior to conventional methods of producing particles?

EPILOGUE
LITERATURE ABOUT
MACROMOLECULES

By now, we hope, the reader is convinced of the importance of macromolecular science, is intrigued by the challenges it poses, and most of all, wants to learn more about it. This part of the book provides a guide for further study.

Macromolecular science comprises sizable parts of chemistry, physics, and technological science. The literature devoted to it is therefore extensive; it is virtually impossible to make a fair selection from it. Instead, we present here (1) a short list of textbooks that cover the same material as the present book but from different viewpoints; (2) a list of major series of monographs in the macromolecular field, polymer encyclopedias, and similar works; (3) a list of major reference sources and handbooks; (4) a list of journals that deal exclusively with macromolecules (innumerable other journals publish macromolecular articles together with articles in other fields).

A. TEXTBOOKS ABOUT MACROMOLECULES

Allcock H., and F. Lampe, *Contemporary Polymer Chemistry,* 2nd ed., Prentice Hall, New York, 1990.

Batzer, H., and F. Lohse, *Introduction to Macromolecular Chemistry,* Wiley, New York, 1979.

Billmeyer, F.W., *Textbook of Polymer Chemistry,* 3rd ed., Wiley, New York, 1984.

Bovey, F.A., and F.H. Winslow, Eds., *Macromolecules: An Introduction to Polymer Science,* Academic, New York, 1979.

Carraher, C.E., Jr., *Seymour/Carraher's Polymer Chemistry: An Introduction,* 4th ed., Dekker, New York, 1996.

Chanda, M., *A Problem Solving Guide,* Dekker, New York, 2000.

Cowie, J.M.G., *Polymers: Chemistry and Physics of Modern Materials,* 2nd ed., International Textbook, Aylesbury, UK, 1991.

Elias, H.-G., *Macromolecules,* 2nd ed., two vols., Plenum, New York, 1984.

Elias, H.-G., *An Introduction to Polymer Science,* VCH, New York, 1997.

Flory, P.J., *Principles of Polymer Chemistry,* Cornell University Press, Ithaca, NY, 1953.

Fried, J.R., *Polymer Science and Technology,* Prentice Hall PTR, Englewood Cliffs, NJ, 1995.

Mandelkern, L., *An Introduction to Macromolecules,* Springer, New York, 1972.

Mark, H., and A.V. Tobolsky, *Physical Chemistry of High Polymeric Systems* (High Polymers Ser., Vol. 2), Interscience, New York, 1950.

Meyer, K.H., *Natural and Synthetic High Polymers,* 2nd ed. (High Polymers Ser., Vol. 4), Interscience, New York, 1950.

Meyer, K.H., and H. Mark, *Hochpolymere Chemie: Ein Lehr- und Handbuch für Chemiker und Biologen,* Akademie Verlag, Leipzig, 1940.

Misra, G.S., *Introductory Polymer Chemistry,* Wiley, New York, 1993.

Morawetz, H., *Polymers. The Origins and Growth of a Science,* Wiley, New York, 1985.

Rodriguez, F., *Principles of Polymer Systems,* 3rd ed., Hemisphere, New York, 1989.

Rosen, S.L., *Fundamental Principles of Polymer Materials,* 2nd ed., Wiley, New York, 1993.

Rudin, A., *The Elements of Polymer Science and Engineering. An Introductory Text for Engineers and Chemists,* 2nd ed., Academic, New York, 1999.

Seymour, R.B., *Introduction to Polymer Chemistry,* McGraw-Hill, New York, 1971.

Stevens, M.P., *Polymer Chemistry: An Introduction,* 2nd ed., Oxford University Press, New York, 1990.

Vollmert, B., *Polymer Chemistry,* Springer, New York, 1973.

Young, R.J., *Introduction to Polymers,* Chapman and Hall, London, 1981.

B. MONOGRAPH SERIES AND ENCYCLOPEDIAS

Advances in Polymer Technology, Wiley, New York.

Advances in Polymer Science (previously *Fortschritte der Hochpolymeren-Forschung*), Springer, Berlin.

Encyclopedia of Polymer Science and Technology, Interscience, New York.

High Polymers, Wiley, New York.

Kunststoff-Handbuch, Carl Hanser, Munich.

Macromolecular Syntheses, Wiley, New York.

Polymer Science Library, Elsevier, Amsterdam.

C. HANDBOOKS AND REFERENCE SOURCES

Brandrup, J., E.H. Immergut, and E.A. Grulke, Eds., *Polymer Handbook,* 4th ed., Wiley, New York, 1999.

Culbertson, B.M., Ed., *Contemporary Topics in Polymer Science,* Vol. 6, Plenum, New York, 1989.

Jenkins, A.D., Ed., *Polymer Science: A Materials Science Handbook,* North-Holland, Amsterdam, 1972.

W.V. Metanomski, Ed., *Compendium of Macromolecular Nomenclature,* Blackwell, London, 1991.

Pethrick, R.A., Ed., *Polymer Handbook,* Harwood Academic Publisher, 1984–1999.

Staeckert, K., *Kunststoff-Lexikon,* Carl Hanser, Munich.

Van Krevelen, D. W., and P. J. Hoftyzer, *Properties of Polymers: Correlations with Chemical Structure,* Elsevier, Amsterdam, 1972.

D. MACROMOLECULAR JOURNALS

Acta Polymerica
Angewandte Makromolekulare Chemie
Biomacromolecules
Biomaterials
Biomedical Polymers
Biopolymers
Carbohydrate Polymers
Colloid & Polymer Science
European Polymer Journal
High Performance Polymers
International Journal of Biological Macromolecules
International Journal of Polymer Analysis and Characterization
International Journal of Polymeric Materials
Journal of Applied Polymer Science
Journal of Applied Polymer Science: Applied Polymer Symposia
Journal of Elastomers and Plastics
Journal of Macromolecular Science—Physics
Journal of Macromolecular Science—Pure and Applied Chemistry
Journal of Macromolecular Science—Reviews in Macromolecular Chemistry and Physics
Journal of Polymer Engineering
Journal of Polymer Science, Part A, Polymer Chemistry
Journal of Polymer Science, Part B, Polymer Physics
Journal of Polymer Science, Polymer Symposia
Macromolecular Chemistry (London)
Macromolecules
Makromolekulare Chemie
Makromolekulare Chemie, Rapid Communications
Macromolecular Rapid Communications
Macromolecular Theory and Simulations
Makromolekulare Chemie, Supplement
New Polymeric Materials

Polymer
Polymer Bulletin
Polymer Composites
Polymer Degradation and Stabilization
Polymer Engineering and Science
Polymer International
Polymer Journal
Polymer Plastics Technology and Engineering
Polymer Preprints, American Chemical Society
Polymer Process Engineering
Polymer Science and Technology
Polymer Science Series A (Vysokomolekulyarnye Soedineniya)
Polymer Science Series B (Vysokomolekulyarnye Soedineniya)
Polymer Testing
Polymeric Materials Science and Engineering
Reactive and Functional Polymers
Rubber Chemistry and Technology

INDEX

Abrasion resistance, 573
Abstraction, by radicals, 140, 157
Accelerators, 227
Acetal interchange, 120
Acetate fibers, 38
Acrylic fibers, 8. *See also* Polyacrylonitrile
Active sites of proteins, 63
Activity, 250
Activity coefficient, 250
Adenosine triphosphate, 63
Aerosols, 2, 73, 75
Agarose, 432, 436
Aggregates of small molecules, 71
Aging, 9
Agostic, 212
Allylic monomers, 159
Amic acids, 125
Aminoplasts, 35, 570
Amphiphilic, 273
Ampholytes, 434
Amylopectin, 38
Amylose, 15, 38
Anexes, 33
Angular scattering function, 395, 396
Anisotropic solutions, 271
Anisotropy, form, 376, 380
    intrinsic, 376, 380
    molecular, 376
    of statistical segments, 377
Annealing, 481
Antibodies, 42
Anticode, genetic, 67
Anticodon, 70

Antioxidants, 233
Antisolvent, 589
Aramid fibers, 125
Aramid polymers, 11, 125
Archibald method, 359
Assymetry of particles, 365
Attrition, 55
Autocorrelation function, 409
Autocorrelator, 409
Autoxidation, 231
Azo initiators, 146

Backbiting, 20, 161
Backbone, 2
Bakelite, 34, 122
Banbury mixer, 558
Barrier properties, 588
Binodal decomposition, 312
Biodegradation, 233
Biogel, 436
Biomedical applications, 7
Birefringence:
    acoustic, 377
    of crystalline materials, 510
    electric, 377
    flow, 374
    intrinsic, 380
    magnetic, 377
    magnitude of, 375, 385
Blends of polymers, 309, 514, 530
Blob, 342
Blowing agents, 571
Boltzmann relation, 244

Bragg condition, 410, 536
Branched structures, 96, 370
Branching:
    coefficient, 109
    degree of, 162
Branched polymers, 18
Brinell hardness test, 572
Bulk modulus, 477
Buoyancy factor, 301, 307
Buoyant density, 305

Cabannes factor, 392
Cage, molecular, 103
Calendering, 567
Caoutchouc, 5
Capillary waves, 566
Carothers equation, 105
Casting, 559
Catexes, 33
Ceiling temperature, 174, 192, 204, 231, 574
Cellophane, 226, 279
Celluloid, 38
Cellulose, 15, 37, 61, 71, 506, 509
    acetate, 38, 225, 234, 566
    acetate-butyrate, 225
    aging of, 225
    alkali, 225
    carboxymethyl, 38, 226
    derivatives of, 366, 371
    ethers, 226
    ethyl, 226
    fibers, 224
    methyl, 38, 226
    modification of, 224
    nitrate, 38, 225, 234, 279, 493
    reaction with ethylene oxide, 226
    sulfate, 225
    triacetate, 225
    xanthate, 38, 226, 566
Chain transfer, 140, 157, 198, 228
Characteristic pressure, 264
Characteristic temperature, 264
Cotton, 37, 224
Cotton-Mouton effect, 377
Counterions, 184
Coupling, 421
    dipolar, 421
    nuclear, 422
Cracks, 519
Crankshaft motion, 475
Craze, 517
Creep, 494, 526
Critical micelle concentration, 78
Critical point, 312
Critical temperature:
    lower, 316, 516
    upper, 316
Critical transition concentration, 272
Cross-flow filtration, 582

Cross-linked structures, 181
Cross-linking, 10, 32, 226
    agents, 181, 516
    control of, 161
Crotonic acid, copolymerization of, 148
Crystalline polymers, 504
    inverse gas chromatography of, 534
    mechanical properties of, 511
    size of crystallites, 539
Crystallinity, control of, 24
Crystallization exotherm, 529
Curie temperature, 521
Curing agents, 131
Cycle forming condensations, 124
Cyclization:
    constant, 106
    intramolecular, 182
    reactions, 101, 224

Dead-end filtration, 582
Dead polymer, 19
deBroglic relation, 540
Debye-Huckel ionic atmosphere, 57
Decomposition:
    binodal, 312
    spinodal, 312
Deformation of particles in flow, 343
Degradation of polymers, 230
Degree of polymerization, 3, 153
    distribution of, 153
Degrees of freedom, external, 244, 263
Demixing, rate of, 311
Denaturation:
    chemical, 62
    of DNA, 68, 418
    heat, 61
    of proteins, 61
Dendrimers, 21
    convergent synthesis, 113
    dendritic, 21
    dendrons, 21
    divergent synthesis, 111
    hyperbranched polymers, 21
    polyamidoamine, 112
    starburst effect, 111
    treelike structures, 21
Density scattering, 401
Deoxyribonucleic acids, 45, 305, 336, 376, 377,
        385, 418
    denaturation of, 68, 418
    mammalian, 359
    renaturation of, 68
Depolarization, 391
    of fluorescence, 419
Depolymerization, 232
Depropagation, 172, 192
Desalination of seawater, 523
Detergents, 2, 75, 180
Dextran, 38, 436

Diads:
  isotactic, 27, 423
  meso, 27
  syndiotactic, 27, 424
Dialysis, 292
Dianions, 189, 192
Dielectric:
  constant, 375, 399, 520
  complex, 520
  fluctuation of, 399
  relaxation, 521
  tensor, 375
Dielectrics, optical properties of, 374
Diels-Alder, condensation, 133
Die swell, 560, 561
Differential scanning calorimetry, 528
Differential thermal analysis, 530
Diffusimetry, 349
Diffusion, 346, 490
  free, 349
Diffusion coefficient, 57, 347, 410, 449, 523
  in gases, 534
  of gases in polymers, 534
  integral, 351
  interpretation of, 351
  z average of, 411
Diffusion-controlled reaction, 104, 165
Diffusion flow, 347
Dip-coating, 564
Disentanglement, 342
  time, 490
Dispersion forces, 245, 248
Displacement tensor, 478
Disproportionation of radicals, 142, 149
Dissymmetry of scattered light, 405
Distribution coefficient, 532
Distribution function of orientations, 343, 377, 520
Distribution of molecular weights, *see* Molecular
  weight
DNA, *see* Deoxyribonucleic acids
DNA profile analysis, 455
DNA sequencing, 456
Dogbone, 526
Domains, 272, 517
  incompatible, 514
Donnan effect, 289, 430
Double helix of DNA, 66, 305
Downstream, 574
Drawing of polymers, 509, 511, 566
Drug carriers, polymeric, 222
Drug delivery, 588
Dry spinning, 566
Dumbbell model in simple shear flow, 343
Dynamic experiments, 498

Elastic force, 477, 482
Elastic networks, 481
Elastic recovery, 498, 561
Elastic solids, 477

Electrets, 521
Electrical properties, 519
Electrofocusing, 434
Electron density in crystals, 541
Electronic excitations, 417
Electron microscopy, 542
  scanning, 543
  transmission, 543
Electron spectroscopy for chemical analysis
  (ESCA), 547
Electrons, solvated, 187
Electrophoresis, 428
  capillary, 433
  capillary zone, 434
  field inversion, 454
  free, 429
  gel, 432
  orthogonal field, 453
  paper, 430
  of proteins, 435
  pulsed field gel, 453
  SDS, 435
Electrophoretic boundary, 429
Electrophoretic mobility, 428
Elementary chains, 31, 482, 486
Ellipsoids:
  flow birefringence of, 374
  frictional ratio of, 328
  orientation by flow, 343, 374, 379
  viscosity of, 332
Elongation at break, 479, 512, 527
Emulsions, 73, 76, 408
Encounter, molecular, 103
Endothermic mixtures, 251
End-to-end distance, 46, 49, 365
  average square of, 52, 483
  of deformed coils, 382
  vector, 485
Entanglements, 325, 374, 486
Enthalpy of polymerization, 196
Entropy:
  configurational, 244
  of disorientation, 249
  of intermolecular interactions, 244
  of mixing, excess, 266
  of mixing, residual, 266, 316
Enzymes, 42, 62
  proteolytic, 63
  specificity of, 63
Epoxy resins, 33, 130, 558, 559
Equation-of-state:
  effects, 516
  terms, 266
  theories, 261
Equilibrium:
  definition of, 277
  in a density gradient, 305
  methods, 277
  phase, 308

Equilibrium (*cont.*)
  sedimentation, 298
  in the ultracentrifuge, 293
Erosion, 588
Excess contact energy, 247
Excimers, 419
Excluded-volume theory, 254, 260
Exothermic mixtures, 251
Exotherms, reaction, 165
Expansion coefficient, 260
Extension ratio, 484
External degrees of freedom, 244, 263
External potentials, 278
Extinction coefficient, 408
Extruders, 560, 563
Extrusion, 559
  bubble blown film, 561, 562
  cast film, 563
  die, 566
  melt, 559

Fatigue, 572
Ferroelectric polymers, 521
Fiber-reinforced composites, 519
Fibers, 4, 8, 38, 117, 125, 226, 511, 519, 539, 564
Fiber-spinning, 561, 564
Fibrils, 517
Fibrinogen, 65
Fibroin, 41, 59
Fick's laws, 347, 348, 524
Field flow fractionation, 448
Fillers, 516
Flexible spacer, 275
Flocculation, 8
Flory-Huggins equation, 244, 247, 533
  for polydisperse polymers, 319
  for three components, 286
Flory-Huggins theory, 243, 244, 250, 270, 314, 488
Flory's constant, 366
Flow:
  curves, 493
  disturbance of, 333
  index, 492
  plane of, 323, 379
  power dissipated by, 325, 331
  simple shear, 322, 330
  through porous media, 341
  viscometric, 322
  viscous, 322
Flow birefringence, 374
Fluctuations, 400
  in dielectric constant, 399
  in molecular orientation, 402
  theory of, 400
  thermal, 311
Fluorescence, 418
  depolarization of, 419

Flux, 576
Foam:
  fabrication of, 570
  polystyrene, 5
  polyurethane, 11, 117
  structural, 571
  water, 76
Foaming agents, 570
Folding of polymer chains, 508
Fourier transform infrared, 420
Fracture, cohesive brittle, 564
Free-draining coil, 336
Free energy of mixing, 242
Free volume, 263, 475, 480, 502
Freeze-drying, 321
Frequency-temperature superposition, 528
Frictional coefficient, 326, 327, 349, 351, 357, 428
Frictional ratio, 328, 329
Frictional tensor, 327, 476
  principal values of, 327
Fringed micelle model, 508
Functionality of monomers, 96

Gaussian subchains, 53, 335
Gelatin, 41, 73
Gel effect, 164, 165
Gel exclusion chromatography, 435
Gel filtration, 436
Gel permeation chromatography, 435
  calibration curve, 440
  column selection, 445
  critical point GPC, 444
  detectors, 446
  molecular weight, 440
  molecular weight distribution, 444
  principles, 439
  universal calibration, 443
  universal calibration curve, 443
Gel point, 109
Gels, 32, 73, 109
  concentration gradient, 432
  resolving, 433
  stacking, 433
  swelling of, 488
Genes, 69
Genetic coding, 68
Gibbs energy of mixing, 242
Glass:
  organic, 7
  permeability of, 524
Glass transition, 502, 511, 521, 530, 534
  temperature, 481
Glycogen, 38
Glyptal polyester, 108
Grafting, 228
Graphite, 521
Guinier approximation, 414
  plot, 542
Gutta-percha, 5

Hardness, 572
Head-to-tail addition, 147
Height equivalent to one theoretical plate, 534
Helices, 272, 506
α-Helix, 60, 376
Helix, double of DNA, 66
Helix-coil transition, 386
Hemicelluloses, 224
Hemoglobin, 65
Henry's law, 254
    constant, 532
Heterogeneity in density, 305
Heterophase materials, 516
Hildebrand's solubility parameter, 248, 515
Histones, 68
Homolytic decomposition, 135
Homopolymers, 3
Huggins' constant, 372
Hyaluronic acid, 38
Hydrodynamically dilute solutions, 340
Hydrodynamically ideal solutions, 340
Hydrodynamically semidilute solutions, 341
Hydrodynamic interactions, 333
    Oseen tensor of, 333, 338
Hydrodynamic radius, 328, 336, 351, 370
Hydrodynamics:
    effect of concentration, 340
    of macromolecular coils, 335
    molecular, 321
Hydrodynamic volume, 365
Hydrophilic bonds, 75
Hydrophobic bonds, 75
Hydrophobic forces, 61
Hyperbranched polymers, 110, 111
Hyperconjugation, 143, 200
Hypersharpening of boundary, 357, 359
Hypochromicity, 418
Hypothetical standard state, 253

Immersion precipitation, 578
Immobilized catalysts, 222
Immobilized enzymes, 222
Impact-resistant materials, 516
Impact strength, 517, 572
Impermeable coil, 336
Induced dipoles, 245
Infrared dichroism, 420
Infrared spectra, 419
Inhibition, 140, 162
Inhibitors, 162, 233
Initiation, 135, 145, 146
    anionic, 188
    thermal, 146, 156
Initiators, 135
    addition on monomer, 149
    anionic, 188
    cationic, 199

    coordination, 205
    decomposition of, 145, 149
    efficiency of, 150
    lithium, 191
    waste of, 146
    Ziegler-Natta, 205
Insertion reactions, 129, 207
Instron, 527
Insulators, 519
Interaction energy, 245
Interaction parameter:
    B_{12}, 515
    polymer-polymer, 533, 542
    ternary, 534
    χ_{AB}, 247
    χ_{12}, 250, 489, 533
Interactions:
    donor-acceptor, 515
    intersegmental, 257
Interfacial polycondensation, 100
Interference of light waves, 392
Interference fringes, 298, 349
Intermolecular interactions, 46, 75
    attractive, 45
    repulsive, 45
Internal viscosity, 346
Interpenetrating networks, 516
Interplanar distances in crystals, 536
Intramolecular interactions, 49
    long-range, 49, 53
    short-range, 49
Intrinsic viscosity, *see* Viscosity, intrinsic
Inverse gas chromatography, 531
Ion exchanging resins, 33, 222
Ionic pairs, 186
Ionomers, 58
Irradiation, high-energy, 138, 228, 232
Isocyanates, 115
Isoelectric point, 429
Isothermal compressibility, 264, 401, 478
Isotropic melt, 276
Izod notched impact test, 572

Keratins, 41, 59, 71, 509
Kerr effect, 377
Kevlar, 11
Kinetic chain length, 152
Kinetics:
    at high conversion, 164
    of polycondensation, 102
    of radical polymerization, 145

Ladder polymers, 130
Lag time, 523
Lamellae, crystalline, 509
Laminated materials, 33, 131
Lamm equation, 355
Latex, 563, 570
Layered structure, 273

Light scattering, 386
anisotropic, 402
by anisotropic particles, 390
compositional, 403
depolarization of, 391, 402
dissymmetry of, 405
function K, 403
by gases, 401
by large particles, 393
by liquids, 401
low-angle laser, 405, 413
by macroscopic systems, 398
multiple, 409, 519
photometers, 411
polydispersity effect, 406
by polymer solutions, 403
quasi-elastic, 409, 413
by small isotropic particles, 386
Lignin, 224
Lipoyl transacetylase, 65, 66
Liquid crystals, 271, 272
cholesteric, 273, 274
mesophore, 273
nematic, 273, 274
smectic, 273, 274
Liquid crystalline polymers, 275
main-chain thermotropic, 275
side-chain polymers, 275
Loss angle, 500, 521
Loss modulus, 499
Loss tangent, 501
Lower critical temperature, 316, 516
Low molecular weight mixtures, 244
Lucas effect, 377
Lucite, 7. *See also* poly(methyl methacrylate)
Lyotropic, 273

Macromers, 229
Macromolecular coils, *see* Coils, macro
molecular
MALDI, 425, 447. *See also* mass spectrometry
Maleic anhydride, copolymerization of, 148
Maltese cross, 510
Mandrel, 562
Mark-Houwink-Sakurada relation, 365, 369, 370,
371
Markers, 454
Mass spectrometry, 425
matrix assisted laser desorption ionization,
425, 447
time-of-flight, 425, 426, 447
Mastication, 23, 558, 560
Maxwell effect, 377
Maxwell element, 496
Mercerization, 225
Melamine-formaldehyde resins, 35, 121
Melting endotherm, 530
Melting temperature, 511
Melt spinning, 564, 565, 566

Melts of polymers, 489
viscosity of, 491, 495
Membrane clogging, 583
Membrane fouling, 581, 583
Membranes, semipermeable, 280, 292
anisotropic, 577
asymmetric, 577
capillary modules, 576
frame module, 575
heterogeneous, 579
homogeneous, 579
ion exchange, 579, 580
nucleation track, 576
spiral wound, 575, 585
symmetric, 577
tubular, 576, 581
Membrane based separation processes, 580
electrodialysis, 577, 584
gas separation, 577, 583
microfiltration 577, 582
nanofiltration, 582
pervaporation, 585, 586
reverse osmosis, 577
ultrafiltration, 577, 581
Mercurization, 38
Mesogenic units, 277
Metallocene catalysts, 208
Metallocene polymers, 208
Metallocenes, 208
Metastable solutions, 311
Metastable, 311
Micelles, 2, 75, 77, 180
block copolymer, 79, 80
equilibrium, 80
Microcapsules, 588
Microcrystals, 385
Microencapsulation, 589
Microfibrils, 71
Microgels, 385
Microspheres, 588
Microscopy, 542
atomic force, 545, 546
electron, 543, 544
optical, 542
scanning tunneling, 545
transmission electron, 543
Mie's theory, 398
Miscibility window, 316
Mixed solvents, solutions in, 284, 304,
406
Modulus:
bulk, 477, 485
complex, 501
of elasticity, 479, 517, 527
loss, 499
shear, 477, 478, 485
storage, 499
time-dependent, 495
Young, 479, 485

Molding:
  blow, 568, 569
  compression, 122, 181, 569
  injection, 181, 567
  reaction, 118, 568
  transfer, 569
Molecular weight, 80, 281
  in anionic polymerization, 194
  from Archibald method, 359
  differential, 90
  from diffusion, 351
  from diffusion and sedimentation velocity,
    358
  distribution of, 84, 303
  distribution function of, 85, 444
  of elementary chains, 485
  equilibrium, 105
  from equilibrium in density gradient, 305
  integral, 90
  from intrinsic viscosity, 366
  from light scattering, 403, 406
  logarithmic normal, 89
  most probable, 84, 153, 293
  narrow, 193
  number-average, 85, 282, 293, 308
  from osmotic pressure, 281, 282, 290
  Poisson, 89, 194, 359
  Schulz-Zimm, 86, 87, 108, 154
  from SDS electrophoresis, 437
  from sedimentation equilibrium, 301, 303
  from sedimentation velocity, 358
  from sedimentation velocity and viscosity,
    366
  viscosity-average, 85, 371
  weight-average, 85, 303, 406
  z average, 85, 406
Monofilaments, 564
Monolithic, 590
Monomeric unit, 3, 84
Monte Carlo methods, 55, 261
Mooney equation, 486
Morphology of crystalline polymers, 506
Motion, intramolecular, dynamics of, 419
Multienzyme complexes, 65
Myosin, 65

Nanocapsules, 588
Nanoparticles, 588
Nanospheres, 588
Natural macromolecules, 36
Natural rubber, 5, 16, 227, 558
  cyclization of, 224
N-carboxyanhydrides, 197
Necking, 512
Necklace model, 335, 336
Networks -macromolecular, 31
  dense, 32
  loose, 31, 226
  two-dimensional, 35

Neutron scattering, 539
  coherent, 541
  incoherent, 541
Newtonian liquids, 326
Nitrocellulose, 454, *see also* Cellulose
  nitrate
NMR, *see* Nuclear magnetic resonance
  C-13, 424
  magic angle, 424
Nomex, 11
Nomenclature of polymers, 28
  source-based, 28, 29
  structure-based, 28, 29
Non-Newtonian liquids, 326
Non-Newtonian solutions, 342
Normal stresses, 383, 477, 525, 561
Novolacs, 34, 123, 131
Nuclear magnetic resonance, 27, 421
Nucleation, 74, 118, 311, 508, 571
Nucleic acids, 1, 42, 298, 357, 417. *See also*
  Deoxyribonucleic acids; Ribonucleic acids
Nucleosides, 43
Nucleosomes, 68
Nucleotides, 43
Nylons, 11, 58, 99, 102, 511

Occupied volume, 480
Oligomers, 3, 293
Optical:
  activity, 272
  anisotropy, 272
  antipodes, 25
  nearfield scanning microscopy, 543
  properties of bulk polymers, 519
  rotation, 417
  tensor, 382
Orientational correlation, 272
Orientation:
  angle, 378, 379, 384
  distribution function of, 343, 377, 520
  factor, 378, 379
  intrinsic, 380
  of particles in flow, 343
Oriented adsorption of solvent, 385
Orlon 12. *See also* polyacrylonitrile
Orthogonal field gel electrophoresis, 453
Oscillating dipole, 386
Oscillations:
  forced, 527
  free, 527
Oseen interaction tensor, 333, 338
Osmometers, 294
Osmometry, 279
  of polyelectrolytes, 288
  technical aspects, 291
Osmotic flow, 280
  pressure, 281, 285
  shock, 280
Overlap concentration, 341

Parameter X_{12}, 266
Parison, 568
Partition coefficient, 319, 531
Partition function, 262
Pearl-string model, 335
Pentane effect, 48
Peptides, 41, 432
Peptization of colloids, 73
Permeation, 574
Permeability, 279, 522
    hydrodynamical, 336, 490
    of glasses, 524
    for oxygen, 13
Peroxides, decomposition of, 136
Persistence length, 53, 416
Perturbation theories, 54, 243, 257
Phase:
    boundaries, 278
    diagrams, 314
    equilibria, 308
    inversion, 578
    separation, 73, 530
    triangular, 318
Phases, conjugate, 309
Phenol-formaldehyde resins, 34, 122, 558,
    569
Phenomenological theories, 243, 253, 269, 287
Photochemical decomposition, 138
Photodegradation, 232
Photoinitiation, 145
Photoresists, 13
Photostabilization, 233
Pick-up rate, 564
Piezoelectric materials, 521
Pitch of a helix, 59
Plasticizers, 6, 513, 558
Plate and frame module, 583
Plateau region, 353, 355
Pleated sheet structures, 59, 71
Plexiglass, 7, 178. *See also* Poly(methyl
    methacrylate)
Poiseuille equation, 361
Poisson distribution function, 88, 89
Poisson ratio, 478, 485
Polarizability:
    anisotropy of, 391
    average, 391
    molecular, 375, 402
    tensor, 375
Polarization, of dielectrics, 374
Poling of electrets, 521
Polyacenaphthylene, 147
Polyacetals, 120
Polyacetylenes, 522
Poly(N-acetyl-D-glucosamine), 38
Polyacrylamide, 8, 32, 179, 183, 221, 432,
    436
Polyacrylates, 230
Polyacrylhydrazide, 222

Poly(acrylic acid), 8, 57, 58, 179, 290, 291
Polyacrylonitrile, 8, 18, 223, 228, 566
Polyaddition, 95, 98
Polyalkenes, 506
Poly(alkylene polysulfides), 12, 132
Poly(alkylene sulfides), 132
Poly(alkyl vinyl ethers), 200
Polyamic acids, 126
Polyamides, 11, 98, 117, 506, 564
    aromatic, 100, 125, 272
Polyanhydrides, 10
Polybenzimidazoles, 17, 127
Poly(benzyl glutamate), 42, 59, 272, 336, 376,
    418
Polybenzyls, 133
Poly(p-bromostyrene), 222, 229
Polybutadiene, 4, 207, 227, 420, 482, 516
Polybutene, 506
Polybutyrolactam, 197
Polycaprolactam, 3, 11, 196, 232
Polycaprolactone, 10, 234, 515
Polycarbonates, 10, 100, 515, 523, 569
Polychloroprene, 6
Polycondensation, 95
    interfacial, 100
    kinetics of, 102
Polycrystalline materials, 538
Polycyclobutene, 207
Polycyclopentene, 207
Poly(diacetylenes), 5, 36
Poly(dichlorophosphazene), 13, 221
Polydienes, 231
Poly(dimethylphenylene oxide), 15, 134
Poly(dimethyl siloxane), 13, 102, 127, 205, 232,
    233, 482, 491
Polyelectrolytes, 3, 8, 56, 288, 428
Polyepichlorohydrin, 9, 205, 220, 515
Polyesters, 10, 97, 117, 506, 511, 515, 564
Polyethers, 9, 132
Poly(ethyl cyanoacrylate), 8
Polyethylene, 3, 51, 134, 161, 205, 228, 232, 490,
    506, 507, 509, 521, 569
    high-density, 4, 220
    high-pressure, 4
    low-density, 4, 509, 510
    low-pressure, 4
Poly(ethylene adipate), 10
Poly(ethylene glycol), 9
Poly(ethylene oxide), 9, 118, 196, 205
Poly(ethylene terephthalamide), 100
Poly(ethylene terephthalate), 10, 98
Polyformaldehyde, 9, 102. *See also*
    Polyoxymethylene
Poly-HEMA, *see* poly(hydroxyethyl methacrylate)
Poly(hexadecyl methacrylate), 20
Poly(hexamethylene adipamide), 11. *See also*
    Nylons
Poly(hydroxyethyl methacrylate), 7, 32, 183
Polyimidazopyrollone, 18

Polyimides, 11, 125
Polyisobutylene, 4, 200, 283, 482, 506
Polyisocyanates, 12
Polyisoprene, 5, 192, 207, 233, 482
Poly(lactic acid), 234
Poly(*p*-lithio styrene), 222, 229
Poly(maleic imide), 147
Polymer blends, 514
Polymarase chain reation, 451, 452
Polymer melts, 489
Polymerase chain reaction, 451
Polymeric sulfur, 13
Polymerization:
  acyclic diene metathesis, 217
  anionic, 27, 128, 184
  atom transfer radical, 176
  bulk, 178
  cationic, 198
  chain, 95
  coordination, 27, 184, 205
  in crystalline lattice, 36
  degree of, 3, 84, 104
  emulsion, 180, 558
  enthalpy of, 173, 195
  entropy of, 173
  group transfer, GTP, 218
  industrial, 178
  insertion, 129, 207
  ionic, 184
  kinetics of, 145, 201
  living cationic, 203
  living radical, 175
  metathesis, 214
  nitroxide-mediated radical, TEMPO, 177
  precipitation, 179
  radical, 27, 135, 166
  rate of, 149
  ring-opening metathesis, 215
  ring opening, 195, 215
  solution, 178
  stepwise, 95
  suspension, 179, 544, 558
  thermodynamics of, 172
Polymers:
  all-carbon backbone, 3
  atactic, 27, 506
  branched, 18, 108
  bulk, 474
  comblike, 20
  conducting, 521
  crystalline, 504
  cyclolinear, 14
  decomposition of, 529
  derivatized at chain ends, 193
  ferroelectric, 521
  fractionation of, 318
  glassy, 480
  isotactic, 27, 423, 506

  ladder, 14, 17, 130, 223
  linear, 2
  living, 23, 184, 193, 229
  monodisperse, 3
  morphology of, 506
  organo-inorganic, 14
  pearl, 179
  phase-separated, 530
  polyaromatic, 223
  polydisperse, 3, 81, 303, 318, 351, 370
  precipitation of, 321
  star, 20
  stereoblock, 27, 207
  syndiotactic, 27, 423
  thermally resistant, 16
Polymethacrylamide, 222
Polymethacrylates, 230
Poly(methacrylic acid), 8
Poly(methyl acrylate), 7
Polymethylene, *see* Polyethylene
Poly(methyl methacrylate), 7, 157, 232, 233, 377, 422, 423, 480, 559
Poly(α-methyl styrene), 192, 200, 232
Poly(methyl vinyl ketone), 16
Polynorbornene, 207
Poly(*n*-octyl methacrylate), 502, 503
Polyoxadiazoles, 17
Polyoxymethylene, 9, 204, 205, 232, 233
  stabilization of, 204
Polypeptides, 41, 59, 198, 420, 437
Poly(*p*-phenylene), 16
Poly(phenylene oxide), 15
Poly(*m*-phenylene oxide), 15, 132
Poly(phenyl sesquisiloxane), 17, 130, 223
Polyphosphazenes, 13, 221, 232
Polypropylene, 3, 4, 205, 231, 232, 506, 511, 564
  atactic, 4, 207
  isotactic, 4, 207, 539, 540
Poly (propylene oxide), 205, 220
Polypyrroles, 522
Polysaccharides, 37, 61, 224
Polysilanes, 13, 135
Polysiloxanes, 13, 127, 228
Polystyrene, 5, 15, 32, 157, 160, 192, 222, 228, 230, 302, 306, 317, 354, 377, 381, 385, 405, 418, 419, 490, 506, 515, 516, 542, 559, 569, 570
  chlorosulfonated, 222
  cross-linked, 222
  fractionation of, 320
  liquid crystalline, 275
  nitrated, 222
  plasticization, 512
  side chain, 275
  sulfonate, 57
Polysulfones, 12, 132
Polytetrafluoroethylene, 6, 147, 232, 506
Polytetrahydrofuran, 205
Polythiozoles, 17

Polytriazoles, 17
Polytrifluorochloroethylene, 6
Polyureas, 11, 117
Polyurethanes, 11, 117, 118, 229, 559, 568, 570
Polyvalerolactam, 197
Poly(vinyl acetals), 15, 222
Poly(vinyl acetate), 7, 9, 157, 160, 221, 228, 232, 515
Poly(vinyl alcohol), 3, 9, 221, 222, 506
Poly(vinyl butyral), 15
Poly(vinyl butyrate), 7
Poly(vinyl carbazole), 9
Poly(vinyl chloride), 6, 223, 228, 232, 514, 515, 519, 544, 567, 569, 574
Poly(vinylene carbonate), 147
Poly(vinyl ethers), 200, 203
Poly(vinylidene fluoride), 521
Poly(vinyl methyl ether), 515
Poly(vinyl methyl ketone), 223, 232
Poly(vinyl naphthalene), 418
Poly(vinyl pyridine), 418
Poly(vinyl pyrrolidone), 9, 32, 183
Poly(*p*-xylylene), 15, 564
Porous glass beads, 437
Porous media, flow through, 341
Potential:
    centrifugal, 279, 295
    chemical, 278
    electrical, 279
    external, 278
    gradient of, 348
    gravitational, 279, 293
    of mean force, 258
    total, 279, 298, 347
Precipitation, incipient, 314
Preferential adsorption, 285, 307, 406, 587
Prepolymers, 33, 122, 131, 558
Prigogine-Flory:
    equation of state, 261, 264
    theory, 262
Primary structure, 2
Primer, 451
Principal values of tensors, 375
Promoters, 138
Propagation, 135, 145
    anionic, 188
Proteins, 42, 59, 290, 357, 417, 418, 420, 432, 433
    blood, 430
    crystalline, 506
    fibrous, 41
    globular, 41, 59, 61, 70, 376
    multiunit, 64
    muscle, 2, 42
    precipitation of, 320
    storage, 42
    structural, 2
    synthesis of, 69
    transport, 42

Pseudo-asymmetric centers, 26
Pseudocrystalline lattice, 55, 244
Pseudocrystals, 514, 519
Pseudo-ideal solutions, 54, 252, 314
Purple of Cassius, 74
Pyroelectric materials, 521
Pyruvate dehydrogenase, 65

Q-e scheme, 167, 170, 171
Quaternary structure, 64
Quenching of fluorescence, 419

Radial dilution law, 355
Radial distribution function, 258
Radical-anions, 188, 192
Radicals:
    disproportionation of, 142
    reactions of, 138
    recombination of, 141
    stabilization of, 143
    stable, 143, 163
Radius of gyration, 56, 339, 366, 370, 395, 404, 414, 415, 542
    average of, 406
    cross-sectional, 415
Random walk, 53, 409
Raoult's law, 251, 281
Rapid expansion, 589
Rate of strain tensor, 477, 479
Rayleigh ratio, 388, 400, 403
Rayon, 38, 226
Reaction molding, 118
Reactivity ratios, 167
Recombination of radicals, 141, 149
Reduced variables, 264
Refractive tensor, 375
Regular solutions, 244
Rejection, 581
Rejection coefficient, 581
Relaxation:
    dielectric, 520
    spectrum, 497, 502
    time, 496, 502, 520
Reptation, 490
Resins, 33
    alkyd polyester, 33
    epoxy, 33, 130, 558, 559
    ion exchanging, 33, 222
    melamine-formaldehyde, 35,121
    phenol-formaldehyde, 34, 122, 558, 569
    urea-formaldehyde, 35, 122
Resists, 230, 233
Resites, 124
Resoles, 124
Restriction enzymes, 453
Restriction endonucleases, 453
Retardants, 162
Retardation, 162
Retardation spectrum, 498

Retardation time, 498, 502
Retention volume, 532
Rheological equation of state, 477
Ribonucleic acids, 45, 358
  messenger, 45, 69
  ribosomal, 45, 62
  transfer, 45, 62, 70
Ribosomes, 69, 358
Rigidity, 478
Ring equilibria, 106
Ripening of viscose, 226
RNA, *see* Ribonucleic acids
Rockwell hardness test, 572
Rotary diffusion, 344, 411
  coefficient of, 344, 378, 419, 520
Rotary friction coefficient, 378
Rubbers, 481
  crystallinity of, 487
  theory of elasticity of, 482
  thermoplastic, 517

SAXS, *see* X-ray scattering, small
Scaling laws, 341
Scattered intensity, reduced, 414
  Rayleigh ratio
Scattering:
  angle, 389, 413
  cross section, 541
  function, angular, 395, 397, 404, 414, 416
  length, 541
  profile, 389
  vector, 394, 398, 409, 535
Schlieren boundary, 354
Schlieren optics, 298, 349, 354
Schlieren peaks, 354
Schotten-Baumann reaction, 100
Schulz-Zimm distribution function, 86
Scratch resistance, 572
Screening length, 342
Secondary structure, 58
Second virial coefficient, 252
Sedimentation coefficient, 353
  concentration dependence of, 356
  distribution of, 359
  normalized, 358
Sedimentation equilibrium, 298
Sedimentation velocity, 352
Sedimenting boundary, 354
Segments:
  anisotropy of, 376
  molecular, 248, 262
  statistical, 52, 335
Selectivity, 575
Self-beating spectroscopy, 411
Self-initiation, 146
Self-sharpening of boundary, 357
Semidilute solutions, 341
Semipermeable membranes, 280, 292
Separation of DNA fragments, 453

Separation factor, 587
Separation of gases, 522
  liquids, *see* Pervaporation
Sephadex, 436
Shear compliance, 478
Shear flow, 561
Shear modulus, 477, 480, 485
Shear stress, 383, 476
Shell, 79
Shift factor, 502
Shrinking of polymers, 559
Silicones, *see* Polysiloxanes
Silk, 41, 59, 71
Siloxanes, 127
Simple shear flow, 322, 330, 337, 343, 363, 477
Single crystals of polymers, 507, 508
Size-exclusion chromatography, 435
Small-angle, 413
Soaps, 75
Sols, 73, 109, 376
Solubility of gases, 523
Solubility parameter, 248, 515
Solution spinning, 564
Solvated electrons, 187
Solvent evaporation, 588
Solvents:
  complexing, 187
  good, 54
  poor, 54
  theta, 54
Sorption, enthalpy of, 532
Southern transfer, 454
Southern transfer technique, 454
Specific interactions, 248, 515
Specific volume, 304
  of glasses, 481
Spectral methods, 417
Spherulites, 510, 538, 542
Spinneret, 564, 566
Spinning, 564
  dry, 566
  fiber, 564
  melt, 565, 566
  solution, 566
  wet, 566
Spinodal decomposition, 312
Spiroacetals, 120
Spray drying, 588, 589
Stabilization of polymers, 230
Stacks, 432
Staining, 544
Starch, 432
Star macromolecules, 20
Statistical weight matrices, 51
Steady-state:
  assumption, 152
  systems, 277
Stereoisomerism, 25
Steric hindrance, 51

Stilbene, copolymerization of, 148
Stockmeyer-Fixman method, 368, 369
Stokes equation, 328
Storage compliance, 503
Storage modulus, 499
Stored-energy function, 483, 485
Strain, 476, 485
    tensor, 479, 484
Strain in rings, 102
Stress, 476
    internal, 498, 567
    normal, 383, 476, 525, 561
    relaxation, 495
Stress-optical:
    coefficient, 382, 383
    law, 382
Stress tensor, 337, 382, 478
Structure, primary, 2
Styragel, 437
Substrates, 62
Sulfur:
    colloidal, 74
    plastic, 13
Supersaturation, 74
Supercritical fluid, 589
Supercritical fluid chromatography, 450
Supported oxide catalysts, 220
Surface:
    fractions, 250, 265
    molecular, 263
    replicas, 544
    tension, 76
Suspending agents, 179
Svedberg equation, 357
Swelling factor, 365
Swelling of gels, 489
Systems:
    continuous, 278
    heterogeneous, 278
    homogeneous, 278

Tacticity, 24
Teflon, *see* Polytetrafluoroethylene
Tensile failure, 517
Tensile strength, 479, 512, 517, 527
Tensile stress, 477
Tension, 477
    biaxial, 477
Termination, 135, 145
Tertiary structure, 61
Testing:
    mechanical, 571
    thermal, 573
Thermal convections, 297
Thermal diffusion, 449
Thermal expansivity, 264
Thermal fluctuations, 311
Thermal pressure coefficient, 264
Thermal stability, 6, 231

Thermogravimetric analysis, 573
Thermomechanical analysis, 573
Thermoplastics, 32
Thermosets, 32, 569
Thermotropic, 275
Theta conditions, 54
Theta solutions, 252, 314, 367
Theta solvents, 54
Theta temperature, 54
Thickening agents, 9
Thickness factor, 416
Third virial coefficient, 252
Three-dimensional structures, 108
Tie-lines, 318
Time-temperature correlation, 502, 528
Tobacco mosaic virus, 272
Torsion pendulum, 527
Transdermal delivery, 589
Transdermal patches, 588
Transfer:
    degradative, 159
    to initiator, 157
    to monomer, 158
    to polymer, 161
    to solvent, 159
Translational diffusion, 344
Transparency, 519
Transport through polymers, 522
Triads:
    heterotactic, 27, 191, 424
    isotactic, 27, 191, 424
    syndiotactic, 27, 191, 424
Turbidity, 408
Two-parameter theories, 259
Two-roll mill, 558
T cell receptor, 42

Ultracentrifuge, 293, 295
Ultraviolet absorption, 68
    optics, 359
Ultraviolet spectrophotometry, 417
Undisturbed velocity, 333
Unimers, 79
Unit cell, 505
Unperturbed dimensions, 367
Unzipping, 232
Upper critical temperature, 316
Upstream, 574
Urea-formaldehyde resins, 35, 122
Urethanes, 116

van Laar model, 247
Velocity gradient, 322
    Parameter β_0, 373, 380
Vesicles, 35
Vibrations, molecular, 419
Virial coefficients, 252, 256, 259, 282, 287, 290, 301, 404, 542
Viscoelasticity, linear, 499

Viscoelastic materials, 492, 493
  liquids, 494, 501
  solids, 494, 501
Viscometers, 361
  capillary, 361, 525
  cone and plate, 525
  Couette, 363, 525
  Ostwald, 361
  parallel plate, 525
  rotary, 363
  Ubbelohde, 362
  Zimm-Crothers, 363, 364
Viscometric flow, 322
Viscometry, 360
Viscose, 38, 226, 566
Viscosity, 323, 525
  complex, 501
  of ellipsoids, 332
  internal, 346, 380
  intrinsic, 364
  kinematic, 361
  macroscopic, 330, 332
  non-Newtonian, 372
  of polymer melts, 325, 490, 495
  reduced, 364
  relative, 362, 364
  specific, 364
  steady-state, 498, 499
  zero-shear, 492

Viscous flow, 322
Void volume, 531
Voigt element, 497
Volume:
  change in mixing, 267
  fractions, 249
Vulcanization, 5, 32, 226, 227, 571

Weissenberg rheogoniometer, 525
Wet spinning, 566
Williams-Landel-Ferry equation, 492, 502
Wood, 224
Wormlike model, model of coil, 53, 416

Xerography, 9
X-ray diffraction, 535, 537, 539
  of fibers, 538
  of perfect crystals, 537
  of polycrystalline materials, 538
X-ray scattering, 413
X-ray photoelectron spectroscopy, 547

Yarn, 564
Yielding, 73
Young modulus, 479, 485

Ziegler Natta catalyst, 205, 206
Ziegler Natta initiators, 205, 206
Zimm plot, 404, 405, 542